T0189431

Lecture Notes in Computer Science 979

Edited by G. Goos, J. Hartmanis and J. van Leeuwen

Advisory Board: W. Brauer   D. Gries   J. Stoer

Springer

Berlin
Heidelberg
New York
Barcelona
Budapest
Hong Kong
London
Milan
Paris
Tokyo

Paul Spirakis (Ed.)

Algorithms – ESA '95

Third Annual European Symposium
Corfu, Greece, September 25-27, 1995
Proceedings

 Springer

Series Editors

Gerhard Goos
Universität Karlsruhe
Vincenz-Priessnitz-Straße 3, D-76128 Karlsruhe, Germany

Juris Hartmanis
Department of Computer Science, Cornell University
4130 Upson Hall, Ithaca, NY 14853, USA

Jan van Leeuwen
Department of Computer Science, Utrecht University
Padualaan 14, 3584 CH Utrecht, The Netherlands

Volume Editor

Paul Spirakis
Computer Technology Institute
Kolokotrini 3, GR-262 21 Patras, Greece

Cataloging-in-Publication Data applied for

Die Deutsche Bibliothek - CIP-Einheitsaufnahme

Algorithms : third annual European symposium ; proceedings
ESA '95, Corfu, Greece, September 25 - 27, 1995. Paul Spirakis
(ed.). - Berlin ; Heidelberg ; New York : Springer, 1995
  (Lecture notes in computer science ; Vol. 979)
  ISBN 3-540-60313-1
NE: Spirakis, Paul [Hrsg.]; ESA <3, 1995, Kerkyra>; GT

CR Subject Classification (1991): F.2, G.2.4, I.3.5, E.1, E.5, I.7.3, B.4.3,
C.1.2

ISBN 3-540-60313-1 Springer-Verlag Berlin Heidelberg New York

© Springer-Verlag Berlin Heidelberg 1995
Printed in Germany

Typesetting: Camera-ready by author
SPIN 10485634   06/3142 – 5 4 3 2 1 0   Printed on acid-free paper

Preface

The 3rd Annual European Symposium on Algorithms (ESA'95) was held September 25-27, 1995, in Corfu (Greece), in the conference facilities of the Corfu Hilton International Hotel. This volume contains all contributed papers presented at the symposium, together with the extended abstract of the invited lecture by Shimon Even (Technion University).

ESA was established in 1993 as a main annual event for all researchers interested in algorithms, theoretical as well as applied. The international symposium covers all research on algorithms and their analysis as it is carried out in the fields of computer science, discrete applied mathematics, and all other areas of algorithm-oriented research and its application. The proceedings of ESA'93 appeared as Volume 726 in the series Lecture Notes in Computer Science and the proceedings of ESA'94 appeared as Volume 855 in the same series.

In response to the Call for Papers for ESA'95, the program committee received 119 submissions, indicating a strong and growing interest in the symposium. The program committee for ESA'95 consisted of:

Gianfranco Bilardi (Padova)	Mike Paterson (Warwick)
Josep Diaz (Barcelona)	Miklos Santha (Orsay)
Alan Gibbons (Warwick)	Hava Siegelmann (Technion)
Michel Goemans (MIT)	Paul Spirakis (Patras, Chair)
Giuseppe Italiano (Rome)	Ondrej Sykora (Bratislava)
Michael Juenger (Cologne)	Emo Welzl (Berlin)
Marek Karpinski (Bonn)	Laurence Wolsey (Louvain)
Jan van Leeuwen (Utrecht)	

The program committee met May 12-13, 1995, in Athens and selected 42 contributed papers for inclusion in the scientific program of ESA'95. The selection was based on originality, quality, and relevance to the study of algorithms, and reflects many of the current directions in algorithms research. We wish to thank all those who submitted extended abstracts for consideration, and all subreferees and colleagues who helped in the extensive evaluation process.

The scientific program of ESA'95 evidently shows the gradual expansion of algorithms research into new areas of computational endeavor in science and industry, and the program committee hopes that this component will continue to manifest itself as part of the ESA program in future years.

Many thanks are due to the Organizing Committee of ESA'95 for making the symposium happen. The Organizing Committee consisted of:

Paul Spirakis (Chair), Lefteris Kirousis, Christos Bouras, Vaggelis Kapoulas, Ioannis Stamatiou, Antonis Tatakis,

all from the Computer Technology Institute and the Computer Engineering and Informatics Department of Patras University. Also we are grateful to the Local Arrangements Committee for all of its help and cooperation. This committee consisted of:

Rozina Efstathiadou (Chair), Lena Gourdoupi (Secretary), Eleni Patoucha (Treasurer), Sonia Antonakopoulou (Public Relations), Jenny Theodosopoulou (Site Coordinator),

all from the Computer Technology Institute's administrative staff.

ESA'95 was conducted in cooperation with the Association for Computing Machinery (ACM) and the European Association for Theoretical Computer Science (EATCS), and was sponsored by the Computer Technology Institute, the General Secretariat of Research and Technology, the project ALCOM II of the ESPRIT Basic Research, Intracom, Intrasoft, and the Greek Ministry of Education.

Patras, September 1995   Paul Spirakis

List of Referees

David Alberts
Helmut Alt
Matthew Andrews
Boris Aronov
Vincenzo Auletta
Franz Aurenhammer
Ilan Baron
Andras Benczur
Martin Bečka
Sergej Bezrukov
Gianfranco Bilardi
Johannes Blömer
Norbert Blum
Carlo Blunod
John Bond
Stephane Boucheron
Lorenzo Brunetta
Andrea Capitanio
Benny Chor
Bruno Codenotti
Livio Colussi
Bruno Courcelle
Jean-Marc Couveignes
Maxime Crochemore
Josep Diaz
Martin Diehl
Paul Dunne
Uriel Feige
Paolo Ferragina
Alfonso Ferreira
Philippe Flajolet
Jean-Luc Fouquet
Pierre Fraigniand
Hubert de Fraysseix
Ulrich Fuchs
Luisa Gargano
Bernd Gärtner
Ricard Gavalda
Dan Gayger
Ori Gerstel
Dora Giammarresi
Alan Gibbons

Michael Godau
Michel Goemans
Concettina Guerra
Tibor Hegedüs
Kieran Herley
Juraj Hromkovič
Leon Hsu
Ferran Hurtado
Sandy Irani
Alon Itai
Giuseppe Italiano
Birgitt Jenner
Eric Jordan
Michael Juenger
Volker Kaibel
Sampath Kannan
Vaggelis Kapoulas
Stratos Karaivazoglou
David Karger
Marek Karpinski
Jozef Kelemen
Claire Kenyon
Lutz Kettner
Lefteris Kirousis
Jon Kleinberg
Martin Kochol
Klaus Kriegel
Luděk Kučera
Jean-Luc Lambert
Jan Van Leeuwen
Stefano Lonardi
Janos Makowski
Massimo Maresca
Conrado Martinez
Rolf Möhring
S. Muthukrishnan
Shafi Naor
Pekka Orponen
Dana Pardubská
Domenico Parente
Mike Paterson
Marco Pelligrini

Giuseppe Persiano
C. Peyrat
Andrea Pietracaprina
Michel Pocchiola
Teresa Przytycka
Geppino Pucci
Tomasz Radzik
Rajeev Raman
Claus Rick
Peter Ružička
Wojtek Rytter
Miklos Santha
Vittorio Scarano
Petra Scheffler
Sven Schönherr
Oded Shmueli
Assaf Shuster
Maria Serna
Jay Sethuraman
Hava Siegelmann
Roman Smolensky
Paul Spirakis
Ladislav Stacho
Jacques Stern
Peter Stormer
Ravi Sundaram
Ondrej Sýkora
Róbert Szelepcsényi
Vassilis Tampakas
Sovanna Tan
Stefan Thienel
Dimitris Thilikos
Sivan Toledo
Jacobo Toran
Vassilis Triantafillou
Eli Upfal
Ugo Vaccaro
Wenceslas Fernandez de la Vega
Michel las Vergnas
Frederic Voisin
Imrich Vrťo
Frank Wagner
Emo Welzl
Lorenz Wernisch
Laurence Wolsey

Fatos Xhafa
Alex Zelikovski
Lisa Zhang

Contents

Session 10. Chair: *Paul Spirakis*

On Mixed Connectivity Certificates

(Extended Abstract)

Shimon Even[1*], Gene Itkis[1**], Sergio Rajsbaum[2***]

[1] Department of Computer Science, Technion, Haifa, Israel 32000
[2] Instituto de Matemáticas, U.N.A.M., Ciudad Universitaria, D.F. 04510, México

Abstract. Vertex and edge connectivity are special cases of mixed connectivity, in which all edges and a specified set of vertices play a similar role. Certificates of k-connectivity for a graph are obtained by removing a subset of its edges, while preserving its connectivity up to k.

We unify the previous work on connectivity certificates and extend it to handle mixed connectivity and multigraphs. Our treatment contributes a new insight of the pertinent structures, yielding more general results and simpler proofs. Also, we present a communication-optimal distributed algorithm for finding mixed connectivity certificates.

1 Introduction

1.1 Basic concepts

Let $G = (V, E)$ be a finite undirected graph with no self-loops, and $x, y \in V$ be a pair of distinct vertices of G. The *edge connectivity* of x and y in G is the maximum number of edge-disjoint paths connecting x and y. Similarly, their *vertex connectivity* is the maximum number of vertex-disjoint paths connecting x and y.

---

[*] Part of this research was conducted while visiting the Instituto de Matemáticas, U.N.A.M. Supported by the Fund for the Promotion of Research at the Technion and DGAPA Projects, U.N.A.M. E-mail: even@cs.technion.ac.il

[**] Supported by the Israeli Council for Higher Education and NSF grant CCR-90 15276. Part of this research was conducted in Boston University. E-mail: itkis@cs.technion.ac.il

[***] Part of this research was conducted while visiting the MIT Laboratory for Computer Science, and CRL Digital Equipment Corporation. Partly supported by DGAPA Projects, U.N.A.M. E-mail: rajsbaum@theory.lcs.mit.edu

Following [FIN93], we consider a generalization of these two particular types of connectivity. Let $S \subseteq V$. We say that a family of paths connecting vertices x, y is S-*independent* if the paths are edge-disjoint and every element of S appears as an inner vertex in at most one of these paths. The S-*mixed connectivity* $\lambda_S(x, y; G)$ of x and y in G is the maximum number of S-independent paths connecting x and y in G. The cases of $S = \emptyset$ and $S = V$ correspond to edge and vertex connectivity, respectively. For brevity, if G is clear from the context, we omit it. Say that x and y are S-*mixed k-connected* if $\lambda_S(x, y) \geq k$. This is also referred to as *local* connectivity, as opposed to global: the *global connectivity* of graph G is $\lambda_S(G) \stackrel{\text{def}}{=} \min_{x, y \in V} \lambda_S(x, y; G)$. G is S-*mixed k-connected* if $\lambda_S(G) \geq k$.

For each type of connectivity, a certificate of k-connectivity for G is a subgraph preserving the connectivity up to k. Namely, $G' = (V, E')$, $E' \subseteq E$, is a *certificate of local S-mixed k-connectivity* for G if for any two vertices x and y, $\lambda_S(x, y; G') \geq \min\{k, \lambda_S(x, y; G)\}$. Similarly, G' is a *certificate of global S-mixed k-connectivity* for G if $\lambda_S(G') \geq \min\{k, \lambda_S(G)\}$. The *size* of G' is $|E'|$.

Clearly, certificates of local k-connectivity are also certificates of global k-connectivity; the opposite is generally not true. Unless stated otherwise, we will speak about certificates of local connectivity.

For a k-connected G there is a trivial lower bound of $k|V|/2$ on the size of a certificate of k-connectivity, because the degree of every vertex in a k-connected graph is at least k. Results of Mader ([M72], [M73], [M79]) imply that every edge k-connected graph G contains an edge k-connected subgraph with $O(k|V|)$ edges. Also, Mader's results imply a similar result for vertex connectivity of simple graphs. Namely, an edge k-connected graph has a certificate of global edge k-connectivity of size $O(k|V|)$, and a simple graph, which is vertex k-connected, has a certificate of global vertex k-connectivity of size $O(k|V|)$. However, as is shown below, if one following Zykov's point of view [Z69], as we do, if parallel edges are allowed, the statement does not hold for vertex connectivity.

Example. Let $G = (V, E)$ be a complete graph, each pair of vertices connected by p parallel edges. Then $|E| = p|V|(|V| - 1)/2$ and $\lambda_V(G) = p + |V| - 2$ (since any two vertices have $p + |V| - 2$ vertex-disjoint paths between them). At the same time, the removal of any edge reduces the graph's global connectivity. In this example, for $p \geq |V|$, Mader's $O(p|V|)$ bound is off by a factor of $|V|$.

1.2 Applications

Certificates with fewer edges than the original graph are useful for improving the efficiency of a number of graph algorithms. One may perform a preprocessing step to find a sparse certificate, and then run the algorithms on the certificate. For example, this method has been used to improve the sequential time complexity of testing simple undirected graphs for k-connectivity [NI92, CKT93, G91]; to improve the running time for finding three independent spanning trees [CM88]; to design efficient fault tolerant protocols for distributed computer networks [IR88], dynamic and distributed algorithms [EGIN92].

Sparse certificates are of special utility for the *distributed model* of computation. This model consists of a graph G with the vertices representing processors and the edges representing bidirectional communication links. There is no common memory and all communication is through messages sent along the links. The messages take a finite but arbitrary time to traverse a link. The *communication complexity* of the algorithm is the number of messages sent by the algorithm, assuming each message contains $O(\log |E|)$ bits. For the sake of discussing the *time-complexity* of asynchronous distributed algorithms, it is customary to assume that each message is transmitted and processed within one unit of time.

The vertex (edge) connectivity of the graph is related to the number of processors (links) failures that can be tolerated by the distributed system before the network is disconnected. S-mixed connectivity allows to deal with networks where any link can fail but only processors in S are subject to failure. The number of messages sent by the distributed algorithm often depends critically on $|E|$. In such cases, if the algorithm is executed on a sparse certificate then the message complexity is reduced while preserving the number of faults that can be tolerated.

1.3 Previous work

Given a k-(vertex or edge) connected graph, the problem of finding a k-connected spanning subgraph with the minimum number of edges is NP-hard for any fixed $k \geq 2$; ([GJ79] cites personal communication with F. Chung and R. Graham). Indeed, for $k = 2$ the reduction from the Hamiltonian Cycle problem is trivial.

However, finding good approximations is possible for all k: Nagamochi and Ibaraki [NI92] find, in time $O(|E|)$, edge and vertex connectivity certificates of size $< k|V|$ (which is within a factor of 2 from the trivial lower bound described in

4

Section 1.1). Their algorithm, as well as others, consists of finding a sequence of forests $F_1, F_2 \ldots$ in the graph. Each forest F_j is maximal in the remaining graph $G - \cup_{i=1}^{j-1} F_i$ (e.g. each connected component of the remaining graph is spanned by a tree of F_j). The maximality alone suffices to show that $G_k \stackrel{\text{def}}{=} \cup_{i=1}^{k} F_i$ is an edge k-connectivity certificate of size $< k|V|$. Moreover, if G is simple and the forests are grown according to a certain rule (see Section 4), then G_k is a vertex k-connectivity certificate as well.

Graph G is called S-*simple* if it has no parallel edges incident to vertices of S. Frank, Ibaraki and Nagamochi [FIN93] show that for all $S \subseteq V$ the algorithm of [NI92] applied to S-simple graphs produces certificates of S-mixed connectivity.

Cheriyan, Kao and Thurimella [CKT93] introduce a more flexible way of constructing certificates of vertex connectivity of size $< k|V|$, and use it in their distributed and parallel algorithms. They show that for simple graphs, constructing F_i in a scan-first-search manner (see Section 4.3) is sufficient to produce certificates of vertex k-connectivity. Their distributed algorithm uses the synchronizers of [AP90] and runs in time $O(k|V|\log^3|V|)$ using $O(k|E| + k|V|\log^3|V|)$ messages. Every certificate obtained by the algorithm of [NI92] can be obtained using the scan-first-search; but as we show in the sequel, the converse does not hold. Thus, the results of [FIN93] do not apply to the certificates of [CKT93].

A new distributed algorithm by Thurimella [T95] for computing $O(k|V|)$ size certificates has time complexity $O(k(D + |V|^{0.614}))$, where D is the diameter of the network. The communication complexity of the algorithm was not analyzed.

In all previous results quoted above, the graphs are assumed to be S-simple; $S = \emptyset$ and $S = V$ for edge and vertex connectivity, respectively.

1.4 Our results

We present a general scheme for generating S-mixed k-connectivity certificates. It consists of an optimum reduction from general graphs, allowing parallel edges, to S-simple graphs, and a scheme to generate certificates of size $< k|V|$ for S-simple graphs.[3] This scheme includes as special cases the results of [CKT93, NI92, FIN93] and has a simpler correctness proof. We believe that the generality of this scheme as well as the simplicity of the proof contribute towards a better understanding of connectivity certificates.

---

[3] Note that an optimum reduction may not produce an optimum certificate, even if one is given an optimum certificate for the S-simple graph.

For the distributive model, we present a communication-optimal algorithm which is an implementation of the scheme above. For simple graphs it generates certificates of size $< k|V|$ in time $(2k+2)|V|$ using $\leq 4|E|$ messages. In addition to the improved complexity, our algorithm is simpler and works in a more restricted single server model. A *single server* algorithm has the following property. At any time there is exactly one vertex which is active, and all activities in the network, while this vertex is in charge, are restricted to its immediate neighborhood. In the course of the computation the server travels in the network along its edges.

In this extended abstract, some of the proofs are omitted.

2 Notation and Basic Notions

Let $V(E')$ be the set of vertices which are the endpoints of the edges of $E' \subseteq E$. For disjoint sets $X, Y \subseteq V$, let $E(X, Y) \subseteq E$ denote the set of edges of G with one endpoint in X and the other in Y; $E(X)$ is the set of edges with an endpoint in X. We write $E(x, Y)$ and $E(x)$ if $X = \{x\}$. A path *linking* sets X and Y is a path with one endpoint in X and the other endpoint in Y; X and Y are then said to be *linked*. $\Gamma_G(v) \stackrel{\text{def}}{=} V(E(v)) - \{v\}$ is the set of neighbors of v in G.

Following the definitions in [FIN93], let $\{Z, A, B\}$ be a partition of V such that $Z \subseteq S$, $A \neq \emptyset$ and $B \neq \emptyset$. We say that the pair $C = (Z, E(A, B))$ is an *S-mixed cut*. If $\emptyset \neq A' \subseteq A$, $\emptyset \neq B' \subseteq B$ then the cut $C = (Z, E(A, B))$ *separates A' and B'*. The *size* of C is defined to be $|C| \stackrel{\text{def}}{=} |Z| + |E(A, B)|$. As was observed in [FIN93], following [DF56], using the max-flow min-cut theorem, it is easy to derive the following version of Menger's theorem.

Theorem 1 *For any $a, b \in V$, the minimum size of an S-mixed cut separating a, b is $\lambda_S(a, b; G)$.*

Henceforth, unless otherwise specified, we discuss S-mixed cuts and connectivity. So, for brevity, we omit the "S-mixed".

3 Reduction to S-simple graphs

$\widetilde{G}$ is the *S-simplification* of G, $\widetilde{G} = simple_S(G)$, if $\widetilde{G}$ is the maximal S-simple subgraph of G. To obtain $\widetilde{G}$ from G, for each $\{a, b\} \cap S \neq \emptyset$ replace all parallel edges between a and b, if there are any in G, by a single edge a—b in $\widetilde{G}$.

We will reduce the construction of a global connectivity certificate for a general graph G (with arbitrary parallel edges) to finding a local connectivity certificate for $simple_S(G)$. The following lemma will be useful:

Lemma 1 *Let G' be a certificate of local k-connectivity for G. Then any cut C' in G' is either a cut in G or $|C'| \geq k$.*

(Proof omitted.)

3.1 Global connectivity

Define the S-degree of a vertex v, $d_S(v; G)$, to be $|V| - 2 + \min_{w \neq v} |E(v, w)|$ if $\Gamma_G(v) = V - \{v\} \subseteq S$ and $|E(v, V - S)| + |\Gamma_G(v) \cap S|$ otherwise. The S-degree of graph G is $d_S(G) \stackrel{\text{def}}{=} \min_{v \in V} \{d_S(v; G)\}$. The traditional definitions of degree coincide with $d_\emptyset$. It is easy to show that $d_S(G) \geq \lambda_S(G)$.

Let $m = \min\{k, d_S(G)\}$, and $G' = (V, E' \subseteq E)$. The following procedure $Incr\_Deg(G', G, m)$ increases $d_S(G')$ to be $\geq m$ by adding edges of G to G':

Procedure **Incr_Deg(G', G, m):**
1. for every $v \in V$, starting with those $\in V - S$, do
2.   while $d_S(v; G') < m$ do
3.   if $\Gamma_G(v) - \Gamma_{G'}(v) \neq \emptyset$ then add some $e \in E(v, \Gamma_G(v) - \Gamma_{G'}(v))$ to E'
4.   else if $\Gamma_{G'}(v) = V - \{v\} \subseteq S$ then
5.   for every $u \in V - \{v\}$ such that $|E'(v, u)| = \min_{w \neq v} |E'(v, w)|$
6.   add some $e \in E(v, u) - E'$ to E'
7.   else add some $e \in E(v, V - S) - E'$ to E'

Lemma 2 *After executing $Incr\_Deg(G', G, m)$, $d_S(G') \geq m$.*

(Proof omitted.)

Reduction: To obtain a certificate of global S-mixed k-connectivity for G first obtain a certificate G' of local S-mixed m-connectivity for $simple_S(G)$, and then apply $Incr\_Deg(G', G, m)$ to turn G' into the desired certificate.

First, $\lambda_S(G) \leq d_S(G)$ implies $k \geq m \geq \min\{k, \lambda_S(G)\}$, so there is no difference between certificates for global k- and m-connectivity. Now, the following theorem, together with Lemma 2, implies the correctness of the above reduction.

Theorem 2 *Let $G' = (V, E' \subseteq E)$, $d_S(G') \geq m$, and let G' contain as a subgraph a certificate of local m-connectivity for $simple_S(G)$. Then $\lambda_S(G') \geq \min\{m, \lambda_S(G)\}$.*

(Proof omitted.)

If $\lambda_S(G') \geq k$ then $d_S(G') \geq k$. So, at least for k-connected graphs, it is necessary to increase the S-degree of the simplification's certificate (as done by *Incr_Deg*), in order to turn it into a certificate for G. As we showed above, surprisingly, it turns out to be sufficient as well.

In general, $Incr\_Deg(G', G, m)$ adds minimal but not necessarily minimum number of edges to achieve $d_S(G') \geq m$. But when applied to an m-connectivity certificate of $simple_S(G)$, as in our particular reduction, this number is indeed minimum (i.e. adding any smaller set of edges from G to G' than that added by $Incr\_Deg(G', G, m)$, leaves $d_S(G') < m$). This implies the reduction is optimal.

Lemma 3 (Reduction Optimality) *Let $H = (V, E_H)$ be a certificate of local m-connectivity for $simple_S(G)$, and let $G' = (V, E')$ be obtained from H by applying $Incr\_Deg(H, G, m)$, and $G'' = (V, E'')$ be an arbitrary m-connectivity certificate for G such that G'' contains H as a subgraph. Then, $|E'| \leq |E''|$.*

(Proof omitted.)

Incr_Deg can be executed at each vertex almost independently (it should be executed at the vertices of $V - S$ before the vertices of S). So, say, in the distributed model, it can be implemented to work in 3 steps. For the sequential model its time complexity is $O(|V| + |E|)$, and it can even be adjusted to work in $O(|V| + |\tilde{E}|)$, where $(V, \tilde{E}) = simple_S(G)$.

Finally, we observe that it is natural that even for the global connectivity the reduction is to the local one. Indeed, the simplification of a k-connected G may happen to be just 1-connected. Then any spanning tree of the simplification is a satisfactory certificate for it. In fact, it is possible to show that (similarly to the reduction above) given a certificate G' of global k-connectivity for G, a certificate of global k-connectivity for $simple_S(G)$ can be obtained by adding edges from $simple_S(G)$ to $simple_S(G')$ to increase its degree to $d_\emptyset(simple_S(G')) \geq \min\{k, d_\emptyset(simple_S(G))\}$.

3.2 Local connectivity

In general the reduction of sec. 3.1 may not produce certificates of *local* connectivity. However, in some specific cases, obtaining certificates of local connectivity is easy. If $\{x, y\} \cap S = \emptyset$ then $\lambda_S(x, y; G) = \lambda_S(x, y; simple_S(G))$. Therefore, if a certificate of local connectivity between vertices of $V - S$ is required, then a certificate of local connectivity of $simple_S(G)$ can be used. A connectivity certificate for a specific pair s, t can be constructed by defining $S' = S - \{s, t\}$ and obtaining a certificate of local connectivity for $simple_{S'}(G)$.

For general graphs, a certificate of local connectivity for G can be constructed as follows. First, find a certificate of local connectivity for $simple_S(G)$. Next, flesh out each edge $x \overset{e}{-} y$ of the certificate to have $\max\{k, |E(x, y)|\}$ parallel edges between x and y. This could increase the certificate size, unnecessarily, by factor of $k - 1$ above the minimum. The problem of efficiently reducing the task of finding sparse certificates of local connectivity to finding sparse certificates for S-simple graphs remains open.

4 Certificates for S-simple graphs

Let $G = (V, E)$ be an S-simple graph. Let $\{F_i\}$ be a sequence of mutually disjoint non-empty sets of edges partitioning E, and define $E_k \overset{\text{def}}{=} \bigcup_{1 \le i \le k} F_i$, $\overline{E_k} \overset{\text{def}}{=} E - E_k$, $G_k \overset{\text{def}}{=} (V, E_k)$. Note that $\overline{E_0} = E$. For every $i \ge 1$, let F_i be a maximal forest in $(V, \overline{E_{i-1}})$. Then each forest F_i consists of a set of spanning trees, one for each connected component of $(V, \overline{E_{i-1}})$. The next lemma follows directly from the maximality of the forests [NI92].

Lemma 4 *If u and v are connected in F_j, then for each i, $1 \le i < j$, u and v are connected in F_i.*

Lemma 4 is sufficient to prove that G_k is a certificate of local edge (i.e. $\emptyset$-mixed) k-connectivity of size $< k|V|$. We skip the proof since it is a special case of the mixed connectivity results which follow. If $S \ne \emptyset$, G_k may not be a certificate of k-connectivity, even in the global sense.

Next, we define S-greedy forests (which are also maximal) and show that the S-greedy forests yield certificates of local S-mixed connectivity.

4.1 Greedy Forests

The following search procedure produces a maximal forest F of G. Initially $F = \emptyset$. The vertices of G will be visited as specified shortly. When a vertex is visited some of its incident edges may be added to F. The first vertex to be visited can be chosen arbitrarily. Whenever the visitation of a vertex terminates, the next vertex to be visited can be chosen to be any other vertex of $V(F)$, or any vertex of a component of G which has no vertices in $V(F)$. The process ends when all vertices of G have been visited at least once. Upon termination F is a maximal forest of G. The edges are added to F (one at a time) as follows. While visiting a vertex $v \in S$, for *every* neighbor x of v, $x \notin V(F)$, add the edge $v{-}x \in E$ to F. S-simplicity implies that there is only one edge joining v and x. When visiting $v \notin S$, if $v \overset{e}{-} x \in E$ and if $x \notin V(F)$, one is allowed to add e to F. (Clearly, if e is added to F then none of its parallel edges, if any, may be added to F neither in the same nor in any other visitation.) So, no edges incident to $v \in S$ may be added after its first visitation. If $v \notin S$, edges incident to v may be added during several of its visitations.

The forests produced by the above procedure are called S-*greedy*. Next, we define them without referring to any algorithm (as a static counterpart to the above algorithmic construction). Let F be a maximal forest in G, and let $t : V \to \{1, 2, \ldots, |V|\}$ be a 1-1 *numbering* of the vertices. The numbering t induces orientation on edges: $\vec{F}(t) \overset{\text{def}}{=} \{u{\to}v : u{-}v \in F \wedge t(u) < t(v)\}$. If $\vec{T}$ is a tree rooted at r, directed from the root towards the leaves, then for each $v \neq r$, $\mathsf{parent}(v)$ is the unique u such that $u \to v \in \vec{T}$; also $\mathsf{parent}(r) \overset{\text{def}}{=} r$.

Definition 1 *Maximal F is S-greedy in G if there exists a numbering t of the vertices such that*

(1)   For every tree T of F, $\vec{T}(t)$ is a rooted tree.

(2)   If $w{-}v \in E$ and $w \in S$ then $t(\mathsf{parent}(v)) \leq t(w)$.

Intuitively, Definition 1 reflects the above algorithmic construction of greedy forests as follows. The order in which the vertices of G are visited for the first time is specified by t. If $\mathsf{parent}(v) = u \neq v$, then v has been added to $V(F)$ when an edge $u{-}v$ has been added to F, while visiting u. If $\mathsf{parent}(v) = v$ then v is the first vertex in its component to be visited. The second item in the above definition reflects the requirement that on the first visitation of $v \in S$ all its non-forest neighbors join the forest. Obviously, an S-greedy forest F in G is also S'-greedy for all $S' \subseteq S$.

4.2 Certificates

Next, we show that if for each $i \leq k$ the forest F_i is S-greedy in $(V, \overline{E}_{i-1})$, then G_k is a k-connectivity certificate (note: $|E_k| < k|V|$).

Theorem 3 below is a generalization of the main theorem of [FIN93], where it has been stated for a specific subclass of S-greedy forests.

Theorem 3 (Mixed Connectivity Certificate)
$\lambda_S(a, b; G_k) \geq \min\{k, \lambda_S(a, b; G)\}$, *for all* $a, b \in V$.

We need a couple of lemmas before proving the theorem. Let $C = (Z, E(A, B))$ be a cut of G. We use the following obvious fact:

Fact 1 *Let* $v \in Z$, $E(v, B) = \emptyset$, $C' = (Z - \{v\}, E(A \cup \{v\}, B))$. *Then* $|C'| < |C|$.

Say, a cut $C' = (Z', E'(A', B'))$ of $G' = (V, E' \subseteq E)$ *narrows* a cut $C = (Z, E(A, B))$ of G if $|C'| < |C|$ and C' separates A and B in G' ($A \subseteq A'$, $B \subseteq B'$, so $Z' \subseteq Z$). For example, C' narrows C in Fact 1 above.

Lemma 5 *Let* $C = (Z, E(A, B))$ *be a cut,* $|C| > 0$, *and forest* F *be* S-greedy in G. *There is a cut* $C' = (Z', \overline{F}(A', B'))$ *in* $(V, \overline{F})$ *which narrows* C.

Proof: If A and B are not linked in G, then a zero size cut narrows C. Thus, assume there is a path in G which connects some vertex $a \in A$ with some vertex $b \in B$. By the maximality of F, such a path exists in F as well. If $F \cap E(A, B) \neq \emptyset$ then $C' = (Z, \overline{F}(A, B))$ narrows C, since $|\overline{F}(A, B)| < |E(A, B)|$.

Now, suppose $F \cap E(A, B) = \emptyset$. Let t be a numbering of F as in Definition 1. Clearly, there is some tree $T \subseteq F$, in which both a and b appear. Thus, $V(T) \cap Z \neq \emptyset$. Let w be the least vertex (w.r.t. t) in $V(T) \cap Z$. Let r be the root of $\vec{T}(t)$; wlog assume $r \notin B$.

If $\overline{F}(w, B) = \emptyset$, then by Fact 1, there is a cut in $\overline{F}$, narrowing $(Z, \overline{F}(A, B))$. Therefore, this cut narrows C as well, and the Lemma follows.

Suppose $\overline{F}(w, B) \neq \emptyset$ and let $w \overset{e}{\longrightarrow} v \in \overline{F}(w, B)$. Since there is an edge in $\overline{F}$ connecting w and v, by the maximality of F, $v \in V(T)$ as well. Let $u = \mathsf{parent}(v)$

in $\vec{T}(t)$. By Definition 1 item (2), $t(u) \leq t(w)$. Let $\vec{P}$ be the directed path in $\vec{T}(t)$ from r (through u) to v. For every vertex x on $\vec{P}$, from r to u, $t(x) \leq t(u)$. $\vec{P}$ must have a vertex $z \in Z$, since $r \notin B$ and $F \cap E(A, B) = \emptyset$. By $t(z) \leq t(u) \leq t(w)$ and the minimality of $t(w)$, it follows that $z = u = w$. Thus, $w = \mathsf{parent}(v)$ in $\vec{T}(t)$. By the S-simplicity, $e \in T$, in contradiction to $w \xrightarrow{e} v \in \overline{F}(w, B)$. □

Lemma 6 *Let A and B be linked in F_k and separated by a cut C in G. Then $|C| \geq k$.*

Proof: A and B are linked in F_i for all $1 \leq i \leq k$ (by Lemma 4). Use Lemma 5 to construct a sequence of $k+1$ cuts $C_j = (Z_j, \overline{E}_j(A_j, B_j))$, $0 \leq j \leq k$, with $C_k = C$ and C_{j-1} narrowing C_j. $|C_i| > |C_{i-1}| \geq 0$ for all $1 \leq i \leq k$, thus $|C| \geq k$. □

Proof *(of Theorem 3):*
By Theorem 1 there is a cut $C = (Z, E_k(A, B))$ separating a and b in G_k, such that $\lambda_S(a, b; G_k) = |C|$. If $E_k(A, B) = E(A, B)$ then $|C| = |(Z, E(A, B))| \geq \lambda_S(a, b; G)$. Otherwise, if $E(A, B) - E_k \neq \emptyset$ then by Lemma 4, A and B are linked in F_k, and so by Lemma 6 (applied to G_k), $|C| \geq k$. In either case, $\lambda_S(a, b; G_k) = |C| \geq \min\{k, \lambda_S(a, b; G)\}$. ∎

4.3 Sequential Algorithms

A naive use of the greedy search procedure (described in sec. 4.1) to construct k greedy forests one after the other, takes $O(k|E|)$ time.

When $S = V$, this algorithm is called by Cheriyan et al. *scan-first-search* [CKT93]. They prove that the union of these forests constitutes a certificate of vertex k-connectivity (of size $< k|V|$). This is a special case of our Theorem 3.

Nagamochi and Ibaraki [NI92] describe an algorithm, which we call *NI-search*. This algorithm produces a partition of E into (S-greedy) forests $F_1, F_2, \ldots$ in a single search of the graph, thus reducing the complexity to $O(|V|+|E|)$. Frank et al. show that if G is S-simple, then the resulting G_k is a certificate of local S-mixed k-connectivity [FIN93]. As we shall see, each F_i produced by NI-search is V-greedy in $(V, \overline{E}_{i-1})$, and therefore these results are subsumed by our Theorem 3.

However, there are certificates composed of S-greedy forests that cannot be produced by NI-search (e.g. see Figure 1, where $S = V$). Hence, results of

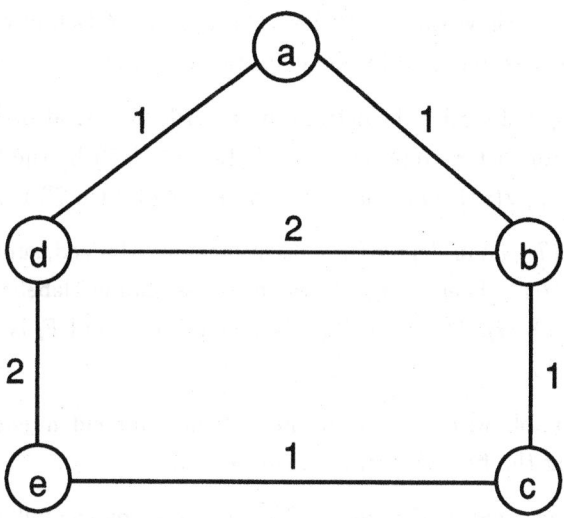

Fig. 1. *A certificate made of V-greedy forests which cannot be obtained by NI-search.*

[FIN93] do not imply that in general S-greedy forests (and in particular scan-first forests) yield mixed connectivity certificates.

NI-search is the only previously published sequential algorithm we know of, to build greedy forests in linear time. Our generic strategy, as presented in sec. 4.1, is more general and provides greater flexibility for other implementations. A case in point is the method used in [CKT93] to design efficient parallel and distributed algorithms that produce vertex connectivity certificates.

We describe NI-search, and prove that it generates S-greedy forests. This is done in detail for self-containment and since our distributed algorithm can be viewed as an implementation of NI-search.

NI-search assigns $\mathsf{rank}(e) > 0$ to each edge e. $F_i \overset{\mathsf{def}}{=} \{e \in E | rank(e) = i\}$. Each vertex v keeps $\mathsf{label}(v) \overset{\mathsf{def}}{=} \max_{e \in E(v)} \{\mathsf{rank}(e)\}$. Initially, $\mathsf{rank}(e) = 0$ (we say, e is *unranked*) for all e, so $\mathsf{label}(v) = 0$ for all v. In each step NI-search visits a yet unvisited v with a highest $\mathsf{label}(v)$ among the unvisited vertices, and assigns $\mathsf{rank}(v \overset{e}{\text{---}} w) = \mathsf{label}(w) + 1$ to each unranked $e \in E(v)$. NI-search terminates when each vertex has been visited. Notice that if $label(v) = i$ then for every $1 \le j \le i$, v has at least one incident edge e for which $rank(e) = j$.

Lemma 7 *For each $i > 0$, F_i produced by NI-search is a V-greedy forest in $(V, \overline{E_{i-1}})$.*

Proof. First we show that conditions (1) and (2) of Definition 1 hold, and then maximality (also required by the definition) is proved.

Let numbering t of vertices be defined by the order of visitation by NI-search. Let $T \subseteq F_i$ be a connected component of F_i. Let $r \in V(T)$ be the first visited in T: $t(r) = \min_{v \in V(T)}\{t(v)\}$. Obviously, the in-degree of r (in $\vec{T}$) is zero.

For any $v \in V(T)$, its in-degree (in $\vec{T}$) is at most one. Indeed, suppose $v \overset{e}{\longrightarrow} u \in T$, and $t(u) < t(w) < t(v)$. Then $v - w \notin T$: when w was scanned $\mathsf{label}(v)$ already was $\geq i$ (due to $\mathsf{rank}(e) = i$). Therefore, $\vec{T}(t)$ is a rooted tree and F_i is a forest. Thus condition (1) holds.

Condition (2) follows directly from the fact that all incident edges of a vertex are ranked during the first visitation of the vertex.

Finally, the maximality of F_i in $(V, \overline{E_{i-1}})$ is shown by the following sequence of three claims. Tree $T \subseteq \overline{E_{i-1}}$ is called *active* if some of its vertices are unvisited.

Claim 1. At any time of NI-search, each F_i contains at most one active tree.

When an edge $v \overset{e}{\longrightarrow} w$ is ranked i, while visiting v either $\mathsf{label}(v) \geq i$ (and the ranking of e creates no new tree in F_i) or $\mathsf{label}(v) = \mathsf{label}(w) = i-1$ (thus creating a new tree in F_i). But in the latter case, there is no unvisited $u \in V$ with $\mathsf{label}(u) \geq i$ (or u would be visited rather than v), and thus there is no other active tree in F_i. $\square$

Claim 2. If $u \overset{e}{\longrightarrow} v \in F_j$, then for each $0 < i < j$, u and v are connected in F_i.

Just before e is ranked, $\mathsf{label}(v), \mathsf{label}(u) \geq j-1$, so $v, u \in V(F_i)$ for each $i < j$. By Claim 1, both u and v are in the unique active tree of F_i. $\square$

Claim 3. Each forest F_i is maximal in $(V, \overline{E_{i-1}})$.

Let $u \overset{e}{\longrightarrow} v \in \overline{E_i}$. Thus, $e \in F_j$ for some $j > i$. Therefore, by Claim 2, u, v are connected in F_i. $\square$

■

Lemma 7 and Theorem 3 yield the following:

Corollary 1 ([FIN93]) *If G is S-simple then, G_k (produced by NI-search) is a certificate of local S-mixed k-connectivity for G of size $< k|V|$.*

4.4 Distributed Algorithm

In this section we present a new distributed algorithm for finding mixed connectivity certificates of size $< k|V|$, for connected S-simple graphs. The algorithm, described in Figure 2, is executed in a network identified with a graph G: each vertex of V corresponds to a node of the network, and each edge of E corresponds to a communication link. Each node v maintains variables corresponding to the ones in NI-search of sec. 4.3: $\mathsf{label}(v)$ and $\mathsf{rank}(e)$, for each incident edge e (all initially 0). In addition, each node v has a boolean variable $first\_time(v)$ initially set to $true$, and a list $unvisited$ initially including all edges incident to v. The algorithm is initiated by sending a VISIT message to any one node (on a nil edge) and terminates with the RETURN message received back (on the same nil edge). A node v receives no messages (i.e. v completes its work) after it sends a RETURN message on an edge of $\mathsf{rank} = 1$. When algorithm halts, $\mathsf{rank}(e) > 0$ for all e, $|E_k| < k|V|$ for any $k > 0$, and if G is S-simple then G_k is a certificate of local S-mixed k-connectivity.

The algorithm is of a restricted form. A single center of activity — we call it the $server$ — travels from vertex to vertex around the network performing all the work. When the server is in a node v, messages are sent only between v and its neighbors. The messages used by the algorithm are: VISIT, RETURN, RANK_EDGE, and EDGE_RANKED(i), $1 \leq i \leq |E|$. The server is said to be in the vertex that has received VISIT, has not yet sent RETURN, and for each VISIT it has sent, a RETURN message has been received. All the unranked edges incident to a vertex are ranked when the server arrives at it for the first time, and in the same way as in the NI-search. Say a node is $visited$ if it has received at least one VISIT message and has completed step (2.2.2). Thus, visited nodes have no unranked edges.

Each (non-nil) edge sees the following sequence of messages: RANK_EDGE, EDGE_RANKED($\cdot$), VISIT, RETURN. All of these are constant size except the EDGE_RANKED(i), which has $\lceil \log i \rceil$ bits, where $i < |E|$. Thus the total number of messages sent by the algorithm is $4|E|$. Since in each step a message is sent by the algorithm the time complexity is $O(|E|)$.

Theorem 4 *Algorithm D implements NI-search, runs in time $4|E|$ and sends $4|E|$ messages.*

(Proof omitted.)

ALGORITHM D (at node v)

(1) **for** RANK_EDGE message arriving at v on edge e **do**

    (1.1) label, rank(e) ← label+1

    (1.2) **send** EDGE_RANKED(rank(e)) on edge e

(2) **for** VISIT message arriving at v on edge e **do**

    (2.1) **drop** e from *unvisited*

    (2.2) **If** *first_time* **then**

        (2.2.1) *first_time*←*false*

        (2.2.2) **for all** edges e', s.t. rank(e') = 0 **do**

            (2.2.2.1) **send** RANK_EDGE on edge e'

            (2.2.2.2) **wait for** EDGE_RANKED(i) to arrive on edge e'

            (2.2.2.3) rank(e') ← i

            (2.2.2.4) **If** label < i **then** label ← i /* OPTIONAL */

    (2.3) **for each** $e' \in$ *unvisited* with rank(e') ≥ rank(e) **in decreasing order** of rank(e') **do**

        (2.3.1) **drop** e' from *unvisited*

        (2.3.2) **send** VISIT message on e'

        (2.3.3) **wait for** RETURN message on e'

    (2.4) **send** RETURN message on e

Fig. 2. Single server algorithm D for S-simple graphs.

Remarks: Line (2.2.2.4) can be omitted without affecting the algorithm correctness, and is included only to guarantee that, analogously to NI-search, if the server is at v then for any unvisited u, label(v) ≥ label(u).

Also, note that the search for an unvisited vertex of maximum label in the whole graph, as required in the original NI-search, is avoided.

$2|E'| - |V|$ steps of the algorithm can be saved by parallelizing step (2.2.2). It is also possible to modify the algorithm to reduce the time and number of messages to $4|E'|$ for $(V, E') = simple_V(G)$. If k-connectivity is desired only for a fixed k, then the algorithm can be modified to run in time complexity $O(k|V|)$ (and each message size is ≤ $\lceil \log k \rceil$).

References

[AP90]   B. Awerbuch and D. Peleg, "Network synchronization with polylogarithmic overhead", in *FOCS*, 1990, pp. 514–522.

[CM88]   J. Cheriyan and S. N. Maheshwari, "Finding nonseparating induced cycles and independent spanning trees in 3-connected graphs", *J. Algorithms*, 9, 1988, pp. 507–537.

[CKT93]   J. Cheriyan, M.-Y. Kao, R. Thurimella, "Scan-first search and sparse certificates: an improved parallel algorithm for k-vertex connectivity", *SIAM J. of Computing*, 22(1), 1993, pp. 157–174.

[DF56]   G. B. Dantzig and D. R. Fulkerson, "On the Max-Flow Min-Cut Theorem of networks", *Linear Inequalities and Related Systems, Annals of Math. Study*, 38, Princeton University Press, 1956, pp.215–221.

[EGIN92]   D. Eppstein, Z. Galil, G. F. Italiano, A. Nissenzweig, "Sparsification — a technique for speeding up dynamic algorithms", *FOCS*, 1992, pp. 60–69.

[FIN93]   A. Frank, T. Ibaraki, H. Nagamochi, "On sparse subgraphs preserving connectivity properties", *J. of Graph Theory*, 17(3), 1993, pp. 275–281.

[G91]   H. Gabow, "A matroid approach to finding edge connectivity and packing arborescences", *STOC*, 1991, pp. 112–132.

[GJ79]   M. R. Garey and D. S. Johnson, *Computers and intractability, a guide to the theory of NP–completeness*, Freeman, San Francisco, 1979, p. 198.

[IR88]   A. Itai and M. Rodeh, "The Multi-Tree Approach to Reliability in Distributed Networks", *Information and Computation*, 79(1), 1988, pp. 3-59.

[M72]   W. Mader, "Über minimal n-fach zusammenhängende, unendliche Graphen und ein Extremalproblem," *Arch. Mat.*, Vol. XXIII, 1972, pp. 553–560.

[M73]   W. Mader, "Grad und lokaler Zusammenhang in endlichen Graphen," *Math. Ann.*, Vol. 205, 1973, pp. 9–11.

[M79]   W. Mader, "Connectivity and edge-connectivity in finite graphs," in *Surveys on Combinatorics*, (B. Bollobas, ed.), London Math. Soc. Lecture Note Series, Vol. 38, 1979, pp. 293–309.

[NI92]   H. Nagamochi and T. Ibaraki, "A Linear-time algorithm for finding a sparse k-connected spanning subgraph of a k-conneceted graph," *Algorithmica*, 7, 1992, pp. 583–596.

[T95]   R. Thurimella, "Sub-linear Distributed Algorithms for sparse certificates and Biconnected Components," to appear in Proc. of the 14th ACM Symposium on Principles of Distributed Computing, August 1995.

[Z69]   A. A. Zykov, *Theory of Finite Graphs,,* (in Russian), Nauka, Novosibirsk, 1969, see pp. 104–105.

Truly Efficient Parallel Algorithms: c-Optimal Multisearch for an Extension of the BSP Model*

(Extended Abstract)

Armin Bäumker, Wolfgang Dittrich
and Friedhelm Meyer auf der Heide

Department of Mathematics and Computer Science
and Heinz Nixdorf Institute, University of Paderborn
33095 Paderborn, Germany

Abstract

In this paper we design and analyse parallel algorithms with the goal to get exact bounds on their speed-ups on real machines. For this purpose we define an extension of Valiant's BSP model, BSP*, that rewards blockwise communication, and uses Valiant's notion of c-optimality. Intuitively a c-optimal parallel algorithm for p processors achieves speed-up close to p/c. We consider the Multisearch problem: Assume a strip in 2D to be partitioned into m segments. Given n query points in the strip, the task is to locate, for each query, its segment. For $m \leq n$ we present a deterministic BSP* algorithm that is 1-optimal, if $n = \Omega(p \log^2 p)$. For $m > n$, we present a randomized BSP* algorithm that is $(1 + \delta)$-optimal for arbitrary $\delta > 0$, $m \leq 2^p$ and $n = \Omega(p \log^2 p)$. Both results hold for a wide range of BSP* parameters where the range becomes larger with growing input sizes m and n. We further report on implementation work in progress. Previous parallel algorithms for Multisearch were far away from being c-optimal in our model and do not consider blockwise communication.

1 Introduction

The theory of efficient parallel algorithms is very successful in developing new original algorithmic ideas and analytic techniques to design and analyse efficient parallel algorithms. For this purpose the PRAM has proven to be a very convenient computation model, because it abstracts from communication problems. On the other hand, the asymptotic results achieved only give limited information about the behaviour of the algorithms on real parallel machines. This is (mainly) due to the following reasons.

- The PRAM cost model (communication is as expensive as computation) is far away from reality, because communication is by far more expensive than internal computation on real parallel machines [6].

- The number of processors p is treated as an unlimited resource, (like time and space in sequential computation) whereas in real machines p is small (a parallel machine (MIMD) with 1000 processors is already a large machine).

There are several approaches to define more realistic computation models and cost measures to overcome the first objection mentioned above: The BSP model due to

* email: {abk,dittrich,fmadh}@uni-paderborn.de, Fax: +49-5251-603514. Supported in part by DFG-Sonderforschungsbereich 1511 "Massive Parallelität: Algorithmen, Entwurfsmethoden, Anwendungen", by DFG Leibniz Grant Me872/6-1, and by the Esprit Basic Research Action Nr 7141 (ALCOM II)

Valiant [14], the LogP model due to Culler et al. [6], the BPRAM of Aggarval et al. [1], and the CGM due to Dehne et al. [7] to name a few. Note that most of the models except the BPRAM neglect the negative effects of communicating small packets.

To deal with the second objection, Kruskal et al. [10] have proposed a complexity theory which considers speed-up. Valiant has proposed a very strong notion of work optimality of algorithms, *c-optimality*. It gives precise information about the possible speed-up on real machines, the speed-up of a c-optimal algorithm should be close to p/c. Besides [8] and [5] there are seemingly no systematic efforts undertaken to design parallel algorithms with respect to this strong optimality criterion.

The computation model used in this paper is the BSP enhanced by a feature that rewards blockwise communication. We design and analyse two algorithms for a basic problem in computational geometry, the Multisearch problem. Our first algorithm is deterministic and 1-optimal. It works for the case of many search queries compared to the number of segments. The second algorithm is designed for the case if only few search queries are asked. It is randomized and proven to be $(1 + \delta)$-optimal with high probability, for $\delta > 0$ arbitary. Both results hold for wide ranges of BSP* parameters.

1.1 The Multisearch Problem

Multisearch is an important basic problem in computational geometry. It is the core of e.g. planar point location algorithms, segment trees and many other data structures.

Given an ordered *universe* U and a partition of U in *segments* $S = \{s_1, \ldots, s_m\}$. The segments are ordered in the sense that, for each $q \in U$ and segment s_i, it can be determined with unit cost whether $q \in s_i, q \in \{s_1 \cup \ldots \cup s_{i-1}\}$, or $q \in \{s_{i+1} \cup \ldots \cup s_m\}$. We assume that, initially, the segments and queries are evenly distributed among the processors. Each processor has a block of at most $\lceil m/p \rceil$ consecutive segments and arbitrary $\lceil n/p \rceil$ queries, as part of the input. The *Multisearch problem* is: Given a set of queries $Q = \{q_1, \ldots, q_n\} \subseteq U$ and a set of segments $S = \{s_1, \ldots, s_m\}$, find, for each q_i, the segment it belongs to (denoted $s(q_i)$). Sequentially this needs time $n \log m$ in the worst case.

An important example is: A strip in 2D is partitioned into segments, and queries are points in the strip, see Figure 1. The task is to determine for each query point which segment it belongs to. Note that sorting the points and merging them with the segments would not solve the problem, as our example shows. In case of $n < m$ we refer to *Multisearch with few queries*, otherwise to *Multisearch with many queries*.

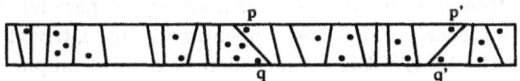

Fig. 1. Strip with segments and query points. Note, that p lies left to q (q' left to p') but $s(p)$ is right to $s(q')$ ($s(q')$ right to $s(p')$).

1.2 BSP, BSP* and c-Optimality

The BSP (Bulk-synchronous parallel) model [14] consists of:

- a number of processor/memory components,
- a router that can deliver messages point to point among the processors, and
- a facility to synchronize all processors in barrier style.

A computation on this model proceeds in a succession of *supersteps* separated by synchronisations. For clarity we distinguish between *communication* and *computation* *supersteps*. In computation supersteps processors perform local computations on data that is available locally at the beginning of the superstep. In communication supersteps all the necessary exchange of data between the processors is done by the router. An instance of the BSP model is characterized by the parameters p, L and g.

- The parameter p is the number of processor/memory components.
- L is the minimum time between successive synchronisation operations. Thus L is the minimum time for a superstep.
- The parameter g is the time the router needs to deliver a packet when in continuous use.

Many routers of real parallel machines support the exchange of large messages and achieve much higher throughput for large messages compared to small packets. Thus good parallel algorithms should try to communicate only large packets, i.e. use *blockwise communication*. Communicating small messages is not significantly faster than communicating messages up to a certain size. To incorporate this crucial property in our model we extend the BSP model to the BSP* model by adding the parameter B, the minimum size the packets must have in order to fully exploit the bandwidth of the router.

For a computation superstep with at most t local operations on each processor we charge $\max\{L, t\}$ time units. For an h-relation, i.e. a routing request where each processor sends and receives at most h messages, we charge $\max\{g \cdot h \cdot \lceil \frac{s}{B} \rceil, L\}$ time units in a communication superstep, if the messages have maximum size s. Thus in the BSP* model it is worthwhile to send messages of size at least B.

Let A^* be the best sequential algorithm on the RAM for the problem under consideration, and let $T(A^*)$ be its worst case runtime. Let c be a constant with $c \geq 1$. In order to be *c-optimal* (with respect to p, L, g and B) a BSP* algorithm A has to fulfill the following requirements [8]:

- The ratio between the time spent for computation supersteps of A and $T(A^*)/p$ has to be in $c + o(1)$.
- The ratio between the time spent for the communication supersteps of A and $T(A^*)/p$ has to be in $o(1)$.

All asymptotic bounds refer to the problem size as $n \to \infty$.

1.3 Known Results

Sequentially Multisearch can be done in time $n \log m$ in the worst case. There are some parallel algorithms for this problem on a variety of parallel models, mainly for the case $n = m$. Multisearch can be solved optimally by a trivial algorithm for the CREW-PRAM. For the EREW-PRAM it is already a very complicated problem. Reif and Sen [13] developed an asymtotically optimal randomized EREW-PRAM algorithm, which works also on the butterfly network. It runs in time $O(\log n)$, with high probability, using n processors. It is easily seen that large constant factors are involved and that it performs badly on the BSP* model. Further it is not obvious how to generalize the algorithm work optimally for the case $n < m$. Ranade [12] has developed a multisearch algorithm for the p processor butterfly network for $n = p \log p$ queries and $m = O(p^c)$ segments. For the case $m \geq n$ this algorithm is asymptotically optimal but not for the case $m < n$. As in the case of the algorithm mentioned above this algorithm has large constant factors and does not consider blockwise communication. Atallah and Fabri [3] achieved (non-optimal, deterministic) time $O(\log n(\log \log n)^3)$ on a n-processor hypercube. A $O(\sqrt{n})$ time algorithm on a $\sqrt{n} \times \sqrt{n}$ mesh network is from Atallah et al. [2]. Dehne et al. [7] have developed some algorithms on the CGM for geometric problems including Mulitsearch. In contrast to the other results quoted above Dehne et al. assume that the machine is much smaller than the problem size, i.e. $p \leq \sqrt{n}$. On their model Multisearch can be solved with p processors in time $O(\frac{n}{p} \log n)$ using a constant number of communication rounds, where constant factors and block-wise communication are not considered. Some 1-optimal algorithms for Sorting and Gauss-Jordan Elimination have been developed by Valiant et al. [8]. McColl [11] has developed some algorithms for matrix problems also on the BSP model.

1.4 New Results

We present and analyse two parallel algorithms for Multisearch. The first algorithm (*ManyQueries*) works for Multisearch with many queries, i.e. $m \leq n$. It is a deterministic BSP* algorithm that is 1-optimal, if $n = \Omega(p \log^2 p)$. The second algorithm (*FewQueries*) works for Multisearch with few queries, i.e. for $2^p \leq m > n$. It is a randomized BSP* algorithm that is $(1 + \delta)$-optimal with probability $1 - p^{-(\log^\beta p)}$ for arbitrary $\delta > 0$, β small enough and $n = \Omega(p \log^2 p)$. These results hold for a wide range of BSP* parameters. E.g. $L \leq n^\eta$, $B \leq n^\xi$ and $g = o(B \log n)$ suffice for ξ and η small enough. Note that p, g and B may grow with the problem size. Therefore we can expect that our algorithms are fast even on machines with relatively slow routers that need large packets. Our algorithms use routines for broadcast, parallel prefix and variants of load balancing as basic routines. BSP* algorithms for these problems are part of our work. Due to page limitations, this extended abstract only sketches most of the algorithms and proofs. A full version of the paper appears as technical report [4].

1.5 Experiments

We have performed preliminary measurements of the Algorithm ManyQueries for the case $n \geq m$ on the GCel from Parsytec. The GCel is a network of T800 transputers as processors, a 2-dimensional mesh as router and Parix as its operation system. In order to implement our algorithm in BSP* style we have realized a library on top of Parix containing the basic routines mentioned above.

First expermiments show, for $n = 1572864$ and $m = 524288$, a speed-up of 49 with 128 processors, where 12288 points and 4096 segments are stored in each processor. Further experiments are in progress, especially with more processors and for the randomized algorithm.

1.6 Organisation of the Paper

In Section 2 we give an outline of the algorithm and introduce some notations, in Section 3 we describe some basic routines which are used by the two Multisearch algorithms presented in Sections 4 and 5.

2 Outline of the Algorithm and Notations

In a preprocessing phase a suitable balanced search tree will be constructed from the input segments. In order to guarantee few communication supersteps and to achieve blockwise communication we choose the search tree to be of high degree (large nodes) and therefore low height.

In the course of the algorithm the queries travel through the search tree along their search paths from the root to their leaves level by level. The algorithm proceeds in rounds, each consisting of a small number of supersteps. The number of rounds corresponds to the height of the search tree. In round i it will be determined which query visits which node on level $i + 1$ of the search tree. In order to obtain an efficient BSP* implementation we have to cope with the following problems.

The first problems concern the initial distribution of the nodes of the search tree among the processors. If $m < n$ an one-to-one-mapping of tree nodes to processors works fine, but if there are more nodes than processors, as in the case of $m > n$ (large search trees), a deterministic distribution causes contention problems: One can always find an input such that only nodes mapped to the same processor are accessed. In order to make contention unlikely we distribute the nodes of the search tree randomly.

This leads to another problem. Random distribution destroys locality and therefore makes it more difficult to communicate in a blockwise fashion. In order to cope with this we use a partially random distribution that preserves some locality properties.

During the travel of the queries through the tree the following problems arise. For each node v and each query q visiting v, the next node to be visited has to be determined. It may occur that some nodes are visited by many queries and that other nodes may only be visited by few queries. Thus a careful load balancing has to be done.

As it might happen that a processor holds many queries visiting different nodes, it can be too expensive to send all the appropriate nodes to that processor. In that case we send the queries to the processors that hold the appropriate nodes.

In order to present the algorithm we need some notation for describing the distribution of the queries among the processors:

For a node v of the search tree we define the *job* at node v to be the set of queries visiting v. *Executing this job* means determining, for each query of the job, which node of the search tree it has to visit next. Let w be a child node of v and let J be the job at node v, then the job at node w is called a *successor job* of J. If J is a job at a node v on level i of the search tree then J is called a *job on level i*.

Let $J_1, \ldots, J_k$ be some jobs on the same level of the search tree and $P_i, \ldots, P_{i+l}$ some consecutive processors. Let the queries from $J_1 \cup \ldots \cup J_k$ be numbered $1, \ldots, n'$, $n' \leq n$, with the first $|J_1|$ jobs numbered $1, \ldots, |J_1|$, and so on. A distribution of the queries of $J_1, \ldots, J_k$ among the processors $P_i, \ldots, P_{i+l}$ is *balanced*, if no P_j gets more than $\frac{n}{p}$ of the queries. It is *ordered*, if the queries in P_j are lower numbered than those in P_{j+1}. The processors holding queries from the same job form a *group*. The first processor of a group is the *group leader*. If the group of a job consists of one processor the job is called *exclusive*, if not it is *shared*. For an example see Figure 2.

Let α, α' be constants with $0 < \alpha' < \alpha < 1$. Further conditions on α, α' can be derived from the analysis of the algorithms. We define *small jobs* to be of size smaller than $(n/p)^\alpha$ and *large jobs* to be of size at least $(n/p)^\alpha$. The nodes of the search tree will have degree $(n/p)^{\alpha'}$, thus a job can have up to $(n/p)^{\alpha'}$ successor jobs.

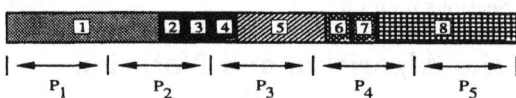

Fig. 2. Jobs $1, \ldots, 8$ are distributed in an ordered and balanced way among processors $P_1, \ldots, P_5$. $1, 5$ and 8 are shared jobs, the others are exclusive ones. Job 1, $(5, 8)$ is held by the processor group $\{P_1, P_2\}$ $(\{P_3, P_4\}, \{P_4, P_5\})$.

3 Communication Routines

In this section we describe and analyse BSP* algorithms for several communication routines: Broadcast, Parallel-Prefix, Distribute, DistributeFQ and Load-Balance. All these are auxiliary routines for the two Multisearch algorithms described in the next two sections. There, these routines will be executed by groups of consecutive processors. These groups can be of any size between two and p processors. We present the communication routines as they are needed for the execution on the group of all p processors $P_1, \ldots, P_p$. Thus the analysis will give an upper bound for the complexity of the routines. The algorithms and their analysis are only briefly sketched in this extended abstract.

3.1 Broadcast

Consider a vector of size d stored in processor P_1. The task is to send this vector to all the other processors.

Algorithm Broadcast: The processors are organized as a binary tree with root P_1. The other processors are called internal processors. P_1 splits the vector in $\lceil d/b \rceil$ packets, each of size at most b. The algorithm proceeds in rounds, each consisting of a communication and a computation superstep. In the i-th round P_1 makes two copies of the i-th packet and sends them to its children processors. Each internal processor that got a packet in the previous round makes two copies of it and sends them two its children. Thus the packets travel through the tree in a pipelined fashion. The algorithm performs $\lfloor \log p \rfloor + \lceil d/b \rceil$ rounds. Each round takes time $\max\{g\lceil b/B \rceil, L\}$ for the communication superstep and time $\max\{\lceil b/B \rceil, L\}$ for the computation supersteps.

Result 1 *Broadcast requires communication time $O((\log p + \lceil d/b \rceil) \cdot \max\{g\lceil b/B \rceil, L\})$ and computation time $O((\log p + \lceil d/b \rceil) \cdot \max\{\lceil b/B \rceil, L\})$. In particular, if $d \geq B \cdot L \log p$ set $b = B \cdot L$ thus communication time $O(g\frac{d}{B})$ and computation time $O(\frac{d}{B})$ is needed.*

3.2 Parallel-Prefix

Let p vectors of size d are given where the i-th vector is stored in the i-th processor. The task of the algorithm *Parallel-Prefix* is to compute, for each $k \in \{1, \ldots, d\}$, the prefix sums of the k-th components of the vectors. The resulting prefix sums are stored in the corresponding processors. As in the broadcast algorithm the processors are organized as a balanced binary tree. We employ the standard parallel prefix algorithm that proceeds in two phases. In the first phase the sums move from the leaves of the tree, in the second from the root to the leaves making some calculations at each level. The vectors are split into $\lceil d/b \rceil$ packets of size at most b in order to proceed in a pipelined fashion as in the algorithm Broadcast.

Result 2 *Parallel-Prefix needs communication time $O((\log p + \lceil d/b \rceil) \cdot \max\{g\lceil b/B \rceil, L\})$ and computation time $O((\log p + \lceil d/b \rceil) \cdot \max\{b, L\})$. In particular, if $d \geq B \cdot L \log p$ set $b = B \cdot L$ thus communication time $O(g\frac{d}{B})$ and computation time $O(d)$ is needed.*

3.3 Distribute and DistributeFQ

Input: A set of at most p large jobs distributed among the p processors in a balanced and ordered way (refer to Section 2 for the notations).

Output: Input jobs of size at most $\frac{n}{p}$ (we have at most p of them) are mapped to different processors. Let r_i be the size of the job which is mapped to processor P_i. Then there is enough space for $\frac{n}{p} - r_i$ additional queries on P_i. We call this value the *gap* of P_i. Input jobs of size larger than $\frac{n}{p}$ are distributed among the processors such that they fill up the gaps. See Figure 3 for an example. Note that afterwards each processor stores queries of at most three different input jobs.

a)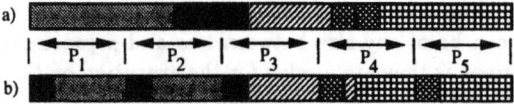

P_1 P_2 P_3 P_4 P_5

b)

Fig. 3. a) shows the situation before and b) after Distribute. Before P_4 holds queries of four jobs. Afterwards P_4 holds queries of 3 jobs.

For multisearch with few queries we need a slightly different version of the algorithm Distribute. We call it *DistributeFQ*.

Input: Small jobs $J_1, \ldots, J_k$ and large jobs $J'_1, \ldots, J'_{k'}$. Each of the small jobs is stored in a processor such that no processor holds more than $(1 + \delta)\frac{n}{p}$ queries of these jobs (δ will be specified later). The large jobs are stored in an ordered way among the processors.

Output: Let r_i be the amount of queries of the small jobs stored in processor P_i. We call the value $(1 + \delta)\frac{n}{p} - r_i$ the gap of P_i. The queries of $J'_1, \ldots, J'_{k'}$ are redistributed such that they fill up the gaps of the processors as in the algorithm Distribute. The small jobs are not moved.

Result 3 *Algorithms Distribute and DistributeFQ need time $O(\frac{n}{p})$ for computation, time $O(g(\frac{n}{pB} + (\frac{n}{p})^{1-\alpha}) + \frac{n}{p})$ for communication, if $\frac{n}{p} \geq \log p \max\{L, g\}$. Further, they need space $O(\frac{n}{p})$ and $O((1 + \delta)\frac{n}{p})$, respectively.*

3.4 Load-Balance

Let J be a large job on a certain level of the search tree and let $J_1, \ldots, J_t$ be the successor jobs of J.

Input: The queries of J are distributed among the processors such that each processor holds at most $\frac{n}{p}$ queries of J. P_1 holds a vector A with the sizes of all successor jobs of J. Further each query is labelled with the number of the successor job it belongs to. A query that belongs to the k-th successor job of J gets the label k, with $1 \leq k \leq (\frac{n}{p})^{\alpha'}$.

Output: The jobs $J_1, \ldots, J_t$ distributed among the processors in an ordered and balanced way. See Figure 4 for an example.

The redistribution specified above can easily be done. If P_1 broadcasts the vector A to the other processors, they have the necessary information in order to calculate the appropriate target processor for each query of J. But if the queries are directly sent to their target processor a problem arises: Many processors could hold only few queries for a certain target procesor, especially less than the block size B. So $\Theta(\frac{n}{p})$ small packets may be sent to a processor. This would require communication time $O(g\frac{n}{p})$ which is too large (if $g = o(B \log n)$). In order to cope with this situation we have to combine these queries to larger packets before we can finally send them to their target processors.

a)

P_1 P_2 P_3 P_4 P_5

b)

Fig. 4. a) shows the situation before and b) after Load-Balance for a group.

Result 4 *Let $0 < \alpha' < 1$, $(\frac{n}{p})^{\alpha'} \geq B \cdot L \log p$, and $B \leq (\frac{n}{p})^{(1-\alpha')/2}$. Then algorithm Load-Balance needs computation time $O(\frac{n}{p})$, communication time $O(g(B(\frac{n}{p})^{\alpha'} + \frac{n}{Bp}))$, and space $O(\frac{n}{p})$.*

4 Multisearch with Many Queries

In this section we show how to do Multisearch for the case $n \geq m$ such that the internal work is almost exactly the same as for the sequential algorithm and the ratio of communication time to computation time is in $o(1)$.

In order to simplify the presentation we consider only the case $m = n$. The case $n > m$ can easily be concluded. The main idea of the search procedure is the following. Construct a balanced search tree over S and let the queries "flow" through the search tree level by level from the root to the leaves. The main problem arises from the fact that many queries can visit a node of the search tree and many nodes are visited per level.

Preprocessing: At most $\lceil n/p \rceil$ consecutive segments lie on each processor as part of the input, they form an *interval*. For each processor P_i we denote the largest segment held by P_i as a *separating segment*. Now we build a balanced binary search tree T

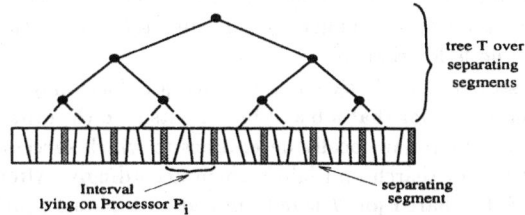

Fig. 5. Tree T above the set of separating segments.

where the leaves are formed by the intervals and the internal nodes are formed by the separating segments, compare Figure 5. We choose the nodes of T to be of degree 2^μ where μ is the largest integer such that $2^\mu - 1 < \lceil (\frac{n}{p})^{\alpha'} \rceil$.

Fact 1 *Let $0 < \alpha' < 1$ be arbitrary, and $B \leq (\frac{n}{p})^{(1-\alpha')/2}$. Then the preprocessing needs computation time $O(\max\{\frac{n}{p}, L\} \log p)$ and communication time $O(\max\{g(\frac{n}{p})^{\alpha'}, L\} \log p)$. The constructed tree T has at most p nodes each of degree $2^\mu \leq \lceil (\frac{n}{p})^{\alpha'} \rceil$. T has $O(\frac{\log p}{\alpha' \log(n/p)})$ levels and each processor stores the segments of at most one interval and one node of T. The data structure needs $O(\frac{n}{p} + (\frac{n}{p})^{\alpha'})$ storage per processor. The root of T is stored on processor P_1.*

Executing Queries: The algorithm has three phases. The first phase is subdivided into rounds. In the i-th round level i of T is considered. During each round only large jobs are executed. Queries of small jobs are directly sent to the processors which hold the appropriate nodes. They are not considered further until Phase 3. The second phase handles jobs at leaf nodes of T after Phase 1. For each such job the correct interval is already computed. Finally, in the last phase the small jobs which have been put aside in the first phase are processed. These can only be small jobs. No two of these need to inspect the same interval, therefore they can be broadcasted to the processors, which store the intervals they have to inspect (in order to find the correct segment) and the processors can compute the correct segments independently.

Algorithm ManyQueries:
Description of Phase 1: Phase 1 proceeds in rounds. In round i the jobs on level i of T will be executed.

Input for round i: Large jobs on level i of the search tree T, distributed in a balanced and ordered way among the processors (For notations compare Section 2).

Output of round i: Large jobs on level $i+1$ of T, distributed in a balanced and ordered way among the processors. Small jobs are directly sent to the processors which hold the appropriate nodes.

Thus we have as input for round i exclusive jobs and we have shared jobs with a group of consecutive processors allocated to each of them. Note that, since we are restricted to large jobs, there are at most $\left(\frac{n}{p}\right)^{1-\alpha}$ jobs mapped exclusively to one processor. This is crucial for the analysis. We now describe how the algorithm executes the input jobs of round i:

Each processor P executes the exclusive jobs which are mapped to it as follows: For each of its exclusive jobs it fetches the nodes from the processors that store them. These can only be $\left(\frac{n}{p}\right)^{1-\alpha}$ many nodes, since there cannot be more exclusive jobs mapped to one processor. Thus, this step is not too expensive. After that each processor determines the successor jobs by means of binary search and stores the large successor jobs ordered in its memory. Successor jobs of smaller size are directly sent to the processors that hold the appropriate node for that job.

A group of processors $P_i, \ldots, P_j$ executes a shared job J as follows: The group leader fetches the appropriate node for that job and broadcasts it to the other group members. Each processor of that group now locally determines to which successor job its queries belong by means of binary search and labels them accordingly. After that the size of each successor job of the shared job J is determined by a parallel prefix computation. With this information the algorithm *Load-Balance* is called which redistributes the queries of J such that afterwards the successor jobs of J are distributed in a balanced and ordered way among the processors $P_i, \ldots P_j$. After that small successor jobs are directly sent to the processors that hold the appropriate node for that job.

We describe the algorithm as if a processor was only involved in either the execution of one shared job or the execution of some exclusive jobs. In fact it can be involved in the execution of up to two shared jobs and up to $\left(\frac{n}{p}\right)^{1-\alpha}$ exclusive ones (see Section 2). It is not difficult to schedule the instructions such that the performance is not affected. The algorithm Load-Balance ensures that each processor has to perform local binary search on at most $\frac{n}{p}$ queries in each round. Here comes the algorithm in detail. The input for the first round is the job at the root node of T consisting of all queries.

1. For each large job (exclusive or shared) the corresponding node of T has to be fetched. Therefore each group leader fetches the node of T and broadcasts it to the processors of its group. Additionally each processor holding exclusive jobs fetches for each exclusive job the corresponding node of T.

2. Each processor determines by means of binary search (on the fetched nodes) for each of its queries which node to visit next. The queries visiting the same node in the next level belong to the same successor job and are marked with the same label.

3. The processors of each group compute the size of the new successor jobs by means of parallel prefix on vectors, where the i-th component of the vector corresponds to the number of queries which visit the i-th child node of the current fetched node of T. The group leader knows the size of each of there successor jobs afterwards.

4. The successor jobs of every shared job are redistributed in a balanced and ordered way among the processors of the group by executing procedure Load-Balance. The processors holding exclusive jobs compute the reorganisation sequentially (by a bucket sort approach).

5. Each processor sends the queries of each small successor job it holds to the processor which holds the node of T that successor job wants to visit next. If a group has reached a leaf of T it leaves this phase. Goto Step 1.

Description of Phase 2: In this phase the large jobs which have reached the leaves of T after Phase 1 are processed. Remember T has p leaves, therefore the queries are partitioned into at most p jobs. Unfortunately a processor can hold queries of up to $(\frac{n}{p})^{1-\alpha}$ large jobs, since a large job can be as small as $(\frac{n}{p})^{\alpha}$.

1. The processors redistribute the queries of the jobs by means of Distribute such that each processor holds queries of at most three jobs.
2. Each processor which has received an exclusive job and each group leader of a shared job fetches the corresponding interval. Additionally each group leader initiates a broadcast of the interval to the processors of its group.
3. Eventually each processor performs binary search for each query on the appropriate interval. Note that a processor has to store at most three intervals.

Description of Phase 3: In this phase the small jobs are considered. Remember that small jobs are sent (during Phase 1) to the processors that hold the nodes of T they have to visit next.

1. Each processor which has received a small job broadcasts it to the processors storing the intervals reachable from the corresponding node of T.
2. Each processor determines for the received queries the correct segment by means of binary search on the segments of its interval.

Theorem 1. *Let $(\frac{n}{p})^{\alpha'} \geq B \cdot L \log p$. For $0 < \alpha' < 1$ there is a constant $\eta > 0$ such that algorithm ManyQueries needs time $(1+o(1))\frac{n}{p}\log n + O(\frac{\log p}{\alpha' \log(n/p)}\frac{n}{p})$ for computation, $O(g\frac{\log p}{\alpha' \log(n/p)}\frac{n}{pB} + g \cdot \frac{n}{pB})$ for communication and $O(\frac{n}{p})$ space per processor, if $B = (\frac{n}{p})^{\eta}$. Thus algorithm ManyQueries is 1-optimal*

- *for $n \geq p^{1+\epsilon}$, $\epsilon > 0$ arbitrary, if $g = o(B \log n)$, and*
- *for $n \geq p \log^2 p$, if $g = o(B \log \log n)$.*

Proof of Theorem 1:
Analysis of Phase 1: Let h be the number of levels of T, i.e. $h = O(\log p/(\alpha' \log(n/p)))$. The Steps 1 to 6 are repeated at most h times.

   Step 1: Each processor holds queries of at most $O((\frac{n}{p})^{1-\alpha})$ large jobs. Therefore each processor has to fetch $O((\frac{n}{p})^{1-\alpha})$ nodes. This can be realized in two communication supersteps. In the first one the requests for nodes will be sent to the processors that hold them. Thus an $O((\frac{n}{p})^{1-\alpha})$-relation with packets of size 1 is realized. In the second superstep the nodes will be sent to the requesting processors. Thus an $O((\frac{n}{p})^{1-\alpha})$-relation with packets of size $O((\frac{n}{p})^{\alpha'})$ is routed. Since $(\frac{n}{p})^{\alpha'} \geq B \cdot L \log p$ these two supersteps need time $O(g \cdot (\frac{n}{p})^{1-\alpha} \cdot (\frac{n}{p})^{\alpha'}/B)$. The broadcast needs time $O(g(\frac{n}{p})^{\alpha'}/B)$ by Result 1.

   Step 2: Binary search is performed during h rounds. For each query at most $\log p + 1$ comparisons are made, $\lceil \frac{n}{p} \rceil$ queries are handled by each processor therefore at most $\lceil \frac{n}{p} \rceil (\log p + 1)$ comparisons are made during Phase 1.

   Step 3: By Result 2, if $(\frac{n}{p})^{\alpha'} \geq B \cdot L \log p$, it needs communication time $O(g(\frac{n}{p})^{\alpha'}/B)$ and computation time $O((\frac{n}{p})^{\alpha'})$ for the parallel prefix on vectors of size $O((\frac{n}{p})^{\alpha'})$.

Step 4: By Result 4, if $(\frac{n}{p})^{\alpha'} \geq B \cdot L \log p$ and $B \leq (\frac{n}{p})^{(1-\alpha')/2}$, Load-Balance needs communication time $O(g(B(\frac{n}{p})^{\alpha'} + \frac{n}{Bp}))$ and computation time $O(\frac{n}{p})$. For exclusive jobs the reorganisation is done sequentially in time $O(s + (\frac{n}{p})^{\alpha'})$, where $s \leq (n/p)$ is the total size of the exclusive jobs on a processor.

Step 5: The queries of small successor jobs are directly sent to the nodes they have to visit next. A job can have at most $\lceil (\frac{n}{p})^{\alpha'} \rceil$ different small successor jobs. A processor has at most $(\frac{n}{p})^{1-\alpha}$ exclusive jobs therefore a processor has to send $O((\frac{n}{p})^{1-\alpha+\alpha'} + \frac{n}{pB})$ packets. A processor receives small jobs of size $O((\frac{n}{p})^{\alpha})$, therefore it has to receive $O((\frac{n}{p})^{\alpha}/B)$ packets. Thus Step 5 can be done in one communication superstep that needs time $(g((\frac{n}{p})^{1-\alpha+\alpha'} + \frac{n}{pB}))$.

Thus Phase 1 needs computation time $\leq \frac{n}{p}\log(p) + O(\frac{\log p}{\alpha' \log(n/p)} \cdot \frac{n}{p} + \frac{n}{p})$ and communication time $O(g \cdot \frac{\log p}{\alpha' \log(n/p)} \cdot (B(\frac{n}{p})^{\alpha'} + \frac{n}{Bp} + (\frac{n}{p})^{1-\alpha+\alpha'}))$.

Analysis of Phase 2: Step 1: By Result 3 it requires time $O(g(\frac{n}{pB} + (\frac{n}{p})^{1-\alpha}) + \frac{n}{p})$ for communication and $O(\frac{n}{p})$ for computation, if $\frac{n}{p} \geq \log p \max\{L, g\}$. *Step 2:* By Result 1 it requires communication time $O(g\frac{n}{Bp})$ and $O(\frac{n}{p})$ computation time, if $\frac{n}{p} \geq B \cdot L \log p$. *Step 3:* It requires computation time $\lceil \frac{n}{p} \rceil \log \lceil \frac{n}{p} \rceil \leq \frac{n}{p}\log n - \frac{n}{p}\log p + O(\frac{n}{p})$.

Analysis of Phase 3: By Result 1 Step 1 needs communication time $O(g(\frac{n}{p})^{\alpha}/B)$ and computation time $O(\frac{n}{p})$, if $(\frac{n}{p})^{\alpha} \geq B \cdot L \log p$. Step 2 needs computation time $\lceil (\frac{n}{p})^{\alpha} \rceil \log \lceil \frac{n}{p} \rceil = O(\frac{n}{p})$

Adding up the communication and computation times for Phases 1, 2, 3 yields the time bounds claimed in Theorem 1. Note that from the analysis we get conditions on the constant α. The value of η in Theorem 1 depends on α and α'. $\qquad\square$

5 Multisearch with Few Queries

In the sequel only few queries are asked: we allow now $n < m \leq 2^p$, rather than $n \geq m$. Recall that the sequential time needed is $n \log m$ in the worst case. It is easily seen that algorithm ManyQueries is far from optimal in this case, e.g. if all queries happen to belong to segments stored in the same processor. Therefore we need a slightly different search tree and a new way to distribute its nodes among the processors.

Preprocessing: In contrast to ManyQueries we now organize the m segments in a search tree T^* of size $m/(\frac{n}{p})^{\alpha'}$ with degree $(\frac{n}{p})^{\alpha'}$ and depth $\log m/\alpha' \log(\frac{n}{p})$, with $\frac{1}{2} < \alpha' < 1$ (see Section 2). The segments within each node are organized as a binary search tree. They will always be processed sequentially.

As m can be very large the number of nodes of T^* can become much larger than p. Thus, for every distribution of its nodes among the processors, there is a bad set of queries which causes high contention. The reason is that many of the queries have to travel through different nodes mapped to the same processor. A standard technique to reduce contention is randomization: If the nodes are randomly distributed among the processors, the above event becomes very unlikely for arbitrary sets of queries. For our purposes, however, a naive random distribution is not suitable, because the neighbours of the nodes stored in a processor are likely to be distributed almost injectively. Thus the communication to be executed when the queries travel through the tree is fine grained, blockwise communication is not possible. Therefore our approach uses a randomized distribution which ensures that children of nodes that are mapped to the same processor are distributed among few processors only. This is done by the following algorithm. It works for any tree T with degree d and depth h.

Algorithm Build-Up:

$i = 0$: The root will be placed on an arbitrary processor.

$i > 0$: If level i of T has at most p nodes, they are mapped to the processors such that each processor gets at most one node. If level i has more than p nodes, they are distributed as follows: Let R be the subset of nodes on level $i - 1$ of T that have been placed on processor P. P chooses a random set of $d^{1+\epsilon}$ processors (neighbour processors) and distributes the children of nodes of R randomly among these neighbour processors. For our algorithm ϵ has to be chosen such that $\alpha' \cdot (1 + \epsilon) < 1$.

The following is easy to check:

Fact 2 *The algorithm Build-Up needs time $O(Lh + g(\frac{m}{p}))$ and $O(\frac{m}{p})$ space per processor.*

The next lemma captures properties of the distribution produced by algorithm Build-Up that are crucial for the use of blockwise communication.

Lemma 2. *a) Consider a tree T with degree d and depth h distributed among p processors by algorithm Build-Up. Then for every $\delta > 0$ and ζ with $0 < \zeta < 1$, there exists an $\eta > 0$ such that the following holds: Fix l, $l \leq M$, nodes $v_1, \ldots, v_l$ of a level of T and give them non-negative weights $g_1, \ldots g_l$ each at most $(\frac{M}{p})^\zeta$, such that the total weight of $v_1, \ldots, v_l$ is at most M. The probability that the total weight of nodes placed on the same processor exceeds $(1 + \delta)\frac{M}{p}$ is less than $4ph/2^{d^\eta}$.*

b) For each processor P_i, the parents of the nodes stored in P_i are distributed among $O(d^{1+\epsilon})$ processors, with probability at least $1 - 2^{-d}$.

The proof can be found in the full version of this paper [4].

Executing Queries:

The algorithm FewQueries proceeds similar to the first phase of algorithm ManyQueries with the following exceptions:

- It works on the data structure generated by algorithm Build-Up that represents T^* rather than on the data structure used by algorithm ManyQueries.
- Many small jobs may be generated in this setting. Fetching nodes for each of them is too expensive. Therefore we send these jobs to the processors that hold the appropriate nodes. The treatment of small jobs is not postponed to a later phase as in algorithm ManyQueries. Small jobs are handled together with large ones.

In the i-th round of algorithm FewQueries we have the following input and output:

Input for round i: The jobs on level i are distributed among the processors as follows: Each small job on level i is placed on the processor that holds the appropriate node for that job. Let r_i be the number of queries of these small jobs that are stored on processor P_i. Lemma 2 guarantees that r_i is not larger than $(1 + \delta)\frac{n}{p}$. The value $\frac{n}{p} - r_i$ is called the gap of P_i. The large jobs on level i are distributed in an ordered way among the processors such that they fill up the gaps.

Output for round i: Each small job on level $i + 1$ is placed on the processor that holds the appropriate node for that job. Large jobs on level $i + 1$ are distributed in an ordered way among the processors such that they fill up the gaps.

In each round i the small jobs of level i are sent to processors holding the appropriate nodes for these jobs. The new data structure guarantees that a processor has to send the small jobs only to $(\frac{n}{p})^{\alpha'(1+\epsilon)}$ neighbour processors. Remember that ϵ is chosen such that $\alpha'(1 + \epsilon) < 1$. Thus jobs can be combined in order to form large messages. This is crucial in order to achieve blockwise communication.

As with the algorithm ManyQueries in each round exclusive and shared input jobs have to be executed. Here we may have much more exclusive jobs mapped to one processor, but the nodes for small exclusive jobs are already stored on the same processor. They need not to be fetched from other processors.

Algorithm FewQueries:

If h is the depth of T^* then the algorithm makes h iterations. In each iteration it executes the following 6 steps:

1. For each shared job the group leader fetches the appropriate node of T^*. Each processor fetches for each of its large exclusive jobs the appropriate node of T^* (Small jobs are already placed together with their appropriate nodes on a processor). Each group leader broadcasts the fetched node to the other group members.

2. Each processor P_i determines by means of binary search for each of its queries which node to visit next. The queries visiting the same node in the next level belong to the same successor job and are marked with the same label. It is guaranteed by Step 6 that P_i has to perform binary search for at most $(1+\delta)\frac{n}{p}$ queries.

3-5. These steps are the same as Step 3-5 in Phase 1 of algorithm ManyQueries.

6. The processors execute algorithm *DistributeFQ*. This guarantees that the queries of large jobs of level $i+1$ are distributed ordered among the processors such that they fill up the gaps that have been left by the small jobs of level $i+1$.

Theorem 3. *Let* $m \le 2^p$, $\frac{n}{p} \ge \log^2 p$ *and* $(\frac{n}{p})^{\alpha'} \ge B \cdot L \log p$. *For* $\frac{1}{2} < \alpha' < 1$ *and* $\delta > 0$ *there are constants* $\beta, \eta > 0$, *such that algorithm FewQueries needs time* $(1 + \delta + o(1))\frac{n}{p}\log m + O(\frac{\log m}{\alpha' \log(n/p)}\frac{n}{p})$ *for computation,* $O(g\frac{\log m}{\alpha' \log(n/p)}(\frac{n}{Bp}))$ *for communication and* $O(\frac{n}{p} + \frac{m}{p})$ *space per processor with probability at least* $1 - p^{-(\log^\beta p)}$, *if* $B = (\frac{n}{p})^\eta$. *Thus algorithm FewQueries is* $(1 + \delta)$-*optimal*

- *for* $m \le 2^p$, $n = p^{1+\zeta}$ *with* $\zeta > 0$, *if* $g = o(B \log n)$ *and*
- *for* $m \le 2^p$ *and* $n = p \log^2 p$, *if* $g = o(B \log \log n)$.

Proof of Theorem 3:

In the following we analyse the time bounds for each round i:

Step 1: Each processor fetches at most $(\frac{n}{p})^{1-\alpha}$ nodes. Mark each of these nodes with weight $(\frac{n}{p})^\alpha$. Lemma 2a) guarantees that Build-Up distributes these nodes such that the total weight of nodes placed on each processor is at most $(1+\delta)\frac{n}{p}$ with probability $1 - p^{-(\log p)^\beta}$ for a certain $\beta > 0$. Thus each processor gets $O((\frac{n}{p})^{1-\alpha})$ requests for nodes of T^*. Therefore the nodes can be fetched in two communication supersteps. In the first one requests will be sent to the processors. In the second superstep the nodes are sent to the requesting processors. The first superstep realizes an $(\frac{n}{p})^{1-\alpha}$-relation of packets of size 1, the second realizes a $(\frac{n}{p})^{1-\alpha}$-relation of packets of size $(\frac{n}{p})^\alpha$. Thus this step has the same time bounds as Step 1 of Phase 1 of algorithm ManyQueries.

Step 2: At the start of each round each processor holds at most $(1+\delta)(\frac{n}{p})$ queries. This is guaranteed by Lemma 2a) and the execution of the routine DistributeFQ at the end of each round. Therefore in Step 2 of each round every processor performs at most $(1+\delta)\frac{n}{p}\alpha' \log(\frac{n}{p})$ comparisons.

Step 3-4: The analysis for these steps is the same as for Steps 3-4 of Phase 1 of algorithm ManyQueries.

Step 5: Each processor sends small jobs with total weight at most $(1+\delta)(\frac{n}{p})$ to at most $(\frac{n}{p})^{\alpha'(1+\epsilon)}$ neighbour processors which hold the appropriate nodes. Mark these nodes with the size of the respective jobs. By Lemma 2a) and b) we know that each processor receives jobs of total weight at most $(1 + \delta)(\frac{n}{p})$ from $O(d^{1+\epsilon})$ neighbour processors with probability $1 - p^{-(\log p)^{\beta}}$ for an appropriate $\beta > 0$. Note that in order to achieve this probability we need $\alpha' > \frac{1}{2}$. Each processor needs computation time $O(\frac{n}{p})$ to combine the queries to large packets. Each processor sends and receives $O(\frac{n}{pB} + (\frac{n}{p})^{\alpha'(1+\epsilon)})$ packets of size B in one communication superstep. Thus the communication time is $O((\frac{n}{p \cdot B})g)$ for Step 5, if $B \leq (\frac{n}{p})^{1-\alpha'(1+\epsilon)}$.

Step 6: The complexity is the same as for the algorithm DistributeFQ (see Section 3).

Since the depth of T^* and therefore the number of iterations is $\frac{\log m}{\alpha \log(n/p)}$, we reach the resource bounds stated by Theorem 3. Note that from the analysis the conditions on α can be derived. The values of β and η in Theorem 3 depend on the values of α and α'. □

References

1. A. Aggarwal, A.K. Chandra and M. Snir, On communication latency in PRAM computations, Proc. ACM Symp. on Parallel Algorithms and Architectures, 1989, 11-21.
2. M.J. Atallah, F. Dehne, R. Miller, A. Rau-Chaplin and J.-J. Tsay, Multisearch Techniques for Implementing Data Structures on a Mesh-Connected Computer, Proc. ACM Symp. on Parallel Algorithms and Architectures, 1991, 204–214.
3. M.J. Atallah and A. Fabri, On the Multisearching Problem for Hypercubes, Parallel Architectures and Languages Europe, 1994.
4. A. Bäumker, W. Dittrich, F. Meyer auf der Heide, Truly efficient parallel algorithms: c-optimal multisearch for an extension of the BSP model, Technical Report, Universität-Gesamthochschule Paderborn, Department of Mathematics and Computer Science, to appear.
5. R. H. Bisseling, W. F. McColl, Scientific computing on bulk synchronous parallel architectures, Proc. 13th IFIP World Computer Congress, Volume 1, 1994.
6. D. Culler, R. Karp, D. Patterson, A. Sahay, K.E. Schauser, E. Santos, R. Subramonian and T. von Eicken, LogP: Towards a Realistic Model of Parallel Computation, Proc. ACM SIGPLAN Symposium on Principles and Practice of Parallel Programming, 1993.
7. F. Dehne, A. Fabri, A. Rau-Chaplin, Scalable Parallel Computational Geometry for Coarse Grained Multicomputers, Proc. ACM Conf. on Comp. Geometry, 1993.
8. A.V. Gerbessiotis and L. Valiant, Direct Bulk-Synchronous Parallel Algorithms, Journal of Parallel and Distributed Computing, 1994.
9. W. Hoeffding, Probability inequalities for sums of bounded random variables, American Statistical Association Journal, 1963, 13–30.
10. C.P. Kruskal, L. Rudolph and M. Snir, A complexity theory of efficient parallel algorithms, Proc. 15th Int. Coll. on Automata, Languages, and Programming, 1988, 333–346.
11. W F McColl, The BSP Approach to Architecture Independent Parallel Programming, To appear in CACM on General Purpose Practical Models of Parallel Computation, 1995.
12. Abhiram Ranade, Maintaining dynamic ordered sets on processor networks, Proc. of the 4th ACM Symp. on Parallel Algorithms and Architectures, 1992, 127–137.
13. J.H. Reif and S. Sen, Randomized Algorithms for Binary Search and Load Balancing on Fixed Connection Networks with Geometric Applications, SIAM J. Comput., Vol. 23, No. 3, June 1994, 633–651.
14. L. Valiant, A Bridging Model for parallel Computation, Communications of the ACM, August 1994, Vol. 33, No. 8.

Optimal Parallel Shortest Paths
in Small Treewidth Digraphs *

SHIVA CHAUDHURI and CHRISTOS D. ZAROLIAGIS

Max-Planck-Institut für Informatik, Im Stadtwald, D-66123 Saarbrücken, Germany
E-mail: {shiva, zaro}@mpi-sb.mpg.de

Abstract. We consider the problem of preprocessing an n-vertex digraph with real edge weights so that subsequent queries for the shortest path or distance between any two vertices can be efficiently answered. We give parallel algorithms for the EREW PRAM model of computation that depend on the *treewidth* of the input graph. When the treewidth is a constant, our algorithms can answer distance queries in $O(\alpha(n))$ time using a single processor, after a preprocessing of $O(\log^2 n)$ time and $O(n)$ work, where $\alpha(n)$ is the inverse of Ackermann's function. The class of constant treewidth graphs contains outerplanar graphs and series-parallel graphs, among others. To the best of our knowledge, these are the first parallel algorithms which achieve these bounds for any class of graphs except trees. We also give a dynamic algorithm which, after a change in an edge weight, updates our data structures in $O(\log n)$ time using $O(n^\beta)$ work, for any constant $0 < \beta < 1$. Moreover, we give an algorithm of independent interest: computing a shortest path tree, or finding a negative cycle in $O(\log^2 n)$ time using $O(n)$ work.

1 Introduction

Finding shortest paths in digraphs is a fundamental problem in network optimization [3]. Given an n-vertex digraph G with real edge weights, the shortest paths problem asks for paths of minimum weight between vertices in G. In the single-source problem we seek such paths from a specific vertex to all other vertices and in the all-pairs shortest paths (apsp) problem we seek such paths between every pair [18].

For general digraphs the best parallel algorithm for the apsp problem takes $O(\log^2 n)$ time using $O(n^3)$ work on an EREW PRAM [12]. In the case of planar digraphs there is an $O(\log^4 n)$-time, $O(n^2)$-work EREW PRAM algorithm [10]. An apsp algorithm must output paths between $\Omega(n^2)$ vertex pairs and thus requires this much work and space. For sparse digraphs a more efficient approach is to preprocess the digraph so that subsequently, *queries* can be efficiently answered. A query specifies two vertices and a *shortest path query* asks for a minimum weight path between them, while a *distance query* only asks for

* This work was partially supported by the EU ESPRIT Basic Research Action No. 7141 (ALCOM II).

the weight of such a path. For example, for *outerplanar* digraphs, it was shown in [11] that after preprocessing requiring $O(\log n)$ time and $O(n \log n)$ work on a CREW PRAM, a distance query is answered in $O(\log n)$ time using a single processor and a shortest path query in $O(\log n)$ time using $O(L + \log n)$ work (where L is the number of edges of the reported path). In [11] it is also shown how distance queries in planar digraphs can be answered in $O(\log n + \log^2 q)$ time using $O(\log n + q)$ work, after polylog-time and $O(n \log n \log^* n + q^{1.5})$-work preprocessing on a CREW PRAM. These latter bounds are given in terms of a minimum number of faces q that collectively cover all vertices of the planar digraph. Note that q varies from 1 (outerplanar digraph) up to $\Theta(n)$.

The study of graphs using the *treewidth* as a parameter was pioneered by Robertson and Seymour [16, 17] and continued by many others (see e.g. [5, 7]). Informally, the treewidth is a measure of how close the structure of the graph is to a tree (see Section 2 for a formal definition). Graphs of treewidth t are also known as partial t-trees. These graphs have at most tn edges. Classifying graphs based on treewidth is useful because diverse properties of graphs can be captured by a single parameter. For instance, the class of graphs of bounded treewidth includes outerplanar graphs, series-parallel graphs, graphs with bounded bandwidth and cutwidth and many other classes [5, 7]. Thus, giving efficient algorithms parameterized by treewidth is an important step in the development of better algorithms for many natural classes of sparse graphs.

In this paper we consider the problem of preprocessing a digraph of small treewidth in parallel, so that afterwards, queries can be efficiently answered using a single processor. We also consider the dynamic version of the problem, where edge weights may change. In [9] sequential algorithms are given that, for digraphs of constant treewidth, after $O(n)$ time preprocessing answer a distance (resp. shortest path) query in $O(\alpha(n))$ (resp. $O(L\alpha(n))$) time[2]. After a change in an edge weight, the algorithm updates the data structure in $O(n^\beta)$ time, for any constant $0 < \beta < 1$. The main contribution of this paper is an algorithm that achieves optimal parallelization, on the EREW PRAM, of the above results. For digraphs of constant treewidth, after $O(\log^2 n)$ time and $O(n)$ work preprocessing our algorithm answers a distance (resp. shortest path) query in $O(\alpha(n))$ (resp. $O(L\alpha(n))$) time using a single processor. Updates can be performed in $O(\log n)$ time using $O(n^\beta)$ work for any constant $0 < \beta < 1$. This improves all previous parallel results for this class of graphs. Moreover, it improves the results in [11] for outerplanar digraphs in many ways: it improves the preprocessing and distance query bounds, it runs on the weakest PRAM model and it applies to a larger class of graphs. We note that the time bottleneck in preprocessing is the computation of the *tree-decomposition* (see Section 2) of the input graph. If an explicit tree-decomposition of the graph is also provided with the input, then the preprocessing time is $O(\log n)$.

As in [9], we give a tradeoff between the preprocessing work and the query time. For bounded treewidth digraphs, after $O(n I_k(n))$ preprocessing, we can answer queries in $O(k)$ time, for $1 \leq k \leq \alpha(n)$. $I_k(n)$ is a function that decreases

---

[2] $\alpha(n)$ is the inverse of Ackermann's function [1] and is a very slowly growing function.

rapidly with k (see Section 3). In particular $I_1(n) = \lceil \log n \rceil$ and $I_2(n) = \log^* n$.

A solution to the single-source problem consists of a *shortest path tree* rooted at a given vertex. A shortest path tree exists iff there is no negative weight cycle in the graph. In parallel computation, the best algorithm for constructing a shortest path tree (or finding a negative cycle) in a general digraph G takes as much time as computing apsp in G [12]. Some improvements have been made for outerplanar [11] and planar digraphs [10] with no negative cycles. In those papers, a shortest path tree can be computed in $O(\log^2 n)$ time, after a preprocessing of the input digraph. The preprocessing work of [10] is $O(n^{1.5})$ on an EREW PRAM, while the preprocessing work in [11] is $O(n \log n)$ on a CREW PRAM. Even with randomization allowed, and the weights restricted to being positive integers, for planar digraphs, the best polylog-time algorithm uses n processors (and hence $\Omega(n \log n)$ work) on an EREW PRAM. Although, on a CREW PRAM, a negative cycle in an outerplanar digraph can be found in $O(\log n \log^* n)$ time and $O(n)$ work, this algorithm does not construct the shortest path tree [14]. Hence, the work for finding a shortest path tree in polylog-time is $\Omega(n \log n)$, even for the case of outerplanar digraphs.

We give an algorithm to construct a shortest path tree (or find a negative cycle) in digraphs of constant treewidth that runs on an EREW PRAM in $O(\log^2 n)$ time using $O(n)$ work (Section 3). If a tree-decomposition is also provided with the input, then the algorithm runs in $O(\log n)$ time. To the best of our knowledge, this is the first deterministic parallel algorithm for the shortest path tree problem that achieves $O(n)$ work.

Our algorithms start by computing a tree-decomposition of the input digraph G. The tree decomposition of a graph with constant treewidth can be computed in $O(\log^2 n)$ time using $O(n)$ work on an EREW PRAM [8]. The main idea behind our algorithms is the following. We define a certain value for each node of the tree-decomposition of G, and an associative operator on these values. We then show that the shortest path problem reduces to computing the product of these values along paths in the tree-decomposition. (A similar idea was used in [2], to show that computing shortest paths reduces to computing the product of certain elements in a closed semiring.) Parallel algorithms to compute the product of node values along paths in a tree are given in [4]. Our preprocessing vs. query bound trade-off arises from a similar trade-off in [4]. The dynamization of our data structures is based on the above ideas and on a graph equipartitioning result which is of independent interest. We note that a similar approach is used in [9], however, the parallel algorithms presented in this paper require substantially different techniques.

2 Preliminaries

In this paper, we will be concerned with finding shortest paths or distances between vertices of a directed graph. Thus, we assume that we are given an *n-vertex weighted digraph* G, i.e. a digraph $G = (V(G), E(G))$ and a weight function $wt : E(G) \longrightarrow \mathbb{R}$. We call $wt(u, v)$ the *weight* of the edge $\langle u, v \rangle$. The weight of a path in G is the sum of the weights of the edges on the path. For $u, v \in V(G)$, a *shortest path* in G from u to v is a path whose weight is minimum among all paths from u to v. The *distance* from u to v, written as $\delta(u, v)$ or $\delta_G(u, v)$, is the weight of a shortest path from u to v in G. A cycle in G is a (simple) path starting and ending at the same vertex. If the weight of a cycle in G is less than zero, then we will say that G contains a *negative cycle*. It is well-known [18] that shortest paths exist in G, iff G does not contain a negative cycle.

For a subgraph H of G, and vertices $x, y \in V(H)$, we shall denote by $\delta_H(x, y)$ the distance of a shortest path from x to y in H. A *shortest path tree* rooted at $v \in V(G)$, is a spanning tree such that $\forall w \in V(G)$, the tree path from v to w is a shortest path in G from v to w.

Let G be a (directed or undirected) graph and let $W \subseteq V(G)$. Then by $G[W]$ we shall denote the subgraph of G induced by W. Let V_1, V_2 and S be disjoint subsets of $V(G)$. We say that S is a *separator for V_1 and V_2*, or that S *separates V_1 from V_2*, iff every path from a vertex in V_1 (resp. V_2) to a vertex in V_2 (resp. V_1) passes through a vertex in S. Let H be a subgraph of G. A *cut-set for H* is a set of vertices $C(H) \subseteq V(H)$, whose removal separates H from the rest of the graph.

Often, we will want to focus on a subgraph induced by a subset of the vertices of a graph, however, we would like the distances between vertices in this subgraph to be the same as in the original graph. Let H be a digraph, with V_1, V_2 and U a partition of $V(H)$ such that U is a separator for V_1 and V_2. Let H_1 and H_2 be subgraphs of H such that $V(H_1) = V_1 \cup U$, $V(H_2) = V_2 \cup U$ and $E(H_1) \cup E(H_2) = E(H)$. We say that H_1' is a graph obtained by *absorbing H_2 into H_1*, if H_1' is obtained from H_1 by adding edges $\langle u, v \rangle$, with weight $\delta_{H_2}(u, v)$ or $\delta_H(u, v)$, for each pair $u, v \in U$. (In case of multiple edges, retain the one with minimum weight.) The following lemma, proved in [9], shows that absorbing a subgraph into another preserves distances.

Lemma 1. *Let H_1 and H_2 be subgraphs of H and let H_1' be obtained by absorbing H_2 into H_1. Then, for all $x, y \in V(H_1')$, $\delta_{H_1'}(x, y) = \delta_H(x, y)$.*

A *tree-decomposition* of a (directed or undirected) graph G is a pair (X, T) where $T = (V(T), E(T))$ is a tree and X is a family $\{X_i | i \in V(T)\}$ of subsets of $V(G)$, such that $\cup_{i \in V(T)} X_i = V(G)$ and also the following conditions hold:

- (*edge mapping*) $\forall (v, w) \in E(G)$, there exists an $i \in V(T)$ with $v \in X_i$ and $w \in X_i$.
- (*continuity*) $\forall i, j, k \in V(T)$, if j lies on the path from i to k in T, then $X_i \cap X_k \subseteq X_j$, or equivalently: $\forall v \in V(G)$, the nodes $\{i \in V(T) | v \in X_i\}$ induce a connected subtree of T.

The *treewidth* of a tree-decomposition is $\max_{i \in V(T)} |X_i| - 1$. The treewidth of G is the minimum treewidth over all possible tree-decompositions of G.

Fact 2. *[8] Given a constant $t \in \mathbb{N}$ and an n-vertex graph G, there exists an EREW PRAM algorithm, running in $O(\log^2 n)$ time using $O(n)$ work, which tests whether G has treewidth at most t and if so, outputs a tree-decomposition (X, T) of G with treewidth at most t.*

Fact 3. *[6, 8] Given a constant $t \in \mathbb{N}$ and a tree-decomposition of treewidth t of an n-vertex graph G, we can compute a rooted, binary tree-decomposition of G with depth $O(\log n)$ and treewidth at most $3t + 2$, in $O(\log n)$ time using $O(n)$ work on an EREW PRAM.*

We shall call the tree-decomposition found in Fact 3 *balanced*. Given a tree-decomposition of G, we can easily find separators in G, as the following proposition shows.

Proposition 4. *[17] Let G be a graph, (X, T), its tree-decomposition, $e = (i, j) \in E(T)$ and T_1 and T_2 the two subtrees obtained by removing e from T. Then $X_i \cap X_j$ separates $\cup_{m \in V(T_1)} X_m$ from $\cup_{m \in V(T_2)} X_m$.*

3   The Static Data Structures

For a function f let $f^{(1)}(n) = f(n)$; $f^{(i)}(n) = f(f^{(i-1)}(n))$, $i > 1$. Define $I_0(n) = \lceil \frac{n}{2} \rceil$ and $I_k(n) = \min\{j \mid I_{k-1}^{(j)}(n) \leq 1\}$, $k \geq 1$. The functions $I_k(n)$ decrease rapidly as k increases, in particular, I_1 behaves like $\log n$ and I_2 like $\log^* n$. Define $\alpha(n) = \min\{j \mid I_j(n) \leq 1\}$. The following was proved in [4].

Fact 5. *[4] Let $\bullet$ be an associative operator defined on a set S, such that for $x, y \in S$, $x \bullet y$ can be computed in $O(m)$ time and $O(w)$ work. Let T be a tree with n nodes such that each node is labelled with an element from S. Then: (i) for each $k \geq 1$, after $O(m \log n)$-time and $O(wnI_k(n))$-work preprocessing on an EREW PRAM, the composition of labels along any path in the tree can be computed in $O(wk)$ time by a single processor; and (ii) after $O(m \log n)$-time and $O(wn)$-work preprocessing on an EREW PRAM, the composition of labels along any path in the tree can be computed in $O(w\alpha(n))$ time by a single processor.*

The main idea of our algorithm is, as in [9], to reduce shortest path computations to the above problem. We define a certain value for each node of the tree-decomposition of G, as well as an associative operator on these values. We then show that shortest path computation reduces to computing products of those values along paths in the tree-decomposition. Then the rest follows by the above Fact.

Call a tuple (a, b, c) a *distance tuple* if a, b are arbitrary symbols and $c \in \mathbb{R}$. For two distance tuples, $(a_1, b_1, c_1), (a_2, b_2, c_2)$, define their product $(a_1, b_1, c_1) \otimes (a_2, b_2, c_2) = (a_1, b_2, c_1 + c_2)$ if $b_1 = a_2$ and as nonexistent otherwise.

For a set of distance tuples, M, define $\text{minmap}(M)$ to be the set $\{(a, b, c) : (a, b, c) \in M$ and $\forall (a', b', c') \in M$ if $a' = a, b' = b$, then $c \leq c'\}$, i.e. among

all tuples with the same first and second components, minmap retains only the
tuples with the smallest third component.

Let M_1 and M_2 be sets of distance tuples. Define the operator $\circ$ by $M_1 \circ M_2 = \text{minmap}(M)$, where $M = \{x \otimes y : \ x \in M_1, \ y \in M_2\}$. It is not hard to show that
$\circ$ is an associative operator.

Let G be a digraph with real edge weights. Note that in the above definition,
if M_1 and M_2 have tuples of the form (a, b, x) where $a, b \in V(G)$ and x is the
weight of a path from a to b, then $M_1 \circ M_2$ computes tuples (a, b, y) where y is
the (shortest) distance from a to b using only the paths represented in M_1 and
M_2.

For $X, Y \subseteq V(G)$, not necessarily distinct, define $P(X, Y) = \{(a, b, \delta_G(a, b)) : a \in X, b \in Y\}$. We will write $S(X)$ for $P(X, X)$ (by definition, $S(X)$ contains
tuples $(x, x, 0)$, $\forall x \in X$).

The following lemma, proved in [9], establishes the desired connection between computing shortest paths and products along tree paths of the operator
$\circ$ defined above.

Lemma 6. *Let G be a weighted digraph and (X, T) its tree decomposition. For
$i \in V(T)$, define $\gamma(i) = S(X_i)$. Let $v_1, \ldots, v_p$ be a path in T. Then $\gamma(v_1) \circ \ldots \circ \gamma(v_p) = P(X_{v_1}, X_{v_p})$.*

Therefore, it only remains to show how the γ values can be efficiently computed in parallel for each node of a tree-decomposition. This is shown in the
next lemma. The following algorithm first converts the given tree-decomposition
into a balanced one, and then repeatedly shrinks the tree, by absorbing the subgraphs corresponding to leaves. When the tree is reduced to a single node, the
algorithm computes γ using brute force, for this node. The distances computed
are the distances in the original graph, since distances are preserved during absorption. Then, it reverses the shrinking process and expands the tree, using the
γ values already computed to compute γ values for the newly expanded nodes.

Lemma 7. *Let G be an n-vertex weighted digraph and let (X, T) be the tree-
decomposition of G, of treewidth t. For each pair u, v such that $u, v \in X_i$ for some
$i \in V(T)$, let $Dist(u, v) = \delta(u, v)$ and $Int(u, v) = x$ where x is some intermediate
node (neither u nor v) on a shortest path from u to v. (If $wt(u, v) = \delta(u, v)$,
then $Int(u, v) = null$.) Then in $O(\log n \log^2 t)$ time using $O(t^3 n)$ work on an
EREW PRAM, we can either find a negative cycle in G, or compute the values
$Dist(u, v)$ and $Int(u, v)$ for each such pair u, v.*

Proof. Initially all the values $Dist(u, v)$ are set to ∞ and $Int(u, v)$ to null.
We give an inductive algorithm. First, convert (X, T) into a balanced tree-
decomposition of G using Fact 3. Then, for each vertex of T, we compute its
level number, which is one more than the level of its parent, with the root having
level number 1. This computation can be done in $O(\log n)$ time and $O(n)$ work
([13], Theorem 3.4).

We use induction on the number of levels of T. Let d be the depth of T
and N_d be the set of tree nodes at level d. For all nodes $z \in N_d$, we run the

algorithm of [12] to solve the apsp problem in $G[X_z]$. This will take $O(\log^2 t)$ time and $O(|N_d|t^3)$ work. If there is a negative cycle in some $G[X_z]$, it will be found by the algorithm of [12]; so henceforth assume that there is no $G[X_z]$ containing a negative cycle. Update the values for pairs $u, v \in X_z$ as follows: if the weight of the shortest path found is less than the current value of $Dist(u, v)$, then set $Dist(u, v)$ to the new value and $Int(u, v)$ to any intermediate vertex on the shortest path found. If $wt(u, v)$ is equal to the weight of the shortest path found, then set $Int(u, v) = $ null.

If $d = 1$ (which implies that $|V(T)| = 1$), we are done. Otherwise remove all nodes $z \in N_d$ from T and call the resulting tree T'. Let $V' = \cup_{i \in V(T')} X_i$ and construct G' by absorbing every $G[X_z]$ into $G[V']$, where the weight of each added edge $\langle u, v \rangle$ is $\delta_{G[X_z]}(u, v)$. Then, for any vertices $u, v \in V'$, $\delta_{G'}(u, v) = \delta_G(u, v)$, by Lemma 1. In particular, if G contains a negative cycle, so does G'. Note that $(X - Y, T')$ is a tree-decomposition for G', where $Y = \cup_{z \in N_d} X_z$.

Inductively run the algorithm on G'. If a negative cycle is found in G', then a negative cycle in G can be found by replacing any edges added during the absorption by their corresponding paths in the subgraphs $G[X_z]$. Hence, we may assume that G' does not contain a negative cycle.

For $a, b \in V'$, $Dist(a, b) = \delta_{G'}(a, b) = \delta_G(a, b)$, as desired. If $Int(a, b) = x \neq$ null, then x is an intermediate vertex on a shortest a to b path in G' and hence also in G, as desired. If $Int(a, b) = $ null, then $\langle a, b \rangle$ is a shortest path in G'. If $wt(a, b) > Dist(a, b)$, then this edge must have been added during the absorption. Correct the value $Int(a, b)$ by setting it to some intermediate vertex on the corresponding a to b shortest path found in $G[X_z]$. After this, all Int values are correct for $a, b \in V'$.

Construct a digraph G'' by absorbing $G[V']$ into every $G[X_z]$, with each added edge $\langle u, v \rangle$ having weight $\delta_G(u, v)$. By Lemma 1, $\delta_{G''}(x, y) = \delta_G(x, y)$, $\forall x, y \in X_z$. Run the algorithm of [12] on G'' to recompute all pairs shortest paths. Update the values $Dist(a, b)$ and $Int(a, b)$ for $a, b \in X_z$ as before. Now for each $z \in N_d$ we have: For $a, b \in X_z$, $Dist(a, b) = \delta_{G''}(a, b) = \delta_G(a, b)$ as desired. For $a, b \in V' \cap X_z$, $Int(a, b)$ is not changed since $Dist(a, b)$ is already $\delta_G(a, b)$. If either a or b does not belong to $V' \cap X_z$, $Int(a, b) = $ an intermediate vertex on a shortest path in G'' and hence in G, or $Int(a, b) = $ null in which case $wt(a, b) = \delta_G(a, b)$. Thus, the values computed are correct for all pairs a, b which completes the induction.

Concerning the resource bounds, it suffices to notice that the algorithm performs a bottom-up and a top-down traversal of T by processing the tree level-by-level and visiting every tree node at most twice. At each level, the algorithm takes $O(\log^2 t)$ time using $O(t^3)$ work per node. Hence, in total it takes $O(\log n \log^2 t)$ time and $O(t^3 n)$ work on an EREW PRAM. □

We are now ready for our main theorem.

Theorem 8. *For any integer t and any $k \geq 1$, let G be an n-vertex weighted digraph of treewidth at most t, whose tree-decomposition can be found in $T(n, t)$ parallel time using $W(n, t)$ work on an EREW PRAM. Then, the following hold* on an EREW PRAM: (i) After $O(\log n \log^2 t + T(n,t))$ time and $O(t^3 n I_k(n) + W(n,t))$ work and space preprocessing, distance queries in G can be answered in $O(t^3 k)$ time using a single processor. (ii) After $O(\log n \log^2 t + T(n,t))$ time and $O(t^3 n + W(n,t))$ work and space preprocessing, distance queries in G can be answered in $O(t^3 \alpha(n))$ time using a single processor.

Proof. Using Fact 2, we first compute a tree-decomposition (X, T) of G. By Lemma 7, we either find a negative cycle (and in such a case we stop), or compute values $Dist(u, v)$ for u, v such that $u, v \in X_i$ for some $i \in V(T)$. From these values, we can easily compute $\gamma(i)$, $\forall i \in V(T)$. By Fact 5 we preprocess T so that product queries on γ can be answered. Given a query, $u, v \in V(G)$, let i, j be vertices of T such that $u \in X_i$ and $v \in X_j$. We ask for the product of the γ values on the path between i and j. By Lemma 6, the answer to this query contains the information about $\delta(u, v)$. The bounds follow easily by the ones given in Fact 5 and by the fact that the composition of any two γ values can be computed in $O(\log^2 t)$ time using $O(t^3)$ work. □

As in [9], a distance query time of Q yields a path query time of $O(t^3 LQ)$, where L is the length of the reported path. (We omit the details for lack of space.) Hence, we can summarize all the preceding results as follows.

Theorem 9. *Let G be an n-vertex weighted digraph of constant treewidth and let $k \geq 1$ be any constant integer. Then, the following hold on an EREW PRAM: (i) After $O(\log^2 n)$ time and $O(n I_k(n))$ work and space preprocessing, distance (resp. shortest path) queries in G can be answered in $O(k)$ (resp. $O(kL)$) time using a single processor (where L is the length of the reported path). (ii) After $O(\log^2 n)$ time and $O(n)$ work and space preprocessing, distance (resp. shortest path) queries in G can be answered in $O(\alpha(n))$ (resp. $O(L\alpha(n))$) time using a single processor (where L is the length of the reported path).*

For the single source problem, we have the following theorem.

Theorem 10. *Let G be an n-vertex weighted digraph of constant treewidth and let $s \in V(G)$. Then, in $O(\log^2 n)$ time using $O(n)$ work on an EREW PRAM, we can either compute a shortest path tree rooted at s, or find a negative cycle in G (if exists). If the tree-decomposition of G is also provided with the input, then the computation takes $O(\log n)$ time.*

Proof. We first compute a tree-decomposition (X, T) of G of treewidth t, using Fact 2. Using Lemma 7, we either compute $Dist(u, v)$, for u, v such that $u, v \in X_i$, for some $i \in V(T)$, or find a negative cycle in G. If there is no negative cycle, we can easily compute $\gamma(i)$, $\forall i \in V(T)$. Let $i \in V(T)$ such that $s \in X_i$. Root T at node i and make it balanced, using Fact 3. Starting at i, perform a top-down traversal of T by visiting all nodes of T level-by-level. (This can be done in parallel as follows: each processor associated with some node of T forks two other processors and associates them with the children of its node.) At each node $j \in V(T)$ store the product of the γ values on the path from i to j. Since the composition of two γ values can be computed in $O(\log^2 t)$ time using $O(t^3)$ work on an EREW PRAM and each node is visited exactly once, the whole process takes $O(\log n \log^2 t)$ time using $O(t^3 n)$ work.

Let $y \in V(G)$ and let $j \in V(T)$ such that $y \in X_j$. By Lemma 6, the value stored at vertex j during the above mentioned top-down traversal, is $P(X_i, X_j)$ which contains the tuple $(s, y, \delta(s, y))$. Thus, we may assume that for each $y \in V(G)$, we have the value $\delta(s, y)$. For an edge $\langle v, u \rangle$ of G, define $h(v, u)$ to be the node z of T such that $v, u \in X_z$ and z is the closest such node to the root. (By the continuity condition, $h(v, u)$ is unique.) It is not hard to see that during the previous top-down traversal of T, we can found such nodes $h(v, u)$ for each edge $\langle v, u \rangle$ in G.

To construct the shortest path tree ST, we do the following. Starting at the root node i, we perform a second, level-by-level, top-down traversal of T. For a node $j \in V(T)$ at level $\ell \geq 1$, we check (sequentially) edges $\langle v, u \rangle$, where $v, u \in X_{h(v,u)}$ and v belongs to the shortest path tree constructed so far. (Initially, $j = i$ and $v = s$.) If $\delta(s, u) = \delta(s, v) + wt(v, u)$, then make v the parent of u in ST. If v, u belong also to any child of $X_{h(v,u)}$, then mark the edge $\langle v, u \rangle$ as being "examined" in the local memory of the processor associated with this child. Note that this last operation is needed in order to avoid concurrent access conflicts in the shared memory, in the case where there is another node $k \in V(T)$ at the same level with j for which $v, u \in X_k$.

A simple induction argument shows that the above procedure creates a shortest path tree rooted at s. It is again easy to see that each tree node is visited exactly once and that we need $O(t)$ time (using a single processor) in such a node. Hence, in total, ST can be constructed in $O(t \log n)$ time using $O(t^3 n)$ work. Since t is constant, the bounds claimed in the theorem follow. □

4   The Dynamic Algorithm

In this section we shall give our dynamic data structures and algorithms. The approach follows the one in [9], but the parallel implementation is rather different. We divide the digraph into subgraphs with disjoint edge sets and small cut-sets, and construct another (smaller) digraph – the reduced digraph – by absorbing each subgraph. The sizes of the subgraphs are chosen so that the subgraphs and the reduced digraph have size $O(\sqrt{n})$. We then construct a query data structure for each subgraph and for the reduced digraph. Queries can be efficiently answered by querying these data structures. Since the edge sets are disjoint, a change in the weight of an edge affects the data structure for only one subgraph. Then we update the data structure of this subgraph. This may result in new distances between vertices in its cut-set, which appear in the reduced digraph as changes in the weights of edges between these cut-set vertices. Since the cut-set is small, the weights of only a few edges in the reduced digraph change. The data structure for the reduced digraph is updated to reflect these changes. Thus an update in the original digraph is accomplished by a constant number of updates in subgraphs of size $O(\sqrt{n})$, which yields $O(\sqrt{n})$ update work. By recursively applying this idea, we get an update work of $O(n^\beta)$, for any constant $0 < \beta < 1$. In the following, we first give the graph partitioning results and then give the details of our algorithms.

4.1 Graph Equipartitions

Lemma 11. *Let T be a rooted binary tree on n nodes and let $1 \le m \le n$. Then, in $O(\log n)$ time and $O(n)$ work on an EREW PRAM, we can partition the nodes of T into at least n/m and at most $8n/m$ groups such that each group: (i) is a connected subtree; (ii) is connected to the rest of the tree through at most 3 edges; and (iii) has at most m nodes.*

Proof. The algorithm is a variant of the usual parallel tree contraction algorithm [13]. By adding a leaf as a child to each node that has one child, we obtain a tree in which each node is a leaf or has two children. Assign a weight of 1 to each node in the tree. Number the leaves of the tree from left to right using the Euler tour technique [13]. From now on assume that we have a tree with weights on the nodes adding up to n, in which each internal node has two children, and in which some of the leaves are numbered from left to right.

In parallel, for each odd numbered leaf that is a left child, if the sum of the weights of the leaf, its parent and its sibling is at most m, then shrink the edges connecting the leaf and its sibling to their parent. Assign the parent a weight equal to the sum of the weights of the three nodes. If the sibling is a leaf, it is even numbered. Assign this number to the parent (which is now a leaf in the modified tree). If the sum of the weights exceeds m, then delete the numbers (if they exist) from the leaf and the sibling.

Repeat the above for each odd numbered leaf that is a right child. After these two steps, all the numbered leaves in the tree have an even number. Divide each of these numbers by 2.

We repeat these two steps $\log n$ times. It is not hard to see that after the ith iteration, at most $l/2^i$ leaves have numbers, where l is the initial number of leaves. Thus, at the end, there are no numbered leaves. Throughout, the following invariant is maintained: if a leaf does not have a number, then the weights of the leaf, its parent and sibling add up to more than m. Call such a triple of leaf, parent and sibling an *overweight* group.

Each non-numbered leaf is contained in some overweight group, and no node can belong to more than two overweight groups. Thus, the sum of the weights of all the overweight groups is at most $2n$, hence the number of overweight groups is at most $2n/m$. Since each overweight group contains at most two non-numbered nodes, the total number of non-numbered leaves at the end is $4n/m$. Since each internal node has two children, the total number of nodes remaining in the tree is at most $8n/m$.

Each node v in the remaining tree is associated with the connected subtree induced by the nodes that were shrunk into v in the above process. These are the required groups. It is easy to see that v has a weight equal to the number of nodes in the associated subtree. Since this weight is at most m, there are at least n/m such connected subtrees. Also, as shown above, there are no more than $8n/m$ connected subtrees. It follows from the construction that each subtree is connected to the rest of the tree through at most 3 edges. □

Input-Output Conventions: We assume that the above algorithm has its input tree specified as a linked structure in n contiguous memory cells. The output it produces is in $O(n)$ contiguous memory cells, divided into contiguous blocks, each block containing one of the connected components in the same linked format, and one final block containing the compressed tree (i.e. the tree at the end of the shrinking process) in a linked format. This can be accomplished using standard EREW PRAM methods using $O(\log n)$ time and $O(n)$ work, which we now describe briefly.

By assigning the preorder number to each node in the compressed tree, we can assign a unique number between 1 and q (where q is the number of nodes in the compressed tree) to each connected subtree. Then, by solving a prefix summation problem on q elements, where the ith element is the number of nodes in subtree i, we can allocate contiguous memory blocks for the various subtrees. It remains to copy the subtrees into the appropriate blocks.

Since each node in the compressed tree knows the memory addresses allocated for its subtree, reversing the shrinking process, we can assign a unique memory address in the appropriate block to each node in a subtree. Now it is a simple matter for each node to copy itself into this address, and duplicate its link structure.

Definition 12. Let λ, δ be positive integer constants and let $1 \leq m \leq n$. Then, given an n-vertex digraph G as well as its balanced tree-decomposition of treewidth t, we define an (λ, δ, m)-*equipartition of G* to be a partition of G into q subgraphs $H_1, \ldots, H_q$, where $n/m \leq q \leq \delta n/m$, along with the construction of another subgraph H' such that: (i) H_i has at most tm vertices and a cut-set $C(H_i)$ of size at most λt; (ii) H' is the induced subgraph on vertices $\cup_{i=1}^{q} C(H_i)$, augmented with edges $\langle x, y \rangle$, $x, y \in C(H_i)$ for each $1 \leq i \leq q$; and (iii) we have a tree-decomposition of treewidth t for each H_i and a tree-decomposition for H' of treewidth $3t$.

The following lemma shows that an $(3, 8, m)$-equipartition can be efficiently computed.

Lemma 13. *Given an n-vertex digraph G along with its balanced tree-decomposition of treewidth t, we can compute an $(3, 8, m)$-equipartition of G in $O(\log n)$ time using $O(t^2 n)$ work on an EREW PRAM, where $1 \leq m \leq n$.*

Proof. Let (X, T) be the balanced tree decomposition of G. Then, by Fact 3, T has at most $2n$ nodes. Partition the nodes of T into $n/m \leq q \leq 8n/m$ connected components using Lemma 11. For each component T_i, $1 \leq i \leq q$, create a subgraph H_i, that is the induced subgraph of G on the vertices in $\cup_{v \in V(T_i)} X_v$. Note that T_i is a tree decomposition of H_i. The number of vertices in H_i is at most $t|V(T_i)| = tm$. Let v_1, v_2 and v_3 be the nodes through which T_i is connected to the other components. Then, $C(H_i) = X_{v_1} \cup X_{v_2} \cup X_{v_3}$, and $C(H_i)$ has at most $3t$ vertices. H' is constructed by constructing a clique on $C(H_i)$ for each $1 \leq i \leq q$. The tree decomposition for H' is constructed by shrinking each component T_i into a single node u and assigning $X_u = C(H_i)$. It is easily verified that this is a tree decomposition of H' of width $3t$. Also, it is not hard to see that the work required for the above constructions is bounded by $O(t^2 n)$ and the time by $O(\log(tn))$. The EREW PRAM implementation can be easily done using the data structures described after the proof of Lemma 11. □

4.2 Data Structures and Algorithms

Let $PD(G, \{P_W, P_T\}, \{U_W, U_T\}, Q)$ be a parallel dynamic data stucture for a digraph G, where $O(P_W)$ (resp. $O(P_T)$) is the preprocessing work and space (resp. time) to be set up, $O(Q)$ is the time to answer a distance query using a single processor and $O(U_W)$ (resp. $O(U_T)$) is the work (resp. time) to update it after the modification of an edge-weight.

Theorem 14. *Assume that we are given an n-vertex weighted digraph G and its balanced tree decomposition of treewidth t. Then, for $r > 0$, we can construct, on an EREW PRAM, the following (with $A = 5t^3 11^{3r}$):*

(i) $PD(G, \{A^r n, 2^r \log n\}, \{A^r n^{(1/2)^{r-1}}, A^r \log n\}, A^r \alpha(n))$; *and*

(ii) $PD(G, \{A^r n I_k(n), 2^r \log n\}, \{A^r n^{(1/2)^{r-1}}, A^r \log n\}, A^r k)$, *for* $k \geq 1$.

Proof. We shall prove part (i). Part (ii) can be proved similarly. We use induction on r. If $r = 1$, then, the work and time allowed for updates exceeds the preprocessing, and the static data structure of Theorem 8 suffices, with updates implemented by simply recomputing the whole data structure. We use the notation $D(G, n, r, t)$ for $PD(G, \{A^r n, 2^r \log n\}, \{A^r n^{(1/2)^{r-1}}, A^r \log n\}, A^r \alpha(n))$.

Assume the theorem holds for $r' < r$. We show how to construct $D(G, n, r, t)$.

We first construct an $(3, 8, \sqrt{n})$-equipartition of G using Lemma 13, yielding H' and $H_1, \ldots, H_q$, $\sqrt{n} \leq q \leq 8\sqrt{n}$.

Define G_i to be H_i with all edges joining pairs of vertices in its cut-set deleted. Define G' to be H' with edges $\langle x, y \rangle$ weighted $\delta_{G_i}(x, y)$ for each pair $x, y \in C(G_i)$, $1 \leq i \leq q$. Replace multiple edges by the edge of minimum weight. Note that G' is exactly the graph obtained by absorbing $G_1, G_2, \ldots, G_q$ into the rest of the graph. By Lemma 1, it follows that $\delta_{G'}(x, y) = \delta_G(x, y)$, $\forall x, y \in V(G')$.

Let $u \in V(G_i), v \in V(G_j) - V(G_i)$. Then, any path from u to v must pass through a vertex in each of the cut-sets of G_i and G_j. Then we have $\delta_G(u, v) = \min\{\delta_{G_i}(u, x) + \delta_{G'}(x, y) + \delta_{G_j}(y, v) : x \in C(G_i), y \in C(G_j)\}$. Similarly, for $u, v \in V(G_i)$, we have $\delta_G(u, v) = \min\{\delta_{G_i}(u, v), \min\{\delta_{G_i}(u, x) + \delta_{G'}(x, y) + \delta_{G_i}(y, v) : x, y \in C(G_i)\}\}$. If we are able to make queries of the form $\delta_{G_i}(x, y)$ and $\delta_{G'}(x, y)$, the above directly yields a query algorithm for any pair of vertices x, y.

In the following, let $n_i = |V(G_i)|$ and $n' = |V(G')|$. The $(3, 8, \sqrt{n})$-equipartition of G gives us a tree-decomposition of treewidth t for each subgraph G_i, and a tree-decomposition of treewidth $3t$ for G'. We balance these tree-decompositions, yielding tree-decompositions for each G_i with treewidth at most $3t + 2 \leq 5t$ and for G' with treewidth at most $9t + 2 \leq 11t$. Inductively, we construct in parallel, $D(G_i, n_i, r - 1, 5t)$ for each $1 \leq i \leq q$, which enables us to answer queries of the form $\delta_{G_i}(x, y)$, and $D(G', n', r - 1, 11t)$ which enables us to answer queries of the form $\delta_{G'}(x, y)$.

The update procedure is the following: note that $E(G_i) \cap E(G_j) = \emptyset$, $i \neq j$ and $E(G_i) \cap E(G') = \emptyset$, i.e. each edge of G belongs to exactly one of the G_i's or to G'. Suppose the cost of an edge belonging to G_i is changed. Then, we update the data structure for G_i. This may result in new values for $\delta_{G_i}(x, y)$, $x, y \in C(G_i)$. We query the updated data structure for $\delta_{G_i}(x, y)$, $x, y \in C(G_i)$ and change the weights of the corresponding edges of G', updating the data structure for G' after each change. That the procedure is correct follows from the fact that changing the cost of an edge in G_i does not change $\delta_{G_j}(x, y)$, $x, y \in C(G_j)$ when $j \neq i$. Thus, after we change, in G', the cost of edges $\langle x, y \rangle$, $x, y \in C(G_i)$, we have $\delta_{G'}(u, v) = \delta_G(u, v)$, $u, v \in V(G')$, again, by repeated applications of Lemma 1. After the last update, the data structure for G' yields correct distances in G, between vertices in $V(G')$.

Now suppose we change the cost of an edge belonging to G'. Then the distances $\delta_{G_i}(x, y)$ do not change. Thus, in this case, we simply update the data structure for G'. This completes the description of the preprocessing and update algorithms.

The time and work required to set up this data structure is the time and work required to construct (1) the equipartitions of G_i's and G', and (2) the data structures of G_i's and G' inductively. By Lemma 13, (1) requires $O(\log n)$ time and $O(t^2 n)$ work. Then, writing $PW(r, t)n$ and $PT(r, t) \log n$ for the preprocessing work and time respectively, we have

$$PW(r, t)n \leq t^2 n + \sum_{i=1}^{q} PW(r - 1, 5t)n_i + PW(r - 1, 11t)n'$$

$$PT(r, t) \log n \leq \log n + \max\{PT(r - 1, 11t) \log n', PT(r - 1, 5t) \log N\}$$

where $N = \max\{n_1, \ldots, n_q\}$.

Querying involves taking the minimum of the results of the sub-queries specified in the query algorithm previously. Writing $Q(r, t)\alpha(n)$ for the query time, we have

$$Q(r, t)\alpha(n) \leq (5t)^2 [2Q(r - 1, 5t)\alpha(N) + Q(r - 1, 11t)\alpha(n')]$$

During updates, in the worst case, there is one update in a graph G_i and then, at most $(5t)^2$ queries in G_i and updates in graph G'. Thus, with $UW(r, t)n^{(1/2)^{r-1}}$ and $UT(r, t) \log n$ representing the work and time respectively, we have

$$UW(r, t)n^{(1/2)^{r-1}} \leq UW(r - 1, 5t)N^{(1/2)^{r-2}}$$
$$+ (5t)^2 [Q(r - 1, 5t)\alpha(N) + UW(r - 1, 11t)(n')^{(1/2)^{r-2}}]$$

$$UT(r, t) \log n \leq UT(r - 1, 5t) \log N + (5t)^2 [Q(r - 1, 5t)\alpha(N) + UT(r - 1, 11t) \log n']$$

It is easy to show that $n' \leq 88tn^{1/2}$, $\sum_{i=1}^{q} n_i \leq 5tn$ and $N = 5tn^{1/2}$. Using these facts and easy estimates, we obtain the following recurrences.

$$PW(r, t)n \leq 2t^2 PW(r - 1, 11t)$$
$$PT(r, t) \leq 2PT(r - 1, 11t)$$
$$Q(r, t) \leq (5t)^3 Q(r - 1, 11t)$$
$$UW(r, t) \leq (5t)^3 UW(r - 1, 11t)$$
$$UT(r, t) \leq (5t)^3 UW(r - 1, 11t)$$

from which the claimed bounds follow.

Thus we can construct $D(G, n, r, t)$, completing the induction. $\square$

The following theorem shows how to obtain an update work of $O(n^\beta)$, for any constant $0 < \beta < 1$, in a digraph of constant treewidth.

Theorem 15. *Let $k \geq 1$ be any constant integer and let $0 < \beta < 1$ be any constant. Given an n-vertex weighted digraph G of constant treewidth, we can construct on an EREW PRAM: (i) $PD(G, \{n, \log^2 n\}, \{n^\beta, \log n\}, \alpha(n))$; and (ii) $PD(G, \{nI_k(n), \log^2 n\}, \{n^\beta, \log n\}, k)$.*

Proof. Using Facts 2 and 3, we can compute a balanced tree-decomposition of G in $O(\log^2 n)$ time and $O(n)$ work on an EREW PRAM. The rest of the proof follows now by Theorem 14, if we set $r = 1 - \log \beta$. $\square$

The algorithms described above give answers to distance queries only. They can be modified to answer path queries as well. Also, before running our update procedure after a change in the weight of an edge, we have to assure that this change does not create a negative cycle in the digraph G. This can be easily tested as follows. Let $\langle u, v \rangle$ be an edge with weight $wt(u, v)$ and let $wt'(u, v)$ be its new weight. Clearly, the new weight $wt'(u, v)$ creates a negative cycle in G iff $\delta_G(v, u) + wt'(u, v) < 0$. This test takes time proportional to that of finding $\delta_G(v, u)$ and hence does not affect our update bound.

References

1. W. Ackermann, "Zum Hilbertschen Aufbau der reellen Zahlen", *Math. Ann.*, 99(1928), pp.118-133.
2. A. Aho, J. Hopcroft and J. Ullman, "The Design and Analysis of Computer Algorithms", Addison-Wesley, 1974.
3. R. Ahuja, T. Magnanti and J. Orlin, "Network Flows", Prentice-Hall, 1993.
4. N. Alon and B. Schieber, "Optimal Preprocessing for Answering On-line Product Queries", Tech. Rep. No. 71/87, Tel-Aviv University, 1987.
5. S. Arnborg, "Efficient Algorithms for Combinatorial Problems on Graphs with Bounded Decomposability - A Survey", *BIT*, 25 (1985), pp.2-23.
6. H. Bodlaender, "NC-algorithms for Graphs with Small Treewidth", *Proc. 14th WG'88*, LNCS 344, Springer-Verlag, pp.1-10, 1989.

7. H. Bodlaender, "A Tourist Guide through Treewidth", *Acta Cybernetica*, Vol.11, No.1-2, pp.1-21, 1993.

8. H. Bodlaender and T. Hagerup, "Parallel Algorithms with Optimal Speedup for Bounded Treewidth", *Proc. 22nd ICALP'95*, LNCS, Springer-Verlag, to appear.

9. S. Chaudhuri and C. Zaroliagis, "Shortest Path Queries in Digraphs of Small Treewidth", *Proc. 22nd ICALP'95*, LNCS, Springer-Verlag, to appear.

10. E. Cohen, "Efficient Parallel Shortest-paths in Digraphs with a Separator Decomposition", *Proc. 5th ACM SPAA*, 1993, pp.57-67.

11. H. Djidjev, G. Pantziou and C. Zaroliagis, "On-line and Dynamic Algorithms for Shortest Path Problems", *Proc. 12th STACS'95*, LNCS 900, pp. 193-204, Springer-Verlag, 1995.

12. Y. Han, V. Pan and J. Reif, "Efficient Parallel Algorithms for Computing All Pair Shortest Paths in Directed Graphs", *Proc. 4th ACM SPAA*, 1992, pp.353-362.

13. J. JáJá, "An Introduction to Parallel Algorithms", Addison-Wesley, 1992.

14. D. Kavvadias, G. Pantziou, P. Spirakis and C. Zaroliagis, "Efficient Sequential and Parallel Algorithms for the Negative Cycle Problem", *Proc. 5th ISAAC*, 1994, LNCS 834, pp.270-278, Springer-Verlag.

15. P. Klein and S. Subramanian, "A linear-processor polylog-time algorithm for shortest paths in planar graphs", *Proc. 34th IEEE Symp. on FOCS*, 1993, pp.259-270.

16. N. Robertson and P. Seymour, "Graph Minors I: Excluding a Forest", *J. Comb. Theory Series B*, 35 (1983), pp.39-61.

17. N. Robertson and P. Seymour, "Graph Minors II: Algorithmic Aspects of Treewidth", *J. Algorithms*, 7 (1986), pp.309-322.

18. R.E. Tarjan, "Data Structures and Network Algorithms", SIAM, 1983.

Shared Memory Simulations
with Triple-Logarithmic Delay [*]

(Extended Abstract)

Artur Czumaj[1], Friedhelm Meyer auf der Heide[2], and Volker Stemann[1]

[1] Heinz Nixdorf Institute, University of Paderborn,
D-33095 Paderborn, Germany
[2] Heinz Nixdorf Institute and Department of Computer Science,
University of Paderborn, D-33095 Paderborn, Germany

Abstract. We consider the problem of simulating a PRAM on a distributed memory machine (DMM). Our main result is a randomized algorithm that simulates each step of an n-processor CRCW PRAM on an n-processor DMM with $\mathcal{O}(\log \log \log n \log^* n)$ delay, with high probability. This is an exponential improvement on all previously known simulations. It can be extended to a simulation of an $(n \log \log \log n \log^* n)$-processor EREW PRAM on an n-processor DMM with optimal delay $\mathcal{O}(\log \log \log n \log^* n)$, with high probability. Finally a lower bound of $\Omega(\log \log \log n / \log \log \log \log n)$ expected time is proved for a large class of randomized simulations that includes all known simulations.

1 Introduction

Parallel machines that communicate via a shared memory (*Parallel Random Access Machines, PRAMs*) are the most commonly used machine model for describing parallel algorithms (see e.g. [J92]). The PRAM is relatively comfortable to program, because the programmer does not have to deal with the hardware limitations, like e.g. synchronization, data locality or interprocessor communication, and can only focus on the combinatorial properties of the problem at hand. On the other hand shared memory machines are very unrealistic from the technological point of view, because, for example, on large machines a parallel shared memory access can only be realized at the cost of a significant time delay. A more realistic model which tries to overcome this unrealistic assumption, is the *Distributed Memory Machine (DMM)*, in which the memory is divided into a limited number of memory modules, one module per processor. Each such module can respond to only one access at a time. Thus DMMs exhibit the phenomenon of *memory contention*, in which an access request is delayed because of concurrent requests to the same module.

[*] Supported in part by DFG-Graduiertenkolleg "Parallele Rechnernetzwerke in der Produktionstechnik", ME 872/4-1, by DFG-Sonderforschungsbereich 1511 "Massive Parallelität: Algorithmen, Entwurfsmethoden, Anwendungen", by DFG Leibniz Grant Me872/6-1, and by the Esprit Basic Research Action Nr 7141 (ALCOM II)

In an effort to understand the effects of memory contention on the performance of parallel computers, several authors have investigated the simulation of shared memory machines on DMMs. Often the authors assumed that processors and modules are connected by a bounded degree network, and packet routing is used to access the modules [R91, L92a, L92b, U84, KU86]. In this paper we study DMMs with a complete interconnection between processors and modules.

Simulations based on hashing distribute the shared memory cells U among the modules using one or more hash functions $h_i : U \to [n]$, [3] $i \in [a]$; cell $u \in U$ is stored in the modules $M_{h_1(u)}, \ldots, M_{h_a(u)}$. All such simulations assume that $h_1, \ldots, h_a$ are randomly chosen from a *high performance universal class* of hash functions as e.g. presented by Dietzfelbinger and Meyer auf der Heide [DM90] or Siegel [S89]. The *delay* of a simulation is the time needed to simulate a parallel memory access of a PRAM. We say a randomized simulation of a p-processor PRAM on an n-processor DMM is *time-processor optimal* if the delay is $\mathcal{O}(p/n)$ with high probability (w.h.p.)[4]. It is easily seen that a simulation of an n-processor EREW PRAM on an n-processor DMM using one hash function has contention $\Omega(\log n / \log \log n)$, w.h.p., even if the hash function behaves like a random function. Karp et al. [KLM92] invent the idea of performing a simulation based on more than one hash function. They present a simulation of an n-processor EREW PRAM on an n-processor DMM with delay $\mathcal{O}(\log \log n)$, w.h.p., with three hash functions. Thus, at the expense of increasing the total storage requirement by a constant factor, the running time of the simulation is exponentially decreased. They also obtain a time-processor optimal simulation of an $(n \log \log n \log^* n)$-processor EREW PRAM on an n-processor DMM. Using the majority technique due to Upfal and Wigderson [UW87], Dietzfelbinger and Meyer auf der Heide [DM93] extend this result to a much simpler schedule for an $\mathcal{O}(\log \log n)$ simulation on the weaker c-collision DMM. In this model a module can only answer, if it gets less than c requests; otherwise it sends a collision symbol. Goldberg et al. [GMR94] show that one can perform a time-processor optimal simulation with delay $\mathcal{O}(\log \log n)$, w.h.p., even on a 1-collision model (called also OCPC). Meyer auf der Heide et al. [MSS95] extend the simulation on a DMM to more hash functions which yields a delay of $\mathcal{O}(\log \log n / \log \log \log n)$, w.h.p.. As it uses a non-constant number of hash functions it cannot be turned into a time-processor optimal simulation.

The techniques used in these papers seem not to yield simulations with smaller delay. In particular, MacKenzie et al. [MPR94] and independently Meyer auf der Heide et al. [MSS95] show lower bounds for classes of algorithms that capture all these algorithms. To break the $\Omega(\log \log n)$ lower bound of Meyer auf der Heide et al. [MSS95] for simulations with constant memory redundancy one has to use non-oblivious techniques. Using information about the number of accesses at a module, Czumaj et al. [CMS95] present a randomized simulation of an EREW PRAM with delay $\mathcal{O}(\log \log n / \log \log \log n)$ and constant memory redundancy, w.h.p.. They also extend this result to a time-processor optimal

---

[3] In this paper $[n]$ will always denote the set $\{1, 2, \ldots, n\}$.

[4] W.h.p. means "with probability at least $1 - n^l$ for any constant l".

simulation of an $(n \log \log n \log^* n / \log \log \log n)$-processor EREW PRAM on an n-processor DMM.

In this paper we design new shared memory simulations that improve all previously known results by an exponential decrease of the delay. We design a simulation of an n-processor EREW PRAM on an n-processor DMM with delay $\mathcal{O}(\log \log \log n \log^* n)$, w.h.p.. We model access requests by an access graph, in which the nodes correspond to the modules and the edges correspond to the accesses of the processors. In all previously known simulations each module essentially decides which request to answer based only on the structure of the corresponding node, its incident edges, and its neighbors. We show that using as much information about the structure of the neighborhood of each node as available during the time of the simulation may drastically improve simulations. A clever access protocol ensures that our access graph will decay in small independent subgraphs. Within these small components we can perform a constant time protocol to answer all remaining accesses of the processors. The algorithms are based on sophisticated log-star techniques that explore the neighborhood of each node and an analysis of the random structure of the access graph. In a full paper we present a simple protocol which enables us to turn any simulation with constant storage overhead of an n-processor EREW PRAM on an n-processor DMM into a time-processor optimal one. Finally we show that our implementations are almost optimal. We present an $\Omega(\log \log \log n / \log \log \log \log n)$ lower bound for a large class of randomized simulations that includes all the known simulations.

Note that our simulations of an n-processor EREW PRAM on an n-processor DMM can be easily extended to simulations of an n-processor CRCW PRAM without increasing the delay. We are not able, however, to obtain a time-processor optimal simulation of a CRCW PRAM. Although the techniques of Karp et al. [KLM92] can be used to get a simulation of an $\mathcal{O}(n \log \log \log n \log^* n)$-processor CRCW PRAM on an n-processor DMM with delay $\mathcal{O}(\log \log \log n (\log^* n)^2)$, which is only a factor of $\log^* n$ away from optimality.

The paper is organized as follows. In Section 2 we proceed with the definition of the computation models and state a graph-theoretic lemma that is the basis of all our analysis. In Section 3 we elaborate some algorithmic utilities for a random graph, especially how to explore the k-neighborhood of a node, which is essential for our algorithms. In Section 4 we present the $\mathcal{O}(\log \log \log n \log^* n)$ simulation. Finally Section 5 presents the $\Omega(\log \log \log n / \log \log \log \log n)$ lower bound for simulations on the DMM.

Because of space limitations some details and proofs are omitted in this extended abstract.

2 Preliminaries

A *parallel random access machine (PRAM)* consists of p processors $P_1, \ldots, P_p$ and a shared memory with cells $U = [m]$. The processors work synchronously and have random access to the shared memory cells, each of which can store an integer. We consider two models of a PRAM, an *exclusive read exclusive write (EREW)* PRAM, in which concurrent reads and writes are forbidden, and a *concurrent read concurrent write (CRCW)* PRAM, which allows concurrent reads and writes. We only deal with an ARBITRARY CRCW PRAM, in which if several processors want to write to the same memory cell simultaneously an arbitrary one succeeds.

A *distributed memory machine (DMM)* has n processors, $Q_1, \ldots, Q_n$, which communicate via a distributed memory consisting of n modules, $M_1, \ldots, M_n$. Each module has a communication window. A module can read from or write into its window. From the point of view of the processors, a window acts like a shared memory cell, where concurrent accesses are allowed. If more than one processor want to access the same module simultaneously then an arbitrary one succeeds. The following lemma can easily be obtained.

Lemma 1. *An n-processor* ARBITRARY *CRCW PRAM with $\mathcal{O}(n)$ shared memory cells and an n-processor DMM can simulate each other with constant delay.*

Our PRAM simulations follow the ideas of Upfal and Wigderson [UW87], and Dietzfelbinger and Meyer auf der Heide [DM93] (see also [M92]). The memory of a PRAM is hashed using three hash functions h_1, h_2, and h_3. That means, each memory cell $u \in U$ of a PRAM will be stored in the modules $M_{h_1(u)}, M_{h_2(u)}$, and $M_{h_3(u)}$ of a DMM. We will call the representations of u in the $M_{h_i(u)}$'s the copies of u. For simplicity of presentation we assume that all the hash functions used are random functions. Universal classes of functions developed by Siegel [S89] are sufficient for our purposes, however (see e.g. [KLM92, MSS95, GMR94]). For the simulation of a PRAM step we use the technique of Upfal and Wigderson [UW87] which ensures that it suffices to access arbitrary two out of the three copies of a shared memory to guarantee a correct simulation. To write to a memory cell a processor of the DMM accesses at least two of the copies and adds a time stamp to them indicating the (PRAM-) time of the update. To read a memory cell a processor has to access two of the copies and takes the one with the latest time stamp.

As in [CMS95], we modify the two-out-of-three idea and split this schedule into three steps of trying to access one out of two copies with a different pair of hash functions in each step. Clearly in this way we always access at least two copies. Therefore in the following we will analyze how to simulate an access of the shared memory that uses two hash functions h_1 and h_2.

For technical reasons, we do not perform all n accesses to the shared memory simultaneously but split the requests into batches of size n/c, for some constant $c \geq 1$, which will be specified in Lemma 2. Since we only have a constant number of batches, this will slow down our algorithm only by a constant factor.

Let S denote such a batch. Let us call a schedule where one has to access for each $u \in S$ at least one of the two possible copies a *one-out-of-two schedule*. Let $G = ([n], E)$ be the labeled directed graph, defined by h_1, h_2 and the set of requests S, that has an edge $(h_1(u), h_2(u))$ labeled u for each $u \in S$. Note that parallel edges and self-loops are allowed in G. Let H be the labeled graph obtained from G by removing all directions from the edges.

Our simulations rely on the properties of the random graph H. The following lemma was proved partially by Karp et al. [KLM92] and Czumaj et al. [CMS95].

Lemma 2. *For each constant l there is a constant c such that for the graph H, consisting of n nodes and n/c edges, the following conditions hold with probability at least $1 - 1/n^l$.*

(a) H has no connected component of size larger than $\log n$.

(b) For each pair of nodes v and w there are only $O(1)$ simple paths from v to w.

(c) There are only $O(1)$ cycles in H.

(d) For each $\xi < \log c/4l$

$$\sum_{\text{connected components } C} |C| \cdot 2^{|C| \xi} = O(n) \ .$$

3 Algorithmic Utilities

All the algorithms we describe in this section are designed on an ARBITRARY CRCW PRAM with $\mathcal{O}(n)$ space. Lemma 1 shows that they run on a DMM as well, with constant delay.

3.1 Log-Star Techniques

The following are the main tools used by our algorithms. Let $b > 1$ be any small constant, e.g. $b = 2$. If an array of size bn contains at least n objects we will call it *padded-consecutive*.

Given n integers $x_1, x_2, \ldots, x_n$, the *strong semisorting* problem [BH93] is to store them in a padded-consecutive array, such that all variables with the same value occur in a padded-consecutive subarray. Given n bits $x_1, x_2, \ldots, x_n$, the *chaining* problem [R93, BV93] is to find for each x_i, the nearest 1's both to its left and to its right. The *processor allocation* problem [GMV91] is to redistribute m tasks among n processors, so that each processor gets $O(1+m/n)$ tasks. For a given sequence of integers, $x_1, \cdots, x_n$, $x_i \in [n]$, the *approximate parallel prefix sums* problem [GMV94] is to find a sequence $y_0 = 0, y_1, \cdots, y_n$, $y_i \in [n]$, such that for $i \in [n]$, $x_i \leq y_i - y_{i-1} \leq bx_i$. We combine results of [R93, BV93, GMV91, GMV94, BH93] in the following lemma.

Lemma 3. *The strong semisorting problem, the chaining problem, the processor allocation problem, and the approximate parallel prefix sums can be solved on an ARBITRARY CRCW PRAM in $\mathcal{O}(\log^* n)$ time with linear total work and linear space, with probability at least $1 - 2^{-n^\varepsilon}$ for some constant $\varepsilon > 0$.*

3.2 Algorithms on the Random Graph H

Throughout this section H will always denote a graph which fulfills the conditions of Lemma 2. Our algorithms need a data structure, called *path-access-structure*, that allows fast access to all paths of length at most k in H. Note that, by the properties of H from Lemma 2, the total length of all the paths is $O(n)$. We store the paths in an array S of length $O(n)$, which consists of padded consecutive subarrays S_v, for each node v of H. Each S_v contains each path starting in node v as consecutive cells. In order to access the paths we build up an array P which consists of padded consecutive subarrays P_v, for each node v of H. P_v contains a pointer to the header of each path starting at node v together with the length of this path.

We first show how to build up the path-access-structure for paths of length at most k, and then how to update the structure under edge deletions.

Lemma 4. *The path-access-structure can be built on an n-processor DMM for all paths in H of length at most k w.h.p. in time $\mathcal{O}(\log k \log^* n)$, with linear total work.*

For the proof of Lemma 4 we need a lemma that was essentially proved in [CMS95].

Lemma 5. *In time $\mathcal{O}(\log^* n)$ and with linear total work, w.h.p., an n-processor DMM can compute the degree of each node of H and can store its adjacency list in a consecutive subarray of an array of length $O(n)$, evenly distributed among the modules.*

Proof of Lemma 4. The algorithm is based on the standard doubling technique (see e.g. [J92]). We perform $\log k + 1$ iterations and ensure the following invariant after r iterations, $0 \leq r \leq \log k$: Each node v has already found all simple paths of length at most 2^r that start at v, stored them in a padded-consecutive subarray S_v and stored the pointers to each path together with its length in a padded-consecutive subarray P_v. Additionally, if a given node has already found all simple paths then we call it *inactive*. Otherwise it is *active*. In each iteration of the algorithm only active nodes participate.

When $r = 0$ then all the paths of length 1 are exactly the edges incident to v. Thus, Lemma 5 can be used to find the adjacency list in $\mathcal{O}(\log^* n)$ time with linear total work, w.h.p.. Additionally we inactivate all isolated nodes.

We perform the $(r+1)$-st iteration, for $r \geq 0$, by pointer jumping. Let v be any active node and C_v be its connected component. Note that since v is active $|C_v| \geq 2^r$. First, v computes how many simple paths of length at most 2^r start at v. Because it is hard to compute this value exactly, each node v computes a value $\tilde{\delta}_r(v)$ which is not smaller and at least b times larger than the number of paths that start at v. Since the subarray P_v containing the pointers to all simple paths starting at v is padded-consecutive, simply finding the first and the last such paths enables to compute $\tilde{\delta}_r(v)$ after performing the chaining algorithm. Now we compute approximately the total length, $\tilde{\gamma}_r(v)$, of all simple paths stored at S_v. Since the lengths of all such paths are stored at P_v, we compute $\tilde{\gamma}_r(v)$ using the approximate prefix sums algorithm, for all v. The difference between the last path-length in two consecutive subarrays P_v and $P_{v'}$ gives us $\tilde{\gamma}_r(v)$.

Let (x, y) be the last edge of any simple path p of length 2^r which starts at v. To find all paths of length l, $2^r < l \leq 2^{r+1}$, starting with p, we must combine p with all paths of length at most 2^r starting at y. Then we remove their anomalies, i.e., the paths that create cycles. Observe that using the values $\tilde{\delta}_r(y)$ and $\tilde{\gamma}_r(y)$ we know how big the new arrays P_v and S_v have to be (for each path p, P_v has to be extended by $\tilde{\delta}_r(y)$ and S_v by $2^r \cdot \tilde{\delta}_r(y) + \tilde{\gamma}_r(y)$). This space allocation can be done using global approximate prefix sums. Observe that $\tilde{\delta}_r(v) = O(|C_v|)$ and $\tilde{\gamma}_r(v) = O(|C_v| \cdot 2^r) = O(|C_v|^2)$. Hence the size of the new P_v is $O(|C_v|^2)$, and the size of the new S_v is $O(|C_v|^3)$. It is easy to compute the length of each new created path to maintain P_v. To update S_v we only have to copy the old paths from S_v and concatenate the paths from v to y with simple paths from y. Hence these operations can be performed in constant time with the total work proportional to the sizes of the new P_v and S_v. This means that the total work for all nodes at all is $\sum_{C:|C| \geq 2^r} O(|C|^4)$, and the running time in each iteration is $\mathcal{O}(\log^* n)$, that is needed for computing $\tilde{\gamma}_r(v)$, $\tilde{\delta}_r(y)$ and allocate P_v's and S_v's.

Finally we have to remove the obtained paths that are not simple. We identify each path in S_v with the position of the first node. Then we perform strong semisorting within all arrays S_v with respect to the pairs [path, a node on the path]. Now, if there is more than one pair [p,y], which can be easily verified, then the path p is not simple and we eliminate it. Strong semisorting within all arrays P_v can be used to remove non-simple paths from these arrays. Using the lengths of the paths in the P_v's, approximate prefix sums enables to remove all non simple paths in the arrays S_v. Hence we can maintain the padded-consecutivety of the P_v's and S_v's. Now we inactivate a node v if the new P_v contains no path of length 2^{r+1}.

The running time of iteration r is $\mathcal{O}(\log^* n)$ with total work $\mathcal{O}(\sum_{C:|C| \geq 2^r} |C|^4)$. Therefore, the total work of the algorithm is

$$\mathcal{O}(\sum_{r=0}^{\log k} \sum_{C:|C| \geq 2^r} |C|^4) = \mathcal{O}(\sum_C |C|^4 \log |C|)$$

and by Lemma 2, this is $\mathcal{O}(n)$.

Now we show how to maintain the path-access-structure when we allow removing edges from the graph. The proof of the following lemma can be obtained using techniques from the proof of Lemma 4.

Lemma 6. *Assume that the path-access-structure is already built. Then after removing some edges from the graph it can be updated in time $\mathcal{O}(\log^* n)$ with linear work on an n-processor DMM, w.h.p..*

The next lemma shows how to use the path-access-structure to count the number of all simple paths of length at most k.

Lemma 7. *Suppose that the path-access-structure is given. Then for all edges e and indices $r \in [k]$, the number of simple paths of length exactly r that start with edge e can be computed in $\mathcal{O}(\log^* n)$ time with linear work on an n-processor DMM, w.h.p..*

Proof. For each simple path in all S_v's we consider the pairs [starting edge of the path, length of the path]. Now we perform strong semisorting with respect to these keys. Hence all the simple paths (in fact their representatives) that start with the same edge e and are of the same length r are stored in a padded-consecutive subarray, which we call $X_{e,r}$. If $y_{e,r}$ denotes the number of such paths, then $y_{e,r} \leq |X_{e,r}| \leq by_{e,r} = O(y_{e,r})$. We allocate $|X_{e,r}| \cdot 2^{|X_{e,r}|}$ processors to the pair $[e, r]$ and compute $y_{e,r}$ in the same way as in the proof of Lemma 5 in constant time. Now we have to show that we only use $\mathcal{O}(n)$ processors. Let C_e be the connected component e belongs to. By Lemma 2, $y_{e,r} = O(|C_v|)$ and hence

$$\sum_{e \in E} \sum_{r=0}^{k} |X_{e,r}| \cdot 2^{|X_{e,r}|} \leq \sum_{C} \sum_{e \in C} |C| \cdot O(|C|) 2^{O(|C|)} \leq O(\sum_{C} |C|^3 2^{O(|C|)})$$

Lemma 2 ensures that this is bounded by $O(n)$.

4 Triple-Logarithmic Simulation

One can view the one-out-of-two schedule as the following process on the graph H. Each processor that wants to access a shared memory cell $u \in U$ asks in each step either $M_{h_1(u)}$ or $M_{h_2(u)}$. This corresponds to directing the edge u in H to $M_{h_1(u)}$ or $M_{h_2(u)}$, respectively. Then, if a module M_j answers the request to cell u (that is, either the content of u is shown in the window of M_j during the reading phase, or the content of u is changed according to the processor that wants to write into u) then the edge labeled u is removed from H. That is, we direct every edge in H and then every node removes one edge (if any) that points to it. Before the next step starts, the orientations from the remaining edges are erased. The simulation ends when all the edges from H are removed.

The $\mathcal{O}(\log \log /\log \log \log n)$-time simulation from [CMS95] is based on very local information. Each node looks only at its neighbors in the access graph and, based on their degrees, chooses one incident edge. The main observation leading to improvements of that result is to look more globally and try to use information on as many nodes and edges as possible. The notion of the *k-neighborhood* plays the crucial role in our paper. The *k-neighborhood* of a node v is the subgraph of H containing the nodes and the edges that are reachable from v by a path of length at most k. Instead of looking only at the neighbors, now each node v will base the decision which incident edge to remove on the structure of its k-neighborhood. We will explore the k-neighborhood of each node in H. As we show in Section 3, essentially all information on the k-neighborhood can be computed in $\mathcal{O}(\log k \log^* n)$ time with linear total work. In this section we will first show a process that removes all edges from a connected component with diameter δ in time $\mathcal{O}(\log \delta \log^* n)$. Then we present an algorithm that essentially breaks large connected components into smaller ones and then use the process on graphs with small diameter.

4.1 Cleaning up the Neighborhood

The access schedules described in previous papers (e.g. in [KLM92]) show how to remove all edges of a connected component C of H in time $\mathcal{O}(\log(|C|))$. The following lemma describes how to achieve a time $\mathcal{O}(\log(\text{diameter of } C))$. Note that this does not give fast simulations on its own, because H has a connected component of diameter $\Omega(\log n / \log \log n)$, with constant probability (see Lemma 12).

Lemma 8. (Cleaning up connected components)
Let δ be the maximal diameter of the connected components in H. Then one can remove all edges in H in time $\mathcal{O}(\log \delta \log^* n)$ with linear total work, w.h.p..

Proof. As it is shown in Lemma 4, one can find, for each node v in H, all simple paths that start at v (and of course are of length at most δ) in time $\mathcal{O}(\log \delta \log^* n)$ with linear total work, w.h.p.. All these paths are stored in a padded-consecutive subarray S_v. Thus S_v contains exactly the nodes of v's connected component. Now each node v can find the node w with the minimal identifier in its component. This can be easily done in $\mathcal{O}(\log^* n)$ time. If $v \neq w$, then v finds all nodes $u_1, u_2, \cdots, u_r$, such that (v, u_i) is the first edge of a simple path from v to w. v "directs" the edges (v, u_i), $1 \leq i \leq r$, to v, that is, the processor assigned to the edge (v, u_i) will try to access the module corresponding to the node v. Because of Lemma 2 we have $r = O(1)$, w.h.p.. Therefore, after $O(1)$ steps each processor will get the answer on its request.

4.2 An $\mathcal{O}(\log \log \log n \log^* n)$-Time Simulation

In this section we describe a simulation of an n-processor EREW-PRAM on an n-processor DMM that improves all the previously known simulations exponentially. Essentially we show how to reduce the diameter of H to $(\log \log n)^2$ efficiently.

A *k-branch* of a node v in the access graph H is the set of all different simple paths of length at most k that start with the same edge incident to v. Clearly every node v has $deg(v)$ many k-branches. Define LEVEL(r) of a k-branch of a node v to be the set of all simple paths of length r, $0 < r \leq k$, of this k-branch. The *weight* of a k-branch of a node v is the bit-vector $\overline{w} = (w_1, \cdots, w_k)$. The value of w_r is 1 if and only if the number of simple paths in LEVEL(r) of the branch is at least 2^{r-1}. If the weight of a k-branch satisfies $w_1 = w_2 = \cdots = w_r = 1$, then we call it *r-complete*. We order the weights with respect to the lexicographical ordering. Informally, a k-branch is lexicographically larger than another k-branch, if it is more similar to a complete binary tree with respect to the number of nodes in each level.

SIMULATION

- *Each node v removes the incident edge which is the beginning of the k-branch with the maximal weight.*

- *Clean up all connected components.*

Using the path-access-structure, Lemma 4 and Lemma 7 we can easily derive an algorithm for the first step of SIMULATION, that needs time $\mathcal{O}(\log k \log^* n)$ and linear total work, w.h.p.. Now we prove that at the beginning of the clean up procedure the maximal diameter of each connected component in H is at most $O(k^2)$, w.h.p., for $k \geq \log \log n$.

Lemma 9. *Let $k \geq \log \log n$ and ζ denote the number of cycles in H. After the first step of the algorithm, w.h.p., H does not contain any simple path of length at least $(\zeta + 1) \cdot (2k + 1)^2$.*

Proof. Assume in the contrary, that at the beginning of the clean up procedure a simple path p of length $(\zeta + 1) \cdot (2k + 1)^2$ survives. Let us call each branch that starts with an edge from p the *path-branch* and we call any k-branch chosen in the first step of the algorithm by a node from p the *side-branch* of this node. Since no node of the path p has been removed in the first step, for each node of p the side-branch differs from the path-branch. Lemma 2 ensures that $\zeta = O(1)$. Hence there exists a subpath $\tilde{p} = (v_0, v_1, \ldots, v_{2k})$ of p, such that all vertices of the side-branches of nodes from $\tilde{p}$ are not contained in any cycle of length smaller than $2k$.

We show that for each node from $\tilde{p}$ the side-branch must be r-complete, for all nodes v_i, $0 \leq i \leq 2k - r$, the right path-branch (starting from the edge (v_i, v_{i+1})) is r-complete, and for all nodes v_i, $r \leq i \leq 2k$, the left path-branch (starting from the edge (v_i, v_{i-1})) is also r-complete.

We prove the desired properties by induction on levels.

LEVEL(1) Because no node of a simple path $\tilde{p}$ of length $2k$ was removed in the first step of the algorithm, each node from $\tilde{p}$ had to remove an incident edge not belonging to $\tilde{p}$. Since for each path-branch of a node from $\tilde{p}$ we have $w_1 = 1$, all the desired path-branches and side-branches are 1-complete.

LEVEL(r) Now assume that $r > 1$ and for each node of $\tilde{p}$ the side-branch and the respective path-branches are $(r - 1)$-complete. Consider a node v_i, $0 \leq i \leq 2k - r$, and the edge (v_i, v_{i+1}). Since the side-branch and the right path-branch of v_{i+1} are both $(r - 1)$-complete and they are disjoint and have no cycle of length smaller than or equal to k, the right path-branch of v_i also must be r-complete. The nodes v_i, $r \leq i \leq 2k$, can be treated in a similar way.

This implies that we need a connected structure of at least $2k \cdot 2^{k-1}$ nodes in H for a simple path of length $\zeta \cdot (2k+1)^2$ to survive the first step of the algorithm. For $k \geq \log \log n$ this contradicts to Lemma 2.

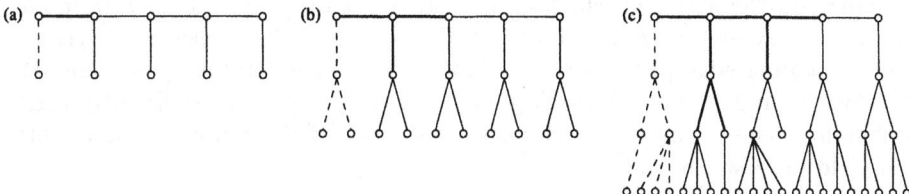

Fig. 1. (a) the path $\tilde{p}$ with the edges that are removed by the nodes of $\tilde{p}$; (b) the path $\tilde{p}$ with the edges required by the elimination rule on LEVEL(2); (c) the path $\tilde{p}$ with the edges required by the elimination rule on LEVEL(3).

The following theorem follows from Lemma 4, Lemma 7, Lemma 8, and Lemma 9.

Theorem 10. SIMULATION *simulates a step of an n-processor EREW PRAM on an n-processor DMM in* $\mathcal{O}(\log\log\log n \log^* n)$ *time with linear total work, w.h.p..*

5 A Lower Bound for PRAM Simulations

In this section we show a lower bound for shared memory simulations based on hashing and performing the *one-out-of-two schedule*. Again, we view the one-out-of-two schedule as a process of removing edges from the graph H as defined in Section 2. We consider the following class of *topological algorithms*:

– Each node of H knows its k-neighborhood; we charge for this $\mathcal{O}(\log k)$ time.

– All nodes repeat in parallel the following iteration until all edges are removed:

  • Each node that has degree one removes all edges in its k-neighborhood.

  • Each other node removes an incident edge basing its decision which one to remove only on the topology of its k-neighborhood.

Observe that this class of algorithms covers all previously known algorithms for randomized shared memory simulations (which essentially use at most the 1-neighborhood) and the algorithms presented in this paper. This also means that our simulation is almost optimal within the class of topological algorithms. The *cleaning up connected components* procedure (Lemma 8) is covered by the capability of the nodes of degree one to eliminate their whole k-neighborhood. In the proof of the lower bound we assume that we can without loss of generality delete edges from the graph H, i.e., with this operation we do not prejudice any simulation. The capability of removing edges has positive effects for all known algorithms. Finally, a node can only get the knowledge of its k-neighborhood if the simulation needs time $\log k$. This maximal knowledge is given in the class of topological algorithms in advance.

The main idea of the lower bound is to focus only on completely symmetric structures in the access graph. Each node that has distance at least k from all leaves has a symmetric k-neighborhood. Hence it randomly makes the decision which outgoing edge to remove. We show that after performing these random decisions a smaller symmetric subgraph will still be left, with sufficiently high probability. The bound for the decrease of the size of the symmetric subgraph will yield the lower bound.

Definition 11. A (d_i, T_i)-tree is a complete d_i-ary tree of depth T_i.

Define the values of d_i and T_i for $0 \le i < \frac{\log\log\log n}{8\log\log\log\log n}$ as follows:

$$d_i = \frac{\log\log n}{(\log\log\log n)^{4(i+1)}} \text{ and } T_i = \frac{\log\log n}{(\log\log\log n)^{4(i+1)-2}}$$

Fix an algorithm A that belongs to the class of topological algorithms. We want to maintain a (d_i, T_i)-tree in the access graph remaining after performing i iterations of the algorithm, with sufficiently high probability. The proof of this invariant is done by induction, which is based on the following two lemmas. Their proofs are omitted in this extended abstract. Let H be a random graph defined in Section 2.

Lemma 12. *Let $c < \log\log n$ and let T be a fixed tree with q nodes. For $q \le \frac{\log n}{9\log\log n}$ the probability that T is a subgraph of H is at least $1/2$.*

Lemma 13. *Consider a (d_i, T_i)-tree, for $0 \le i < \frac{\log\log\log n}{8\log\log\log\log n}$, $k \le \sqrt{\log\log n}$. If every node randomly removes an incident edge then, with probability at least $1 - 1/\log n$, at most $d_i^{T_i - T_{i+1} - 2k}$ of the nodes have degree smaller than d_{i+1}.*

Using these two lemmas we can show that after iteration i of algorithm A a (d_i, T_i)-tree is left, with sufficiently high probability.

Lemma 14. *After performing i iterations of the algorithm A, a (d_i, T_i)-tree is a subgraph of the remaining access graph with probability at least $(1 - 1/\log n)^i$, for $1 \le i < \frac{\log\log\log n}{8\log\log\log\log n}$ and $k \le \sqrt{\log\log n}$.*

Proof. The proof is done by induction on i. For $i = 0$ we use Lemma 12. Assume that the lemma holds for $i < \frac{\log\log\log n}{8\log\log\log\log n}$. From the induction hypothesis we know that a (d_i, T_i)-tree is a subgraph of the access graph at the beginning of round $i + 1$. Without loss of generality, we only consider the edges from this tree and remove all other edges. Because of the definition of our class of topological algorithms we remove all nodes that have distance at most k from any leaf of this tree.

For the remaining nodes the topology of their k-neighborhood is fully symmetric, so it is not possible for them to distinguish between the incident edges because every permutation of the labeling of the edges and modules is equally likely. Therefore each decision based on the topology of the k-neighborhood made by these nodes is random and is independent on decisions of other nodes.

Hence, using Lemma 13 at most $d_i^{T_i - T_{i+1} - 2k}$ nodes have degree smaller than d_{i+1}, and a (d_i, T_i)-tree is a subgraph of the remaining graph.

The invariant of Lemma 14 holds in each iteration i, for $1 \leq i < \frac{\log\log\log n}{8\log\log\log\log n}$, and k small enough, even if we start only with $\frac{n}{\log\log\log n}$ edges. This implies the following theorem.

Theorem 15. *For any algorithm of the class of topological algorithms the expected number of iterations until all edges of the access graph H will be removed is $\Omega(\frac{\log\log\log n}{\log\log\log\log n})$.*

Proof. In time $\mathcal{O}(\frac{\log\log\log n}{\log\log\log\log n})$ it is only possible to see a k-neighborhood for

$$k \leq 2^{O(\frac{\log\log\log n}{\log\log\log\log n})} \leq \sqrt{\log\log n}.$$

Therefore using Lemma 14, after $i \leq \frac{\log\log\log n}{8\log\log\log\log n}$ iterations some edges will be left. The probability for this event can be bounded by

$$\left(1 - \frac{1}{\log n}\right)^{\frac{\log\log\log n}{8\log\log\log\log n}} \geq 1/e.$$

This yields the desired expected number of iterations.

References

[BH93] H. Bast and T. Hagerup. Fast parallel space allocation, estimation and integer sorting (revised). Technical Report MPI-I-93-123, Max-Planck-Institut für Informatik, Im Stadtwald 66123 Saarbrücken, June 1993. A preliminary version appeared in *Proceedings of the 23rd Annual ACM Symposium on Theory of Computing* under the title "Constant-Time Parallel Integer Sorting", 1991.

[BV93] O. Berkman and U. Vishkin. Recursive star-tree parallel data structure. *SIAM Journal on Computing*, 22(2):221–242, 1993. A preliminary version appeared in *Proceedings of the 30th IEEE Symposium on Foundations of Computer Science*, pages 196–202, 1989.

[CMS95] A. Czumaj, F. Meyer auf der Heide, and V. Stemann. Improved optimal shared memory simulations, and the power of reconfiguration. To appear in *Proceedings of the 3rd Israel Symposium on Theory of Computing and Systems*, 1995.

[DM90] M. Dietzfelbinger and F. Meyer auf der Heide. A new universal class of hash functions and dynamic hashing in real time. In *Proceedings of the 17th Annual International Colloquium on Automata, Languages and Programming*, pages 6–19, 1990.

[DM93] M. Dietzfelbinger and F. Meyer auf der Heide. Simple, efficient shared memory simulations. In *Proceedings of the 5th Annual ACM Symposium on Parallel Algorithms and Architectures*, pages 110–119, 1993.

[GMR94] L. A. Goldberg, Y. Matias, and S. Rao. An optical simulation of shared memory. In *Proceedings of the 6th Annual ACM Symposium on Parallel Algorithms and Architectures*, pages 257–267, 1994.

[GMV91] J. Gil, Y. Matias, and U. Vishkin. Towards a theory of nearly constant time parallel algorithms. In *Proceedings of the 32nd IEEE Symposium on Foundations of Computer Science*, pages 698–710, 1991.

[GMV94] M. T. Goodrich, Y. Matias, and U. Vishkin. Optimal parallel approximation algorithms for prefix sums and integer sorting. In *Proceedings of the 5th Annual ACM-SIAM Symposium on Discrete Algorithms*, pages 241–250, 1994.

[J92] J. JáJá. *An Introduction to Parallel Algorithms*. Addison Wesley, 1992.

[KLM92] R. M. Karp, M. Luby, and F. Meyer auf der Heide. Efficient PRAM simulation on a distributed memory machine. Technical Report tr-ri-93-134, University of Paderborn, 1993, to appear in *Algorithmica*. A preliminary version appeared in *Proceedings of the 24th Annual ACM Symposium on Theory of Computing*, pages 318–326, 1992.

[KU86] A. Karlin and E. Upfal. Parallel hashing - an efficient implementation of shared memory. In *Proceedings of the 18th Annual ACM Symposium on Theory of Computing*, pages 160–168, 1986.

[L92a] F. T. Leighton. *Introduction to Parallel Algorithms and Architectures: Arrays, Trees, Hypercubes*. Morgan Kaufmann Publishers, San Mateo, 1992.

[L92b] F. T. Leighton. Methods for packet routing in parallel machines. In *Proceedings of the 24th Annual ACM Symposium on Theory of Computing*, pages 77–96, 1992.

[M92] F. Meyer auf der Heide. Hashing strategies for simulating shared memory on distributed memory machines. In *Proceedings of the Heinz Nixdorf Symposium on Parallel Architectures and Their Efficient Use*, pages 20–29, 1992.

[MPR94] P. D. MacKenzie, C. G. Plaxton, and R. Rajaraman. On contention resolution protocols and associated probabilistic phenomena. In *Proceedings of the 26th Annual ACM Symposium on Theory of Computing*, pages 153–162, 1994.

[MSS95] F. Meyer auf der Heide, C. Scheideler, and V. Stemann. Exploiting storage redundancy to speed up randomized shared memory simulations. To appear in *Proceedings of the 12th Annual Symposium on Theoretical Aspects of Computer Science*, 1995.

[R93] P. Ragde. The parallel simplicity of compaction and chaining. *Journal of Algorithms*, 14:371–380, 1993.

[R91] A. G. Ranade. How to simulate shared memory. *Journal of Computer and System Sciences*, 42:307–326, 1991.

[S89] A. Siegel. On universal classes of fast high performance hash functions, their time-space tradeoff, and their applications. In *Proceedings of the 30th IEEE Symposium on Foundations of Computer Science*, pages 20–25, 1989.

[U84] E. Upfal. Efficient schemes for parallel communication. *Journal of the ACM*, 31:507–517, 1984.

[UW87] E. Upfal and A. Wigderson. How to share memory in a distributed system. *Journal of the ACM*, 34:116–127, 1987.

Implementing Shared Memory on Multi-Dimensional Meshes and on the Fat-Tree*

(Extended Abstract)

Kieran T. Herley[1], Andrea Pietracaprina[2], Geppino Pucci[3]

[1] Department of Computer Science, University College Cork, Cork, Ireland
[2] Dipartimento di Matematica Pura e Applicata, Università di Padova, Padova, Italy
[3] Dipartimento di Elettronica e Informatica, Università di Padova, Padova, Italy

Abstract. We present deterministic upper and lower bounds on the slowdown required to simulate an (n, m)-PRAM on a variety of networks. The upper bounds are based on a novel scheme that exploits the splitting and combining of messages. Such a scheme can be implemented on an n-node d-dimensional mesh, with d constant, and on an n-leaf pruned butterfly, attaining the best worst-case slowdowns to date for such interconnections. Moreover, the one for the pruned butterfly is the first PRAM simulation scheme on an area-universal network. Finally, under the standard point-to-point assumption, we prove a bandwidth-based lower bound on the slowdown of any deterministic PRAM simulation on an arbitrary network, formulated in terms of its decomposition tree.

1 Introduction

The problem of implementing shared memory on distributed-memory architectures has been intensively studied over the last decade. Generally, this problem has been referred to as the PRAM simulation problem and involves representing the m cells of the PRAM shared memory (called *variables*) among the n processor-memory nodes of the simulating machine, in such a way that an arbitrary PRAM *step*, in which any n-tuple of variables is read or written, can be efficiently executed. The time required to simulate a PRAM step is known as the *slowdown* of the simulation. A number of approaches to this problem, both probabilistic and deterministic, have been investigated for a variety of well-known architectures such as the complete interconnection, the mesh of trees, the butterfly, as well as a variety of expander-based architectures, among others.

We will not attempt to summarize the entire literature on this problem here but only quote those results that directly relate to our work, and refer the interested reader to [11] for a recent and comprehensive summary of previous work on this topic. In [1], Alt *et al.*, building on the work of [14], presented a deterministic scheme to simulate a PRAM with n processors and m variables (called an (n, m)-PRAM) on an n-node *Module Parallel Computer* (MPC), an

---

* This research was supported, in part, by the ESPRIT III Basic Research Programme of the EC under contract No. 9072 (project GEPPCOM).

architecture in which each node includes both a processor and a private memory module accessible only to that processor, and in which the nodes are connected by a crossbar that allows each node to transmit one message to, and receive one message from any other node per step. Their scheme employs the following copy-based method for the representation of the PRAM variables, which most of the deterministic simulation algorithms, including this present work, adopt. Specifically, each variable is represented by $2c - 1 = O(\log m)$ copies, each consisting of a value and a timestamp, and distributed carefully among the memory modules of the simulating machine. To write a variable, at least c of its copies are overwritten to reflect the intended value and the time of writing. To read a value, at least c copies are inspected. This set of copies must contain at least one of the copies most recently written, which is readily identifiable by virtue of its timestamp. They show that for a suitable distribution of the copies among the nodes of the machine, any n-tuple of variables may be accessed (read or written) in $O(\log m)$ time.

The above scheme can be ported to an arbitrary network by simulating each MPC step using standard techniques. In particular, this approach yields a simulation with slowdown $O(n^{1/d} \log m)$ on an n-node d-dimensional mesh, $d = O(1)$, and a simulation with slowdown $O(\sqrt{n} \log m)$ on an n-leaf pruned butterfly, which are the interconnections that we consider in this paper. In [1], it was also observed that a simple PRAM simulation for the 2-dimensional mesh with an optimal slowdown of $O(\sqrt{n})$, is indeed possible. Unfortunately, this simulation requires up to n copies per variable, resulting in an unacceptable memory blow-up. Moreover, the method does not appear to extend easily to higher dimensional meshes or to the pruned butterfly.

Most of the deterministic simulations that appear in the literature, including those of this paper, rely on certain expander-based graphs whose existence can be proved, but for which no efficient construction is known. Recently, Pietracaprina et al. [11, 12] have studied deterministic simulations based entirely on explicitly constructible structures. By resorting to a complex hierarchical arrangement of constructible, mildly-expanding graphs, they achieve $O(\sqrt{n} \log n)$ slowdown on an n-node mesh for memories of $O(n^{1.5})$ size, using $O(\log^{1.59} n)$ copies per variable. In this paper, our focus is on slowdown rather than constructivity. By employing more powerful expander-based structures, we achieve a better slowdown than that of [12] at a lower level of redundancy.

It might appear, at least at first glance, that updating c copies apiece for n variables would require the physical movement of cn distinct packets across the entire network, resulting in an overly expensive access protocol. In order to circumvent this difficulty, we devise a novel splitting and combining technique based on the following idea. If a processor p wishes to send the same packet to two nodes x and y that are distant from p but close to one another, then rather than dispatch a separate packet for each, it may be more efficient to dispatch a single message to some node z close to both x and y. At z, the message is then split into two packets which are forwarded to x and y. A careful implementation of this idea yields the result stated in the following theorem, which constitutes the fastest deterministic PRAM simulation on multi-dimensional meshes to date.

Theorem 1. *There exists a scheme to simulate an (n, m)-PRAM on a n-node d-dimensional mesh, $d = O(1)$, with worst-case slowdown $O\left(n^{1/d}(\log(m/n))^{1-1/d}\right)$, using $O(\log(m/n))$ copies per variable and $O((m/n)\log^3(m/n))$ storage per node.*

In order to implement the strategy outlined above, the scheme relies on a recursive decomposition of the mesh and on efficient algorithms for k-sorting, where each processor initially holds k packets, and for (k, k)-routing, where each processor sends and receives at most k packets. In order to achieve the storage bound, special techniques are needed, as reported in the appendix.

The n-leaf pruned butterfly [2] is an area-universal network that is a variant of Leiserson's fat-tree [9]. Although quite different from the 2-dimensional mesh in terms of the details of its structure, it is sufficiently similar in its bandwidth characteristics to support the key operations upon which our simulations rely, with comparable efficiency. By providing novel sorting and routing primitives for this network, and by using its natural decomposition into subtrees, we are able to implement the above simulation scheme with the same slowdown achieved for the 2-dimensional mesh, thereby obtaining the first PRAM simulation on an area-universal network. The result is stated in the following theorem.

Theorem 2. *There exists a scheme to simulate an (n, m)-PRAM on an n-leaf pruned butterfly with worst-case slowdown $O(\sqrt{n}\log(m/n))$, using $O(\log(m/n))$ copies per variable and $O((m/n)\log^3(m/n))$ storage per node.*

Lower bounds on the slowdown of PRAM simulations on bounded degree networks have been presented in a number of studies [1, 8, 5]. All such bounds, however, apply to the entire class of bounded degree networks, among which there are networks with low diameter and high bandwidth. For example, in [5] the authors show an $\Omega(\log^2(m/n)/\log\log(m/n))$ lower bound, which is too weak for our purposes, since a trivial $\Omega(n^{1/d})$ (resp., $\Omega(\sqrt{n})$) lower bound holds for d-dimensional meshes (resp., pruned butterfly), due to diameter (resp., bandwidth) limitations. In this paper, we present the first lower bound argument that takes into account the characteristics of the individual network. To capture the properties of the network topology, the bound exploits the notion of decomposition tree [3, 9], which provides a partition of the network into disjoint regions of limited bandwidth. We consider only *on-line* simulations, in which the simulation of each step does not commence until the simulation of all preceding read steps has concluded. Furthermore, as in all previous works, our lower bound is proved under the *point-to-point* assumption, which requires that a processor updating a number of copies of a variable dispatch a separate message for each copy.

Assume that the simulating network has a $[w_0, w_1, \ldots, w_{\log n}]$ *decomposition tree* [9], that is, for any i, $0 \le i \le \log n$, the network can be partitioned into 2^i disjoint regions, each connected to the rest of the network by at most w_i edges. The following theorem states our general lower bound.

Theorem 3. *Let $m \geq 2n^2$ and let $T = \Omega(m/n)$. The worst-case running time of any on-line, point-to-point simulation of a T-step (n,m)-PRAM computation on an n-processor network with a $[w_0, w_1, \ldots, w_{\log n}]$ decomposition tree, is*

$$\Omega\left(T \min_{r \geq 0}\left\{\max_{1 \leq h,k \leq \log n}\left\{g_{h,k}(r) + r\frac{n}{2^h w_h}\right\}\right\}\right), \tag{1}$$

where $g_{h,k}(r) = \min\left\{n/w_h, n/(rw_k), n\left(m/2n^2\right)^{1/2^r}/(2^k w_k)\right\}$.

By specializing the lower bound to d-dimensional meshes, with d constant, and to the pruned butterfly, we obtain the following result.

Corollary 4. *Under the same hypotheses as Theorem 3, any simulation of a T-step (n,m)-PRAM computation requires time*

$$\Omega\left(Tn^{\frac{1}{d}}\left(\frac{\log(m/2n^2)}{\log\log(m/2n^2)}\right)^{1-\frac{1}{d}}\right) \quad \left[resp., \ \Omega\left(T\sqrt{\frac{n\log(m/2n^2)}{\log\log(m/2n^2)}}\right)\right]$$

on an n-node d-dimensional mesh [resp., an n-leaf pruned butterfly].

Note that the point-to-point assumption precludes the splitting and combining of messages, therefore the above lower bounds do not apply to our simulations. However, we also prove similar bounds in the unrestricted model by limiting the total level of redundancy used to represent the variables. Such bounds show that our simulations use an amount of redundancy which is only a doubly logarithmic factor higher than the minimum redundancy needed to achieve the same slowdown. Specifically, we have:

Theorem 5. *For any constant $\alpha \geq 1$ there exists a constant $\beta > 0$ such that any simulation of an (n,m)-PRAM that uses a total of at most $\beta m \log(m/2n^2)/\log\log(m/2n^2)$ copies for the variables requires slowdown*

$$\Omega\left(n^{\frac{1}{d}}\left(\frac{\log(m/2n^2)}{\log\log(m/2n^2)}\right)^{(1-1/d)\alpha}\right) \quad \left[resp., \ \Omega\left(\sqrt{n}\left(\frac{\log(m/2n^2)}{\log\log(m/2n^2)}\right)^{\alpha/2}\right)\right]$$

on an n-node d-dimensional mesh [resp., an n-leaf pruned butterfly].

The rest of the paper is organized as follows. Section 2 discusses the memory map that describes the distribution of the copies among the memory modules. In Section 3, the mesh simulation algorithm is presented. The algorithm consists of two phases, copy-selection and routing, which are described in Subsections 3.2 and 3.1, respectively. In Section 4, the scheme is ported to the pruned butterfly. The lower bounds are presented in Section 5. Finally, the techniques needed to meet the space bounds of the simulations are discussed in the appendix.

2 Memory Organization

Consider the simulation of an (n, m)-PRAM on an n-node machine and suppose that each variable be replicated into $2c - 1$ copies, for a suitable integer c. It is convenient to model the distribution of the copies of the variables among the nodes of the machine by means of a bipartite graph $G = (U, V; E)$, where U represents the set of variables, V the set of nodes, and $2c - 1$ edges connect each variable to the distinct nodes storing its copies. Note that there is a one-to-one correspondence between E and the set of all copies.

Let $S \subseteq U$ and let $F \subseteq E$ contain exactly k edges incident on every $s \in S$. We call F a k-*bundle* for S. We say that F has *degree* q if at most q edges of F are incident on any node of $\Gamma_F(S)$, where $\Gamma_F(S)$ denotes the subset of V touched by the edges in F. Our simulations adopt the standard majority protocol requiring that the access to a set of variables S be performed by accessing the copies corresponding to a c-bundle for S. In order to achieve fast simulation times, it is desirable to access a c-bundle of low degree, since the degree of a c-bundle is a measure of the congestion at the nodes of the underlying machine, that is, the number of physical copies that ultimately have to be accessed from some individual node.

The existence of c-bundles of low degree is intimately related to the expansion properties of the graph G. This motivates the following definition [5] that characterizes a class of graphs that make good memory organizations.

Definition 6. A bipartite graph $G = (U, V; E)$ with $|U| = m$, $|V| = n$, and input degree d is a (λ, d, c, σ)-generalized expander if, for every $S \subseteq U$ such that $|S| \le \sigma n$ and for every c-bundle F of S, $|\Gamma_F(S)| \ge \lambda c |S|$.

The memory organization of our simulations is governed by a $(\lambda, 2c - 1, c, 1/gc)$-generalized expander $G = (U, V; E)$, where $\lambda < 1$ is some arbitrary constant, $g = (4e^{1+\lambda})^{2/(1-\lambda)}$ and $c = (2/(1 - \lambda)) \log(3\lambda m/n) = \Theta(\log(m/n))$. We will also assume when referring to such graphs that the degree of each vertex in V is $O((m/n)c)$. The existence of such a graph can be established with a minor modification to a result presented in [5], but no efficient construction for graphs of this type is yet known.

It is assumed for the moment that each node in the network holds a copy of a read-only table that encodes the structure of the memory organization. In the appendix, we will show how this table may be represented in a distributed fashion, requiring $O((m/n) \log^3(m/n))$ storage per node. Note that the total storage of the simulating machine will then be only a polylogarithmic factor away from the size of the PRAM memory.

Let $S \subseteq U$ be the set of variables to be accessed. The construction of a c-bundle for S starts with the set $E(S) \subseteq E$ of all edges incident on S, and proceeds as a sequence of *whittling steps*. Each whittling "prunes" the set of edges by selecting c edges apiece for some of the variables in S, and discarding the remaining $c - 1$. At the beginning of a whittling step, a variable is said to be *alive* if the c edges for the variable have not been selected yet, and *dead* otherwise. The sequence terminates when all variables are dead, at which point we are left with the desired c-bundle. The variables to whittle at each step are chosen to ensure that the degree of the c-bundle will not exceed a fixed threshold q, whose value will be specified later.

Let $S_i \subseteq S$ denote the set of live variables and $F_i \subseteq E(S)$ the residual set of edges at the beginning of the i-th whittling step. Initially, $S_1 = S$ and $F_1 = E(S)$. Conceptually, the i-th whittling step identifies a set W_i of congested nodes and selects c edges apiece for as many live variables as possible without involving W_i. We say that $x \in S_i$ is *confined* to W_i if x has c or more copies stored in nodes of W_i. In the i-th whittling step, for each $x \in S_i$ which is not confined to W_i, we select c edges not incident on W_i and remove the remaining $c - 1$ from F_i. This operation will be referred to as *whittling of S_i with respect to W_i*.

Let $F \subseteq E$. We say that a node $v \in V$ is *q-congested* with respect to F if at least q edges of F are incident on v. The following definition formalizes the notion of a sequence of whittling steps that construct a c-bundle.

Definition 7. Let $S \subseteq U$, $|S| \leq n$. A *q-whittling sequence* of length k for S is a sequence (S_1, F_1, W_1), (S_2, F_2, W_2), ..., (S_k, F_k, W_k) such that

- $S_1 = S$, $F_1 = E(S)$, and $W_1 \subseteq V$ is a set of at most $(2c - 1)n/q$ nodes including all nodes that are q-congested with respect to $E(S)$;
- For $i > 1$, S_i and F_i are, respectively, the set of live variables and the set of residual edges obtained after whittling S_{i-1} with respect to W_{i-1}. W_i is the set of q-congested nodes with respect to F_i;
- $S_k = W_k = \emptyset$ and F_k is a c-bundle for S.

Note that each W_i, with $i > 1$, contains only the q-congested nodes, while W_1 may include an additional small number of noncongested nodes. The rationale behind this asymmetry will become clear later in the paper. Note also that F_k, the final c-bundle for S, has degree at most q. The following lemma characterizes the rate at which the variables "die" during a whittling sequence.

Lemma 8. *Let S be a set of at most n variables, and $q = (2c - 1)gn/(n - g)$ $= \Theta(\log(m/n))$. Then, for any q-whittling sequence of length k, (S_1, F_1, W_1), (S_2, F_2, W_2), ..., (S_k, F_k, W_k), we have $|S_i| \leq n/(\lambda gc)^{i-1}$ for $i \geq 1$. Therefore $k = O(\log n / \log \log(m/n))$.*

Proof (sketch). The proof proceeds by induction on i. The basis $i = 1$ is trivial. Now consider the case $i = 2$ and suppose, for a contradiction, that $|S_2| > n/\lambda gc$. Then we can select an arbitrary subset $S' \subseteq S_2$ of $n/\lambda gc$ variables, and a c-bundle F' for S', with $\Gamma_{F'}(S') \subseteq W_1$. By the expansion properties of the memory organization, this would imply $|W_1| \geq n/g$, which contradicts the hypothesis $|W_1| \leq (2c - 1)n/q < n/g$. Finally, suppose that $i \geq 3$. Since no edge in F_{i-1} relating to variables in $S - S_{i-1}$ can be incident on nodes of W_{i-1}, we have that $|W_{i-1}| \leq (2c-1)|S_{i-1}|/q$. On the other hand, all variables in S_i are confined to W_{i-1} and $|S_i| \leq |S_2| \leq n/(\lambda cg)$, therefore $|S_i| \leq |W_{i-1}|/(\lambda c)$. The bound for S_i follows by combining the two inequalities.

3 Simulation on Meshes

We restrict ourselves to describing the simulation of an (n, m)-EREW PRAM step on a $\sqrt{n} \times \sqrt{n}$ 2-dimensional mesh. The extension to higher dimensions or more powerful PRAM variants requires minor technical modifications which will be provided in the full version of this extended abstract.

Suppose that each processor-memory node is connected to its immediate neighbours by means of bidirectional wires capable of transmitting a single $O(\log m)$-bit packet per step, and that the PRAM variables are distributed among the mesh nodes according to the memory organization described in the previous section. Consider the simulation of a write step, and let S be the set of n variables to be written. The set S will be represented by means of a set of *variable packets*, each bearing the name of a referenced variable, the value to be written to that variable, and space for a *bitvector* of length $2c - 1$ to be explained later. We will use the symbol S to refer both to the set of variables and the set of packets that represent it. The simulation consists of two phases: *Copy Selection*, during which a c-bundle of degree less than q for S is determined and encoded in the bitvectors of the variable packets; and *Access*, where the actual updates on the selected copies take place. (The simulation of a read step is virtually identical, except that Access will also involve routing the values back to the reading processors.)

Our algorithms will also employ a set of *copy packets* representing the set of copies of S, with one packet per copy. Note that there are up to $(2c - 1)|S|$ distinct copy packets, and any naive manipulation of such a large set is too expensive for our purposes. A crucial feature of our simulation is to use this explicit representation sparingly.

The simulation is based on two algorithmic primitives: *k-sorting* and (k, k)-*routing*. Given a set of packets distributed among the n nodes so that each node holds at most k packets, k-sorting is used to rearrange the packets so that node 1 holds the packets with the k smallest keys, node 2 holds the next k smallest keys, and so on. For any value of k, this can be accomplished in $O\left(k(\sqrt{n} + \log k)\right)$ time on the mesh. The (k, k)-routing problem involves routing a set of packets subject to the constraint that no node is the source (resp., destination) of more than k packets. This can be done on the mesh in $O(k\sqrt{n})$ time. Both algorithms can be found in [7].

For ease of presentation, we will discuss the implementation of *Access* first and then discuss that of *Copy Selection*, which embodies some of the same ideas.

3.1 Access

Recall that following the completion of *Copy Selection*, the bitvectors of S encode a c-bundle of degree less than q for S. (The i-th bit of the bitvector for $u \in S$ indicates whether or not the i-th copy of u belongs to the c-bundle.) The mesh is conceptually partitioned into submeshes of size $\sqrt{n/s} \times \sqrt{n/s}$, called *cells*. Updating the selected copies of u will involve routing a copy of u's variable packet to each node holding a selected copy of u. This is accomplished in three steps.

Firstly, the set S is replicated in each cell of the mesh. Secondly, within each cell a copy packet is generated for each copy of a referenced variable that resides in that cell. (A copy packet is effectively a copy of the appropriate variable packet labelled with the location of the copy in question (destination).) Thirdly, each copy packet is routed locally within its cell to its destination. Note that delaying the generation of the copy packets to the second step is the crucial feature of the algorithm, since in this way all the copy packets within a cell will "share" the cost of the routing from the requesting processor to the cell.

For brevity, we omit the discussion of the first step and only mention that it can be accomplished in $O(\sqrt{ns})$ time using an appropriate combination of routing and splitting.

The second step involves the generation of the appropriate copy packets from the copy of S held locally within each individual cell. Although the copy selection guarantees that there will be at most q copy packets to be sent to each destination, a node of the cell may initially hold up to s variable packets, each of which may generate from 0 to c copy packets. In order to avoid congestion in routing the copy packets, great care has to be exercised to ensure that each node of the cell generate approximately the same number of such packets. Within each cell C, we partition S into *degree classes* $S_C^{(i)}$, $0 \le i \le \lceil \log(2c-1) \rceil$, where $S_C^{(i)}$ contains those variables with between 2^i and 2^{i-1} copies in C. The variables of each class are distributed (using $\lceil \log(2c-1) \rceil$ prefix computations and one execution of (s,s)-routing), so that each node receives at most $\lceil |S_C^{(i)}|/(n/s) \rceil$ of them, which ensures that each node generates $O(q)$ copy packets. The complexity of this step is $O(\log c \sqrt{n/s} + \sqrt{ns})$.

The third step is effectively an instance of (q,q)-routing within each cell and can be completed in $O(q\sqrt{n/s})$ time. Summing the contribution of the three steps and selecting q as in Lemma 8 and $s = q$, we have:

Lemma 9. *The copies corresponding to a c-bundle for S of degree at most $q = O(\log(m/n))$ can be accessed in $O\left(\sqrt{n \log(m/n)} \right)$ time.*

3.2 Copy Selection

In this subsection we describe how we select a c-bundle of low degree for S by performing a sequence of whittling steps as described in Section 2. We break the algorithm into two stages for reasons of efficiency. The first stage performs the first whittling step and applies to n variables, while the second stage performs the rest of the whittling sequence and applies to the variables that are left alive by the first stage.

Stage 1 The first stage whittles S with respect to a set W_1 of nodes that contains (i) all nodes q-congested with respect to $E(S)$; (ii) all nodes belonging to cells storing at least $q(n/s)$ copies of variables in S. Note that $|W_1| \le (2c-1)n/q$, therefore, by Lemma 8, this stage selects c copies apiece, stored outside W_1, for all but $n/(\lambda g c)$ variables in S. Appropriate values for λ and g can be chosen so that $n/(\lambda gc) \le n/(2c-1)$. The implementation of this stage, omitted here for brevity, involves identifying the nodes in W_1 and determining for each variable in S which of its copies are incident on this set. In essence, the algorithm is similar to that for the *Access*, though more involved, and has the same running time.

Stage 2 Lemma 8 implies that the end of Stage 1 we have determined a c-bundle for all but $n/(2c-1)$ variables. The i-th whittling step $(i \ge 2)$ involves whittling S_i with respect to the set W_i of nodes q-congested with respect to F_i. By representing the edges of S_i by a set of $r_i = (2c-1)|S_i|$ copy packets, this whittling may be completed efficiently within a submesh of size $\sqrt{r_i} \times \sqrt{r_i}$ in time $O(\sqrt{r_i})$ by means of sorting and prefix operations. By Lemma 8, $r_{i+1} \le r_i/(\lambda gc)$, therefore the cost of the sequence of whittling steps performed during the second stage is $O\left(\sum_{i=1}^{\infty} \sqrt{r_i}\right) = O(\sqrt{n})$, which is always dominated by the cost of Stage 1. Combining the costs of the two stages, we obtain the following.

Lemma 10. *Given a set S of n distinct variables, a c-bundle for S of congestion at most $q = \Theta(\log(m/n))$ can be constructed in $O\left(\sqrt{n \log(m/n)}\right)$ time.*

The combination of Lemma 10 with Lemma 9 shows that our simulation achieves the running time stipulated in Theorem 1.

4 Simulation on the Pruned Butterfly

A closer look at the PRAM simulation devised for the mesh reveals that the copy selection and access phases rely upon the recursive decomposition of the network into subnetworks, and upon the primitives of sorting and routing. In fact, the scheme can be ported to any network topology with an appropriate decomposition into subnetworks (cells) and for which an implementation of the above primitives is available. In this section, we provide efficient algorithms for k-sorting and (k, k)-routing on the pruned butterfly, a universal network belonging to the class of fat-trees. As a consequence, we obtain the first efficient PRAM simulation on a universal network.

The n-leaf *pruned butterfly*, introduced in [2], is a variant of Leiserson's fat-tree [9]. Its coarse structure may be interpreted as a n-leaf complete binary tree where the leaves represent the processor-memory nodes of the machine, the internal nodes represent clusters of routing switches, and where the edges represent channels of bandwidth that doubles every other level from the leaves to the root. More precisely, each subtree of n' leaves is connected to its parent through a channel of capacity $\Theta\left(\sqrt{n'}\right)$. A 16-leaf pruned butterfly is illustrated in Figure 1, where each ellipse indicates a cluster of routing switches that collectively constitute a single internal node of the underlying binary tree. A formal definition of the network can be found in [2]. The pruned butterfly is *area-universal* in the sense that it can route any set of messages almost as efficiently as any circuit of similar area. The following lemma summarizes the results on sorting and routing for the pruned butterfly.

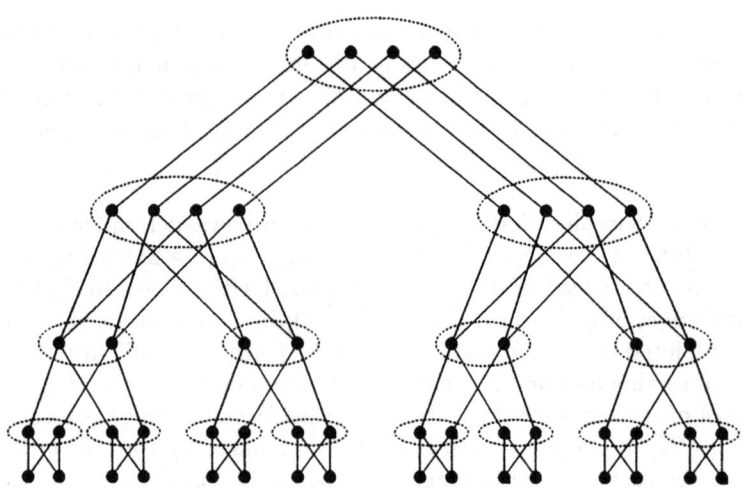

Fig. 1. A 16-leaf pruned butterfly

Lemma 11. *Any instance of k-sorting and (k,k)-routing can be executed on the n-leaf pruned butterfly in time $O\left(k(\sqrt{n}+\log k)\right)$.*

Proof (sketch). Both sorting and routing algorithms rely upon the following primitive. Consider a collection of packets $p_1, p_2, \cdots, p_h$, $h \leq n$, stored one per node among the leaves of an n-leaf pruned butterfly, where the i-th packet has source s_i and destination d_i. We refer to the collection as a *wave* if (i) $s_1 < s_2 < \cdots < s_h$, (ii) $d_1 < d_2 < \cdots < d_h$, and (iii) h is no greater than $\sqrt{n}$. In [2] it is argued that any wave can be routed in $O(\log n)$ time. Moreover, a sequence of t waves may be routed in a pipeline fashion in $O(t + \log n)$ time.

For sorting, we adapt Batcher's bitonic sorting paradigm to handle the case where $k = 1$, and invoke the techniques of Baudet and Stevenson [4] to provide the generalization to larger values of k. Namely, the algorithm is structured as a sequence of $\log n$ merging phases, where the i-th phase merges pairs of sequences of length 2^{i-1} into sequences of length 2^i. The i-th phase is executed in parallel as a pipelined sequence of waves in distinct subtrees of the pruned butterfly, and it is completed in $O\left(\sqrt{2^i}\right)$ time.

As for the routing, we adapt the structure of Peleg and Upfal's algorithm [13], in which the (k,k)-routing is accomplished by means of balancing, sorting, and parallel prefix techniques. In particular, we develop a novel algorithm for balancing, which evenly redistributes any set of packets among the leaves of the pruned butterfly in $O(k\sqrt{n})$ time, provided that no leaf holds more than k packets initially. The algorithm works by progressively balancing the packets within larger and larger subtrees of the pruned butterfly. More specifically, pipelined waves of packets are moved from heavier to lighter subtrees in a bottom-up fashion.

By using the above primitives we can port the PRAM simulation scheme devised for the mesh to the pruned butterfly, thereby obtaining the result stated in Theorem 2.

5 Lower Bound

In this section, we develop a novel lower bound for deterministic PRAM simulation on processor networks. The lower bound argument is similar in spirit to the one used in [1, 8, 5], but embodies a number of critical enhancements that make it sensitive to the bandwidth characteristics of the interconnection, whereas previous techniqies yielded significant results only when applied to networks of high bandwidth. The bound relies on the notion of decomposition tree [3, 9], which provides a partition of the network into disjoint regions of limited bandwidth. We first formulate the general lower bound in terms of such a decomposition, and then specialize it to multi-dimensional meshes and to the pruned butterfly.

Consider the simulation of an (n, m)-PRAM on an n-processor network $\mathcal{N}$. Each PRAM step involves the reading (*read step*) or writing (*write step*) of some n-tuple of variables. It is assumed that the simulation is *on-line*, in the sense that the simulation of a PRAM step begins only after the simulation of previous read steps has been completed. Furthermore, we only consider simulations that are *point-to-point*, by which we mean that a processor that wants to write a variable must dispatch a distinct message for each copy of the variable it wants to update. This is not the case in the simulations presented in this paper. However, at the end of the section we observe that part of the argument does not rely on the point-to-point assumption. This enables us to derive a non point-to-point lower bound formulated in terms of the global space used to represent the PRAM memory, which applies to our upper bounds.

We assume that the simulating network $\mathcal{N}$ has a $[w_0, w_1, \ldots, w_{\log n}]$ *decomposition tree*, as defined in [9], that is, for any i, $0 \leq i \leq \log n$, $\mathcal{N}$ can be partitioned into 2^i disjoint i-*regions*, $R_1^{(i)}, \ldots, R_{2^i}^{(i)}$, where each i-region is connected to the rest of the network by at most w_i edges. For any two indices h and k, with $1 \leq h, k \leq \log n$ and any variable u, let $r_{h,k}^t(u)$ be the minimum, taken over all h-regions $R_j^{(h)}$, of the number of k-regions containing valid (i.e., most recently updated) copies of u that lie outside $R_j^{(h)}$. The *average redundancy* at time t is $r_{h,k}^t = \sum_{u \in U} r_{h,k}^t(u)/m$. Intuitively, the lower bound hinges on a tradeoff between the cost of the write and the cost of the read steps. Unless the average redundancy is high, and therefore writing steps are expensive, an adversary is always able to generate read steps that will be relatively expensive to simulate.

Lemma 12. *Let* $m \geq 2n^2$. *If* $r = r_{h,k}^t$ *is the average redundancy at time* t, *then there exists an n-tuple of variables which requires time*

$$g_{h,k}(r) = \Omega \left(\min \left\{ \frac{n}{w_h}, \frac{n}{r w_k}, \frac{n \left(m/2n^2 \right)^{1/2r}}{2^k w_k} \right\} \right)$$

to be read.

Proof (sketch). We identify a set of n variables all of whose valid copies are confined within a low-bandwidth portion of the network. Let $U' = \{u \in U : r_{h,k}^t(u) \leq 2r\}$. Clearly, $|U'| \geq m/2$ and there exists an h-region $R_{j_0}^{(h)}$ for which there are at least $m/2^{h+1} \geq m/2n \geq n$ variables in U' achieving their minimum redundancy with respect to $R_{j_0}^{(h)}$. We distinguish between two cases. If $r < 1/2$, then there exists an n-tuple of variables whose valid copies are all within $R_{j_0}^{(h)}$ and therefore $g_{h,k}(r) \geq n/w_h$. If instead $r \geq 1/2$, a combinatorial argument [10] shows that there exist n variables whose updated copies are contained in the union of $R_{j_0}^{(h)}$ and at most

$$\max\left\{4r, 2^k \left(\frac{2n^2}{m}\right)^{\frac{1}{2r}}\right\}$$

k-regions. Reading these variables would take time

$$g_{h,k}(r) \geq \frac{n}{w_h + w_k \left(4r + 2^k \left(2n^2/m\right)^{1/2r}\right)}.$$

The lemma is proved by taking the minimum between the two cases.

In order to prove Theorem 3, we construct a PRAM program made of batches of m/n write steps that update all the variables, followed by m/n read steps suitably chosen by the adversary. Consider the simulation of one such batch. If at time t the average redundancy is r, then mr packets have crossed boundaries of h-regions. In particular, there must be an h-region $R_j^{(h)}$ such that at least $mr/2^h$ packets have crossed its boundary, requiring time at least $mr/(2^h w_h) = (m/n)rn/(2^h w_h)$. The theorem follows by combining the cost of the writes with the cost of the reads established in Lemma 12, by maximizing over all choices of h and k, and by minimizing over all possible values of r.

The lower bound specializes to d-dimensional meshes and to the pruned butterfly as follows. For d-dimensional meshes, we choose the natural decomposition into submeshes, which yields a $[w_0, w_1, \ldots, w_{\log n}]$ balanced decomposition tree with $w_i = \Theta\left((n/2^i)^{1-1/d}\right)$. For the pruned butterfly, the natural decomposition into subtrees yields a $[w_0, w_1, \ldots, w_{\log n}]$ balanced decomposition tree with $w_i = \Theta\left((n/2^i)^{1/2}\right)$. We choose h and k such that $2^h = \log(m/2n^2)/\log\log(m/2n^2)$ and $2^k = r(m/2n^2)^{1/2r}$. It can be shown that for both networks, the minimum in Equation (1) is attained for $r = \Theta\left(\log(m/2n^2)/\log\log(m/2n^2)\right)$, yielding the result of Corollary 4.

Finally, we observe that Lemma 12 is proved without resorting to the point-to-point assumption, therefore it can be used to obtain an unrestricted lower bound (based only on the difficulty of the read operations), parametric in the total space used by the simulation. More precisely, suppose we want to simulate an (n, m)-PRAM using a total of at most mr copies of the variables. Then, r provides a bound on the average redundancy at any time during the simulation, and therefore $g_{h,k}(r)$ is a lower bound on the slowdown of any such simulation. By specializing this argument for the case of multi-dimensional meshes and the case of the pruned butterfly we obtain the result stated in Theorem 5, which implies that the simulations presented in this paper use an amount of redundancy which is only a factor $O\left(\log\log(m/2n^2)\right)$ away from the best redundancy needed to attain the same running times.

References

1. H. Alt, T. Hagerup, K. Mehlhorn, and F. P. Preparata. Deterministic simulation of idealized parallel computers on more realistic ones. *SIAM Journal on Computing*, 16(5):808–835, Oct 1987.
2. P. Bay and G. Bilardi. Deterministic on-line routing on area-universal networks. In *Proceedings of the 31st Annual Symposium on Foundation of Computer Science*, pages 297–306, 1990.
3. S.N. Bhatt and F.T. Leighton. A framework for solving VLSI graph layout problems. *Journal of Computers and System Sci.*, 28(2):300–343, Apr 1984.
4. G. Baudet and D. Stevenson. Optimal sorting algorithms for parallel computers. *IEEE Transactions on Computers*, c-27(1):84–87, Jan 1978.
5. K. T. Herley and G. Bilardi. Deterministic simulations of PRAMs on bounded-degree networks. *SIAM Journal on Computing*, 23(2):276–292, Apr 1994.
6. K.T. Herley. Representing shared data on distributed-memory parallel computers. *Mathematical Systems Theory*, 1995. To appear.
7. M. Kaufmann, J. F. Sibeyn and T. Suel. Derandomizing Algorithms for Routing and Sorting on Meshes. In *Proceedings of 5th ACM-SIAM Symp. on Discrete Algorithms*, pages 669–679, 1994.
8. A.R. Karlin and E. Upfal. Parallel hashing: An efficient implementation of shared memory. *Journal of the ACM*, 35(4):876–892, Oct 1988.
9. C. E. Leiserson. Fat-trees: Universal networks for hardware-efficient supercomputing. *IEEE Transactions on Computers*, c-34(10):892–900, Oct 1985.
10. A. Pietracaprina, G. Pucci. Work-Efficient Deterministic PRAM Simulations are Impossible. GEPPCOM Report 17-94. Dipartimento di Elettronica e Informatica, Università di Padova, Italy, Oct 1994.
11. A. Pietracaprina, G. Pucci, and J. F. Sibeyn. Constructive deterministic PRAM simulations on a mesh-connected computer. In *Proceedings of 6th Annual ACM Symposium on Parallel Algorithms and Architectures*, pages 248–256, Jun 1994.
12. A. Pietracaprina, G. Pucci. Improved deterministic PRAM simulation on the mesh. In *Proceedings of 22nd International Colloquium on Automata, Languages and Programming*, Jul 1995. To appear.
13. D. Peleg and E. Upfal. The token distribution problem. *SIAM Journal on Computing*, 18(2):229–243, Apr 1989.
14. E. Upfal and A. Wigderson. How to share memory in a distributed system. *Journal of the ACM*, 34(1):116–127, Jan 1987.

Appendix: Space-Efficient Simulations

The simulations presented in the paper are based on a memory organization whose structure is modelled by a bipartite graph $G = (U, V; E)$, with $|U| = m$,

$|V| = n$, and where every vertex in U has degree $2c - 1 = \Theta\left(\log(m/n)\right)$. This graph may be represented by means of a read-only table $T_G = [t_1, t_2, \cdots, t_m]$ consisting of m entries, where the i-th entry $t_i = (t_i(1), t_i(2), \cdots, t_i(2c-1))$ contains the addresses of the copies of the i-th variable (items). In this appendix, we show that such a table may be represented in a distributed fashion among the nodes of the simulating network, so that the maximum number of items stored per node is $O((m/n)\log^3(m/n))$ and that any n-tuple of entries (corresponding to the variables to be accessed in the PRAM step) may be read in time proportional to the slowdown of the simulation step. We sketch the required techniques for the 2-dimensional mesh, which are akin to those presented in [6]. The result extends immediately to the other interconnections considered in the paper.

Partition the $\sqrt{n} \times \sqrt{n}$ mesh into n/n' tiles of size $\sqrt{n'} \times \sqrt{n'}$. Each tile will contain a distinct copy of T_G distributed as follows. Partition T_G into m/b pages of b entries each, and partition each tile into n'/b blocks each of size $\sqrt{b} \times \sqrt{b}$. Within each tile, replicate and distribute the m/b pages among the n'/b blocks that make up that tile according to a $(\lambda, 2d - 1, d, 1/gc)$-generalized expander $H = (U_H, V_H, E_H)$ such that $|U_H| = m/b$, $|V_H| = n'/b$, $d = 2c - 1$, and where parameters $\lambda < 1$, $c = \Theta(\log(m/n))$, and $g = O(1)$ are as defined in Section 2. The maximum number of pages mapped to any individual block is $O((m/n')d))$ which amounts to $O((m/n')bd(2c - 1))$ items in all. The items mapped to a particular block are distributed evenly (without replication) among the nodes of the block, $O((m/n')d(2c-1))$ per node. Within each node, the individual items are held in a static dictionary in order to facilitate retrieval.

Note that there are a total of $(n/n')(2d - 1)$ copies of each entry and that to read an entry it suffices to read any one copy. Note also that the structure of the graph H can be represented by means of a read-only table T_H of m/b entries. This latter table is replicated and represented in every block in the network, with each node of each block holding m/b^2 entries of T_H.

To read an n-tuple of entries of T_G, each tile deals locally with the reads relating to its own nodes, independently of other tiles, by executing the following steps.

1. Generate a set S that contains $2d-1$ numbered request packets $r_1(x), r_2(x)$, $\ldots, r_{2d-1}(x)$ for each referenced entry x. Packet $r_i(x)$ bears the name of the processor that generated it, the entry to which it refers, and the name of the block that contains the i-th copy of that entry within the tile in question.
2. Select a subset S' of the packets that contains d packets $r_1'(x), \cdots, r_d'(x)$ per referenced entry such that the number of selected packets relating to any individual block is $O(bd)$.
3. Route each packet in S' to the appropriate block, ensuring that the number of packets routed to any individual node is $O(d)$.
4. Within each block, circulate the packets around a tour that visits each node exactly once (Hamiltonian circuit). As a packet, say $r_i'(x)$, visits a node, check whether that node contains a copy of entry t_x and if so load a copy of item $t_x(i)$ into the packet.
5. Route each packet back to the node that generated it.

Notice that the $d = 2c - 1$ selected packets relating to entry x are ultimately returned to the node that generated them, each bearing the value of a distinct item of that entry.

In order to discover the locations of the various copies of the entries, which are needed to generate packets during Step 1, the nodes need to query T_H. Since each block maintains a private copy of this table and each block generates $b(2d-1)$ request packets, this operation can be accomplished in the same fashion as that outlined for Step 4 and has the same $O(bd)$ running time. Steps 3 and 5 involve (d, d)-routing within an n'-node tile so these contribute $O(d\sqrt{n'})$ to the running time.

As for Step 2, note that for each page of T_G the number of entries referenced may be up to b, the page size. For a particular tile, let P_i denote the set of pages where the number of referenced entries lies in the interval $[2^i, 2^{i+1})$. Clearly, $\sum_{i=0}^{\log b} 2^i |P_i| \leq n'$. Since H is a generalized expander, it is possible to construct a d-bundle for the pages in each P_i that has degree $O((|P_i|/(n'/b))d)$. Each edge in such a bundle corresponds to at most 2^{i+1} request packets, and so the total number of selected packets over all the P_i is $\sum_{i=0}^{\log b} 2^{i+1}(|P_i|/(n'/b))d = O(bd)$. The algorithmic techniques required to perform the selection include straightforward combinations of sorting and parallel prefix akin to those employed during the second stage of the copy selection process of Section 3, and this step also has a running time of $O(d\sqrt{n'})$.

Thus, the overall running time is $O((b + \sqrt{n'})d)$ which is $O(\sqrt{n}\log(m/n))$ for $b = \sqrt{n'} = \sqrt{n/\log(m/n)}$ and $d = O(\log(m/n))$. We can see that the distributed representation of the memory map T_G requires $O((m/n')d(2c - 1)) = O((m/n)\log^3(m/n))$ storage per node, while the representation of T_H contributes a further $O((m/b^2)d) = O((m/n)\log^2(m/n))$ per node. The total storage requirement per node is $O((m/n)\log^3(m/n))$.

Beyond the Worst-Case Bisection Bound: Fast Sorting and Ranking on Meshes*

Michael Kaufmann[1], Jop F. Sibeyn[2], Torsten Suel[3]

[1] Wilhelm-Schickard-Institut für Informatik, Universität Tübingen, Sand 13, 72076 Tübingen, Germany. Email: mk@informatik.uni-tuebingen.de.
[2] Max-Planck-Institut für Informatik, Im Stadtwald, 66123 Saarbrücken, Germany. Email: jopsi@mpi-sb.mpg.de.
[3] NEC Research Institute, 4 Independence Way, Princeton, NJ 08540, USA. Email: torsten@research.nj.nec.com.

Abstract. Sorting is an important subroutine in many parallel algorithms and has been studied extensively on meshes and related networks. If every processor of an $n \times n$ mesh is the source and destination of at most k elements, then sorting requires at least $k \cdot n/2$ steps in the worst-case, and simple algorithms have recently been proposed that nearly match this bound. However, this lower bound does not extend to non-worst-case inputs, or weaker definitions of sorting that are sufficient in many applications. In this paper, we give algorithms and lower bounds for several such problems.

We first present a very simple scheme for k-k routing that performs optimally under both average-case and worst-case inputs. As an application of this scheme, we describe a simple k-k sorting algorithm based on sample sort that nearly matches this bound.

The main part of the paper considers several 'sorting-like' problems. In the ranking problem, the ranks of all elements have to be determined, but there is no requirement about their final positions. We describe an algorithm running in time $(1+o(1)) \cdot k \cdot n/4$ steps, which is nearly optimal under the considered model of the mesh. We show that integer versions of the sorting and ranking problems, where keys are drawn from $\{0, \ldots, m-1\}$, can be solved asymptotically faster than the general problems for small values of m. A related problem, the excess counting problem, can be solved in $O(n)$ steps in many interesting cases.

1 Introduction

One of the most thoroughly investigated interconnection schemes for parallel computation is the $n \times n$ *mesh*, in which a set of n^2 processing units (PUs) is connected by a two-dimensional grid of communication links. While the mesh has a large diameter in comparison to the various hypercubic networks, it is nonetheless of great importance due to its simple structure and efficient layout. A number of parallel machines with mesh topology have been built, and a variety of algorithmic problems have been analyzed on theoretical models of the mesh.

1.1 Routing and Sorting

Routing and sorting are probably the two most extensively studied algorithmic problems on fixed-connection networks. In a routing problem, a set of *packets* has to be redistributed in the network such that every packet ends up at the PU specified in its destination address. Here, the address of a PU is determined by some fixed numbering of the PUs called an *indexing scheme*. A routing problem in which each PU is the source and destination of at most k packets is called a k-k routing problem.

---

* Part of this work was done while the third author was visiting the Max-Planck-Institut.

In the k-k sorting problem, instead of a destination address each packet contains a key from a totally ordered set, and the packets have to be rearranged such that the packet of rank i ends up at the PU with index $\lfloor i/k \rfloor$, for all i. Thus, in a routing problem the destinations of the packets are given as part of the input, while in a sorting problem, the destinations have to be computed from the given key values. For an introduction into the problems of routing and sorting, and a survey of basic results, we refer the reader to [12].

There is a trivial lower bound of $2 \cdot n - 2$ for routing and sorting on the two-dimensional mesh due to the diameter of the network. For large k, this bound is dominated by the lower bound of $k \cdot n/2$ for k-k routing and sorting due to the bisection width of the network, and several algorithms running in $(1 + o(1)) \cdot k \cdot n/2$ steps have recently been presented [7, 9, 8, 23].

Note that, while the diameter lower bound of $2 \cdot n - 2$ (or $n - 1$ for algorithms that output the result at a single PU) actually applies to most non-trivial algorithmic problems on the mesh, the bisection bound of $k \cdot n/2$ does not seem to extend to any general class of problems with k items per PU. As a trivial example, we point out that computing the prefix sums takes $2 \cdot n - 2$ steps with one item per PU, and $2 \cdot n + O(k)$ steps with k items per PU. On the other hand, the bisection lower bound of $k \cdot n/2$ for k-k sorting relies on the assumption that the final positions of the packets are given by a fixed indexing scheme, and that we can thus force all $k \cdot n^2/2$ packets initially located in one half of the network to move to the other half. If we drop this assumption, and only require the algorithm to compute the ranks of the packets, then this simple lower bound no longer applies.

1.2 Sorting-Like Problems.

In this paper, we study several problems that are closely related to sorting, but that are not covered by the above worst-case bisection bound for k-k sorting. These problems are motivated by the observation that in many applications of parallel sorting it is actually not necessary to perform a sorting operation in its most general form, but it suffices to solve a slightly weaker problem.

One example of such a problem is the *ranking* problem, already alluded to above, in which the ranks of all packets have to be computed, but the final positions of the packets are not fixed. Many applications, e.g., in scientific computing, use ranking as a subroutine. In fact, the 'Integer Sort' of the NAS Parallel Benchmarks [1], which is probably the most widely used set of benchmark problems for parallel machines, requires only a ranking, rather than sorting, of the keys.

Other problems considered in this paper are the *integer sorting*, *integer ranking*, and *excess counting* problems. Among other things, they have applications in PRAM emulations, list ranking, and unbalanced communication problems on fixed-connection networks. In the following, we describe and motivate these problems in more detail.

Ranking. Let $\mathcal{P} = \{p_0, \ldots, p_{m-1}\}$ be a set of packets, where each packet p_i contains a value x_i drawn from some totally ordered set X. The ranking problem is the problem of computing, for each packet p_i, the function $Rank(p_i, \mathcal{P}) = |\{j \mid x_j < x_i \vee (x_j = x_i \wedge j < i)\}|$.

Note that a sorting problem can be solved by first ranking the packets, and then routing each packet to the destination corresponding to its rank. In the case of the sequential RAM model, as well as in the shared-memory PRAM, the time for permuting the data in the second step is dominated by that for computing the ranks. However, this is not the case for fixed-connection networks of bounded degree such as the mesh, where routing can take a significant portion of the time for sorting. On such networks, ranking can sometimes be performed more efficiently than sorting. Given the close relation between routing, sorting, and ranking described above, we can also think about the ranking problem as capturing some of those features that distinguish sorting from routing.

As we will show, ranking takes only about half as much time as sorting on the mesh. While the definition of ranking allows the final positions of the packets to be arbitrary, our algorithm in fact arranges the packets in a fairly regular pattern. More precisely, the packets are arranged in blocks of side length $o(n)$, such that each block contains only packets with consecutive ranks; we refer to this as a *blocked ranking*. As it turns out, a blocked ranking is advantageous in some applications. For example, a blocked ranking is required as part of the list ranking algorithm for fixed-connection networks in [25].

Integer Sorting and Ranking. The *integer sorting* and *integer ranking* problems are special cases of the sorting and ranking problems where the values of the keys are restricted to the set $\{0, \ldots, m-1\}$. These problems are motivated by the observation that in many applications the keys are drawn from a fairly small set of values, often not much larger than the input size. (In fact, this assumption is part of the definition of the 'Integer Sort' in the NAS Parallel Benchmarks [1].) For example, each key could be the index of a PU, or a pointer into an array or other data structure.

In the *integer ranking* problem, each PU has to compute the ranks of the packets initially located in it. In the *integer sorting* problem, at the end of the computation the PU with index i has to contain the key value of rank $\lfloor i/k \rfloor$. Note that this definition of integer sorting is fairly weak in that it does not require any way of determining the origins of the key values in the final output. In fact, it turns out that integer sorting can be done faster than integer ranking. We will show that both problems can be solved asymptotically faster than general sorting as long as m is not significantly larger than $k \cdot n^2$.

Excess Counting. A problem related to integer sorting is the *excess counting* problem, which asks to mark all packets whose key values appear more than t times. Excess counting can be used to detect imbalances, say, in irregular communication problems, or in the distribution of key values. The application that originally motivated our interest in this problem arises in the context of a deterministic PRAM emulation on the mesh [17]. The algorithm maintains $2 \cdot c - 1$ copies of each memory location, for some constant c, that are distributed over the network. In order to access a memory location, it suffices to access c of the $2 \cdot c - 1$ copies [16]. In a simulation of the algorithm of [17], requests for all $2 \cdot c - 1$ copies are generated, and then excess counting is used to identify PUs that receive a large number of requests and eliminate some of these requests, thus speeding up the routing of the remaining packets [15].

The running time of our algorithm for excess counting only depends on the threshold t, and is independent of m, the size of the set of possible key values. If $t = \Omega(n^\epsilon)$, for some $\epsilon > 0$, then excess counting can be solved in time proportional to the diameter of the network. This is significantly faster than the corresponding integer sorting operation (which could also be used to solve the problem).

1.3 Standard vs. Arbitrary Model

Before we give an overview of the contributions of this paper, a few remarks about our model of computation are in order. A common assumption in the literature on routing and sorting is that packets are 'atomic entities' whose contents can only be accessed and manipulated in a very restricted fashion. More precisely, each packet consists of a destination address or key value plus some additional data, and the algorithm can read the destination or key but cannot read or alter the data. In each step, a single packet plus $O(\log n)$ bits of auxiliary information can be transmitted across any edge.

In the case of sorting, it is also commonly assumed that the only way to access a key value is by means of a comparison with another key located in the same (or a neighboring) PU. This model, which we will refer to as the *standard model*, is the one assumed in our results for general sorting and ranking.

However, the standard model does not seem appropriate for problems such as integer sorting and excess counting, which are more efficiently solved by non-comparison-based methods. For these problems, we assume a less restrictive model, called the *arbitrary model*, which allows unrestricted access and manipulation of the input. If the keys are chosen from $\{0, \ldots, m - 1\}$, then we assume that $\log m + O(\log n)$ bits can be transmitted across a communication link in a single step; this allows us to compare the running times with those for general ranking under the standard model. (All results can also be transformed into a 'bit model', by multiplying the stated bounds by a factor of $\log m + O(\log n)$.)

Many lower bounds for the standard model, say the $k \cdot n/2$ worst-case lower bound for k-k routing and sorting, follow directly from the observation that only a bounded number of packets can cross a given bisection of the network in a single step; we call such a lower bound a *standard bisection bound*. However, when proving bisection-based lower bounds in a less restrictive model, we have to be more careful, as it may not be immediately clear how much information has to be exchanged between different areas of the network in order to compute the given function. In the case of the arbitrary model, lower bounds can be established with combinatorial arguments similar to those used in VLSI theory [13]; we call such a bound an *information-theoretic bisection bound*.

1.4 Overview of this Paper

In this paper, we study several problems related to routing, sorting, ranking, and counting on the mesh. In addition to providing algorithms and lower bounds, we also attempt to motivate and discuss these problems in a more general context.

In Section 2 we review some recent algorithms for k-k routing and sorting on meshes, and describe their relation to the total-exchange operation, an important communication primitive in parallel computation. We present a modified scheme for k-k routing that nearly matches the bisection bound for both average-case and worst-case inputs. Finally, we discuss an application of this scheme to the efficient implementation of sorting algorithms on parallel machines.

Section 3 considers the ranking problem with unrestricted key length. We prove a lower bound of $k \cdot n/4$ in the standard model (where k is the number of items per PU), and provide nearly matching randomized and deterministic upper bounds.

In Sections 4 and 5 we consider the integer sorting, integer ranking, and excess counting problems under the arbitrary model. We give tight or close to tight bounds for all of these problems, and show that in most cases they can be solved asymptotically faster than the general sorting problem.

1.5 Model of Computation

Our model of computation is the d-dimensional *mesh*. It consists of $N = n^d$ *processing units* (PUs) laid out in a d-dimensional grid of side length n, where every PU is connected to each of its (at most) $2 \cdot d$ immediate neighbors by a bidirectional communication link. We assume that in a single step a PU can exchange a bounded amount of data with each of its neighbors, as defined in the following paragraph. For the sake of simplicity, we assume that a PU can perform an unbounded amount of internal computation in each step. Unless stated otherwise, all presented algorithms can be adapted to run in the same time bounds on a model with bounded internal computation, provided the input size is at most polynomial in n.

In the *standard model*, assumed in Sections 2 and 3, a single packet, plus $O(\log n)$ bits of auxiliary information, can be transmitted across a communication link in each step. The only way of accessing a key value in the standard model is by performing a comparison with another key located in the same PU. In the *arbitrary model*, assumed in Sections 4 and

5, the key values can be manipulated in an arbitrary fashion, and up to $\log m + O(\log n)$ bits of data can be transmitted across a directed communication link in each step, where the key values are chosen from $\{0, \ldots, m-1\}$.

Throughout the paper, we focus on the case $d = 2$. For most results, generalizations to meshes of higher dimension, as well as toroidal networks, follow immediately. We also assume that $k = \omega(1)$, which is the most interesting case, and which allows for a more succinct statement of many of our upper bounds.

2 Routing, Sorting, and the Total-Exchange Operation

In this section, we briefly review the basic ideas underlying the recent optimal algorithms for k-k routing and sorting on the mesh [7, 9, 8]. In particular, we focus on the relation of these algorithms to the total-exchange operation [2], a communication primitive used in many parallel algorithms. We then present a very simple routing scheme based on the total-exchange operation which achieves nearly optimal performance on both average-case and worst-case inputs, and which we believe to be of practical interest. Finally, as an application of our routing scheme, we describe an efficient implementation of a sorting algorithm based on sample sort. In the description of the algorithms in this section, we omit most details about their exact implementation on the mesh, in order to keep the presentation as simple and general as possible.

2.1 Total-Exchange Operation

The *total-exchange* operation (also sometimes called *personalized all-to-all* or *gossiping*) is a communication primitive in which each PU of a parallel machine sends a different message of fixed length to each other PU. If each message consists of w packets of data, then we say that the total-exchange has size w. The total-exchange operation arises in many parallel applications [2], and its simple and regular structure allows for a very efficient implementation (relative to the available bandwidth) [5]. As a consequence, the total-exchange operation has also been studied with respect to its ability to efficiently perform other, more general communication patterns (e.g., see [5, 19]).

Given a partition of a mesh into blocks, we can also define the *blocked total-exchange* operation, in which the blocks play the role of the PUs in the total-exchange, that is, each block sends a fixed amount of data to each other block.

Several optimal randomized [7] and deterministic [8, 9] schemes for k-k routing and sorting on the mesh have recently been proposed. The randomized schemes are based on work by Valiant [26] and Reif and Valiant [20], while the deterministic schemes can be viewed as efficient implementations of Leighton's Columnsort [11]. The basic structure of these algorithms consists of two phases, with each phase containing a blocked total-exchange. This is most explicitly described in the deterministic algorithms in [8, 9], where the network is partitioned into m sufficiently large blocks, and each block B_i is subdivided into m *buckets*, $b_{i,0}, \ldots, b_{i,m-1}$. In every blocked total-exchange, the content of bucket $b_{i,j}$ in B_i is sent to bucket $b_{j,i}$ in B_j. Following is a high-level description of the algorithms.

1. In the first phase, each block distributes its packets 'approximately evenly' over its buckets. Then it calls a blocked total-exchange of size $w = k \cdot N/m^2$. The distribution of the packets over the buckets can either be done deterministically by sorting the packets inside each block B_i and placing the packet of rank j into bucket $b_{i,j \bmod m}$, or we can use randomization.

2. In the second phase, each block B_i sorts its packets, and places the packet of rank j into bucket $b_{i,\lfloor j/w \rfloor}$. Then another blocked total-exchange of size w is performed. Afterwards, local operations are used to bring the packets to their destinations.

It can be shown that after the second blocked total-exchange, each packet is within one block of its final destination (assuming that the blocks are sufficiently large, and that blocks with consecutive indices are adjacent in the network). Note that in the case of routing, we can simply place each packet into the bucket corresponding to its destination block during the second phase. This results in an approximately equal number of packets in each bucket, and the size of the second total-exchange is then determined by (an upper bound on) the maximum number of packets in any bucket.

The running time of this scheme is dominated by the time for the two blocked total-exchange operations. In [8, 9] it is shown that a blocked total-exchange of size $k \cdot n^2/m^2$ can be performed in $(1+o(1)) \cdot k \cdot n/4$ steps on the $n \times n$ mesh. This is also clearly optimal, and leads to a running time of $(1 + o(1)) \cdot k \cdot n/2$ for k-k routing and sorting.

2.2 Routing with Good Average-Case and Worst-Case Performance

The approach described in the previous subsection is nearly optimal with respect to worst-case inputs, in which all $k \cdot N$ packets have to cross the bisection of the network. However, the running time is a factor of 2 away from the lower bound with respect to average-case inputs, in which only half of the packets have to cross the bisection.

It is not difficult to modify the scheme such that it runs in optimal time on average-case inputs, by essentially omitting the first phase [10]. (This is because the sole purpose of the first phase is to distribute the packets evenly over the network, thus reducing a worst-case to an average-case problem.) Unfortunately, this one-phase algorithm has a miserable worst-case performance. Of course, one could add an additional step to the algorithm that scans the input to decide whether it should be treated as worst case or average case. While this approach may solve the problem from a purely theoretical point of view, it seems unlikely that it would lead to any practical and elegant routing schemes.

In the following, we describe a simple refinement of the above two-phase routing scheme that has both good average-case *and* good worst-case performance. A somewhat similar idea, although in a different context, is described in [19]. For the remainder of this subsection, we restrict our attention to routing problems. Our scheme consists of the following two phases.

1. In the first phase, each block B_i computes w_i, the minimum, over all blocks B_j, of the number of packets that have to be sent from B_i to B_j. It then places w_i packets with destination B_j into each bucket $b_{i,j}$, and distributes all remaining packets evenly over the buckets, either by sorting with respect to destination blocks, or through randomization. Then a blocked total-exchange of size $w = k \cdot N/m^2$ is performed. This delivers (at least) $m \cdot w_i$ packets from each B_i to their destination blocks, and distributes the remaining $k \cdot N - m \cdot \sum_i w_i$ packets evenly over the network.

2. In the second phase, each B_i places all remaining packets with destination in B_j into bucket $b_{i,j}$. Then it calls a blocked total-exchange operation of size approximately $w' = k \cdot N/m^2 - \sum_i w_i/m$. Afterwards, local operations can be used to bring the packets to their final destinations.

It can be shown that for a random input we have $w' = o(w)$ with high probability. Hence, the total running time for this case is dominated by the cost of the first total-exchange, and we obtain the following result.

Theorem 1 *The described scheme performs k-k routing on meshes in $(1 + o(1)) \cdot k \cdot n/2$ steps for all distributions, and in $(1 + o(1)) \cdot k \cdot n/4$ steps on the average.*

Thus, we achieve optimality for both average-case and worst-case inputs. In addition, the algorithm also seems to perform well on inputs that are 'between average and worst case', although it does not achieve optimality on every possible input,

2.3 Application to Sorting

As a simple application of our routing scheme, we can obtain a sorting algorithm based on sample sort [21, 20, 3] that performs optimally on both worst-case and average-case inputs. Informally speaking, sample sort uses randomly [21, 20, 3] or deterministically [8] selected splitters in order to reduce a sorting problem to a routing problem. In many cases, the time for solving this routing problem dominates the time for selecting and handling the splitters. (Although for realistic problem sizes the latter may also be significant, see [24] for a discussion.) The routing problem resulting from this reduction can be either worst case or average case (or somewhere in between), depending on the input to the sorting problem. By applying the routing scheme from the previous subsection, we obtain the following result.

Theorem 2 *There are simple randomized and deterministic algorithms that perform k-k sorting on meshes in $(1 + o(1)) \cdot k \cdot n/2$ steps for all distributions, and in $(1 + o(1)) \cdot k \cdot n/4$ steps on the average.*

While in our algorithm the routing times for the average case and worst case differ by only a factor of two, this gap can be significantly larger in actual implementations of sample sort on parallel machines (e.g., see [3, 4]), due to the algorithm used in the routing phase (and also due to the router). Because most parallel machines can efficiently implement highly regular communication patterns such as the total-exchange, we believe that the simple routing scheme in Section 2.2 may provide a practical solution to the problem of worst-case key distributions in sample sort and related algorithms.

3 Ranking

In this section, we give nearly tight bounds for ranking in the standard model. We first prove a lower bound, and then present an algorithmic scheme that leads to nearly optimal randomized and deterministic solutions. We assume that every PU initially holds k packets.

3.1 Lower Bound

We prove a simple lower bound for ranking in the standard model, where we can only compare two key values located in the same PU.

Theorem 3 *Any randomized or deterministic algorithm for ranking requires at least $k \cdot n/4$ steps on the standard model of the mesh.*

Proof: Partition the packets into pairs (l_i, r_i), $0 \le i < k \cdot n^d/2$, where each l_i is initially located in the left half, and each r_i is initially located in the right half of the network. Consider all assignments of key values to the packets such that $val(l_i) = 3 \cdot i + 1$ and $val(r_i) \in \{3 \cdot i, 3 \cdot i + 2\}$, for all $0 \le i < k \cdot n^d/2$. Thus, in order to compute the rank of l_i, it is necessary to perform a comparison between l_i and r_i. Since a comparison can only be performed between packets located in the same PU, at least one packet in each of the $k \cdot n^d/2$ pairs has to cross the bisection. As at most $2 \cdot n^{d-1}$ packets can cross the bisection in a single step, the theorem follows. $\square$

In the arbitrary model, we have to be more careful, since an algorithm might try to perform a comparison between packets in different areas of the network with only partial knowledge of their key values. In fact, there are well-known randomized protocols that compare two κ-bit values located in different processors by communicating $o(\kappa)$ bits between the processors [27]. If the range of key values is sufficiently large, then they can be used to obtain asymptotically faster algorithms for ranking in the arbitrary model.

3.2 Basic Scheme

We now present our basic algorithmic scheme for ranking, which can be seen as an extension of the well-known sample sort algorithm [3, 20, 21]. We will later use this scheme to obtain nearly optimal randomized and deterministic solutions.

Partition the mesh into g square blocks of equal size, called *G-blocks*, where $g = \omega(1)$. Let $f = \omega(g)$ be a multiple of g. (Suitable choices for g and f are discussed further below.) Our basic scheme consists of the following steps.

1. Select a global set of $s = \omega(f)$ approximately evenly spaced splitter elements, and compute the exact global ranks of the splitters. Then broadcast the splitters and their ranks to all G-blocks.

2. Use the splitters to estimate for every packet p_i its ranks r_i, and define $R_i = \lfloor f \cdot r_i/(k \cdot n^2) \rfloor$. Each R_i, $0 \le R_i < f$, is the index of the *destination interval* of p_i.

3. Compute a suitable assignment of destination intervals to G-blocks, such that f/g intervals are assigned to each G-block.

4. Route each packet to the G-block to which its destination interval was assigned.

5. Complete the ranking by locally sorting the packets in the G-blocks.

Lemma 1 *Step 1, 2 and 5 can be performed in $o(k \cdot n)$ steps. At most $(1 + o(1)) \cdot k \cdot n^2/f$ packets are allocated to any destination interval.*

Proof: The selection of the splitters and computation of their global ranks in Step 1 can be performed in $o(k \cdot n)$ steps using either randomized [7, 20, 21] or deterministic [8] sampling techniques. Step 2 can be performed in time $o(k \cdot n)$ by sorting the packets together with the splitters in the G-blocks.

Every destination interval is assigned approximately $k \cdot n^2/f$ packets, up to a small inaccuracy due to packets that lie between two splitters that are in different destination intervals. This inaccuracy is bounded by the maximum number of packets between any two splitters. If s is chosen sufficiently larger then f, then this is number is lower-order compared to $k \cdot n^2/f$. □

It remains to show how we can assign the destination intervals to the G-blocks such that the routing in Step 4 can be performed efficiently. The assignment must be such that the routing is as 'balanced' as possible. More precisely, we want to minimize the maximum number of packets that have to be sent from any G-block to any other G-block. This goal is inspired by the following lemma.

Lemma 2 *If at most $(1 + o(1)) \cdot k \cdot n^2/g^2$ packets have to be routed between any two G-blocks, then the routing in Step 4 can be performed in $(1 + o(1)) \cdot k \cdot n/4$ steps.*

Proof: Apply a blocked total-exchange of size $(1 + o(1)) \cdot k \cdot n^2/g^2$ with respect to the G-blocks, as defined in Section 2.1. □

Thus, if we can efficiently find a good assignment, then we immediately obtain a ranking algorithm running in $(1 + o(1)) \cdot k \cdot n/4$ steps. We formalize the problem as follows. Let $M = k \cdot N = k \cdot n^2$ be the total number of packets, and let $A = (a_{i,j})$ be a $g \times f$ matrix, where entry $a_{i,j}$ gives the number of packets in the ith G-block belonging to the jth destination interval. It follows that

$$0 \le a_{i,j} \le M/f, \text{ for all } 0 \le i < g, 0 \le j < f,$$
$$\sum_j a_{i,j} = M/g, \text{ for all } 0 \le i < g, \text{ and}$$
$$\sum_i a_{i,j} = M/f, \text{ for all } 0 \le j < f.$$

We have to partition the set of columns into g disjoint subsets of f/g columns each, such that every row sum in every subset is at most $(1 + o(1))$ times the average value. More formally, we have to construct an $f \times g$ zero-one matrix $S = (s_{i,j})$, with

$$\sum_j s_{i,j} = 1, \text{ for all } 0 \le i < g, \qquad \sum_i s_{i,j} = f/g, \text{ for all } 0 \le j < f, \qquad (1)$$

satisfying

$$(A \cdot S)_{i,j} = (1 + o(1)) \cdot M/g^2, \text{ for all } 0 \le i, j < g. \qquad (2)$$

3.3 Randomized Solution

We now show that a randomly chosen matrix S satisfies (2) with high probability (i.e., with failure probability at most $M^{-\epsilon}$, for some $\epsilon > 0$). A similar idea was mentioned in [3] in the context of an implementation of sample sort on the CM-5, where the authors propose to 'randomize the locations of the buckets' in order to make the performance of the algorithm independent of the key distribution.

Thus, we randomly select a matrix S satisfying (1) and prove that (2) holds with high probability given an appropriate choice of f and g. Note that (2) imposes g^2 conditions that must be satisfied simultaneously. If any single condition is violated with probability at most $M^{-\epsilon}/g^2$, then the probability that any of them is violated is at most $M^{-\epsilon}$.

Lemma 3 *If $f = \omega(\ln^{1/3} M)$, then a randomly selected zero-one matrix S that satisfies (1) also satisfies (2) with high probability.*

Proof: We can bound the probability for a single condition by analyzing the following situation. Given a multi-set T of f numbers between 0 and M/f whose sum is equal to M/g, we have to bound the probability that the sum of the elements of a random subset S of size f/g exceeds $M/g^2 + t$, for $t = o(M/g^2)$. The expected value of this sum is M/g^2.

A majorization argument shows that the probability that the sum of the values of the elements in S exceeds $M/g^2 + t$ is smaller than the probability that the sum of f/g independently selected elements X_i, with $0 \le X_i \le M/f$ and with expected value M/g^2, exceeds $M/g^2 + t$ (see [22] and the references therein). Applying Hoeffding's Inequality [6] this probability can be estimated as follows.

$$Pr(\sum_i X_i \ge M/g^2 + t) \le \exp(-2 \cdot f^3 \cdot t^2/(g \cdot M^2)).$$

So, we should take

$$t(f, g, M) = c \cdot M \cdot \ln^{1/2} M \cdot g^{1/2}/f^{3/2},$$

for some constant c, to get the desired bound on the probability. In order that $t(f, g, M) = o(M/g^2)$, we must have $\ln^{1/2} M \cdot g^{5/2} = o(f^{3/2})$. Since the only condition on g is $g = \omega(1)$, the lemma follows. □

3.4 Deterministic Solution

We now derive a deterministic solution by performing a (very simple) derandomization [14] of the above algorithm. To do so, we show that the parameters f and g in the randomized solution can be chosen such that the sample space is of size $O(N)$ and the success probability is non-zero. We can then search the entire sample space by assigning a constant number of samples to each of the N PUs.

Lemma 4 *Let k be polynomial in N. If $f = \log N / \log\log N$ then it suffices to test at most N matrices satisfying (1) in order to find one that also satisfies (2).*

Proof: Lemma 3 implies that for $f = \omega(\ln^{1/3} M)$ a randomly selected S satisfies (2) with high probability. Hence, there *exists* an S that satisfies (2). The size $s(f, g)$ of the sample space can be estimated as

$$s(f, g) = \prod_{i=0}^{g-1} \binom{f - i \cdot f/g}{f/g} = \frac{f!}{((f/g)!)^g} < f^f.$$

Thus, if we choose $f \le \log N / \log\log N$, then $s(f, g) < N$. Assuming that k is polynomial in N, the condition $f = \omega(\ln^{1/3} M)$ is also satisfied. $\square$

Since the matrices that have to be tested can be generated in a systematic way, the tests can be performed in a distributed fashion. Thus, Lemma 4 implies that every PU has to generate and test only one $f \times g$ zero-one matrix. This takes $\mathcal{O}(f \cdot g^2)$ time. Since $f < \log N$, this time is negligible in comparison to the routing time. Combining all results from Lemma 1 to 4, we obtain the main result of this section.

Theorem 4 *If every PU initially holds k packets, then the ranking problem can be solved deterministically in time $(1 + o(1)) \cdot k \cdot k/4$ on the $n \times n$ mesh.*

It follows from Lemma 1 that at most $(1 + o(1)) \cdot k \cdot n^2/g$ packets are routed to any G-block. If desired, we can redistribute the packets in $o(k \cdot n)$ steps such that every PU holds exactly k packets at the end.

The large lower-order terms make the deterministic algorithm impractical for reasonable values of n and k.[4] The probabilistic algorithm, on the other hand, has a fairly simple structure and may be of practical interest.

4 Integer Ranking and Sorting

In this section, we consider the integer ranking and integer sorting problems, where the keys are restricted to the set $\{0, \ldots, m - 1\}$. We will show that both problems can be solved asymptotically faster than general sorting, provided that m is not much larger than $k \cdot n/2$. Throughout this section, we assume the arbitrary model of the mesh, which is more useful than the comparison-based standard model for problems with restricted key size. We begin with the following lower bound.

Theorem 5 *If $m = O(k \cdot n^2)$, then any deterministic algorithm for integer sorting requires $\Omega(n + \sqrt{m \cdot k} / \log n)$ steps on the arbitrary model of the mesh.*

Proof: (Sketch) Consider any two square blocks B_0 and B_1 of side length $\sqrt{m/(2 \cdot k)}$ in the mesh. Let X be the set of those inputs where every possible key value appears exactly once in $B_0 \cup B_1$, while all keys in the rest of the network have value 0. Note that every input in X results in the same output. If we do not distinguish between inputs that can be obtained from each other by only permuting keys within B_0 or within B_1, then X contains $\binom{m}{m/2}$ distinct inputs.

A simple crossing-sequence argument (e.g., see [13]) shows that at least $\log(|X|) = \Omega(m)$ bits of data have to be communicated between B_0 and B_1, since otherwise there is an input π that coincides with some $\pi_0 \in X$ in B_0, and with some $\pi_1 \in X$, $\pi_0 \ne \pi_1$, in the rest of the network, and that results in the same output as π_0 and π_1. The lower bound then follows from the fact that B_0 and B_1 have only $O(\sqrt{m/k})$ outgoing edges, each of which can transmit at most $m + O(\log n)$ bits in each step. $\square$

---

[4] Going through the proofs of the above lemmas, we find that for $f = \log N / \log\log N$, we can take $g = \log^{1/3} N$. On the mesh this gives lower-order terms bounded by $\mathcal{O}(k \cdot n / \log^{1/3} n)$.

For the case of integer ranking, the above lower bound can be strenghtened slightly. We obtain the following result, the proof of which is omitted.

Theorem 6 *If $m = O(k \cdot n^2)$, then any deterministic algorithm for integer ranking requires $\Omega(n + \sqrt{m \cdot k} \cdot \frac{\log(k \cdot n^2/m)}{\log n})$ steps on the arbitrary model of the mesh.*

We now describe algorithms for integer sorting and ranking whose running times nearly match the lower bounds. Our algorithms are based on counting sort, and rely on an efficient implementation of a restricted form of the 'multiprefix' operation described in [18]. We first present a simple, but non-optimal, algorithm that solves both integer sorting and ranking, and then explain how the algorithm can be modified into optimal solutions for each problem.

In the following, we assume that $k \cdot n^2 > m > k$. (If $m \le k$, then the problem can be solved in $O(n + k)$ steps by a simple pipelined prefix operation.) The algorithm consists of the following steps.

1. Partition the mesh into blocks of side length $\sqrt{m/k}$, and determine in each block the number a_i of occurrences of key value i, for $0 \le i < m$. Arrange the numbers inside each block such that a_i is contained in PU $P_{i \bmod k}$ of the block.

2. Combine the values a_i from all blocks in the mesh by repeatedly 'merging' groups of four adjacent blocks. That is, add the corresponding a_i values from the four blocks, and distribute the resulting values over the larger block.

3. Use a simple prefix operation to compute the ranks of the key values.

4. In the case of integer sorting, simple routing and segmented broadcast operations can be used to bring the key values to the correct PUs. For integer ranking, we have to deliver the rank information to the PUs initially holding the keys. To do so, we reverse the 'merging' in Step 2 by performing a downwards pass in the merging tree (as in standard prefix algorithms).

Step 1 of the algorithm can be performed in $O(\sqrt{m \cdot k})$ steps using local sorting. It can be shown that the running time for Steps 2 and 4 is dominated by the time needed for the lowest level of the merging tree, which is also $O(\sqrt{m \cdot k})$. Thus, the algorithm runs in time $O(\sqrt{m \cdot k})$.

The non-optimality of this simple algorithm is due to the inefficient representations of the key values as a collection of values a_i in the upward pass in Step 2, and the rank values in the downward pass in Step 4. To improve the running time of the algorithm, we assume that the keys and ranks in Steps 2 and 4 are given as sorted lists. Note that a sorted list of ν key values from $\{0, \ldots, \mu - 1\}$ can be represented with $O((1 + \log(\mu/\nu)) \cdot \nu)$ bits. Using such a representation, and starting with blocks of side length $\sqrt{m/k}/\log n$ in Step 1, we can reduce the time for Steps 2 and 4 to $O(\sqrt{m \cdot k}/\log n)$ and $O(\sqrt{m \cdot k} \cdot \frac{\log(k \cdot n^2/m)}{\log n})$, respectively. This establishes the following result.

Theorem 7 *If $m = O(k \cdot n^2)$, then integer sorting and integer ranking can be performed in time $O(n + \sqrt{m \cdot k}/\log m)$ and $O(\sqrt{m \cdot k} \cdot \frac{\log(k \cdot n^2/m)}{\log n})$, respectively, on the arbitrary model of the mesh.*

Thus, the results asymptotically match the lower bounds. The algorithms can be adapted to the case where $m = o((k \cdot n^2)^{1+\epsilon})$ for any constant $\epsilon > 0$, for which they still achieve an asymptotic improvement over general sorting.

5 Excess Counting

In this section, we consider the excess counting problem, where each PU holds k colored packets, and we want to mark those packets whose colors occur more than some threshold t times. We start with a lower bound. The proof uses an information-theoretic bisection bound applied to a corner section, similar to that in Section 4.

Lemma 5 *The excess counting problem with threshold t requires that some connections transfer $\Omega(k \cdot n/\sqrt{t})$ bits.*

Proof: Divide the mesh in a corner C and the remainder of the network $\mathcal{R}$. C has size $(n/(2 \cdot t)^{1/2}) \times (n/(2 \cdot t)^{1/2})$, and contains $x = k \cdot N/(2 \cdot t)$ packets. All colors occurring in C are unique. In $\mathcal{R}$ there are $k \cdot N \cdot (1 - 1/(2 \cdot t)) > 2 \cdot x \cdot (t - 1)$ packets.

Suppose that $2 \cdot x$ colors occur exactly $t - 1$ times in $\mathcal{R}$, and that the x colors occurring in C are chosen from these $2 \cdot x$ colors. Then clearly the complete information on the colors occurring in C has to be sent into $\mathcal{R}$. There are $\binom{2 \cdot x}{x}$ possible choices of these x colors. Thus, at least $\log \binom{2 \cdot x}{x} = \Omega(x)$ bits have to be transferred over the $2 \cdot n/(2 \cdot t)^{1/2}$ connections between C to $\mathcal{R}$, and the result follows. $\square$

In the following, we derive an algorithm whose running time comes quite close to the above lower bound. Note that a trivial possibility is to sort the packets on their colors, and then count the packets of each color; this can be done in $(1 + o(1)) \cdot k \cdot n/2 + 2 \cdot n$ steps. A factor of two can be gained by applying the ranking algorithm of Section 3 (using the fact that the algorithm produces a *blocked* ranking). If the number of possible colors is very large, and t small, then this is the best algorithm we know. If the number of possible colors is not too large, then we can apply the techniques of Section 4.

In the following, we present an alternative scheme that runs in time independent of the number of possible colors, and that is considerably faster than sorting for sufficiently large values of t. Its structure differs significantly from the algorithm for integer sorting. The algorithm is inspired by the following observation.

Observation 1 *Consider a network that is partitioned into x subnetworks. If a color occurs more than t times in total, then in some subnetwork it occurs at least $\lceil t/x \rceil$ times.*

We assume the arbitrary model. Initially the network is partitioned into t subnetworks with N/t PUs each, called *blocks* (assume that t divides N). The packets in each subnetwork are sorted on their colors and the frequency of each color is determined. All occurring colors are *candidates*. Iteratively, the candidates in pairs of subnetworks are merged together. Inductively we assume that the following invariant holds at all times:

Invariant 1 *After i merges, and for any subnetwork S with $2^i \cdot N/t$ PUs, a color c is a candidate in S iff it occurs at least 2^i times. There are at most $k \cdot N/t$ candidates in S, and the candidates and their frequencies are known in all blocks in S.*

Initially Invariant 1 holds. Finally, for $i = \log t$, the invariant states that a packet is a candidate iff it occurs more than t times in the whole network, and that the candidates are known in every block. In that case, one more local operation suffices to mark all packets with colors that occur more than t times. Assume that Invariant 1 holds after merge i. Then the following steps are performed to merge two subnetworks S_1 and S_2 of 2^i blocks each (we only describe the operations in S_2, as the situation in S_1 is symmetric):

Algorithm EXCESS-MERGE

1. The candidates from S_1 and their frequencies are broadcast to all blocks of S_2.

2. In each block of S_2, determine the frequency of the candidates from S_1.

3. The frequencies of the candidates of S_1 in the blocks of S_2 are added up, and made available in every block. These numbers are then added to their frequencies in S_1.

4. All old candidates from S_1 and S_2 that occur more than 2^{i+1} times are selected as candidates.

Theorem 8 *By iterating* EXCESS-MERGE *for* $i = 0, 1, \ldots, \log t - 1$, *excess counting with threshold* t *can be performed in time* $\mathcal{O}(n + \log t \cdot k \cdot n/\sqrt{t})$ *on an* $n \times n$ *mesh.*

Proof: The correctness follows from the invariant, which is obviously restored after every merge. To analyze the running time, we consider the described merging of two subnetworks consisting of 2^i blocks each. The blocks have size $n/\sqrt{t} \times n/\sqrt{t}$ and the subnetworks have size $(2^{\lfloor i/2 \rfloor} \cdot n/\sqrt{t}) \times (2^{\lceil i/2 \rceil} \cdot n/\sqrt{t})$. Step 1 takes $\mathcal{O}((2^{i/2} + k) \cdot n/\sqrt{t})$ steps (distance plus bisection bound). Step 2 can be implemented by sorting in every block S, and Step 3 is similar to a broadcast. Thus, this application of EXCESS-MERGE takes $\mathcal{O}((2^{i/2} + k) \cdot n/\sqrt{t})$ steps. By summing over all $\log t$ iterations we obtain the result. $\square$

The result can be generalized to a bound of $O(d \cdot n + \log t \cdot k \cdot n/t^{1/d})$ for excess counting on a d-dimensional mesh. Note that our scheme outperforms sorting for all $t = \omega(1)$. To the best of our knowledge, the deterministic sequential complexity of this problem is the same as that of sorting. (Randomizedly, better performance can be achieved by applying hashing and bucket-sort.) So our algorithm gives an interesting example of a problem for which an adaptation of a sequential algorithm does not lead to a good parallel algorithm for the mesh. Also, we see here that problems which are apparently of about the same complexity sequentially, might have a substantially different complexity on the mesh.

Acknowledgement

We thank Greg Plaxton for helpful discussions about the material in Section 2.

References

1. Bailey et al., 'The NAS Parallel Benchmarks,' *Tech. Rep. RNR-94-007*, NASA Ames Research Center, 1994.

2. Bertsekas, D. P., J. N. Tsitsiklis, *Parallel and Distributed Computation: Numerical Methods*, Prentice-Hall, 1989.

3. Blelloch, G. E., C. E. Leiserson, B. M. Maggs, C. G. Plaxton, S. J. Smith, M. Zagha, 'A Comparison of Sorting Algorithms for the Connection Machine CM-2,' *Proc. 3rd Symp. on Parallel Algorithms and Architectures*, pp. 3–16, ACM 1991.

4. Dusseau, A. C., 'Modeling Parallel Sorts with LogP on the CM-5,' *Tech. Rep. CSD-94-829*, University of California at Berkeley, 1994.

5. Hinrichs, S., C. Kosak, D. R. O'Hallaron, T. M. Stricker, R. Take, 'An Architecture for Optimal All-to-All Personalized Communication,' *Proc. 6th Symp. on Parallel Algorithms and Architectures*, pp. 310–319, ACM, 1994.

6. Hofri, M., *Probabilistic Analysis of Algorithms*, Springer, 1987.

7. Kaufmann, M., S. Rajasekaran, J. F. Sibeyn, 'Matching the Bisection Bound for Routing and Sorting on the Mesh,' *Proc. 4th Symp. on Parallel Algorithms and Architectures*, pp. 31–40, ACM, 1992.

8. Kaufmann, M., J. F. Sibeyn, T. Suel, 'Derandomizing Algorithms for Routing and Sorting on Meshes,' *Proc. 5th Symp. on Discrete Algorithms*, pp. 669–679 ACM-SIAM, 1994.

9. Kunde, M., 'Block Gossiping on Grids and Tori: Deterministic Sorting and Routing Match the Bisection Bound,' *Proc. 1st European Symp. on Algorithms*, LNCS 726, pp. 272–283, Springer, 1993.

10. Kunde, M., R. Niedermeier, K. Reinhardt, P. Rossmanith, 'Optimal Average Case Sorting on Arrays,' *Proc. 12th Symp. on Theoretical Aspects of Computer Science*, pp. 503–513, Springer, 1995.

11. Leighton, F. T., 'Tight Bounds on the Complexity of Parallel Sorting,' *IEEE Transactions on Computers*, C-34(4), pp. 344–354, 1985.

12. Leighton, F. T., *Introduction to Parallel Algorithms and Architectures: Arrays, Trees and Hypercubes*, Morgan Kaufmann, 1991.

13. T. Lengauer, 'VLSI Theory,' in *Handbook of Theoretical Computer Science, Volume A: Algorithms and Complexity*, J. van Leeuwen (ed.), pp. 805–833, Elsevier/MIT Press, 1990.

14. Luby, M., 'A Simple Parallel Algorithm for the Maximal Independent Set Problem,' *SIAM Journal on Computing*, 15, pp. 1036–1053, 1986.

15. Meyer, U., J.F. Sibeyn, 'Simulating the Simulator: Deterministic PRAM Simulation on a Mesh Simulator,' *Proc. Eurosim '95*, F. Breitenecker and I. Husinsky (Eds), Elsevier, 1995, to appear.

16. Mehlhorn, K., U. Vishkin, 'Randomized and Deterministic Simulations of PRAMs by Parallel Machines with Restricted Granularity of Parallel Memories,' *Acta Informatica*, 9(1), pp. 29–59, 1984.

17. Pietracaprina, A., G. Pucci, J. F. Sibeyn, 'Constructive Deterministic PRAM Simulation on a Mesh-Connected Computer,' *Proc. 6th Symp. on Parallel Algorithms and Architectures*, pp. 248–256, ACM, 1994.

18. Ranade, A., S. N. Bhatt, S. L. Johnsson, 'The Fluent Abstract Machine', *Advanced Research in VLSI: Proc. 5th MIT Conference*, pp. 71–94, MIT Press, 1988.

19. Rao, S. B., T. Suel, Th. Tsantilas, M. Goudreau, 'Efficient Communication Using Total-Exchange', *Proc. 9th International Parallel Processing Symposium*, pp. 544–550, IEEE, 1995.

20. Reif, J. H., L. G. Valiant, 'A logarithmic time sort for linear size networks,' *Journal of the ACM*, 34, pp. 68–76, 1987.

21. Reischuk, R., 'Probabilistic Parallel Algorithms for Sorting and Selection,' *SIAM Journal of Computing*, 14, pp. 396–411, 1985.

22. Schmidt, J.P., A. Siegel, A. Srinivasan, 'Chernoff-Hoeffding Bounds for Applications with Limited Independence,' *Proc. 4th Symp. on Discrete Algorithms*, pp. 331–340, ACM-SIAM, 1993.

23. Sibeyn, J. F., 'Desnakification of Mesh Sorting Algorithms,' *Proc. 2nd European Symp. on Algorithms*, LNCS 855, pp. 377–390, Springer, 1994.

24. Sibeyn, J.F., 'Sample Sort on Meshes,' *Tech. Rep. MPI-I-95-1012*, Max-Planck Institut für Informatik, Saarbrücken, Germany, 1995.

25. Sibeyn, J. F., 'List Ranking on Interconnection Networks,' *Tech. Rep. MPI-I-95*, Max-Planck-Institut für Informatik, Saarbrücken, Germany, 1995, to appear.

26. Valiant, L. G., 'A Scheme for Fast Parallel Communication,' *SIAM Journal on Computing*, 11, pp. 350–361, 1982.

27. Yao, A. C., 'Some Complexity Questions Related to Distributive Computing,' *Proc. 11th Symp. on the Theory of Computing*, pp. 209–213, ACM 1979.

Fast Deterministic Simulation of Computations on Faulty Parallel Machines

Bogdan S. Chlebus[1] Leszek Gąsieniec[1] * Andrzej Pelc[2] **

[1] Instytut Informatyki, Uniwersytet Warszawski,
Banacha 2, 02–097 Warszawa, Poland.
[2] Département d'Informatique, Université du Québec à Hull,
Hull, Québec J8X 3X7, Canada.

Abstract. A method of deterministic simulation of fully operational parallel machines on the analogous machines prone to errors is developed. The simulation is presented for the exclusive-read exclusive-write (EREW) PRAM and the Optical Communication Parallel Computer (OCPC), but it applies to a large class of parallel computers. It is shown that simulations of operational multiprocessor machines on faulty ones can be performed with logarithmic slowdown in the worst case. More precisely, we prove that both a PRAM with a bounded fraction of faulty processors and memory cells and an OCPC with a bounded fraction of faulty processors can simulate deterministically their fault-free counterparts with $\mathcal{O}(\log n)$ slowdown and preprocessing done in time $\mathcal{O}(\log^2 n)$. The fault model is as follows. The faults are deterministic (worst-case distribution) and static (do not change in the course of a computation). If a processor attempts to communicate with some other processor (in the case of an OCPC) or read a memory word (in the case of a PRAM) then it is immediately notified whether the operation was successful (fault-free addressee) or failed (faulty addressee).

This is for the first time that a general fast deterministic simulation technique is designed for the EREW PRAM with the worst-case fault distribution. The simulation is designed in such a way that it relies only on a fraction of all the operational processors. During preprocessing, the active processors retrieve the original input provided to all processors before the simulation started. This is accomplished by adapting the information-dispersal method.

1 Introduction

We witness the size of multiprocessor computers grow. The result of the increase of the number of processing elements is that the machines become prone to hardware failures. This poses the task for algorithm designers to make algorithms resilient to faults encountered in the course of a computation.

---

* This work was done during the author's stay at the Université du Québec à Hull as a postdoctoral fellow, E-mail: lechu@mimuw.edu.pl
** Research supported in part by NSERC grant OGP 0008136, E-mail: pelc@uqah.uquebec.ca

One of the most popular models of parallel multiprocessor computers is the *Parallel Random-Access Machine* (PRAM) (see [5, 7, 10]). It consists of a number of processors, each with its local control and memory, and a global shared memory. In a step of computation, each processor can access an arbitrary memory cell for both reading or writing. The following three variants of the PRAM are considered, depending on the allowed concurrent memory access: *concurrent-read, concurrent-write* (CRCW), in which many processors can simultaneously read from a cell or write to it, *concurrent-read, exclusive-write* (CREW), in which reading can be done concurrently but only one processor may attempt to write to a given cell at a time, and *exclusive read, exclusive write* (EREW), in which at most one processor may try to reach a given cell at a time. In the case of CREW and EREW PRAM, violating the exclusivity restrictions results in a run-time computation error.

Another model attracting growing interest is the *Optical Communication Parallel Computer* (OCPC), also called *Distributed Memory Machine* [16], *Direct Connection Machine* [13] or S^*PRAM [20]. It models a completely connected optical network (see [1, 4, 6, 15, 18]). All processors have their local control and memory modules attached. Every processor A can attempt to communicate directly with any processor B in a unit of time. If at the moment of communication A is the only processor attempting to communicate with B, then the communication is successful. If several processors attempt to communicate with B at the same time then all these attempts are unsuccessful; moreover, all the involved processors receive collision messages, and the computation proceeds with no run-time errors.

There has been a lot of research done recently to develop simulation techniques transforming PRAM algorithms designed to operate in a fault-free environment into algorithms that are reliable on fault-prone machines. Many approaches have been applied, depending on the nature of faults (static versus dynamic, deterministic versus stochastic, fail-stop versus restartable), the efficiency criteria (time versus work), the properties and capabilities of the underlying model (synchronous versus asynchronous, concurrent read and write versus exclusive read and write), or the types of simulations (randomized versus deterministic, definite versus tentative). Kanellakis and Shvartsman [8] introduced the fail-stop PRAM and developed many deterministic and robust (that is, work efficient) algorithms. This approach to fault tolerance was continued in a series of papers. Kedem, Palem and Spirakis [12] and Shvartsman [19] developed robust general PRAM simulations for dynamic fail-stop errors. Kedem *et al.* [11] produced robust randomized tentative simulations (without per-step synchronization) which have constant expected slowdown, under some assumptions on the stochastic faults. Kanellakis and Shvartsman [9] developed deterministic robust PRAM emulations under the model of restart failures. Chlebus, Gambin and Indyk [2] studied a CRCW PRAM with deterministic memory faults, and designed efficient PRAM simulations for both static and dynamic errors. Diks an Pelc [3] developed reliable and efficient algorithms for the EREW PRAM, under the stochastic processor-error model.

In this paper we present a general deterministic method of simulating parallel algorithms in a faulty environment. It can be applied in synchronous models of computation in which any two processors can communicate in constant time. The models of PRAM and OCPC fall in that category. The communication may be via global shared memory in the case of a PRAM, or by beams of light in the case of an optical computer.

The simulations presented are of the fully operational PRAM on a PRAM with faulty memory and faulty processors, and of the fully operational optical computer on such a computer with faulty processors. The fault model is as follows: all the faults are deterministic and static. Here *deterministic* means that the simulation performance is with respect to any distribution of faults, and *static* means that the faulty/operational status of error-prone components does not change in the course of a computation. It is assumed that at most a bounded fraction of processors or memory cells (in the case of a PRAM) are faulty. If an operational processor attempts to communicate with some other processor (in the case of the OCPC) or read a memory cell (in the case of the PRAM) then it is immediately notified whether the operation was successful (fault-free addressee) or failed (faulty addressee).

The main contribution of this paper is showing that one can obtain a *logarithmic slowdown in the worst case* in simulations of operational multiprocessor machines on faulty ones. More precisely, we prove the following main result:

Theorem 1. *Both a PRAM with a bounded fraction of faulty processors and memory cells and an OCPC with a bounded fraction of faulty processors can simulate deterministically its fault-free counterpart in a step-by-step fashion with $\mathcal{O}(\log n)$ slowdown and preprocessing done in time $\mathcal{O}(\log^2 n)$.* □

The presented fault-free PRAM simulation on a PRAM with faulty shared memory is related to the results obtained in [2]. There are two simulations developed in [2] for static memory errors: one with a $\mathcal{O}(\log n)$ slowdown on the optimal number of $n/\log n$ processors, the other operating in real time on $n \cdot \log n$ processors. To compare the results notice that the simulations developed in [2] are *randomized* and the performance bounds are *expected*, and the model is a *CRCW PRAM*, whereas the simulation presented in this paper is *deterministic* and applicable to the whole PRAM family, in particular to the weakest model, the *EREW PRAM*.

This is for the first time that a general deterministic simulation technique is designed for the EREW PRAM with the worst-case fault distribution. At first glance, it could seem that *no* deterministic and efficient simulation of a PRAM with faulty memory cells is possible, under the worst-case fault distribution. Namely, if the simulation is deterministic then the memory cells accessed by each processor are predetermined until the first operational cell is encountered. This implies that some operational processors may *never* (within the time bounds of a fast simulation) find a fault-free memory cell and hence may never be able to communicate with other processors. To circumvent this obstacle, the developed simulation is designed in such a way that it relies only on a fraction of the operational processors. Then the question arises: what with the input supplied to processors that do not participate in the simulation? We solve this problem by adapting the method of information dispersal of Rabin [17] and show how all the original input can be retrieved from the information available to the processors being active in the simulation.

The paper is organized as follows. In section 2 we describe the simulation problem and reduce it to two abstract tasks: construction of a list of fault-free processors and communication along this list. In sections 3 and 4 we show how these two tasks can be efficiently performed in a faulty environment. Finally, in section 5 we discuss the problem of how to efficiently supply the input to a faulty PRAM.

2 Simulations on faulty machines

In this section we describe the problem of simulating fault-free PRAM and OCPC on their fault-prone counterparts. Then we reduce the simulation to two abstract problems which are solved in the following sections.

2.1 Simulation on a faulty PRAM

Two kinds of faults in the PRAM are considered: faulty processors and faulty cells in the global shared memory. The shared memory cells are simply called *cells* in what follows. Let n denote the number of processors and suppose they are labeled $1, ..., n$. For the sake of simplicity of exposition, let us assume that the number of cells in the simulated fault-free machine is also n. Assume also that the fraction of faulty cells and the fraction of faulty processors are bounded by constants q_1 and q_2 respectively such that $q = q_1 + q_2 < 1$. Faults are *static*, in the sense that the faulty/operational status of the processors and the memory cells is the same during a computation, in particular, no new faults appear in the course of a computation. Each operational processor has a fully reliable local memory of size $O(1)$. A fault-free processor accessing a cell learns immediately the fault status of the cell; this mechanism is an extension of the standard PRAM model: we assume that there is a special one-bit register at each processor which is set to the value corresponding to the fault status of the accessed cell. Faulty processors do not perform any attempts of memory access. The fault-prone machine used for the simulation has n processors and cn cells. The consecutive segments of c cells each are assigned to consecutive processors. The constant integer c (depending on q_1 and q_2 but not on n) is chosen in such a way that at least an operational processors have each sufficiently many fault-free cells assigned to it, for some positive constant α. (This can be done in view of $q < 1$.) These processors are called *active*. Notice that each fault-free processor can learn in time $O(1)$ if itself or another processor is active.

We describe the simulation in the case of the EREW PRAM, but it can be adapted directly to other variants of the model. A step of computation consists of accessing a cell and local computation. In a specific step, the processor i needs to access the cell corresponding to processor $f(i)$, where f is a certain partial one-to-one function. In order to simulate such a computation on a faulty machine with at least αn active processors, we first perform a preprocessing and then simulate the computation step by step.

During the preprocessing, a list of all the active processors is built. When the list is constructed, every active processor knows the length l of the list, its own rank in it and the label of its successor. In section 3 it is shown how this preprocessing can be done in time $O(log^2 n)$. For simplicity, let us assume that l divides n and let $m = n/l$. The ith active processor will simulate the actions of processors with labels in the set $S_i = \{(i-1)m + 1, ..., im\}$. Suppose that, in a given computation step of the simulated algorithm, processor j_1 reaches the cell corresponding to processor j_2. Let $j_1 \in S_{i_1}$ and $j_2 \in S_{i_2}$, $j_2 = (i_2 - 1)m + x$. In the faulty machine, the active processor A with rank i_1 in the list knows that it simulates the action of j_1 and can compute i_2 and x. The main component of the simulation of a step of computation consists in accessing the xth fault-free cell corresponding to processor B with rank i_2 in the list. The only information that A is missing is the physical identification (label) of processor B. Hence the simulation would be complete if every active processor could learn the label of another active processor knowing its rank in the list. We call this task *communication along the list*. In section 4 we prove that this task can be accomplished by a procedure working in logarithmic time, thus yielding logarithmic slowdown of the simulated algorithm.

2.2 Simulation on a faulty OCPC

Consider an OCPC with processors labeled $1, ..., n$. At most qn processors are faulty, for some constant $0 \leq q < 1$. Faults are static in the same sense as above. Faulty processors do not attempt to communicate and if a fault-free processor A attempts to communicate with processor B then A receives back a message determining whether B is fault-free or faulty. Such an attempt takes a unit of time. We keep the notion of active processor: in the present context it simply means fault-free, thus $\alpha = 1 - q$.

One computation step in a fault-free OCPC includes attempts of some processors to communicate with other processors: processor i attempts to communicate with $f(i)$ for some function f, not necessarily one-to one. In order to simulate a fault-free computation on a faulty OCPC we first perform the same preprocessing as in the case of the PRAM, creating the list of active (fault-free) processors. Then the computation is simulated step by step. We keep the notation introduced in the previous case. Suppose that, in a given computation step σ of the simulated algorithm, processor $j_1 \in S_{i_1}$ communicates with processor $j_2 = (i_2 - 1)m + x$. In the faulty OCPC, the active processor A with rank i_1 in the list knows that it simulates the action of j_1 and can compute i_2 and x. As before we need to accomplish *communication along the list*, for A to learn the identification of the processor B with which it needs to communicate (A knows that B has rank i_2 in the list). The rest of the simulation of step σ is divided into m synchronous phases. Processor A first checks if more than one processor in S_{i_1} attempts to communicate with j_2 in step σ. If this is the case, there is a collision in the original algorithm and B does not get any message. Thus in this case A is passive. Otherwise A waits till phase x and attempts to communicate with B.

Hence, for both cases of a faulty PRAM and faulty OCPC, we reduced the simulation to performing two procedures: building a list of all the active processors and communication along the list. In the next two sections we describe these procedures and evaluate their performance.

3   Building a list

There are n processors p_1, p_2, ..., p_n. They are in two categories: *active* and *dormant*. We assume that there is a constant $0 < \alpha \leq 1$ such that the number of active processors is at least $\alpha \cdot n$. The dormant processors do not participate in the computations, except that when an active processor inquires a dormant processor about its status, it is immediately notified that it attempted to communicate with a dormant processor.

A set S of processors is *good* if the number of active elements of S is at least $\alpha \cdot |S|$, where $|S|$ denotes the size of S. The *partition tree* (PT$_n$), for the given n processors, is defined as follows. It is a full binary tree with the nodes labeled by intervals of integers. The notation $[a..b]$ means the interval consisting of the integers x such that $a \leq x \leq b$. The root is labeled by $[1..n]$. If a node v is labeled by $[a..b]$ and v is an internal node then its left child is labeled by $[a..\lfloor(a+b-1)/2\rfloor]$ and the right child is labeled by $[\lceil(a+b)/2\rceil..b]$. The leaves are labeled by intervals $[a..b]$ of length $\lceil\frac{n}{2^h}\rceil$ or $\lfloor\frac{n}{2^h}\rfloor$, for $h = \lfloor\log(\alpha n)\rfloor$. A processor p_i is *associated* with node v if the index i belongs to the label of v. The processors associated with a node of a tree PT$_n$ are referred to as the *group* of the node. A node is *good* if its group is good. Notice the following:
a) the root of PT$_n$ is good;
b) if a node of PT$_n$ is good then one of its children is good.
From this the following key property of partition trees follows:

Lemma 2. *There is a good leaf in PT$_n$ such that the path from v to the root consists of good nodes.*   □

We design an algorithm BUILD_LIST organizing all the active processors in a list. The algorithm works in phases corresponding to the levels of the tree PT$_n$. During a phase the processors are partitioned into groups of processors associated with nodes at the corresponding level of the tree. A group is either *busy* or not. If a group is busy then:
a) each processor in the group knows this;
b) the active processors in the group are organized in a list, and each element of the list knows its *rank*, that is, the distance from the head of the list;
c) each element knows the number of elements in the list.

During the first phase of the algorithm the groups correspond to the leaves of PT$_n$. The processors in each group check each other and if the group is good then the processors arrange themselves in a list, compute ranks and the group becomes busy. In general, once a phase has been completed, the computation proceeds to the next one corresponding to the next level of PT_n towards the root. Consider such a phase. It consists of three parts. During the first part the busy groups corresponding to nodes being left children perform computations, while the other groups pause, in the second part the busy groups corresponding to right children perform the computations, but only if their left-sibling group was idle during part one. Consider the first part. Each processor in a busy group G_1 is assigned a set of processors in the sibling group G_2 as follows. The processors in G_2 are partitioned into as many subsets as there are active elements in G_1, the sizes of subsets are constant (G_1 is good) and differing by at most one. The active processor with rank k in the list in G_1 is assigned the kth subset of G_2. Each active processor of G_1 checks on the assigned subset and includes all its active processors into the list. This completes part one. The second part is similar, but the roles between sibling groups are reversed. For each pair of sibling groups, if at least one of the groups is busy then after part two all the active processors in these groups are connected in a common list. This list is processed during the third part. The ranks and the size of the list are computed and broadcast to all the elements of the list by the standard procedure of pointer jumping. If the size of the list is large enough for the group of the parent node to be good then the group of the parent node becomes busy in the next phase.

In this description of the algorithm one detail is missing, namely how to control the duration of the first two parts of a phase, when the processors check on the assigned elements of sibling groups. Notice that this time can be bounded by a common constant since all the busy groups are good.

Theorem 3. *The algorithm* BUILD_LIST *produces a list connecting all the active processors, and operates in time* $\mathcal{O}(\log^2 n)$.

Proof. By Lemma 2, there is at least one busy group at each phase. The correctness follows by induction on the height of PT_n. There are $\mathcal{O}(\log n)$ phases. The first two parts of a phase take time $\mathcal{O}(1)$, but the third one takes time proportional to the logarithm of the size of the list, which is $\mathcal{O}(\log n)$.

4 Communication along the list

In this section the word *processor* means *active* processor. Let us consider a specific step of computation performed by all (active) processors. The processors are assumed to have already been connected in a list. The word *list* means the underlying directed list of (active) processors. The indices of processors used in this section are their ranks in the list. Each processor $P(i)$ knows the index of another processor $P(j)$ that it wants to communicate with. The processor $P(j)$ is the *target processor* for $P(i)$, in this step of computation. The goal of each processor is to learn the identification of its target processor. Once this is done, processors may communicate with the targets and simulate the required step of computation, as described in section 2.

The list is assumed to be cyclic. The distance $dist(i,j)$ from processor $P(i)$ to its target $P(j)$ is the number of links that need to be traversed in order to get from $P(i)$ to $P(j)$. Thus $dist(i,j) = (j - i) \bmod l$, where l is the length of the list.

Communication along the list uses the idea of pointer jumping. We use two kinds of links. The links of the underlying cyclic list are referred to as the *primary links*. They are never modified in the course of computation, once having been established. There are the *secondary links* that are actually doubled. When communication starts, the secondary links are initialized to the primary ones. By the doubling instruction the new link value $L(i)$, for the ith node of the list, becomes $L(L(i))$.

We show the advantage of link doubling in the following example. Suppose that just *one* processor $P(i)$ needs to communicate with some other processor $P(j)$, so there is no concurrency involved. Let $d = dist(i,j)$ and let $b_0 \ldots b_{s-1} b_s$ be the binary representation of d, where b_0 is the least significant bit. The value b_k is called the kth *distance bit* of $P(i)$. The distance to the target is $\sum_{k=0}^{s} b_k \cdot 2^k$. Processor $P(i)$ covers this distance in at most s steps. In step 0 it traverses the primary link iff $b_0 = 1$. Further, in the kth step $P(i)$ traverses the appropriate secondary link iff $b_k = 1$. By saying that processor $P(i)$ traverses a link pointing to processor $P(j)$ we mean that $P(i)$ scans the shared memory segment assigned to $P(j)$ (and copies appropriate information to its own memory segment).

Next we show how to extend this approach so that all processors may concurrently move to their targets. The main problem that arises is that there may be many processors which have currently reached a node of the list, and just reading the value of the link could create access conflicts. We are interested in a solution without any such conflicts, having in view the application to the EREW PRAM.

The computation proceeds in $\lceil \log l \rceil$ phases, where l is the length of the list. The kth phase begins with the kth doubling of the secondary links. At each phase k there is a set of processors $\mathcal{P}(v,k)$ associated with the list node v: these are the processors traversing the secondary links that are currently at the node v. A phase is performed in three parts. In part one, just after link doubling, the processors in set $\mathcal{P}(v,k)$ are partitioned into two groups (each one possibly empty) depending on the kth distance bit. In part two processors with the bit equal to 1 are moved to the next node along the appropriate secondary link, those with bit 0 stay at the node. Simultaneously, some new processors arrive at node v. During the third part of the phase, the old and the new processors at a node are merged into a new set $\mathcal{P}(v, k + 1)$.

The processors in every set $\mathcal{P}(v,k)$ are organized in a *compressed distance tree* defined below. The *distance tree* is a binary tree whose edges are labeled by ones and zeros. Every branch in a distance tree represents a suffix $d|k = b_k b_{k+1} \ldots b_s$, where d is the distance of some processor in set $\mathcal{P}(v,k)$ from its target. The distance tree is a prefix tree, i.e. two branches split at their longest common prefix. The *compressed distance tree* is a distance tree whose all chains (sequences of nodes of out-degree one) are compressed into single edges. The length of an edge is the number of bits in the corresponding chain. The processors from set $\mathcal{P}(v, k)$ are placed in internal nodes of the compressed distance tree as follows. Every processor $P(i) \in \mathcal{P}(v, k)$ is placed in some internal node in the branch $d_i|k$, where d_i is (the binary representation of) the distance of $P(i)$ from its target. Every internal node, except for the root, has exactly one processor placed in it. The root of degree two has two processors placed in it. A processor placed in the root of the compressed distance tree is called the *leader* of the set $\mathcal{P}(v, k)$. If a root has two processors placed in it, then the set $\mathcal{P}(v, k)$ has two leaders, they are called *partners* in this case. Only leaders of the group $\mathcal{P}(v, k)$ have links to the node v in the list, all other processors in this group are involved in holding the structure of the tree.

In phase k, the trees which represent sets $\mathcal{P}(v, k)$ are split into two temporary trees if the root has two children. The first one, the *waiting tree*, contains all the processors that do not use the current secondary link, i.e. their kth distance bit is 0. The second tree, called the *traversing tree*, contains all the processors which use the current secondary link. It is used in the construction of the distance tree for the set $\mathcal{P}(u, k + 1)$, where u is the node of the list reachable from v by the current secondary link. Namely, the traversing tree derived from $\mathcal{P}(v, k)$ is merged with the waiting tree derived from $\mathcal{P}(u, k)$. When the tree is split, only the leader of the set of processors in the traversing tree uses the current secondary link in the list and moves to the new destination u. When the traversing part of phase k is finished, every leader is either replaced from the root to its only child z copying information about the list to its new partner at z, or deletes the first bit in the edge directed to the child z, if the length of the edge is at least two. The waiting and traversing trees, which are associated with the same node in the list, are then merged starting from the their roots. The merging process continues towards the leaves recursively. At a node, the merging consists of combining the subtrees in such a way that the tree structure is as specified in the definition, and can be performed in time $O(1)$. The key idea behind the performance of the algorithm is to allow the root (leader) to traverse the secondary links just after the operations needed for splitting and merging the trees regarding it and its children have been completed. Below the root, other processors are still busy reorganizing the tree, but this does not affect the root, hence the latter can follow the current secondary link. Since the operation is done in parallel on many levels, the recursive merging is actually a pipelined operation. If we waited for the whole structure of the tree to be updated, there would be another logarithmic factor in the time performance. A detailed description of the merging process will appear in the full version of the paper.

We assume that processors have a sufficiently large menu of standard bit operations that allows them to perform the needed operations on strings of bits in time $\mathcal{O}(1)$. There is a logarithmic number of phases in the communication process, thus we get the following theorem.

Theorem 4. *The communication along the list is performed deterministically in time $\mathcal{O}(\log n)$.* □

Now theorem 1 follows immediately from theorems 3 and 4.

5 Input for faulty machines

The aspect of input/output operations for a PRAM has never been much discussed, due mainly to the fact that the model itself is theoretical and disregards the real costs of communication between the processors and the memory modules. Usually it is simply assumed that the input has been stored somehow either in the processors' local memories or in the shared memory prior to the start of the computation. We adopt the approach to provide the input directly to the processors. A processor may stay idle during the computation because either it is faulty or its assigned segment of shared memory cells does not contain sufficiently many fault-free elements. The categorizing of processors into active and dormant is done during preprocessing. The question arises, how are the preprocessing and input operations related. It would be hardly acceptable to assume that only *after* preprocessing the input operation starts because that would require a very flexible (and hence complicated and costly) input hardware. Therefore we work under the assumption that the input is provided to all processors regardless of their faults or faults in the shared memory.

As before we assume that there is a constant $\alpha > 0$ such that there are at least $\alpha \cdot n$ active processors arranged in a list after preprocessing. We want to provide the input to the first $\alpha \cdot n$ processors in the list, by resorting to the general input mechanism which distributes the input among all the n processors. A possible solution is to *encode* the input, in such a way that it can be retrieved from the fraction of information known to the active processors. This adds some extra burden on the input hardware, but some price has to be paid anyway.

Methods of encoding and then retrieving the original message in a situation when only a part of the transmitted information is available are known as error-correcting codes. We apply a specific simple and efficient method popularized by Rabin [17] under the name of *information dispersal*.

There are $u_1, \ldots, u_{\alpha n}$ original input strings, each comprised of some t bits; it is required that $2^t \geq n$. They are encoded as a sequence $v_1, \ldots, v_n$, where $v_i = g(\omega^i)$ and g is the polynomial

$$g(x) = u_1 + u_2 x + \ldots + u_{\alpha n} x^{\alpha n - 1}$$

The arithmetic is in the field $GF(2^t)$. The element ω is a 2^tth primitive root of unity in $GF(2^t)$. The task of encoding is equivalent to computing the Discrete Fourier Transform and can be implemented efficiently by the FFT algorithm. Suppose that $v_{i_1}, \ldots, v_{i_{\alpha n}}$ are the strings stored by the active processors in the list (in any order). This gives $\alpha \cdot n$ values at distinct points of a polynomial of degree $\alpha \cdot n - 1$. The task of retrieving the input $u_1, \ldots, u_{\alpha n}$ is equivalent to obtaining the coefficients of a polynomial from its values, that is, to *interpolating* the polynomial. There is an algorithm for this problem that runs in time $O(\log^3 n)$ on an EREW PRAM with n processors (see [7] and the references therein). The algorithm computes the Lagrange interpolation formula and is reduced to polynomial evaluation at $O(n)$ points and then to the FFT algorithm. This algorithm specialized to our problem can be implemented to run faster, even on the butterfly (see [14] for the description and properties of the butterfly). By a *normal butterfly algorithm* we mean an algorithm in which a step of computation is performed by the nodes on one level, and the consecutive levels are used in a cyclic fashion.

Lemma 5. *The decoding of information dispersal from the fraction of $\alpha \cdot n$ strings can be performed on the $\mathcal{O}(\log n)$-dimensional butterfly in time $\mathcal{O}(\log^2 n)$ by a normal algorithm.*

Proof. The general interpolation algorithm resorts to the algorithm of evaluating a polynomial at $\mathcal{O}(n)$ points, which runs in time $\mathcal{O}(\log^2 n)$. This algorithm can be replaced by the algorithm to evaluate polynomials at the powers of a root of unity, because the arithmetic is in a finite field. Notice that we need to evaluate the polynomials at some bounded fraction of *all* the elements of the field, which can be performed in time $\mathcal{O}(\log n)$ by the FFT algorithm; in this way we gain the $\log n$-factor. The communication of processors is that needed for the FFT algorithm and of a full-binary-tree pattern, and can be implemented on the butterfly as a normal algorithm.

We can adapt this algorithm, due to the following:

Lemma 6. *A normal butterfly algorithm can be implemented on a list of active processors with delay $\mathcal{O}(1)$, provided the size of the list is at least equal to the number of nodes in one level of the butterfly.*

Proof. We need a two-directional cyclic list of length equal exactly to the size of a level of the butterfly, but this can be assumed to have been taken care of during preprocessing. Each processor simulates a row in the butterfly. The connections to other rows can be obtained dynamically by doubling the list links in both directions. A processor chooses the particular link depending on the bit representation of the row number.

After decoding, the input strings are stored in the order corresponding to physical ordering of processors and not in the order of the list. They can be rearranged by one application of the operation of communication along the list. As a conclusion of this discussion we obtain the following result:

Theorem 7. *The input encoded by the information dispersal method can be retrieved by the active processors of an EREW PRAM in time $\mathcal{O}(\log^2 n)$.*   □

Providing the input in the case of OCPC is handled similarly: in this case all fault-free processors are active and retrieve original input from information available to them.

References

1. R.J. Anderson, and G.L. Miller, Optical communication for pointer based algorithms, *Tech. Rep. CRI 88-14*, Comp. Sci. Dpt., University of Southern California, Los Angeles, 1988.

2. B.S. Chlebus, A. Gambin, and P. Indyk, PRAM computations resilient to memory faults, in *Proceedings of the 2nd Annual European Symposium on Algorithms*, ed. J. van Leeuwen, Utrecht, The Netherlands, 1994, Springer LNCS 855, pp. 401–412.

3. K. Diks and A. Pelc, Reliable computations on faulty EREW PRAM, *Theoretical Computer Science*, to appear.

4. M. Geréb-Graus and T. Tsantilas, Efficient optical communication in parallel computers, in *Proceedings of the 4th Ann. ACM Symposium on Parallel Algorithms and Architectures*, 1992, pp. 41–48.

5. A. Gibbons, and W. Rytter, *"Efficient Parallel Algorithms,"* Cambridge University Press, 1988.

6. L.A. Goldberg, M. Jerrum, F.T. Leighton, and S.B. Rao, A doubly logarithmic communication algorithm for the completely connected optical communication parallel computer, in *Proceedings of the 5th Ann. ACM Symposium on Parallel Algorithms and Architectures*, 1993, pp. 300–309.

7. J. JáJá, *"An Introduction to Parallel Algorithms"*, Addison Wesley, Reading, MA, 1992.

8. P.C. Kanellakis, and A.A. Shvartsman, Efficient parallel algorithms can be made robust, *Distributed Computing*, 5 (1992) 201-217.

9. P.C. Kanellakis, and A.A. Shvartsman, Efficient parallel algorithms on restartable fail-stop processors, in *Proceedings of the 10th Annual ACM Symposium on Principles of Distributed Computing*, 1991, pp. 23–36.

10. R.M. Karp, and V. Ramachandran, Parallel algorithms for shared-memory machines, in *"Handbook of Theoretical Computer Science,"* ed. J. van Leeuwen, Elsevier, 1990, vol. A, pp. 869–941.

11. Z. M. Kedem, K. V. Palem, A. Raghunathan, and P. Spirakis, Combining tentative and definite executions for very fast dependable parallel computing, in *Proceedings of the 23rd Annual ACM Symposium on Theory of Computing*, 1991, pp. 381-390.

12. Z. M. Kedem, K. V. Palem, and P. Spirakis, Efficient robust parallel computations, in *Proceedings of the 22nd Annual ACM Symposium on Theory of Computing*, 1990, pp. 138-148.

13. C.P. Kruskal, L. Rudolph, and M. Snir, A complexity theory of efficient parallel algorithms, *Theoretical Computer Science* 71 (1990) 95–132.

14. F.T. Leighton, *"Introduction to Parallel Algorithms and Architectures: Arrays, Trees, Hypercubes,"* Morgan Kaufman Publishers, San Mateo, California, 1991.

15. W.F. McColl, General purpose parallel computing, in *"Lectures on Parallel Computation"*, ed. A. Gibbons and P. Spirakis, Cambridge University Press, 1993, pp. 337–391.

16. F. Meyer auf der Heide, Hashing strategies for simulating shared memory on distributed memory machines, in *Proceedings of the 1st Heinz Nixdorf Symposium "Parallel Architectures and their Efficient Use,"* ed. F. Meyer auf der Heide, B. Monien, A.L. Rosenberg, Paderborn, Germany, 1992, Springer LNCS 678, pp. 20–29.

17. M.O. Rabin, Efficient dispersal of information for security, load balancing, and fault tolerance, *Journal of ACM*, 36 (1989), 335–348.

18. S.B. Rao, Properties of an interconnection architecture based on wavelength division multiplexing, *Technical Report TR-92-009-3-0054-2*, NEC Research Institute, Princeton, 1992.

19. A. A. Shvartsman, Achieving optimal CRCW PRAM fault-tolerance, *Information Processing Letters*, 39 (1991) 59-66.

20. L.G. Valiant, General purpose parallel architectures, in *"Handbook of Theoretical Computer Science,"* ed. J. van Leeuwen, Elsevier, 1990, vol. A, pp. 869-941.

Average Circuit Depth and Average Communication Complexity

Bruno Codenotti[1]* and Peter Gemmell[2]** Janos Simon[3]***

[1] Istituto di Matematica Computazionale del CNR, Pisa, Italy.
[2] Sandia National Labs.
[3] Department of Computer Science, The University of Chicago.

Abstract. We use the techniques of Karchmer and Widgerson [KW90] to derive strong lower bounds on the expected parallel time to compute boolean functions by circuits. By average time, we mean the time needed on a self-timed circuit, a model introduced recently by Jakoby, Reischuk, and Schindelhauer, [JRS94] in which gates compute their output as soon as it is determined (possibly by a subset of the inputs to the gate).
More precisely, we show that the average time needed to compute a boolean function on a circuit is always greater than or equal to the average number of rounds required in Karchmer and Widgerson's communication game. We also prove a similar lower bound for the monotone case. We then use these techniques to show that, for a large subset of the inputs, the average time needed to compute $s - t$ connectivity by monotone boolean circuits is $\Omega(\log^2 n)$.
We show, that, unlike the situation for worst case bounds, where the number of rounds characterize circuit depth, in the average case the Karchmer-Widgerson game is only a lower bound. We construct a function g and a set of minterms and maxterms such that on this set the average time needed for any monotone circuit to compute g is polynomial, while the average number of rounds needed in Karchmer and Widgerson's monotone communication game for g is a constant. Related work by Raz and Widgerson [RW89] shows that the monotone probabilistic communication complexity (a model weaker than ours) of the s-t connectivity problem is $\Omega(\log^2 n)$.

Keywords: circuit complexity, parallel time, communication complexity, lower bounds

1 Introduction

A recent paper of Jakoby, Reischuk, and Schindelhauer, [JRS94] formalizes the important and useful notion of self-timed circuits. In a self-timed circuit a gate produces its output as soon as a large enough subset of the inputs to it becomes

* Partially supported by ESPRIT Basic Research Action, Project 9072 'GEPPCOM'.
** Portions of this work were done while visiting IMC-CNR in Pisa, partially supported by GNIM-CNR
*** Portions of this work were done while visiting IMC-CNR in Pisa, sponsored by a grant from CNR

available. More precisely, if at time t a subset S of the inputs of gate g is defined, and the output of g is determined by the inputs in S, then at time $t + \Delta$ the output of g is determined (and becomes available to the gates that it is an input to.) Here Δ is the *gate delay*, which we will suppose to be 1. Thus, if a 0 appears at the input of an AND gate, by the next clock tick the gate will produce a 0 output, without waiting for the other input value to be defined. For some inputs, some circuits may produce an output much earlier than the worst-case bound given by the depth of the circuit.

Self-timed circuits are used in some fast computer architectures, [4] and [JRS94] makes a compelling argument for their theoretical study.

We will examine the *expected parallel time* to compute boolean functions by self-timed circuits. [JRS94] discuss the average time needed to compute some simple functions like addition and parity, showing that they can be done much faster on the average than in the worst case.

There are several possible definitions for expected or average behavior. A simple and natural choice is to compute the expected delay under the assumption of uniform distribution of the inputs. We will call this measure the *average parallel time*. Unfortunately, average time is not a very robust measure: it is possible to change it radically by simple padding (for example Wilf exhibits an NP-complete problem that can be solved in constant average time [W]). We refer to the more complete discussion of these problems in [JRS94], [RS93], [L86]. A more robust, and more complicated measure was introduced by Levin [L86]. We will call the distribution given by Levin's definition the *Levin measure*. Our results hold for both. In the interest of clarity and brevity, we postpone the definitions and proofs related to the Levin measure to the full paper.

A beautiful technique, due to Karchmer and Widgerson [KW90], relates circuit depth to the communication complexity of boolean relations. Given a boolean function f, consider the following *communication game*: There are two players. Player 0 has an input $x \in f^{-1}(0)$ and player 1 has an input $y \in f^{-1}(1)$.

The objective of the game is for player 1 to learn an index i such that $x_i \neq y_i$.

The game consists of messages exchanged by the two players according to some protocol, agreed upon in advance by the two players. The protocol is such that the messages sent by player 1 depend only on x and on the messages previously received; similarly the messages sent by player 2 depend only on y and on the previous messages. The communication complexity of f (more precisely, of the relation $f^{-1}(1) \times f^{-1}(0) \times i$ such that $x_i \neq y_i$) is the number of bits exchanged by the best protocol in the worst case. We refer to [KW90] for precise definitions.

Karchmer and Widgerson showed that the number of bits of communication necessary, in the worst case, in this game is exactly the minimum depth of a bounded fan-in boolean circuit that computes f. A similar characterization holds for monotone circuits: one of the players has a minterm of the circuit, the

---

[4] In actual circuits it is also possible to take advantage of varying gate delays and varying delays in different subcircuits: we will ignore these, as we are interested in asymptotics.

other a maxterm. They exchange messages until they can agree on an index that belongs to both the maxterm and the minterm. (To see how the monotone game relates to the general game, recall that a minterm is a minimal set of variables that set to 0 make the function have the value 0. Similarly, setting the variables of maxterm to 1 will force the function to have the value 1. So finding an index i that belongs to the minterm and to the maxterm means that there is an input x extending the maxterm – and thus $f(x) = 1$ – an input y extending the minterm – and thus $f(y) = 0$ – with $x_i = 1$ and $y_i = 0$).

Communication complexity of boolean relations is of intrinsic interest, and has been extended to probabilistic versions by Raz and Widgerson [RW89]. Average case communication complexity has also been previously studied [O91], [FKN91], but in a different context.

Lower bounds on the communication complexity for explicit functions can be used to derive lower bounds on worst case parallel time (circuit depth). While the notoriously hard problem of proving superlogarithmic parallel time bounds for explicit boolean functions remains open, there are some strong results for monotone circuits. Karchmer and Widgerson were able to prove that s-t connectivity (given a directed graph and two distiguished vertices s and t, is there a path from s to t) requires monotone circuits of depth $\Omega(\log^2 n)$ [KW90]. An even stronger, linear depth lower bound was obtained by Raz and Widgerson [RW92] for matching, a problem known to require nonpolynomial size monotone circuits [R85].

In this paper we study the average communication complexity of monotone boolean functions, in both the average and the Levin measure. We show that in both measures, the average number of rounds (bits) required in the Karchmer-Widgerson game is a lower bound on the average parallel time. We can then adapt the Karchmer-Widgerson lower bound proof to show an $\Omega(\log^2 n)$ lower bound on parallel average time for s-t connectivity on a certain natural set of inputs.

We also show, that unlike the deterministic case, the complexity of the game is not a characterization of average time – in fact the difference between the two bounds may be unbounded. More precisely, we construct a function g and a set of minterms and maxterms such that, on this set, the average time needed for any monotone circuit to compute g is a polynomial (n^α), while the average number of rounds needed in Karchmer and Widgerson's monotone communication game for g is $O(1)$.

Our results are somewhat related to those in [RW89]. Raz and Widgerson show that the monotone probabilistic communication complexity of the s-t connectivity problem is $\Omega(\log^2 n)$. Since probabilistic complexity is asymptotically not less than average complexity, our lower bounds imply theirs. They also show that in the general (nonmonotone) model, every relation has an $O(\log n)$ complexity protocol. This implies a similar bound in our model. It also shows that probabilistic communication complexity is not a characterization for probabilistic parallel time, and the difference can be from a polynomial to a logarithm. In our model, we can show an even bigger gap, from a polynomial to a constant.

2 Definitions

We introduce some general notation as well as the definitions that we will use for the computing time of functions on circuits and the communication complexity of functions.

First, some simple notation: For an integer n, let $[n]$ denote $\{1, 2 \ldots n\}$. $x \in_U S$, means that element x is chosen uniformly at random from set S.

We now consider definitions to do with the depth and the computing time of circuits.

Given a circuit C and input x, let $C(x)$ be the output of C on x.

Given a function f on domain D, let $depth_D(f)$ be the minimum depth of any fan-in 2 circuit that computes f.

Given a monotone function f on domain D, let $depth_D^m(f)$ be the minimum depth of any monotone fan-in 2 circuit that computes f.

In the following sense, the output of a circuit may be often determined in time less than the depth of the circuit:

Definition 1. Let C be a circuit with NOT gates only on the first level. Let x be an input for C. We will say that a subcircuit, C_x, of C **determines** $C(x)$ if:

C_x contains the output gate of C.

If $C(x) = 1$ then:

- All the gates or inputs that feed into the AND gates of C_x are also contained in C_x.

- At least one of the gates or inputs, with value equal to 1, that feed into every OR gate of C_x is contained in C_x.

If $C(x) = 0$ then:

- All the gates or inputs that feed into the OR gates of C_x are also contained in C_x.

- At least one of the gates or inputs, with value equal to 0, that feed into every AND gate of C_x is contained in C_x.

Definition 2. $time_C(x)$, or the time required on circuit C to compute on x, is equal to the minimum depth of any subcircuit C_x that determines $C(x)$.

We define the *average computing time* of a function f on domain D:

Definition 3. Given a function f on domain D,

$$AT_D(f) = min_C : C \text{ computes } f \text{ on } D \left[\sum_{x \in D} time_C(x)/|D| \right]$$

If f is monotone, then

$$AT_D^m(f) = min_C : C \text{ monotone, computes } f \text{ on } D \left[\sum_{x \in D} time_C(x)/|D| \right]$$

We now make definitions for communication games, where the goal is for two players to learn one index where their inputs differ.

Definition 4. Let f be a boolean function. Given two players, P_0 who knows string $x \in f^{-1}(0)$ and P_1 who knows string $y \in f^{-1}(1)$, we define a communication game G for f on domain D to be a sequence of interactions where P_0 and P_1 exchange bits back and forth until both know the same value i such that $x_i \neq y_i$. We refer to each bit communicated as a *round* and to the number of rounds communicated by the players using a game strategy G as $Rounds_G(x, y)$.

We now define **average communication complexity** for a function f on domain D.

Definition 5.

$$AC_D(f) = min_{G: \text{game for } f \text{ on } D} \left[\sum_{(x,y) \in (D \cap f^{-1}(0) \times D \cap f^{-1}(1))} \frac{Rounds_G(x,y)}{|(D \cap f^{-1}(0)) \times (D \cap f^{-1}(1))|} \right]$$

Definition 6. Let f be a monotone function.

Let MAX_f be the set of *maxterms* for f, ie $MAX_f =$ all minimal subsets, q, of $[n]$ such that $(\forall i \in q, x_i = 0) \Rightarrow f(x) = 0$.

Let MIN_f be the set of *minterms* for f, ie $MIN_f =$ all minimal subsets, p, of $[n]$ such that $(\forall i \in p, x_i = 1) \Rightarrow f(x) = 1$.

Given sets of minterms B_1 and maxterms B_0, we define extensions of the minterms/maxterms to the input set $\{0,1\}^n$:

$$B_0^* = \{x | i \in q \Rightarrow x_i = 0, i \in \bar{q} \Rightarrow x_i = 1\}_{q \in B_0}$$

$$B_1^* = \{y | i \in p \Rightarrow y_i = 1, i \in \bar{p} \Rightarrow y_i = 0\}_{p \in B_1}$$

Given a boolean function, f, and two players, P_0 who knows a maxterm $q \in B_0$ and P_1 who knows a minterm $p \in B_1$, we define a monotone communication game G for f on $B_0 \cup B_1$ to be a sequence of interactions where P_0 and P_1 exchange bits back and forth until both know the same value i such that $i \in p \cap q$. We refer to each bit communicated as a *round* and to the number of rounds communicated by the players using a game strategy G as $Rounds_G^m(x,y)$.

Definition 7. We now define *average monotone communication complexity*. Let D be a set of minterms and maxterms:

$$AC_D^m(f) = min_{G: \text{monotone game for } f \text{ on } D} \left[\frac{\sum_{(x,y) \in (D \cap MAX_f \times D \cap MIN_f)} Rounds_G^m(x,y)}{|(D \cap MAX_f) \times (D \cap MIN_f)|} \right]$$

3 Average Communication Complexity Lower Bounds Average Computing Time

Theorem 8. *For any boolean function f on domain D, we have:*

$$AC_D(f) \leq AT_D(f)$$

Proof. The proof is similar to the corresponding worst-case proof in [KW90] except that the path that the two players take as they "walk up" the circuit is determined by the quickest way to determine the output of the gates.

Let C be a circuit such that $AT(f) = \left[\sum_{x \in D} time_C(x)/|D| \right]$

Let $(x,y) \in_u (f^{-1}(0) \cap D \times f^{-1}(1) \cap D)$.

From the fact that $x \in f^{-1}(0)$ and by the definition of *time*, we know there is a subcircuit, C_x, of C, with the same output gate as C, with inputs that are a subset of C's inputs, and with maximum depth $time(x)$. Furthermore, each of C_x's AND gates has one input and one output and each of C_x's OR gates has two inputs and one output.

Similarly, we know there is a subcircuit C_y of C with the same output gate as C, with inputs that are a subset of C's inputs, and with maximum depth $time(y)$. Furthermore, each of C_y's OR gates has one input and one output and each of C_y's AND gates has two inputs and one output.

The game that players P_0 and P_1 will play is to start at the output of C and move up their subcircuits as follows:

- If the gate is an AND gate, it is P_0's move, and (s)he sends an R or an L to P_1 according to whether the single branch that C_x takes at this point is the right or left input of this gate in C.
- If the gate is an OR gate, it is P_1's move, and (s)he sends an R or an L to P_0 according to whether the single branch that C_y takes at this point is the right or left input of this gate in C.
- If the players have reached an input, then the players have arrived at an input bit where they disagree.

We note that, in the course of the game, the two players take a walk in $C_x \cap C_y$ and so the total number of rounds is at most $min(time(x), time(y))$.

The expected communication time is:

$$E_{(x,y)\in_u((D\cap f^{-1}(0))\times(D\cap f^{-1}(1)))}min(time_C(x), time_C(y)) \leq E_{z\in_u D}(time_C(z))$$

We now discuss a corresponding theorem for average monotone time and communication complexity.

Theorem 9. *For any boolean function f and sets of maxterms and minterms, B_0 and B_1, we have:*

$$AC^m_{B_0 \cup B_1}(f) \leq AT^m_{B_0^* \cup B_1^*}(f)$$

The proof is identical to the one above, except that we may have the two parties extend their inputs. P_0 extends his/her maxterm input with 1's. P_1 extends his/her minterm input with 0's. The index i that the communication protocol will find is one where $x_i = 0$ and $y_i = 1$. Therefore, i is a desired element of $p \cap q$.

4 An $\Omega(\log^2 n)$ Lower Bound on Average Time on Monotone Circuits for STCONN

We now show that, on the set of input graphs, D, described in [KW90], $AT^m_D(STCON) = \Omega(\log^2 n)$. The overall strategy of the proof is very similar, but the fact that we must use averages instead of individual graphs is responsible for some added difficulty.

First, we change the requirement of a lower bound on the number of bits exchanged into a requirement for a lower bound on the number of rounds. Our lemma is analogous to [KW90]'s Theorem 2.3:

Lemma 10. *For any function f and domain D, there exists a communication game G for f on D where at each round P_0 sends 2^a bits while P_1 responds with a bits and such that the* average *number of rounds k satisfies:*

$$k \le \frac{AT_D(f)}{a}$$

Similarly, for any monotone function f and set of minterms and maxterms D, there exists a monotone communication game G for f on D where at each round P_0 sends 2^a bits while P_1 responds with a bits and such that the average *number of rounds k satisfies:*

$$k \le \frac{AT_D^m(f)}{a}$$

Proof. Let C be the best average case circuit for f on domain D. At each round, the two players traverse (away from the output gate) a levels of the circuit C. There are 2^a inputs to the subcircuit that is rooted at the round's starting gate. The function computed by this subtree can be expressed in CNF using at most 2^{2^n} clauses of the 2^n input variables to the subcircuit. P_0 sends to P_1 the index of a clause for which all his/her variables are most quickly determined to be 0. P_1 then returns the index of the variable within the clause that s/he most quickly determines is equal to 1.

We will restrict ourselves to the following domain: $D = B_0 \cup B_1$, where B_0 is the set of those graphs with exactly one $s-t$ cut and all other edges present, and B_1 is the set of those graphs with one path connecting s and t, with length at most $n^{1/10}$ and all other edges absent.

We now state the main result of the section:

Theorem 11.
$$AT_D^m(STCONN) = \Omega(\log^2 n)$$

To prove the theorem, we first note that by lemma 10 it suffices to prove an $\Omega(\log n)$ lower bound on the number of rounds needed in the game to obtain an $\Omega(\log^2 n)$ lower bound on the number of bits exchanged. The theorem then follows from theorem 9.

We need some rather technical definitions, borrowed from [KW90].

Definition 12. Let $S' \subset S$.

$$\mu_S(S') = \frac{|S'|}{|S|}$$

Let $\epsilon = 1/10$. Let $l = n^\epsilon$.

Let $U_p = [n]^l$, the set of all paths of length l and let $U_c = \{0, 1\}^l$, the set of all 2-colorings of l vertices.

Assume that there is a monotone communication game G that can be used by players P_0 and P_1, for inputs (x, y) from $(B_0 \times B_1)$, to find an index $i : i \in x \cap y$. Assume that G is of the form described in lemma 10 with $a = \epsilon \log(n)$ and that the expected number of rounds $k_{max} = \frac{\epsilon}{2} \log(n)$.

Define $t_{max} = \log(l) - 1, n_0 = n, l_0 = l$. For $j \in [t_{max} - 1]$, let $n_{t+1} = n_t - 4\sqrt{n_t}, l_{t+1} = l_t/2$.

Definition 13. We define the property $\mathbf{H}(t, k)$:

There exists a collection of vectors $P^t \subset [n_t]^{l_t}$ and a collection of colorings $Q^t \subset \{0,1\}^{n_t}$, with $\mu_{[n_t]^{l_t}}(P^t) \geq \frac{1}{8}n^{-\epsilon}$ and $\mu_{\{0,1\}^{n_t}}(Q^t) \geq 2^{-2tn^\epsilon}$ such that there is a monotone communication game G^t for $STCONN$ on (P^t, Q^t) with $E_{(q,p) \in Q^t \times P^t}(Rounds_{G^t}^m(q, p)) \leq k$.

Note that $\overline{H(0, k_{max})}$ implies that the protocol needs $\Omega(\log n)$ rounds, and this suffices to prove our theorem.

We will prove the following two claims that imply (by induction) that $\overline{H(0, k_{max})}$, concluding the proof of Theorem 4.

Claim 14 *For* $t \leq t_{max}, \overline{H(t, 0)}$.

Proof. Because we have $\mu_{[n_t]^{l_t}}(P^t) \geq \frac{1}{8}n^{-\epsilon}$, and because the fraction of all paths of length t, $[n_t]^{l_t}$, that contain any one node is $1 - (1 - \frac{1}{n_t})^{l_t} \ll n^{-\epsilon}$, we know that there is no one node that is contained in all of P_1's paths. Therefore, P_0 can not know a vertex adjacent to a bichromatic edge.

Claim 15 *For* $t \leq t_{max}, H(t, k) \Rightarrow H(t + 1, k - \frac{1}{2})$.

Proof. We assume $H(t, k)$ and construct P^{t+1}, Q^{t+1}, and G^{t+1}.

Divide P^t into equivalence classes so that for every member of each equivalence class, P_1 sends the same message to P_0 in the first round of G^t. We know that the combined weight of those equivalence classes such that G^t requires at most expected $k + 1/4$ rounds (when that equivalence class is paired with Q^t) is at least $\frac{1}{4k}$.

By the pigeon-hole principle, there exists a subset $P \subset P^t$ such that $\mu_{[n_t]^{l_t}}(P) \geq \frac{1}{8}\frac{1}{4k}n^{-2\epsilon}$, for every $p \in P$, P_1 sends the same message in the first round of G^t, and the average expected time that G^t takes on $P \times Q^t$ is at most $k + \frac{1}{4}$.

Now divide Q^t into equivalence classes so that for every member of each equivalence class, P_0 sends the same message to P_1 in the first round of G^t. We know that the combined weight of those equivalence classes such that G^t requires at most expected $k + 1/2$ rounds (when that equivalence class is paired with P) is at least $\frac{1}{4k}$. By the pigeon-hole principle, there exists a subset $Q \subset Q^t$ such that $\mu_{\{0,1\}^{n_t}}(Q) \geq \frac{1}{4k}2^{-(2t+1)n^\epsilon}$, for every $q \in Q$, P_0 sends the same message in the first round of G^t, and the average expected time that G^t takes on $P \times Q$ is at most $k + \frac{1}{2}$.

We now prune the sets P and Q down further so that they will meet the requirements of claim 15.

We must introduce several further pieces of notation, borrowed from [KW90].

Definition 16. Given an interval $I \subset [l_t]$ of consecutive integers and a vector $p \in [n_t]^{l_t}$, let p_I be the projection of p into I. For $P \subset [n_t]^{l_t}$, let $P_I = \{p_I : p \in P\}$.

Given a set of paths, P, a set of vertices, $I \subset [n_t]$, and a vector $p \in I$, let $\mathbf{Ext}_{P,I}(\mathbf{p}) = \{\tilde{p} \in P : \tilde{p}_I = p\}$

Similarly, given a subset $T \subset [l_t]$ and a coloring $q \in \{0,1\}^{n_t}$, let q_T be the projection of q into T. For a set of colorings, Q, let $Q_T = \{q_T : q \in Q\}$.

Given a restriction function $\rho : [n_t] \rightarrow \{0, 1, *\}$, let $Q^\rho = \{q \in Q : \forall i \in [n_t], \rho(i) \neq * \Rightarrow q_i = \rho(i)\}$.

Definition 17. Let $L = [l_t/2]$, $R = [l_t] - L$.

We say that P is L (or R) good if many left (right) projections of P each have many extensions to the right (left):

In particular, P is **L good** if:

$$\mu_{P_L}(\{p_L : \mu_P(Ext_{P,L}(p_L)) \geq n^{-2\epsilon}/32\}) \geq \frac{1}{4}n^{-\epsilon}$$

P is **R good** if:

$$\mu_{P_R}(\{p_R : \mu_P(Ext_{P,R}(p_R)) \geq n^{-2\epsilon}/32\}) \geq \frac{1}{4}n^{-\epsilon}$$

Lemma 18. P is either L good or R good.

Proof. The proof is exactly as in [KW90] and follows from a lemma due to Hastad.

Without loss of generality, we will assume that P is L good.

Lemma 19. There exists a restriction $\rho : [n_t] \rightarrow \{0, 1, *\}$ with $|\rho^{-1}(*)| = n_{t+1}$ such that the following properties hold:

G1: $\mu_{\{0,1\}^{n_{t+1}}}(Q^\rho_{\rho^{-1}(*)}) \geq 2^{-2(t+1)n^\epsilon}$.

G2: $\exists \tilde{P} \subset P$ such that :
- $\forall p \in \tilde{P}, p_L \in \rho^{-1}(*)$ and $p_R \in \rho^{-1}(1)$
- $\forall p, p' \in \tilde{P}, p_L \neq p'_L$
- $\mu_{[n_t]^{l_t}}(\tilde{P}_L) \geq \frac{1}{8}n^{-\epsilon}$.

Proof. This lemma is almost straight out of [KW90]

Let $Q^{t+1} = Q^\rho_{\rho^{-1}(*)}$ and $P^{t+1} = \tilde{P}_L$ and rename the coordinates so that $[n_{t+1}] = \rho^{-1}(*)$.

There is a natural 1-1 correspondence between Q^{t+1} and Q^ρ and between P^{t+1} and $\tilde{P}$.

Also, for any $q \in Q$ and $p \in P$, any bichromatic edge lies in the interval L.

The game G^{t+1} simulates the remaining rounds of G^t on vector in P^{t+1} and coloring in Q^{t+1} by following G_t on the associated vector in P and coloring in Q.

5 Average Communication Complexity May be Much Lower than Average Circuit Time

We now construct an example that shows that, unlike the situation with the worst-case complexity of communication games and circuits, the reverse of theorem 8 does not hold.

Theorem 20. For all constants $\alpha < 1$, there exists a function f and domain $D \subset \{0, 1\}^n$ such that
$$AC_D(f) = O(1), \text{ whereas } AT_D(f) = \Omega(n^\alpha).$$

Proof. We let g be a random boolean function such that $|g^{-1}(1)| = 2^{n^{\alpha}}$.

Most such functions must be such that $AT_D(g) = \Omega(n^{\alpha})$ and we let g be one such function.

We now create f by padding with one extra bit to include a larger set of values for which $f = 1$:

We define $f : \{0,1\}^n \times \{0,1\} \to \{0,1\}$:

$$f(x,y) = \begin{cases} g(x) & \text{if } y = 0 \\ 1 & \text{otherwise} \end{cases}$$

We have:

$$f^{-1}(0) = g^{-1}(0) \times \{0\}$$

$$f^{-1}(1) = (g^{-1}(1) \times \{0\}) \cup (\{0,1\}^n \times \{1\})$$

Note that $|f^{-1}(0)| = 2^n - 2^{n^{\alpha}}$, and $|f^{-1}(1)| = 2^n + 2^{n^{\alpha}}$.
The proof of theorem 8 gives us:

$$AC_D(f) \leq E_{(x,y) \in_* f^{-1}(0) \times f^{-1}(1)} min(time_C(x), time_C(y)) \leq E_{y \in f^{-1}(1)}(time_C(y)),$$

for any circuit C that computes f.

Because $2^n >> 2^{n^{\alpha}}$, and because the last bit can be determined in unit time, we have $AC_D(f) \leq E_{y \in f^{-1}(1)}(time_C(y)) = O(1)$ and then the thesis follows by noting that $AT_D(f) = AT_D(g) = \Omega(n^{\alpha})$.

References

[FKN91] T. Feder, E. Kushilevitz, M. Naor, *Amortized Communication Complexity*, Proc. 32nd Symposium on the Foundations of Computer Science, October 1991, pp. 239 - 248.

[JRS94] A. Jakoby, R. Reischuk, C. Schindelhauer, *Circuit Complexity: from the Worst Case to the Average Case*, Proc. 26th ACM Symposium on the Theory of Computing, May 1994, pp.58 - 67.

[KW90] M. Karchmer and A. Wigderson, *Monotone Circuits for Connectivity Require Superlogarithmic Depth*, Siam Journal of Discrete Math, vol. 3, no 2, May 1990, pp.255 - 265.

[L86] L. Levin, *Average Case Complexity Problems*, SIAM J. Computing 15, 1986, pp. 285-286.

[O91] A. Orlitsky, *Interactive Communication: Balanced Distributions, Correlated Files, and Average-Case Complexity*, Proc. 32nd Symposium on the Foundations of Computer Science, October 1991, pp. 228 - 238.

[RW89] R. Raz and A. Wigderson, *Probabilistic Communication Complexity of Boolean Relations*, Proc. 30th Symposium on the Foundations of Computer Science, October 1989, pp. 562 - 567.

[RW92] R. Raz and A. Wigderson, *Monotone Circuits for Matching Require Linear Depth*, Journal of the ACM v.39 1992.

[R85] A. A. Razborov, *A lower bound on the monotone network complexity of the logical permanent* Mat. Zametki v.37 n.6 1985 pp. 887 - 900.

[RS93] R. Reischuk, C. Schindelhauer, *Precise Average Case Complexity*, Proc. 10th Symposium on Theoretical Aspects of Computer Science, Feb. 1993, pp.650 - 661.

[W] H. S. Wilf, *Backtrack: An $O(1)$ expected time algorithm for the graph coloring problem*, Information Processing Letters v.18 n.3 1984 pp. 119-121.

Packing Trees

Joseph Gil and Alon Itai

Department of Computer Science
Technion – Israel Institute of Technology
Technion City; Haifa; Israel 3200

Abstract. In a virtual memory system, the address space is partitioned into pages, and main memory serves as a cache to the disk. The problem we address is: Given a tree T, find a packing, an allocation of its nodes to pages, which optimizes the cache performance. We investigate a model for tree access in which a node is accessed only via the path leading to it from the root. Two cost functions are considered: the total number of different pages visited in the search, and the number of page faults incurred. It is shown that both functions can be optimized simultaneously. An efficient dynamic programming algorithm to find an optimal packing is presented. The problem of finding an optimal packing which also uses the minimum number of pages, is shown to be NP-complete. However, an efficient approximation algorithm is presented. This algorithm finds a packing that uses the minimum number of pages, and requires at most one extra page fault per search. Finally, we study dynamic trees which allow insertions and deletions.

1 Introduction

1.1 Motivation

The simplicity and the elegance of the Random Access Machine model have made it the model of choice for many algorithm designers. However, it should be recognized that memory is hardly ever "random access". The rapid decrease in the cost-performance ratio of CPUs has not been matched by that of storage units. As a result, true random access machines are still economically infeasible. Most modern computer systems feature a memory hierarchy of several levels: the topmost level is small, fast, and expensive; each successive level is larger, slower and cheaper. This model is very appealing since by caching, it offers an expected access time close to that of the fastest level while keeping the average cost per memory cell near the cost of the cheapest level.

We need therefore to pay more attention to questions pertaining to the cost of memory reference in real life computers. One such work is that of Aggarwal, Alpernm, Chandra, and Snir [1] who suggested a computation model in which a charge of $\lg i$ is associated with the access to the ith memory cell. A less theoretical approach was pursued by researchers from the compiler optimization and numerical algorithms communities. A significant body of research (see [9] for a brief survey) demonstrated performance enhancement in many numerical algorithms operating on matrices. The key idea behind these results was the alignment of the algorithm and the data structures to achieve better cache performance.

The performance gains were obtained by non-trivial loop transformations and judicious layouts of the matrices with the aim of reducing the number of *cache-misses* and *page-faults*. Although a more precise definition is required, the basic principle is simple: since a "locality of reference" assumption underlies every hierarchical memory system, a program coerced to exhibit this kind of behavior, will perform better. Finding algorithms that enhance the effectiveness of caches is in a sense a problem dual to that of cache management policies that attempt to optimize the performance of algorithms.

Further, an increase in the locality of memory references can improve parallel performance by reducing communication overheads. Indeed, we see reports [10] of these optimizations leading to almost doubling the measured speedup on a multi-processor system, bringing it close to the theoretical maximum.

A natural question arising here is whether combinatorial computing can capitalize on similar techniques. In this paper, we try to give a partial answer to this problem by studying the issue of memory-level sensitive representation of trees in memory. Why trees? Most importantly, large trees are a ubiquitous data representation. They occur in a wide range of application domains including AI (Lisp functions and data), text processing (e.g., SGML [8] documents), and geometric modeling in CAD systems [5]. Also, trees are more interesting than matrices in that they have more different topologies and since the pointer structure allows for many different layouts in memory.

1.2 Definitions

To continue the discussion in a more rigorous manner we need the following notations and definitions. Let T be a given input tree of n nodes, whose root is r. We say that a node v belongs to a tree T, and by abuse the notation $v \in T$. For a node $v \in T$, T_v denotes the sub-tree of T whose root is v. If $v \neq r$, then parent(v) denotes its parent. Also, if T is a binary tree then the left and right children of v are denoted by left(v) and right(v). A *descent* is a sequence $v_1, \ldots, v_k$ such that $v_i = \text{parent}(v_{i+1})$ for $i = 1, \ldots, k-1$. We say that the descent leads from v_1 to v_k. The *path* to a node v is a descent leading from r to it; $\ell(v)$ denotes the number of edges in this path.

The nodes of T are to be represented in memory as records and its edges as pointers. Memory is partitioned into *pages*, each of which can contain p nodes. A *packing of T* is a function τ which maps its nodes to memory pages; formally $\tau : T \to Z^+$ such that $\left|\tau^{-1}(i)\right| \leq p$. A page is full if $\left|\tau^{-1}(i)\right| = p$. The *space* of a packing is the number of pages it uses, i.e., $|\tau(T)|$. We say that a packing is *compact* if its space consumption achieves the minimum, $N = \lceil n/p \rceil$; it is *k-compact* if its space usage is kN.

1.3 The problem

Operating systems' jargon includes the idiom "working set of a process"—a loosely defined term which refers to the set of pages a process will reference in the "near future". Virtual memory systems work only because in practice the working set is small, and most of it is stored in memory. [1] If a large tree is packed randomly in pages, then the working set is large and its processing may lead to extensive disk activity and even to thrashing. Our overall goal is therefore to find a packing that minimizes the working set.

The model we use for patterns of access to the tree is that of paths. A node is accessed only by traversing the path leading to it. With each node v we associate a positive weight $w(v)$ that can be thought of as the probability of accessing it (although weights are not necessarily normalized). We also use the notation $w(T_v) = \sum_{u \in T_v} w(u)$. The model is quite general and corresponds to a large family of tree algorithms ranging from ordinary search trees to ray tracing [4].

Consider a packing τ and the path to the node v: $r = v_0, v_1, \ldots, v_{\ell(v)} = v$. There are two extremal cases:

1. Suppose that the cache can accommodate only one page which is initially used for some other purpose. Then, the number of page faults that occur along a path is exactly the number of times *a page boundary is crossed*. More formally, we define for every node u

$$\Delta_\tau(u) = \begin{cases} 1 \ u = r \\ 0 \ \text{if} \ \tau(\text{parent}(u)) = \tau(u) \\ 1 \ \text{otherwise} \ . \end{cases}$$

The cost associated with the path, $c_\tau(v)$ is defined as

$$\sum_{i=0}^{\ell_v - 1} \Delta_\tau(v_i) \ .$$

2. Conversly, suppose that the size cache is unlimited. The appropriate charge for a path in this case is $s_\tau(v)$, the number of *distinct pages accessed* along it, defined formally as

$$s_\tau(v) = |\{\tau(v_i) \mid 0 \le i \le \ell(v)\}| \ .$$

Two functions which should be used as targets for our global optimizations are therefore $c_\tau(T)$ and $s_\tau(T)$, defined as

$$c_\tau(T) = \sum_{v \in T} w(v) c_\tau(v) \ ,$$

$$s_\tau(T) = \sum_{v \in T} w(v) s_\tau(v) \ .$$

---

[1] For the sake of concreteness, we chose to relate to disk-caching in memory. However, the discussion equally applies to caching of main memory in a separate or on-chip CPU cache.

A packing τ is said to be *page-fault-optimal* if it minimizes $c_\tau(T)$, and *working-set-optimal* if it minimizes $s_\tau(T)$.

Henceforth, the subscript τ is omitted whenever it can be understood from the context. For a page P and a node v, $\tau(v) = P$, we say that v belongs to P, and by abuse of notation, we write $v \in P$.

2   Results

In this section we present our main results. Proofs are postponed to later sections.

We first observe that a trivial instance of the problem occurs when $|T| \leq p$. This case is tacitly excluded from our consideration.

It is natural to conjecture that the page that contains the root must be full.

Lemma 1. *Let τ be a page-fault-optimal packing, and $P = \tau(r)$. Then, P is full.*

Proof. Suppose to the contrary that P is not full. Then, since we assumed $|T| > p$, there must exist a node v, $v \notin P$, parent$(v) \in P$. Moving v to P decreases $c(v)$ without increasing $c(u)$ for any other node u.   ∎

Intuition may also lead us to believe that in an optimal packing a path never visits the same page twice. Indeed, this is the case.

Definition 2. A packing is *convex* if for every descent $u_1, \ldots, u_k$ if $\tau(u_1) = \tau(u_k)$ then for $i = 1, \ldots, n$, $\tau(u_i) = \tau(u_1)$.

Theorem 3. *All page-fault-optimal packings are convex.*

Proof. See Section 3.   ∎

How about working-set-optimal packings? Perhaps surprisingly, we can show that both of our, seemingly different, complexity measures, c and s, can in fact be optimized simultaneously.

Theorem 4. *All page-fault-optimal packings are also working-set-optimal.*

Proof. See Section 3.   ∎

We therefore may use the term *optimal* to refer to both *page-fault-optimal* and *working-set-optimal*. As it turns out, an optimal packing can be computed efficiently by using dynamic programming in a bottom-up traversal of the tree.

Theorem 5. *Let T be a tree of n nodes and degree d. Then, an optimal packing can be computed in time $O(np^2 \lg d)$ while using $O(p \lg n)$ space.*

Proof. See Section 4.   ∎

The packing obtained by this algorithm is 2-compact. Is it also possible to simultaneously optimize space and cache performance? Unfortunately, the answer is (most likely) negative.

Theorem 6. *The problem of computing an optimal compact packing is NP-complete.*

Proof. See Section 5. ■

On the other hand, it is possible to achieve a compact packing in which the average cost is not much worse than that of the optimal.

Theorem 7. *Let τ be an optimal packing. Then, in $O(n \lg n)$ time and using $O(p)$ space, it is possible to compute from it a compact packing τ' such that*

$$\frac{c_{\tau'}}{n} \leq \frac{c_\tau}{n} + \frac{1}{n} \sum_{v \in T} w(v) \ .$$

Furthermore, if all $w(v) = 1$ then

$$c_{\tau'}/n \leq c_\tau/n + 0.5 \ .$$

Proof. See Section 6. ■

Most of this paper is devoted to the question of packing fixed trees. The dynamic version of this problem, that is, efficiently updating the packing as the tree changes due to, say, insertions and deletions, seems to be much more difficult. In Section 7 we discuss this further and give some partial results that show how to maintain a reasonably good packing for two specific schemes of tree updates: B-trees and a variation of weight-balanced trees.

3 Equivalence of the two Complexity Measures

In order to show that the page-fault and the working-set complexity measures are equivalent, we show that the optimal packings for either measures are convex.

3.1 Proof of Theorem 3

Let τ, τ' be packings and let $v \in T$ be a node. Then, we define

$$\Delta_{\tau',\tau}(v) = \Delta_{\tau'}(v) - \Delta_\tau(v) \ .$$

Since $\Delta_\tau(v), \Delta_{\tau'}(v) \in \{0,1\}$, we have $\Delta_{\tau',\tau}(v) \in \{-1,0,1\}$. It is easy to see that

$$c_{\tau'}(v_i) - c_{\tau'}(v_i) = \sum_{j=0}^{i} \Delta_{\tau',\tau}(v_j) \ .$$

Also

$$c_{\tau'}(v) - c_\tau(v) = c_{\tau'}(\text{parent}(v)) - c_\tau(\text{parent}(v)) + \Delta_{\tau',\tau}(v) \ .$$

It follows that if $c_{\tau'}(\text{parent}(v)) \leq c_\tau(v)$ and $\Delta_{\tau',\tau}(v) \leq 0$ then $c_{\tau'}(v) \leq c_\tau(v)$.

Lemma 8. *Suppose that for all nodes v for which $\Delta_{\tau',\tau}(v) = 1$ holds, $c_{\tau'}(v) \leq c_\tau(v)$ also holds. Then, for every node $u \in T$, $c_{\tau'}(u) \leq c_\tau(u)$.*

Proof. Consider a path $r = v_0, \ldots, v_{\ell(v)} = v$. We show by induction on i that $c_{\tau'}(v_i) \leq c_\tau(v_i)$. The induction base, $i = 0$ is trivial: $c_\tau(v_0) = 1 = c_{\tau'}(v_0)$. By the inductive hypothesis and the definition of $\Delta_{\tau',\tau}$,

$$c_{\tau'}(v_i) = c_{\tau'}(v_{i-1}) + \Delta_{\tau'}(v_{i-1}) \leq c_\tau(v_i) + \Delta_{\tau',\tau}(v_i) \ .$$

If $\Delta_{\tau',\tau}(v_i) \leq 0$, we are done. If $\Delta_{\tau',\tau}(v_i) = 1$, the lemma follows from the hypothesis. ∎

Lemma 9. *There exists a page-optimal tree which is convex.*

Proof. Let τ be a non-convex packing. We will construct a packing τ' of lower cost.

Since τ is non-convex, T contains a descent $u_0, \ldots, u_k$, such that $\tau(u_0) = \tau(u_k) = P$, while for some $i < k$, $\tau(u_i) \neq P$. Without loss of generality, assume that $u_0, \ldots, u_k$ is the shortest non-convex descent for τ. Therefore,

$$\tau(u_1), \tau(u_2), \ldots, \tau(u_{i-1}) \neq P \ .$$

Let i be the maximal integer for which $\tau(u_1) = \tau(u_2) = \ldots = \tau(u_i)$.

Let T_k be the maximal subtree of T which contains u_k and all its nodes are assigned to page P, i.e., $v \in T_k$ implies $\tau(v) = P$. Note that u_k is the root of this tree since $\tau(\text{parent}(u_i)) \neq P$.

There are two cases to consider, $i \geq |T_k|$ and $i < |T_k|$. In each case we modify τ as in Figure 1. In the full paper we apply Lemma 8 to show that for every node $v \in T$, the new packing τ' satisfies $c_{\tau'}(v) \leq c_\tau(v)$. ∎

The proof of Theorem 3 is completed by noting that in the above construction, $c_{\tau'}(u_1)$ is strictly less than $c_\tau(u_1)$.

3.2 Proof of Theorem 4

We first show that working-set-optimal packings enjoy the property of Theorem 3.

Theorem 10. *All working-set-optimal packings are convex.*

Proof. The proof is similar and simpler than that of Theorem 3. ∎

Theorem 4 is immediately implied from the following easy lemma:

Lemma 11. *Let τ be a convex packing. Then for every $v \in T$*

$$c_\tau(v) = s_\tau(v) \ .$$

4 An Algorithm for Finding an Optimal Packing

In this section we give the dynamic programming algorithm that lies behind Theorem 5. We begin by deriving the additional properties of optimal packings on which the algorithm is based. Next, the basic algorithm for binary trees is presented. This algorithm uses relatively large space for its computation. We then show how space consumption can be reduced. Finally, the extension to trees of arbitrary degree is described.

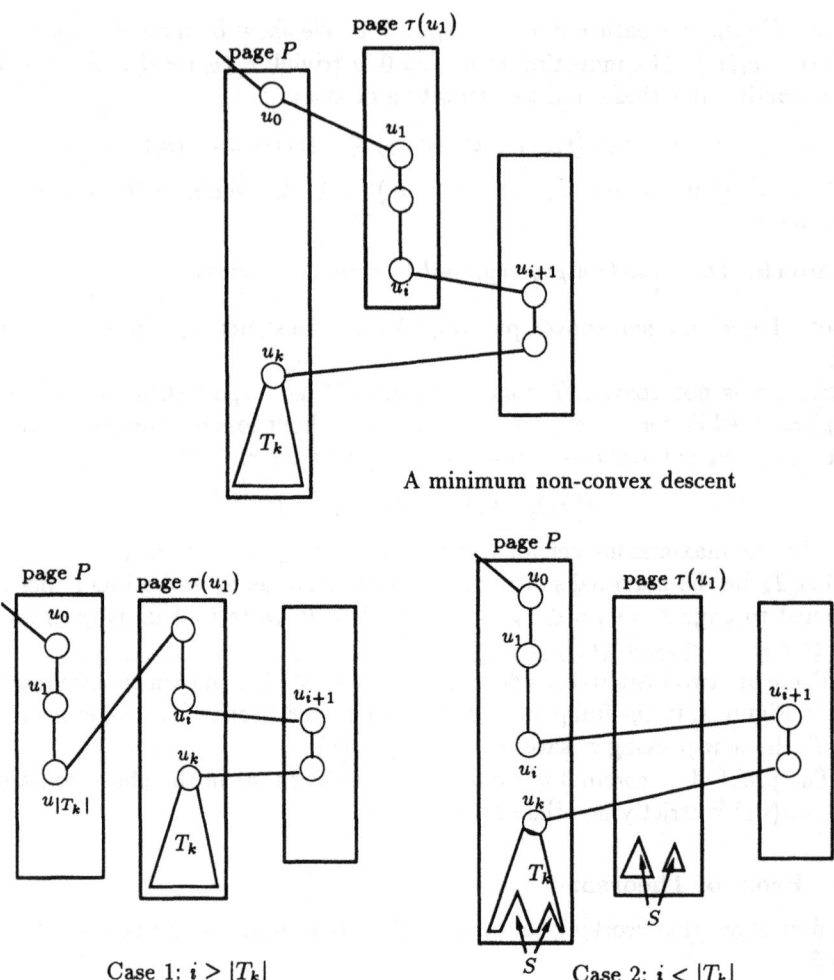

A minimum non-convex descent

Case 1: $i \geq |T_k|$

Case 2: $i < |T_k|$

Fig. 1. Proof of Lemma 9

4.1 Properties of optimal packings

Lemma 12. *Let τ be an optimal packing of T. Let P be a non-full page of τ. Then P must contain a leaf of T, and furthermore, if $v \in P$, then $T_v \subseteq P$.*

Let $\mathbf{T}_\tau$ be the graph whose nodes are the pages of τ and all edges (P, P') for which there exist $v \in T$ and $\tau(\mathrm{parent}(v)) = P$ and $\tau(v) = P'$.

Lemma 13. *There exists an optimal packing τ such that $\mathbf{T}_\tau$ is a tree.*

Proof. Connectivity of $\mathbf{T}_\tau$ follows from the connectivity of T. If $\mathbf{T}_\tau$ is not a tree then it contains a cycle. Let τ be an optimal packing such that $\mathbf{T}_\tau$ has the least number of cycles. In view of Lemma 9, no path from the root to a leaf can contain such a cycle. Therefore, there exists a page P that contains two non-root nodes v_1, v_2, for which $\Delta(v_1) = \Delta(v_2) = 1$. By Lemma 12, $T_{v_2} \subseteq P$. Let τ' be the packing which is identical to τ, except for the nodes in T_{v_2} which are mapped by τ' to a new page $P' \notin \tau(T)$. Then, $c_\tau(T) = c_{\tau'}(T)$ and τ' has less cycles than τ. ∎

Corollary 14. *Let τ be an optimal packing and $v \in T \setminus \{r\}$ a node for which $\tau(v) \neq \tau(\text{parent}(v))$. Then, the pages assigned by τ to the nodes of T_v are disjoint from those assigned to the remainder of the tree. Further, the restriction of τ to T_v is an optimal packing of T_v.*

4.2 Algorithm for Binary trees

Since initially all pages are isomorphic, we may assume that the root r is always mapped to a fixed page R. Consider the set

$$V = \{v \in T \mid \tau(v) \notin R, \tau(\text{parent}(v)) = R\} \ .$$

By Corollary 14, τ is an optimal packing of T_v for all $v \in V$. Thus, in order to find an optimal packing, it may be possible to first determine which other nodes reside in R and then continue recursively with all trees T_v, $v \in V$. Dynamic programming provides a more efficient implementation of the idea presented above. However, in order to implement this technique, we consider the following slightly more general problem: For $i = 1, \ldots, p$ an *i-confined* packing of a tree T is a packing of T in which R contains i nodes.

Suppose that in an optimal packing i nodes of $T_{\text{left}(r)}$ are mapped to R. Then τ, restricted appropriately, is both an optimal i-confined packing of $T_{\text{left}(r)}$ and an optimal $(p-i-1)$-confined packing of $T_{\text{right}(r)}$. This property is the basis of our bottom-up dynamic programming algorithm.

Let $A[v, i]$ be the cost of an optimal i-confined packing of T_v. If exactly $1 \leq i \leq p-2$ of the nodes of $T_{\text{left}(r)}$ are mapped to R, then exactly $p-1-i \geq 1$ nodes of $T_{\text{right}(r)}$) are mapped to $\tau(r)$. Accounting for the cost of accessing r itself we have

$$A[r, p] = w(r) + A[\text{left}(r), i] + A[\text{right}(r), p - 1 - i] \ . \tag{1}$$

What about the cases $i = 0$ and $i = p - 1$? If $T_{\text{left}(r)} \cap R = \emptyset$, then all the nodes of $T_{\text{left}(r)}$ incur a page fault when going from r to $\text{left}(r)$. Hence, the cost of T in this case is

$$A[r, p] = w(r) + w(T_{\text{left}(r)}) + A[\text{left}(v), p] + A[\text{right}(r), p - 1] \ . \tag{2}$$

Similarly, in the case $T_{\text{right}(r)} \cap R = \emptyset$, we have

$$A[r, p] = w(r) + A[\text{left}(v), p - 1] + w(T_{\text{right}(r)}) + A[\text{right}(r), p - 1] \ . \tag{3}$$

Generalizing over equations (1), (2), and (3) for every node v and $i = 2, \ldots, p$ we obtain:

$$A[v, i] = w(v) + \min \left(\begin{array}{l} A[\text{left}(v), i-1] + w(T_{\text{right}(v)}) + A[\text{right}(v), p], \\ w(T_{\text{left}(v)}) + A[\text{left}(v), p] + A[\text{right}(v), i-1], \\ \min_{1 < j < i-1} (A[\text{left}(v), j] + A[\text{right}(v), i-j-1]) \end{array} \right) ,$$

$$(4)$$

where, $A[\text{left}(v), i]$ is defined, for notational convenience, as 0 whenever $\text{left}(v)$ does not exist (and similarly for $\text{right}(v)$). The case $i = 1$ is special and is given by

$$A[v, 1] = w(T_v) + A[\text{left}(v), p] + A[\text{right}(v), p] . \qquad (5)$$

This gives us the desired dynamic programming algorithm. The algorithm is detailed in Figure 2. It is easy to check that the algorithm runs in $O(np^2)$ time.

```
For all leaves v of T do
    begin
        mark v;
        for i := 1 to p do
            A[v, i] := w(v);
    end;
While there exists an unmarked node v do
    begin
        mark v;
        for i := 2 to p do
            update A[v, i] according to equation (4);
        update A[v, 1] according to equation (5);
    end;
```

Fig. 2. Dynamic programming algorithm for computing an optimal packing of a binary tree T.

4.3   Reducing the space requirement of the algorithm

The algorithm computes finding np values of $A[v, i]$. Therefore a naive implementation would use $O(np)$ space. In this section we show how to save space without compromising the time complexity.

In examining the algorithm we find that in order to calculate $A[v, \cdot]$, we need only the values of $A[\text{left}(v), \cdot]$ and $A[\text{right}(v), \cdot]$. Moreover, after $A[v, \cdot]$ has been calculated, the values of $A[\text{left}(v), \cdot]$ and $A[\text{right}(v), \cdot]$ can be discarded. Let $\sigma(t)$ denote the number of nodes v for which at time t the value of $A[v, \cdot]$ has been calculated, but $A[\text{parent}(v), \cdot]$ has not yet been calculated. Then, at time t we need $p\sigma(t)$ space. The total space requirement is $p \max_t \{\sigma(t)\}$. We wish to organize the computation so as to minimize $\max_t \{\sigma(t)\}$.

This problem has been studied before in a different context: finding the smallest number of registers required to calculate an arithmetic expression [2]. If T denotes a binary expression tree where the leaves are data elements and the internal nodes are operators, then $\sigma(v)$, the minimum number of registers required to calculate the value of a node v, can be found by the following recursive formula:

$$\sigma(v) = \begin{cases} 1 & v \text{ is a leaf,} \\ \max\{\sigma(\text{left}(v)), \sigma(\text{right}(v))\} & \text{if } \sigma(\text{left}(v)) \neq \sigma(\text{right}(v)) \\ 1 + \text{left}(v) & \text{otherwise .} \end{cases}$$

The maximum value of $\sigma(r)$ over all n-node trees occurs for the complete binary tree. In this case, $\sigma(T) = 2 + \lg(n)$.

By using $\sigma(r)$ registers, each of size $O(p)$, to store the values of $A[v, \cdot]$ of "active" nodes, we can compute $A[r, \cdot]$ and hence also the minimal value of $c_\tau(T)$. However, since all the temporary values are discarded, the value of $\tau(v)$ for $v \in T$ is not yielded. It is possible to retrace the evaluation in order to compute the actual packing: Since we know the number of nodes in $T_{\text{left}(r)}$ that were mapped R we may may calculate $\tau(r)$, $\tau(\text{left}(r))$ and $\tau(\text{right}(r))$. However, to calculate the optimal mapping of the grandchildren of r we must rerun the algorithm on $T_{\text{left}(r)}$ and $T_{\text{left}(r)}$. The running time of this modified version is $O(p \sum_{v_T} |T_v|)$. For a full tree this is $O(pn \lg n)$. However, for trees of depth $\Omega(n)$, the running time is in the order of n^2.

We now show a better tradeoff between space and time. For all $k > 0$ we compute the optimal packing τ of any binary tree in $O(p^2 kn)$ time and $O(pkn^{1/k})$ space. In particular, for $k = \lg n$ this is $O(p^2 n \lg n)$ time and $O(p^2 \lg n)$ space.

We first consider $k = 2$. Recall that every n-node tree T has a vertex v such that both T'_v and $T \setminus T'_v$ have at least $n/3$ vertices. A simple generalization of this argument shows that T can be partitioned into subtrees $T'_1, \ldots, T'_{\sqrt{n}}$, each having between $\sqrt{n}$ and $2\sqrt{n}$ nodes.

Let $r = r_1, \ldots, r_m$ ($m \leq \sqrt{n}$) be the roots of these trees. To save space, we compute $A[r, p]$ using our space saving scheme, but retain the values of $A[r_j, \cdot]$ ($j = 2, \ldots, m$).

Now consider the subtree T'_1 whose root r_1 is r, the root of T. Its leaves are either one of the r_j's or a leaf of T. Thus, we may recompute the values of $A[v, \cdot]$ for all $v \in T'_1$ using $O(p\sqrt{n})$ space. We now compute the optimal packing for $v \in T'_1$. Also, for every leaf of T'_1 which belongs to $\{r_2, \ldots, r_m\}$ we know the number of nodes of its subtree that belong to the same page as that leaf. We store this information, and free all the space used to compute $A[v, \cdot]$, $v \in T'_1$.

In the general step, let $\mathcal{T} \subseteq \{T'_2, \ldots, T'_m\}$ be the set of trees for which we have not yet calculated the packing, but whose root is a leaf of a subtree for which the packing has been calculated. Pick $T'_j \in \mathcal{T}$, recompute $A[v, \cdot]$ for all $v \in T'_j$, and update its leaves as we did for T'_1.

At any time we need the values of $A[v, \cdot]$ only for $v \in \{r_1, \ldots, r_m\}$ and for one of the subtrees T'_j. The working space requirements are hence at most $p(m + \sqrt{n}) = 3p\sqrt{n}$.

The values of $A[v, \cdot]$ are calculated once for $v \in \{r_1, \ldots, r_{\sqrt{n}}\}$ and twice for the remaining nodes. Thus, the running time is multiplied by a factor of two.

To reduce the working space requirements even further to $O(pkn^{1/k})$, partition T into $n^{1/k}$ subtrees each of size between $n^{(k-1)/k}$ and $2n^{(k-1)/k}$. Then compute $A[r, \cdot]$, and store the values of $A[r_j, \cdot]$ for the roots of these subtrees. Finally, apply the procedure recursively on all these subtrees going from the subtree containing the root down.

The depth of the recursion is k. For each level of recursion we store $\leq 2n^{1/k}$ tables, making the total storage requirements $O(pkn^{1/k})$. The value of $A[v, \cdot]$ is calculated at most once at each level of the recursion. Thus, the running time is multiplied by a factor of k.

To prove of Theorem 5 we set $k = \lg n$.

4.4 Arbitrary degree trees

Our results can be extended to trees of arbitrary degrees: Let v be a vertex with d children. Construct a full binary tree $\tilde{T}^v$ with d leaves. Then, $\tilde{T}^v$ has $2d-1$ nodes and is of height $\lg_2 d$. We identify v with the root of $\tilde{T}^v$, and the children of v with its leaves, and replace the edges from v to its children by $\tilde{T}^v$. Repeating this procedure for all internal vertices v yields a binary tree $\tilde{T}$, which has at most twice as many vertices as T.

To find an optimal packing of T, we run a modified version of the binary tree algorithm which accounts for the fact that the new vertices have zero weight and do not occupy any space.

5 Optimal Compact Packing is NP-Complete

In the previous section, we looked for a packing which minimized the number of page misses, and disregarded the number of pages required for the packing. We now show that if we insist on a compact packing, one that uses the minimum number of pages, then finding the time-optimal packing becomes NP-complete.

Consider the following decision problem:

TREE-PACKING:
Instance: An integer p, a binary tree T of $n = mp$ nodes, and a number C.
Question: Does there exist a compact packing of cost C, i.e., a packing τ such that $c_\tau(T) \leq C$ and $|\tau(T)|m$.

Theorem 15. *TREE-PACKING is NP-complete.*

Proof. The problem is obviously in NP. In the full paper we show a reduction from the 3-PARTITION problem, which Garey and Johnson proved is strongly NP-complete [7]. ∎

6 Approximating an Optimal Compact Packing

The NP-Completeness result is due to our requirement that the packing be compact and that the cost be optimal. Here we show that if the cost is allowed to to increase by an additive factor, then a compact packing can be found efficiently.

We start with the Dynamic Programming solution. Let $P_1, ..., P_q$ be its non-full pages. Because of the optimality of the solution, each P_i is a leaf-page (the children of all the nodes of P_i are also mapped to the same page). Since the Dynamic Programming solution did not attempt to save space, P_i consists of a single tree (call it T_i).

We now apply an iterative process to repack these pages: In the i'th step, let P_i be the first page that is neither full nor empty and P_j the next such page. (We are done if no such P_j exists.)

If $|P_j| + |P_i| \leq p$, we move all of P_j to P_i and free P_j. This change does not affect the cost of the packing. Otherwise, we split T_j (the tree residing in P_j) into two pieces: one of size $p - |P_i|$ and the other consisting of the remaining nodes. The split is done so that the piece moved to P_i is a connected graph containing T_j's root. (This can be accomplished by conducting a breadth-first search or a depth-first search from T_j's root.) We move the first piece to P_i, thus filling it, and P_j remains with $|P_j| - (p - |P_i|)$ nodes. In this case, after the move, P_i is full and P_j is the next page to be filled. In this manner, no tree of P_j will be split again. The important point to notice is that each tree is split at most once and hence each path is split at most once. Therefore, the number of page faults of each path from the root to a node increased by at most 1. Hence, the increase in cost is less than $\sum_{v \in \bigcup T_j} w(v) < w(T)$. If the $w(v)$ are probabilities, then $w(T) = 1$ and the cost of our packing is at most 1 greater than that of the optimal tight solution.

If all the nodes have the same weight, say $1/n$, we can improve on this by observing that there are several ways to split s nodes from the tree T_j. If $s > |T_j|/2$, we split the tree so that the larger piece contains the root, and we move the root to P_i. Otherwise, we keep the root in P_j and move the other piece to P_i. Hence, at least half the vertices of T_j are mapped to the same page as the root of T_j, and therefore their cost did not increase. The cost increased for at most half the nodes. Since every path was split at most once, the increase is by 1 for these. This leads to a bound that exceeds the optimal cost by an additive factor of at most $1/2$.

This improvement cannot be guaranteed when the weights are not equal. Let T_j consist of two nodes: u_0—the root, and its child u_1. Suppose exactly one node of T_j should be moved to P_i. Regardless whether u_0 or u_1 is moved, a page fault is incurred between u_0 and u_1. Therefore, the number of page faults on the path to u_1 increases by 1. Thus the expected path length has increased by $w(u_1) \cdot 1$. When $w(u_1) > 1/2$, the added cost is greater than $1/2$.

7 Packing of Dynamically Changing Trees

The schemes we described for packing trees assumed that the tree T was given and does not change. Now we consider dynamic trees, i.e., trees in which nodes can be inserted and deleted. Often algorithms maintain "well-balanced" trees—trees whose height is $O(\lg n)$. We assume that the reader has some familiarity with B-trees and weight balanced trees (see [3] for a textbook presentation). These schemes perform local operations on trees such as rotations and node splitting (replacing a node v by two siblings v_1, v_2, and attaching each of v's children to either v_1 or v_2). We will examine how an optimal packing is affected by such operations, and show how to maintain a near optimal packing for B-trees and a variant of weight balanced trees.

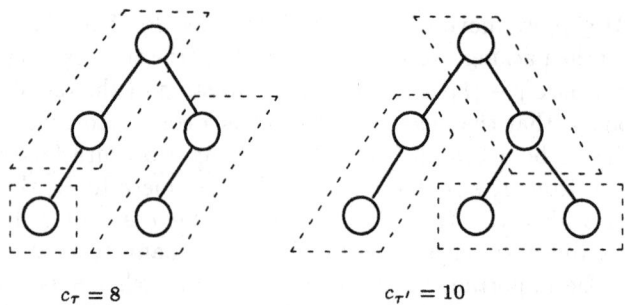

$$c_T = 8 \qquad\qquad c_{T'} = 10$$

Fig. 3. Adding a node completely changes the packing ($w(v) = 1$).

As can be seen from Figure 3, the insertion of a single node may affect the contents of all the pages of an optimal mapping. Therefore, if an efficient update algorithm is desired we can can only hope to approximate the optimal packings. More formally, let $o_1, \ldots, o_m$ be a series of inserts, deletions and rotations, T_0 the initial tree, and T_i the tree obtained after applying the operations $o_1, \ldots, o_i$. Our aim is a *dynamic packing scheme* that, given T_i, the tree obtained after applying the operations $o_1, \ldots, o_i$, the packing τ_{i+1} which was produced for it, and the next update operation o_{i+1}, will efficiently compute τ_{i+1}, the packing for T_{i+1}. It is required that

$$c_{\tau_i}(T_i) = O\left(c_{\tau_{\text{opt}}}(T_i)\right) \qquad i = 0, \ldots, m , \tag{6}$$

where $\tau_{\text{opt}}(T_i)$ is the optimal packing of T_i.

Requirement (6) will follow from a stronger condition

$$c_{\tau_i}(v) = O\left(c_{\tau_{\text{opt}}}(v)\right) \qquad v \in T_i \text{ and } i = 0, \ldots, m. \tag{7}$$

We also wish that the size of $\tau_i(T_i)$ does not exceed that of a compact packing by more than a small factor.

A general solution to this problem still remains to be found. However, we can provide one for two families of trees. The main ideas behind our constructions are sketched below.

B-trees: B-trees are the most common data structure used to store databases. It is a tree for which the degree of every node is between $d/2$ and d, and all leaves are at the same distance from the root. If the node size of the B-tree is equal to the page size, then every packing maps each node to a unique page and thus is optimal. Suppose therefore that several nodes can be stored in a page. Let h be the height of the B-tree. Let η be such that a full binary tree of degree d and of height η can be stored in a page. Specifically, we set

$$\eta = \lfloor \lg_d(1 + p(d-1)) \rfloor \ .$$

To get a packing we first cut the edges emanating from the nodes at height (distance from the leaves) $\eta, 2\eta, \ldots, \lfloor h/\eta \rfloor \, \eta$, whereby producing subtrees, each with at most p nodes. We then map each such subtree to a separate page. This is illustrated in Figure 4.

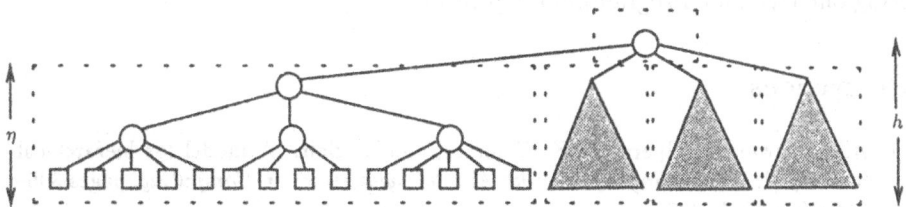

Fig. 4. Packing a B-tree of degree 5, $\eta = 3$.

For a leaf v the cost of the packing is $c_\tau(v) = \lceil h/\eta \rceil$. Thus the cost of the packing is at most $n \lceil h/\eta \rceil$, while the cost of an optimal packing is at least $\frac{n}{2} \lceil h/\eta' \rceil$, where

$$\eta' = \left\lfloor \lg_{d/2}(1 + p(d/2 - 1)) \right\rfloor \ .$$

Since $\eta' = \theta(\eta)$, $c_\tau(T) = \theta(c_{\text{opt}}(T))$.

Insertions to a B-tree may add nodes only to leaf pages. With our selection of η, the page is large enough to accommodate all insertions, unless the root of the subtree stored in it splits. If this happens, we store the two resultant subtrees in a separate page. If the root of the entire tree is split and η divides h, then the root is moved to a separate page. It follows that that the updating of τ requires changing at most two pages. It can be shown that the amortized time for updating the packing is constant. Deletions are carried out similarly within the same complexity.

This dynamic packing scheme might require an excessive number of pages. The space consumption may be reduced by storing several trees in a single memory page. Using a rather standard memory management policy we can guarantee that at least a constant fraction of each page is used. Although more page overflows may occur with this policy, we can show that incurred overhead per operation is, in an amortization sense, constant.

Weight Balanced Trees We follow the scheme suggested by G. Varghese [3, Problem 18-3, p. 376]: Let $1/2 < \alpha < 1$. We say that T is α-balanced if for every node $v \in T$

$$|T_{\text{left}(v)}| \leq \alpha |T_v| \text{ and } |T_{\text{right}(v)}| \leq \alpha |T_v| \ . \tag{8}$$

Whenever, as a result of insertions and deletions, condition (8) is violated for a node v but not for any of its ancestors, the entire subtree T_v is reorganized: it is replaced by a full binary tree with $|T_v|$ nodes. This reorganization requires $O(|T_v|)$ time. One can show that the amortized time for insertion/deletion is $O(\lg n)$.

To maintain a near optimal packing, every page P_i consists of a subtree of T. Since only leaves are added, if there is no room for a new node in its parent's page, it is allocated to a new page. Whenever T_v is reorganized we also repack T_v using our Dynamic Programming algorithm.

References

1. A. Aggarwal, B. Alpern, A. K. Chandra, and M. Snir. A model for hierarchical memory. In *Proceedings of the 19th Annual Symposium on Computing*, pages 305–314, New York, 1987.
2. A. V. Aho, R. Sethi, and J. D. Ullman. *Compilers: principles, techniques, and tools.* Addison-Wesley, Reading, Massachusetts, 1988.
3. T. Cormen, C. Leiserson, and R. Rivest. *Introduction to Algorithms.* McGraw Hill and The MIT Press, 1990.
4. B. Farizon. Local and global memory for the implementation of ray tracing on a parallel machine. Master's thesis, Computer Science Dept., Technion – Israel Institute of Technology, 1995.
5. J. D. Foley and A. V. Dam. *Fundamental of Interactive Computer Graphics.* The Systems Programming Series. Addison-Wesley, Reading, Massachusetts, 1984.
6. S. Gal, Y. Hollander, and A. Itai. Optimal mappings in a direct mapped cache environment. *Math. Programming B*, 63:371–387, 1994.
7. M. R. Garey and D. S. Johnson. *Computers and Intractability, A Guide to the Theory of NP-Completeness.* W.H. Freeman and Company, 1979.
8. C. F. Goldfarb. *The SGML Handbook.* Clarendon Press, Oxford, 1990.
9. M. S. Lam, E. Rothberg, and M. E. Wolf. The cache performance and optimizations of blocked algorithms. In *Fourth International Conference on Architectural Support for Programming Languages and Operation Systems (ASPLOS IV)*, 1991.
10. M. E. Wolf and M. S. Lam. A data locality optimizing algorithm. In *Proceedings of the ACM SIGPLAN'91 Conference on Programming Language Design and Implementation*, 1991.

Sometimes Travelling is Easy: The Master Tour Problem *

Vladimir G. Deĭneko[1], Rüdiger Rudolf[1] and Gerhard J. Woeginger[2]

[1] TU Graz, Institut für Mathematik B
A-8010 Graz, Austria.
[2] Department of Mathematics and Computer Science
Eindhoven University of Technology
P.O. Box 513, 5600 MB Eindhoven, The Netherlands.

Abstract. In 1975, Kalmanson proved that in case the distance matrix in the Travelling Salesman Problem (TSP) fulfills certain combinatorial conditions (nowadays called the *Kalmanson conditions*) then the TSP is solvable in polynomial time.

We deal with the problem of deciding for a given instance of the TSP, whether there is a renumbering of the cities such that the corresponding renumbered distance matrix fulfills the Kalmanson conditions. Two results are derived: First, it is shown that such a renumbering can be found in polynomial time (in case it exists). Secondly, it is proved that such a renumbering exists if and only if the instance possesses the so-called *master tour* property. Thereby a recently posed question by Papadimitriou is answered in the negative.

1 Introduction

The *travelling salesman problem* (TSP) is defined as follows. Given an $n \times n$ distance matrix $C = (c_{ij})$, find a permutation $\pi \in S_n$ that minimizes the sum $\sum_{i=1}^{n-1} c_{\pi(i)\pi(i+1)} + c_{\pi(n)\pi(1)}$ (the salesman must visit the cities 1 to n in arbitrary order and wants to minimize the total travel length). This problem is one of the fundamental problems in combinatorial optimization and known to be NP hard. For more specific information on the TSP, the reader is referred to the book by Lawler, Lenstra, Rinnooy Kan and Shmoys [7].

In this paper, we are interested in a special case of the TSP where — due to special combinatorial structures in the distance matrix — the problem is solvable in polynomial time: The case of *Kalmanson* distance matrices. A symmetric $n \times n$ matrix C is called a *Kalmanson* matrix if it fulfills the conditions

$$c_{ij} + c_{k\ell} \leq c_{ik} + c_{j\ell} \quad \text{for all } 1 \leq i < j < k < \ell \leq n \tag{1}$$

$$c_{i\ell} + c_{jk} \leq c_{ik} + c_{j\ell} \quad \text{for all } 1 \leq i < j < k < \ell \leq n. \tag{2}$$

* This research has been supported by the Spezialforschungsbereich F 003 "Optimierung und Kontrolle", Projektbereich Diskrete Optimierung.

Note that these conditions do not involve any diagonal entries c_{ii}. Since every city is visited only once, diagonal entries are of no relevance for the TSP and may as well be considered to be 'undefined' or to be zero. Originally, Kalmanson introduced these conditions in order to generalize the concept of *convexity* of finite point sets in the plane: For some convex planar point set, let $p_1, \ldots, p_n$ denote its clockwise ordering around the convex hull. Then the Euclidean distance matrix $c_{ij} = d(p_i, p_j)$ fulfills all conditions (1) and (2) (Proof: In a convex quadrangle, the total length of the diagonals is greater or equal to the total length of two opposite sides). Moreover, if we 'rotate' the ordering by one point, the distance matrix of the resulting *rotated* point sequence $p_2, p_3, \ldots, p_n, p_1$ also is a Kalmanson matrix. It is easy to verify that this 'rotation property' does not result from special Euclidean features, but solely from inequalities (1) and (2). Hence, if one removes the first row and first column from a Kalmanson matrix and appends them after the last row and column, the result of this operation is another Kalmanson matrix. Similarly, reversing the ordering of the rows and columns transforms Kalmanson matrices into Kalmanson matrices.

Kalmanson [6] proved that for the TSP with a Kalmanson distance matrix, the identity permutation $(1, 2, 3, \ldots, n)$ always constitutes an optimal tour and thus, the TSP is easily solved for this special case. Observe that the length of the optimum TSP tour is not changed when the cities are renumbered, i.e. when the rows and columns of the distance matrix are permuted according to some common permutation (however, such a renumbering will usually destroy the Kalmanson conditions). Intuition tells us that a renumbered instance is still a rather trivial special case of the TSP, since it is just a Kalmanson instance in disguise, but it is by no means obvious, how to recognize this disguise. Hence, the problem arises of finding a permutation that transforms the distance matrix back into a Kalmanson matrix.

Another problem related to the TSP is the detection of a master tour, motivated by the following observation. Suppose that all cities in a Euclidean instance of the TSP are the vertices of a convex polygon. Then the optimum tour is not only easy to find (it is the perimeter of the polygon), but the instance also fulfills the much stronger *master tour* property: There is an optimum TSP tour π such that the optimum TSP tour of any subset of cities can be obtained by simply omitting from the tour π the cities that are not in the subset. Such a tour π is called a master tour.

The concept of a master tour was first formulated by Papadimitriou [8, 9]. It is easy to prove that deciding whether a given instance of the TSP has the master tour property is in the complexity class $\Sigma_2 \mathbf{P}$. Papadimitriou also considered the corresponding decision problem as a "good candidate for a natural $\Sigma_2 \mathbf{P}$-complete problem".

The results presented in this paper are the following.

(1) For a symmetric $n \times n$ matrix C, it can be decided in $O(n^2 \log n)$ time whether C is a permuted Kalmanson matrix.

(2) A distance matrix allows a master tour if and only if it is a permuted Kalmanson matrix.

Combining results (1) and (2) yields a polynomial time algorithm for the master tour problem. Hence, unless $\Sigma_2 P = P$, the conjecture of Papadimitriou is false.

Organization of the paper. Section 2 summarizes elementary definitions and results on permutations and matrices. In Section 3, several lemmas on the combinatorial structure of Kalmanson matrices are collected. These lemmas are used in Section 4 to derive an $O(n^2 \log n)$ time algorithm for recognizing permuted $n \times n$ Kalmanson matrices. Section 5 explains the connection between permuted Kalmanson matrices and master tours and shows that a master tour can be detected in polynomial time. Finally, Section 6 closes with a short discussion.

2 Definitions and Preliminaries

In this section, several basic definitions for permutations and matrices are summarized.

For an $n \times n$ matrix C, denote by $I = \{1, \ldots, n\}$ the set of rows (columns). A row i *precedes* a row j in C ($i \prec j$ for short), if row i occurs before row j in C. For two sets K_1 and K_2 of rows, we write $K_1 \prec K_2$ if and only if $k_1 \prec k_2$ for all $k_1 \in K_1$ and $k_2 \in K_2$.

Let $V = \{v_1, v_2, \ldots, v_r\}$ and $W = \{w_1, w_2, \ldots, w_s\}$ be two subsets of I. We denote by $C[V, W]$ the $r \times s$ submatrix of C which is obtained by deleting all rows not contained in V and all columns not in W.

For permutations, we adopt the notation $\pi = \langle x_1, x_2, \ldots, x_n \rangle$ for "$\pi(i) = x_i$ for $1 \leq i \leq n$". The *concatenation* of permutations $\langle x_1, \ldots, x_n \rangle$ and $\langle y_1, \ldots, y_m \rangle$ is $\langle z_1, \ldots, z_{n+m} \rangle$, where $z_i = x_i$ for $1 \leq i \leq n$ and $z_{n+j} = y_j$ for $1 \leq j \leq m$. The identity permutation is denoted by ε, i.e. $\varepsilon(i) = i$ for all $i \in I$. For a permutation ϕ, the permutation ϕ^- defined by $\phi^-(i) = \phi(n - i + 1)$ is called the *reverse permutation* of ϕ. Permutation ϕ is called a *cyclic shift* or a *rotation* if there exists a $k \in I$ such that $\phi = \langle k, k+1, \ldots, n, 1, \ldots, k-1 \rangle$.

By $C_{\phi, \pi}$ we denote the matrix which is obtained from matrix C by permuting its rows according to ϕ and its columns according to π, i.e. $C_{\phi, \pi} = (c_{\phi(i), \pi(j)})$. For $C_{\phi, \phi}$, we usually write C_ϕ. A permutation ϕ is called a *Kalmanson permutation for some matrix* C if C_ϕ is a Kalmanson matrix. A matrix C is called a *permuted Kalmanson matrix*, if there exists a Kalmanson permutation for C.

For a partition $V = \langle V_1, \ldots, V_v \rangle$ of I into v subsets, the set $\text{STR}(V_1, \ldots, V_v)$ is defined to contain all permutations ϕ that fulfill $\phi(v_i) \prec \phi(v_j)$ for all $v_i \in V_i$ and $v_j \in V_j$ with $1 \leq i < j \leq v$. $\text{STR}(V_1, \ldots, V_v)$ is called the set of permutations induced by the sequence of *stripes* $V_1, \ldots, V_v$. An appropriate data structure for storing, manipulating and intersecting such sets of permutations are *PQ-trees* as introduced by Booth and Lueker [1] (in fact, PQ-trees of *height two* suffice to represent these permutations).

Proposition 1. *(Booth and Lueker [1])*
For two partitions $\langle U_1, \ldots, U_u \rangle$ and $\langle V_1, \ldots, V_v \rangle$ of I, the set $\mathrm{STR}(U_1, \ldots, U_u) \cap \mathrm{STR}(V_1, \ldots, V_v)$ either equals $\mathrm{STR}(W_1, \ldots, W_w)$ for an appropriate partition $W = \langle W_1, \ldots, W_w \rangle$ of I or it is empty. The partition W can be computed in $O(|I|)$ time.

An $m \times n$ matrix C is called a *sum-matrix* if there exist numbers $x_1, \ldots, x_m$ and $y_1, \ldots, y_n$ such that $c_{ij} = x_i + y_j$ for all i and j. Note that this implies $c_{ij} + c_{rs} = c_{is} + c_{rj}$ for $1 \leq i < r \leq m$ and $1 \leq j < s \leq n$ (i.e. in any two by two submatrix, both diagonals have equal sums). For convenience, single rows and columns are also considered to be sum-matrices.

An $m \times n$ matrix C is called a *Contra Monge matrix* if $c_{ij} + c_{rs} \geq c_{is} + c_{rj}$ holds for $1 \leq i < r \leq m$ and $1 \leq j < s \leq n$. The combinatorial structure of Contra Monge matrices and of permuted Contra Monge matrices is well-understood (see the original paper by Deineko and Filonenko [4] or the survey paper by Burkard, Klinz and Rudolf [2]). The known main results are summarized in the following proposition.

Proposition 2. *Let $X = (x_{ij})$ be an $m \times n$ matrix. Let $\Pi \subseteq S_m \times S_n$ denote the set of all pairs of permutations (π, ϕ) such that $X_{\pi, \phi}$ is a Contra Monge matrix.*

(i) *Then either Π is the empty set, or there exists an appropriate partition $R_1, \ldots, R_r$ of the set R of rows and an appropriate partition $C_1, \ldots, C_c$ of the set C of columns of X such that*

$$\Pi = \{(\pi, \phi) | \pi \in \Pi_R, \phi \in \Pi_C\} \cup \{(\pi, \phi) | \pi^- \in \Pi_R, \phi^- \in \Pi_C\}$$

where $\Pi_R = \mathrm{STR}(R_1, \ldots, R_r)$ and $\Pi_C = \mathrm{STR}(C_1, \ldots, C_c)$.
(ii) *The partitions $R_1, \ldots, R_r$ and $C_1, \ldots, C_c$ can be computed in $O(mn + m \log m + n \log n)$ time (in case they exist).*
(iii) *Every submatrix $X[R_i, C]$ and every submatrix $X[R, C_j]$ is a sum-matrix. These sum-matrices are maximal sum-matrices in X (i.e. neither rows nor columns may be added without destroying the sum-matrix property).*
(iv) *In case Π is not empty, either $r = c = 1$ holds (and X is a sum-matrix), or the numbers r and c are both at least two (and X is horizontally and vertically divided into several stripes by Π).*
(v) *Matrix X is a Contra Monge matrix if and only if for all pairs of indices $1 \leq i \leq m - 1$ and $1 \leq j \leq n - 1$, the inequality*

$$x_{ij} + x_{i+1,j+1} \geq x_{i,j+1} + x_{i+1,j} \tag{3}$$

is fulfilled.

Similarly to the above alternate characterization (3) of Contra Monge matrices, an alternate characterization of Kalmanson matrices can be given.

Proposition 3. *An $n \times n$ matrix C is a Kalmanson matrix if*

$$c_{i,j+1} + c_{i+1,j} \leq c_{ij} + c_{i+1,j+1} \quad \text{for all } 1 \leq i \leq n - 3, \ i + 2 \leq j \leq n - 1 \tag{4}$$

$$c_{i,1} + c_{i+1,n} \leq c_{in} + c_{i+1,1} \quad \text{for all } 2 \leq i \leq n - 2. \tag{5}$$

Observe that the conditions (4) and (5) can be verified in $O(n^2)$ time. This yields the following proposition.

Proposition 4. *For a symmetric $n \times n$ matrix C, it can be decided in $O(n^2)$ time whether C is a Kalmanson matrix.*

Proposition 5. *Let C be an $n \times n$ Kalmanson matrix, let $I' \subseteq I$ and let $\phi \in S_n$ be a cyclic shift. Then (i) $C_{e^-} \in \mathbb{K}$, (ii) $C[I', I'] \in \mathbb{K}$ and (iii) $C_\phi \in \mathbb{K}$ holds.*

3 Combinatorial Properties of Kalmanson Matrices

In this section, we derive several technical lemmas on the combinatorial structure of Kalmanson matrices.

Lemma 6. *Let C be an $n \times n$ symmetric Kalmanson matrix, $2 \leq m \leq n-1$, let $V = \langle 1, 2, \ldots, m \rangle$ and $W = \langle m+1, \ldots, n \rangle$. Then $C[V, W]$ is a Contra Monge matrix.*

Proof. This is a consequence of condition (4) in Proposition 3.

Lemma 7. *Let C be a symmetric $n \times n$ matrix. Let V and W be a partition of I with $|V| = r \geq 2$, $|W| = s \geq 2$ such that $C[V, W]$ is a sum-matrix. Let $q \in W$ and $p \in V$ be arbitrary. Let $D = C[V \cup \{q\}, V \cup \{q\}]$ and $E = C[\{p\} \cup W, \{p\} \cup W]$. Assume that there is a permutation $\psi = \langle v_1, \ldots, v_r, q \rangle$ of $V \cup \{q\}$ and a permutation $\pi = \langle p, w_1, \ldots, w_s \rangle$ of $W \cup \{p\}$ such that $D_\psi \in \mathbb{K}$ and $E_\pi \in \mathbb{K}$.*

Under these conditions either $C_\phi \in \mathbb{K}$ for $\phi = \langle v_1, \ldots, v_r, w_1, \ldots, w_s \rangle$ or there does not exist any permutation $\sigma \in \text{STR}(V, W)$ with $C_\sigma \in \mathbb{K}$.

Proof. We prove the stronger statement that under the conditions in the lemma either the inequality

$$c_{v_1 v_r} + c_{w_1 w_s} \leq c_{v_1 w_1} + c_{v_r w_s} \tag{6}$$

is fulfilled and $C_\phi \in \mathbb{K}$ for $\phi = \langle v_1, \ldots, v_r, w_1, \ldots, w_s \rangle$ or inequality (6) is not fulfilled and there does not exist any permutation $\sigma \in \text{STR}(V, W)$ with $C_\sigma \in \mathbb{K}$.

First assume that inequality (6) is fulfilled. We prove that $C_\phi \in \mathbb{K}$ according to Proposition 3 by verifying conditions (4) and (5). Consider two indices i and j in C_ϕ, $1 \leq i \leq n-3$ and $i+2 \leq j \leq n-1$. In case $i \prec i+1 \prec j \prec j+1$ are all in V or are all in W, condition (4) holds for C_ϕ since $D_\psi \in \mathbb{K}$ and $E_\pi \in \mathbb{K}$. If i and $i+1$ are in V and j and $j+1$ are in W, the four elements $c_{i,j+1}$, $c_{i+1,j}$, c_{ij} and $c_{i+1,j+1}$ lie in the sum-matrix $C[V, W]$ and thus trivially fulfill (4). Next, if i is in V and $i+1 \prec j \prec j+1$ are in W then $i = v_r$ and $i+1 = w_1$ holds. The relations $p \prec w_1 \prec j \prec j+1$ in $E_\pi \in \mathbb{K}$ yield $c_{p,j+1} + c_{w_1,j} \leq c_{pj} + c_{w_1,j+1}$. Since $p, v_r \in V$, $j, j+1 \in W$ and $C[V, W]$ is a sum-matrix, $c_{pj} + c_{v_r,j+1} = c_{p,j+1} + c_{v_r,j}$. Adding this equality to the previous inequality yields (4). The last case where $i \prec i+1 \prec j$ are in V and $j+1$ is in W is handled symmetrically. Summarizing, (4) is true in any case.

Next, consider an index $2 \leq i \leq n-2$. In case $i \neq v_r$, (5) is true since $D_\psi \in \mathbb{K}$ (respectively, $E_\pi \in \mathbb{K}$) holds. In case $i = v_r$, (5) is exactly (6). Hence, (6) implies (5), and the first half of the lemma is proven.

To prove the remaining half, assume that for some $\sigma \in \mathrm{STR}(V, W)$, $C_\sigma \in \mathbb{K}$ holds. We show how to derive inequality (6) from this. Since V precedes W in σ, only two cases arise:

(i) $v_1 \prec v_r \prec w_1 \prec w_s$ or $v_r \prec v_1 \prec w_s \prec w_1$ in σ. Then condition (1) yields (6).

(ii) $v_1 \prec v_r \prec w_s \prec w_1$ or $v_r \prec v_1 \prec w_1 \prec w_s$ in σ. Then condition (1) yields $c_{v_1 v_r} + c_{w_1 w_s} \leq c_{v_r w_1} + c_{v_1 w_s}$. Since $C[V, W]$ is a sum–matrix, $c_{v_1 w_s} + c_{v_r w_1} = c_{v_1 w_1} + c_{v_r w_s}$. Adding these two inequalities gives (6).

Lemma 8. *Let C be a symmetric $n \times n$ matrix. Let $U_1, \ldots, U_m$ be a partition of I such that $C[U_i, I \setminus U_i]$ is a sum–matrix for $1 \leq i \leq m$. Let u_i be an arbitrary element in U_i. Let π_i be a Kalmanson permutation for $C[U_i \cup \{u_{i+1}\}, U_i \cup \{u_{i+1}\}]$ (indices are taken modulo m, i.e. $u_{m+1} = u_1$) that has u_{i+1} as last element. Let ϕ_i denote the permutation of U_i induced by π_i.*

Under these conditions either $C_\phi \in \mathbb{K}$ where ϕ is the concatenation of $\phi_1, \ldots, \phi_m$ or there does not exist any Kalmanson permutation for C in $\mathrm{STR}(U_1, \ldots, U_m)$.

Proof. The proof is done by induction on the number t of stripes U_i with cardinality at least two. If $t = 0$, the statement trivially holds. Otherwise if $t \geq 1$, we may assume without loss of generality that $|U_1| \geq 2$. Moreover, we assume that there exists a Kalmanson permutation for C in $\mathrm{STR}(U_1, \ldots, U_m)$, since otherwise there is nothing to show.

Set $W = U_2 \cup \ldots \cup U_m$ and consider the sets U_i' where $U_1' = \{u_1\}$ and $U_i' = U_i$ for $2 \leq i \leq m$. By the induction assumption, the concatenation of $\langle u_1 \rangle, \phi_2, \ldots, \phi_m$ is a Kalmanson permutation for $C[\{u_1\} \cup W, \{u_1\} \cup W]$. Set $V = U_1$. By the conditions of the lemma, the concatenation π_1 of ϕ_1 and $\langle u_2 \rangle$ is a Kalmanson permutation for $C[V \cup \{u_2\}, V \cup \{u_2\}]$. Moreover, the matrix $C[V, W]$ is a sum–matrix. Summarizing, all conditions for applying Lemma 7 with $p = u_1$ and $q = u_2$ are fulfilled. The statement in Lemma 7 yields that the concatenation ϕ is indeed a Kalmanson permutation and the inductive proof is complete.

The following notation is convenient. For two rows i and j of a matrix C, define the set

$$\mathcal{M}(i, j) = \left\{ k \in I \setminus \{i, j\} \mid c_{ik} - c_{jk} = \min_{\ell \neq i, j} \{ c_{i\ell} - c_{j\ell} \} \right\}.$$

Note that $C[\{i, j\}, \mathcal{M}(i, j)]$ is a sum matrix. In case $|\mathcal{M}(i, j)| = n - 2$ holds, $c_{ik} - c_{jk} = const$ for all $k \in I \setminus \{i, j\}$. Such a pair of rows is called *equivalent*, and is denoted by $i \sim j$. For the sake of completeness, we define that every row is equivalent to itself.

Lemma 9. *Let C be a symmetric $n \times n$ matrix with $i, j_1, j_2 \in I$. If $i \sim j_1$ and $i \sim j_2$ holds, then $j_1 \sim j_2$.*

Hence, the relation $\sim$ is symmetric, reflexive and transitive.

Proof. The goal is to show that for any $k, \ell \in I$, $\{j_1, j_2\} \cap \{k, \ell\} = \emptyset$ the equality
(∗) $c_{j_1 k} - c_{j_2 k} = c_{j_1 \ell} - c_{j_2 \ell}$ holds. If $i \notin \{k, \ell\}$ holds, we use $j_1 \sim i$ and $j_2 \sim i$:
Subtract the equalities $c_{ik} - c_{j_1 k} = c_{i\ell} - c_{j_1 \ell}$ and $c_{ik} - c_{j_2 k} = c_{i\ell} - c_{j_2 \ell}$ from each
other, and derive (∗).

Otherwise, assume that $k = i$. Use $i \sim j_2$ to obtain $c_{ij_1} - c_{j_2 j_1} = c_{i\ell} - c_{j_2 \ell}$
and use $i \sim j_1$ to obtain $c_{ij_2} - c_{j_1 j_2} = c_{i\ell} - c_{j_1 \ell}$. Subtracting these equations
yields (∗).

Lemma 10. *Let C be a symmetric $n \times n$ matrix. If $1 \sim i$ for all $i \in I$, then $C \in \mathbb{K}$.*

Proof. By Lemma 9 above, all inequalities (1) and (2) are fulfilled with equality.

Lemma 11. *Let C be a symmetric $n \times n$ Kalmanson matrix. Let i and j be two rows of C with $i \prec j$, let $K_1 = \mathcal{M}(i,j) \cup \{i\}$ and $K_2 = I \setminus K_1$. Then there exists a cyclic shift ϕ such that $C_\phi \in \mathbb{K}$ and $K_1 \prec K_2$ in C_ϕ.*

Proof. By definition $i \in K_1$ and $j \in K_2$. Consider any $k \in \mathcal{M}(i,j)$. Then
$c_{ik} - c_{jk} = c_{i\ell} - c_{j\ell}$ for all $\ell \in K_1 \setminus \{i\}$ and $c_{ik} - c_{jk} < c_{i\ell} - c_{j\ell}$ for all $\ell \in K_2 \setminus \{j\}$.
We distinguish the following three cases on the relative position of i, j and k in
C.

(i) $k \prec i \prec j$. Then condition (2) yields $c_{ip} - c_{jp} \leq c_{ik} - c_{jk}$ for any p with
$k \prec p \prec i \prec j$. Hence, $p \in K_1$ for all $p \in I$ with $k \prec p \prec i$.
(ii) $i \prec k \prec j$. Analogously to the argument in (i), condition (1) implies $p \in K_1$
for all $p \in I$ with $i \prec p \prec k$.
(iii) $i \prec j \prec k$. Analogously to the argument in (i), conditions (2) and (2) imply
$p \in K_1$ for all $p \in I$ with $p \prec i$ or $k \prec p$.

Summarizing, we conclude that there exist two elements r and s such that either
$K_1 = \{r, \ldots, i, \ldots, s\}$ or $K_2 = \{s+1, \ldots, j, \ldots, r-1\}$. By Proposition 5(iii),
every cyclic shift of C again yields a Kalmanson matrix. Hence, choosing $\phi = \langle r, \ldots, s, \ldots, n, 1, \ldots, r-1 \rangle$ or choosing $\phi = \langle r, \ldots, n, 1, \ldots, s, s+1, \ldots, r-1 \rangle$
completes the argument.

4 Recognition of Permuted Kalmanson Matrices

This section shows how to recognize permuted Kalmanson matrices in polynomial time. The recognition algorithm is described in two steps: First, in Subsection 4.1, we give a rough outline of the algorithm. We sketch a divide and
conquer approach that is based on the Lemmas derived in the preceding section.
Then in Subsection 4.2, we describe a fast implementation of the algorithm that
runs in $O(n^2 \log n)$ time.

4.1 Outline of the Algorithm

Given an $n \times n$ matrix C, we want to decide whether there exists a permutation σ such that $C_\sigma \in \textbf{K}$ and we want to compute σ in case it exists. Our solution algorithm follows a divide and conquer strategy. The main goal is to find in polynomial time $D(n)$ a so-called *nice bipartition* of the set I of rows into two sets V and W fulfilling the following three properties.

(i) $|V|, |W| \geq 2$.
(ii) $C[V, W]$ is a sum–matrix.
(iii) The matrix C is a permuted Kalmanson matrix if and only if there exists a permutation $\sigma \in \text{STR}(V, W)$ with $C_\sigma \in \textbf{K}$.

If we have found some nice bipartition, we choose rows $q \in W$ and $p \in V$ and recursively compute Kalmanson permutations ψ and π for the two matrices $C[V \cup \{q\}, V \cup \{q\}]$ and $C[\{p\} \cup W, \{p\} \cup W]$. According to Lemma 7 and property (iii) above, either the concatenation of ψ and π is a Kalmanson permutation for C or C cannot be a permuted Kalmanson matrix. By Proposition 4, it can be decided in $O(n^2)$ time whether the concatenation of ψ and π indeed yields a Kalmanson permutation. Summarizing, this results in a recursive algorithm with time complexity

$$T(n) \leq \max_{2 \leq k \leq n-2} \left\{ T(k+1) + T(n-k+1) \right\} + D(n) + O(n^2).$$

It is easy to verify that $T(n) = O(nD(n) + n^3)$ and hence the algorithm runs in polynomial time. It remains to explain how to find a nice bipartition.

First, we find a row k that is *not* equivalent to row 1 (if such a row k does not exist, if follows from Lemma 10 that the identity permutation is a Kalmanson permutation). Compute $\mathcal{M}(1, k)$ and define the sets $K_1 = \mathcal{M}(1, k) \cup \{1\}$ and $K_2 = I \setminus K_1$. Since $1 \not\sim k$, $|K_1|, |K_2| \geq 2$ holds. By Lemma 11, it is sufficient to deal with permutations ϕ for which $K_1 \prec K_2$ holds, i.e. with permutations $\phi \in \text{STR}(K_1, K_2)$. Now, if $C[K_1, K_2]$ is a sum–matrix, K_1 and K_2 form a nice bipartition and we are done. Otherwise, by Lemma 6, it is necessary to deal with permutations ϕ for which the matrix $C_\phi[K_1, K_2]$ is a Contra Monge matrix. According to Proposition 2, these permutations can be described by $\phi \in \text{STR}_1 \cup \text{STR}_1^*$ where $\text{STR}_1 = \text{STR}(K_{11}, \ldots, K_{1r}, K_{21}, \ldots, K_{2s})$ and $\text{STR}_1^* = \text{STR}(K_{1r}, \ldots, K_{11}, K_{2s}, \ldots, K_{21})$ for appropriate partitions $K_{11}, \ldots, K_{1r}$ of K_1 and $K_{21}, \ldots, K_{2s}$ of K_2. It is easy to see that by rotation and reversion, every $\phi^* \in \text{STR}_1^*$ can be transformed into some $\phi \in \text{STR}_1$. By Proposition 5 we conclude that in case C can be permuted into a Kalmanson matrix, this can also be reached by some $\phi \in \text{STR}_1$ and thus it is sufficient to consider permutations in STR_1.

In case all stripes in STR_1 have cardinality one, there remains just a single potential Kalmanson permutation (and it can be checked in $O(n^2)$ time whether the permutation indeed is a Kalmanson permutation). Otherwise, there is some stripe K_{ij} of cardinality at least two. We rotate the sequence of stripes in STR_1 in such a way that K_{ij} becomes the first stripe, and rename the stripes into

$L_1, \ldots, L_\ell$ with $L_1 = K_{ij}$. In case $C[L_1, I \setminus L_1]$ is a sum–matrix, $V = L_1$ and $W = I \setminus L_1$ forms a nice bipartition. Otherwise, we observe that any Kalmanson permutation in $\mathrm{STR}(L_1, \ldots, L_\ell)$ must transform $C[L_1, I \setminus L_1]$ into a Contra Monge matrix. According to Proposition 2, we compute an appropriate partition of L_1 and an appropriate partition of $I \setminus L_1$ that encodes all permutations that transform $C[L_1, I \setminus L_1]$ into a Contra Monge matrix. This either results in a refinement of the stripes in $L_1, \ldots, L_\ell$ (cf. Proposition 1) or the set of potential permutations becomes empty (and C is not a permuted Kalmanson matrix).

This procedure is repeated over and over again as long as there are stripes of cardinality at least two. Either we find a nice bipartition of I, or we may refine the stripes, or eventually all stripes are of cardinality one. Since the stripes can be refined at most $(n-1)$ times, a conservative estimation yields $D(n) = O(n^3)$. According to the above arguments, the recognition algorithm runs in polynomial time $O(n^4)$.

4.2 Implementation of the Algorithm

In this subsection, we explain how to implement the divide and conquer algorithm described in the preceding section in $O(n^2 \log n)$ time. Our main tools are advanced data structures (PQ-trees and union–find structures) and a slight modification of the divide step. Let us start with three simple but important statements.

• All through the algorithm we will derive and exploit sufficient conditions on the matrix for being Kalmanson. These conditions will restrict and cut down the set of potentially feasible permutations. We will not verify at every single step, whether we are indeed dealing with Kalmanson permutations (this would be too time consuming). Hence, the output of the algorithm will be some $\sigma \in S_n$ with the following property: "In case C is a permuted Kalmanson matrix, then $C_\sigma \in \mathbb{K}$." The verification whether C_σ is indeed in $\mathbb{K}$ is postponed to a single $O(n^2)$ check in the end, *after* the algorithm.

• All sets of permutations induced by stripes are stored in PQ-trees of height two (as already stated in Section 2). For a set of permutations Π over I stored in a PQ-tree of constant height and a subset $J \subseteq I$, the following operation can be performed in $O(|J|)$ time: "Restructure the PQ-tree in such a way that the restructured PQ-tree stores exactly those permutations $\pi \in \Pi$ in which the objects in J are consecutive" (cf. Booth and Lueker [1]). Note that this operation might lead to an empty set of permutations.

• The algorithm represents its knowledge on the equivalence of rows in a union–find data structure in order to answer questions of the form "Are the rows i and j already known to be equivalent?" in constant time. In case it receives new information on the equivalence of two rows i and j, the corresponding two equivalence classes have to be combined into a single class. We choose an implementation of this data structure that supports the FIND operation in constant time and that supports the UNION operation in time that is linear in the size of the merged classes (This can be done e.g. via pointers from every element to the name of the corresponding class; see e.g. Cormen, Leiserson and Rivest [3]).

Next, we give a precise low-level description of the algorithm. The algorithm performs the following five steps (S0)–(S4).

(S0) If $n \leq 3$, output any permutation $\sigma \in S_n$.

(S1) Row 1 is compared to rows k, $2 \leq k \leq n$, until some row k is found that is not equivalent to 1 or all rows are known to be equivalent to row 1. If $1 \not\sim k$, the algorithm moves on to (S2). If all rows are found to be equivalent to row 1, the algorithm outputs the identity permutation and stops.

Observing that any $n \times n$ matrix with $n \leq 3$ is a Kalmanson matrix justifies Step (S0). In step (S1), a comparison between rows 1 and k is performed as follows: First, the algorithm checks whether 1 is already known to be equivalent to k (from some higher level of the recursion). In case it is, the algorithm immediately moves on to the next row. Otherwise, it scans row k in linear time. If it turns out that $1 \sim k$, this information is handled over to the union–find structure and the next row is investigated.

The following step (S2) is a kind of initialization for step (S3).

(S2) Let $k \not\sim 1$ be the result of (S1). Compute $\mathcal{M}(1, k)$ and define the sets $K_1 = \mathcal{M}(1, k) \cup \{1\}$ and $K_2 = I \setminus K_1$. Compute for $C[K_1, K_2]$ the partitions $K_{11}, \ldots, K_{1r}$ of K_1 and $K_{21}, \ldots, K_{2s}$ of K_2 that encode all permutations that transform $C[K_1, K_2]$ into a Contra Monge matrix (this is done according to Proposition 2).
Set $\mathrm{STR}_0 = \mathrm{STR}(K_1, K_2)$ and $\mathrm{STR}_1 = \mathrm{STR}(K_{11}, \ldots, K_{1r}, K_{21}, \ldots, K_{2s})$.

By the discussion in Subsection 4.1, it is sufficient to search for Kalmanson permutations in STR_1. Note that from the submatrix $C[K_1, K_2]$, we will not receive any further information on refining the stripes: By Proposition 2(iii) for every stripe K_{1i}, the matrix $C[K_{1i}, K_2]$ is a sum–matrix and for every stripe K_{2j}, the matrix $C[K_1, K_{2j}]$ is a sum–matrix. However, we may receive informations from $C[K_{1i}, K_1]$ and $C[K_{2j}, K_2]$.

In the following "refinement step" (S3), a sequence of sets of permutations $\mathrm{STR}_1, \mathrm{STR}_2, \ldots$ is constructed with $\mathrm{STR}_i = \mathrm{STR}(Q_1^{(i)}, \ldots, Q_{m_i}^{(i)})$. The partition in STR_{i+1} is always a refinement of the partition in STR_i. The refinement procedure stops as soon as for every stripe $Q = Q_j^{(i)}$, the submatrix $C[Q, I \setminus Q]$ is a sum–matrix.

(S3) Starting with $j = 1$, perform the following refinement procedure for every stripe $Q = Q_j^{(i)}$ in STR_i: Rotate STR_i in such a way that Q becomes the first stripe in STR_i. Let Q^* be the *mother stripe* of Q in STR_{i-1}, i.e. the stripe that contains set Q. Let q be any column that is *not* in Q^*. In case $Q \neq Q^*$ holds, consider the matrix $D = C[Q, (Q^* \setminus Q) \cup \{q\}]$ and compute according to Proposition 2 the partition of the rows and columns of D that implicitly describe all permutations that transform D into a Contra Monge matrix. If $Q = Q^*$ holds, consider the next set $Q_j^{(i)}$.

The set STR_{i+1} results from refining STR_i according to all these new partitions. Step (S3) is repeated until $\text{STR}_{i+1} = \text{STR}_i$, i.e. the partition has not been refined.

Consider the matrices $E = [Q, I \setminus Q]$ and $F = [Q, I \setminus Q^*]$. Since Q was obtained as a stripe from Q^*, Proposition 2(iii) ensures that the matrix F is a sum–matrix. By Lemma 6, we may restrict our attention to permutations that transform matrix E into a Contra Monge matrix. E essentially decomposes into D and F (where D and F have the common column q). Proposition 2 states that in Contra Monge matrices, sum–submatrices may as well be represented by a single column (in our case by column q). This simplifies matrix E down to matrix D.

If no more refinement of a stripe Q is possible, it follows from Proposition 2 that the corresponding matrix D is a sum–matrix. Since F is a sum–matrix, too, and the sum–matrices D and F have a common column q, this implies that the matrix E itself is a sum–matrix. Hence at the end of (S3), all matrices $C[Q, I\setminus Q]$ are sum–matrices.

Note that some of the derived refinements may contradict each other: We receive constraints (i.e. subsets of I that must be consecutive) from every single stripe Q. The constraints arise from the rows and from the columns of the Contra Monge matrices and they also concern the other stripes within Q^*. Another type of contradiction arises if the Contra Monge conditions force q to become an interior column of D. Hence, if some constraint cannot be fed into the PQ-tree, there does not exist a consistent refinement for Q^* and matrix C cannot be a permuted Kalmanson matrix. In this case, the algorithm returns any permutation and stops.

(S4) Recursion. Let $\text{STR}_i = \text{STR}(Q_1^{(i)}, \ldots, Q_m^{(i)})$ be the resulting set of permutations as derived in step (S3). For every stripe $Q_j^{(i)}$, select any row q_{j+1} from $Q_{j+1}^{(i)}$. Set $X_j = Q_j^{(i)} \cup \{q_{j+1}\}$ and determine the restriction of the union-find structure to the elements in X_j. Compute recursively a Kalmanson permutation π_j for $C[X_j, X_j]$ under this union-find structure. Afterwards, rotate π_j such that q_{j+1} becomes the last element and remove q_{j+1} from the rotated permutation. This yields permutation σ_j.

The output consists of the concatenation of the permutations $\sigma_1, \sigma_2, \ldots, \sigma_m$ in exactly this order.

Step (S4) essentially applies Lemma 8 to the stripes in STR_i.

Lemma 12. *Either matrix C is not a permuted Kalmanson matrix, or $C_\sigma \in \mathbb{K}$ holds for the output permutation $\sigma \in S_n$ of the above algorithm.*

The correctness of this claim follows from the lemmas derived in Section 3. It remains to investigate the time complexity of the algorithm.

Lemma 13. *For any $n \times n$ matrix C, the above algorithm runs in $O(n^2 \log n)$ time.*

Proof. It is convenient to make an analysis for the main part of the algorithm and a separate analysis for the UNION operations.

Time complexity of the algorithm. Let $\text{STR}_i = \text{STR}(Q_1^{(i)}, \ldots, Q_m^{(i)})$ be the final set of permutations computed in step (S3) and let a_j denote the cardinality of $Q_j^{(i)}$. Clearly, $a_j \leq n - 2$ for all j and $\sum_{j=1}^m a_j = n$. Define $S_a = \sum_{j=1}^m a_j^2$. For every j, the algorithm treats the submatrix corresponding to stripe $Q_j^{(i)}$ recursively. The remaining area of size $n^2 - S_a$ is covered by small, almost disjoint submatrices (they are disjoint with the exception of the negligible columns used to represent the sum–submatrices). Such a covering submatrix with dimensions $x \times y$ is handled in $O(xy + x \log x + y \log y)$ time according to Proposition 2. Hence, handling a matrix of area A is done in $O(A \log A)$ time and this complexity is superlinear in the concerned area. Thus the total cost for handling all these matrices is at most $O((n^2 - S_a) \log(n^2 - S_a))$. Storing, refining and modifying the partitions with the help of PQ-trees costs time that is linear in the size of the concerned set (i.e. proportional to the sidelengths of the submatrices) and thus is dominated by the cost for handling the submatrices. The overall cost for the FIND operations in step (S1) is $O(n)$. Summarizing, the time $T(n)$ for treating a matrix of sidelength n obeys

$$T(n) \leq \max_{a_j} \left\{ \sum_{j=1}^m T(a_j + 1) + c_1 \left(n^2 - \sum_{j=1}^m a_j^2\right) \log\left(n^2 - \sum_{j=1}^m a_j^2\right) + c_2 n \right\},$$

where the maximum is taken over all integers a_j with $1 \leq a_j \leq n - 2$ and $\sum_{j=1}^m a_j = n$ and where c_1 and c_2 are appropriate positive constants. By setting $T(n) = F(n-1)$, by applying $\log(a_j) \leq \log(n-2)$ and by observing $n \leq \sum_{j=1}^m a_j^2 \leq n^2 - 4n + 8$, one easily proves by induction that $F(n) \leq c_3 n^2 \log(n+1)$ where $c_3 = 10 \max\{c_1, c_2\}$.

Time complexity for manipulating the union-find structure. The union-find structure is only used in step (S1). The cost for the FIND operations was already investigated above and it remains to analyze the UNION operations. We represent the recursive process in the standard way by a tree: The root of the tree represents the original problem. The sons of a vertex v in the tree represent the subproblems originating in step (S4) when the problem corresponding to v is treated. Every vertex v is labelled by two numbers a_v and b_v, where a_v equals the number of rows (i.e. the *size*) of the corresponding subproblem and b_v denotes the number of UNION operations that result from treating the subproblem. Clearly, $b_v \leq a_v$ holds.

Since every UNION operation decreases the number of equivalence classes by one and since in every leaf of the tree there remains at least one equivalence class, on every branch going from the root to some leaf the overall sum of all values b_v is at most $n - 1$. Moreover, the sum of the a-labels of all sons of vertex v is linear in the size a_v. The overall cost of all UNION operations is $O(\sum a_v b_v)$ where the sum is taken over all vertices in the tree. With the help of the above inequalities, it is straightforward to show that this overall cost is dominated by $O(n^2)$. This completes the proof of the claim.

Theorem 14. *For a symmetric $n \times n$ matrix C, it can be decided in $O(n^2 \log n)$ time whether C is a permuted Kalmanson matrix.*

Proof. From Lemmas 12 and 13, we get the correctness and the time complexity of the algorithm. In the end, we permute C according to the output permutation σ and check whether the permuted matrix C_σ indeed is Kalmanson. This is done in $O(n^2)$ time as described in Proposition 4.

5 Master Tours in Polynomial Time

A *master tour* π for a set V of cities fulfills the following property: For every $V' \subseteq V$, an optimum travelling salesman tour for V' is obtained from π by removing from it the cities that are not in V'. Given the distance matrix C for a set of cities, the *master tour problem* consists in deciding whether this set of cities possesses a master tour. In this section, we prove that the master tour problem is closely related to permuted Kalmanson matrices and hence solvable in polynomial time.

Theorem 15. *For an $n \times n$ symmetric distance matrix C, the permutation $\langle 1, 2, \ldots, n \rangle$ is a master tour if and only if C is a Kalmanson matrix.*

Proof. (Only if): Assume that $\langle 1, 2, \ldots, n \rangle$ is a master tour for the distance matrix C. Then by definition, for each subset of four cities, $\{i, j, k, \ell\}$, with $1 \leq i < j < k < \ell \leq n$, the tour $\langle i, j, k, \ell \rangle$ is an optimal TSP tour. Since C is symmetric, there are only three combinatorially different tours through those cities: (i) $\langle i, j, k, \ell \rangle$, (ii) $\langle i, j, \ell, k \rangle$ and (iii) $\langle i, k, j, \ell \rangle$. The optimality of tour (i) implies that $c_{ij} + c_{jk} + c_{k\ell} + c_{\ell i} \leq c_{ij} + c_{j\ell} + c_{\ell k} + c_{ki}$ and $c_{ij} + c_{jk} + c_{k\ell} + c_{\ell i} \leq c_{ik} + c_{kj} + c_{j\ell} + c_{\ell i}$. By exploiting the symmetry of C and simplifying, the above inequalities turn into

$$c_{jk} + c_{i\ell} \leq c_{ik} + c_{j\ell} \quad \text{and} \quad c_{ij} + c_{k\ell} \leq c_{ik} + c_{j\ell},$$

which are exactly the conditions (2) and (1). Hence, C is a Kalmanson matrix.

(If): Let $K = \{x_1, \ldots, x_k\}$ be a subsequence of $\langle 1, 2, \ldots, n \rangle$. Then by Proposition 5(ii), the matrix $C[K, K]$ is again a Kalmanson matrix and by Kalmanson's result [6] the tour $\langle x_1, \ldots, x_k \rangle$ is an optimal tour for K. Consequently, $\langle 1, 2, \ldots, n \rangle$ is a master tour.

Theorem 16. *For a symmetric $n \times n$ matrix C, it can be decided in $O(n^2 \log n)$ time whether C possesses a master tour.*

Proof. By Theorem 15, a symmetric distance matrix has a master tour if and only if it is a permuted Kalmanson matrix. By Theorem 14, permuted Kalmanson matrices can be recognized in $O(n^2 \log n)$ time.

6 Discussion

In this paper we have developed an algorithm for recognizing permuted $n \times n$ Kalmanson matrices in $O(n^2 \log n)$ time and we showed that this problem is equivalent to detecting master tours. Since the input is of size n^2, the derived time complexity is close to optimal. Two questions remain open.

(1) We would like to know whether the $\log n$ factor in the time complexity can be shaved off in the *Random Access Machine* model of computation.

(2) The second question concerns characterizing all Kalmanson permutations for some given input matrix C. Our algorithm just outputs a single Kalmanson permutation. However, we would like to have a complete and concise description of *all* Kalmanson permutation similar to the concise description of all Contra Monge permutations in Proposition 2. One of the main obstacles in deriving such a description is that we do not fully understand the structure of equivalent columns. E.g. it is *not true* that equivalent columns must stick together in Kalmanson permutations. Consider the following two matrices.

$$A = \begin{pmatrix} * & 0 & 0 & 0 \\ 0 & * & 0 & 1 \\ 0 & 0 & * & 0 \\ 0 & 1 & 0 & * \end{pmatrix} \qquad A_\sigma = \begin{pmatrix} * & 0 & 0 & 0 \\ 0 & * & 0 & 0 \\ 0 & 0 & * & 1 \\ 0 & 0 & 1 & * \end{pmatrix}$$

Matrix A is a Kalmanson matrix where rows 1 and 3 are equivalent. However, its permutation A_σ is not a Kalmanson matrix and *no* permutation of A which makes rows 1 and 3 neighboring rows yields a Kalmanson matrix.

References

1. K.S. Booth and G.S. Lueker, Testing for the consecutive ones property, interval graphs and graph planarity using PQ-tree algorithms, *Journal of Computer and System Sciences* **13**, 1976, 335–379.
2. R.E. Burkard, B. Klinz and R. Rudolf, Perspectives of Monge Properties in Optimization, SFB Report S02, Institut fuer Mathematik, TU Graz, Austria.
3. T.H. Cormen, C.E. Leiserson and R.L. Rivest, *Introduction to Algorithms*, MIT Press, 1990.
4. V.G. Deĭneko and V.L. Filonenko, On the reconstruction of specially structured matrices, *Aktualnyje Problemy EVM i programmirovanije*, Dnepropetrovsk, DGU, 1979, (in Russian).
5. P.C. Gilmore, E.L. Lawler and D.B. Shmoys, Well-solved special cases, Chapter 4 in [7], 87–143.
6. K. Kalmanson, Edgeconvex circuits and the travelling salesman problem, *Canadian Journal of Mathematics* **27**, 1975, 1000–1010.
7. E.L. Lawler, J.K. Lenstra, A.H.G. Rinnooy Kan and D.B. Shmoys, *The Travelling Salesman Problem*, Wiley, Chichester, 1985.
8. C.H. Papadimitriou, Lecture at the Maastricht Summerschool on Combinatorial Optimization, August 1993.
9. C.H. Papadimitriou, Computational Complexity, *Addison-Wesley*, 1994.

Interval Graphs with Side (and Size) Constraints

Itsik Pe'er and Ron Shamir*

Department of Computer Science,
Sackler faculty of Exact Sciences,
Tel Aviv University, Tel-Aviv, 69978 Israel
email: {izik,shamir}@math.tau.ac.il.

Abstract. We study problems of determining whether a given interval graph has a realization which satisfies additional given constraints. Such problems occur frequently in applications where entities are modeled as intervals along a line (events along a time line, DNA segments along a chromosome, etc.). When the additional information is order constraints on pairs of disjoint intervals, we give a linear time algorithm. Extant algorithms for this problem (known also as *seriation with side constraints*) required quadratic time. When the constraints are bounds on distances between endpoints, and the graph admits a unique clique order, we show that the problem is polynomial. However, we show that even when the lengths of all intervals are precisely predetermined, the problem is NP-complete. We also study unit interval satisfiability problems, which are concerned with the realizability of a set of unit intervals along a line, subject to precedence and intersection constraints. For all possible restrictions on the types of constraints, we either give polynomial algorithms or prove their NP-completeness.

1 Introduction

A graph $G(V, E)$ is an *interval graph* if one can assign to each vertex v an interval I_v on the real line, so that two intervals have a non-empty intersection if and only if their vertices are adjacent. The set of intervals $\{I_v\}_{v \in V}$ is then called a *realization* of G. Interval graphs have been intensively studied, due to their central role in many applications (cf. [32, 16, 10]). They arise in numerous practical problems which require the construction of a time line where each particular event or phenomenon corresponds to an interval representing its duration. Among the applications are planning [3], scheduling [19, 30], archaeology [24], and circuit design [33]. There are also non-temporal applications in genetics [5] and behavioral psychology [8]. In the Human Genome Project, a central problem which bears directly on interval graphs is the physical mapping of DNA [7, 23]: It calls for the reconstruction of a map (a realization) for a collection of DNA segments, based on partial information on the relations between pairs of segments.

---

* Research supported in part by a grant from the Ministry of Science and the Arts, Israel.

The question whether a given graph has an interval realization (interval graph recognition) can be answered in linear time [6, 26, 20]. Surprisingly little is known about realization problems when the input contains also additional constraints on the realization. The problems which we study here are concerned with the realizability of a set of intervals along a line, subject to various types of constraints. The constraints may be order constraints, distances between interval endpoints or sizes of the intervals. Such problems arise naturally in many of the applications mentioned above, and we shall mention additional motivation for specific problems.

1.1 Interval Graphs with Side Order Constraints

When two intervals x, y on the real line intersect, we denote their relation by $x \cap y$ (read 'x intersects y'). If x is completely to the left of y we write $x \prec y$ or, equivalently, $y \succ x$. Suppose we are given a graph $G(V, E)$ and a set C of ordered pairs from $\overline{E}$. Is there an interval realization of G which satisfies $I_x \prec I_y$ whenever $(x, y) \in C$? This problem contains as special cases the interval graph and the interval order recognition problems. The problem (called also *seriation with side constraints*) was originally motivated by problems in archaeology and in consecutive retrieval with access priorities [25]. It arises also in temporal reasoning [17], and in physical mapping, when order information on some pairs of segments is available (see, e.g., [27]).

Golumbic and Shamir [17] give a simple $O(n^3)$ algorithm for the problem on a graph with n vertices, and Korte and Möhring [25] give an $O(n^2)$ algorithm using MPQ-trees. We give here a new linear-time algorithm of complexity $O(n + |E| + |C|)$. This improvement is substantial when the interval graph is sparse. Such sparse graphs are typical to many applications, including physical mapping (see, e.g., [22]). Our algorithm builds on a new interval graph recognition algorithm by Hsu and Ma [20].

1.2 Interval Graphs with Difference Constraints

In this problem the input is an interval graph and a set of difference constraints on the endpoints of the intervals. A difference constraint has the form $x - y < c$ or $x - y \leq c$. The goal is to determine if there is a realization in which the endpoints of the intervals satisfy all the constraints. Note that this problem contains as special cases the previous one, and the problem in which every constraint involves two endpoints of the same interval, which we shall discuss later. Fishburn and Graham[11] studied a special case of the problem in which the lengths of all intervals are bounded by *the same* pair of integers $p \leq |I_v| \leq q$. Their characterization implicitly gives a $n^{O(pq)}$ time algorithm for the realization problem. Isaak [21] gave a polynomial algorithm for the problem when the bounds are integers, the input is an interval order, and the realization must be discrete, i.e., all endpoints must be integers. He also posed the more general question which we study here.

We prove here that the problem is polynomial when the interval graph has a unique clique order, up to complete reversal (we call such graphs *UCO graphs*). This class of graphs properly contains the prime interval graphs, and is recognizable in linear time. We also show that the problem of solving a system of difference constraints is *linearly equivalent* to solving the realization problem on UCO interval graphs with difference constraints. On the other hand, the problem is NP-complete in general, as will follow from the next result.

1.3 Interval Graphs with Exact Given Sizes

Suppose we are given an interval graph and for each interval its exact length in the realization is prescribed. How hard is it to decide if a realization exists? This problem arises naturally when both intersection and size data is given. See [18] for an example of a physical mapping experiment in which such information is generated. If all lengths are the same then the graph is a unit-interval graph [32] which can be recognized in linear time [9]. We prove that the problem is NP-complete. This implies that the realizability problems for interval graphs with difference constraints, and for interval graphs with individual upper and lower bounds on interval sizes, are also NP-complete. On the other hand, the problem is polynomial on UCO graphs.

1.4 Unit Interval Satisfiability Problems

In an *Interval Satisfiability* (ISAT) problem we are given a set of *events*, and for each pair of events a set of *permitted relations* between them. The set is any subset of $\{\prec, \cap, \succ\}$. The question is whether one can assign to each event an interval on the real line so that for each pair of events, their intervals satisfy one of their permitted relations. We denote the permitted relations between a pair of intervals by their concatenation. Hence, $x \prec\cap y$ is short for "$x \prec y$ or $x \cap y$", $x \prec\succ z$ is short for "$x \prec z$ or $x \succ z$", etc. In addition to the applications discussed above, ISAT problems arise in artificial intelligence, in the context of temporal reasoning [2] and medical diagnosis [28].

An important notion in studying interval satisfiability problems is the *domain*: It is the collection of possible sets of permitted relations in the input. For example, in the domain $\{\cap, \prec\succ\}$, for every two events either the input requires that they intersect or it requires that they should not intersect. These are the *only* types of input restrictions. Clearly, ISAT restricted to this domain is equivalent to the interval graph recognition problem. On the domain $\{\cap, \prec, \succ\}$, ISAT is equivalent to the interval order recognition problem [10]. By allowing other combinations of permitted relations, one gets various interesting generalizations of interval graphs and interval orders. Golumbic and Shamir [17] and Webber [35] have recently determined the polynomiality or NP-completeness of ISAT on all possible restricted domains.

We study here ISAT problems *with the additional constraint that all intervals must be of the same length*. We call this problem the *Unit Interval Satisfiability*

(UISAT) problem. We provide a complexity classification for all possible domains. While we build on the techniques developed by [17] and [35], the known results for ISAT do not imply anything for UISAT, due to the additional length restrictions on intervals.

The motivation to studying UISAT is similar to that for ISAT, as in many applications equal length (or, equivalently, non-containment [31]) of intervals is a natural condition. For example, in some planning and scheduling situations all events must have the same length. In many experiments of physical mapping, all the DNA fragments involved have nearly identical length, due to the recombination technique used to generate them (cf. [34, 15]). In these cases, partial information on intersection and precedence of events gives rise to UISAT problems.

Section 2 describes the linear algorithm for interval graph with side order constraints. Section 3 describes the polynomial algorithm for UCO graphs with constraints on differences between endpoints. Section 4 outlines the NP-completeness proof of interval graphs with exact given sizes. Section 5 provides a complexity classification of unit interval satisfiability problems on all possible domains. For space reasons, many details in the proofs and algorithms are necessarily omitted. Basic terminology on graphs and orders can be found in [32, 16, 10].

2 Interval Graphs with Side Order Constraints

The input for this problem is $J = (G, C)$ where $G = (V, E)$ is a graph and $C = \{(x, y) | x \prec y\}$ is a set of additional order constraints on pairs of non-adjacent vertices. We assume without loss of generality that all the intervals are open, and that G is connected. We denote by l_v, r_v the left and right endpoints of interval I_v.

Our algorithm builds on the new linear time algorithm for recognizing interval graphs of Hsu and Ma [20]. We decompose G into modules and develop new ideas to handle the order constraints. We first note some properties of the decomposition and introduce some definitions:

A set $M \subseteq V$ is called a *module* in $G = (V, E)$ if for each $x, y \in M$ and $u \notin M$: $xu \in E \Leftrightarrow yu \in E$. G is called *prime* if its only modules are the singleton vertices and V. For $M \subseteq V$, $G[M] = (M, E[M])$ is the subgraph of G induced on M. Define $N(M) = \{x \notin M | ux \in E, u \in M\}$. For a module M in the graph G, denote by $G[v/M]$ the graph $G' = (V', E')$, where $V' = V \setminus M \cup \{v\}$, and $E' = E[V \setminus M] \cup \{uv | u \in N(M)\}$. We say that $G[v/M]$ is obtained from G by *contracting M to v*. Following [20], we classify non-singleton modules into three categories: *type I* modules, which are cliques, *type II* modules, which are connected, but not cliques, and *type III* modules, which are not connected.

The key to the algorithm is the fact that modules behave to some extent like single vertices (see Fig. 1): We show that for certain modules, order constraints between all module vertices and any non-module vertex must be consistent.

Moreover, order constraints among vertices in the same module must be consistent. These properties facilitate recursive reduction of the problem, by contracting appropriate modules and passing over the constraints to the new instance. In order to do this efficiently, we handle each type of modules separately.

Lemma 1. *([20]) If S is a module in a chordal graph then either S is a clique or $N(S)$ is a clique.*

Lemma 2. *Let $G(V, E)$ be an interval graph and let $U \subseteq V$ be a set of vertices inducing a connected subgraph $G[U]$. In a realization $\{(l_v, r_v)\}_{v \in V}$, if $v, u \in U$ and $l_v < r_u$, then every point $x \in (l_v, r_u)$ is contained in some interval of a vertex in U.*

Let S be a set of vertices in an interval graph G, and let $\{I_x\}_{x \in V}$ be an open realization of G. Define $I^{out}(S) = I^{out} = (\min_{x \in S} l_x, \max_{x \in S} r_x)$. Define $p = \min_{x \in S} r_x$ and $q = \max_{x \in S} l_x$. Note that $p > q$ if and only if S is a clique. When $p \le q$ we define $I^{in}(S) = I^{in} = [p, q]$. Pick a vertex $t \notin S$. If S is a clique we say that t *shatters* S if I_t intersects I^{out} but does not intersect $\cap_{x \in S} I_x$. If S is not a clique we say that t *shatters* S if either I_t has at least one endpoint in I^{in}, or $I_t \cap I^{out}$ and $I_t \diamondsuit I^{in}$.

Lemma 3. *In a realization of a graph, no module which is maximal or connected is shattered.*

Proof. Suppose t shatters the module M. If M is a clique then t in adjacent to some but not all of the vertices in S so M cannot be a module. Hence, M is not a clique. If $I_t \cap I^{out}$ and $I_t \diamondsuit I^{in}$, or I_t has exactly one endpoint in I^{in}, then again we get a contradiction to M being a module.

The only remaining case is when $I_t \subseteq I^{in}$. If M is connected then by Lemma 2 I_t meets some interval from M but not all, a contradiction. If M is not connected, define $S = \{x \in V | I_x \subseteq I^{in}\}$, and let $M' = M \cup S$. Clearly $t \in M' \setminus M$. It can be shown that M' is a module, contradicting the maximality of M. $\quad\square$

A module M of G is called *compliant* in J if there exist no $v_1, v_2 \in M$ and $u \in V \setminus M$ such that C contains the constraints $v_1 \prec u$ and $u \prec v_2$. By Lemma 3, we can conclude:

Corollary 4. *In a realizable instance, every maximal or connected module is compliant.* $\quad\square$

For a compliant module M in J, denote by $C[M]$ the set of constraints between vertices in M. Denote by $C[v/M]$ the set of constraints obtained from $C \setminus C[M]$, by substituting v instead of every $u \in M$ and deleting repetitions. The compliance of M assures us that the new set will contain no contradicting pairs of constraints (say, $x \prec y$ and $y \prec x$). Denote by $J[v/M]$ and $J[M]$ the instances $(G[v/M], C[v/M])$ and $(G[M], C[M])$ respectively (see Fig. 1). We say that $J[v/M]$ is obtained from J by *contracting M to v*.

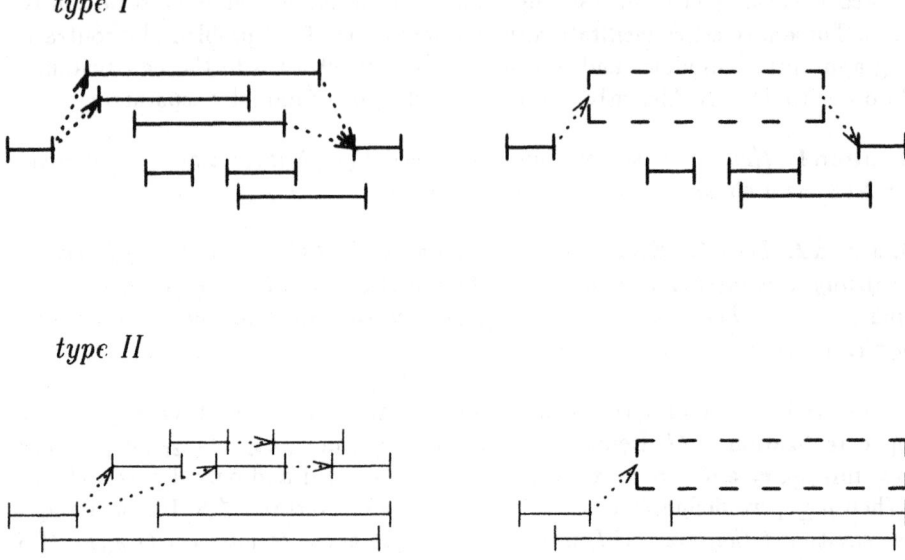

Fig. 1. Compliant modules and their contraction: An instance is represented by a realization of the interval graph, plus dotted arcs denoting the side order constraints. The boxes on the right are single intervals which replace the respective modules in the contraction and inherit the module's constraints. For type I and type II modules, the constraints between all module vertices and each non-module vertex (left) are consistent, therefore the compliant module can be contracted (right). For a type II module, order constraints within the module must also be consistent (this is verified in the sub-instance of that module in a lower level of the decomposition tree).

The algorithm will perform several phases of contractions and decompositions. Below we show that a single contraction of an appropriate module preserves the realizability of an instance. For instances $J, J_1, \ldots, J_k$ we say that J is *equivalent* to $\{J_1, \ldots, J_k\}$ if J is has a realization if and only if every J_i, $1 \leq i \leq k$, has one.

Proposition 5. *If M is a compliant module in J, which is connected or maximal, then J is equivalent to $\{J[v/M], J[M]\}$.*

Proof. ($\Rightarrow$): Suppose that $S = \{I_{v_1}, \ldots, I_{v_n}\}$ is a realization of J where for $1 \leq j \leq n$: $I_{v_j} = (l_j, r_j)$. The set of intervals $S_M = \{I_u | u \in M\}$ is an interval realization of $G[M]$. Moreover, S_M satisfies every constraint in $C[M]$, because S satisfies C. Hence, $J[M]$ is realizable.

Define the interval for the vertex v (which is a contraction of M) to be $I_v = I^{out}(M)$. The set of intervals $(S \setminus S_M) \cup \{I_v\}$ is a realization of $J[v/M]$: The intervals in $S \setminus M$ are left unchanged, hence, they satisfy both the appropriate intersection properties and the appropriate order constraints. It can also be shown, that I_v satisfies the intersection and order properties required from v by the definition of $I^{out}(M)$.

($\Leftarrow$): Suppose that $S = \{I_{v_i}|v_i \in V \setminus M\} \cup \{I_v\}$ and $S_M = \{I_{v_i}|v_i \in M\}$ are sets of intervals realizing $J[v/M]$ and $J[M]$, respectively. If M is a clique, let S' be a set of intervals containing $|M|$ copies of I_v (with the same endpoints). Otherwise, by Lemma 1, $N(M)$ is a clique, so the set $I = \cap_{u \in N(M)} I_u$ is an interval. Moreover, since M is a module, $I \cap I_v \neq \emptyset$. Construct the set of intervals S' by mapping each interval $I_u \in S_M$ to I'_u using the monotone linear transformation that maps $I^{out}(M)$ to $I \cap I_v$. In both cases, it can be proven that $S \setminus \{I_v\} \cup S'$ realizes J.   □

By the above proposition, the following scheme solves our problem: Detect a module M which is connected or maximal. If it is not compliant in the instance J, answer 'no'. Otherwise continue recursively on each of the two instances $J[v/M]$ and $J[M]$. Upon reaching an instance with a prime or empty graph, solving the problem is immediate. Answer "yes" if and only if the answers to all the sub-instances were "yes". In order to implement this scheme in linear time, the algorithm proceeds in phases, as shown schematically below:

Do the following including checking compliance of the relevant modules in each step:
1. Contract all clique modules.
2. Form a decomposition tree for the Type-II modules and construct a sub-instance for each node.
3. For every sub-instance, contract all Type-III modules, solving realizability problems induced on each module and on the contracted instance.
4. Return $TRUE$ if and only if all sub-instances were true.

Finding the clique modules in G can be done in linear time [20]. Given these modules, contracting them in J is simple. Constructing a decomposition tree for the type II modules is possible in $O(m + n)$ time [20]. It can be shown that each node x in the type II decomposition tree gives rise to a sub-instance of our problem, the resulting set of sub-instances is equivalent to the original instance, and the total size of the resulting sub-instances can only decrease. Moreover, the graph of each sub-instance can be obtained from the type II decomposition, and the constraints in it are those between pairs of vertices whose lowest common ancestor is x. Each of these sub-instances contains no type II modules. The assignment of constraints to sub-instances can also be done in linear time, by answering the appropriate lowest common ancestor queries.

For an instance with only type III modules, these modules can be identified in $O(n+m)$ time [20]. Given these modules, they are simple to contract. Contracting these modules yields instances with prime or empty graph. These simple cases are easily solvable in linear time. In summary:

Theorem 6. *Deciding if a graph has a realization which satisfies additional order constraints can be done in linear time.*   □

3 UCO Interval Graphs with Metric Constraints

We call an interval graph *uniquely clique-orderable* (UCO) if its maximal cliques have the same linear order, up to complete reversal, in every realization. UCO graphs can be recognized in linear time by applying the PQ-tree algorithm of Booth and Lueker [6], and noting that G is UCO if and only if the final tree contains a single Q-node. In this section we consider the following problem: Let the input be a UCO graph $G = (V, E)$, and a set S of linear inequalities on the variables $\{l_v, r_v\}_{v \in V}$ (which are the intervals endpoints), in which every inequality has the either the form $x - y < C_{xy}$, or the form $x - y \leq C_{xy}$, for variables x, y and constant C_{xy}. The question asked is whether there is a realization $\{[l_v, r_v]\}_{v \in V}$ for G, whose intervals endpoints solve S.

We reduce this problem to deciding whether there is a feasible solution for a system of inequalities, each of which is either of the form $x - y < C$ or $x - y \leq C$. Such a system is called *a system of difference constraints*.

Let $\prec_G \subseteq E(\bar{G})$ be the unique interval order admitted by G. It is known how to determine if G is UCO, and if so, how to compute $\prec_G$ in $O(m + n)$ time [6, 26, 20]. We construct two systems T and $\bar{T}$ of difference constraints on the variables $\{l_v, r_v\}_{v \in V}$, as follows: Both systems include all inequalities in S. In addition, for each $x, y \in V$: If $x \prec_G y$ then T contains an inequality $r_x < l_y$, and $\bar{T}$ contains an inequality $r_y < l_x$. If $xy \in E$ then both T and $\bar{T}$ contain an inequality $r_x \geq l_y$ (and $r_y \leq l_x$). With these definitions one can prove the following:

Lemma 7. *P has a realization if and only if either T or $\bar{T}$ have a feasible solution.*

Hence, we can solve our problem by deciding whether system T or $\bar{T}$ is feasible. We shall prove now that a system S of weak and strict difference constraints on n variables is reducible in linear time to a system S' of weak difference constraints, with numbers only $O(n)$ larger. (Standard transformation techniques [13] would give numbers $O(2^L)$ larger for binary input length L.) Assume all constants in S to be integral, and fix $\epsilon < \frac{1}{n}$. Define S' to include every weak inequality $x - y \leq c$ in S, and a weak inequality $x - y \leq c - \epsilon$ for every strict inequality $x - y < c$ in S.

Lemma 8. *S has a feasible solution if and only if S' has one.*

Proof. The 'if' direction is trivial, since a feasible solution to S' also satisfies S. The proof of the 'only if' uses the notion of the distance graph for the system of difference constraints (cf. [1, p. 103]), generalized to handle both strict and weak inequalities: For such a system T, construct a directed weighted graph $D(T)$ with vertex set V. For every constraint $x - y \leq C_{xy}$ or $x - y < C_{xy}$ add an arc (y, x) to $D(T)$ with weight C_{xy} and label the arc $\leq$ or $<$, respectively. Bellman [4] has shown that when all inequalities in T are weak, T is feasible if and only if $D(T)$ contains no negative cycles (see also [1, p. 103]).

Assume S' is not feasible. Then $D(S')$ contains a negative-weight cycle c. Let $w(c)$ and $w'(c)$ be the total weight of c in $D(S)$ and $D(S')$, respectively. Suppose $w(c) = w'(c) < 0$. Summing all inequalities along c in $D(S')$ we get $w'(c) \geq 0$, a contradiction. Otherwise, c contains an arc labeled $<$, $w'(c) < 0$, and since $|w(c) - w'(c)| \leq n\epsilon < 1$, it follows that $w(c) < 1$. Hence, $w(c) \leq 0$, since $w(c)$ is integer. Summing all inequalities along c in $D(S)$ we get $w(c) > 0$, a contradiction. □

By Lemma 7, and Lemma 8, solving an instance of our problem linearly reduces into determining if at least one of two systems of difference constraints is feasible. The feasibility of such a system with M weak inequalities on N variables, with sum of constants C, can be decided in $O(\min(NM, \sqrt{N}M \log NC))$ time [29, 12]. In our instance (G, S) there are n vertices, so $N = 2n, M = \Theta(n^2)$. Hence:

Theorem 9. *The realization problem on UCO interval graph with difference constraints can be solved in* $O(\min(n^3, n^{2\frac{1}{2}} \log nC))$ *time.* □

Interestingly, we cannot do any better: We can show a linear time reduction in the opposite direction, which enables us to decide the feasibility of a system of difference constraints, by determining whether a given UCO interval graph has a realization with intervals whose lengths match given bounds, and this is a special case of UCO interval graphs with metric constraints.

4 Interval graphs with exact given sizes

We consider here the most restricted version of difference constraints, in which an exact size is prescribed for each interval. Formally the problem is the following:

METRIC INTERVAL GRAPH RECOGNITION (MIG):
INSTANCE: A graph $G = (V, E)$ and function $s : V \to \mathbb{Q}^+$, assigning a *size* $s(v)$ for each vertex.
QUESTION: Is there a realization $\{I_v\}_{v \in V}$ of G with $|I_v| = s(v)$ for every $v \in V$?

The problem arises naturally whenever a combination of intersection data and specific interval (event) lengths are known. Previous solved special cases were mentioned in the introduction.

Theorem 10. *MIG is NP-complete.*

The proof is by reduction from 3-COLORING. Unfortunately, the current proof we have is rather lengthy (15 pages) so we cannot describe it here. It is interesting to note that in this problem, there *is* loss of generality by assuming that all intervals are closed or open. We first construct a reduction to the problem with the additional designation of each interval as closed or open, and then show that although in general there may not be an all-closed realization whenever there is one with a prespecified designation, this is true for our reduction. Interestingly, the interval graph generated by our reduction has a decomposition tree of depth 2 only, so the problem is hard even in that case. Moreover, the length of all intervals involved is polynomial in $|V|$, so the problem is also Strongly NP-complete. By the results of the previous section, MIG is polynomial on UCO and hence, on prime interval graphs.

5 Unit Interval Satisfiability

In this section we study UISAT restricted to each possible domain. Recall that every domain is a collection of subsets of $\{\prec, \cap, \succ\}$. In a restricted problem only relations from the domain are allowed in the input. The problem restricted to domain Δ is denoted UISAT(Δ). Previously studied cases include UISAT($\{\diamond, \cap\}$), which is equivalent to unit interval graph recognition (a linear-time solvable problem) [31, 9], and UISAT(Δ_0) with $\Delta_0 = \{\diamond, \cap, \prec\cap\succ\}$, which is the NP-complete unit interval graph sandwich problem [15]. The version of UISAT(Δ_0) with the additional restriction that the maximum clique has size $\leq k$ was recently shown to be polynomial for fixed k but hard when k is a parameter [22].

We use the names of domains from [17] and [35]. We denote by N^D the number of occurrences of relation set D in the input. We assume here without loss of generality that all interval are closed.

5.1 The Domain $\Delta_1 = \{\cap, \prec, \succ, \prec\cap, \cap\succ, \prec\cap\succ\}$

Golumbic and Shamir [17] have shown that ISAT(Δ_1) reduces to deciding whether a system of difference constraints (with variables corresponding to interval endpoints) has a feasible solution. The requirement of unit length to all intervals is expressible as a set of additional linear constraints on the endpoints, and substituting for the variables of all right endpoints of intervals, one gets a system of difference constraints. Using arguments similar to those in Sect. 3, the problem reduces to deciding if a distance graph (in which all arc weights are integers between $-2n-1$ and $2n$) contains no negative cycle. Let $m = N^{\prec\cap} + N^{\prec} + 2N^{\cap}$. Using the algorithm of [29] to detect a negative cycle, we conclude:

Theorem 11. *UISAT(Δ_1) can be solved in $O(\sqrt{n}m\log n)$ time.* □

5.2 The Domain $\Delta_2 = \{\cap, \prec, \succ, \diamond\}$

Let J be an instance of UISAT(Δ_2). We first apply the linear algorithm for ISAT(Δ_2) which was described in Sect. 2. If the algorithm returns *FALSE*, then J has no proper interval realization. Otherwise, the algorithm gives a realization of J and, implicitly, an interval order $P = (V, <)$ which respects all the order constraints, but is not guaranteed to be a proper interval order. Form the interval graph $G = (V, E)$, where $uv \in E$ if and only if $v \cap u$ in J. Fishburn and Graham [11] have shown that an interval order is a proper interval order if and only if its incomparability graph G does not contain an induced $K_{1,3}$. By that result, as we have already established that P is an interval order, it suffices to check if G contains an induced $K_{1,3}$. To check this in linear time we recall some properties of interval graphs: Let $C_1, \ldots, C_k$ be the maximal cliques in a graph $G = (V, E)$, where $V = \{v_1, \ldots, v_n\}$. The *clique matrix* of G is the $n \times k$ zero-one matrix $C(G) = (m_{ij})$ where $m_{ij} = 1$ if and only if $v_i \in C_j$. If the columns in $C(G)$ can be permuted so that the ones in each row are consecutive, then we say that $C(G)$ has the *consecutive ones* property, and we call such a permutation of the columns a *consecutive (clique) order*. According to Gilmore and Hoffman [14], G is an interval graph if and only if $C(G)$ has the consecutive ones property. For $C(G)$ in a consecutive order, let L_i and R_i be the column indices of the leftmost and rightmost 1's, respectively, in the i-th row of $C(G)$. Clearly, $\{[L_i, R_i]\}_{v_i \in V}$ is a realization of G. Such realization is called *canonical*. One can now prove the following:

Lemma 12. *An interval graph $G = (V, E)$ is a unit interval graph if and only if no canonical realization $\{[L_i, R_i]\}_{v_i \in V}$ for G satisfies $L_i < L_j$ and $R_i > R_j$ for any adjacent vertices $v_i, v_j \in V$.*

Let $m = |E| = N^{\cap}$. The clique matrix $C(G)$ of an interval graph and a consecutive clique order can be computed in $O(n + m)$ time (see, e.g., [6]). The consecutive permutation on the columns can also be performed in $O(n + m)$ time. The indices L_i and R_i can be computed, for all vertices v_i, in a total time of $O(n + m)$, and the check, for each edge $v_i v_j$ whether $L_i < L_j$ and $R_i > R_j$, takes $O(m)$ time for all edges. Therefore:

Theorem 13. *UISAT(Δ_2) can be solved in $O(n + N^{\cap} + N^{\prec})$ time.* □

5.3 The Domain $\Delta_3 = \{\leftdiamond, \prec, \succ, \prec\cap\succ\}$

It is easy to see that an instance of UISAT(Δ_3) is satisfiable if and only if it is satisfiable as an instance of ISAT(Δ_3). The results of [17] on ISAT(Δ_3), together with the above observation imply:

Theorem 14. *UISAT(Δ_3) can be solved in $O(n + N^{\prec})$ time.* □

5.4 The Domain $\Delta_5 = \{\cap\succ, \prec\cap, \leftdiamond\}$

This problem appears to be technically the hardest in this section. In fact, the corresponding problem for ISAT was the only problem left open in the classification of ISAT domains of [17]. Weber [35] subsequently proved it is NP-complete.

Theorem 15. *UISAT(Δ_5) is NP-complete.*

We omit the proof which is based on a reduction from Not-All-Equal Satisfiability. Unlike Webber's proof for ISAT, we use a symmetric six-interval gadget for each clause. It is interesting to note that the construction generates a directed graph whose diameter is at most four, irrespective of the size of the formula.

5.5 Summary

Every possible domain Δ is either contained in one of $\Delta_1, \Delta_2, \Delta_3$, and in this case UISAT(Δ) is polynomial, or it contains one of Δ_0, Δ_5, in which case UISAT(Δ) is clearly NP-hard. Thus, we have resolved the complexity of UISAT(Δ) for all possible domains Δ. In the final analysis, there are 21 polynomial domains and 10 NP-complete ones. (The symmetry of some relation sets reduces the number of non-isomorphic nonempty domains to 31.) As noted by [17], one of the byproducts of this classification is the ability to speed up enumerative solution of problems on NP-complete domains, by finding the polynomial domain which contains the largest number of relations in the input, thereby minimizing the size of enumeration needed.

References

1. R. K. Ahuja, T. A. Magnanti, and J. B. Orlin. *Network Flows: Theory, Algorithms and Applications*. Prentice Hall, Englewood Cliffs, NJ, 1993.
2. J. F. Allen. Maintaining knowledge about temporal intervals. *Comm. ACM*, 26:832–843, 1983.
3. J. F. Allen. *Reasoning about plans*. Morgan Kaufman, 1991.
4. R. Bellman. On a routing problem. *Quarterly of Applied Mathematics*, 16:87–90, 1958.
5. S. Benzer. On the topology of the genetic fine structure. *Proc. Nat. Acad. Sci. USA*, 45:1607–1620, 1959.
6. K. S. Booth and G. S. Lueker. Testing for the consecutive ones property, interval graphs, and planarity using PQ-tree algorithms. *J. Comput. Sys. Sci.*, 13:335–379, 1976.
7. A. V. Carrano. Establishing the order of human chromosome-specific DNA fragments. In A. D. Woodhead and B. J. Barnhart, editors, *Biotechnology and the Human Genome*, pages 37–50. Plenum Press, 1988.
8. C. H. Coombs and J. E. K. Smith. On the detection of structures in attitudes and developmental processes. *Psych. Rev.*, 80:337–351, 1973.
9. X. Deng, P. Hell, and J. Huang. Linear time representation algorithms for proper circular arc graphs and proper interval graphs. Technical report, School of Computing Science, Simon Fraser University, 1993.
10. P. Fishburn. *Interval Orders and Interval Graphs*. Wiley, New York, 1985.
11. P. Fishburn and R. L. Graham. Classes of interval graphs under expanding length restrictions. *J. Graph Theory*, 9:459–472, 1985.
12. H. Gabow and R. E. Tarjan. Faster scaling algorithms for network problems. *SIAM J. Comput.*, 18:1013–1036, 1989.
13. P. Gács and L. Lovász. Khachiyan's algorithm for linear programming. *Math. Prog. Study*, 14:61–68, 1981.
14. P. C. Gilmore and A. J. Hoffman. A characterization of comparability graphs and of interval graphs. *Canad. J. Math.*, 16:539–548, 1964.
15. P. W. Goldberg, M. C. Golumbic, H. Kaplan, and R. Shamir. Four strikes against physical mapping of DNA. Technical report, Computer Science Dept., Tel Aviv University, April 1993. To appear in *Journal of Computational Biology*.

16. M. C. Golumbic. *Algorithmic Graph Theory and Perfect Graphs*. Academic Press, New York, 1980.

17. M. C. Golumbic and R. Shamir. Complexity and algorithms for reasoning about time: A graph-theoretic approach. *J. ACM*, 40:1108–1133, 1993.

18. E. D. Green and M. V. Olson. Chromosomal region of the Cystic Fibrosis gene in yeast artificial chromosomes: a model for human genome mapping. *Science*, 250:94–98, 1990.

19. G. Hajös. Über eine art von graphen. *Intern. Math. Nachr.*, 11, 1957. problem 65.

20. W-L. Hsu and T-H Ma. Substitution decomposition on chordal graphs and applications. In W-L. Hsu and R. C. T. Lee, editors, *Proc. 2nd Int. Symp on Algorithms (ISA '91)*, pages 52–60. Springer-Verlag, 1991. LNCS 557.

21. G. Isaak. Discrete interval graphs with bounded representation. *Discrete Applied Mathematics*, 33:157–183, 1993.

22. H. Kaplan and R. Shamir. Pathwidth, bandwidth and completion problems to proper interval graphs with small cliques. Technical report, CS Department, Tel Aviv University, November 1993. To appear in *SIAM J. Comput.*

23. R. M. Karp. Mapping the genome: some combinatorial problems arising in molecular biology. In *Proc. 25th STOC*, pages 278–285. ACM Press, 1993.

24. D. G. Kendall. Incidence matrices, interval graphs, and seriation in archaeology. *Pacific J. Math.*, 28:565–570, 1969.

25. N. Korte and R. H. Möhring. Transitive orientation of graphs with side constraints. In H. Noltemeier, editor, *Proc. of the International workshop on Graph-theoretic concepts in Computer Science (WG '85)*, pages 143–160, Linz, 1985. Universitätsverlag Rudolf Trauner.

26. N. Korte and R. H. Möhring. An incremental linear time algorithm for recognizing interval graphs. *SIAM J. Comput.*, 18:68–81, 1989.

27. P. Lichter et al. High-resolution mapping of human chromosome 11 by in situ hybridization with cosmid clones. *Science*, 247:64–69, 1990.

28. K. Nökel. *Temporally Distributed Symptoms in Technical Diagnosis*. Lecture Notes in Artificial Intelligence 517. Springer Verlag, 1991.

29. J. B. Orlin and R. K. Ahuja. New scaling algorithms for the assignment and minimum cycle mean problems. *Mathematical Programming*, 54:41–56, 1992.

30. C. Papadimitriou and M. Yannakakis. Scheduling interval ordered tasks. *SIAM J. Comput.*, 8:405–409, 1979.

31. F. S. Roberts. Indifference graphs. In F. Harary, editor, *Proof Techniques in Graph Theory*, pages 139–146. Academic Press, New York, 1969.

32. F. S. Roberts. *Discrete Mathematical Models, with Applications to Social Biological and Environmental Problems*. Prentice-Hall, Englewood Cliffs, New Jersey, 1976.

33. S. A. Ward and R. H. Halstead. *Computation Structures*. MIT Press, 1990.

34. J.D. Watson, M. Gilman, J. Witkowski, and M. Zoller. *Recombinant DNA*. W.H. Freeman, New York, 2nd edition, 1992.

35. A.B. Webber. Proof of an interval satisfiability conjecture. manuscript, Wesstern Illinois University, 1994.

Maximum Skew-Symmetric Flows

Andrew V. Goldberg[1] and Alexander V. Karzanov[2]

[1] Computer Science Department, Stanford University, Stanford, CA 94305, USA. (Current address: NEC Research Inst., 4 Independence Way, Princeton, NJ 08540.)
[2] Inst. for Systems Analysis, 9, Prospect 60 Let Oktyabrya, 117312 Moscow, Russia

Abstract. We introduce the maximum skew-symmetric flow problem which generalizes flow and matching problems. We develop a theory of skew-symmetric flows that is parallel to the classical flow theory. We use the newly developed theory to extend, in a natural way, the blocking flow method of Dinitz to the skew-symmetric flow case. In the special case of the skew-symmetric flow problem that corresponds to cardinality matching, our algorithm is simpler and more efficient than the corresponding matching algorithm.

1 Introduction

Flow and matching problems are classical problems in combinatorial optimization [1, 8, 13, 23] and have many practical applications. These problems are closely related; in particular, the bipartite matching problem can be viewed as a special case of the maximum flow problem (see *e.g.* [1, 22]). In general, however, the combinatorial structure of matchings is more complicated than the combinatorial structure of flows. Although matching algorithms are similar to and often motivated by flow algorithms, the former are much more complicated.

This paper continues a systematical study of skew-symmetric graphs and their applications started in our previous paper [15, 16]. That paper, devoted to problems on regular paths, extends to the skew-symmetric graphs the usual path reachability and shortest path problems, as well as certain problems on alternating paths in matching theory. The present paper deals with the *maximum integral skew-symmetric flow (maximum IS-flow)* problem. We study combinatorial and linear programming structure of this problem and develop algorithms for it.

As the flow and matching problems are closely related to path problems, the maximum IS-flow problem is closely related to regular path problems in skew-symmetric graphs. For example, duals of the shortest regular path and maximum IS-flow problems are similar, and the shortest regular path algorithm developed in [15] is used as a subroutine in the algorithms developed in the present paper.

The theory of IS-flows, developed in the current paper, is parallel to that for usual maximum flows [13]. The maximum IS-flow problem generalizes both the maximum flow and maximum matching problems. This appears to be a right generalization since combinatorial and linear programming theorems and algorithms for flows extend in a natural way to IS-flows. The implied results for matching are often simpler and better motivated.

The fundamental results in the flow theory are the decomposition theorem, the augmenting path theorem, and the max-flow min-cut theorem. Next we state these theorems and their skew-symmetric equivalents. Definitions of the concepts

---

* The first author was supported in part by NSF Grant CCR-9307045.

we use appear in Sections 2 and 3, and proofs of the new theorems appear in Section 3.

The flow decomposition theorem says that a flow can be decomposed into a collection of source-to-sink paths and cycles; the IS-flow decomposition theorem says that an IS-flow can be decomposed into a collection of symmetric pairs of source-to-sink paths and symmetric pairs of cycles. The augmenting path theorem says that a flow is maximum if and only if it admits no augmenting path. The IS-flow augmenting path theorem says that an IS-flow is maximum if and only it it admits no regular augmenting path. The max-flow min-cut theorem says that the maximum flow value is equal to the minimum cut capacity. Its skew-symmetric equivalent is that the maximum IS-flow value is equal to the minimum odd-barrier capacity.

Linear programming duality, in particular complementary slackness, plays an important role in understanding the structure of network flow problems and in developing algorithms for these problems. We study linear programming formulation of the maximum IS-flow problem and the complementary slackness conditions for it in Section 4.

The augmenting path theorem is the basis for the augmenting path algorithm of Ford and Fulkerson [13] and the shortest augmenting path algorithm of Edmonds and Karp [10]. A more efficient version of the latter algorithm is the blocking flow algorithm of Dinitz [7]. These algorithmic results can be extended to IS-flows using the combinatorial and linear programming results of Sections 3 and 4. In Section 5 we introduce a skew-symmetric version of the blocking flow algorithm. The resulting algorithm has the same running time bound as Dinitz' algorithm on general networks [7] as well as on networks with unit arc or node capacities [11, 19, 20].

Modifications of our algorithm achieve better time bounds. Let n and m denote the number of nodes and arcs in the input network, respectively. For the special case arising from the cardinality matching problem, we use the graph compression techniques of Feder and Motwani [12] to get an $O\left(\sqrt{n}m\frac{\log(n^2/m)}{\log n}\right)$ time bound.[3] This improves the previous bound of $O(\sqrt{n}m)$ for cardinality matching [14, 24, 25].

For the general problem, we can use any integral maximum flow algorithm to find a good initial solution for our skew-symmetric blocking flow algorithm. This yields an $O(M(n,m)+nm)$ time bound, where $M(n,m)$ is the time to find an integral maximum flow in a network with with n nodes, m arcs, and integral capacities. Note that, with one exception, all fastest currently known maximum flow algorithms [2, 5, 17, 21] find integral solutions and run in $\Omega(nm)$ time. The only exception is the algorithm of Cheriyan et. al. [5] that runs in $O(n^3/\log n)$ time (taking advantage of bit operations in the uniform-cost RAM model of computation). Thus except for dense graphs, our bound for the maximum IS-flow problem matches the best bounds for the maximum flow problem.

This paper is organized as follows. Section 2 gives basic definitions and reviews results related to the r-path problems. Sections 3 and 4 deal with the combinatorial and linear programming aspects of the skew-symmetric flow theory, respectively. Section 5 develops the general blocking flow method for regular flows and Section 6 deals with improved performance of the method on special classes of networks. Section 7 gives a reductions from the b-matching problem to the skew-symmetric flow problem. In Section 8 we give concluding remarks.

---

[3] This bound was conjectured in [12].

2 Background

A digraph $G = (V, E)$ is *skew-symmetric* if there is a mapping σ of $V \cup E$ onto itself such that: σ is an involution (*i.e.*, $\sigma(x) \neq x$ and $\sigma(\sigma(x)) = x$ for any $x \in V \cup E$); for every $v \in V$, $\sigma(v) \in V$; and for every $a = (v, w) \in E$, $\sigma(a) = (\sigma(w), \sigma(v))$. We assume that a description of G includes σ. In this paper, we shall usually use the term *symmetric* instead of skew-symmetric. We say that the node $\sigma(v)$ is symmetric to v, and the arc $\sigma(a)$ is symmetric to a; symmetric elements are also called *mates*. The mate $\sigma(x)$ of an element x will often be denoted by x'. We extend the symmetry σ to paths (cycles) in a natural way by saying that two paths (cycles) are symmetric if the elements of one of them are symmetric to those of the other and go in the reverse order. Note that G cannot contain non-trivial self-symmetric paths or cycles. Indeed, if $P = (x_0, a_1, x_1, \ldots, a_k, x_k)$ is such a path (cycle), choose arcs a_i and a_j such that $i \leq j$, $a_j = \sigma(a_i)$ and $j - i$ is minimum. Then $j > i + 1$ (as $j = i$ would imply $\sigma(a_i) = a_i$ and $j = i + 1$ would imply $\sigma(x_i) = x_{j-1} = x_i$). Now $\sigma(a_{i+1}) = a_{j-1}$ contradicts the minimality of $j - i$.

Note also that if $v' = \sigma(v)$ and $(v, v') \in E$, then there are even number of copies of (v, v') which are partitioned into pairs of mates. A function h on arcs is called *symmetric* if $h(a) = h(\sigma(a))$ for all $a \in E$.

Given a *source* $s \in V$, a *sink* $t \in V$, and a capacity function $u : E \to \mathbf{Z}_+$, a function $f : E \to \mathbf{R}_+$ is a *flow* if it obeys the capacity constraints $f(a) \leq u(a) \quad \forall a \in E$ and the conservation constraints

$$div_f(v) := \sum_{(u,v) \in E} f(u, v) - \sum_{(v,w) \in E} f(v, w) = 0 \quad \forall v \in V - \{s, t\}.$$

The *value* of a flow f is given by $|f| = div_f(s)$. An *IS-flow* abbreviates a *symmetric integer flow*, the main object that we study in this paper.

An input to the *maximum skew-symmetric flow problem (MSFP)* is a skew-symmetric graph G, a symmetric integer capacity function $u : E \to \mathbf{Z}_+$, a source s, and a sink $s' = \sigma(s)$. The goal is to find an IS-flow of maximum value. W.l.o.g., we may assume that no arc of G enters s; otherwise we can delete such arcs and the arcs leaving s', which produces an equivalent problem in a skew-symmetric graph.

Note that the integrality requirement is important: if we do not require f to be integral, then for any maximum flow f, the flow $(f + \sigma(f))/2$ is a symmetric maximum flow. For the rest of this paper, we consider only integral flows.

In what follows we will utilize some results on regular paths in skew-symmetric graphs. A *regular path (r-path)* is a path in G that does not contain a pair of symmetric arcs. Similarly, an *r-cycle* is a cycle that does not contain a pair of symmetric arcs. Suppose we are given two symmetric nodes s and s'. The *r-reachability problem* is to find an r-path from s to s' or a proof that there is none. Given a symmetric function of arc lengths, the *shortest r-paths problem* to find the shortest r-path from s to s' or a proof that there is none. The r-reachability problem can be solved in $O(m)$ time [3, 15]. For nonnegative lengths, the shortest r-path problem can be solved in $O(m \log n)$ time; for unit lengths, in $O(m)$ time [15].

2.1 Buds and Trimming Operation

Buds, introduced in the context of skew-symmetric paths [15], play an important role in our algorithms. A bud is a triple $\tau = (V_\tau, E_\tau, e_\tau = (v, w))$ such that

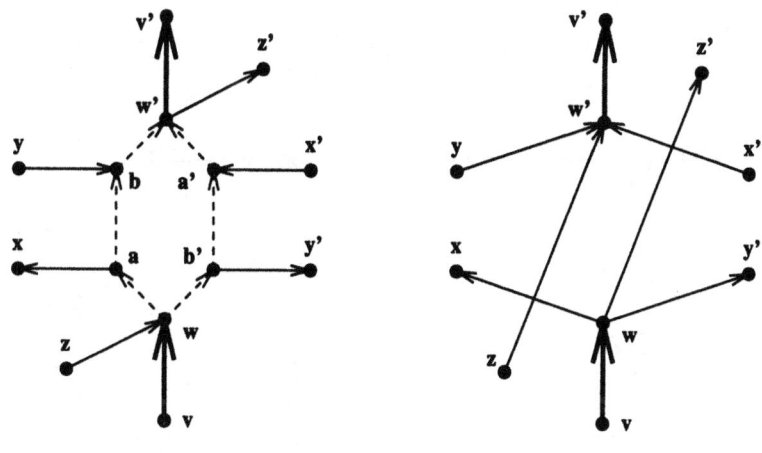

Before trimming   After trimming

Fig. 1. *Bud trimming example.*

(1) (V_τ, E_τ) is a symmetric subgraph of G with $s \notin V_\tau$.
(2) e_τ is an arc of G entering V_τ, i.e., $v \notin V_\tau \ni w$.
(3) For every node $x \in V_\tau$ there is an r-path from w to x in (V_τ, E_τ) (and therefore an r-path from x to $\sigma(w)$).
(4) There is an r-path from s to v which is node-disjoint from V_τ.

The node w is called the *base node* of the bud and the arc (v, w) is called the *base arc*. The node $\sigma(w)$ is called the *anti-base node* of the bud and the arc $\sigma(v, w)$ is called the *anti-base arc*.

Given a symmetric. graph $G = (V, E)$ with a bud $(V_\tau, E_\tau, e_\tau = (v, w))$, let $v' = \sigma(v)$, $w' = \sigma(w)$, and $e'_\tau = \sigma(e_\tau)$. The *trimming operation* transforms G into $\overline{G}$ with the node set $\overline{V} = V - (V_\tau - \{w, w'\})$ and arc set $\overline{E}$ constructed as follows.

1. Each arc $a = (x, y) \in E$ such that either $x, y \in V - V_\tau$, or $a = e_\tau$, or $a = e'_\tau$ remains in $\overline{E}$.
2. Each arc $(x, y) \in E - \{e'_\tau\}$ that leaves V_τ is replaced by an arc from w to y.
3. Each arc $(x, y) \in E - \{e_\tau\}$ that enters $V\tau$ is replaced by an arc from x to w'.

See Figure 1 for an example of bud trimming. Clearly $\overline{G}$ has a natural skew-symmetry.

A useful property of buds is as follows.

Lemma 1. *[15] There is an r-path from s to s' in G if and only if there is an r-path from s to s' in $\overline{G}$.*

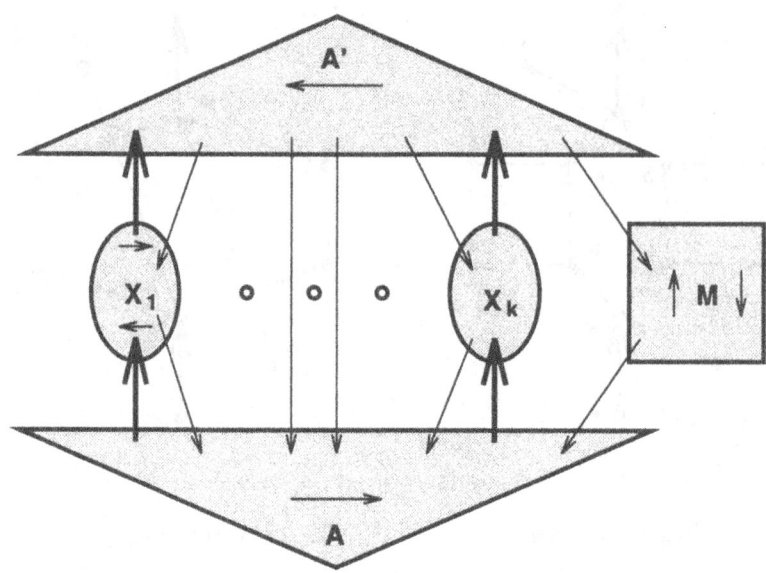

Fig. 2. *A barrier.*

2.2 Barriers

Barriers are dual to r-paths in a sense that either there is an r-path from s to $\sigma(s)$ or there is an s-barrier. Given a node $s \in V$, we say that $\mathcal{B} = (A; X_1, \ldots, X_k)$ is an *s-barrier* if the following conditions hold.

(B1) $A, X_1, \ldots, X_k$ are pairwise disjoint subsets of V, and $s \in A$.

(B2) For $A' = \sigma(A)$, $A \cap A' = \emptyset$.

(B3) For $i = 1, \ldots, k$, X_i is symmetric, *i.e.*, $\sigma(X_i) = X_i$.

(B4) For $i = 1, \ldots, k$, there is a unique arc, a_i, from A to X_i.

(B5) For $i, j = 1, \ldots, k$ and $i \neq j$, no arc connects X_i and X_j.

(B6) For $M = V - (A \cup X_1 \cup \ldots \cup X_k)$ and $i = 1, \ldots, k$, no arc connects X_i and M.

(B7) No arc goes from A to $A' \cup M$.

(Note that arcs from A' to A, from X_i to A, and from M to A are possible.) Figure 2 illustrates the definition. Unless mentioned otherwise, in this paper s and $s' = \sigma(s)$ are the source and the sink, respectively, and we refer to an s-barrier as barrier.

Theorem 2. *[15] There is an r-path from s to s' if and only if there is no barrier.*

3 Skew-Symmetric Flow Theory

In this section we extend the classic flow decomposition, augmenting path, and max-flow min-cut theorems of Ford and Fulkerson [13] to the skew-symmetric case. First we need a few definitions. Let u be an integral symmetric capacity function, and f be an IS-flow in G.

It is convenient to consider the graph $G^+ = (V, E^+)$ formed by adding a reverse arc to each arc of G. For $a \in E^+$, a^R denotes the corresponding reverse arc. The symmetry σ is extended to G^+ in a natural way. An arc $a \in E^+$ has the *residual capacity* $u_f(a)$ equal to $u(a) - f(a)$ if $a \in E$ and $f(a)$ otherwise. An arc $a \in E^+$ is called *residual* if $u_f(a) > 0$. We denote the set of residual arcs by E_f and form the *residual graph* $G_f = (V, E_f)$. We say that an arc a is *saturated* by f if $u_f(a) = 0$.

Given a symmetric function $h : E \to \mathbf{Z}_+$ (or $h : E^+ \to \mathbf{Z}_+$), we define the *split-graph* $G^{[h]} = (V, E^{[h]})$ by splitting all arcs $a \in E$ (or $a \in E^+$) into two arcs a_1 and a_2 with capacities $\lceil h(a)/2 \rceil$ and $\lfloor h(a)/2 \rfloor$, respectively, and then by deleting the arcs with zero capacity. We extend σ to $E^{[h]}$ in a natural way by defining $\sigma(a_1) = (\sigma(a))_1$ and $\sigma(a_2) = (\sigma(a))_2$.

We make extensive use of the graph $G^{[u_f]}$ generated by the residual capacity function u_f. The capacity function in $G^{[u_f]}$ is denoted by u'_f. An *augmenting r-path* for f is an s-s' r-path in $G^{[u_f]}$. If P is an augmenting r-path and δ is the minimum capacity of its arcs, then we can push δ units of flow through the path in G corresponding to P and then δ units through the path corresponding to $\sigma(P)$. Such an augmentation produces a feasible IS-flow in G and increases the value of f by 2δ.

An *elementary flow* is a triple (P, P', δ), where P is a simple cycle or a simple path from s to s' (not necessarily r-cycle or r-path), $P' = \sigma(P)$, and $\delta \in \mathbf{N}$. A set D of elementary flows is a *decomposition* of an IS-flow f if for all $a \in E$ we have

$$f(a) = \sum_{(P,P',\delta)\in D:\ a\in P} \delta + \sum_{(P,P',\delta)\in D:\ a\in P'} \delta.$$

The *symmetric decomposition theorem* is as follows.

Theorem 3. *An IS-flow f can be decomposed into at most m elementary flows.*

The proof of this theorem (omitted due to the lack of space) also gives a polynomial time algorithm for symmetric decomposition.

The decomposition theorem and the fact that the network has no self-symmetric paths imply the following lemma.

Lemma 4. *For any symmetric set $S \subseteq V$ and any IS-flow f, the total flow on the arcs entering S, as well as the total flow on the arcs leaving S, is even.*

The following symmetric augmenting paths theorem is easy to prove using the symmetric decomposition theorem.

Theorem 5. *An IS-flow f is maximum if and only if there is no augmenting r-path.*

The classical max-flow min-cut theorem states that the maximum flow value is equal to the minimum cut capacity. To state a skew-symmetric version of this theorem we need the notion of *odd s-barrier*. Given $s \in V$, we say that $\mathcal{B} = (A; X_1, \ldots, X_k)$ is an *odd s-barrier* if the following conditions hold.

(O1) $A, X_1, \ldots, X_k$ are pairwise disjoint subsets of V, and $s \in A$.
(O2) For $A' = \sigma(A)$, $A \cap A' = \emptyset$.
(O3) For $i = 1, \ldots, k$, X_i is symmetric, i.e., $\sigma(X_i) = X_i$.
(O4) For $i = 1, \ldots, k$, $u(A, X_i)$ is odd.
(O5) For $i, j = 1, \ldots, k$ and $i \neq j$, no arc connects X_i and X_j.
(O6) For $M = V - (A \cup X_1 \cup \ldots \cup X_k)$ and $i = 1, \ldots, k$, no arc connects X_i and M.

We refer to an odd s-barrier as odd barrier. We define *capacity* of an odd barrier $B = (A; X_1, \ldots, X_k)$ by $u(B) = u(A, V - A) - k$.

The following is the symmetric max-flow min-cut theorem.

Theorem 6. *The maximum IS-flow value is equal to the minimum odd barrier capacity.*

Proof. First we show that the capacity of any odd barrier is an upper bound on the value of an IS-flow f. By Lemma 4, the value of the flow going through each set X_i is even. In view of (O5)–(O6), this easily implies $f(A, X_i) - f(X_i, A)$ is at most $u(A, X_i) - 1$, and therefore the total value of f is at most $u(A, A - V) - k$.

Next we show that the two values are equal. Let f be a maximum IS-flow. By Theorem 5, $G^{[u_f]}$ contains no s-s' r-path, so it must contain a barrier $B = (A; X_1, \ldots, X_k)$.

Let a_i be the arc from A to X_i in $G^{[u_f]}$ defined as in (B4) (see Section 2). From (B4) and the construction of $G^{[u_f]}$ it follows that the residual capacity u_f of every arc from A to X in G^+ is equal to zero except for the arc corresponding to a_i, whose residual capacity is one. Hence,

(i) if a_i was formed by splitting an arc $\tilde{a}_i$ in G then $\tilde{a}_i$ goes from A to X_i, and $f(\tilde{a}_i) = u(\tilde{a}_i) - 1$;

(ii) if a_i was formed by splitting $\tilde{a}_i^R$ for some $\tilde{a}_i \in E$, then $\tilde{a}_i$ goes from X_i to A, and $f(\tilde{a}_i) = 1$;

(iii) all arcs from A to X_i in G except $\tilde{a}_i$ (in case (ii)) are saturated by f;

(iv) all arcs from X_i to A in G except $\tilde{a}_i$ (in case (iii)) are free of flow.

Furthermore, comparing arcs in $G^{[u_f]}$ and G, we observe that:

(v) property (B7) implies that the arcs from A to $A' \cup M$ are saturated and the arcs from $A' \cup M$ to A are free of flow;

(vi) property (B5) implies (O5) and (B6) implies (O6).

From (i)-(vi) it follows that B is an odd s-barrier in G and $|f| = u(B)$. $\blacksquare$

4 Integer and Linear Programming Formulations

We can state MSFP as an integer program in a straightforward way. To be consistent with the rest of the paper, we use function rather than vector notation. We also use the dot notation for inner products: given two functions g and h on a set S, we define $g \cdot h = \sum_{x \in S} g(x) h(x)$.

The integer program corresponding to MSFP is as follows.

$$\text{maximize } |f| = \sum_{(s,v) \in E} f(s, v) \qquad (4\text{-}1)$$

$$\text{subject to}$$

$$f(a) \geq 0 \qquad \forall a \in E \qquad (4\text{-}2)$$

$$f(a) \leq u(a) \qquad \forall a \in E \qquad (4\text{-}3)$$

$$\sum_{(u,v) \in E} f(u, v) - \sum_{(v,w) \in E} f(v, w) = 0 \qquad \forall v \in V - \{s, t\} \qquad (4\text{-}4)$$

$$f(a) - f(\sigma(a)) = 0 \qquad \forall a \in E \qquad (4\text{-}5)$$

$$f(a) \text{ integral} \qquad \forall a \in E \qquad (4\text{-}6)$$

(Recall that no arc of G enters s.)

We obtain an alternate linear programming formulation for MSFP by replacing the integrality condition (4-6) by certain linear constraints related to so-called fragments. This linear program and its dual (discussed below) are analogous to, but somewhat more complicated than, those for the usual maximum flow problem and its dual in [13].

A *fragment* is a pair $\tau = (V_\tau, E_\tau)$, where V_τ is a symmetric set of nodes with $s \notin V_\tau$, and E_τ is a subset of arcs entering V_τ such that the total capacity $u(E_\tau)$ is odd. The *characteristic function* χ_τ of τ is the function on E defined by

$$\chi_\tau(a) = \begin{cases} 1 \text{ if } a \in E_\tau \cup \sigma(E_\tau), \\ -1 \text{ if } a \in \delta(V_\tau) - (E_\tau \cup \sigma(E_\tau)), \\ 0 \text{ otherwise.} \end{cases}$$

Here $\delta(V_\tau)$ is the set of arcs with one end in V_τ and the other in $V - V_\tau$. We denote the set of fragments by $\mathcal{T}$.

Let f be a (feasible) IS-flow and $\tau \in \mathcal{T}$. Since u is symmetric, the definition of χ_τ shows that $f \cdot \chi_\tau \leq 2u(E_\tau)$. An elementary but important property of a fragment is that $f \cdot \chi_\tau$ is at most $2u(E_\tau) - 2$; it follows immediately from Lemma 4 and the fact that $u(E_\tau)$ is odd. This gives additional linear constraints for MSFP:

$$f \cdot \chi_\tau \leq 2u(E_\tau) - 2 \quad \forall \tau \in \mathcal{T}. \tag{4-7}$$

Note that if we add these constraints, we can drop the symmetry constraints (4-5) without changing the optimum value of the linear program. This fact follows from the proof of the following theorem.

Theorem 7. *Every maximum IS-flow is an optimal solution to the linear program (4-1)–(4-4), (4-7).*

Proof. Assign a dual variable $\pi(v) \in \mathbf{R}$ (a *potential*) to each node $v \in V$, $\gamma(a) \in \mathbf{R}_+$ (a *length*) to each arc $a \in E$, and $\rho(\tau) \in \mathbf{R}_+$ to each fragment $\tau \in \mathcal{T}$. For an arc $a = (v, w)$ define $\Delta_\pi(a) = \pi(w) - \pi(v)$ (the *potential difference*), and $q_{\gamma,\tau}(a) = \gamma(a) + \sum_{\mathcal{T}} \chi_\tau(a)\rho(\tau)$.

Consider the linear program:

$$\text{minimize } \psi(\pi, \gamma, \rho) = \sum_E u(a)\gamma(a) + \sum_{\mathcal{T}} (2u(E_\tau) - 2)\rho(\tau) \tag{4-8}$$

$$\text{subject to}$$

$$\gamma(a) \geq 0 \qquad \forall a \in E \tag{4-9}$$

$$\rho(\tau) \geq 0 \qquad \forall \tau \in \mathcal{T} \tag{4-10}$$

$$\pi(s) = 0 \tag{4-11}$$

$$\pi(s') = 1 \tag{4-12}$$

$$q_{\gamma,\tau}(a) - \Delta_\pi(a) \geq 0 \qquad \forall a \in E. \tag{4-13}$$

We observe that (4-1)-(4-4),(4-7) and (4-8)-(4-13) are mutually dual linear programs (up to taking each potential with the opposite sign and shifting it by the constant $1/2$). Therefore,

$$\max |f| = \min \psi(\pi, \gamma, \rho), \tag{4-14}$$

where the maximum and minimum range over the feasible solutions to (4-1)-(4-4),(4-7) and (4-8)-(4-13), respectively.

Next we show that every maximum IS-flow f achieves the maximum in (4-14). To see this, choose an odd barrier $\mathcal{B} = (A; X_1, \ldots, X_k)$ of minimum capacity $u(\mathcal{B})$. For $i = 1, \ldots, k$, let E_i be the set of arcs from A to X_i; then $\tau_i = (X_i, E_i)$ be a fragment for G, u. Define $\pi(v)$ to be 0 for $v \in A$, 1 for $v \in A'$, and 1/2 otherwise. Define $\gamma(a)$ to be 1 for $a \in (A, A')$, 1/2 for $a \in (A, M) \cup (M, A')$, and 0 otherwise. Define $\rho(\tau_i) = 1/2$ for $i = 1, \ldots, k$, and $\rho(\tau) = 0$ for the other fragments in G.

One can check that (4-13) holds for all a (e.g., $q_{\gamma,\rho}(a) = \Delta_\pi(a) = 1$ for $a \in (A, A')$ and $q_{\gamma,\rho}(a) = \Delta_\pi(a) = 1/2$ for $a \in (A, L) \cup (L, A')$, where $L = X_1 \cup \ldots \cup X_k \cup M$). Thus π, γ, ρ are feasible.

Finally, we observe that $u \cdot \gamma = u(A, A') + u(A, M)$ (using the fact that $u(A, M) = u(M, A')$) and that

$$\sum_{\mathcal{T}} (2u(E_\tau) - 2)\rho(\tau) = \sum_{i=1}^{k} \frac{1}{2}(2u(E_i) - 2) = \left(\sum_{i=1}^{k} u(A, X_k) \right) - k.$$

This implies $\psi(\pi, \gamma, \rho) = u(\mathcal{B})$, and now the result follows from Theorem 6. ∎

Given potentials $\pi(v)$, $v \in V$, and a function $\rho : \mathcal{T} \to \mathbf{R}_+$, we define the *reduced cost* function $c_\pi^\rho : E \to \mathbf{R}$ by

$$c_\pi^\rho(a) = \pi(w) - \pi(v) - \sum_{\mathcal{T}} \rho(\tau)\chi_\tau(a) \quad \text{for } a = (v, w) \in E.$$

Using reduced costs, we can get rid of the dual variables γ in (4-8)–(4-13) by doing the substitution $\gamma(a) = \max\{0, c_\pi^\rho(a)\}$. This gives an optimality criterion for MSFP as follows.

Theorem 8. *An IS-flow f is maximum if and only if there are potentials $\pi(v) \in \mathbf{R}$, $v \in V$, with $\pi(s) = 0$ and $\pi(s') = 1$, and a function $\rho : \mathcal{T} \to \mathbf{R}_+$, such that the following "complementary slackness" conditions hold:*

(CS1) $\forall \tau \in \mathcal{T}$, $\rho(\tau) > 0$ implies $f \cdot \chi_\tau = 2u(E_\tau) - 2$;
(CS2) $\forall a \in E$, $c_\pi^\rho(a) > 0$ implies $f(a) = u(a)$;
(CS3) $\forall a \in E$, $c_\pi^\rho(a) < 0$ implies $f(a) = 0$.

5 R-Blocking Flow Method

Given a network H, we say that a flow f in H is *blocking* if every path from s to s' in H contains a saturated arc. If H is symmetric, we say that an IS-flow f is *r-blocking* if every r-path from s to s' contains a saturated arc. In this section we describe our r-blocking flow algorithm for the MSFP (called RBFM). The algorithm is a generalization of the blocking flow algorithm of Dinitz [7], and the complexity of these algorithms is the same. At the end of the section we show how the time bound for the MSFP problem can be improved further.

The RBFM algorithm maintains an IS-flow f. Initially f is the zero flow. The algorithm applies the following two steps until there is no augmenting r-path with respect to f:

1. Construct the auxiliary network $H_f = (V, E(H_f))$.
2. Find an r-blocking IS-flow g in H and let g' be the corresponding function on E. Set $f \leftarrow f + g'$.

In the Dinitz' algorithm, the auxiliary network can be defined as a subgraph of G_f induced by zero reduced cost arcs (with respect to the shortest path distances from s for unit arc lengths). This network is acyclic, which is crucial for the correctness and the running time of the algorithm. If we use the same definition in the IS-flow case, however, the resulting network may have cycles. We define the auxiliary network H_f as a subgraph of the trimmed graph $\overline{G_f}$ computed by the shortest r-path algorithm. The construction of H_f is described in detail in Section 5.2, where we also show how an IS-flow g in H_f can be extended to the corresponding function g' in G. We will prove two key properties of H_f: (i) H_f is acyclic and (ii) if g' corresponds to an r-blocking flow g in H_f, then the length of a shortest augmenting r-path for $f + g'$ is greater than the length of a shortest augmenting r-path for f. Note that the latter property implies a bound on the number of iterations of the algorithm.

Lemma 9. *The RBFM algorithm terminates in $O(n)$ iterations.*

An algorithm for finding an r-blocking IS-flow in a symmetric acyclic network is described in Section 5.3. This algorithm is based on Dinitz' algorithm and the r-reachability algorithm [15].

5.1 Buds in Split-Graph

Our implementation of steps 1 and 2 of the RBFM algorithm use the r-reachability and the shortest r-path algorithms of [15] on split-graphs. We take advantage of the structure of these graphs to maintain the *unit base invariant:* the residual capacity of base arcs of all buds trimmed by these algorithms is one. This guarantees that an IS-flow in the trimmed graph can be extended to an IS-flow in the original graph.

Given a split-graph $G^{[h]}$, we maintain G and h, which implicitly represent $G^{[h]}$. Note that if P is a simple path in G^+, then every arc a on P with $h(a) > 1$ corresponds to two parallel arcs a_1 and a_2 in $G^{[h]}$. Even if $a' = \sigma(a)$ is on P, we can put arcs a_1 and $a_2' \neq \sigma(a_1)$ on the corresponding path in $G^{[h]}$. The only way P does not correspond to an r-path in $G^{[h]}$ is when P contains arcs $a, a' = \sigma(a)$ with $h(a) = 1$. Using this observation, one can easily modify the r-path algorithms to maintain the unit base invariant. We discuss such a modification of the r-reachability algorithm in Section 5.3; the modification of the shortest r-path algorithm is similar.

5.2 The Auxiliary Network

To build the auxiliary network $H_f = (V, E(H_f))$, we use the shortest r-paths algorithm from [15] on $G^{[u_f]}$ with the unit length function ℓ. Because of the unit input lengths, the implementation of the algorithm that uses buckets runs in linear time.

The shortest paths algorithm finds a path P and a dual solution ℓ_π^ρ, where π is a potential function and ρ is a function on shortest path fragments (which are similar to, but not the same as, the flow fragments of Section 4) proving optimality of the path in the graph $\overline{G^{[u_f]}}$ with buds corresponding to positive values of ρ trimmed. In the postprocessing stage the algorithm extends P and ℓ_π^ρ to $G^{[u_f]}$.

We construct H_f from the graph $\overline{G^{[u_f]}}$: $E(H_f)$ consists of all arcs a of $\overline{G^{[u_f]}}$ with $\ell_\pi^\rho(a) = 0$. Note that $E(H_f)$ contains all arcs which are on shortest r-paths from s to s' and that any s–s' r-path in H_f is a shortest r-path.

Lemma 10. *The auxiliary graph H_f is acyclic.*

Proof. Consider a cycle Γ in $\overline{G^{[u_f]}}$. We claim that $\ell_\pi^\rho(\Gamma) = \ell(\Gamma)$. This implies the lemma.

To see the claim, first note that the changes in ℓ_π^ρ due to π cancel out as we go around the cycle. Next consider the changes due to ρ. Note that only trimmed bud arcs are affected by these changes, and each time a cycle intersects a trimmed bud, it either enters through the base arc and exits through an arc which is not the base and not the anti-base arc, or enters through such an arc and exits through the anti-base arc. In both cases, the contributions due to ρ cancel out. ∎

Let g be a blocking flow in H_f. Since the buds trimmed by the shortest r-path algorithm are nested, each bud contains an r-path from its base w to $\sigma(w)$, and the unit base invariant is maintained, the corresponding IS-flow $\bar{g}$ in G_f can be found in linear time using techniques similar to those used in [15] for extending the shortest path in the trimmed graph to the shortest path in the original graph. The IS-flow $\bar{g}$ gives the desired function g' in G.

The proof of the next lemma is based on the understanding of how the residual graph and the function ℓ_π^ρ change when f is augmented.

Lemma 11. *If g' corresponds to an r-blocking flow in H_f and $f' = f + g'$, then the shortest augmenting r-path for f' is longer than the shortest augmenting r-path for f.*

5.3 Finding R-Blocking Flows

In this section we describe an algorithm *RBF* for finding an r-blocking IS-flow in an acyclic skew-symmetric network $H = (V, E(H))$ with a symmetric capacity function $h : E(H) \rightarrow \mathbf{Z}_+$. The algorithm is based on Dinitz' blocking flow algorithm [7] and the r-reachability algorithm [15]. Let g denote the flow being constructed in H. Initially g is the zero flow.

A simple $O(m^2)$ RBF algorithm works by successive augmentations. Given g, consider the "forward" residual graph $F = (V, E(F))$, where $E(F) = \{a \in E(H) | h(a) - g(a) > 0\}$. If $F^{[h-g]}$ does not contain an augmenting r-path with respect to g, then g is a blocking IS-flow and the algorithm terminates.

Otherwise we augment g in a way similar to that in the proof of Theorem 3 Let P be an augmenting r-path in $F^{[h-g]}$ and let $P' = \sigma(P)$. Let $\tilde{P}$ and $\tilde{P}'$ be the images of P and P' in G. For an arc $a \in \tilde{P} \cup \tilde{P}'$, let $\epsilon(a)$ be $f(a)$ if a occurs in exactly one of $\tilde{P}, \tilde{P}'$, and $\lfloor f(a)/2 \rfloor$ otherwise. Define $\delta = \min\{\epsilon(a) : a \in \tilde{P} \cup \tilde{P}'\}$. The augmentation increases g on each arc of $\tilde{P}$ by δ, and then increases g on each arc of $\tilde{P}'$ by δ. It is easy to see that $\delta > 0$ and that the resulting g is a feasible IS-flow.

Note that an augmentation either saturates an arc of H or reduces the residual capacity of an arc of H to one. The same holds for the symmetric arc. Therefore the simple RBF algorithm does at most m augmentations. Since the path P can be found in $O(m)$ time, the algorithm runs in $O(m^2)$ time.

Next we describe an RBF algorithm that runs in $O(nm)$ time. We start with a review of the r-reachability algorithm [15]. The algorithm maintains a set A of nodes reachable from s by r-paths, with the r-paths represented by a spanning tree $T \subseteq E$ of A rooted at s. The algorithm also maintains $A' = \sigma(A)$ and

$T' = \sigma(T)$. By symmetry, from every node $v' \in A'$ there is an r-path to s' in T'. The invariant $A \cap A' = \emptyset$ always holds. Initially $A = \{s\}$ and $T = \emptyset$.

The algorithm searches the graph as follows. At each step it scans an arc (v, w) with $v \in A$ and $w \notin A$. If $w \notin A'$, then w is added to A, (v, w) to T, $\sigma(w)$ to A', and $\sigma(v, w)$ to T'. Now suppose $w \in A'$ and let P be the concatenation of the s to v path P_1 in T, (v, w), and the w to s' path P_2 in T'. If P is regular, then the algorithm terminates and returns P. If P is not regular, then there is a bud τ containing a subpath of P; the algorithm trims τ and updates A, A' and T, T'. If there is no arc (v, w) with $v \in A$ and $w \notin A$, the algorithm concludes that s' is not reachable and terminates.

The following modifications transform the r-reachability algorithm into the r-blocking flow algorithm.

The first modification maintains the unit base invariant. We search the split-graph $H^{[h]}$ represented by H and h. The algorithm scans arcs in $E(H)$. Suppose the arc a with head node w is scanned. Recall that if $h(a) > 1$ then there are two arcs, a_1 and a_2, corresponding to a in the split-graph, and if $h(a) = 1$ then there is just one arc, a_1, and that $\sigma(a_i) = \sigma(a)_i$. If w is added to A and $h(a) > 1$, then a_1 is added to T and $\sigma(a)_2$ is added to T'; otherwise a_1 is added to T and $\sigma(a)_1$ is added to T'. (Note that T and T' are no longer symmetric, but their images in H are symmetric.) The modification works correctly because of a is split into a_1 and a_2 and there is an s to s' r-path using a_2, the symmetric r-path uses a_1 and any r-path using a_1 does not use $\sigma(a)_1$. Since the only arcs in T with symmetric arcs in T' have unit capacities, the modification maintains the unit base invariant.

The second modification is to search the graph in the depth-first manner. (The r-reachability algorithm works in linear time regardless of the search strategy as long as the selection is efficient.) We maintain a stack that initially contains s. Let v be the node on top of the stack. If v has no outgoing arcs left, v is popped from the stack and deleted from the graph. The symmetric node v' is also deleted. If the stack is empty, there is no augmenting r-path.

If v has outgoing arcs, we examine such an arc (v, w). Note that $w \notin A$ since the graph is acyclic and we delete nodes after depth-first search processes them. If $w \notin A'$, we add w to the stack and start scanning the arcs out of w. Otherwise $w \in A'$; we either discover an augmenting r-path or trim a bud. In the former case we do an augmentation and decrease h. In the latter case, a group of nodes on top of the stack which the trimming operation merges into the base of the bud is replaced by the base node and the search continues.

If an augmenting r-path Γ is discovered, an augmentation is performed similar to the simple RBF algorithm above. The only difference is that we are working in the trimmed graph. After the RBF algorithm terminates, the blocking flow it constructs is in the trimmed graph. At the end of the algorithm, we expand the buds contracted by the algorithm and extend the blocking flow to H.

Correctness of the RBF algorithm follows from the unit base invariant, Lemma 1, and the observation that when a node v is popped from the stack, there is no s to s' r-path through v (and by symmetry no r-path through $\sigma(v)$).

Next we analyze the algorithm's running time.

The following lemma follows from the fact that the RBF algorithm runs in $O(m + A)$ time, where A is the total number of arcs on all augmenting paths used by the algorithm.

Lemma 12. *The RBF algorithm runs in $O(nm)$ time.*

Lemmas 9 and 12 yield the following theorem.

Theorem 13. *The RBFM algorithm solves MSFP in $O(n^2 m)$ time.*

5.4 Improved Bounds

Next we show that the $O(n^2m)$ bound can be improved to $O(M(n,m) + nm)$, where $M(n,m)$ is the time needed to solve the integer maximum flow problem on a network with n nodes and m arcs. The improvement is achieved by running the RBFM algorithm with a good initial solution.

To obtain the initial solution, we round the arc capacities down to the nearest multiple of two and find a maximum flow g in the resulting network such that all values of g are even. Define $f = (g + \sigma(g))/2$. Clearly f is an IS-flow in the original network, and the value of f is at most m units away from the optimal value.

Suppose we run the RBFM algorithm with f as the initial solution. Because of our choice of f, the number of augmentations done by the algorithm in all r-blocking flow computations is at most $m/2$, since each augmentation increases the flow value by at least two. By the proof of Lemma 12, the total work involved in the r-blocking flow computations is $O(nm)$. Since the auxiliary network construction takes linear time, we have the desired result.

Theorem 14. *The MSFP problem can be solved in $O(M(n,m) + nm)$ time.*

6 Special Networks

The $O(n^2m)$ time bound for Dinitz' maximum flow algorithm can be improved for several important classes of networks [11, 12, 19, 20]. In this section we use similar techniques to obtain the same improvements for the RBFM algorithm.

Proof of the following lemma is based on the symmetric decomposition theorem.

Lemma 15. *On networks with unit arc capacities, the RBFM algorithm runs in $O(m^{1.5})$ time.*

We define *capacity of a node* y by

$$u(y) = \min \left(\sum_{(x,y) \in E} u(x,y), \sum_{(y,z) \in E} u(y,z) \right).$$

A network with *unit node capacities* is a network in which all nodes in $V - \{s, s'\}$ have capacity of one.

The proof of the following lemma is similar to that of the previous lemma.

Lemma 16. *On networks with unit node capacities, the RBFM algorithm runs in $O(\sqrt{n}m)$ time.*

We say that a network is *bipartite* if $V - \{s, s'\}$ can be partitioned into two sets X and Y such that every arc in E goes from s to X, from X to Y, or from Y to s'.

Next we show how to use the graph compression techniques of Feder and Motwani [12] to improve the bound of Lemma 16 to $O\left(\sqrt{n}m \log(n^2/m)/\log n\right)$ for bipartite networks with unit capacities and no arcs between symmetric nodes.

Let $V = \{s, t\} \cup X \cup Y$, where arcs of G go from s to X, from X to Y, or from Y to t. Suppose that the subgraph induced by $X \cup Y$ has a bipartite clique Q induced by L, R where $L \subseteq X$, $R \subseteq Y$. Since the network has no arcs between symmetric nodes, the clique $Q' = \sigma(Q)$ is disjoint from Q. We compress the cliques as follows. We delete the arcs of Q and Q', add two symmetric nodes q and q', and add symmetric arcs (x, q), $(q', \sigma(x))$ for each $x \in L$, and symmetric arcs (q, y), $(\sigma(y), q')$ for each $y \in R$. Note that there is a one-to-one correspondence between IS-flows in the original and the transformed networks. If the input graph is dense enough, this transformation can be used to reduce the number of arcs.

It is shown in [12] how to transform a bipartite graph with n nodes and m arcs into a graph with $m^* = O(m \log(n^2/m)/\log n)$ arcs in $o(\sqrt{n}m \log(n^2/m)/\log n)$ time by compressing large enough cliques as described above. The compression algorithm can be easily modified so that when a clique is found and compressed, the symmetric clique is compressed as well.

Let Z be the set of nodes added while compressing cliques. The results of [12] imply that $|Z| = o(m/\sqrt{n})$. Thus we can have an asymptotic increase in the number of nodes but not in the graph size. An iteration of the RBFM algorithm on the compressed graph takes $O(m \log(n^2/m)/\log n)$ time. To bound the number of iterations, note that for any IS-flow f in the compressed graph, augmenting r-paths cannot share nodes in $X \cup Y$ and an augmenting r-path with k arcs must pass through at least $(k-1)/2$ nodes in $X \cup Y$. An argument similar to that of Theorem 16 shows that the number of iterations is $O(\sqrt{n})$.

Thus we have the following result.

Lemma 17. *On bipartite networks with unit node capacities and no arcs between symmetric nodes, MSFP can be solved in* $O\left(\sqrt{n}m \log(n^2/m)/\log n\right)$ *time.*

7 Relationship to Matching

Given an undirected graph $G' = (V', E')$ with no self-loops, a *matching* M is a subset of edges such that no two edges share a node. The *maximum matching problem* is to find a matching M whose cardinality $|M|$ is as large as possible.

A more general problem is the *capacitated b-matching problem* [9, 23]. The input to this problem includes a supply function $b : V' \to \mathbf{Z}_+$ and a capacity function $u' : E' \to \mathbf{Z}_+$. A *b-matching* is a function $h : E' \to \mathbf{Z}_+$ that satisfies the supply constraints

$$\sum_{\{v', w'\} \in E'} h(v', w') \leq b(v') \qquad \forall v' \in V'$$

and the capacity constraints $h(e) \leq u'(e) \qquad \forall a \in E'$. The value of a b-matching h is $h(E) = \sum_{E'} h(e)$. The goal is to find a b-matching of the maximum value. The maximum matching problem is a special case of the capacitated b-matching problem with all supplies and capacities equal to one.

The capacitated b-matching problem and the skew-symmetric flow problem can be reduced to each other without increasing the problem size by more than a constant factor.

The capacitated b-matching problem can be reduced to the skew-symmetric flow problem in a natural way as described below (in fact, such a reduction is well-known for special cases; see *e.g.* [1, 3, 22]).

Given an instance of the capacitated b-matching problem, we construct a digraph $G = (V, E)$ and a capacity function u on E as follows.

- For every $v \in V'$, V contains two symmetric nodes v_1 and v_2.
- In addition, V contains two symmetric nodes s and s' (the source and the sink).

- For every $\{v, w\} \in E'$, E contains two symmetric arcs (v_1, w_2) and (w_1, v_2) with $u(v_1, w_2) = u(v_2, w_1) = u'(v, w)$.
- For every $v \in V'$, E contains two symmetric arcs (s, v_1) and (v_2, s') with capacity $u(s, v_1) = u(v_2, s') = b(v)$.

Obviously, there is a one-to-one correspondence between b-matchings in G' and IS-flows in G, and the value of an IS-flow is twice the value of the corresponding b-matching. Thus a solution to the MSFP yields a solution to the b-matching problem.

It is easy to see that the reduction from the maximum matching problem yields the problem with unit node capacities. Since G' has no self-loops, G has no arcs between symmetric nodes. Thus, Lemma 16 implies the following result which proves the conjecture of Feder and Motwani [12] and improves the $O(\sqrt{n}m)$ bound on the cardinality matching problem.

Theorem 18. *The cardinality matching problem can be solved in* $O(\sqrt{n}m \log(n^2/m)/\log n)$ *time.*

The reduction from the skew-symmetric flow problem to the capacitated b-matching problem is also very natural.

Given an instance of the skew-symmetric flow problem, $(G = (V, E)$, $s, s' = \sigma(s)$, and $u)$ we construct an instance $(G' = (V', E')$, u', *and* b$)$ of the capacitated b-matching problem as follows.

- For each pair of mates $\{v_1, v_2 = \sigma(v)\}$ such that $v_1 \in V$ $\{v_1, v_2\}! = \{s, s'\}$ we add a node v to V'.
- For each pair of arcs (v_1, w_1) and $\sigma(v_1, w_1)$ such that $v_1, w_1 \notin \{s, s'\}$ we add an edge $\{v, w\}$ to E' and define $u'(v, w) = u(v_i, w_1)$.
- We extend u to $V \times V$ by defining $u(v_1, w_1) = 0$ if $(v_1, w_1) \notin E$. For each $v' \in V'$ we define $b(v')$ to be $|u(s, v_1) - u(v_1, s')|$.

It is easy to see that a maximum b-matching in this network gives a maximum skew-symmetric flow in the original network.

8 Concluding Remarks

The cardinality matching algorithm of Blum [3, 4] is very similar to the special case of our RBFM algorithm on networks with unit node capacities. However, the algorithm does not explicitly maintain buds, and the paper [3, 4] does not address the maximum IS-flow problem.

We conjecture that the dynamic tree data structure [6] can be used to find an r-blocking IS-flow in $O(m \log(n^2/m))$ time, matching the blocking flow bound of [18]. However, the results of Section 5.4 allow us to get the corresponding bound for MSFP using the dynamic tree data structure indirectly (in a maximum flow algorithm).

Finally, we would like to mention that the results of this paper can be extended to the minimum-cost skew-symmetric flows (which generalize minimum-cost flows and weighted matchings).

References

1. G. M. Adel'son-Vel'ski, E. A. Dinits, and A. V. Karzanov. *Flow Algorithms.* Nauka, Moscow, 1975. In Russian.

2. R. K. Ahuja, J. B. Orlin, and R. E. Tarjan. Improved Time Bounds for the Maximum Flow Problem. *SIAM J. Comput.*, 18:939–954, 1989.
3. N. Blum. A New Approach to Maximum Matching in General Graphs. In *Proc. ICALP*, pages 586–597, 1990.
4. N. Blum. A New Approach to Maximum Matching in General Graphs. Technical report, Institut für Informatik der Universität Bonn, 1990.
5. J. Cheriyan, T. Hagerup, and K. Mehlhorn. Can a Maximum Flow be Computed in $o(nm)$ Time? In *Proc. ICALP*, 1990.
6. D. D. Sleator and R. E. Tarjan. Self-adjusting binary search trees. *J. Assoc. Comput. Mach.*, 32:652–686, 1985.
7. E. A. Dinic. Algorithm for Solution of a Problem of Maximum Flow in Networks with Power Estimation. *Soviet Math. Dokl.*, 11:1277–1280, 1970.
8. J. Edmonds. Paths, Trees and Flowers. *Canada J. Math.*, 17:449–467, 1965.
9. J. Edmonds and E. L. Johnson. Matching, a Well-Solved Class of Integer Linear Programs. In R. Guy, H. Haneni, and J. Schönhein, editors, *Combinatorial Structures and Their Applications*, pages 89–92. Gordon and Breach, NY, 1970.
10. J. Edmonds and R. M. Karp. Theoretical Improvements in Algorithmic Efficiency for Network Flow Problems. *J. Assoc. Comput. Mach.*, 19:248–264, 1972.
11. S. Even and R. E. Tarjan. Network Flow and Testing Graph Connectivity. *SIAM J. Comput.*, 4:507–518, 1975.
12. T. Feder and R. Motwani. Clique Partitions, Graph Compression and Speeding-up Algorithms. In *Proc. 23st Annual ACM Symposium on Theory of Computing*, pages 123–133, 1991.
13. L. R. Ford, Jr. and D. R. Fulkerson. *Flows in Networks*. Princeton Univ. Press, Princeton, NJ, 1962.
14. H. N. Gabow and R. E. Tarjan. Faster scaling algorithms for general graph-matching problems. *J. Assoc. Comput. Mach.*, 38:815–853, 1991.
15. A. V. Goldberg and A. V. Karzanov. Path Problems in Skew-Symmetric Graphs. Technical Report STAN-CS-93-1489, Department of Computer Science, Stanford University, 1993.
16. A. V. Goldberg and A. V. Karzanov. Path Problems in Skew-Symmetric Graphs. In *Proc. 5th ACM-SIAM Symposium on Discrete Algorithms*, pages 526–535, 1994.
17. A. V. Goldberg and R. E. Tarjan. A New Approach to the Maximum Flow Problem. *J. Assoc. Comput. Mach.*, 35:921–940, 1988.
18. A. V. Goldberg and R. E. Tarjan. Finding Minimum-Cost Circulations by Successive Approximation. *Math. of Oper. Res.*, 15:430–466, 1990.
19. A. V. Karzanov. O nakhozhdenii maksimal'nogo potoka v setyakh spetsial'nogo vida i nekotorykh prilozheniyakh. In *Matematicheskie Voprosy Upravleniya Proizvodstvom*, volume 5. Moscow State University Press, Moscow, 1973. In Russian; title translation: On Finding Maximum Flows in Network with Special Structure and Some Applications.
20. A. V. Karzanov. Tochnaya otzenka algoritma nakhojdeniya maksimalnogo potoka, primenennogo k zadache "o predstavitelyakh". In *Problems in Cibernetics*, volume 5, pages 66–70. Nauka, Moscow, 1973. In Russian; title translation: The exact time bound for a maximum flow algorithm applied to the set representatives problem.
21. V. King, S. Rao, and R. Tarjan. A Faster Deterministic Maximum Flow Algorithm. *J. Algorithms*, 17:447–474, 1994.
22. E. L. Lawler. *Combinatorial Optimization: Networks and Matroids*. Holt, Reinhart, and Winston, New York, NY., 1976.
23. L. Lovász and M. D. Plummer. *Matching Theory*. Akadémiai Kiadó, Budapest, 1986.
24. S. Micali and V. V. Vazirani. An $O(\sqrt{|V|}|E|)$ algorithm for finding maximum matching in general graphs. In *Proc. 21th IEEE Annual Symposium on Foundations of Computer Science*, pages 17–27, 1980.
25. V. V. Vazirani. A Theory of Alternating Paths and Blossoms for Proving Correctness of the $O(\sqrt{V}E)$ General Graph Maximum Matching Algorithm. *Combinatorica*, to appear.

Certificates and Fast Algorithms for Biconnectivity in Fully-Dynamic Graphs

Monika R. Henzinger[1] * and Han La Poutré[2] **

[1] Department of Computer Science, Cornell University, Ithaca, NY 14853, USA.
[2] Department of Computer Science, Utrecht University, Utrecht, The Netherlands.

Abstract. In this paper, we present sparse certificates for biconnectivity together with algorithms for updating these certificates. We thus obtain fully-dynamic algorithms for biconnectivity in graphs that run in $O(\sqrt{n \log n} \log \lceil \frac{m}{n} \rceil)$ amortized time per operation, where m is the number of edges and n is the number of nodes in the graph. This improves upon the results in [11], in which algorithms were presented running in $O(\sqrt{m})$ amortized time, and solves the open problem to find certificates to speed up biconnectivity, as stated in [2].

1 Introduction

The field of dynamic graph algorithms has become an important field in algorithmic research in recent years. Currently, several results exist for incremental and fully-dynamic graph problems, like for maintaining spanning trees, the 2-edge- or the 2-vertex-connected components of a graph, or the planarity of a graph under the insertions and/or deletions of edges and vertices [3, 4, 5, 6, 7, 8, 9, 10, 11, 12].

In [4, 5, 11], algorithms for maintaining minimum spanning trees and the connectivity, 2-edge-connectivity and the 2-vertex-connectivity relations in fully-dynamic graphs were presented that run in $O(\sqrt{m})$ time per operation (amortized time for 2-vertex-connectivity). (In this paper, n is the number of nodes and m is the number of edges.) In the meantime, in [2], the concept of certificates and sparsification trees (for definitions, see Section 2) was introduced to speed up several fully-dynamic graph algorithms. In particular, sparse certificates could be used for speeding up fully-dynamic algorithms for maintaining minimum spanning trees and the connectivity and 2-edge-connectivity relations in graphs [4, 5], to $O(\sqrt{n})$ time per operation. Basically, these sparse certificates were defined in terms of (successive, minimum) spanning trees, and were maintained by applying Fredericksons minimum spanning trees data structure [4]. The algorithm to be speeded up was thus used on the resulting certificate for the whole graph (viz., in the root of the sparsification tree).

---

* Maiden name: Monika H. Rauch. This research was supported in part by the NSF Career Award 27813.
** Full name: Johannes A. La Poutré. The research of the author has been made possible by a fellowship of the Royal Netherlands Academy of Arts and Sciences (KNAW).

However, this approach seems not to work for certificates for biconnectivity. Like for many other problems involving k-connectivity, such as designing static, parallel, incremental, or fully-dynamic algorithms, 2-vertex-connectivity appears to be substantially harder to deal with than 2-edge-connectivity. An example of this can also be observed in [1], considering parallel algorithms for k-connectivity, where sparse *static* certificates for biconnectivity are defined in terms of breadth-first trees, whereas those for 2-edge-connectivity can consist of any kind of spanning trees. Up to now, no efficient fully-dynamic algorithms for maintaining breadth-first trees are known, and it is commonly felt that these trees are hard to maintain indeed. So, designing certificates for fully-dynamic biconnectivity and thus speeding up the fully-dynamic algorithms for it is an appealing open problem [1, 2, 11].

In this paper, we present new, sparse certificates for biconnectivity that can efficiently be maintained under insertion and deletion of edges. We thus obtain $O(\sqrt{n \log n} \log \lceil \frac{m}{n} \rceil)$ amortized time algorithms for maintaining the biconnectivity relation in fully-dynamic graphs, and therefore show that fully-dynamic biconnectivity falls in the same (current) time complexity class as fully-dynamic 2-edge-connectivity.

We introduce sparse certificates that are determined in a relaxed, history-dependent way, and that thus are not defined in a more mathematical, "static" way, like in [2]. In particular, the certificates we use are not "stable", as defined and used throughout in [2] (see also Definition 1). Also, we develop extensions of the algorithms that must be speeded up themselves [11] as well, to be able to maintain the certificates. Thus, our approach also appears to be the first combination of certificates and maintenance algorithms that also uses the algorithm that must be "speeded up" itself to maintain the certificates, and not just Fredericksons minimum spanning tree algorithms.

In our solutions, the (on-line) sequence of update operations is split into two parts, which are treated separately: one concerning deletions and one concerning insertions. Viz., we typically start with subgraphs of which we only want to maintain the certificates in a *decremental* way: after any (delete) operation, roughly either the number of biconnected components increases, or we find some edges to "repair" the biconnectivity relation in the certificate of the subgraph (with some additional constraints). Insertions of edges are then temporarily processed outside the sparsification paradigm, and from time to time handled further in a batch-like way.

Thus, our relaxation of (non-stable) certificates, usage of sparsification trees, and the separation of decremental subproblems are new. We conjecture, that our approach may lead to fast algorithms for various other fully-dynamic graph problems as well, like k-vertex connectivity for $k \geq 3$.

The paper is organized as follows. Section 2 contains the preliminaries, including a description of sparsification trees. In Section 3, we define certificates for biconnectivity, and give algorithms for maintaining them. In Section 4, the replacement data structure, used in Section 3, is described. In this extended abstract, we omit many details and special cases.

2 Preliminaries

For a graph G, two nodes x and y are called *biconnected* (or 2-vertex-connected) if the deletion of a node from G does not separate x and y in G (i.e., x and y are still connected).

For a tree T, we denote by $\pi_T(x, y)$ the (unique) path between nodes x and y in T.

We give some definitions concerning certificates, as occurring in [1, 2]. For any graph property P, and graph G, a *certificate* for G is a graph G' such that G has property P if and only if G' has property P. A *strong* certificate for G is a graph G' such that, for any graph H, $G \cup H$ has property P if and only if $G' \cup H$ has property P.

A property is said to have *sparse* certificates if there is some constant c such that for every graph G on an n-vertex set, we can find a strong certificate for G with at most cn edges.

In [2], some sparse certificates are given for k-edge-connectivity, minimum spanning trees, and bipartiteness, all depending on the data structures for minimum spanning trees presented by Frederickson [4, 5]. Also, the concept of stable certificate is defined, which is important for the use of certificates in sparsification trees, as follows.

Definition 1 *Let A be a function mapping graphs to strong certificates. Then A is* stable *if it has the following properties:*

1. For any graphs G and H, $A(G \cup H) = A(A(G) \cup H)$.

2. For any graph G and edge e in G, $A(G - e)$ differs from $A(G)$ by $O(1)$ edges.

However, for fully-dynamic biconnectivity, this definition seems to be too strict. Amongst others, it supposes that for each graph, a (unique) certificate can be chosen, which then has to be maintained by the dynamic algorithms with only $O(1)$ changes (see [2]). We seem to need a more liberal concept of certification.

2.1 Sparsification tree

We sketch how certificates can be used in sparsification trees [2] to maintain a property P. Our final strategies will be somewhat different, though.

As in [2], maintain a partition of the graph edges in $\lceil \frac{m}{n} \rceil$ groups, all but one containing exactly n edges. The remaining group is called the small group.

Insertion of an edge in the graph is always done in the small group. Deletion of an edge is performed in the proper group, after which an edge from the small group is transferred to this group (via a deletion and insertion in these groups, respectively). If the small group becomes empty, we delete it; if it contains n edges and we want to insert a new edge, we start a new group.

We form and maintain a *sparsification tree*, which is a binary tree of height $O(\log \lceil \frac{m}{n} \rceil)$, with $\lceil \frac{m}{n} \rceil$ leaves corresponding to the $\lceil \frac{m}{n} \rceil$ groups. Each node x in the sparsification tree corresponds to a subgraph $G(x)$ of G formed by the edges in the groups at the leaves that are the descendants of that node x. Also, to node x, a subgraph $S(x)$ of $G(x)$ is related, such that $S(x)$ has property P iff $G(x)$ has property P ($S(x)$ is actually a sparse certificate for $G(x)$). (If x is a leaf, then $S(x) = G(x)$.) We say that the edges in $S(x)$ are the *edges related to* x. Furthermore, a sparse certificate $C(x)$ (for $G(x)$ or $S(x)$) is maintained. The subgraph $S(x)$ related to node x is found by forming the union of the certificates $C(y)$ and $C(z)$ of the two child nodes y and z. The certificate $C(x)$ is found and maintained in $S(x)$. Thus, the certificate at the root is the certificate of the whole graph. For further details and elaboration, we refer to [2].

3 Certificates for biconnectivity

In this section, we present sparse certificates for biconnectivity together with algorithms for updating these certificates, and handling sparsification trees.

3.1 Monotone sparse certificates and their maintenance

Consider a graph G of n nodes. For a spanning tree T of graph G and a sequence of (ordered) edges e_i ($1 \le i \le l$), the extension graph TT_j ($0 \le j \le l$) consists of all the edges e_i with $1 \le i \le j$. We call the sequence of edges an *add-on sequence* if for every j, $0 \le j < l$, the graphs $T \cup TT_j$ and $T \cup TT_{j+1}$ have a different number of biconnected components. Hence, the number of components in TT_{j+1} is at least one smaller than the number of components in $T \cup TT_j$. We call $T \cup TT_l$ the add-on graph.

Lemma 2 *Let G be a graph, T a spanning tree of G, and S an add-on sequence for biconnectivity. Then S contains at most $n - 1$ edges.*

Proof. $TT_0 = T$ contains exactly n different biconnected components. Since any graph contains at least one biconnected component, it follows by the definition of add-on sequence that S contains at most $n - 1$ edges. □

We first describe a certificate and the maintenance of it for some appropriate graph G.

A *dynamic colour partition* of G colours the edges of G from three colours viz. blue, red, and green.

For convenience in this abstract, we assume in this subsection that G always is a connected graph (this is not essential, though). Blue and red edges have a *cost* related to them, which is 0 for blue edges and which is a natural number otherwise. No two red edges have the same cost.

At any time, the blue edges form a spanning tree of G, and the red edges are $O(n)$ non-tree edges that form an add-on sequence if ordered according to their cost. In addition, this add-on sequence is *maximal*, i.e., such that the add-on graph contains the same (number of) biconnected components as G. The blue and red edges together form the certificate for G. Hence, this is a *sparse certificate* for G.

We give algorithms for maintaining the sparse certificate of G under deletions and certain insertions of edges. The edge insertions are only allowed if they do not change the biconnectivity relation. Therefore, the changes in the biconnectivity relation are monotone for a sequence of these operations, i.e., biconnected components are only split up and are never joined. Thus, we call this sparse certificate a *monotone sparse certificate*. (We want to point out that the restriction on insertions only applies to maintaining the certificates themselves, and not to the overall algorithms presented in Subsection 3.2.)

Here and in the sequel, we use (implicitly) an operation that turns a given green edge into a red edge, while simultaneously labelling this edge with a cost which is the number of preceding operations.

Initialisation. We can construct a sparse certificate for G as follows. Initially, all edges are of G are green. First, a breadth-first search tree of G is constructed, and all the edges in it are made blue. This blue tree is denoted by T_b. The graph G_r consisting of red edges is then incrementally constructed as follows. First, a breadth-first search forest B of $G \setminus T_b$ is made and the edges are enumerated in some way. A (green) edge of B is converted into a red edge only if it changes the number of components of $T_b \cup G_r$. This can be done by maintaining the biconnected components dynamically [7, 12], thus yielding an add-on sequence indeed. As was shown in [1], the graph $T_b \cup B$ is a certificate for biconnectivity, hence, so is $T_b \cup G_r$. (We do not really need the breadth-first forest, but can do this with all the existing edges once they are ordered as well.)

Insertions and Deletions. The algorithms for maintaining the certificate processes deletions and insertions as follows.

Edge insertions into G are only allowed if they do not change the biconnectivity relation. The algorithm inserts a new edge as a green edge.

When an edge is deleted, the following may happen to the colour of edges. After the deletion of one red edge, some green edges may become red. After the deletion of a blue edge, one red edge may become blue and some green edges may become red. And finally, the deletion of a green edge is done without consequences for other edges and their colours.

For the deletion of an edge, we have the following more detailed strategy. Whenever a blue edge e is deleted, it is replaced by the minimum cost red edge e' which restores a blue spanning tree of G. This is processed by first interchanging the colours of e and e', and subsequently deleting the (now) red edge e, as in the case below.

Whenever a red edge (x, y) is deleted, a node a that now is a new articulation point for x and y in the *current* $T_b \cup G_r$ is generated. Such a node is called a *potential* articulation point. Then, for potential articulation point a, it is tested whether it is also an articulation point of G: If it is, we call it a *definite* articulation point. Otherwise, (i.e., if not,) a green edge (x, y) is turned red such that a is not an articulation point in $T_b \cup G_r$ any more, and G_r is updated accordingly. Successively, the next potential articulation point is generated for the current (possibly updated) $T_b \cup G_r$ (if any), and the above process is repeated until no unprocessed potential articulation points for x and y exist.

Lemma 3 *At any moment, the current red edges (in the order of their costs) form an add-on sequence for biconnectivity.*

Proof. We distinguish the three basic changes with respect to red and blue edges.

1. Obviously, if a red edge is deleted from the add-on sequence, the remaining sequence still is an add-one sequence.

2. When a red edge $e = (x, y)$ and a blue edge $d = (u, v)$ interchange colours, then $d = (u, v) \in \pi_{T_b}(x, y)$. We show that replacing e by d in the add-on sequence (while interchanging the numbers related to these edges as well) yields another add-on sequence.

Let $S = e_1, ..., e_k$ be the red add-on sequence before deletion of d, where $e = e_j$. Let T_1 and T_2 be the two subtrees remaining after deleting d from T_b, and let T_b' be $T_1 \cup T_2 \cup \{(x, y)\}$ (the new spanning tree).

Let the sequence S' equal S where e is replaced by d. Then S' is an add-on sequence for T_b', which is seen as follows. Note that $T_b \cup TT_j$ equals $T_b' \cup TT_j'$. Hence, $T_b \cup TT_i = T_b' \cup TT_i'$ for $i \geq j$, and hence the graphs $T_b' \cup TT_i'$ and $T_b' \cup TT_{i+1}'$ have a different number of biconnected components, for $j \leq i < k$. Furthermore, for $i < j$, an edge e_i must have both its end points in either T_1 or T_2, because $e = e_j$ was the minimum cost edge connecting T_1 and T_2. Therefore, since $T_b \cup TT_i$ and $T_b \cup TT_{i+1}$ have a different number of biconnected components for $0 \leq i < j$, and since T_b and T_b' equal on their subtrees T_1 and T_2 and differ only in the edges e and d, it follows that the graphs $T_b' \cup TT_i'$ and $T_b' \cup TT_{i+1}'$ have a different number of biconnected components, for $0 \leq i < j$.

Hence, the (new) sequence of red edges is an add-on sequence for T_b'.

3. When a green edge $e = (x, y)$ is turned red, the thus extended new sequence of red edges obviously is an add-on sequence again, since otherwise this edge would not have been added. □

Corollary 4. *During a sequence of d deletions of edges, at most $d + 2n - 2$ edges have been blue or red for some time.*

Proof. The total number of blue or red edges that have existed (as blue or red edge) is at most $2(n - 1)$ (the final number of blue and red edges) plus d (the number of deleted edges). □

We refer to the above algorithms for maintaining the certificates as *certificate algorithms*.

Data structures for monotone sparse certificates. To determine which edges change colour, as described above, we use the following dynamic data structures, which allow deletions, restricted insertions, and colour changes of edges as described above.

1. We keep the graph $T_b \cup G_r$ in a dynamic minimum spanning tree data structure of Frederickson [4] with all blue edges having cost 0. Whenever a blue edge is deleted, this data structure provides us with the minimum cost red edge. This edge becomes blue.
2. We keep $T_b \cup G_r$ in a dynamic biconnectivity data structure of Rauch [11]. Whenever a blue or red edge is deleted, this data structure provides us with the new potential articulation points generated one at a time, in time proportional to the number of actually generated potential articulation points.
3. We keep G in a dynamic biconnectivity data structure of Rauch [11]. This allows to test for every new potential articulation point a of $T_b \cup G_r$ whether it is also an articulation point of G. If not, some green edge (x, y) such that $a \in \pi_{T_b}(x, y)$ should become red, as above.
4. We keep a *replacement data structure* that provides such a green edge. It is described in the next section.

By [11], the operations on the data structures 1 and 2 can be performed in $O(\sqrt{n})$ time per returned edge or node, since $T_b \cup G_r$ has $O(n)$ edges, and the operations on data structure 3 can be performed in $O(\sqrt{m})$ time, while the operations on data structure 4 can be performed in $O(\sqrt{m \log n})$ time, by Theorem 9.

In the following, a certificate operation denotes the insertion, deletion or colour change of an edge, the generation of a new potential articulation point or the other queries on the data structures as described above.

Lemma 5 *A certificate operation can be performed in $O(\sqrt{m \log n})$ time.*

3.2 The overall algorithms and data structures

We can now combine the above certificate and certificate algorithms with sparsification as follows. (For terminology, we refer to Subsection 2.1.)

The edges in the graph are partitioned into yellow and non-yellow. The sparsification tree "contains" only non-yellow edges. (A non-yellow edge has (local) colour(s) defined at relevant sparsification nodes.) Each leaf corresponds to a group of at most n non-yellow edges. At *each* node x of the sparsification tree, the above certificate algorithms are executed locally on the graph $S(x)$, to maintain the certificate $C(x)$ in $S(x)$. Note that, thus, $S(x)$ has a colour partition, with just the colours blue, red, and green, where the colours are defined "locally" in $S(x)$ (independent of the actual colours red and blue in e.g. $C(y)$ and $C(z)$ in the children y and z, if any), and where thus $C(x)$ consists of the blue and red edges of $S(x)$.

Whenever a new edge e is inserted in the graph, keep it as a yellow edge in the "yellow delay set" Y. If the total number of yellow edges exceeds $O(n)$, we create a new leaf x with $S(x)$ consisting of these yellow edges, then colour these edges green and run the certificate initialization algorithm on $S(x)$. Subsequently, we "build/rebuild" every node on the root path of x.

Whenever an edge e is deleted, it is deleted from the proper leaf x, i.e., from $S(x)$. The deletion from x might cause that at most $n-1$ green edges turn red at x. Let y be the parent of x in the sparsification tree. First add all the new red edges of x as green edges to y, called "forced insertions". (Note that forced insertions do not change the biconnectivity relation of $S(y)$.) Subsequently, if y contains e, delete e from y (where again, green edges may turn red) and (recursively) repeat this procedure for the parent of node y, until e is not deleted from a tree node any more or until we reach the root.

Whenever there are more than $2\lceil \frac{m}{n} \rceil$ leaves, rebuild the entire sparsification tree.

Thus, the certificate $C(G)$ for G is given by $C(root) \cup Y$, i.e., the certificate related to the root, joined with the existing yellow edges in the yellow delay set. This obviously is a sparse certificate again. We run the biconnectivity algorithms of [11] on this for maintaining $C(G)$ and for computing the queries asked.

Lemma 6 *Let x be a node of the sparsification tree. At any time during a sequence S of operations, the number of edges related to x is at most $4(n-1)$, and the number of red and blue edges is at most $2(n-1)$. During sequence S, at most $d + 2n - 2$ edges have been blue or red for some time, and at most $d + 4n - 4$ edges have been related to x, where d is the number of edge deletions in the node x.*

Note that (new) green edges are added to a sparsification node only if they do not change the biconnectivity relation indeed (at the very moment of their insertion), as mentioned before.

Lemma 7 *For a sparsification node x, since its last (re)building, the total number of new, definite articulation points in $S(x)$ is $O(n)$, and the total number of certificate operations in $S(x)$ is at most $O(n+d)$, where d is the number of edge deletions in node x.*

Proof. The first statement follows since definite articulation points do not vanish any more. For the second statement, we charge the cost of obtaining a potential articulation point that vanishes again by making a green edge red, to this colour change; We observe that colour changes only occur from green to red to blue; and we use Lemma 6. □

Theorem 8. *There exists a fully-dynamic algorithm for biconnectivity in graphs such that a sequence of s operations starting from the empty graph takes $O(s.\sqrt{n \log n} \log \lceil \frac{m}{n} \rceil)$ time, while each query takes $O(1)$ time, and where n is the current number of nodes. Thus, each operation takes $O(\sqrt{n \log n} \log \lceil \frac{m}{n} \rceil)$ amortized time, and each query takes $O(1)$ worst-case time.*

Proof. (Sketch) The number of certificate operations in a sparsification node x since its last (re)building is $O(n+d)$, where d is the number of (local) deletions in that node. Charge the $O(n)$ operations to the rebuilding (initialisation) at x, and charge each occurring local deletion for one operation. Thus, each "global" deletion is charged for $O(\log \frac{m}{n})$ certificate operations in total (viz., for the sparsification nodes that all lie on a root path). Furthermore, maintaining the graph $S(root) \cup Y$ (of $O(n)$ edges) takes time proportional to maintaining $S(root)$ and $O(1)$ amortized certificate operations per yellow edge. We charge the cost of the former to maintaining $S(root)$ and the cost of the latter to the insertion of a yellow edge, viz., $O(1)$ certificate operations per operation. Finally, each insertion of an edge is now charged for $O(\log \frac{m}{n})$ amortized operations in total, by the delay-build/rebuild strategy. Each such certificate operation takes $O(\sqrt{n \log n})$, since the occurring graphs have $O(n)$ edges, and by [11] and Theorem 9.

Adding up the number of certificate operations and using [11], yields the theorem. □

4 The replacement data structure

In this section, we describe the replacement data structure, as mentioned in Subsection 3.1.

We are given a graph G of blue, red, and green edges such that the blue edges form a spanning tree T_b of G (for convenience, we assume that G is connected). We refer to an edge of T_b just as *tree edge* and denote the tree path $\pi_{T_b}(x, y)$ just by $\pi(x, y)$.

Let P be a path in T_b and let b_1, a, and b_2 be three consecutive vertices on P such that a is an articulation point in $T_b \cup G_r$ that separates b_1 and b_2. We say a green edge e *covers* a on P iff a does not separate b_1 and b_2 in $T_b \cup G_r \cup e$.

We describe the functionality of the replacement data structure. On the one hand, it is able to perform the deletions, restricted insertions, and colour changes of edges as described in Subsection 3.1. On the other hand, it can return the appropriate green edges as described in Subsection 3.1. We describe this operation in more detail.

Let the red edge (u, v) have to be deleted. The replacement data structure is given a (newly generated) potential articulation point a of $\pi(u, v)$ on the current $T_b \cup G_r$ that is *not* an articulation point on $\pi(u, v)$ in G (called *candidate*). Then for candidate a, it outputs a green edge covering a on $\pi(u, v)$.

In this section, we take that each blue or red edge has cost 0 and each green edge has cost 1.

Now, first assume we would maintain the following (too costly) data structure, namely, for each node x the minimum spanning tree $F(x)$ of $G \setminus x$. Note that two neighbours of x are biconnected iff they are connected in $F(x)$. Since the blue and red edges form a sparse certificate of G before an edge deletion, all edges in F are blue or red. The deletion of a blue or red edge removes at most one edge from F and adds at most one, potentially green, edge. This edge is a green edge covering a, i.e. it is an edge that the replacement data structure is looking for.

However, it is too expensive to maintain for each node x the minimum spanning tree $F(x)$ of $G \setminus x$. Thus, we decompose the graph into subgraphs, called *clusters*, which are connected by blue edges, and maintain for each cluster a data structure similar to the one described above. Since the nodes in a cluster are connected by a spanning tree containing only blue edges, we have to focus on edges between clusters.

4.1 Graph decomposition

We expand every node of G with degree $d > 3$ into d nodes that are connected by a chain of $d - 1$ *dashed* edges. We naturally expand T_b to be a spanning tree of the expanded graph G' (where all dashed edges thus are in T_b).

We decompose G' as in [11]. A *cluster* is a set of vertices that induces a connected subgraph of T. An edge is *incident* to a cluster if exactly one of its endpoints is in the cluster. A *restricted partition of order k* with respect to T is a partition of the vertices so that

1. Each set in the partition is a cluster that is incident to ≤ 3 tree edges and contains $\leq k$ vertices.
2. A cluster that is incident to 3 tree edges contains exactly one vertex.
3. If a cluster is incident to a dashed edge, then all tree edges incident to the cluster are incident to the same vertex of G.
4. No two adjacent clusters can be combined and still satisfy 1 to 3.

The partition splits G into $O(m/k)$ clusters of size $\leq k$ and is found in time $O(m+n)$ [5]. We denote by $C(x)$ a cluster containing a representative of a vertex x and say that $C(x)$ *contains x*. If the representatives of x are contained in > 1 clusters, then x is a *shared vertex* and all clusters $C(x)$ are called *x-clusters* and *share x*. By condition 3 every cluster shares ≤ 1 vertex. Since each dashed edge between two clusters is a tree edge, there are $O(m/k)$ shared vertices.

This decomposition induces the following *graph H of clusters*. Two vertices C and C' of H are connected by an edge (respectively blue edge) if and only if there is an edge (respectively blue edge) between a vertex of C and a vertex of C'. If there is no blue edge between a vertex of C and C', but a red edge, then C and C' are connected by a red edge. If there is neither a blue nor a red, but a green edge between them, they are connected by a green edge.

We denote by H_{br} the subgraph of H induced by blue and red edges. The blue edges form a spanning tree of H and of H_{br}. Two clusters that are adjacent in this spanning tree are called *tree neighbours*.

4.2 Outline of the data structure

We maintain both H and H_{br} dynamically in the high-level data structure of [11]. For a graph H' (with H' either H or H_{br}), the high-level data structure of [11] maintains for each cluster C in H' the following graph $H'(C)$: $H'(C)$ contains a node for each tree neighbour of C. There is an edge between two tree neighbours L_1 and L_2 of C iff there is an edge in $H' \setminus C$ between the subtree containing L_1 and the subtree containing L_2. The data structure in [11] can be extended to label an edge (L_1, L_2) in $H'(C)$ with an edge in $H' \setminus C$ connecting the subtree of L_1 with the subtree of L_2.

As shown in [11], all graphs $H'(C)$ can be maintained in time $O(k + m/k)$ per edge insertion in G or edge deletion in G. Augmenting the data structure to label edges of $H'(C)$ with edges of H' does not increase the running time. A connectivity query in $H'(C)$ can be answered in constant time. Note that two tree neighbours of C are connected in $H'(C)$ iff they are biconnected in H.

We use the high-level data structures of H' to

1. test in constant time if two tree neighbours of a cluster C are biconnected in H',
2. output all biconnected tree neighbours in H' of a cluster C in time linear in their number and a spanning forest connecting them in $H'(C)$,
3. for a shared vertex s, test in constant time if two non s-clusters that are tree neighbours of s-clusters are biconnected in H', and
4. for a shared vertex s, output all biconnected non s-clusters that are tree neighbours in H' of a s-cluster, together with a spanning forest connecting them in $H' \setminus \{C, s \in C\}$.

If there is an edge between L_1 and L_2 in $H(C)$, but not in $H_{br}(C)$, then all edges in $H \setminus C$ between the subtree containing L_1 and the subtree containing L_2 are green. Thus, using the high-level data structure of H and of H_{br}, we can determine the minimum cost edge in H connecting the subtree containing L_1 and the subtree containing L_2. Given the spanning forests connecting the tree neighbours of C in $H(C)$ and $H_{br}(C)$, we can compute a minimum spanning forest connecting the tree neighbours of C in $H(C)$ in time linear in the number of tree neighbours. A query to determine this minimum spanning forest is called a *minimum spanning query for C*.

We state some observations. An edge deletion can disconnect the edges in a cluster and thus force the split of a cluster. Furthermore, if a blue tree edge is deleted and a non-blue edge becomes blue, the tree degree of a cluster which contains more than one node can become three: Thus the cluster has to be split into up to five clusters, up to four of with tree degree 2 and one with tree degree 3. This can also create one *new* shared vertex. Finally, if an edge is inserted, this can give the above changes in the cluster partition as well.

To guarantee that the number of clusters is $O(m/k)$, we rebuild the data structure every m/k deletions or insertions. This takes time $O(m)$ and adds an amortized cost of $O(k)$ to each operation.

As mentioned above, new shared vertices can be created during the sequence of operations. To distinguish them from the others, we call all shared vertices that are not new, *old*.

In the next subsections, we will describe how we find a green covering edge for different types of candidates. Again, we have the high-level data structures of [11] as the base, but we modify and extend it. We will consider a candidate a, where b_1 and b_2 are the neighbours of a on $\pi(u, v)$. We distinguish between a being a "non-shared vertex", a "new shared vertex", or an "old shared vertex."

4.3 Non-shared candidates

To cover a non-shared candidate a with a green edge e, we first build a graph $G(a)$ of size $O(k)$ and then we determine e using $G(a)$. Both steps take time $O(k + m/k)$. We do this as follows.

Let $C(a)$ be incident to $j \leq 3$ tree edges. The graph $G(a)$ contains as nodes all the nodes of $C(a)$ and one additional node for each tree neighbour L_j of $C(a)$, called *cluster-node*. The graph $G(a)$ contains as edges (1) all the edges between two nodes of $C(a)$, (2) an edge (x, L_j) for each edge (x, y) incident to a node $x \in C(a)$ if $\pi(x, y)$ contains the tree edge between $C(a)$ and L_j, and (3) the edges of a minimum spanning tree in $H(C(a))$ (between the tree neighbours of $C(a)$).

The graph $G(a)$ can be built as follows: The edges of (1) can be found by inspecting the edges incident to nodes of $C(a)$ in time $O(k)$. The edges of (3) can be found in time $O(1)$ using a minimum spanning query for C. To determine the edges of (2), we first compute a prefix and a postfix numbering of the clusters in H using the blue spanning tree in time $O(m/k)$. Then we scan all edges incident to nodes in $C(a)$ to find every edge (x, y) with $x \in C(a)$ and $y \notin C(a)$. For each such edge we compare the prefix and postfix number of $C(y)$ with the prefix and postfix numbers of the tree neighbours of $C(a)$. This determines in constant time the edge (x, L_j) that belongs to $G(a)$. Since $O(k)$ edges have at least one endpoint in $C(a)$, the total time is $O(k + m/k)$.

Note that b_1 and b_2 are either contained in $G(a)$ or represented by nodes L_j. Since a is a candidate, b_1 and b_2 or their representatives are disconnected in the graph induced by the blue and red edges of $G(a) \setminus a$, but connected in $G(a) \setminus a$. Thus, there exists a cut in $G(a) \setminus a$ separating b_1 and b_2 or their representatives that only contains green edges. Any one edge e of the cut covers a in $G(a)$. The cut can be found in time $O(k)$ by executing a BFS from b_1 or its representative. If e is an edge between two neighbour clusters L_1 and L_2, the corresponding "original" green edge is the label of e as given by the high-level data structure.

4.4 New shared candidates

This case is similar to the case of non-shared candidates, with the appropriate generalizations. Afterwards, we keep and maintain $G(a)$ as in Subsection 4.5. We refer to the full paper.

4.5 Old shared vertices

To cover a candidate that is an old shared vertex a, we would like to build the same graph $G(a)$ as for a non-shared vertex. However, an old shared vertex can be shared by all clusters and, thus, $G(a)$ can consist of m edges. Thus we cannot afford to construct $G(a) \setminus a$ whenever we have to cover a, but we have to maintain dynamically a data structure that represents $G(a) \setminus a$ instead.

The data structure consists of two *shared graphs*, $G_1(a)$ and $G_2(a)$. (They correspond to $G(a)$ and $\tilde{G}(a)$ in [11].) We maintain the minimum spanning tree of $G_1(a)$ dynamically and we build $G_2(a)$ whenever a is a candidate.

Basically, $G_1(a)$ corresponds to $G(a) \setminus a$ with some edges between cluster-nodes missing. Thus, some cluster-nodes are not connected in $G_1(a)$, even though they are connected in $G(a) \setminus a$. To "add" the missing edges, we build the graph $G_2(a)$ whenever we answer a query.

The data structure and algorithms basically follow ideas from [11]. We will not give a detailed description of the data structure and the corresponding algorithms here, but we refer to the full paper.

4.6 Complexity and Correctness

In the full paper, we prove that the data structure determines in time $O(m/k \cdot \log n)$ an edge that covers an old shared candidate; that it can be updated in amortized time $O(m/k + k)$ after each update operation in G; and that it can be built in time $O(m)$. By choosing $k = \sqrt{m \log n}$, we obtain the following theorem.

Theorem 9. *The given data structure determines in time $O(\sqrt{m \log n})$ an edge that covers a candidate. It can be updated in amortized time $O(\sqrt{m \log n})$ after each update operation in G and it can be built in time $O(m)$.*

5 Conclusion

We have presented certificates for biconnectivity together with algorithms for updating these certificates. We thus obtained fully-dynamic algorithms for biconnectivity in graphs that run in $O(\sqrt{n \log n} \log \frac{m}{n})$ amortized time per operation. We used novel techniques and approaches to handle sparsification.

We conjecture that our liberalization of certificates and their maintenance, and the usage of monotone subproblems, are important for other dynamization problems, like k-vertex-connectivity for $k \geq 3$.

References

1. J. Cheriyan and R. Thurimella, "Algorithms for parallel k-vertex connectivity and sparse certificates" *Proc. 23rd Annual Symp. on Theory of Computing*, 1991, 391-401.
2. D. Eppstein, Z. Galil, G. F. Italiano, A. Nissenzweig, "Sparsification - A technique for speeding up dynamic graph algorithms" *Proc. 33nd Annual Symp. on Foundations of Computer Science*, 1992, 60–69.
3. D. Eppstein, Z. Galil, G. F. Italiano, and T. Spencer. "Separator based sparsification for dynamic planar graph algorithms". *Proc. 25th Annual Symp. on Theory of Computing*, 1993, 208–217.
4. G. N. Frederickson, "Data Structures for On-line Updating of Minimum Spanning Trees" *SIAM J. Comput.* 14 (1985), 781–798.
5. G. N. Frederickson, "Ambivalent Data Structures for Dynamic 2-edge-connectivity and k smallest spanning trees" *Proc. 32nd Annual IEEE Symp. on Foundation of Comput. Sci.*, 1991, 632–641.

6. M.H. Rauch Henzinger and V. King, "Randomized Dynamic Graph Algorithms with Polylogarithmic Time per Operation" *Proc. 27th Annual Symp. on Theory of Computing*, 1995, 519-527.

7. J.A. La Poutré, "Dynamic Graph Algorithms and Data Structures" *Ph.D. Thesis*, Utrecht University, 1991.

8. J.A. La Poutré, "Alpha-Algorithms for Incremental Planarity Testing" *Proc. 26 Annual Symp. on Theory of Computing*, 1994, 706-715.

9. J.A. La Poutré and J. Westbrook, "Dynamic Two-Connectivity with Backtracking" *Proc. 5th Annual ACM-SIAM Symp. on Discrete Algorithms*, 1994, 204-212.

10. M. H. Rauch. "Fully Dynamic Biconnectivity in Graphs" *Proc. 33nd Annual Symp. on Foundations of Computer Science*, 1992, 50–59.

11. M. H. Rauch. "Improved Data Structures for Fully Dynamic Biconnectivity" *Proc. 26 Annual Symp. on Theory of Computing*, 1994, 686–695. An improved version is published as Technical Report 94-1412, Department of Computer Science, Cornell University, Ithaca, NY.

12. J. Westbrook, R. E. Tarjan, "Maintaining bridge-connected and biconnected components on-line" *Algorithmica* 7 (1992), 433–464.

On the All-Pairs Shortest Path Algorithm of Moffat and Takaoka[*]

Kurt Mehlhorn and Volker Priebe[**]

Max-Planck-Institut für Informatik, Im Stadtwald, 66123 Saarbrücken, Germany

Abstract. We review how to solve the all-pairs shortest path problem in a non-negatively weighted digraph with n vertices in expected time $O(n^2 \log n)$. This bound is shown to hold with high probability for a wide class of probability distributions on non-negatively weighted digraphs. We also prove that for a large class of probability distributions $\Omega(n \log n)$ time is necessary with high probability to compute shortest path distances with respect to a single source.

1 Introduction

Given a complete digraph in which all the edges have non-negative length, we want to compute the shortest path distance between each pair of vertices. This is one of the most basic questions in graph algorithms, since a variety of combinatorial optimization problems can be expressed in these terms. As far as worst-case complexity is concerned, we can solve an n-vertex problem in time $O(n^3)$ by either Floyd's algorithm [3] or by n calls of Dijkstra's algorithm [2]. Fredman's algorithm [4] uses efficient distance matrix multiplication techniques and results in a running time of $O(n^3((\log \log n)/\log n)^{1/3})$ (slightly improved to $O(n^3((\log \log n)/\log n)^{1/2})$ by Takaoka [14]). Recently, Karger, Koller, and Phillips [8] presented an algorithm that runs in time $O(nm^* + n^2 \log n)$, where m^* denotes the number of edges that are a shortest path from their source to their target.

However, worst-case analysis sometimes fails to cover the advantages of algorithms that perform well in practice; average-case analysis has turned out to be more appropriate for these purposes. We are not only interested in algorithms with good expected running time but in algorithms that finish their computations within a certain time bound *with high probability* (and might therefore be called reliable).

Two kinds of probability distributions on non-negatively weighted complete digraphs have been considered in the literature. In the so-called *uniform model*, the edge lengths are independent, identically distributed random variables. In

---

[*] This work is supported by the ESPRIT II Basic Research Actions Program of the EC under contract no. 7141 (project ALCOM II) and the BMFT-project "Softwareökonomie und Softwaresicherheit" ITS 9103

[**] Research supported by a Graduiertenkolleg graduate fellowship of the Deutsche Forschungsgemeinschaft. E-mail: priebe@mpi-sb.mpg.de

the so-called *endpoint-independent model*, a sequence c_{vj}, $1 \leq j \leq n$, of n non-negative weights is fixed for each vertex v of K_n arbitrarily. These weights are assigned randomly to the n edges with source v, i.e., a random injective mapping π_v from $[1..n]$ to V is chosen and c_{vj} is made the weight of edge $(v, \pi_v(j))$ for all j, $1 \leq j \leq n$.

Frieze and Grimmett [6] gave an algorithm with $O(n^2 \log n)$ expected running time in the uniform model when the common distribution function F of the edge weights satisfies $F(0) = 0$, $F'(0)$ exists, and $F'(0) > 0$. Under these assumptions, $m^* = O(n \log n)$ with high probability and so the algorithm of Karger et al. also achieves running time $O(n^2 \log n)$ with high probability.

The endpoint-independent model is much more general and therefore harder to analyze. Spira [12] proved an expected time bound of $O(n^2 (\log n)^2)$, which was later improved by Bloniarz [1] to $O(n^2 \log n \log^* n)$. (We use log to denote logarithms to base 2 and ln to denote natural logarithms; $\log^* x := 1$ for $x \leq 2$ and $\log^* x := 1 + \log^* \log x$ for $x > 2$.) In [10] and [11], Moffat and Takaoka describe two algorithms with an expected time bound of $O(n^2 \log n)$. The algorithm in [11] is a simplified version of [10]. In this paper,

- we present an even simpler version of the algorithm in [11] and also correct a small oversight in the analysis given by Moffat and Takaoka.
- Moreover, we prove that the running time of the modified version is $O(n^2 \log n)$ with high probability.
- We show that under modest assumptions $\Omega(n \log n)$ edges need to be inspected to compute the shortest path distances with respect to a single source.

2 Preliminaries

We mention some results from discrete probability theory. Suppose that in a sequence of independent trials, the probability of success is $\geq p$ for each of the trials. Then the expected number of trials until the first successful one is $\leq 1/p$. In the so-called *coupon collector's problem*, we are given a set of n distinct coupons. In each trial, a coupon is drawn (with replacement) uniformly and independently at random. Let X denote the number of trials required to have seen at least one copy of each coupon. By the above argument, $E[X] = \sum_{0 \leq i < n} \frac{n}{n-i} \sim n \ln n$. Actually, it is rather unlikely that we deviate from the expected number of trials by more than a constant multiplicative factor, since the probability that a particular coupon has not been collected after r trials equals $(1 - \frac{1}{n})^r$. Hence, for any $\beta > 1$,

$$\Pr(X > \beta n \ln n) \leq n \left(1 - \frac{1}{n}\right)^{\beta n \ln n} \leq n e^{-\beta \ln n} = n^{-(\beta-1)} . \qquad (1)$$

For a problem of size n, we will say that an event occurs *with high probability*, if it occurs with probability $\geq 1 - O(n^{-C})$ for an arbitrary but fixed constant C and large enough n. For example, (1) tells us that the number of trials in the coupon collector's problem is $O(n \ln n)$ with high probability.

2.1 A Probabilistic Experiment

We will refer to the following probabilistic experiment: An urn contains n balls that are either red or blue; let m be the number of red balls. The balls are repeatedly drawn from the urn (without replacement) uniformly and independently at random. For $1 \leq k \leq m$, let the random variable W_k denote the waiting time for the k-th red ball. In addition, we define the random variables Y_i, $1 \leq i \leq m$, by $Y_1 := W_1$ and $Y_i := W_i - W_{i-1}$ for $2 \leq i \leq m$. Note that both the W_k's and the Y_i's are not independent, e.g., $W_m = \sum_{1 \leq i \leq m} Y_i \leq n$ whereas each Y_i can take values in $\{1, \ldots, n - m + 1\}$. It is shown in the appendix that the Y_i's are exchangeable random variables; in particular, for any i with $1 \leq i \leq m$,

$$E[Y_i] = \frac{n+1}{m+1} \ . \tag{2}$$

It will prove convenient to normalize the Y_i's. For $1 \leq i \leq m$, define random variables $Z_i := (Y_i - 1)/(n - m)$. The Z_i's take values in $[0, 1]$; for any j, $1 \leq j \leq n - m + 1$, $\Pr\left(Z_i = \frac{j-1}{n-m}\right) = \Pr(Y_i = j)$. By linearity of expectation,

$$E[Z_i] = \frac{E[Y_i] - 1}{n - m} = \frac{1}{m+1} \quad \text{for} \quad 1 \leq i \leq m \ . \tag{3}$$

The Z_i's are dependent as well, therefore, we do not expect the relation $E[\prod Z_i] = \prod E[Z_i]$ to hold. However, considering the underlying experiment, we may conjecture that if Z_1 is 'large', then it is less likely to occur that Z_2 is 'large' as well. The following lemma proves that the Z_i's are indeed negatively correlated. (A proof is given in the appendix.)

Lemma 1. For any $I \subseteq \{1, \ldots, m\}$ with $|I| = k \geq 1$,

$$E\left[\prod_{i \in I} Z_i\right] = \frac{[n-m]_k}{(n-m)^k} \cdot \frac{1}{[m+k]_k} \leq \frac{1}{(m+1)^k} = \prod_{i \in I} E[Z_i] \ ,$$

where $[x]_k := x \cdot (x-1) \cdots (x - k + 1)$.

Lemma 1 suffices to establish large deviation estimates of the Chernoff-Hoeffding kind for $Z := \sum_i Z_i$. Let $X_1, \ldots, X_n$ be random variables that take values in $[0, 1]$ and let $X := \sum_{1 \leq i \leq n} X_i$. Following [13], we call the X_i's 1-correlated if for all non-empty $I \subseteq \{1, \ldots, n\}$

$$E\left[\prod_{i \in I} X_i\right] \leq \prod_{i \in I} E[X_i] \ .$$

Note that if the random variables $X_1, \ldots, X_n$ and $Y_1, \ldots, Y_m$ are both 1-correlated and if the X_i's are independent of the Y_j's, then the whole set of random variables $X_1, \ldots, X_n, Y_1, \ldots, Y_m$ is 1-correlated as well.

Lemma 2 ([13]). Let X be the sum of 1-correlated random variables $X_1, \ldots, X_n$ with values in $[0, 1]$. Then for any $\varepsilon > 0$,

$$\Pr(X > (1+\varepsilon)E[X]) \leq \left(\frac{e^\varepsilon}{(1+\varepsilon)^{(1+\varepsilon)}}\right)^{E[X]} \ . \tag{4}$$

Note that for $\varepsilon \geq 2e - 1$, (4) implies that

$$\Pr(X > (1 + \varepsilon)E[X]) \leq 2^{-(1+\varepsilon)E[X]} .$$

We will also use *Azuma's inequality*. The following formulation appears in [9].

Lemma 3. *Let $X_1, \ldots, X_n$ be independent random variables, with X_k taking values in a set A_k for each k. Suppose that the function $f : \prod A_k \to \mathbb{R}$ satisfies $|f(x) - f(y)| \leq c$ whenever the vectors x and y differ only in a single coordinate. Let Y be the random variable $f(X_1, \ldots, X_n)$. Then for any $t > 0$,*

$$\Pr(|Y - E[Y]| \geq t) \leq 2e^{-2t^2/(nc^2)} .$$

We use the following terminology for weighted digraphs. For an edge $e = (u, v)$, we call u the *source* and v the *target* or *endpoint* of e. The weight of an edge is denoted by $c(e)$. We will interpret an entry in the adjacency list of a vertex u either as the endpoint v of an edge with source u or as the edge (u, v) itself, as is convenient.

3   The Algorithm of Moffat and Takaoka

We will now review the algorithm of Moffat and Takaoka in [11] and their analysis. We are given a complete digraph on n vertices with non-negative edge weights. The algorithm first sorts all adjacency lists in order of increasing weight (total time $O(n^2 \log n)$) and then solves n single source shortest path problems, one for each vertex of G. The single source shortest path problem, say, with source $s \in V$, is solved in two phases. Both phases are variants of Dijkstra's algorithm [2] and only differ in the way the priority queue is handled.

Dijkstra's algorithm labels the vertices in order of increasing distance from the source. We use S to denote the set of labeled vertices and $U = V - S$ to denote the set of unlabeled vertices. Initially, only the source vertex is labeled, i.e., $S = \{s\}$. For each labeled vertex v, its exact distance $d(v)$ from the source is known. For the source node s, we have $d(s) = 0$. For each labeled vertex v, one of its outgoing edges is called its *current edge* and is denoted $ce(v)$. We maintain the invariant that all edges preceding the current edge $ce(v)$ in v's (sorted) adjacency list have their endpoint already labeled. Both phases use a priority queue. A priority queue stores a set of pairs (x, k) where k is a real number and is called the key of the pair. We assume that priority queues are implemented as Fibonacci heaps [5]. Fibonacci heaps support the insertion of a new pair (x, k) in constant time and the deletion of a pair with minimum key (*delete min* operation) in amortized time $O(\log p)$ where p is the number of pairs in the priority queue. They also support an operation *decrease key* in constant amortized time. A *decrease key* operation takes a pointer to a node in a Fibonacci heap containing, say, the pair (x, k), and allows the replacement of k by a smaller key k'.

Phase I. In Phase I, the additional invariant is maintained that the targets of all current edges are unlabeled. For each vertex $u \in U$, we also maintain a list $L(u)$ of all vertices $v \in S$ whose current edge ends in u. The priority queue contains all vertices in U. The key of a vertex $u \in U$ is $\min_{v \in L(u)} d(v) + c(v, u)$. In each iteration of Phase I, the vertex $u \in U$ with minimal key value $d(u)$ is selected and is deleted from the priority queue by a *delete min* operation. The vertex u is added to S and for each vertex $v \in \{u\} \cup L(u)$, the current edge $ce(v)$ is advanced to the first edge in v's (sorted) adjacency list whose target is in $V - S$. For any $v \in \{u\} \cup L(u)$, let $ce(v) = (v, w)$ be the new current edge of v and denote w's current key by d_w. We add v to $L(w)$, and if $d(v) + c(v, w) < d_w$, we decrease d_w appropriately. This implies a *decrease key* operation on the priority queue. By our assumption on the implementation of the priority queue, the cost of an iteration of Phase I is $O(\log n)$ plus the number of edges scanned. Phase I ends when $|U|$ becomes $n/\log n$.

Remark: Moffat and Takaoka use a binary heap instead of a Fibonacci heap to realize the priority queue; Fibonacci heaps did not exist at that time. Since a *decrease key* operation in a binary heap takes logarithmic time, they use a slightly different strategy for Phase I. They keep the vertices in S in the priority queue. The key of vertex $v \in S$ is $d(v) + c(ce(v))$. In each iteration, the vertex of minimum key, say, vertex $v \in S$, is selected from the heap. Let w be the target of the current edge of v. The current edge $ce(z)$ is advanced for all $z \in \{w\} \cup L(w)$. Then w is inserted into the priority queue with key $d(w) + c(ce(w))$, and for each $z \in L(w)$, the key of z is increased (since the weight of the new current edge is greater than the weight of the old current edge of z). Moffat and Takaoka show that the expected cost of an *increase key* operation is constant in a binary heap. We believe that the implementation using Fibonacci heaps is slightly simpler since it does not use a non-standard operation on priority queues.

What is the total number of edges scanned in Phase I? In [11], Moffat and Takaoka argue as follows: Let U_0 be the set of unlabeled vertices at the end of Phase I. Then $|U_0| = n/\log n$. Since for every vertex v the endpoints of the edges out of v form a random permutation of V, we should expect to scan about $\log n$ edges in each adjacency list during Phase I and hence about $n \log n$ edges altogether. This argument is incorrect as U_0 is determined by the orderings of the adjacency lists and cannot be fixed independently. The following example makes this fact obvious. Assume that all edges out of the source have length one and all other edges have length two. Then Phase I scans $n - n/\log n$ edges out of the source vertex and U_0 is determined by the last $n/\log n$ edges in the adjacency list of the source. Thus the conclusion that one scans about $\log n$ edges in each adjacency list is wrong. However, the derived conclusion that the expected total number of scanned edges is $O(n \log n)$ is true, as the following argument shows.

Consider Phase I', the following modification of Phase I (similar to Spira's algorithm [12]). In this modification, the target of a current edge may be labeled and the vertices $v \in S$ are kept in a priority queue with keys $d(v) + c(ce(v))$. When a vertex v with minimum key is selected, let w be the target of $ce(v)$. If w does not belong to S, then w is added to S and to the priority queue. In any case, $ce(v)$ is advanced to the *next* edge and v's key is increased appropriately. The modified algorithm finishes Phase I$'$ when $|V - S| = n/\log n$. Let S_0 be the set of vertices that have been labeled in Phase I$'$. For every vertex $v \in S_0$, denote by $A(v)$ the set of edges out of v that have been scanned by the modified algorithm and denote by $S(v)$ the targets of the edges in $A(v)$.

The analysis of the coupon collector's problem implies that $E[\sum_v |A(v)|] = O(n\log n)$. Indeed, let X_i be the number of edges scanned when $|S| = i$. Since the targets of the edges are random, we have $E[X_i] \leq n/(n - i)$ and hence

$$E\left[\sum_{i \leq n-n/\log n} X_i\right] \leq n\sum_{i \leq n} 1/i \leq n(\ln n + 1) \ .$$

This proves that $E[\sum_v |A(v)|] = O(n\log n)$.

It turns out that Phase I of the algorithm by Moffat and Takaoka shows basically the same behavior. In fact, at the end of Phase I, their algorithm has labeled exactly the vertices in S_0, and all the edges in $\bigcup_v A(v)$ have been scanned by the Moffat and Takaoka algorithm as well. However, for the purpose of maintaining the invariant, the current edge pointer of each vertex $v \in S_0$ has been advanced to the first vertex in U_0 in v's adjacency list. For every vertex $v \in S_0$, let e_v and e_v' be the current edge of v at the end of Phase I in the algorithm by Moffat and Takaoka and at the end of Phase I$'$ in the algorithm by Spira, respectively. e_v' precedes e_v in v's adjacency list and the edge e_v is the first edge after e_v' with target in U_0. We can imagine scanning the edges after e_v' only when Phase I$'$ has terminated. Due to the endpoint-independent distribution of edge weights, the targets of the edges after e_v' in the adjacency list form a random permutation of $V - S(v) \supseteq V - S_0 = U_0$. This is the setting of the probabilistic experiment in Sect. 2.1, where the number of edges between e_v' and e_v corresponds to Y_1. Since $|V - S(v)| \leq n$ and $m := |U_0| = n/\log n$, we deduce from (2) in Sect. 2.1 that the expected number of edges between e_v' and e_v is $O(\log n)$. This completes the analysis of Phase I.

Phase II. In Phase II, the weaker additional invariant is maintained that the endpoint of every current edge belongs to U_0. We now keep the vertices $v \in S$ in the queue with key $d(v) + c(ce(v))$.

In each iteration of Phase II, a vertex with minimum key is selected from the queue. Say that vertex v is selected and that w is the endpoint of $ce(v)$. The vertex w is a random vertex in U_0 but it is not necessarily unlabeled. If w is unlabeled, it will be labeled, $d(w)$ is set to $d(v) + c(ce(v))$, and $ce(v)$ and $ce(w)$ are advanced to the next edge whose endpoint is in U_0. If w is already labeled, only $ce(v)$ is advanced. In either case the heap is updated appropriately. All of this takes time $O(\log n + \#\text{edges scanned})$.

We have already stated that w is a random element of U_0 and hence, when $|U| = i < n/\log n$, an expected number of $(n/\log n)/i$ iterations is required to decrease the cardinality of U by one. The expected number of iterations in Phase II is therefore bounded by

$$\frac{n}{\log n} \sum_{1 \leq i \leq n/\log n} \frac{1}{i} = O(n) \ .$$

Moreover, whenever the current edge of a vertex is advanced, it is advanced to the next edge having its endpoint in U_0. As argued above, for any vertex $v \in S$, the vertices in U_0 are distributed randomly in $V - S(v)$, and (2) in Sect. 2.1 shows that whenever $ce(v)$ is advanced in Phase II, it is advanced by an expected number of $O(\log n)$ edges. We conclude that the expected cost of one iteration in Phase II is $O(\log n)$ and, given that we do k iterations in Phase II, the expected cost of Phase II is $O(k \log n)$. Hence, the total expected cost of Phase II is $O(n \log n)$.

The above discussion is summarized in the following theorem.

Theorem 1. *For endpoint-independent distributions the algorithm of Moffat and Takaoka runs in expected time $O(n^2 \log n)$.*

We will next prove that the algorithm by Moffat and Takaoka is reliable, i.e., that, with high probability, its running time does not exceed its expectation by more than a constant multiplicative factor.

Theorem 2. *The running time of the all-pairs shortest path algorithm by Moffat and Takaoka is $O(n^2 \log n)$ with high probability.*

Proof. It is sufficient to prove that solving a single source shortest path problem takes time $O(n \log n)$ with high probability. As in the proof of Theorem 1, we analyze Phase I′, the remaining part of Phase I, and Phase II separately, and we prove that each of them takes time $O(n \log n)$ with high probability. We use the notation that has been introduced for the proof of Theorem 1.

Recall that the running time of Phase I′ is $O(n \log n)$ plus $\sum_{v \in S_0} |A(v)|$, the number of edges scanned. The tail estimate for the coupon collector's problem, (1) in Sect. 2, implies that $\sum_{v \in S_0} |A(v)|$ is $O(n \log n)$ with high probability.

For the analysis of the remaining part of Phase I, for any $v \in S_0$, define the random variable Y_v as being the number of edges between e'_v and e_v. With $m := n/\log n$ and $Z_v := (Y_v - 1)/(n - m)$, (2) and (3) in Sect. 2.1 imply that for $Y_I := \sum_{v \in S_0} Y_v$ and $Z_I := \sum_{v \in S_0} Z_v$, the expected values are $E[Y_I] = |S_0|\frac{n+1}{m+1} = \Theta(n \log n)$ and $E[Z_I] = |S_0|/(m + 1) = \Theta(\log n)$. Since the Z_v's are independent random variables, we get from the usual Chernoff-Hoeffding bound (which is subsumed in Lemma 2) that

$$\Pr(Y_I > (1 + \varepsilon)E[Y_I]) \leq \Pr(Z_I > (1 + \varepsilon)E[Z_I]) \leq 2^{-(1+\varepsilon)E[Z_I]}$$

for large enough ε. This proves that $Y_I = O(n \log n)$ with high probability.

We now turn to the analysis of Phase II. Let the random variable Y_{II} denote the total number of edges scanned in Phase II; we know that $E[Y_{II}] = O(n \log n)$ from the proof of Theorem 1. Suppose that we perform k iterations in Phase II; then we can express Y_{II} as the sum of random variables Y_i, $1 \leq i \leq k$, where Y_i denotes the number of advances of the current edge pointer in the i-th iteration. With $m := n/\log n$, we introduce the normalized random variables $Z_i := (Y_i - 1)/(n - m)$, $1 \leq i \leq k$. Since some of the Z_i's might refer to the adjacency list of the same vertex, the Z_i's are not necessarily independent random variables.

However, Lemma 1 tells us that $Z_1, \ldots, Z_k$ are 1-correlated random variables. Lemma 2 provides a tail estimate for $Z^{(k)} := \sum_{i=1}^{k} Z_i$;

$$\Pr\left(Z^{(k)} > (1+\varepsilon)E[Z^{(k)}]\right) \leq 2^{-(1+\varepsilon)E[Z^{(k)}]}$$

for large enough ε. For $k = O(n)$, we set $(1+\varepsilon) = \Theta(n/k)$ to obtain that $Z^{(k)} = O(\log n)$ with high probability. If we abbreviate by I_k the event that we perform k iterations in Phase II, then

$$\Pr(Y_{II} > (1+\varepsilon)E[Y_{II}]) = \sum_k \Pr(Y_{II} > (1+\varepsilon)E[Y_{II}] \mid I_k) \cdot \Pr(I_k)$$
$$\leq \sum_k \Pr\left(Z^{(k)} > (1+\varepsilon)E[Z^{(k)}]\right) \cdot \Pr(I_k) \ .$$

By the tail estimate for the coupon collector's problem, the number of iterations in Phase II is $O(n)$ with high probability. Hence, $Y_{II} = O(n \log n)$ with high probability. Again, because the number of iterations is $O(n)$ with high probability, the total time needed for updating the heap in Phase II is $O(n \log n)$ with high probability.

Thus we have proved that the running time of the algorithm is $O(n^2 \log n)$ with high probability. □

4 A Lower Bound for the Single Source Problem

Can we achieve running time $o(n^2 \log n)$ for solving the all-pairs shortest path problem? In certain situations we certainly can, e.g., if all edge weights are equal to one. However, in the general case of endpoint-independent distributions, it takes expected time $\Omega(n \log n)$ to compute the shortest path distances with respect to a single source, as we now argue.

Our underlying graph is $\tilde{K}_n = (V, E)$, the complete digraph on n vertices with loops. We restrict ourselves to the case of *simple* weight functions on the edges, i.e., for every vertex v and each integer k, $1 \leq k \leq n$, there is exactly one edge with weight k and source v. A single source shortest path algorithm gets as its input the problem size n, a source vertex s, and a simple weight function c. We assume that c is provided by means of an oracle that answers questions of the following kind:

(1) What is the weight $c(e)$ of a given edge e?
(2) Given a vertex $v \in V$ and an integer $k \in \{1, \ldots, n\}$, what is the target of the edge with weight k and source v?

The algorithm is supposed to compute the function d of shortest distances from s. It is allowed to ask the oracle questions of type (1) and (2), thereby gaining partial information on c. The complexity of the algorithm on a fixed simple weight function c is defined to be the number of questions the algorithm asked in order to compute the distance function d with respect to c.

For simple weight functions, the distance function d maps the set of vertices into $\mathbb{N}_0$. Define $D := \max\{d(v) \ ; \ v \in V\}$ and for all i, $0 \leq i \leq D$, let $V_i :=$

$\{v \; ; \; d(v) = i\}$. We call V_i the i-th layer with respect to d. For all i, $0 \le i \le D$, let $\ell(i) := |\{j \; ; \; j > i \text{ and } V_j \ne \varnothing\}|$ be the number of non-empty layers above layer i. Clearly, D, the sets V_i, and the function ℓ depend on c; for ease of notation, we do not make this dependence visible in the notation.

We first argue intuitively how to provide a lower bound on the complexity of a single source shortest path algorithm in terms of ℓ. Consider any vertex v of distance $d(v)$ from the source and suppose that the algorithm has not inquired about one of v's outgoing edges, say e, of length $c(e) < D - d(v)$. By omitting the check of e, the algorithm cannot exclude that $d(v) + c(e)$ is smaller than the distance label of the target of e, in which case the distance function computed by the algorithm would be incorrect.

Lemma 4. *Let c be a simple weight function and let d be the distance function with respect to c. Then any shortest path algorithm has complexity at least*

$$\sum_{u \in V} (\ell(d(u)) - 1) \; .$$

Proof. Let E' be the set of edges queried by the algorithm by a question of either type (1) or type (2). For an arbitrary but fixed vertex $u \in V$, let $E(u)$ be the set of edges with source u and let $E'(u) := E' \cap E(u)$. We prove that $|E'(u)| \ge \ell(d(u)) - 1$. This is clear if E' contains edge $e \in E(u)$ of weight $c(e) = j$ for all j, $1 \le j < \ell(d(u))$. If there is an edge $e_i \in E(u) - E'$ with weight $c(e_i) = i < \ell(d(u))$, then every non-empty layer V_j above layer $d(u) + i$ must contain the target of an edge in $E'(u)$. Assume otherwise, then there is an edge $e_j = (u, v) \notin E'$ with $v \in V_j$ for a $j > d(u) + i$. Define the simple weight function c' by

$$c'(e) := \begin{cases} c(e), & \text{if } e \notin \{e_i, e_j\} \; ; \\ c(e_j), & \text{if } e = e_i \; ; \\ c(e_i), & \text{if } e = e_j \; . \end{cases}$$

Then $c'(e) = c(e)$ for all $e \in E'$, and therefore the algorithm will output d, the distance function with respect to c, on input c' as well. However, $d(v) = j > d(u) + i = d(u) + c'(e_j)$ for $e_j = (u, v)$, which shows that d is incorrect with respect to c'.

We choose $i = \min\{c(e) \; ; \; e \in E(u) - E'\}$. Note that all edges in $E(u)$ with targets in layer V_j, $j > d(u) + i$, must have weight at least $j - d(u) > i$ by the correctness of the algorithm. Hence, $|E'(u)| \ge i - 1 + \ell(d(u) + i) \ge \ell(d(u)) - 1$. $\quad\square$

Table 1 shows the distribution of vertices over distances for a (typical) simple weight function on a graph of $n = 10000$ vertices. Most vertices have distance about 14 ($\approx \log n$) from the source but there are vertices that have distance as much as 24 ($\approx 2 \log n$). By the argument of Lemma 4, we can guess that any (correct) algorithm must inquire about $\Omega(n \log n)$ edges.

In the remainder of this section, we make this argument more precise. We derive a lower bound of $\Omega(n \log n)$ on the expected value of $\sum_{u \in V} \ell(d(u))$ for random simple weight functions c. More generally, we show that any algorithm has to ask $\Omega(n \log n)$ questions with high probability.

194

Table 1. A typical distribution of vertices over distances for $n = 10000$

distance d	0	1	2	3	4	5	6	7	8	9	10	11	12
# vertices	1	1	2	4	8	16	32	64	120	237	449	796	1306

distance d	13	14	15	16	17	18	19	20	21	22	23	24
# vertices	1845	1952	1562	910	415	181	58	20	16	2	1	2

Our proof strategy is as follows. The lower bound given by Lemma 4 depends only on the distance function d. For random simple weight functions, we re-interpret the calculation of d and the construction of the layers V_i as the outcome of a random labeling process. Note that a random simple weight function is given by n independent permutations of V, one for each vertex. The i-th vertex on the permutation for vertex v is the target of the edge with weight i and source v. The labeling process proceeds in stages. In the 0-th stage, V_0 is set to $\{s\}$ and $d(s)$ is set to 0. In the i-th stage, $i \geq 1$, each vertex $v \in S^{(i)} = \bigcup_{0 \leq j < i} V_j$ picks the $(i - d(v))$-th vertex in its adjacency list. Note that each vertex that v has not yet seen is equally likely to occur. The newly reached vertices are put into V_i and their d-value is set to i. Instead of fixing the n permutations beforehand, we may also view them as being fixed on-line (this is sometimes called the principle of deferred decisions). This leads to the following re-interpretation of the random labeling process: In the i-th stage, each vertex in $S^{(i)} = \bigcup_{0 \leq j < i} V_j$ chooses a vertex uniformly and independently at random from the set of vertices it has not yet seen. The labeling process stops if $S^{(k)} = V$ for some k.

A related process was considered by Frieze and Grimmett in [6]. They assumed that each vertex in $S^{(i)}$ chooses a vertex uniformly and independently at random from the set of *all* vertices. If D_A denotes the number of stages taken by this version of the process, then it is clear that D_A stochastically dominates D, i.e., for all m, $\Pr(D > m) \leq \Pr(D_A > m)$. Frieze and Grimmett prove in [6] that D_A (and hence D) is $O(\log n)$ with high probability. However, we need a lower bound on D and hence their result is of no use to us. (Nevertheless, our proof strategy was inspired by theirs.)

The random labeling process is said to be in state j, if $|S^{(i)}| = j$. We call stage i of the labeling process *central*, if $n/e \leq |S^{(i)}| \leq n - \sqrt{n}$. Layers constructed in central stages are called central.

Our proof will proceed in two steps. First, we show in Lemma 5 that there are $\Omega(\log n)$ central stages with high probability. Second, we prove in Lemma 6 that each central stage gives rise to a non-empty layer with high probability.

Lemma 5. *With high probability, the labeling process has $\Omega(\log n)$ central stages.*

Proof. For a random simple weight function c, let i_0 be the first central stage with respect to c. Then $n/e \leq |S^{(i_0)}| \leq 2n/e$, since $|S^{(i+1)}| \leq 2|S^{(i)}|$ for any $i \geq 0$. We will show that $|S^{(i_0+k)}| \leq n - \sqrt{n}$ with high probability for $k = (\ln n)/17$. Let $U = V - S^{(i_0)}$ be the set of vertices that are still unlabeled after stage i_0. Note that $|U| \geq (e - 2)n/e \geq n/4$.

Let us condition on $m = |U|$. Construct an $n \times m$ matrix A with 0-1 entries as follows. The rows correspond to the vertices in V and the columns correspond to the vertices in U; entry a_{vu} is 1 if and only if the edge (v, u) is among the k shortest edges in v's adjacency list whose head is an element of U. Let $f(A)$ be the number of all-zero columns in A. Then $|S^{(i_0+k)}| \leq n - f(A)$ because no vertex in U corresponding to an all-zero column will be labeled in the k stages following stage i_0. Since A models a process in which all vertices (and not only those that are currently labeled) are allowed to label new vertices, and in which each vertex is prevented from choosing vertices that have been labeled by other vertices before stage i_0, $f(A)$ may seem to be a rather crude lower bound on $|V - S^{(i_0+k)}|$. However, we will now prove that even $f(A) \geq \sqrt{n}$ with high probability.

A row of A is a random 0-1 vector of length m with exactly k ones. Moreover, the row entries $A_i.$, $1 \leq i \leq n$, are independent random variables, and if A, A' differ only in a single row, then $|f(A) - f(A')| \leq k$. Hence, by Azuma's inequality (Lemma 3), we get the following tail estimate for $f(A) = f(A_1., \ldots, A_n.)$,

$$\Pr(f(A) \leq E[f(A)]/2) \leq 2\exp(-E[f(A)]^2/(2nk^2)) .$$

The probability that a fixed column is all-zero is $(1 - k/m)^n$; therefore,

$$E[f(A)] = m\left(1 - \frac{k}{m}\right)^n . \tag{5}$$

Remember that $m = |U| \geq n/4$ and $k = (\ln n)/17$; since $(1 - 1/x)^x \geq e^{-2}$ for large enough x, we get from (5) that

$$E[f(A)] \geq me^{-2kn/m} \geq \tfrac{1}{4}n^{1-8/17} > 2\sqrt{n} \tag{6}$$

for large enough n, where $E[f(A)]$ is conditioned on m. However, the lower bound in (6) is independent of m. Hence,

$$\Pr(f(A) < \sqrt{n}) \leq 2\exp(-\Theta(n^{1/17})/(\ln n)^2) = O(n^{-C})$$

for any fixed $C > 0$ and large enough n. Since $|S^{(i_0+k)}|$ is increasing in k, we have thus proved that, with high probability, it will take $\Omega(\ln n) = \Omega(\log n)$ stages to label all but $\sqrt{n}$ vertices. $\qquad \square$

Lemma 6. *With high probability, each central layer contains at least one vertex.*

Proof. Suppose the process is in state j at the beginning of stage i. For any vertex in $S^{(i)}$, the probability of selecting a vertex in $S^{(i)}$ during this stage is $\leq j/n$. Therefore, the next layer will remain empty with probability $\leq (j/n)^j$. Note that $x \mapsto (x/n)^x$ is an increasing function for $x > n/e$.

Let B denote the event that at least one central layer remains empty. By the estimates provided in the preceding paragraph,

$$\Pr(B) \leq \sum_{j=n/e}^{n-\sqrt{n}} \left(\frac{j}{n}\right)^j \leq n\left(\frac{n-\sqrt{n}}{n}\right)^{n-\sqrt{n}} \leq ne^{-\sqrt{n}+1} = O(n^{-C})$$

for sufficiently large n. $\qquad \square$

Theorem 3. *Any algorithm for the single source shortest path problem has complexity* $\Omega(n \log n)$ *with high probability on random simple weight functions.*

Proof. Suppose that i is the first central stage of the labeling process; as before, let $S^{(i)}$ denote the set of vertices that have already been labeled up to this stage. By Lemma 5, with high probability, the process has $\Omega(\log n)$ central layers. Lemma 6 tells us that all these layers will be non-empty with high probability. With the notation introduced in the discussion of the labeling process, this reads

$$\sum_{u \in S^{(i)}} (\ell(d(u)) - 1) = \Omega(n \log n) \quad \text{with high probability.}$$

By Lemma 4, the left-hand side term is a lower bound on the complexity of any shortest path algorithm. □

Acknowledgements

We learned from discussions with Paul Spirakis that analyzing the Moffat and Takaoka algorithm [11] is not as easy as it might appear at first glance. The remarks of an anonymous referee for ICALP'94 allowed considerable simplification of our proofs of Theorems 1 and 2. Rudolf Fleischer suggested the use of Fibonacci heaps in the implementation of the algorithm. Finally, numerous discussions with Hannah Bast were particularly insightful, as conversations with Devdatt Dubhashi and Torben Hagerup helped to clarify our ideas.

References

1. P.A. Bloniarz, A shortest-path algorithm with expected time $O(n^2 \log n \log^* n)$, *SIAM J. Comput.* **12** (1983) 588–600
2. E.W. Dijkstra, A note on two problems in connexion with graphs, *Numer. Math.* **1** (1959) 269–271
3. R.W. Floyd, Algorithm 97: Shortest path, *Comm. ACM* **5** (1962) 345
4. M.L. Fredman, New bounds on the complexity of the shortest path problem, *SIAM J. Comput.* **5** (1976) 83–89
5. M.L. Fredman and R.E. Tarjan, Fibonacci heaps and their uses in improved network optimization algorithms, *J. ACM* **34** (1987) 596–615
6. A.M. Frieze and G.R. Grimmett, The shortest-path problem for graphs with random arc-lengths, *Discrete Appl. Math.* **10** (1985) 57–77
7. R.L. Graham, D.E. Knuth, and O. Patashnik, *Concrete Mathematics* (2nd ed.), Addison-Wesley, Reading, MA, 1994
8. D.R. Karger, D. Koller, and S.J. Phillips, Finding the hidden path: Time bounds for all-pairs shortest paths, *SIAM J. Comput.* **22** (1993) 1199–1217
9. C. McDiarmid, On the method of bounded differences, in: J. Siemons (Ed.), *Surveys in Combinatorics, 1989* (London Mathematical Society Lecture Notes Series; 141), Cambridge University Press, Cambridge, 1989
10. A. Moffat and T. Takaoka, An all pairs shortest path algorithm with expected time $O(n^2 \log n)$, *Proc. of the 26th Annual Symposium on Foundations of Computer Science*, Portland, OR, 1985, 101–105

11. A. Moffat and T. Takaoka, An all pairs shortest path algorithm with expected time $O(n^2 \log n)$, *SIAM J. Comput.* **16** (1987) 1023–1031
12. P.M. Spira, A new algorithm for finding all shortest path in a graph of positive arcs in average time $O(n^2 \log^2 n)$, *SIAM J. Comput.* **2** (1973) 28–32
13. A. Srinivasan, Techniques for probabilistic analysis and randomness-efficient computation, Ph.D. Thesis, Cornell University, Ithaca, NY, Technical Report 93-1378, August 1993
14. T. Takaoka, A new upper bound on the complexity of the all pairs shortest path problem, *Inform. Process. Lett.* **43** (1992) 195–199

Appendix

Recall the probabilistic experiment from Sect. 2.1: An urn contains n balls that are either red or blue; let m be the number of red balls. The balls are repeatedly drawn from the urn (without replacement) uniformly and independently at random. For $1 \leq k \leq m$, let the random variable W_k denote the waiting time for the k-th red ball. In addition, we define the random variables Y_i, $1 \leq i \leq m$, by $Y_1 := W_1$ and $Y_i := W_i - W_{i-1}$ for $2 \leq i \leq m$. The W_k's are distributed according to the negative hypergeometric distribution, i.e., for k, r with $1 \leq k \leq m$ and $k \leq r \leq n - m + k$,

$$\Pr(W_k = r) = \binom{r-1}{k-1}\binom{n-r}{m-k} \bigg/ \binom{n}{m}.$$

The waiting time for the k-th red ball equals r if and only if there is a k-tuple $(j_1, \ldots, j_k)$ of positive integers with $j_1 + \cdots + j_k = r$ and $Y_i = j_i$ for all i, $1 \leq i \leq k$. Hence, for $j_1, \ldots, j_k \geq 1$ with $j_1 + \cdots + j_k = r$,

$$\Pr\left(\bigwedge_{1 \leq i \leq k} Y_i = j_i\right) = \binom{r-1}{k-1}^{-1} \Pr(W_k = r) = \binom{n-(j_1 + \cdots + j_k)}{m-k} \bigg/ \binom{n}{m}.$$

By using the well-known convolution identity

$$\sum_{0 \leq k \leq l} \binom{l-k}{m}\binom{q+k}{n} = \binom{l+q+1}{m+n+1} \tag{A.1}$$

for integers $l, m, n, q \geq 0$, $n \geq q$ (see [7] for a proof), it is easy to see that

$$\begin{aligned}
\Pr\left(\bigwedge_{2 \leq i \leq k} Y_i = j_i\right) &= \sum_{1 \leq j_1 \leq n-m+1} \Pr\left(\bigwedge_{1 \leq i \leq k} Y_i = j_i\right) \\
&= \sum_{1 \leq j_1 \leq n-m+1} \binom{n-(j_2 + \cdots + j_k) - j_1}{m-k} \bigg/ \binom{n}{m} \\
&= \binom{n-(j_2 + \cdots + j_k)}{m-k+1} \bigg/ \binom{n}{m}
\end{aligned}$$

and, more generally, for any non-empty $I \subseteq \{1, \ldots, m\}$ and positive integers j_i, $i \in I$,

$$\Pr\left(\bigwedge_{i \in I} Y_i = j_i\right) = \binom{n - \sum_{i \in I} j_i}{m - |I|} \Big/ \binom{n}{m}, \qquad (A.2)$$

i.e., the Y_i's are exchangeable random variables. Making use of (A.1), we conclude that for any i, $1 \le i \le m$,

$$E[Y_i] = \sum_{j=1}^{n-m+1} j\binom{n-j}{m-1} \Big/ \binom{n}{m} = \binom{n+1}{m+1} \Big/ \binom{n}{m} = \frac{n+1}{m+1}.$$

This proves (2) in Sect. 2.1.

For $1 \le i \le m$, we introduced normalized random variables $Z_i := (Y_i - 1)/(n - m)$. The Z_i's take values in $[0, 1]$; for any j, $1 \le j \le n - m + 1$, $\Pr\left(Z_i = \frac{j-1}{n-m}\right) = \Pr(Y_i = j)$. By linearity of expectation, $E[Z_i] = 1/(m+1)$ for any i, $1 \le i \le m$. Lemma 1 proves that the Z_i's are negatively correlated.

Lemma 1. *For any $I \subseteq \{1, \ldots, m\}$ with $|I| = k \ge 1$,*

$$E\left[\prod_{i \in I} Z_i\right] = \frac{[n-m]_k}{(n-m)^k} \cdot \frac{1}{[m+k]_k} \le \frac{1}{(m+1)^k} = \prod_{i \in I} E[Z_i],$$

where $[x]_k := x \cdot (x-1) \cdots (x-k+1)$.

Proof. Only the first equation has to be proved and because of (A.2), we can restrict ourselves to the case $I = \{1, \ldots, k\}$. Using (A.1), one can prove by induction on k that

$$\sum_{\substack{j_1, \ldots, j_k \ge 1 \\ j_1 + \cdots + j_k = r}} (j_1 - 1) \cdots (j_k - 1) = \binom{r-1}{2k-1}.$$

Therefore, by (A.2) and (A.1),

$$(n-m)^k E\left[\prod_{1 \le i \le k} Z_i\right]$$

$$= \sum_{k \le r \le n-m+k} \sum_{\substack{j_1, \ldots, j_k \ge 1 \\ j_1 + \cdots + j_k = r}} (j_1 - 1) \cdots (j_k - 1) \cdot \Pr\left(\bigwedge_{1 \le i \le k} Y_i = j_i\right)$$

$$= \binom{n}{m}^{-1} \sum_{k \le r \le n-m+k} \binom{n-r}{m-k}\binom{r-1}{2k-1} = \binom{n}{m}^{-1}\binom{n}{m+k} = \frac{[n-m]_k}{[m+k]_k}.$$

$\square$

Fully Dynamic Transitive Closure in Plane Dags with One Source and One Sink*

Thore Husfeldt

BRICS,** Department of Computer Science, University of Aarhus,
Ny Munkegade, DK–8000 Århus C, Denmark

Abstract. We give an algorithm for the Dynamic Transitive Closure
Problem for planar directed acyclic graphs with one source and one sink.
The graph can be updated in logarithmic time under arbitrary edge
insertions and deletions that preserve the embedding. Queries of the
form 'is there a directed path from u to v?' for arbitrary vertices u and v
can be answered in logarithmic time. The size of the data structure and
the initialisation time are linear in the number of edges.

We also give a lower bound of $\Omega(\log n/ \log\log n)$ on the amortised complexity of the problem in the cell probe model with logarithmic word
size.

1   Introduction

1.1   Dynamic Graph Algorithms

The strategy behind *dynamic algorithms*—algorithms and data structures for
problems whose instances change over time—is to incrementally recompute the
answer to the present instance. Thus, using already computed information about
previous instances, we can hope to do better than the simple-minded who bluntly
solves the whole problem from scratch after each change. This is easier said than
done, since even for very basic problems like graph connectivity, fast dynamic
algorithms seem to be difficult to come by.

The task is to answer connectivity queries in a graph while e.g. edges are inserted and deleted. Much work has gone into these problems, and for undirected
graphs, these endeavours have led to impressive results: The sparsification tech-
nique by Eppstein et al. [5] yields an algorithm that runs in time $O(|V|^{1/2})$. Re-
cently, Henzinger and King [10] showed that with randomisation, the problem

---

* A full version of this paper, including all proofs, can be found as RS–94–31 on the
BRICS World Wide Web server at URL http://www.daimi.aau.dk/BRICS/. This
work was partially supported by the ESPRIT II Basic Research Actions Program of
the EC under contract no. 7141 (project ALCOM II). Part of this work was done
while the author was at the Hebrew University of Jerusalem, Israel, supported by a
grant from the Danish Research Academy.

** Basic Research in Computer Science, Centre of the Danish National Research
Foundation.

can be solved in amortised, expected polylogarithmic time. For planar, undirected graphs we know deterministic algorithms that run in logarithmic time per operation in the worst case [7, 9], see also [6].

In contrast, the same problem on *directed* graphs—where we will use *reachability* or *transitive closure* for connectivity—is notoriously hard. Indeed, no known algorithms beat the trivial one that recomputes the solution after every update, using the sequential algorithm in time $O(|E|)$. (Embarrassingly, not even a logarithmic time lower bound in some strong model for the problem is known.) Even the restriction to planar graphs does not seem to make the problem much easier: the algorithm by Subramanian for this problem, which runs in amortised time $O(|V|^{2/3} \log n)$, gets nowhere near the coveted polylogarithmic execution time. The contribution of the present paper is a logarithmic time algorithm for a restricted class of planar digraphs.

1.2 Sketch of Result

We give an algorithm for the Transitive Closure Problem on directed acyclic graphs that are drawn in the plane without intersecting edges and have exactly one source and one sink, see Fig. 1. The algorithm handles queries of the form 'is there a directed path from vertex u to vertex v?' and updates that add or remove arbitrary edges, as long as the topology and embedding of the graph are not violated. Updates and queries are processed in time logarithmic in the size of the graph. The data structure can be initialised in linear time and uses linear space.

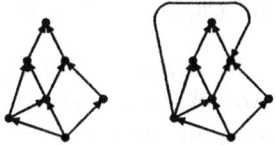

Fig. 1. Plane graphs with one source and one sink

Together with an easily proved lower bound, this characterises the complexity of the Dynamic Transitive Closure Problem on this class of graphs to within an $O(\log \log n)$ factor. The algorithm is pleasantly simple and should be easy to implement efficiently. (The most complicated part is the dynamic tree data structure from [17], where implementation issues are discussed.)

1.3 Outline

This extended abstract is organised as follows: Below, we give some preliminary definitions and state the problem precisely. We also mention a lower bound for the problem. In Sect. 3, we derive a characterisation of the transitive closure in *st*-graphs and re-prove the result of [20] on a restricted version of the problem. Section 4 gives an algorithm for the general case that performs well in the *amortised* sense. We sketch how to remove the amortisation in Sect. 4.3 to get worst-case bounds.

All proofs have been omitted due to lack of space. A full version of the paper, including the proofs and a full description of the worst-case algorithm, is available; see the title page.

2 Preliminaries

2.1 Graphs

A graph is *embeddable* on a surface if it can be drawn on the surface such that the edges do not intersect except at their endpoints. A graph is *planar* if it is embeddable in the plane. Using the stereographic projection, it is easily shown that a graph is planar if and only if it is embeddable on the sphere. For a more thorough coverage of planar graphs, see any text on graph algorithms, e.g. [22].

For vertex v of a digraph we let $\deg^+(v)$ and $\deg^-(v)$ denote its out- and indegree, respectively. A vertex v is a *source* if $\deg^-(v) = 0$, and a *sink* if $\deg^+(v) = 0$. We are now ready to define the class of graphs studied in this paper. The terminology is somewhat awkward (but standard).

Definition 1. A directed acyclic graph is a *source–sink graph* (or short *st-graph*) if it has exactly one source s and one sink t. A *spherical st-graph* is a planar *st*-graph that is embedded in the plane. If in that embedding the source and the sink are on the same face, the graph is a *plane st-graph*.

We require *st*-graphs to be acyclic, which agrees with [21] and disagrees with [19]. Figure 1 shows two spherical *st*-graphs, the left of which is also a plane *st*-graph. The following properties of spherical *st*-graphs can be shown; the last two items explain why we used *spherical* and *plane* the above definition.

1. Every vertex is on a simple directed path from s to t, called an *st-path*.
2. In every embedding, the incoming edges to any vertex appear consecutively around the vertex, and so do the outgoing edges; this determines the *left face* left(v) and the *right face* right(v) of a vertex, see Fig. 2. This implicitly defines an order of the edges appearing around v, say, from the leftmost outgoing edge to the leftmost incoming edge in the clockwise direction. We will sometimes refer to this order as the *ordering of the edges around v*.
3. The boundary of every face consists of two directed paths with common origin and terminus vertices, see Fig. 2.
4. Every spherical *st*-graph can be embedded on the sphere such that all edges are directed upward (i.e., their projection on some fixed direction in the plane is positive). For example, we could embed the graph from Fig. 1 by placing the curved arc on the opposite side of the sphere.

5. Every plane st-graph can be embedded in the plane such that all edges are directed upward.

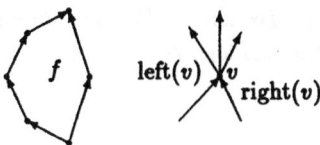

Fig. 2. A vertex and a face in a spherical st-graph

In the rest of this paper, $G = (V, E)$ will denote a spherical st-graph with source s and sink t, vertices V and edges E. We let n denote the size of the problem, i.e. the number of edges in the graph. For brevity, we will sometimes use the notation $u \prec v$ if there is a path from u to v. We will write $u \parallel v$ if neither $u \prec v$ nor $v \prec u$.

2.2 Dynamic Transitive Closure

Operations. We consider the *Dynamic Transitive Closure Problem* for spherical st-graphs. Namely, we present a data structure that handles the following operations (for clarity, we have spelt out the embedding-preserving restrictions on the update operations):

insert(u, v): insert an edge from vertex u to vertex v if they are on the same face and the new edge does not induce a directed cycle,

delete(u, v): delete the edge from u to v provided $\deg^+(u) \geq 2$ and $\deg^-(v) \geq 2$,

query(u, v): answer 'yes' if and only if there is a path from u to v.

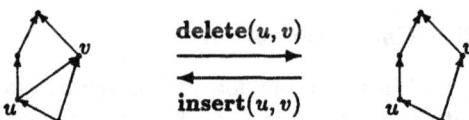

Other operations. Alternatively, we could allow all insertion and deletion operations and let the data structure decide which updates violate the restrictions. To this end, we could use the planarity testing data structure of Tamassia [19] to decide if u and v are on the same face. The acyclicity condition is of course easily checked using our own data structure: edge (u, v) induces a cycle if and only if there is a path from v to u. The restriction on the deletion operation is easily checked by maintaining the in- and outdegree with each vertex.

It is easy to extend the data structure to cope with a **report** operation that outputs a path from u to v if it exists in time $O(\log n) + r$, where r denotes the length of the path.

Lower bound. Our update operations are sufficiently versatile to admit a lower bound proof for the problem in the *cell probe model* with logarithmic word size. The proof is a reduction to the Parity Prefix Problem of [8] in the same fashion as [12] and [16].

Theorem 2. *The Dynamic Transitive Closure Problem on spherical st-graphs requires amortised time $\Omega(\log n / \log \log n)$ in the cell probe model with logarithmic word size.*

2.3 Related Work

Restricted versions of the present problem have been studied by Tamassia and Preparata [20], who consider the case where the source and the sink remain on the same face, and Tamassia and Tollis [21], who allow the source and the sink to be on different faces but modify the repertory of update operations.

Italiano et al. [11] present a dynamic reachability algorithm for *series–parallel* digraphs. Papers by Bodlaender [3] and Cohen et al. [4] extend this to graphs of tree-width two and three, respectively. Apart from these and the class studied in the present paper, no other class of digraphs is known to the author that allows fully dynamic reachability algorithms within polylogarithmic time bounds.

It is easy to see that the $\Omega(\log n / \log \log n)$ lower bound applies to the problems studied in [3, 4, 11, 18]. For the problems in [20, 21], no better bound than $\Omega(\log \log n / \log \log \log n)$ is known to the author; this bound can be proved using techniques from [2, 13, 23], see also [14].

Other dynamic problems on planar *st*-graphs are studied in [1] and [19]. Reference [20] contains pointers to a vast number of applications of these graphs within visibility representations, graph drawing and embedding, motion planning, computational geometry, lattice theory, and VLSI design.

3 Properties of Planar Source–Sink Graphs

3.1 A Reachability Characterisation

We employ an idea used in many polylog-time dynamic graph algorithms: decompose the graph into a number of trees such that all the necessary information can also be derived from the trees.

Definition 3. The tree S is the subgraph of G constructed by removing all edges that are not the leftmost *incoming* edge of any vertex. Similarly, the tree T is constructed by removing all edges that are not the leftmost *outgoing* edge to any vertex.

Observe the following facts:

1. S and T are indeed trees,
2. S is divergent and rooted at s, while T is convergent and rooted at t (hence the names),

3. no subpath of T can ever leave another path to the right, and no subpath of S can ever enter another path from the right.

We need some notation. Let u and v be vertices in V. There are two unique paths in S from s to u and v, respectively. Let s' denote the last vertex that is on both these paths. Similarly, let t' denote the first vertex that is on both paths in T from u and v to t.

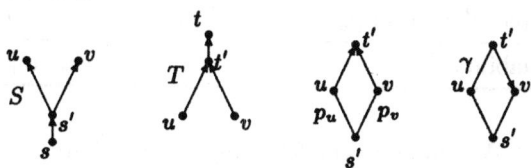

Let p_u (p_v) denote the path that starts in s', follows the unique S-path to u (v) and reaches t' on the unique T-path. Whenever it seems convenient, we will also refer to the two paths as p_l and p_r, such that p_l is the path leaving s' to the left and p_r is the other path.

We will confuse the edges of G with their embedding to alleviate notation. Namely, we introduce the curve γ which is the concatenation of (the embeddings of) p_l and p_r. The orientation of γ will be such that it agrees with the direction of p_l and the reversed direction of p_r. Recall that a curve is *closed* if its endpoints coincide, it is *simple* if it does not intersect itself except at its endpoints. Note that γ is closed and not necessarily simple.

The next lemma is the *crux* of our algorithm. It captures the following fact about reachability in spherical st-graphs: to get from vertex u to vertex v one can always choose a path whose first half stays in T and whose last half stays in S. The proof is not difficult.

Lemma 4. *Let $\leq_S$ and $\leq_T$ denote the predecessor relation in S and T, respectively. Then $u \prec v$ if and only if $\exists w \in V : u \leq_T w \wedge w \leq_S v$.*

3.2 The Plane Case

To see some of the present machinery in motion and to get our hands dirty before we study the full problem, let us derive an algorithm for the case of *plane st-graph*.

We must handle the existential quantifier of the last lemma without searching all of V. One can show that in the plane case, the curve γ is very well-behaved: either it is simple or it intersects exactly once. This boils down to proving that once a path crosses the other, it cannot return. The figure below shows all four possible cases.

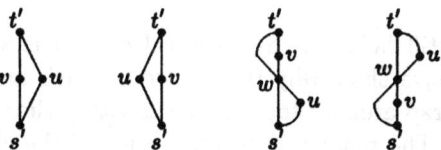

But this means that the existence of w 'between u and v' is uniquely determined by the appearance of p_u and p_v around s' and t'. Table 1 summarises the result.

Table 1: Reachability in the plane case.

p_u leaves s' right of p_v	y	y	n	n
p_u enters t' right of p_v	y	n	y	n
Reachability	$u \parallel v$	$u \prec v$	$v \prec u$	$u \parallel v$

Data Structures. We maintain the following information:

1. With every vertex v, two sequences of the incoming and outgoing edges of v, respectively, ordered according to the cyclic ordering around v (see the remarks after Def. 1). We can used balanced search trees for this.
2. The trees S and T using the *dynamic tree* data structure of Sleator and Tarjan [17].

Updates. After each insertion or deletion we must reorganise our data structures. An edge can be inserted into or deleted from the edge list around a vertex in time $O(\log n)$; maintaining the two dynamic trees is a standard technique.

Queries. Evert u and v in S to find their nearest common ancestor s', see [17]. Evert u and v in T to find their nearest common ancestor t'. From the edge lists around s' and t' we see which of p_u and p_v appears rightmost. By Table 1, this yields the reachability information.

In summary, we have re-proved the following theorem due to Tamassia and Preparata [20], using a different characterisation.

Theorem 5. *The Dynamic Transitive Closure Problem for plane st-graphs can be solved in time $O(\log n)$, where n denotes the number of edges. The data structure uses linear space and can be initialised in linear time.*

3.3 Additional Concepts for Spherical Graphs

The gist of the last section was that

1. if u and v are connected, then p_u and p_v intersect, and
2. if p_u and p_v intersect, then they 'switch sides,' i.e., they appear around s' in another order than they do around t'.

The first item still holds in the spherical case. The second does not. The first two figures above show why the sphere is much more difficult than the plane: Paths can wrap around; the reader can easily check that both examples contradict Table 1. The remedy is to keep track of the globe-trotting of γ by

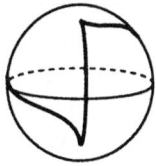 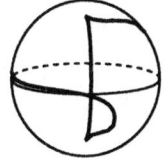 

Fig. 3. The sphere: problems (left) and remedy (right).

maintaining a chain of faces between the poles, as indicated in the third figure; it is helpful to view this chain of faces as a path μ in the dual of the graph. The chain we will call *the meridian* and define in Sect. 4. First, we introduce some additional concepts to be able to formalise what we just sketched.

Definition 6. A *region* is a maximal topologically connected subset in the complement of γ. A curve is *proper* if it intersects γ only at points where γ does not intersect itself. We define the function Ind that maps points to integers as follows: For x in a region the *index* $\mathrm{Ind}(x)$ is the minimum number of intersections between γ and μ over all proper curves μ from s to x. Note that Ind is constant on every region, vanishes on the region of s, and in the plane case, also on the region of t.

For $\mathrm{Ind}(t) > 0$, we define the *orientation of* t as follows: Let x be a point in a region incident to the region of t such that $\mathrm{Ind}(x) = \mathrm{Ind}(t) - 1$. Let μ be a proper curve from x to t that crosses γ only once. Then the orientation of t is *positive* if μ crosses γ from left to right, and *negative* otherwise.

In other words, the orientation of t is the direction of the closed curve that separates the region of t from its neighbouring region with lower index. If this curve is oriented clockwise, the orientation of t is positive. The figure below shows some examples where $\mathrm{Ind}(t) = 2$ and the orientation of t is positive.

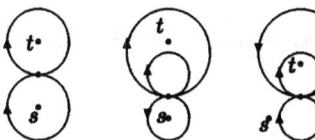

It can be shown that the reachability information between u and v is uniquely determined by (i) the index of t, (ii) the orientation of t, and (iii) the appearance of p_u and p_v around s' and t'.

Table 2: Reachability in the spherical case.

p_r right of p_l at t'	y	n	n	y	y	n	n	n	y	y	n	n
p_u right of p_v at s'	*	*	y	n	y	n	y	n	y	n	y	n
Index of t	0	1	0	1+	1+	2+	1+	0	1+	1+	2+	1+
Orientation of t	*	+	*	+	−	+	−	*	+	−	+	−
Reachability	$u \parallel v$		$u \prec v$					$v \prec u$				

As Table 1 did in the plane case, Table 2 shows the precise connection (aster-isks denote arbitrary or undefined entries). The proof that this table is correct is considerably more involved than for plane graphs, where there were only two cases to consider (not counting symmetries). In contrast, Fig. 4 shows how p_u and p_v can behave on the sphere. The formal proof that the information in Table 2 is correct is long and somewhat tedious.

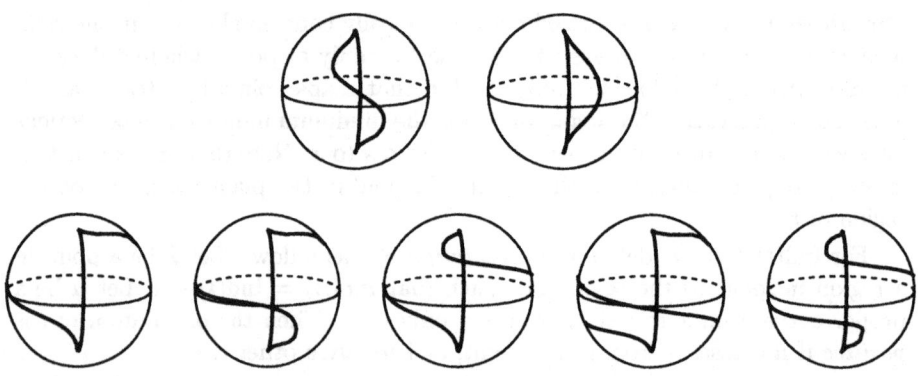

Fig. 4. Examples of the behaviour of p_u and p_r on the sphere. The two topmost cases appear also in the plane, while the five other cases exploit the possibility to travel around the sphere. The bottom examples each represent an infinite number of cases in which the paths cross any number of times; in all those cases, the orientation and the reachability information is the same.

4 Algorithm for Sequences of Updates

4.1 The Meridian

We use the results of the last section to construct an algorithm that performs well in the amortised sense, i.e., a sequence of m updates and queries takes time $O(m \log n)$. We will define a chain of faces between the poles, from which we are able to derive the index and orientation introduced above.

Definition 7. A *meridian* (F^0, E^0) consists of a sequence of *meridian faces* $F^0 = \langle f_1, \ldots, f_m \rangle$ and *meridian edges* $E^0 = \langle e_1, \ldots, e_{m-1} \rangle$ such that

1. for $i = 1, \ldots, m - 1$, edge e_i is on the boundaries of f_i and f_{i+1},
2. $f_i \neq f_j$ for $i \neq j$ (this implies $e_i \neq e_j$).

Moreover, $f_1 = \text{left}(s)$ and $f_m = \text{left}(t)$.

It is easy to see that the meridian corresponds to a *proper* curve μ in the sense of Def. 6.

For curves α and β we let $\phi_r(\alpha, \beta)$ denote the number of times α crosses β from right to left. Symmetrically, $\phi_l(\alpha, \beta)$ denotes the number of times α crosses β from left to right.

Note that ϕ_l and ϕ_r have the nice property that if we decompose α into proper curves $\alpha_1, \ldots, \alpha_k$ then we have, e.g.,

$$\phi_l(\alpha, \beta) = \sum_{i=1}^{k} \phi_l(\alpha_i, \beta).$$

If α is a closed curve and β is a proper curve (with respect to α) whose endpoints are on the same region (with respect to α), then β must leave the region bounded by α as often as it enters it, so

$$\phi_l(\alpha, \beta) - \phi_r(\alpha, \beta) = \phi_l(\beta, \alpha) - \phi_r(\beta, \alpha) = 0.$$

These properties are exploited in the proof of the following lemma.

Lemma 8. *The index and the orientation of t are given by the absolute value and the sign of*

$$\phi_r(\mu, p_l) + \phi_l(\mu, p_r) - \phi_l(\mu, p_l) - \phi_r(\mu, p_r),$$

respectively.

4.2 Data Structure

We extend the data structure of Sect. 3.2, keeping the sequences of outgoing and incoming edges around every vertex and the dynamic trees for S and T. The extensions are:

1. We maintain the sequences of meridian faces F^0 and edges E^0 under insertion and deletion of subsequences, e.g., using balanced trees.
2. With every edge e that is in either S or T, we store

$$\phi_r(\mu, e) = \begin{cases} 1, & \text{if } e = e_i \text{ for some } e_i \in E^0 \text{ and right}(e) = f_i, \\ 0, & \text{otherwise,} \end{cases}$$

   which tells us if e is crossed by the meridian from right to left. Symmetrically, we store $\phi_l(\mu, e)$, which can be derived analogously. Using (4.1) above, we can now in time $O(\log |E|)$ calculate the value of $\phi_r(\mu, p)$ and $\phi_l(\mu, p)$ for every *dynamic path* p of S or T; see [17] for the details and terminology.
3. With every face, we keep a topologically ordered sequence of the edges on the two paths that bound the face.

209

Queries. For the query operation, we again evert u and v in S and T to find their order around s' and t'. Using Lemma 8 and the data structure above, we find the index and orientation of t. Finally, we refer to Table 2 for the answer.

Insertions. Consider the case where a new edge e is inserted into face f, splitting it into f' and f''. The edge lists around f' and f'' are easily derived from the edge lists around f. The meridian is unaffected if $f \notin F^0$. Otherwise, one or both of f' and f'' may become part of the updated meridian, depending on where the meridian edges appear around f (we use the edge list around f to decide which case we are in). For example, if there is a meridian edge on both f' and f'', they both become part of the meridian and e becomes a new meridian edge. In any case, there are only a constant number of updates to the meridian lists.

A straightforward analysis shows that all operations can be performed in logarithmic time, including the updates to the values of ϕ_l and ϕ_r stored in S and T.

Deletions. Consider the case where deletion of the edge e between faces f' and f'' creates a new face f. Creating the edge list around f is handled as above.

In contrast, the meridian may change drastically. The change occurs when both f' and f'' are meridian faces: We cannot just merge them into one, as that would violate the second condition of Def. 7—in other words, the meridian curve μ would no longer be simple.

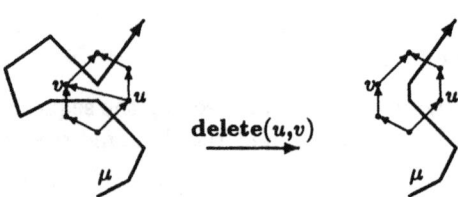

Fig. 5. Deletion of an edge that separates two meridian faces

To remedy this, we must remove everything between f' and f'' from the meridian, as shown in Fig. 5. Even though the data structure for the meridian face and edge lists can be updated in logarithmic time, the values $\phi_l(\mu, e)$ and $\phi_r(\mu, e)$ at every removed meridian edge e also have to be changed, which takes time $O(n \log n)$ in the worst case. However, an easy amortisation argument (store a credit with each meridian edge) shows that a sequence of m updates and queries can be executed in time $O(m \log n)$.

In summary, we have the following theorem:

210

Theorem 9. *The Dynamic Transitive Closure Problem on spherical st-graphs can be solved in amortised time* $O(\log n)$, *where n denotes the number of edges. The data structure uses linear space and can be initialised in linear time.*

4.3 Worst-Case Time Bounds

We will roughly sketch how to remove the amortisation. Obviously, the problem is that we do not have time to remove the meridian cycles arising from a delete operation. However, it is not hard to show that such meridian cycles do not influence the proof of Lemma 8: In a nutshell, whenever a path crosses a such a meridian *cycle*, it most re-cross the same cycle later in the other direction (meridian cycles cannot separate s from t). The effects on the value stated in the lemma cancel. Hence we let sleeping dogs lie—we do *not* remove the meridian cycles but instead just make sure that they remain cycles as the graph undergoes further changes.

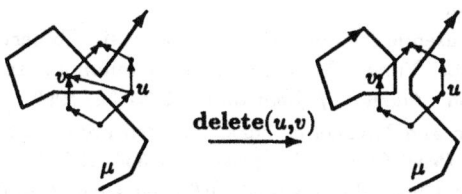

Fig. 6. An edge deletion in the algorithm for worst-case time bounds.

A remaining minor problem is that this results in more and more meridian cycles as we go, so we use *global rebuilding* [15] to construct an unpolluted data structure in the background.

Theorem 10. *The Dynamic Transitive Closure Problem on spherical st-graphs can be solved in time* $O(\log n)$. *The data structure uses linear space and can be initialised in linear time.*

5 Acknowledgements

I am indebted to Gerth Stølting Brodal for several discussions that led to this work and for his insightful comments on an earlier version of this paper. I also thank Lars Arge, Peter Binderup, Peter Bro Miltersen, and Gudmund Frandsen for comments that considerably improved the presentation.

References

1. Giuseppe Di Battista and Roberto Tamassia. Algorithms for plane representations of acyclic digraphs. *Theoretical Computer Science*, 61:175–198, 1988.

2. Paul Beame and Faith Fich, 1994. Personal communication, reported by Peter Bro Miltersen.

3. Hans L. Bodlaender. Dynamic algorithms for graphs with treewidth 2. In *19th International Workshop on Graph Theoretic Concepts in Computer Science (WG)*, volume 790 of *Lecture Notes in Computer Science*, pages 112–124. Springer Verlag, Berlin, 1993.

4. Robert F. Cohen, S. Sairam, Roberto Tamassia, and Jeffrey S. Vitter. Dynamic algorithms for optimization problems in bounded tree-width graphs. In *Proceedings of the 3rd Conference on Integer Programming and Combinatorial Optimization*, 1993.

5. D. Eppstein, Z. Galil, G. F. Italiano, and A. Nissenzweig. Sparsification—A technique for speeding up dynamic graph algorithms. In *Proc. 33rd FOCS*, pages 60–69, 1992.

6. David Eppstein, Zvi Galil, Giuseppe F. Italiano, and Thomas H. Spencer. Seperator based sparsification for dynamic planar graph algorithms. In *Proc. 25th STOC*, pages 208–217, 1993.

7. David Eppstein, Giuseppe Italiano, Roberto Tamassia, Robert E. Tarjan, Jeffery Westbrook, and Moti Yung. Maintenance of a minimum spanning forest in a dyamic planar graph. *Journal of Algorithms*, 13:33–54, 1992.

8. Michael L. Fredman and Michael E. Saks. The cell probe complexity of dynamic data structures. In *Proc. 21st STOC*, pages 345–354, 1989.

9. Harold N. Gabow and Matthias Stallman. Efficient algorithms for graphic matroid intersection and parity. In *Proc. 12th ICALP*, volume 194 of *Lecture Notes in Computer Science*, pages 210–220. Springer Verlag, Berlin, 1985.

10. Monika Rauch Henzinger and Valerie King. Randomized dynamic graph algorithms with polylogarithmic time per operation. In *27th STOC*, pages 519–527. ACM, 1995.

11. Giuseppe F. Italiano, Alberto Marchetti Spaccamela, and Umberto Nanni. Dynamic data structures for series parallel digraphs. In *Proc. First Workshop on Algorithms and Data Structures (WADS)*, volume 382 of *Lecture Notes in Computer Science*, pages 352–373. Springer Verlag, Berlin, 1989.

12. P. B. Miltersen, S. Subramanian, J. S. Vitter, and R. Tamassia. Complexity models for incremental computation. *Theoretical Computer Science*, 130:203–236, 1994.

13. Peter Bro Miltersen. Lower bounds for union-split-find related problems on random access machines. In *Proc. 26th STOC*, pages 625–634. ACM, 1994.

14. Peter Bro Miltersen, Noam Nisan, Shmuel Safra, and Avi Wigderson. On data structures and asymmetric communication complexity. In *Proc. 27th STOC*, pages 103–111. ACM, 1995.

15. Mark H. Overmars. *The design of dynamic data structures*, volume 156 of *Lecture Notes in Computer Science*. Springer Verlag, Berlin, 1983.

16. Monika Rauch. Improved data structures for fully dynamic biconnectivity. In *26th STOC*, pages 686–695. ACM, 1994.

17. Daniel D. Sleator and Robert Endre Tarjan. A data structure for dynamic trees. *Journal of Computer and Systems Sciences*, 26:362–391, 1983.

18. Sairam Subramanian. A fully dynamic data structure for reachability in planar digraphs. In *Proc. 1st Ann. European Symp. on Algorithms (ESA)*, volume 726 of *Lecture Notes in Computer Science*, pages 372–383. Springer Verlag, Berlin, 1993.

19. Roberto Tamassia. A dynamic data structure for planar graph embedding. In *Proc. 15th ICALP*, volume 317 of *Lecture Notes in Computer Science*, pages 576–590. Springer Verlag, Berlin, 1988.

20. Roberto Tamassia and Franco P. Preparata. Dynamic maintenance of planar digraphs, with applications. *Algorithmica*, 5:509–527, 1990.

21. Roberto Tamassia and Ioannis G. Tollis. Dynamic reachability in planar digraphs with one source and one sink. *Theoretical Computer Science*, 119:331–343, 1993.

22. J. van Leeuwen. Graph algorithms. In J. van Leeuwen, editor, *Algorithms and complexity*, volume A of *Handbook of theoretical computer science*, chapter 10, pages 525–631. Elsevier, Amsterdam, 1990.

23. B. Xiao. *New bounds in cell probe model.* Doctoral dissertation, University of California, San Diego, 1992.

Planarity for Clustered Graphs

Qing-Wen Feng   Robert F. Cohen   Peter Eades

Department of Computer Science, University of Newcastle
University Drive, Callaghan NSW 2308, AUSTRALIA
Email: qwfeng, rfc, eades@cs.newcastle.edu.au

(extended abstract)

Abstract. In this paper, we introduce a new graph model known as *clustered graphs*, i.e. graphs with recursive clustering structures. This graph model has many applications in informational and mathematical sciences. In particular, we study C-planarity of clustered graphs. Given a clustered graph, the C-planarity testing problem is to determine whether the clustered graph can be drawn without edge crossings, or edge-region crossings. In this paper, we present efficient algorithms for testing C-planarity and finding C-planar embeddings of clustered graphs.

1   Introduction

Many systems, particularly those which present relational information, include a graph drawing function. Examples include CASE tools, idea organizing systems, reverse engineering systems and software design systems. Such systems have motivated a great deal of research on algorithms for drawing graphs; the survey [2] contains over 250 references.

As the amount of information that we want to visualize becomes larger and more complicated, classical graph models tend to be insufficient. Some more powerful graph formalisms for representing information have been introduced, e.g. hypergraphs [3], compound digraphs [17], cigraphs [13] and higraphs [9]. The problem of automatically drawing these types of graphs appears difficult. To date only heuristic algorithms for hierarchical layout of compound digraphs have been presented [17, 16].

In this paper, we introduce a practical and simple graph model called *clustered graphs*. A clustered graph consists of a graph G and a recursive partitioning of the vertices of G (see Fig. 1). Each partition is known as a *cluster* of a subset of the vertices of G. Clustering appears in the diagrams produced in a wide number of applications areas, such as software engineering [19], knowledge representation [11], software visualization [18], idea organization [12], VLSI design [9], and general divide and conquer problem solving methodologies.

Planarity is a much studied area for classical graphs. For example, the problem of minimizing edge crossings is proved to be NP-hard [8]. However, efficient algorithms for testing whether a graph is planar (i.e. can be drawn without edge crossings) exist [10, 14, 4, 6]. Planarity issues relating to the more powerful graph

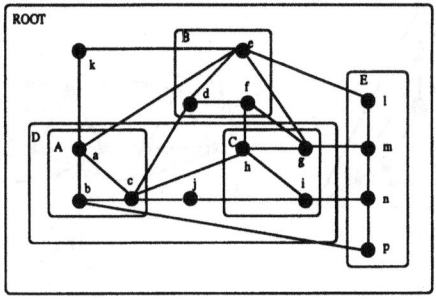

Fig. 1. An Example of a Clustered Graph

models mentioned above have not been studied. In this paper, we introduce *C-planarity*, the planarity of clustered graphs. In a drawing of a clustered graph, vertices and edges are drawn as points and curves as usual. Clusters are drawn as simple closed curves that define closed regions of the plane. The region for each cluster contains the drawing of the subgraph induced by its vertices and no other vertices. A region for a cluster contains the regions for all its subclusters and does not intersect the region for any other cluster. A clustered graph is *C-planar* if it has a drawing with no crossings between distinct edges, or crossings between an edge and a region. Note that the planarity of the underlying graph does not imply the existence of a C-planar drawing of a clustered graph. For example, in Fig. 2, two edges cross a region to which they do not belong.

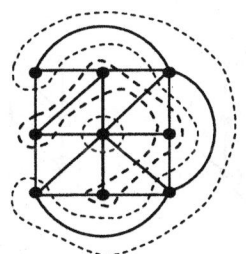

Fig. 2. A Non C-planar Clustered Graph with Planar Underlying Graph

It appears that C-planarity testing is not a trivial extension of planarity testing of classical graphs. For example, consider the clustered graph in Fig. 3. Suppose that the vertices on three triangles belong to three separate clusters. It is obvious that the graph is planar in the usual sense. The graph induced by the vertices of each cluster is planar; and the graph obtained by collapsing any cluster to a vertex is also planar. However, this clustered graph is not C-planar.

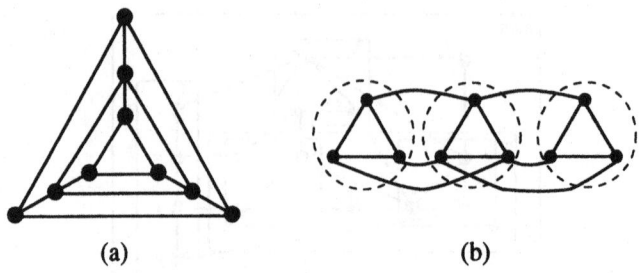

(a)   (b)

Fig. 3. A Planar but Not C-Planar Clustered Graph

Based on analysis of other examples, it appears to us that the st-numbering based planarity testing algorithms in [14] and [4] also cannot be easily adapted to C-planarity testing.

In a clustered graph, the subgraph induced by the vertices of a cluster may not be connected even if the entire underlying graph is connected. This non-connectedness of clusters makes things more complicated. There can be many possible ways to form the regions for clusters even with a fixed embedding of the underlying graph. Fig. 4 gives an example. In Fig. 4(a) and (b), the embeddings of the underlying graph are the same, while the formations of the regions are different. Only the example in (b) gives a C-planar drawing. In the rest of the paper, we will concentrate on clustered graphs where the graph induced by each cluster is connected. We call such graphs *connected clustered graphs*.

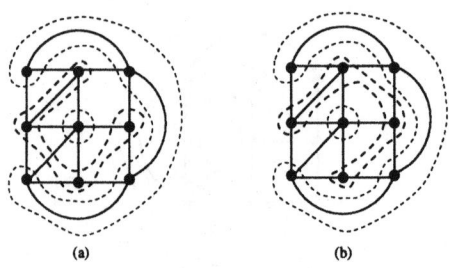

(a)   (b)

Fig. 4. Different Region Formations on the Same Graph

In this paper, we develop algorithms for testing C-planarity and finding C-planar embeddings of connected clustered graphs in $O(n^2)$ time. In Section 2, we present some terminology and some useful characterizations of C-planar clustered graphs. In Section 3, we present an algorithm for testing C-planarity in connected clustered graphs. We also extend the testing algorithm to a C-planar embedding algorithm. We conclude in Section 4 with some interesting open problems.

Due to space limitations of the extended abstract, the proofs are either omitted or presented as sketches, and the algorithms are also described as sketches.

2 Preliminaries

A *clustered graph* $C = (G, T)$ consists of an undirected graph G and a rooted tree T such that the leaves of T are exactly the vertices of G. Each node ν of T represents a *cluster* $V(\nu)$ of the vertices of G that are leaves of the subtree rooted at ν. Note that the tree T describes an inclusion relation between clusters. The tree T is called the *inclusion tree* of C. The graph G is called the *underlying graph* of C. We let $T(\nu)$ represent the subtree of T rooted at node ν and $G(\nu)$ denote the subgraph of G induced by the cluster associated with node ν. We define $C(\nu) = (G(\nu), T(\nu))$ to be the *sub-clustered graph* associated with node ν. When necessary to avoid confusion, we refer to edges of G as *adjacency edges* and edges of T as *inclusion edges*. For the purposes of this paper, we can assume that each node in T has at least two children except for leaf nodes.

A *drawing* of a clustered graph $C = (G, T)$ is a representation of the clustered graph in the plane. Each vertex of G is represented by a point. Each edge of G is represented by a simple curve between the drawing of its endpoints. For each node ν of T, the cluster $V(\nu)$ is drawn as a simple closed region R defined by a simple closed curve in the plane such that:

- the regions for all sub-clusters of R are completely contained in the interior of R;
- the regions for all other clusters are completely contained in the exterior of R;
- if there is an edge e between two vertices of $V(\nu)$ then the drawing of e is completely contained in R.

Given a drawing $\mathcal{D}$ of $C = (G, T)$, we produce a *consistent drawing* $\mathcal{D}'$ of G by removing the region boundary curves from $\mathcal{D}$.

We say that the drawing of edge e and region R have an *edge-region crossing* if the drawing of e crosses the boundary of R more than once. A drawing of a clustered graph is *C-planar* if there are no edge crossings or edge-region crossings. If a clustered graph has a C-planar drawing then we say it is *C-planar* (Fig. 1 gives an example of a C-planar clustered graph). Note that $C = (G, T)$ is C-planar only if G is planar. A C-planar drawing also contains a planar drawing of the underlying graph.

An edge is said to be *incident* with a cluster $V(\nu)$ if one end of the edge is a vertex of $V(\nu)$ but the other end is not in $V(\nu)$. An *embedding* of $C = (G, T)$ includes an embedding of G plus the circular ordering of edges crossing the boundary of the region of each non trivial cluster (a cluster which is not a single vertex). In other words, an embedding of a clustered graph consists of the circular ordering of edges around each cluster which are incident to that cluster.

A clustered graph $C = (G, T)$ is a *connected clustered graph* if each cluster induces a connected subgraph of G. The following theorem gives a necessary and sufficient condition for the C-planarity of connected clustered graphs.

Theorem 1. *A connected clustered graph $C = (G, T)$ is C-planar if and only if graph G is planar and there exists a planar drawing $\mathcal{D}$ of G, such that for each node ν of T, all the vertices and edges of $G - G(\nu)$ are in the outer face of the drawing of $G(\nu)$.*

Proof. Note that since each $G(\nu)$ is connected, the boundary of its outer face in any planar drawing of $G(\nu)$ consists of a connected cycle.

Consider a clustered graph $C = (G, T)$ with a C-planar drawing $\mathcal{D}$, let $\mathcal{D}'$ be the consistent drawing of the underlying graph G. Suppose that there is a node ν of T such that $G - G(\nu)$ are not all in the outer face of the drawing of $G(\nu)$ in $\mathcal{D}'$. Then there must exist a vertex v in $G - G(\nu)$ which is drawn in the interior of the outer facial cycle of the drawing of $G(\nu)$. Then any simple region that contains the drawing of $G(\nu)$ must also contain v. This contradicts the assumption that $\mathcal{D}$ is a C-planar drawing of C.

Now, consider a planar drawing $\mathcal{D}'$ of the underlying graph G, such that for each node ν of T, $G - G(\nu)$ is drawn in the outer face of the drawing of $G(\nu)$. We produce a drawing $\mathcal{D}$ of clustered graph $G = (C, T)$ by adding cluster boundaries to $\mathcal{D}'$ recursively up tree T. For each node ν of T, we make the boundary for cluster of ν by drawing a simple closed curve in the outer face of $G(\nu)$ along its outer facial cycle, $\epsilon > 0$ distance away from the outer facial cycle of $G(\nu)$ or the boundary of the included regions. In this construction, each region is simple, and the region inclusion convention is followed. There are no edge crossings, since D' is a planar drawing of G. By construction, the only edges that cross the boundary of the region for a node ν of T are edges connecting vertices of $G(\nu)$ with vertices of $G - G(\nu)$. Consequently, there are no edge-region crossings. Clearly, this construction produces a C-planar drawing of C. $\qquad\square$

Next, we present a characterization of C-planarity of general clustered graphs. We need some more terminology here. Suppose that $C_1 = (G_1, T_1)$ and $C_2 = (G_2, T_2)$ are two clustered graphs, T_1 is a subtree of T_2, and for each node ν of T_1, $G_1(\nu)$ is a subgraph of $G_2(\nu)$. Then we say C_1 is a *sub-clustered graph* of C_2, and C_2 is a *super-clustered graph* of C_1.

Theorem 2. *A clustered graph $C = (G, T)$ is C-planar if and only if it is a sub-clustered graph of a connected and C-planar clustered graph.*

Proof. Suppose that clustered graph $C = (G, T)$ is C-planar, where $G = (V, E)$; and $\mathcal{D}$ is a C-planar drawing of C. Let ν be a node of T, and let $w_1, w_2, \ldots, w_k$ be the points in circular order where edges cross the region boundary of cluster ν. Let v_i and v_{i+1} be the vertices of $G(\nu)$ which connect to adjacent points w_i and w_{i+1} respectively. In this C-planar drawing $\mathcal{D}$ of C, vertices w_i, w_{i+1}, v_{i+1} and v_i are on boundary of some face f (See Fig. 5(a)), since v_i and v_{i+1} are on the same side of the region boundary of cluster ν. We add edge (v_i, v_{i+1}) to $G(\nu)$

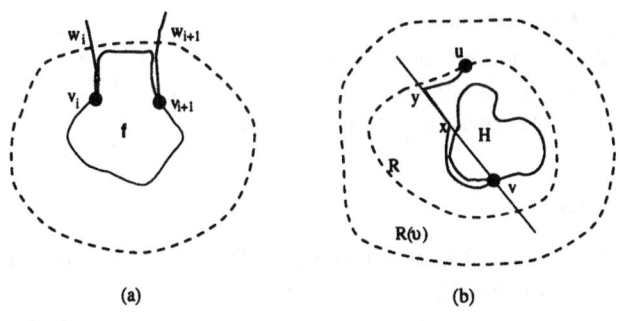

Fig. 5. The Example for the Proof of Theorem 2

if v_i and v_{i+1} are not connected previously. We draw it by making a curve from v_i to v_{i+1} along the curve $(v_i, w_i, w_{i+1}, v_{i+1})$ $\epsilon > 0$ distance away from it, inside face f. This does not produce any crossings since the curve we draw is totally inside face f. Suppose that there is a connected component H of $G(\nu)$ which has no connection with $G - G(\nu)$. In the C-planar drawing $\mathcal{D}$ of C, let region R be the smallest subregion of $R(\nu)$ (the region for cluster ν) which contains the drawing of H. Let $\bar{R}$ be the subregion of R bounded by the the bounding cycle of R and the outer facial cycle of H. Suppose that v is a vertex on the outer facial cycle of H and u is a vertex of $G(\nu)$ on the boundary of region R. Any straight line through v must contain a segment (x, y) which is contained in $\bar{R}$, with x on the outer facial cycle of H and y on the boundary of R (see Fig. 5(b)). We add an edge (v, u) to $G(\nu)$. We draw edge (v, u) in $\bar{R}$ in the following manner. First, we draw a curve from v to x along the outer facial cycle of H. Then we continue along the segment (x, y) to point y. Finally, we follow the boundary of R to u. This does not introduce any crossings since this curve is formed all inside region $\bar{R}$. By the operations above, a super-clustered graph $C'(G', T)$ of C is obtained, where $G' = (V, E')$; a C-planar drawing of C' is formed; and C' is a connected clustered graph.

Suppose that C is a sub-clustered graph of C', where C' is connected and C-planar. A C-planar drawing of C can be obtained by restricting a C-planar drawing of C' to C. Therefore, C is C-planar. $\qquad\square$

3   C-planarity Testing

In this section, we describe an efficient algorithm for testing C-planarity in connected clustered graphs.

Our algorithm is based on Theorem 1. For a clustered graph $C = (G, T)$, we test whether there is a planar embedding of G such that for each node ν of T, $G - G(\nu)$ is embedded in the same face (the outer face) of $G(\nu)$. We try to embed the subgraph induced by each cluster one by one, following a traversal of T from bottom to top. For each node ν of T, we test whether $G(\nu)$ has any planar embeddings that satisfy the conditions of Theorem 1 for $C(\nu)$. If we proceed to the root cluster, and such embeddings exist for the root of T, then the clustered graph is C-planar, otherwise it is not C-planar.

3.1 Background

We apply the well known PQ-tree technique [4] in our algorithm. The following is a brief definition of the PQ-tree data structure.

The *PQ-trees* over a set U are trees whose *leaves* are elements of U and whose *internal* nodes are distinguished as being either *P-nodes* or *Q-nodes*. Reading the leaves of a tree from left to right yields its *frontier*. We can make two types of transformations on a PQ-tree:

1. arbitrarily *permute* the children of a P-node;
2. *reverse* the children of a Q-node.

By making such transformations on a PQ-tree, its frontiers form a set of permutations of the leaves. We say the structure of a PQ-tree *expresses* a set of permutations of its leaves.

In the PQ-tree planarity testing algorithm, graphs are decomposed into biconnected components, and each biconnected component is tested for planarity. Each vertex of a biconnected component is labeled by its *st-number* and added in the st-number order. The st-numbering is calculated in the following manner. An st-numbering consists of a biconnected graph G with n vertices and an arbitrary edge (s, t). The vertices of G can be numbered from 1 to n such that vertex s receives number 1, vertex t receives number n, and every vertex except s and t is adjacent both to a lower-numbered and a higher-numbered vertex. Vertices s and t are called the *source* and the *sink* respectively.

We need the following lemma to understand the PQ-tree planarity testing algorithm and also to show the correctness of our algorithm.

Lemma 3. *[15] Suppose that a graph G is a biconnected and st-numbered planar graph. Let $G_k = (V_k, E_k)$ be the subgraph of G induced by vertices $V_k = \{1, 2, \ldots, k\}$, $1 \leq k \leq n$. If edge (s, t) is drawn on the boundary of the outer face in an embedding of G, then all the vertices and edges of $G - G_k$ are drawn in the outer face of the plane subgraph G_k of G.*

Using the notation of the above Lemma, a planar drawing of G with all the vertices and edges of $G - G_k$ drawn in the outer face of G_k is called a *planar st-drawing* of G.

The PQ-tree planarity testing algorithm maintains a PQ-tree throughout the algorithm. Whenever a vertex v_i is added, an appropriate operation (called *reduction*) on the PQ-tree is made. After each reduction step, the PQ-tree exactly expresses the set of possible permutations of the edges that connect to G_k along the outer face of planar st-drawing of G_k.

The efficient implementation of the PQ-tree technique is fully described in [4].

We use the concept of *virtual edge* and *virtual vertex* in our algorithm. For a graph $G(V, E)$ with subgraph $G'(V', E')$, those edges with one end in V' and the other end in $V - V'$ are called *virtual edges* of G', and those ends of the virtual edges in $V - V'$ are called *virtual vertices* of G'.

3.2 A Testing Algorithm

We test C-planarity based on Theorem 1. We determine whether there is a planar embedding of G such that for each node ν of T, $G - G(\nu)$ is embedded in the same face (the outer face) of $G(\nu)$.

We try to embed the subgraph induced by each cluster recursively, following a traversal of T from bottom to top. For a node ν of T with children $\mu_1, \ldots, \mu_d$, we test whether $G(\nu)$ has any planar embeddings that satisfy the conditions of Theorem 1 for $C(\nu)$. We find such embeddings for $C(\nu)$ by combining the possible embeddings of each child cluster μ_i which are found recursively. Then we record such embeddings of $G(\nu)$ for later testing of the parent cluster of ν. We construct a *representative graph* that represents all the possible orderings of edges that are incident to cluster ν around the outer face of $G(\nu)$; then replace $G(\nu)$ in G with the representative graph. Graph G is changed every time we process a node of tree T. At the time when the algorithm proceeds to cluster ν, planar embeddings of G reflect all planar embeddings of the children of ν that satisfy the conditions of Theorem 1.

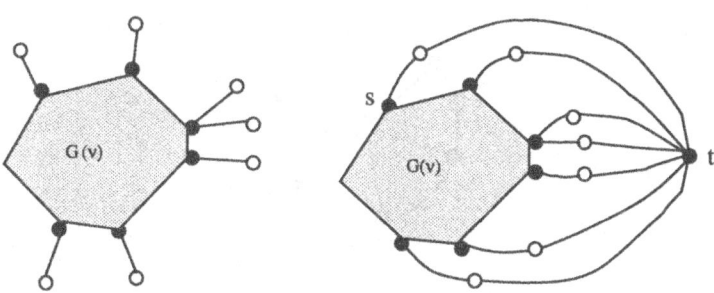

Fig. 6. Illustration of Choosing s and t for Each Cluster

At cluster ν, we not only test whether $G(\nu)$ is planar, but also test whether the edges that are incident to cluster ν can be drawn in the outer face of $G(\nu)$. Therefore, we have to take into account the virtual edges of $G(\nu)$. We form a graph $G'(\nu)$ by adding virtual edges to $G(\nu)$, and apply the PQ-tree planarity testing algorithm to $G'(\nu)$. We add a vertex on each virtual edge of $G(\nu)$ to distinguish them from each other; and let $G'(\nu)$ be the graph resulted from connecting the virtual edges of $G(\nu)$ to a single virtual vertex (see Fig. 6).

The PQ-tree algorithm decomposes a graph into biconnected components and tests each of them for planarity respectively. The following lemma facilitates the application of the PQ-tree algorithm to $G'(\nu)$.

Lemma 4. *Suppose that F is a connected subgraph of G. Let F' be the graph constructed by adding virtual edges to F, and connecting each virtual edge to a single virtual vertex. If there are at least two virtual edges in F', then all the virtual edges belong to the same biconnected component of F'.*

We apply the PQ-tree testing algorithm to $G'(\nu)$. For the biconnected component B that contains the virtual edges, we compute the st-numbering by choosing the single virtual vertex as the sink and any vertex of $G(\nu)$ that connects to the virtual vertex as the source. If the planarity testing on $G'(\nu)$ returns TRUE, then $G'(\nu)$ is planar, and by Lemma 3, all the edges incident to cluster ν can be drawn in the outer face of $G(\nu)$. Let T_{PQ} be the nonempty PQ-tree that results when the planarity testing on biconnected component B is completed. The tree T_{PQ} expresses all the possible orderings of the edges that are incident to cluster of ν along the outer face of $G(\nu)$. We associate T_{PQ} with cluster of ν.

At each cluster ν, we need to determine whether we can combine the planar embeddings of each of its child cluster μ_i that satisfy the conditions of Theorem 1 for $C(\mu_i)$ into planar embeddings of $G(\nu)$ that satisfy the conditions of Theorem 1 for $C(\nu)$. For each child cluster μ_i of cluster ν, we replace $G(\mu_i)$ with a representative graph which is constructed from *wheel* graphs.

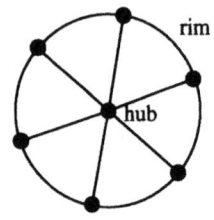

Fig. 7. A "Wheel" Graph

A *wheel* graph consists of a vertex called the *hub* of the wheel and a cycle called the *rim* of the wheel, such that the *hub* is connected to every vertex on the *rim*. (see Fig. 7). Every face of a wheel is a triangle except the face bounded by the rim. We call this face the *rim face*. If the rim face of a wheel graph is drawn as the outer face, then we say the drawing is a *canonical drawing* of the wheel.

The following lemma shows that wheel graphs have certain properties that can be exploited in construction of representative graphs.

Lemma 5. *Suppose that G is a planar graph with subgraph F. Let F' be the subgraph constructed by adding virtual edges to F. Let F_1, F_2, ..., F_k be a collection of wheel subgraphs of F, such that each wheel graph has a distinguished hub, the hubs only connect to vertices on the corresponding rim in F, and every two wheel graphs have at most one common vertex. If there is a planar drawing $\mathcal{D}$ of F' with virtual edges drawn in the outer face of F, then there must also exist a planar drawing $\mathcal{D}'$ of F' such that :*

- *The circular ordering of the virtual edges along the outer face of F is preserved.*
- *Every wheel graph F_i is drawn* canonically.
- *$F - F_i$ is drawn in the outer face of F_i.*

In a PQ-tree, a P-node corresponds to a cut vertex in the graph the PQ-tree represents; a Q-node corresponds to a biconnected component of the graph; and the leaves correspond to the virtual edges. Given a PQ-tree associated with a graph G, we construct a representative graph G_{PQ} in the following manner. For each Q-node, we construct a wheel graph; for each P-node, we construct a vertex which serves as a cut vertex connecting the wheels (see Fig. 8). The constructed G_{PQ} has the following properties:

- The ordering of the virtual edges is the same as the ordering of the leaves of the PQ-tree.
- Biconnected components in G correspond to wheels in G_{PQ}.
- Cut vertices in G correspond to cut vertices in G_{PQ}.
- Every vertex in G_{PQ} has its counterpart in G except the vertices constructed as the hubs of wheels.

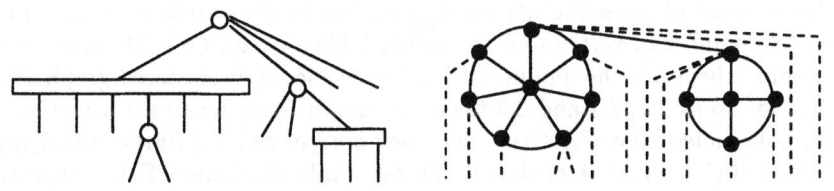

Fig. 8. The Construction of Representative Graphs

The algorithm for testing the C-planarity of a connected clustered graph contains a main loop which is a post order traversal of tree T. For each node ν of T, we test the planarity of $G'(\nu)$, construct a representative graph for $G(\nu)$, and replace $G(\nu)$ with the representative graph in G.

Algorithm 1 CPT
    Input: a connected clustered graph $C = (G, T)$;
    Output: a boolean value indicating whether C is C-planar.

(1) Use the PQ-tree planarity testing algorithm PT to determine whether G is planar. If G is not planar, then return FALSE and exit.

(2) We proceed on T from bottom to top. For each non-leaf node ν of T, perform the following:

(2.1) Form graph $G'(\nu)$ from $G(\nu)$.

Apply the PQ-tree planarity testing algorithm PT to $G'(\nu)$. For the biconnected component that contains the virtual edges, choose the single virtual vertex as the sink t; choose any vertex of $G(\nu)$ that connects to vertex t as the source to compute st-numbering. Let T_{PQ} be the resulted PQ-tree when the testing of this component is completed.

If any biconnected component is non-planar, then return FALSE and exit.

(2.2) Construct representative graph $G(\nu)_{PQ}$ based on T_{PQ}.

(2.3) Replace subgraph $G(\nu)$ in G with $G(\nu)_{PQ}$ and update G.

(3) When we proceed to the root of T, test the planarity of the updated G using algorithm PT. If graph G is not planar then return FALSE otherwise return TRUE.

$\square$

The correctness of our algorithm follows immediately from Theorem 1 and Lemmas 3, 4, and 5.

It is shown in [4] that for a given graph G with n vertices and m edges, the PQ-tree planarity testing algorithm requires at most $O(n)$ steps. There are algorithms requiring $O(n + m)$ steps to find biconnected components of a graph [1], and to generate st-numbering for each biconnected component [7]. Note that graph G is updated throughout our algorithm. At node ν of T, the vertices in G that serve as hubs of wheels have no connection with the rest of the graph except the vertices on the corresponding rim. Therefore, they do not appear in the updated G when the algorithm proceeds to the parent node of ν in T. Thus, the number of vertices of the updated G is always $O(n)$ throughout the algorithm, where n is the number of vertices of the input clustered graph. Since each of the steps 2.1, 2.2 and 2.3 takes linear time in terms of the size of $G'(\nu)$ which is bounded above by $O(n)$, and they are iterated $|T|$ times, step 2 takes $O(|T| \cdot n)$ time in all. Both of step 1 and step 3 take $O(n)$ time. Hence algorithm CPT takes $O(|T| \cdot n)$ time. We have assumed that each node in T has at least two children except for leaf nodes. Thus T has at most $2n$ nodes. Therefore algorithm CPT takes $O(n^2)$ time. The following theorem summarizes the performance of algorithm CPT.

Theorem 6. *Algorithm CPT tests C-planarity of an n vertex, connected clustered graph $C = (G, T)$ in $O(n^2)$ time.*

3.3 An Embedding Algorithm

In this section, we show how to extend the C-planarity testing algorithm CPT to a C-planar embedding algorithm $CPEmbed$. The input to algorithm $CPEmbed$

is a connected clustered graph. The algorithm returns a C-planar embedding if the input clustered graph is C-planar; otherwise, returns an empty embedding.

The algorithm *PEmbed* [5] replaces algorithm *PT* in our algorithm *CPEmbed*. The algorithm *PEmbed* tests the planarity of a graph and finds a planar embedding if the graph is planar. It uses the same PQ-tree technique as algorithm *PT*. If a graph is planar, *PEmbed* records a partial planar embedding of the graph and returns a PQ-tree associated with each biconnected component of the graph. By choosing an ordering of the leaves of each PQ-tree that the PQ-tree accepts, together with the partial planar embedding, a complete planar embedding of the graph can be obtained in linear time [5].

In our algorithm *CPEmbed*, we find a circular ordering of the edges incident to each cluster recursively, following a traversal of tree T from top to bottom. We modify algorithm *CPT* to *CPT'* by replacing the primitive PQ-tree planarity testing algorithm *PT* with algorithm *PEmbed*. We use a stack S to record the partial embeddings and the PQ-trees obtained by algorithm *CPT'* at each node ν of T. Visiting the stack from top to bottom forms a traversal of the inclusion tree T from top to bottom.

The algorithm *CPEmbed* is described as follows.

Algorithm 2 CPEmbed

Input: a connected clustered graph $C = (G, T)$;

Output: an C-planar embedding $\mathcal{E}$ of C which consists of a circular ordering of edges incident to each cluster of C.

(1) Perform algorithm *CPT'* on clustered graph $C = (G, T)$. At node ν of T, push the partial planar embedding and the PQ-tree associated with cluster of ν onto a stack S. If *CPT'* returns FALSE, then return an empty C-planar embedding and exit.

(2) $\mathcal{E} = \emptyset$.

While S is not empty, perform the following:

(2.1) Pop the partial embedding and the PQ-tree from stack S which corresponds to node ν of T.

Let $ORD(\nu)$ be the circular ordering of edges incident to cluster ν in embedding $\mathcal{E}$. Choose $ORD(\nu)$ as the ordering of the leaves of the PQ-tree associated with $G'(\nu)$. Find a complete planar embedding $\mathcal{H}_\nu$ of $G(\nu)$ according to $ORD(\nu)$ and the partial embedding popped from the stack.

(2.2) Call procedure *Formalize*$(\mathcal{H}_\nu)$ (below) to modify $\mathcal{H}_\nu$, such that for each wheel subgraph F of $G(\nu)$, F is embedded canonically, and the vertices and edges of $G - F$ are embedded in the outer face (the rim face) of F.

(2.3) For each child μ_i of ν, find the circular ordering $ORD(\mu_i)$ of the edges incident to cluster μ_i according to $\mathcal{H}_\nu$; and let $\mathcal{E}$ be $\mathcal{E} \cup ORD(\mu_i)$.

$\square$

Procedure *Formalize* changes a planar embedding of graph G, such that each wheel subgraph F_i is embedded canonically, and $G - F_i$ is embedded in the outer face (the rim face) of F_i. By Lemma 5, this kind of embedding exists. We formalize the embedding of G by moving the part of $G - F_i$ which are not embedded in the rim face of F_i to the other side of the rim of F_i. The following description of procedure *Formalize* completes our description of algorithm *CPEmbed*.

Procedure *Formalize($\mathcal{H}_\nu$)*

For each wheel subgraph F_i of $G(\nu)$ updated in CPT':
    For each vertex x on the rim of the wheel F_i:
        Suppose that h is the hub of the wheel, r_1, r_2 are the two vertices on the rim adjacent to x, and
$$(h, v_1, v_2, \ldots, v_p, r_1, v_{p+1}, \ldots, v_q, r_2, v_{q+1}, \ldots, v_l, h)$$
        is the circular order of the vertices which connect to x defined by $\mathcal{H}_\nu$. We change this circular ordering to
$$(h, r_1, v_1, v_2, \ldots, v_p, v_{p+1}, \ldots, v_q, v_{q+1}, \ldots, v_l, r_2, h).$$
□

The correctness of this algorithm follows from the correctness of *PEmbed* and *CPT*.

To compute the running time of algorithm *CPT'*, we first note that algorithm *PEmbed* takes linear time (see [5]). Then, by a similar argument as for algorithm *CPT*, algorithm *CPT'* takes $O(n^2)$ time. Consequently, step 1 of algorithm *CPEmbed* takes $O(n^2)$ time. According to [5], step 2.1 takes linear time in terms of the size of $G'(\nu)$. Step 2.2, and 2.3 also take linear time in terms of the size of $G(\nu)$. Therefore, step 2 requires time $O(n \cdot |T|) = O(n^2)$ in all. Thus, algorithm *CPEmbed* require $O(n^2)$ time. The following theorem summarizes the performance of algorithm *CPEmbed*.

Theorem 7. *Algorithm CPEmbed finds a C-planar embedding of an n vertex, connected clustered graph $C = (G, T)$ in $O(n^2)$ time.*

4 Conclusion and Open Problems

In this paper, we have introduced a graph model known as *clustered graphs* and investigated the planarity of clustered graphs. We have presented an efficient algorithm for testing C-planarity in connected clustered graphs, and also extend the C-planarity testing algorithm to a C-planar embedding algorithm.

Some interesting open problems include:

- Can we improve the performance of the proposed algorithms to linear time?
- For non-connected clustered graphs, with a given embedding of the underlying graph, how do we test whether the embedding admits a C-planar drawing?
- Can we find a polynomial time algorithm that tests C-planarity of non-connected clustered graphs, or show that the problem is NP-hard?

References

1. A. V. Aho, J. E. Hopcroft, and J. D. Ullman. *The Design and Analysis of Computer Algorithms*. Addison-Wesley, Reading, Mass., 1974.
2. G. Di Battista, P. Eades, R. Tamassia, and I. Tollis. Algorithms for drawing graphs: An annotated bibliography. Technical report, Department of Computer Science, Brown University, 1993. To appear in Computational Geometry and Applications, currently available from wilma.cs.brown.edu by ftp.
3. Claude Berge. *Hypergraphs*. North-Holland, 1989.
4. K. Booth and G. Lueker. Testing for the consecutive ones property, interval graphs, and graph planarity using PQ-tree algorithms. *Journal of Computer and System Sciences*, 13:335–379, 1976.
5. N. Chiba, T. Nishizeki, S. Abe, and T. Ozawa. A linear algorithm for embedding planar graphs using PQ-trees. *J. of Computer and Sytem Sciences*, 30(1):54–76, 1985.
6. H. de Fraysseix and P. Rosenstiehl. A depth-first-search characterization of planarity. *Annals of Discrete Mathematics*, 13:75–80, 1982.
7. S. Even and R. E. Tarjan. Computing an st-numbering. *Theoretical Computer Science*, 2:339–344, 1976.
8. M.R. Garey and D.S. Johnson. Crossing number is NP-complete. *SIAM J. Algebraic and Discrete Methods*, 4(3):312–316, 1983.
9. D. Harel. On visual formalisms. *Communications of the ACM*, 31(5):514–530, 1988.
10. J. Hopcroft and R. E. Tarjan. Efficient planarity testing. *Journal of ACM*, 21(4):549–568, 1974.
11. T. Kamada. *Visualizing Abstract Objects and Relations*. World Scientific Series in Computer Science, 1989.
12. J. Kawakita. The KJ method – a scientific approach to problem solving. Technical report, Kawakita Research Institute, Tokyo, 1975.
13. Wei Lai. *Building Interactive Digram Applications*. PhD thesis, Department of Computer Science, University of Newcastle, Callaghan, New South Wales, Australia, 2308, June 1993.
14. A. Lempel, S. Even, and I. Cederbaum. An algorithm for planarity testing of graphs. In *Theory of Graphs, International Symposium (Rome 1966)*, pages 215–232. Gordon and Breach, New York, 1967.
15. T. Nishizeki and N. Chiba. *Planar Graphs: Theory and Algorithms, Annals of Discrete Mathematics 32*. North-Holland, 1988.
16. S. C. North. Drawing ranked digraphs with recursive clusters. preprint, 1993. Software Systems and Research Center, AT & T Laboratories.
17. K. Sugiyama and K. Misue. Visualization of structural information: Automatic drawing of compound digraphs. *IEEE Transactions on Systems, Man and Cybernetics*, 21(4):876–892, 1991.
18. C. Williams, J. Rasure, and C. Hansen. The state of the art of visual languages for visualization. In *Visualization 92*, pages 202 – 209, 1992.
19. Rebecca Wirfs-Brock, Brian Wilkerson, and Lauren Wiener. *Designing Object-Oriented Software*. P T R Prentics Hall, Englewood Cliffs, NJ 07632, 1990.

A Geometric Approach to Betweenness

Benny Chor* and Madhu Sudan**

Abstract. An input to the *betweenness* problem contains m constraints over n real variables. Each constraint consists of three variables, where one of the variables is specified to lie inside the interval defined by the other two. The order of the other two variables (which one is the largest and which one is the smallest) is not specified. This problem comes up in questions related to physical mapping in computational molecular biology. In 1979, Opatrny has shown that the problem of deciding whether the n variables can be totally ordered while satisfying the m betweenness constraints is NP–complete. Furthermore, the problem is MAX SNP complete. Therefore, there is some $\epsilon > 0$ such that finding a total order which satisfies at least $m(1 - \epsilon)$ of the constraints (even if they are all satisfiable) is NP–hard. It is easy to find an ordering of the variables which satisfies $1/3$ of the m constraints (e.g. by choosing the ordering at random).

In this work we present a polynomial time algorithm which either determines that there is no feasible solution, or finds a total order which satisfies at least $1/2$ of the m constraints. Our algorithm translates the problem into a set of quadratic inequalities, and solves a semidefinite relaxation of them in $\mathcal{R}^n$. The n solution points are then projected on a random line through the origin. Using simple geometric properties of the SDP solution, we prove the claimed performance guarantee.

1 Introduction

An input to the *betweenness* problem consists of a finite set of n real variables $S = \{x_1, \ldots, x_n\}$, and a finite set of m constraints. Each constraint consists of a triplet $(x_i, x_j, x_k) \in S \times S \times S$. A total order $x_{i_1} < x_{i_2} < \ldots < x_{i_n}$ satisfies the constraint (x_i, x_j, x_k) if either $x_i < x_j < x_k$ or $x_k < x_j < x_i$. That is, each constraint forces the second variable x_j to be between the two other variables x_i and x_k, but does not specify the relative order of x_i and x_k. The betweenness problem is to decide if all constraint can be simultaneously satisfied by a total order of the variables.

In 1979, Opatrny [13] has shown that the betweenness problem is NP–complete. This problem naturally arises in the analysis of certain mapping problems in molecular biology. For example, when trying to order markers on a chromosome,

* Dept. of Computer Science, Technion, Haifa 32000, Israel (benny@cs.technion.ac.il)
** IBM Thomas J. watson Research Center, P.O.Box 218, Yorktown Heights, NY 10598 (madhu@watson.ibm.com)

given the results of a radiation hybrid experiment [6, 3]. A computational task of practical significance, in this context, is to find a total ordering of the markers (the x_i in our terminology) which maximizes the number of satisfied constraints.

Opatrny gave two reductions in his proof of NP–completeness. One of these reductions is from 3SAT. Following his construction, we show, in Section 2, an approximation preserving reduction from MAX 3SAT. This implies that there exists an $\epsilon > 0$, such that finding a total order which satisfies at least $m(1 - \epsilon)$ of the constraints (even if they are all satisfiable) is NP–hard. It is easy to find a total order which satisfies one third of the m constraints. Simply arranges the points in a random order along the line. The probability that a specific constraint is satisfied by such a randomly chosen order is $1/3$. Thus the expected number of constraints satisfied by a random order is at least a third of the m constraints. On the other hand, it is easy to construct examples where at most $m/3$ constraints are satisfiable. Thus to achieve better approximation factors, one needs to be able to recognize instances of the betweenness problems which are not satisfiable.

We present a polynomial time algorithm which either determines that there is no feasible solution, or finds a total order which satisfies at least $1/2$ of the m constraints. Our algorithm translates the problem into a set of quadratic inequalities, and solves a semidefinite relaxation of them in $\mathcal{R}^n$. Let $v_1, \ldots, v_n \in \mathcal{R}^n$ be a feasible solution to the SDP, where each v_i corresponds to the real variable x_i. The n solution points are then projected on a random line through the origin. We show that if "x_j between x_i and x_k" is one of the betweenness constraint, then the angle between the lines $v_i v_j$ and $v_k v_j$ (in $\mathcal{R}^n$) is obtuse. Using this property, we prove that the random projection satisfies each constraint with probability at least $1/2$. This gives a randomized algorithm with the claimed performance guarantee. Next, we show how to derandomize the algorithm. In addition, we demonstrate that our analysis of the semidefinite program is tight. There is an infinite family of inputs to the betweenness problem, such that the resulting SDP is feasible, but any total order of the variables satisfies at most $1/2 + o(1)$ of the m constraints.

Our use of semidefinite programming is inspired by the recent success in using this methodology to find improved approximation algorithms for several optimization problems. The applicability of SDP in combinatorial optimization was demonstrated by Grötschel, Lovász and Schrijver [7] to show that the Theta function of Lovász [12] was polynomial time computable. This application was then turned into exact coloring and independent set finding algorithms for perfect graphs. The use of SDP in approximation algorithms was innovated by the work of Goemans and Williamson [5] who break longstanding barriers in the approximability of MAX CUT and MAX 2SAT by their SDP based algorithm. Further evidence of the applicability of the SDP approach is provided by the works of Karger, Motwani and Sudan [10], who use it to approximate graph coloring, Alon and Kahale [1] (independent set approximation) and by Feige and Goemans [4] (improvements to MAX 2SAT).

Thus the semidefinite programming method has now been used successfully to solve many optimization problems — exactly and approximately. However all the cases where SDP has been used to find approximation algorithms seem to be essentially partition problems (MAX CUT, Coloring, Multicut etc.). Our solution seems to be (to the best of our knowledge) the only case where SDP has been used to solve an ordering problem. This syntactic difference between ordered structures and unordered ones, and the ability of SDP to help optimize over both, offers critical additional evidence on the power of the SDP methodology.

The remaining of this paper is organized as following: Section 2 presents the approximation preserving reduction from MAX 3SAT, as well as other observations about the betweenness problem. Semidefinite programming is briefly reviewed in Section 3. The algorithm is presented in Section 4. Section 5 shows the tightness of our analysis. Finally, Section 6 contains some concluding remarks and open problems.

2 Preliminaries

We start this section with some preliminary observations about the betweenness problem. We begin by analyzing the complexity of finding approximate solutions to the problem. Opatrny [13] has shown that it is hard to decide if a given instance of the betweenness problem is satisfiable. Following Opatrny's proof, we present an approximation preserving reduction from MAX 3SAT to the betweenness problem. It is known that there exists a constant $\epsilon > 0$ such that finding a solution which satisfies of $1 - \epsilon$ fraction of all clauses in a satisfiable problem is NP-hard [2]. Therefore, we conclude that there exists a constant $\epsilon' > 0$ such that finding an ordering which satisfies $(1 - \epsilon')$ fraction of the constraints in a satisfiable instance of the betweenness problem is NP-hard.

Proposition 1. *If, for some $\epsilon > 0$, there exists a polytime algorithm that satisfies $1 - \epsilon$ fraction of all constraints in a satisfiable instance of the betweenness problem, then there exists a polytime algorithm that satisfies $1 - 6\epsilon$ fraction of all clauses in every satisfiable 3-cnf formula.*

Proof. Given a 3-CNF formula ϕ, we construct an instance I of the betweenness problem as follows. For each Boolean variable x_i of ϕ, we create a point p_i in I. In addition we create two special points T and F. Given a clause say $C_j = x_1 \vee x_2 \vee \overline{x_3}$, we create three points $q_j^{(12)}$, $q_j^{(123)}$, and $q_j^{(3)}$. Corresponding to the clause C_j we introduce the following betweenness constraints: F is between p_3 and $q_j^{(3)}$; $q_j^{(12)}$ is between p_1 and p_2; $q_j^{(123)}$ is between $q_j^{(12)}$ and $q_j^{(3)}$ and $q_j^{(123)}$ is between T and F.

Thus corresponding to each clause we may have upto 6 betweenness constraints (3 + number of negated literals in the clause). Given a satisfiable formula ϕ with say the assignment $x_1, \ldots, x_k =$ false and $x_{k+1}, \ldots, x_n =$ true being a satisfying assignment — we first lay out the points p_i and T and F as follows:

$$p_1 \cdots p_k \; F \; p_{k+1} \cdots p_n \; T.$$

Then for a clause, say $C_j = x_1 \lor x_2 \lor \overline{x_3}$ we insert the points $q_j^{(\overline{3})}$, $q_j^{(12)}$ and $q_j^{(123)}$ in order into the linear order above placing each point as far to the right as possible without violating any betweenness constraint. Notice that if the clause is satisfiable then such an ordering would automatically place $q_j^{(123)}$ between F and T, thereby satisfying all betweenness constraints associated with the clause C_j.

To show that the reduction preserves the quality of the approximation, we show that given an ordering which leaves at most k betweenness constraints that are not satisfied, there exists an assignment which leaves at most k clauses of ϕ unsatisfied. W.l.o.g. assume that T lies to the right of F in the linear order. For every point p_j which lies to the right of F, we set the variable x_j to false. The remaining variables are assigned true. Notice that if all the betweenness constraints associated with a clause C_j are satisfied it must be the case that at least one of the literals associated with C_j is assigned true by such an assignment - thus proving our claim.

Thus given a 3-CNF formula ϕ with m clauses we have constructed a betweenness instance I with $m' \leq 6m$ constraints. Further more given an ordering satisfying $(1 - \epsilon)m'$ constraints, we can reconstruct an assignment satisfying $m - \epsilon m' \geq m(1 - 6\epsilon)$ clauses of ϕ.

Corollary 2. *There exists an $\epsilon > 0$, such that satisfying $1 - \epsilon$ fraction of the constraints in a satisfiable instance of the betweenness problems is NP-hard.*

Next we show what can achieved by the obvious randomized algorithm for the betweenness problem.

The natural randomized algorithm for the betweenness problem arranges the points in a random order along the line. The probability that a specific constraint is satisfied by such a randomly chosen order is $1/3$. Thus the expected number of constraints satisfied by a random order is at least a third of all the constraints. By the method of conditional probabilities one can find such order in polynomial time. Since this order satisfies $1/3$rd of all constraints, it is within $1/3$rd of the optimal ordering.

Before going on to more sophisticated techniques for solving this problem, let us examine the main weakness of the above algorithm. We first argue that no algorithm can do better than attempting to satisfy a third of all given constraints. Consider an instance of the betweenness problem on three points with three constraints insisting that each point be between the other two. Clearly we can satisfy only one of the above three constraints, proving the claim. Thus the primary weakness of the above algorithm is not in the (absolute) number of constraints it satisfies, but in the fact that it attempts to do so for every instance of the betweenness problem; even those which are obviously not satisfiable. Thus to achieve better approximation factors, one needs to be able to recognize instances of the betweenness problems which are not satisfiable. But this is an NP-hard task. In fact, Corollary 2 indicates that one cannot even distinguish instances which are satisfiable from those for which an ϵ fraction of the constraints remain unsatisfied under any assignment. In what follows we use a semidefinite relaxation of our problem to distinguish cases which are not satisfiable from cases where at least 50% of the given clauses are satisfiable. We then go on to show that using this relaxation we can achieve a better approximation than the naive randomized algorithm.

3 Semidefinite Programming

In this section we briefly introduce the paradigm of semidefinite programming. We describe why it is solvable in polynomial time. A complementary technique to that of semidefinite programming is the incomplete Cholesky decomposition. We describe how the combination allows one to find embeddings of points in infinite-dimensional space subject to certain constraints.

Definition 3. For positive integers m and n, a semidefinite program is defined over a collection of n^2 real variables $\{x_{ij}\}_{i=1,j=1}^{n,n}$. The input consists of a set of mn^2 real numbers $\{a_{ij}^{(k)}\}_{i=1,j=1,k=1}^{n,n,m}$, a vector of m real numbers $\{b^{(k)}\}_{k=1}^{m}$ and a vector of n^2 real numbers $\{c_{ij}\}_{i=1,j=1}^{n,n}$. The objective is to find $\{x_{ij}\}_{i=1,j=1}^{n,n}$ so as to

$$\text{maximize} \quad \sum_{i=1}^{n} \sum_{j=1}^{n} c_{ij} x_{ij}$$

subject to

$$\forall k \in \{1,\ldots,m\} \quad \sum_{i=1}^{n} \sum_{j=1}^{n} a_{ij}^{(k)} x_{ij} \leq b^{(k)}.$$

and   The matrix $X = \{x_{ij}\}$ is symmetric and positive semidefinite.

Recall that the following are equivalent ways of defining when a symmetric matrix X is positive semidefinite.

1. All the eigenvalues of X are non-negative.
2. For all vectors $y \in \mathcal{R}^n$, $y^T X y \geq 0$.
3. There exists a real matrix V such that $V^T \cdot V = X$.

It is well-known that the ellipsoid algorithm of Khaciyan [11] can be used to find an additive ϵ-approximate solution to any semidefinite program in time polynomial in the input size and the logarithm of $1/\epsilon$ (see for instance [8]).

In order to use the semidefinite programming approach for solving combinatorial optimization problems, one more tool is useful. This is the ability to find a matrix V as guaranteed to exist in part 3 of the definition of positive semidefiniteness above. The method which yields such a matrix is the incomplete Cholesky decomposition.

The matrix V can be used to interpret the solution obtained by the semidefinite programming problem geometrically. Interpret the columns of the $n \times n$ matrix V as n vectors $v_1, \ldots, v_n$ in $\mathcal{R}^n$. Now the variables x_{ij} of the matrix X simply correspond to the inner product of v_i and v_j. Thus a linear constraint on the x_{ij}'s is simply a linear constraint on the inner products of the vectors v_i's. And the objective function is simply a linear function on the inner products.

Thus the following provides a geometric interpretation of SDP:

Find n vectors $v_1, \ldots, v_n$ so as to maximize the quantity $\sum_{i,j} c_{ij} < v_i, v_j >$, subject to the constraints $\sum_{i,j} a_{ij}^{(k)} < v_i, v_j > \leq b^{(k)}$.

Alternately one can interpret SDP as solving an optimization problem which attempts to find n points in infinite dimensional Euclidean space subject to linear constraints on the squares of the distance between the points. This is done by observing that the square of the distance between points v_i and v_j (denoted d_{ij}^2) is simply $< (v_i - v_j), (v_i - v_j) > = < v_i, v_i > + < v_j, v_j > -2 < v_i, v_j >$. Thus a linear inequality on the d_{ij}^2's is also a linear inequality on the inner products of the v_i's. (Actually the distance square interpretation is equivalent to SDP since we can express $< v_i, v_j >$ as $(d_{i0}^2 + d_{j0}^2 - d_{ij}^2)/2$.) Hence we can use SDP to solve any problem of the form:

Embed n points in $\mathcal{R}^n$ such that the squares of the distance between the points, denoted d_{ij}, satisfy the constraints $\sum_{i,j} a_{ij}^{(k)} d_{ij}^2 \leq b^{(k)}$ while trying to maximize $\sum_{i,j} c_{ij} d_{ij}^2$.

In what follows we will use the last interpretation of SDP to solve the betweenness problem.

4 The Algorithm

The general idea of our algorithm is to express the betweenness constraints as a set of real quadratic inequalities. By considering an n–dimensional relaxation of the problem, we get an instance of semidefinite programming, and can find a feasible solution in $\mathcal{R}^n$ (if one exists). We study simple geometric properties of this solution set. We use them to argue that a projection of the set on a random line satisfies at least half the betweenness constraints (with high probability). Then we show how to derandomize the algorithm.

Consider a set of m betweenness constraints on n real variables $x_1, \ldots, x_n$. Suppose these constraints are satisfiable, and that $x_1 < x_2 < \ldots < x_n$ is a satisfying linear order. We can clearly embed the points in the unit interval, and assign $x_i = (i-1)/(n-1)$ $(i = 1, \ldots, n)$. Let x_i, x_j, x_k be a triplet such that x_j is required to be between x_i and x_k. For the assignment above, it is readily seen that $(x_i - x_j)^2 + (x_k - x_j)^2 < (x_i - x_k)^2$. The x's are at least $1/(n-1)$ apart and at most 1 apart. Therefore the ratio between the left hand side and the right hand side is maximized when x_i and x_k are extreme points (0 and 1), and x_j is as close as possible to one of them $(1/(n-1)$ or $(n-2)/(n-1))$. For these values, the left hand side is

$$\left(\frac{1}{n-1}\right)^2 + \left(1 - \frac{1}{n-1}\right)^2 = 1 - \frac{2}{n-1} + \frac{2}{(n-1)^2} .$$

Denote this value by α_n. Notice that $\alpha_n = 1 - O(1/n)$ depends only on the number of variables.

To set up the SDP instance, we add the inequality

$$d_{i,j}^2 + d_{k,j}^2 \leq \alpha_n d_{i,k}^2$$

for every betweenness constraint "x_j between x_i and x_k". In addition, we add two inequalities for every pair x_i, x_j. These inequalities force the two variables to be neither too close nor too far apart,

$$\left(\frac{1}{n-1}\right)^2 \leq d_{i,j}^2 \leq 1 .$$

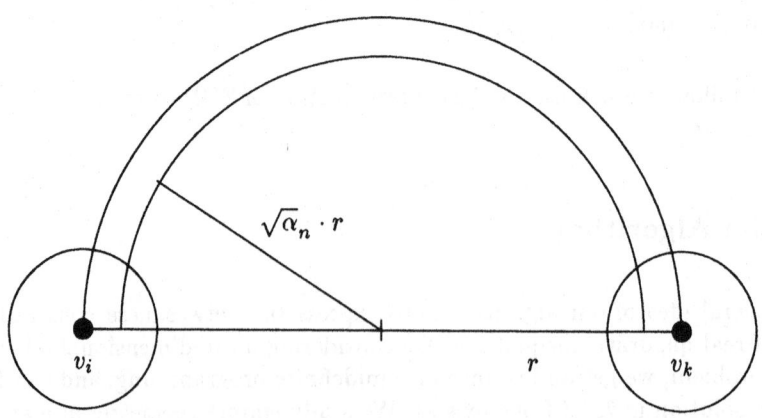

Fig. 1. Possible location for the midpoint v_j

As argued in Section 3, we can use semidefinite programming to test feasibility of these inequalities in $\mathcal{R}^n$, and to find an approximation of a feasible solution (if one exists). Let $v_1, \ldots, v_n \in \mathcal{R}^n$ be a feasible solution, and let $v_i, v_j, v_k \in \mathcal{R}^n$ be a triplet which corresponds to a betweenness constraint. Consider the two dimensional plane through the points v_i, v_j, v_k. Let $2r$ be the distance between v_i and v_k $(1/(n-1) \le 2r \le 1)$. Then v_j, the "midpoint", lies inside the ball of radius $r\sqrt{\alpha_n}$ whose center is the middle $(v_i + v_k)/2$, and outside the two small balls of radius $1/(n-1)$ around v_i and v_k (see figure 1).

This implies that the angle $\theta_{i,j,k} = \angle v_i v_j v_k$ (the angle between the lines $v_i v_j$ and $v_k v_j$) is obtuse. (Right angle correspond to Pythagorean triplets, where equality $d_{i,j}^2 + d_{k,j}^2 = d_{i,k}^2$ holds.) Furthermore, this angle is at least $(1 + O(\sqrt{\alpha_n}))\pi/2 = (1 + O(1/\sqrt{n}))\pi/2$. The algorithm proceeds by picking uniformly at random a line through the origin, and projecting the n points $v_1, \ldots, v_n$ of this random line. Let $x'_1, \ldots, x'_n$ be the n resulting points.

Claim 4. Let $\theta_{i,j,k}$ denote the angle $\angle v_i v_j v_k$. Then the probability that x'_j lies between x'_i and x'_k equals $\theta_{i,j,k}/\pi$.

Proof. Instead of considering an arbitrary line through the origin, we consider a parallel line that goes through the point v_j. This does not change the betweenness relation of the projections. Neither is this relation changed when considering the projection of this line on the two dimensional plane defined by v_i, v_j, v_k. Consider the section of the circle defined by the two lines which go through v_j and are perpendicular to the lines $v_i v_j$ and $v_k v_j$. It is not hard to see that only lines going through this section violate the betweenness constraint of the projections. This section occupies an angle of $\pi - \theta_{i,j,k}$ (see figure 2). The claim follows.

$\square$

Corollary 5. Suppose the SDP has a feasible solution. Then for any of the m constraints, the probability that x'_j lies between x'_i and x'_k is at least $1/2 + O(1/\sqrt{n})$.

As a consequence, the expected number of betweenness constraints satisfied by $x'_1, \ldots, x'_n$ is at least $m(1/2 + O(1/\sqrt{n})) = m(1/2 + o(1))$. Thus we get a randomized polynomial time algorithm which either finds that the constraints are infeasible, or generates a linear order which satisfies at least half the constraints.

We now outline a method for derandomizing our algorithm. Given an embedding of the betweenness problem, we can define a graph and an embedding of the graph in $\mathcal{R}^n$, such that the expected size of the MAX CUT found for this embedding of the graph equals the expected number of betweenness constraints that are satisfied by a random projection.

For every ordered pair of points (v_i, v_j) of the betweenness problem introduce the vertex w_{ij} with embedding $v_i - v_j$. If i, j, k is a betweenness constraint, then put an edge between w_{ij} and w_{kj}. This defines the graph and its embedding.

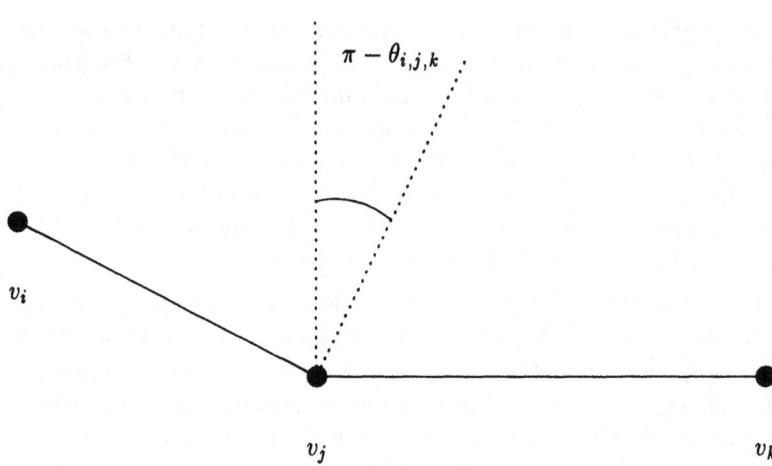

Fig. 2. Lines going through the circular section violate the constraint.

Now consider any hyperplane through the origin that cuts across the edge between w_{ij} and w_{kj}. Let the slope of the normal to the hyperplane be the vector r. Assume w.l.o.g. that $r.w_{ij} < 0$ and $r.w_{jk} < 0$, then $r.v_i < r.v_j$ and $r.v_j < r.v_k$. Thus j lies between i and k. Conversely if projection onto the vector r satisfies the betweenness constraint for i, j, k; then the edge between w_{ij} and w_{jk} must be cut.

Hariharan and Mahajan [9] give a method to deterministically find a vector r whose cut value equals the expected cut value. They use this algorithm to derandomize the MAX CUT and Max 2SAT algorithm of Goemans and Williamson [5]. By using their algorithm we get vector such that projection on to this vector satisfies as many constraints as satisfiable by a random vector.

Remark: Observe that the above reduction is not a generic reduction from betweenness to MAX CUT; and needs to use the fact that the graph produced for the MAX CUT problem has a specified embedding in order to map a solution of the MAX CUT problem to a solution of the betweenness problem.

5   Tightness of our analysis

In this section we show that our analysis of the semidefinite program is tight. We do so by exhibiting a family of instances of the betweenness problem on m constraints for which the optimum value is at most $m(1/2 + o(1))$ but the SDP is nevertheless feasible.

For every integer $d > 1$, we construct the instance I_d as follows. I_d has 2^d points corresponding to the 2^d vertices of the d-dimensional hypercube. I_d has

$m = \binom{d}{2} 2^d$ constraints - one for every simple path of length 2 in the hypercube, with the betweenness constraint expecting the middle vertex of the path to be between the endpoints.

Consider a small perturbation of our SDP, where we have $d_{i,j}^2 + d_{j,k}^2 \leq d_{i,k}^2$ for each betweenness constraint. This SDP is clearly feasible — the natural embedding of the hypercube in d-dimensions (as a hypercube) ensures that every path of length 2 subtends an angle of 90° at the midpoint.

Now consider a linear ordering of the points. Consider any point p and all the paths which have p as the midpoint: The number of such paths is $\binom{d}{2}$. Now let d_1 of the neighbors of p be on its left and d_2 of its neighbors be on its right (where $d_1 + d_2 = d$). The number of betweenness constraints expecting p to be in the middle, that get satisfied, is $d_1 d_2 \leq d^2/4$. Thus the fraction of betweenness constraints associated with any point, that get satisfied, is at most $(d^2/4)/(d(d-1)/2) = d/(2(d-1)) = 1/2 + 1/(2(d-1)) = 1/2 + o(1)$.

6   Concluding Remarks

Our formulation of the problem as SDP only tested for feasibility of the constraints. It is interesting to see if the inclusion of an appropriate objective function, and possibly of additional inequalities, can be used to improve the performance guarantee of the algorithm. Other approaches to the problem, possibly purely combinatorial ones, are also of interest.

Finally, we remark that metric information can be easily incorporated into our algorithm. As a simple example, suppose that for some of the constraints, we know not only that x_j is between x_i and x_k, but that it is exactly in the middle, namely $x_j = (x_i + x_k)/2$. In this case, we add the inequality

$$d_{i,j}^2 + d_{k,j}^2 \leq d_{i,k}^2/4$$

instead of

$$d_{i,j}^2 + d_{k,j}^2 \leq \alpha_n d_{i,k}^2 .$$

Any feasible solution will have v_j exactly in the middle of v_i and v_k, and the same holds with respect to the final projections.

Acknowledgments

Many thanks to Amir Ben-Dor for numerous helpful discussions on the betweenness problem. We would also like to thank Ron Shamir for acquainting us with reference [13] and for useful discussions, Oded Goldreich for his comments on an earlier draft of this manuscript, and Amos Beimel for his expert advice on xfig.

References

1. N. Alon and N. Kahale. "Approximating the independence number via the θ-function", Manuscript, August 1994.
2. S. Arora, C. Lund, R. Motwani, M. Sudan and M. Szegedy. "Proof Verification and Hardness of Approximation Problems", *Proceedings 33rd Annual IEEE Symposium on Foundations of Computer Science*, pp. 14–23, 1992.
3. D. Cox, M. Burmeister, E. Price, S. Kim, R. Myers, "Radiation Hybrid Mapping: A Somatic Cell Genetic Method for Constructing High Resolution Maps of Mammalian Chromosomes", *Science*, Vol. 250, 1990, pp. 245–250.
4. U. Feige and M. Goemans. "Approximating the value of two prover proof systems, with applications to MAX 2SAT and MAX DICUT", *Proceedings of the Third Israel Symposium on Theory and Computing Systems,*, 1995.
5. M. Goemans and D. Williamson. ".878-Approximation Algorithms for MAX CUT and MAX 2SAT", *Proceedings of the Twenty-Sixth ACM Symposium on Theory of Computing*, pp. 422-431, 1994.
6. S. Goss and H. Harris, "New Methods for Mapping Genes in Human Chromosomes", *Nature*, Vol. 255, 1975, pp. 680–684.
7. M. Grötschel, L. Lovász and A. Schrijver. "The ellipsoid method and its consequences in combinatorial optimization", *Combinatorica*, 1:169–197, 1981.
8. M. Grötschel, L. Lovász and A. Schrijver. *Geometric Algorithms and Combinatorial Optimization.* Springer-Verlag, Berlin, 1987.
9. R. Hariharan and S. Mahajan. Derandomization of Semidefinite Programming based algorithms. Manuscript, 1995.
10. D. Karger, R. Motwani and M. Sudan. "Approximate graph coloring via semidefinite programming", Proceedings of the 35th Annual Symposium on Foundations of Computer Science, 1994.
11. L. Khaciyan. "A polynomial algorithm in linear programming", (English translation appears in) *Soviet Mathematics Doklady*, vol. 20, pp. 191–194, 1979.
12. L. Lovász. "On the Shannon capacity of a graph", *IEEE Transactions on Information Theory*, IT-25:1–7, 1979.
13. J. Opatrny, "Total Ordering Problem", *SIAM J. Comput.*, vol. 8 No. 1, February 1979, pp. 111–114.

Efficient Computation of the Geodesic Voronoi Diagram of Points in a Simple Polygon

(Extended Abstract)

Evanthia Papadopoulou and D. T. Lee*

Northwestern University, Evanston, Illinois 60208, USA.

Abstract. We present an $O((n+k)\log(n+k))$ time algorithm for computing the *geodesic Voronoi diagram* of k points in a simple polygon of n vertices improving upon the previously known results. The method introduces a new approach to the construction of geodesic Voronoi diagrams by combining a sweep of the polygon and the merging step of a usual divide-and-conquer strategy.

1 Introduction

Given a simple polygon P with n sides, and a set $S = \{s_1, \ldots, s_k\}$ of k point sites inside P, the *geodesic Voronoi diagram* of S in P is the partitioning of P into k cells $V(s_1), \ldots, V(s_k)$ such that a point $x \in P$ belongs to $V(s_i)$ if and only if the geodesic distance of x from s_i is less than (or equal to) the geodesic distance of x from any other site $s_j \in S$. This diagram is a powerful tool for the solution of several proximity problems involving point sites enclosed in a simple polygon. Examples include the minimum spanning tree, finding the closest pair of sites, and finding the nearest neighbor for every site. Furthermore, it encodes shortest path information from the sites to all points in the polygon, and can thus be used to efficiently answer shortest path queries.

We present an $O((n+k)\log(n+k))$-time algorithm for computing the *geodesic Voronoi diagram* of points in a simple polygon, improving upon Aronov's $O((n+k)\log(n+k)\log n)$ result [1]. An earlier result by Asano and Asano [3] solves the problem in time $O(nk + n\log\log n + k\log k)$. Our method combines a sweep of the polygon and the merging step of a usual divide-and-conquer strategy. It introduces a new approach to the construction of Voronoi diagrams, and can be extended for computing the *geodesic Voronoi diagram* of points in polygons with holes under certain restrictions (see conclusions).

The ordinary Voronoi diagram of point sites in the plane is a classical mathematical object. It is well-known that it can be computed in optimal $O(n\log n)$ time by divide-and-conquer [13], or a sweepline approach [5]. For more information about the Voronoi diagrams, refer to the survey by Aurenhammer [2]. Aronov used a divide-and-conquer algorithm [1] for computing the geodesic Voronoi diagram of point sites in a simple polygon, in which the merge step

---

* Supported in part by the National Science Foundation under the Grant CCR-9309743, and by the Office of Naval Research under the Grant No. N00014-93-1-0272.

is preceded by a procedure that extends a recursively computed diagram of the subset of sites inside half the polygon to the full polygon, which is the reason for the additional $\log(n + k)$ factor in the time complexity. For the special case where the set of sites includes *all* the reflex vertices of the polygon, the extension phase is skipped and thus, the geodesic Voronoi diagram is computed in $O((n + k)\log(n + k))$ time. For the special case where the set of sites coincides with *the vertices* of a simple n-gon, an alternative $O(n \log n)$ algorithm can be obtained from the *generalized Delaunay triangulation* of a simple polygon [4, 10, 14]. In a general polygonal domain of n vertices, there is basically one subquadratic result by Mitchell [12] which computes the geodesic Voronoi diagram of a set of sites in $O((n+k)^{5/3}+\epsilon)$ time and space. This result uses the *continuous Dijkstra paradigm* which simulates the effect of "wavefronts" propagating out of the sites. Very recently, Hershberger and Suri [8] gave an $O(n \log^2 n)$-time and $O(n \log n)$-space algorithm to compute the *shortest path map* of a single source point in a general polygonal domain of n vertices, i.e., $k = 1$. The shortest path map is a planar subdivision that encodes shortest path information from a fixed source to all other points in the plane. Their approach is also based on the continuous Dijkstra paradigm. In a recent manuscript, they improve the complexity to $O(n \log n)$.

In the following, we present an $O((n + k) \log(n + k))$-time algorithm that computes the geodesic Voronoi diagram of a set of k sites in a simple n-gon, using a different approach.

2   An overview of the Algorithm

Consider the dual tree of an arbitrary triangulation of polygon P, where a node corresponds to a triangle, and an edge corresponds to a diagonal of the triangulation. See Figure 1. Assign an arbitrary node of degree one as the *root* of the tree. Then, there is a unique directed path from the root to every triangle which defines a parent-child relation between the triangles of the triangulation. Every triangle Δ has a unique *parent* Δ' which is its predecessor on the dual path from the root to Δ. Let d be the diagonal shared by Δ and Δ'. Diagonal d partitions the polygon into two parts: The part of the polygon "below" d, denoted by $P(d)$, which corresponds to the subtree of the dual tree rooted at Δ, and the part of the polygon "above" d, denoted by $P'(d)$, the complement of $P(d)$. Δ' and Δ are referred to as the triangle *above* d and the triangle *below* d, and are denoted as $\Delta'(d)$ and $\Delta(d)$ respectively.

Our algorithm computes the geodesic Voronoi diagram of S in P in three phases. In the first phase, we sweep the triangulated polygon in the order specified by the postorder traversal of the rooted dual tree. For each diagonal d, we compute the geodesic Voronoi diagram that lies in $\Delta(d)$ of all the sites in $P(d)$. We create an $O(n+k)$ subdivision of P, $SD(P)$, which contains the above information for every triangle. In the second phase, we sweep the triangulated polygon in the order specified by the preorder traversal of the rooted dual tree. We compute the geodesic Voronoi diagram that lies in $\Delta(d)$ of all the sites in

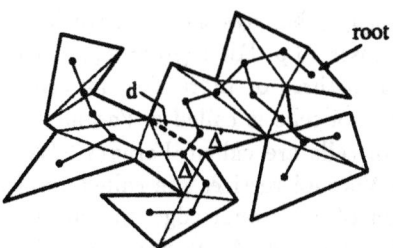

Fig. 1. The dual tree of P.

$P'(d)$, and store it in another $O(n + k)$ subdivision, $SD'(P)$. Finally, we merge the two subdivisions and obtain the geodesic Voronoi diagram of S in P. It is easy to see, that merging in $\Delta(d)$, the Voronoi diagram of sites in $P(d)$ with the Voronoi diagram of sites in $P'(d)$ yields the geodesic Voronoi diagram of S in $\Delta(d)$.

3 Preliminaries

We are given a simple polygon P of n vertices arbitrarily triangulated, and a set of k point sites $S = \{s_1, \ldots, s_k\}$ inside P. Let V denote the vertex set of P. We denote the shortest path between two points $p, q \in P$, that lies totally in P as $\pi(p, q)$. The length of $\pi(p, q)$ is the *geodesic distance* between p and q, and is denoted by $gd(p, q)$. The last vertex on $\pi(p, q)$ before q is called the *anchor* of q (with respect to p). To simplify the presentation, we assume that P and S are in *general position* i.e., no vertex is equidistant from two distinct sites and that no four sites lie on the same geodesic circle. We review some definitions from [1]. See Figure 2 for an illustration.

The Voronoi cell of a site $s_i \in S$ is $V(s_i) = \{x \in P \mid \forall s_j \in S : gd(s_i, x) \leq gd(s_j, x)\}$; s_i is called the *owner* of $V(s_i)$. $V(s_i)$ is partitioned into finer regions by the *shortest path map* from s_i. The *shortest path map* of $V(s_i)$ from s_i is the partition of $V(s_i)$ into maximal sets of points that have the same anchor v with respect to s_i. We refer to such a maximal set of points as the *sub-cell* of $V(s_i)$ induced by vertex v. Vertex v is referred to as the *anchor* of the sub-cell, and it is weighted with its *geodesic distance* from s_i ($w(v) = gd(s_i, v)$). It is well known [6, 1] that the edges of the shortest path map is a collection of *extension segments* emanating from reflex vertices in $V(s_i)$, where the extension segment emanating from a vertex v is the initial part of a half-line collinear with the last edge on $\pi(s_i, v)$, extending in the direction away from s_i, until it hits the boundary of the polygon or the cell $V(s_i)$. The boundary shared by cells $V(s_i)$ and $V(s_j)$ consists of points that are of the same geodesic distance from s_i and s_j, and is the union of portions of bisectors of p and q, where $p, q \in V \cup S$. The bisector of p and q, $p, q \in V \cup S$, is $b(p, q) = \{z \in P \mid d(z, p) + w(p) = d(z, q) + w(q)\}$, where $d(p, q)$ is the Euclidean distance between p and q. $b(p, q)$ belongs to one of the two branches of the hyperbola with foci p, q and eccentricity $d(p, q)/|w(p) - w(q)|$

[11]. If $w(p) < w(q)$, $b(p,q)$ belongs to the branch closer to q, otherwise it belongs to the branch closer to p. For $w(p) = w(q)$, $b(p,q)$ is a straight line. A point which is common to three or more Voronoi cells or on the boundary of P and two or more Voronoi cells is called a *Voronoi vertex*. Bisectors that are common to two Voronoi cells are called *Voronoi edges*. Endpoints of bisector portions which are not Voronoi vertices are called *breakpoints*. A breakpoint is the point of intersection of a Voronoi edge and an edge of the shortest path map of the incident Voronoi cell. As it was shown in [1], the complexity of the geodesic Voronoi diagram is $O(n + k)$. Augmenting Voronoi cells with the their shortest path map yields the *augmented geodesic Voronoi diagram*. Our algorithm computes the *augmented geodesic Voronoi diagram* of S in P. In the following, we always imply the augmented structure when we refer to the geodesic Voronoi diagram of S in P, and we denote it by $Vor_P(S)$ (see Figure 2). Note that given $Vor_P(S)$ and a query point t, a point location query [9], yields in $O(\log(n + k))$-time the site $s_i \in S$ nearest to t, and also retrieves $\pi(s_i, t)$ in additional time proportional to the number of links in the path.

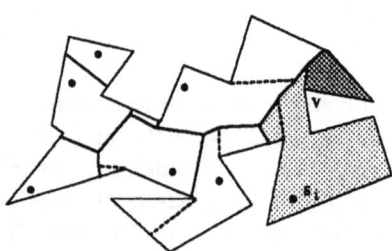

Fig. 2. The geodesic Voronoi diagram of S in P, $Vor_P(S)$. Dashed lines indicate the shortest path map in Voronoi cells; the shaded region is $V(s_i)$, $s_i \in S$; the dark shaded part is the sub-cell of $V(s_i)$ induced by v.

$Vor_P(S)$ can be viewed as a planar graph embedded in P whose vertices are the Voronoi vertices and breakpoints, and edges are the Voronoi edges and shortest path map edges. For notational convenience, we identify the Voronoi diagram with the induced planar graph and use the same notation ($Vor_P(S)$) to denote both. The geodesic Voronoi diagram of $\hat{S} \subseteq S$, restricted to a triangle Δ is denoted as $Vor_\Delta(\hat{S})$. The planar graph induced by $Vor_\Delta(\hat{S})$ consists of the Voronoi vertices and breakpoints of $Vor_P(\hat{S})$ that lie in Δ and the incident Voronoi edges truncated by the sides of Δ. $Vor_\Delta(\hat{S})$ denotes both the diagram and the induced planar graph.

The bisector of two disjoint sets of sites, $S_1, S_2 \subset S$ is $b(S_1, S_2) = \{x \in P \mid gd(x, S_1) = gd(x, S_2)\}$, where $gd(x, S_i) = \min_{s \in S_i}\{gd(x, s)\}$. $b(S_1, S_2)$ is the set of Voronoi edges of $Vor_P(S_1 \cup S_2)$ that are shared by Voronoi cells $V(s_i)$ and $V(s_j)$, for $s_i \in S_1$ and $s_j \in S_2$. As it was shown in [1], for S_1 and S_2 lying on opposite sides of a polygon diagonal, $b(S_1, S_2)$ contains no cycles and all of its components meet the boundary of P. Under the general position assumption, all components of $b(S_1, S_2)$ are simple paths whose vertices (except the endpoints) have degree two.

As defined in section 2, a triangulation diagonal d partitions the polygon into two parts, $P(d)$ and $P'(d)$. We say that $P(d)$ is the part of the polygon "below" d, and that $P'(d)$ is the part of the polygon "above" d. We assume that $d \in P(d)$. Let $S(d) = S \cap P(d)$ and $S'(d) = S \cap P'(d)$. Consider the triangles incident to d. Throughout the paper, we adopt the following convention: $\Delta(d)$ and $\Delta'(d)$ denote the two triangles that are *below* and *above* the common diagonal d respectively, where $d = \overline{v_1 v_2}$. d_1 and d_2 denote the remaining diagonals of $\Delta(d)$, and d_3 and d_4 denote the remaining diagonals of $\Delta'(d)$. We assume that d_4 is the diagonal shared by $\Delta'(d)$ and its parent. We further assume that d_1, d and d_4 are incident to vertex v_1, and that d_2, d and d_3 are incident to vertex v_2. We denote the vertex shared by d_1 and d_2 as $v_{1,2}$ and the vertex shared by d_3 and d_4 by $v_{3,4}$ (see Figure 3). Let the set of sites in $\Delta(d)$ be $S(\Delta(d)) = S(d) - S(d_1) - S(d_2)$.

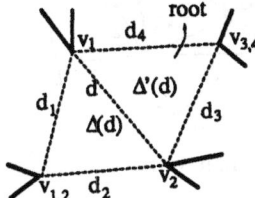

Fig. 3. $d, \Delta(d)$, $\Delta'(d)$

4 The subdivisions

Let $SD(P)$ be the following planar subdivision of P. For every diagonal d, the subdivision contains the Voronoi diagram of $S(d)$ that lies in $\Delta(d)$, $Vor_{\Delta(d)}(S(d))$. See Figure 4. Consider the points of intersection of d with the Voronoi diagrams in $\Delta(d)$ and $\Delta'(d)$ (*i.e.*, $Vor_{\Delta(d)}(S(d))$ and $Vor_{\Delta'(d)}(S(d_4))$). Intersection points that are common to both diagrams belong to the same Voronoi or shortest-path-map edge, and thus are not considered. Intersection points that belong to only one of the diagrams, induce vertices which are referred to as *border* vertices. (In Figure 4c, the border vertex is indicated by an arrow). Any two consecutive border vertices on d (including the endpoints of d) are joined by an edge, if the incident Voronoi cells in $\Delta(d)$ and $\Delta'(d)$ belong to different sites. Edges in $SD(P)$ joining two border vertices or a border vertex and a polygon vertex are called *border* edges. A border edge on d, implies that the owner of the adjacent Voronoi cell in $Vor_{\Delta'(d)}(S(d_4))$ has not been considered in $P(d)$, and that its Voronoi cell should expand in $P(d)$ (see Figures 4c and 5).

$SD'(P)$ is defined similarly. For every diagonal d, $SD'(P)$ contains the Voronoi diagram of $S'(d)$ that lies in $\Delta(d)$, $Vor_{\Delta(d)}(S'(d))$. See Figure 6. The Voronoi diagrams in $\Delta(d)$ and $\Delta'(d)$ (*i.e.*, $Vor_{\Delta(d)}(S'(d))$ *and* $Vor_{\Delta'(d)}(S'(d_4))$) induce

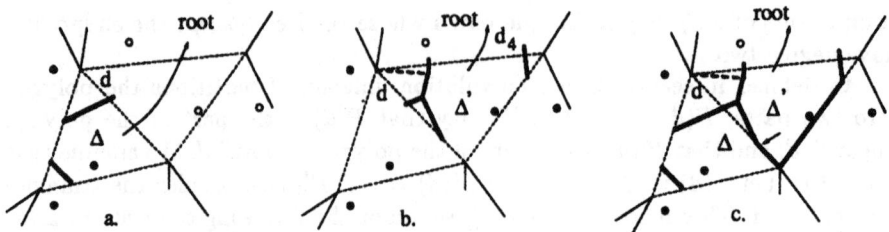

Fig. 4. a. $Vor_{\Delta(d)}(S(d))$, b. $Vor_{\Delta'(d)}(S(d_4))$, c. $SD(P) \cap \{\Delta, \Delta'\}$; border vertices are indicated by arrows.

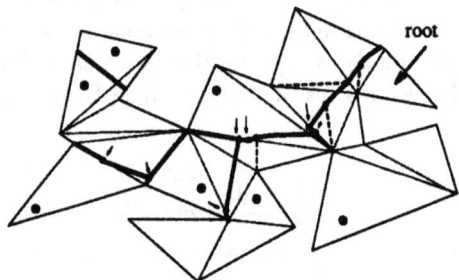

Fig. 5. $SD(P)$: Border vertices are indicated by arrows.

border vertices and *border* edges on d which are defined in exactly the same manner as for $SD(P)$ (see Figure 6c.) In $SD'(P)$, a border edge on a diagonal d implies that the site of the adjacent Voronoi cell in $\Delta(d)$ has not been considered in $P'(d)$ and that its Voronoi cell should expand in $P'(d)$ (see Figures 6c and 7).

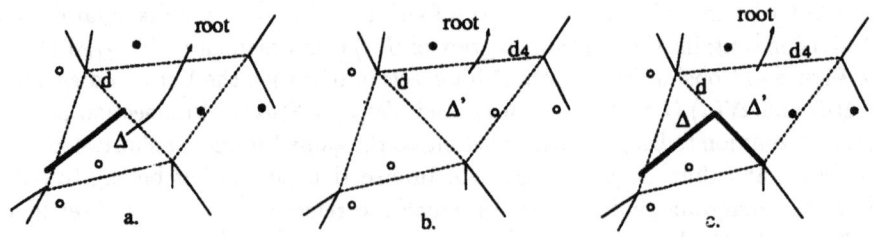

Fig. 6. a. $Vor_{\Delta(d)}(S'(d))$, b. $Vor_{\Delta'(d)}(S'(d_4))$, c. $SD'(P) \cap \{\Delta, \Delta'\}$.

We refer to faces of the two subdivisions as *regions*. Every region of $SD(P)$ and $SD'(P)$ has an *owner* (a site in S) and an *anchor* (a polygon vertex or site in S). Recall that the anchor of a region is the last vertex on the shortest path from the owner to every point in the region, and it is weighted by its geodesic distance from the owner of the region. We denote a region of anchor y, where $y \in V \cup S$, as $r(y)$. The union of regions that have owner $s \in S$ is denoted

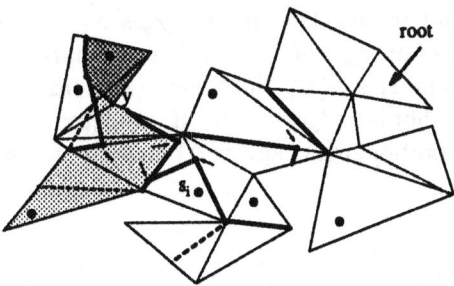

root

Fig. 7. $SD'(P)$. The shaded portion is $R(s_i)$. The face shaded darker is $r(y) \in R(s_i)$

as $R(s)$. In Figure 7, $R(s_i)$ is shown shaded; the face shaded darker indicates $r(y) \in R(s_i)$.

Given a diagonal d, $SD(d)$ denotes the part of $SD(P)$ "below" d excluding border vertices and edges on d i.e., $SD(d) = SD(P) \cap \{P(d) - d\}$. $SD'(d)$ denotes the part of $SD'(P)$ "above" d excluding border vertices and edges on d i.e., $SD'(d) = SD'(P) \cap P'(d)$.

Lemma 1. *The number of border vertices of degree two in $SD(P)$ and $SD'(P)$ is $O(n + k)$.*

By lemma 1, Euler's formula, and the fact that the number of faces is $O(n+k)$, we have the following.

Lemma 2. *$SD(P)$ and $SD'(P)$ are planar subdivions of size $O(n + k)$.*

5 Constructing the subdivisions

Consider a polygon diagonal d, and the incident triangles $\Delta(d)$ and $\Delta'(d)$. For brevity, let Δ and Δ' denote $\Delta(d)$ and $\Delta'(d)$ respectively. The Voronoi diagram of $S(\Delta)$ in Δ, $Vor_\Delta(S(\Delta))$ can be easily computed in the ordinary divide-and-conquer way by ignoring the polygon boundary. Let $S_{d_i}(\Delta)$, $i = 1, 2$, and $S_d(\Delta)$ denote the ordered list of sites whose regions in $Vor_\Delta(S(\Delta))$ intersect d_i and d respectively. Consider the Voronoi diagrams, $Vor_\Delta(S(d_i))$, of $S(d_i)$, $i = 1, 2$ in Δ. $Vor_\Delta(S(d_i))$, $i = 1, 2$ can be obtained from $SD(d_i)$ using the extension procedure of [1].

In order to compute $SD(P)$, we need $Vor_\Delta(S(d))$, the Voronoi diagram of $S(d)$ in Δ. $Vor_\Delta(S(d))$ can be computed by merging $Vor_\Delta(S(d_1))$, $Vor_\Delta(S(\Delta))$, and $Vor_\Delta(S(d_2))$. Let's define a subdivision, $SD_1(d)$, to be used as an intermediate step in the construction of $SD(d)$. $SD_1(d)$ is defined as $SD(d)$ for $S(d_2) = \emptyset$ i.e, $SD_1(d) = SD(d_1) \cup Vor_\Delta(S(d_1) \cup S(\Delta)) \cup \{border\ edges\ on\ d_1\}$. Border edges on d_1 are induced by regions of sites in $S(\Delta)$ that are adjacent to d_1 in $Vor_\Delta(S(d_1)) \cup S(\Delta))$.

Let $\sigma(\Delta)$ denote the set of edges of $SD(d)$ that are shared by regions $R(s_i)$ and $R(s_j)$, for $s_i \in S(d_1) \cup S(\Delta)$ and $s_j \in S(d_2)$. $\sigma(\Delta)$ consists of $\{b(S(d_1) \cup$

$S(\Delta), S(d_2))\} \cap \Delta$ and the incident border edges on d_1 or d_2 (see Figure 8a). Let $\tau(\Delta)$ denote the set of edges of $SD_1(d)$ that are shared by regions $R(s_i)$ and $R(s_j)$, for $s_i \in S(d_1)$ and $s_j \in S(\Delta)$. Note that $\tau(\Delta)$ consists of $\{b(S(d_1), S(\Delta))\}$ $\cap\Delta$ and the incident border edges on d_1 (if any) (see Figure 8b). Under the general position assumption, $\sigma(\Delta)$ and $\tau(\Delta)$ have the following properties (see Figures 8, 9).

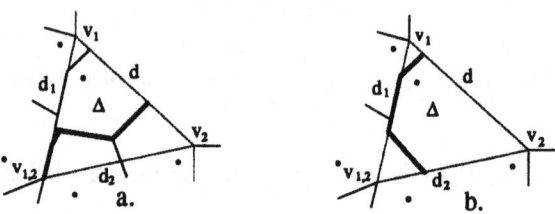

Fig. 8. a. $SD(d) \cap \Delta$; the thick curve is $\sigma(\Delta)$. b. $SD_1(d) \cap \Delta$; the thick curve is $\tau(\Delta)$.

Lemma 3. *If $S(\Delta) = \emptyset$, $\sigma(\Delta)$ is a simple path with one endpoint at vertex $v_{1,2}$ (the vertex incident to d_1 and d_2) and another on diagonal d; $\sigma(\Delta)$ has exactly one common point with d.*

Lemma 4. *If $S(\Delta) \neq \emptyset$, $\sigma(\Delta)$ is a subset of edges of $SD(d)$ with the following properties:*

- *$\sigma(\Delta)$ is a collection of disjoint simple paths, none of which is a cycle.*
- *There is one component of $\sigma(\Delta)$, referred to as the initial component, with one endpoint at vertex $v_{1,2}$ and another on d, which is the only point of the component common with d.*
- *The components of $\sigma(\Delta)$ other than the initial, have both their endpoints on d, and these are their only points common with d.*
- *Each component of $\sigma(\Delta)$ other than the initial, contains an edge which is shared by regions $R(s_i)$ and $R(s_j)$, for $s_i \in S_{d_2}(\Delta)$ and $s_j \in S(d_2)$.*

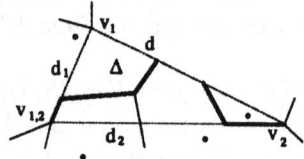

Fig. 9. $\sigma(\Delta)$.

Lemma 5. *$\tau(\Delta)$ is a subset of edges of $SD_1(d)$ with the following properties.*

- *$\tau(\Delta)$ is a collection of disjoint simple paths, none of which is a cycle.*

- *The endpoints of each component of $\tau(\Delta)$ lie on $\{d, d_2\}$, and these are the only points common with $\{d, d_2\}$.*
- *Each component of $\tau(\Delta)$ contains an edge which is shared by regions $R(s_i)$ and $R(s_j)$, for $s_j \in S_{d_1}(\Delta)$ and $s_i \in S(d_1)$.*

Lemmas 3, 4, and 5 imply that the number of connected components of $\sigma(\Delta)$ or $\tau(\Delta)$ is $O(1 + |S(\Delta)|)$.

$SD(P)$ is kept as a doubly-connected-edge-list (DCEL) [13]. For each diagonal d, we compute $SD(d)$ and $T(d)$, the ordered list of regions of $SD(d)$ intersecting d, arranged in a search tree that supports insertions, deletions, splitting, and merging in time logarithmic of its size (for example the red-black tree of [7]). As in [1], every triangle is associated with a bucket to hold bisector intersection points. The algorithm is as follows:

- For every triangle $\Delta = \Delta(d)$, compute the ordinary Voronoi diagram of $S(\Delta)$, $Vor(S(\Delta))$, ignoring the polygon boundary.
- Locate in the triangulation the Voronoi vertices of $Vor(S(\Delta))$ that do not lie in Δ, and store them in the associated buckets. Voronoi vertices lying in triangles of $P'(d)$ are considered for constructing $SD(P)$. Voronoi vertices lying in triangles of $P(d_1)$ and $P(d_2)$ are considered for constructing $SD'(P)$.
- Keep the part of $Vor(S(\Delta))$ that lies in Δ ($Vor_\Delta(S(\Delta))$). Identify $S_{d_1}(\Delta)$, $S_{d_2}(\Delta)$, $S_d(\Delta)$, the ordered list of regions in $Vor_\Delta(S(\Delta))$ intersecting d_1, d_2 and d respectively, and arrange them in search trees [7].
- Call procedure $Construct\text{-}SD(d)$ for d being one of the two boundary edges of the root triangle. Procedure $Construct\text{-}SD(d)$ does a postorder traversal of the dual tree.

Algorithm Construct-$SD(d)$
Input: A triangulated polygon P and a triangulation diagonal d.
Output: $SD(d)$ and $T(d)$.
begin
1. If d is a boundary edge such that there is no triangle *below* d then, if there is no site on d let $T(d) = \emptyset$. If d is a boundary edge and $S \cap d \neq \emptyset$ then initialize $T(d)$ to the ordered list of regions induced by the sites on d. Return $T(d)$, $SD(d) = d$.
2. Let $\Delta = \Delta(d)$.
   (a) Recursively call $Construct\text{-}SD(d_1)$. Let $T(d_1)$ be the ordered list of regions of $SD(d_1)$ intersecting d_1 returned by the recursive call.
   (b) If $T(d_1) \neq \emptyset$, extend $T(d_1)$ into Δ using the algorithm of [1]. Split the extended tree at v_2 (the vertex shared by d and d_2). Let $T_1(d)$ be the part intersecting d, and $T'_1(d_2)$ be the part intersecting d_2. Store $T'_1(d_2)$ in order to be used in the construction of $SD'(P)$. While extending $T(d_1)$, compute $SD(d_1) \cup Vor_\Delta(S(d_1))$.
   (c) Recursively call $Construct\text{-}SD(d_2)$. Let $T(d_2)$ be the ordered list of regions of $SD(d_2)$ intersecting d_2 returned by the recursive call.
   (d) If $T(d_2) \neq \emptyset$, extend $T(d_2)$ into Δ using the algorithm of [1]. Split the extended tree at v_1 (the vertex shared by d and d_1). Let $T_2(d)$ be the

part intersecting d, and $T_2'(d_1)$ be the part intersecting d_1. Store $T_2'(d_1)$ in order to be used in the construction of $SD'(P)$. While extending $T(d_2)$, compute $SD(d_2) \cup Vor_\Delta(S(d_2))$.

(e) If $S(\Delta) = \emptyset$, merge $\{SD(d_1) \cup Vor_\Delta(S(d_1))\}$ and $\{SD(d_2) \cup Vor_\Delta(S(d_2))\}$ to get $SD(d)$. While merging, update $T_1(d)$ and $T_2(d)$ every time a region is trimmed by $\sigma(\Delta)$. At the end, compute the intersection of the bisector of $\sigma(\Delta)$ intersecting d with its neighboring bisectors in $T_1(d)$ and in $T_2(d)$ respectively. Locate the intersections in the triangulation (without checking feasibility) and assign them to the corresponding buckets.

(f) If $S(\Delta) \neq \emptyset$, merge $\{SD(d_1) \cup Vor_\Delta(S(d_1))\}$ and $Vor_\Delta(S(\Delta))$ to get $SD_1(d)$. Merge $SD_1(d)$ and $\{SD(d_2) \cup Vor_\Delta(S(d_2))\}$ to get $SD(d)$. While merging, update $T_1(d)$, $S_d(\Delta)$ and $T_2(d)$ accordingly. Compute the points of intersection of the bisectors of $\tau(\Delta)$ and $\sigma(\Delta)$ intersecting d with the neighboring bisectors in $T_1(d)$, $S_d(\Delta)$, and $T_2(d)$, and locate them in the triangulation.

(g) Merge $T_1(d)$, $S_d(\Delta)$, and $T_2(d)$ to form $T(d)$. Return $T(d)$ and $SD(d)$.

end.

To extend $SD(d_1)$ and $SD(d_2)$ into Δ we use the *extension* procedure of [1], which given a triangulated polygon separated by a triangulation diagonal e into subpolygons P_1 and P_2, and the planar map induced by the Voronoi diagram of a set of sites in P_1, finds the planar map induced on P by the Voronoi diagram of the same set of points. We briefly review here the main ideas of this procedure.

Each triangle in the triangulation of P_2 is associated with a bucket to hold bisector intersection points. From the Voronoi diagram in P_1, extract the ordered list of regions adjacent to e, and arrange them in a search tree T. For each pair of bisectors bounding a single region in T compute their point of intersection (without checking feasibility), locate them in the triangulation of P_2, and store them in the associated buckets. (Here, we are given $T(d_1)$ and $T(d_2)$ thus, we do not extract the regions; intersections of neighboring bisectors have also been computed.) Move from triangle to neighboring triangle starting at the triangle adjacent to e. Upon entering a triangle through edge e, consider the extension segments of the endpoints of e from their owner in the diagram. (Here, we only move into Δ). Compute their point of intersection with the neighboring bisectors and locate them in the triangulation. Process the intersection points in the bucket of the current triangle in order of increasing distance from e. Process only the feasible intersection points. Each feasible intersection point corresponds to a region deletion and the generation of a new bisector. When construction inside the current triangle is complete, split T at the apex of the triangle giving two subtrees, one for each of the remaining edges of the current triangle. For more details see [1]. Slightly modifying the analysis of [1] we have the following.

Lemma 6. *The time to extend $T(d_i)$, $i = 1, 2$ into Δ (and therefore compute $SD(d_i) \cup Vor_\Delta(S(d_i))$) is $O((|I_f(\Delta)|+1) \log |T(d_i)| + |I_{new}(\Delta)| \log n + |I_{inf}(\Delta)|)$, where $I_f(\Delta)$ denotes the feasible intersection points in the bucket associated with Δ, I_{new} denotes the intersection points generated while extending in Δ, and I_{inf} denotes the infeasible intersection points located in the bucket associated with Δ.*

To merge $SD(d_1) \cup Vor_\Delta(S(d_1))$ with $Vor_\Delta(S(\Delta))$, we construct $\tau(\Delta)$, split the diagrams along the components of $\tau(\Delta)$, and clean up dominated regions. Similarly, to merge $SD(d_2) \cup Vor_\Delta(S(d_2))$ with $SD_1(d)$ we construct $\sigma(\Delta)$, split along the components of $\sigma(\Delta)$, and clean up. Lemmas 3, 4, and 5, provide the means to identify a point on every component of the merge curves. The initial component of $\sigma(\Delta)$ has an endpoint at vertex $v_{1,2}$. For the other components of $\sigma(\Delta)$ (resp. $\tau(\Delta)$) we can identify a point as follows: Let s be the first point in $S_{d_2}(\Delta)$ (resp. $S_{d_1}(\Delta)$) that has not been considered while constructing the initial or other components. Let y be the anchor of the region where s lies in $Vor_\Delta(S(d_2))$ i.e., $s \in r(y)$ (resp. $Vor_\Delta(S(d_1))$). The region in $Vor_\Delta(S(d_i))$ where $s \in S(\Delta)$ lies can be easily determined in $O(\log |T(d_i)|)$ time during the extension procedure. Consider the bisector of s and y, $b(s,y)$. If $b(s,y)$ intersects segment $\overline{sy}$ in Δ then obviously the intersection point belongs to a component of $\sigma(\Delta)$ (resp. $\tau(\Delta)$). Otherwise, the point of intersection of $\overline{sy}$ and d_2 (resp. d_1) is closer to $S(\Delta)$ than to $S(d_2)$ (resp. $S(d_1)$) thus, it belongs to a border edge of $\sigma(\Delta)$ (resp. $\tau(\Delta)$). In any case, a point on a component of $\sigma(\Delta)$ (resp. $\tau(\Delta)$) can be identified in $O(\log |T(d_2)|)$ (resp. $O(\log |T(d_1)|)$ time.

Once a point on a component of $\sigma(\Delta)$ (resp. $\tau(\Delta)$) has been identified the component can be traced in a way similar to tracing the merge curve of the ordinary Voronoi diagram. When a component of $\sigma(\Delta)$ (resp. $\tau(\Delta)$) hits $d_i, i = 1, 2$, (resp. d_1) it continues along d_i (resp. d_1) tracing border edges, until it reaches a point on $b(S(d_2), S(d_1) \cup S(\Delta))$, (resp. $b(S(d_1), S(\Delta))$) in which case it continues in the interior of Δ as $b(S(d_2), S(d_1) \cup S(\Delta))$ (resp. $b(S(d_1), S(\Delta))$). Note that for a point x of $\sigma(\Delta)$ on a border edge of d_i, $gd(x, S(d_1) \cup S(\Delta)) \geq gd(x, S(d_2))$, and that for a point x on a border of $\tau(\Delta)$, $gd(x, S(\Delta)) \geq gd(x, S(d_1))$. By lemmas 3, 4, and 5, when $\sigma(\Delta)$ hits d and when $\tau(\Delta)$ hits $\{d, d_2\}$, the tracing of the component ends.

Lemma 7. *Given* $Vor_\Delta(S(d_1))$, $Vor_\Delta(S(\Delta))$, *and a point on every component of* $\tau(\Delta)$, *the time to trace* $\tau(\Delta)$ *is proportional to the size of* $\tau(\Delta)$ *plus* $|S(\Delta)|$.

Lemma 8. *Given* $SD_1(d)$, $Vor_\Delta(S(d_2))$, *and a point on every component of* $\sigma(\Delta)$, *the time to trace sigma*(Δ) *is proportional to the size of* $\sigma(\Delta)$ *plus* $|S(\Delta)|$.

Lemmas 7 and 8, the fact that $\sigma(\Delta)$ and $\tau(\Delta)$ may have at most $O(1 + |S(\Delta)|)$ components, and the fact that the size of $SD(P)$ is $O(n+k)$, imply that that the total time spent for tracing merge curves while constructing $SD(P)$ is $O(n+k)$. Since the number of components of $\sigma(\Delta)$ and $\tau(\Delta)$ is $O(1 + |S(\Delta)|)$, the total time spent to identify a point on components of $\sigma(\Delta)$ and $\tau(\Delta)$ throughout the procedure is $O(k \log(n+k))$. Considering also the time to clean up, generate and locate intersection points, update, split, and merge $T_2(d), T_1(d)$ and $S_d(\Delta)$ we derive the following lemma.

Lemma 9. *The total time spent for merging while constructing $SD(P)$ is $O((n+k) \log(n+k))$.*

The total time spent for extending Voronoi diagrams form triangle to triangle depends (by lemma 6) to the total number of intersection points generated throughout the procedure. Note that bisector intersection points are generated due to $Vor(S(\Delta))$, the bisectors of $\sigma(\Delta)$ and $\tau(\Delta)$ crossing d, and the extension procedure.

Lemma 10. *The total number of bisector intersection points generated during the construction of $SD(P)$ is $O(n + k)$.*

By lemmas 6 and 10 we derive the following.

Lemma 11. *The total time spent for extending during the construction of $SD(P)$ is $O((n + k) \log(n + k))$.*

By the above lemmas, the time complexity to construct $SD(P)$ is $O((n + k) \log(n + k))$.

The construction $SD'(P)$ is similar, only easier, and hence we skip it from this abstract. We compute $SD'(P)$ in a preorder traversal of the rooted dual tree.

6 Merging the two Subdivisions

Once $SD(P)$ and $SD'(P)$ are available it is easy to construct the Voronoi diagram of S in P by merging the two diagrams. Let $\Delta = \Delta(d)$. Let $\phi(\Delta)$ denote the set of Voronoi edges in $Vor_\Delta(S)$ that are shared by pairs of Voronoi cells $V(s_i)$ and $V(s_j)$ for s_i in $S(d)$ and $s_j \in S'(d)$. Let $\phi(P)$ denote the union of $\phi(\Delta)$ for every triangle Δ in P. We have the following lemma.

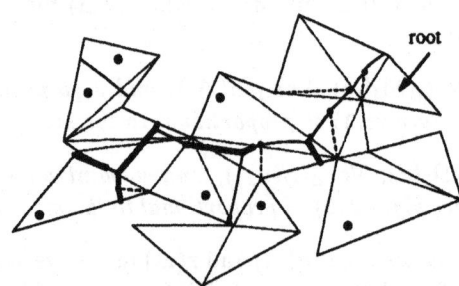

Fig. 10. Merging $SD(P)$ and $SD'(P)$. $\phi(P)$ is shown in thick lines.

Lemma 12. *$\phi(P)$ is a collection of Voronoi edges with the following properties:*
- *$\phi(P)$ consists of disjoint simple paths.*
- *The endpoints of a component of $\phi(P)$ must lie on the polygon boundary or at a border vertex of degree two of $SD(P)$ or $SD'(P)$.*
- *If an endpoint of a component of $\phi(P)$ is at a border vertex u of $SD(P)$ (resp. $SD'(P)$) then the bisector of $\phi(P)$ incident to u and the Voronoi edge in $SD(P)$ (resp. $SD'(P)$) incident to u, belong to the same bisector.*

Lemma 12 implies an algorithm for constructing $Vor_P(S)$ which can be briefly stated as follows: Trace the polygon boundary to identify points which are equidistant from their region owner in $SD(P)$ and their region owner in $SD'(P)$. For every such point, trace the corresponding component of $\phi(P)$ in the usual way [1, 13]. After all components starting at boundary points have been traced, trace the components (if any) with both endpoints at border vertices of degree two. Delete all border edges. Delete all Voronoi edges in $SD(P)$ and $SD'(P)$ incident to border vertices that are not endpoints of $\phi(\Delta)$. Finally, delete all border vertices and merge the two diagrams. It is not hard to see that merging can be done in $O(n + k)$ time.

7 Conclusion

We have presented an $O((n + k) \log(n + k))$-time algorithm for computing the geodesic Voronoi diagram of points in a simple polygon. Our algorithm is generalizable to an $O((n + k) \log(n + k))$-time algorithm for computing the geodesic Voronoi diagram of points in polygons with holes such that there exist *bridges* that partition the polygon into parts of at most one hole (see Figure 11). A *bridge* is a line segment with endpoints on the polygon boundary that is entirely contained in the polygon free space. We first generalize our algorithm for computing the geodesic Voronoi diagram of points in polygons with only one hole as follows. Given a polygon with a hole, transform it into a simple polygon by drawing a diagonal with one endpoint on the polygon boundary and the other on the boundary of the hole, and treating it as ordinary boundary. Compute the Voronoi diagram of points in the obtained simple polygon. Then remove the diagonal and modify the Voronoi diagram accordingly. Given a polygon with more than one hole such that holes are separated by bridges, we triangulate it using the bridges as triangulation diagonals. The dual graph is now a *tree of cycles*, where each cycle corresponds to hole. We call the dual graph *tree of cycles* because it becomes a tree (referred to as the dual *collapsed* tree) if each cycle is collapsed into a single node (referred to as a *collapsed* node). We can now apply the above algorithm using the dual *collapsed* tree as guide, and treating collapsed nodes as polygons with one hole.

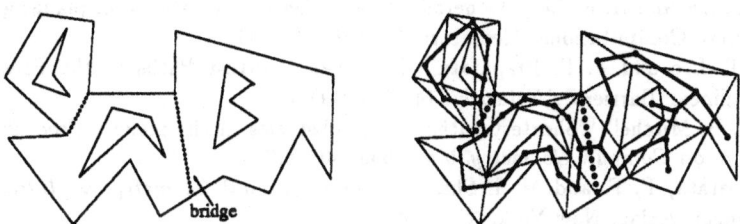

Fig. 11. A polygon with holes that can be partitioned into disjoint parts by *bridges*. The dual graph of the triangulation is a *tree of cycles*.

Furthermore, our approach can be slightly modified to compute the geodesic Voronoi diagram of point sites in polygonal domains where shortest paths are monotone with respect to a constant number of directions. Examples include a polygonal domain of n vertical line segments, where shortest paths are monotone with respect to the horizontal axis, and a polygonal domain of rectangular obstacles under the L_1 metric, where shortest paths are monotone with respect to either the horizontal or the vertical axis. In these cases we construct a subdivision of the plane (e.g., in the case of vertical line segments draw the vertical lines passing through the line segments) such that the dual graph can be transformed to a directed acyclic graph (e.g., in the case of vertical line segments assign a *right* direction to the edges of the dual graph). Sweeping is then guided by the topological sorting of the dual graph. Whether the approach of combining a sweep of the polygonal domain with the merging step of a divide-and-conquer strategy can be used for the geodesic Voronoi diagram of points in a general polygonal domain, is an open problem.

References

1. B. Aronov. "On the geodesic Voronoi diagram of point sites in a simple polygon", *Algorithmica*, 4 (1989), 109-140.
2. F. Aurenhammer, "Voronoi diagrams: A survey of a fundamental geometric data structure," *ACM Comput. Survey*, 23 1991, 345-405.
3. Ta. Asano and Te. Asano, "Voronoi diagrams for points in a polygon," in Discrete Algorithms & Complexity: Perspective in Computing, ed. D. S.Johnson, Academic Press, 1987, 51-64.
4. L. P. Chew, Constrained Delaunay triangulations, *Algorithmica*, 4 (1989), 97-108.
5. S. Fortune, A sweepline algorithm for Voronoi diagrams, *Algorithmica*, 2 (1987), 153-174.
6. L. J. Guibas, J. Hershberger, D. Leven, M. Sharir, and R.E. Tarjan, "Linear-time algorithms for visibility and shortest path problems inside triangulated simple polygons", *Algorithmica*, 2 (1987), 209-233.
7. L. J. Guibas and R. Sedgewick, "A dichromatic framework for balanced trees", *Proc. 19th IEEE Symp. on Foundations of Computer Science*, 1978, 8-21.
8. J. Hershberger and S. Suri, "Efficient Computation of Euclidean Shortest Paths in the Plane", *34th Symp. on Foundations of Computer Science*, 1993, 508-517.
9. D. Kirkpatrick, "Optimal Search in Planar Subdivisions" *SIAM J. Computing*, Vol. 12, No 1, 1983, 28-35.
10. D. T. Lee and A. K. Lin, "Generalized Delaunay triangulations for planar graphs", Discrete Computational Geometry, 1 (1986),201-217.
11. D. T. Lee and F. P. Preparata, "Euclidean Shortest Paths in the Presence of Rectilinear Barriers", *Networks*, 14 1984, 393-410.
12. J. S. B. Mitchell, "Shortest paths among obstacles in the plane", *Proc. 9th ACM Symp. on Comput. Geometry*, May 1993, 308-317.
13. Preparata, F. P. and M. I. Shamos, *Computational Geometry: an Introduction*, Springer-Verlag, New York, NY 1985.
14. C. Wang and L. Schubert, "An optimal algorithm for constructing the Delaunay triangulation of a set of line segments", *Proc. 3rd ACM Symposium on Computational Geometry*, 1987, 223-232.

Linear Size Binary Space Partitions for Fat Objects

Mark de Berg*

Department of Computer Science, Utrecht University,

P.O.Box 80.089, 3508 TB Utrecht, the Netherlands.

Abstract. We describe a new method for constructing binary space partitions for scenes in arbitrary dimensions. If the objects in the scene are fat (that is, they are not extremely long and skinny) then the BSP has linear size and it can be constructed in $O(n \log^2 n)$ time, where n is the number of objects. In fact, the method produces a linear size BSP for a more general class of scenes, namely scenes that satisfy the *bounding-box-fitness condition*—a property that we suspect many realistic scenes exhibit. The method is very simple and should perform well in practice.

1 Introduction

Many geometric problems can be solved more easily if a decomposition of the space of interest into smaller subspaces, or cells, is given. Therefore decompositions of two-, three-, or higher-dimensional scenes play an important role in computer graphics, geographic information systems, and robotics.

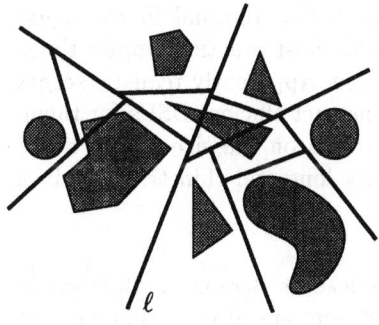

Fig. 1. A BSP in the plane.

There is a variety of schemes available to construct decompositions. Quadtrees and octrees, and *kd*-trees are among the most popular ones [14, 15]. Another popular decomposition scheme is the *binary space partition*, or *BSP*. In this scheme the space is split into two subspaces with a hyperplane, so in $\mathbb{R}^2$ it is split with a line and in $\mathbb{R}^3$ with a plane. These two subspaces are again split with a hyperplane, and so on. The splitting process continues recursively until the subspaces are intersected by only one of the objects in the scene. (We assume that the objects in the scene don't intersect each other, otherwise we cannot require each subspace to contain only one object.) Figure 1 shows a BSP of a two-dimensional scene; the line labeled ℓ is the first splitting line. Observe that the objects in the scene

* Supported by the Dutch Organisation for Scientific Research (N.W.O.) and by ESPRIT Basic Research Action No. 7141 (project ALCOM II: *Algorithms and Complexity*)

can be fragmented by the splitting process. A natural way to model the splitting process is with a binary tree. The root of this tree stores the first splitting hyperplane, its right child stores the splitting hyperplane of the right subspace and its left child stores the splitting hyperplane of the left subspace, and so on. A leaf of the tree corresponds to a cell in the final decomposition; it stores the (fragment of) the object that intersects the cell. Such a tree is called a *binary space partition tree*, or *BSP tree*.

Binary space partitions are used for a large number of purposes. For example, they are used for hidden surface removal with the painter's algorithm [7], for shadow generation [5], for set operations on polyhedra [10, 18], for visibility preprocessing for interactive walkthroughs [17], and for cell decomposition methods in motion planning [2].

The efficiency of algorithms based on BSPs depends crucially on the *size* of the BSP, that is, on the number of cells of the decomposition. Notice that this number is exactly the number of leaves in the corresponding BSP tree, which is linear in the size of the tree. Hence, when constructing a BSP of a given scene, one should choose the splitting hyperplanes carefully, so that the fragmentation of the objects is kept small. In two-dimensional space this is always possible: Paterson and Yao[12] proved that any set of polygons in the plane with n edges in total admits a BSP of size $O(n \log n)$ and that any set of axis-parallel polygons admits a linear size BSP. It is still open whether it is possible to construct a linear size BSP for any set of polygons in the plane. In three-dimensional space— the setting most relevant to computer graphics—the situation is less rosy: the method of Paterson and Yao is only guaranteed to produce a BSP of size $O(n^2)$. (For axis-parallel polyhedra one can obtain a BSP of size $O(n\sqrt{n})$ [13].) They also gave an example of a three-dimensional scene such that *any* BSP must have quadratic size, which shows that their method is optimal in the worst case. A quadratic size BSP is, of course, useless in most practical applications. Nevertheless, BSPs usually perform fine in practice. Apparently realistic scenes have some property that makes it possible to construct efficient BSPs for them. Indeed, the example proving the $\Omega(n^2)$ lower bound on the size of BSPs is a quite artificial construction, which uses extremely long and thin triangles in a grid-like pattern.

The discrepancy between theory and practice lead de Berg et al. [3] to study BSPs for scenes consisting of *fat objects*. Fat objects are objects that are not extremely long and skinny; a formal definition is given in Section 3. Recently, fat objects have attracted a lot of attention in computational geometry [1, 6, 8, 9, 11, 16, 19]. De Berg et al. proved that scenes of fat objects always admit a BSP of linear size. Their algorithm for constructing a BSP runs in $O(n \log n \log \log n)$ time, where n is the number of objects. Unfortunately, their method only works in the plane, so it is not very useful in computer graphics applications. Moreover, it is rather complicated. We propose a new method for constructing BSPs. Our method yields a linear size BSP for fat objects in arbitrary dimensions, and it is quite simple. The running time of the construction algorithm is $O(n \log^2 n)$, which is slightly worse than the running time of the method of de Berg et al.[2]

The method can be viewed as a combination of octree, kd-tree, and BSP tree approaches. Roughly, it works as follows. Consider the set of bounding boxes of the objects in the scene. In the first stage of the algorithm we compute a decomposition guided by the vertices of these bounding boxes. This decomposition uses only axis-parallel splitting planes; these splitting planes are chosen in an octree-like fashion or in a kd-tree like fashion, depending on the distribution of the bounding box vertices within the subspace being subdivided. In the second stage the cells of decomposition are split further, to separate the remaining objects inside each cell; in this stage we use hyperplanes that are not necessarily axis-parallel. Section 2 contains a detailed description of the decomposition, and it gives a simple, yet efficient algorithm for constructing it. For collections of fat objects the method computes a linear size BSP. The construction time is $O(n \log^2 n)$. The method not only works for fat objects; it can be proved to produce a linear size BSP for a more general class of scenes, namely scenes for which the *bounding-box-fitness condition* holds. Intuitively, this condition states that the bounding boxes of (almost all of) the objects provide a fairly good approximation of the object in the following sense: if a region with the shape of a cube is intersected by many objects then it must contain a vertex of a bounding box of one or more of these objects. We believe that most realistic scenes exhibit this property. In Section 3 we analyze our method under the bounding-box-fitness condition and we show that fatness implies the bounding-box-fitness condition.

2 Constructing the BSP

Let S be a set of n non-intersecting objects in $\mathbb{R}^d$, where $d \geqslant 2$. We assume that the scene is polygonal; curved objects will be briefly discussed at the end of Section 3. We also assume that the objects have constant complexity; this is not necessary, but simplifies the notation.

Below we describe our method for constructing a BSP for S, and we give an efficient algorithm for constructing it. Although the method works in arbitrary dimensions we shall mostly use three-dimensional terminology from now on.

Before we start we need a few definitions. The *bounding box* of an object o is the smallest axis-aligned box that contains the object. Let $V = V(S)$ denote the set of vertices of the bounding boxes of the objects in S. For a set S in $\mathbb{R}^3$, the set V contains at most $8n$ points; in $\mathbb{R}^d$, it contains at most $2^d n$ points.

Our algorithm for constructing the BSP for S works in two stages. In the first stage we construct a BSP using axis-parallel planes only. This stage is guided by the set V of vertices of the bounding boxes. The cells of the intermediate BSP that results from the first stage do not contain points from V in their interior. They can still be intersected by a number of objects, however. The remaining

---

[2] Actually, for the planar case our algorithm can be modified slightly so that it runs in the same time as de Berg at al.'s algorithm, namely $O(n \log n \log \log n)$.

objects in each cell are separated in the second stage of the algorithm, which uses splitting planes that are not necessarily axis-parallel.

During the first stage we shall try to ensure that the cells we create are cubes, rather than arbitrary boxes. The reason for this is the following: A cell that is a thin box can easily be intersected by many objects without containing many bounding box vertices; for such a cell the bounding box vertices don't provide sufficient information to control the fragmentation of the objects. Most objects that intersect a cube, however, will have a vertex of their bounding box inside it. Indeed, an object whose bounding box does not have a vertex inside a cube must be relatively big with respect to the cube (its diameter must be at least the edge length of the cube), and there cannot be too many big objects intersecting the cube, otherwise they would start intersecting each other. Hence, we can use the bounding box vertices to control the fragmentation. This intuitive explanation that fragmentation is controlled will be made formal in the next section, after we have described the two stages of our algorithm in more detail. Observe that an object intersecting a cube that does not have one of its own vertices inside the cube can still be small. This is the reason that we cannot use the vertices of the objects themselves to control the fragmentation.

The first stage. To construct the intermediate BSP we proceed as follows. In a generic partitioning step we have cell C, which is a cube, and a non-empty subset $V_C \subset V$ that contains all points from V lying in the interior of C. Initially, C will be a minimal enclosing cube of the set V, and V_C will contain all vertices from V that do not lie on the boundary of C.

We define an *octree split* to be a split of C into 2^d equal-sized sub-cubes. Thus an octree split is performed by taking the planes perpendicular to and bisecting the edges of C. We call an octree split *useless* if all the points in V_C are in the interior of the same sub-cube; otherwise it is called *useful*. The cube C is partitioned according to the following rules, which are illustrated in Figure 2 for the two-dimensional case.

octree split   *kd*-split

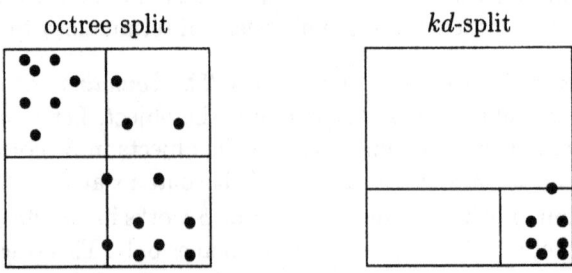

Fig. 2. Splitting a cube.

1. If an octree split of C is useful, it is performed.
2. Otherwise, let $C_1, \ldots, C_{2^d}$ be the sub-cubes that would result from an octree split of C. Suppose that all points in V_C lie in C_j, and let v be the vertex of C that is also a vertex of C_j. Let C'_j be the smallest cube with v as one of its vertices that contains all points from V_C in its closure. Intuitively, C'_j is obtained by shrinking C_j while keeping v as one of its vertices until a point from V_C is hit. Now C is split using planes through the facets of C'_j; the order in which these planes are taken is arbitrary. We call such a split a *kd-split*. Note that it is slightly different from the way splits are usually performed in a *kd*-tree: there one splits the set of points into two equal halves, whereas we have an unbalanced split which guarantees the subspace containing the points to be a cube.

The splitting process is repeated recursively for the resulting subcells that have points in their interior. All subcells resulting from an octree split are cubes, but a *kd*-split can produce cells that are not cubes. The only cell on which we have to recurse after performing a *kd*-split, however, is a cube.

We now analyze the size of the intermediate decomposition.

Lemma 1. *The first stage of the algorithm results in an intermediate BSP consisting of $O(n)$ cells that are boxes and do not contain a vertex from V in their interior.*

Proof. The construction of the partitioning proceeds recursively until each cell is empty, so the second part of the lemma is clearly true. To see that there are $O(n)$ cells we note that any split increases the number of cells by a constant only. More precisely, an octree split results in $2^d - 1$ extra cells, and a *kd*-split results in d extra cells. Furthermore, when a cell C is split either one or more points from the current subset V_C are on the splitting planes, or V_C is partitioned into two or more subsets. The first case can occur at most $|V|$ times, and the second case at most $|V| - 1$ times. Hence, the total number of cells is at most $(2|V| - 1)(2^d - 1) + 1 = O(n)$. $\qquad\square$

The description above specifies which splitting planes are used in the first stage of the algorithm. However, one important detail of the construction process is left open: how do we find the splitting planes for a cell C efficiently? If this is done in a brute-force manner, then the construction time can be high, even though the BSP that is produced may be small. The reason for this is that the splits need not be balanced; in fact, a *kd*-split is always unbalanced. So if we spend linear time on each split and the splits are very unbalanced, then the running time $T(n)$ would satisfy a recurrence like $T(n) = O(n) + T(n-1)$, leading to a quadratic running time. Next we describe a simple technique to avoid this. This technique may also be useful in other decomposition schemes, such as octrees or quadtrees.

Consider a cell C with the non-empty set V_C of bounding box vertices in its interior. Assume that we have three sorted doubly linked lists (in $\mathbb{R}^d$ we have

d lists) available on the points in V_C: a doubly-linked list $\mathcal{L}_x$ that contains the points in V_C sorted on x-coordinate, a doubly-linked list $\mathcal{L}_y$ that contains the points in V_C sorted on y-coordinate, and a doubly-linked list $\mathcal{L}_z$ that contains the points in V_C sorted on z-coordinate. We also assume that we have cross pointers between corresponding points in the three lists. For the initial set V, these lists with the cross pointers can be constructed in $O(n \log n)$ time. In recursive calls, they can be maintained efficiently, as we show below.

The first thing we have to do when splitting C is decide whether we should perform an octree split or a kd-split. Let $C = [c_x : c_x+s] \times [c_y : c_y+s] \times [c_z : c_z+s]$. Let p_{xmin} and p_{xmax} be points in V_C whose x-coordinate is minimal and maximal, respectively. These points are simply the first and last point in the list $\mathcal{L}_x$. Define p_{ymin}, p_{ymax}, p_{zmin}, and p_{zmax} similarly. An octree split is useful if and only if one of the three planes used in an octree split has points on both sides of it, or on it. Thus an octree split is useful if and only if

$$p_{\text{xmin}} \leqslant c_x + s/2 \text{ and } q_{\text{xmax}} \geqslant c_x + s/2, \text{ or}$$
$$p_{\text{ymin}} \leqslant c_y + s/2 \text{ and } q_{\text{ymax}} \geqslant c_y + s/2, \text{ or}$$
$$p_{\text{zmin}} \leqslant c_z + s/2 \text{ and } q_{\text{zmax}} \geqslant c_z + s/2.$$

Hence, using the sorted lists $\mathcal{L}_x$, $\mathcal{L}_y$, and $\mathcal{L}_z$ we can decide in constant time if we should perform an octree split or a kd-split.

Suppose we have to perform a kd-split. The planes we use for this split can be computed in constant time from the the points p_{xmin}, p_{xmax}, p_{ymin}, p_{ymax}, p_{zmin}, and p_{zmax}. After we have computed the planes, we determine the points from V_C that lie on the splitting planes. This can be done by simply walking along the sorted lists $\mathcal{L}_x$, $\mathcal{L}_y$, and $\mathcal{L}_z$, starting at the appropriate end. For example, suppose that we use a plane $h_x : x = x^*$ as a splitting plane, and assume all points in V_C lie to the left of h_x or on h_x. Then we walk along $\mathcal{L}_x$ starting from p_{xmax}. As soon as we encounter a point that does not lie on h_x we can stop; because $\mathcal{L}_x$ is sorted, none of the remaining points lies on h_x. By doing this for each of the three lists, we can identify the points that lie on one of the splitting planes in $O(k)$ time, where k is the number of such points. These points are removed (using the cross pointers) from all three lists. The new lists can be used in the recursive call for the subcell that contains the remaining points.

Now suppose we have to perform an octree split. An octree split can still be very unbalanced: many points from V_C could lie in one subcell and only few in the remaining subcells. If this happens, we are not allowed to spend linear time to distribute the points over the subcells. To overcome this difficulty we use a technique called *tandem search*[4]. Consider the partition of V_C induced by the plane $h_x : x = c_x + s/2$. We have to partition V_C into three subsets: the set V_{left} of points to the left of h_x, the set V_{on} of points on h_x (these points can be excluded from further consideration), and the set V_{right} of points to the right of h_x. The points in V_{left} can be found by walking along $\mathcal{L}_x$ starting from p_{xmin}, and the points in V_{right} can be found by walking along $\mathcal{L}_x$ starting from p_{xmax}. To avoid spending too much time in case of an unbalanced split, we start walking simultaneously from both ends, one step at a time. This way we find the smaller of the subsets V_{left} and V_{right} in time proportional to its size, because the number of steps we need is roughly twice the size of the smaller set. Figure 3 illustrates this. The search started at p_1, then p_8 was visited, then p_2, and so on, until p_6 was found to lie on h_x. Now the subset V_{right} has been found; it consists of p_8 and p_7. The set V_{on} can be found in time proportional to its size by continuing

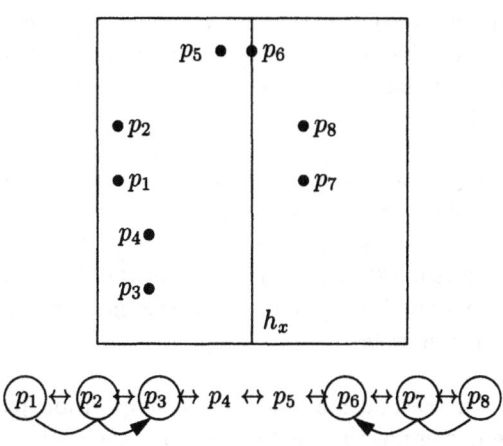

Fig. 3. Tandem search.

to walk from where the smaller set ended, until a point is encountered that is not on h_x. It remains to produce the sorted lists for two subspaces resulting from the split with h_x. For the lists sorted on x-coordinate this is easy, because we can simply split the list we had for C at the appropriate positions. For the lists sorted on y- and z-coordinate this is not possible. These lists are obtained as follows. The lists for the smaller subset are simply computed from scratch, by sorting the points. The lists for the larger subset are generated by deleting the points in the lists we had for C, using the cross pointers. The total time we need to perform the split is $O(l \log l + k)$, where l is the number of points in the smaller subset, and k is the number of points on the plane h_x. Thus, the more unbalanced the split, the faster it is performed.

We have explained how to perform the partitioning for the plane orthogonal to the x-axis; the partitioning for the other two planes in the octree split are performed in a similar fashion.

Lemma 2. *The first stage of the algorithm runs in $O(n \log^2 n)$ time.*

Proof. First, consider a kd-split. The time we need to perform such a split is $O(k)$, where k is the number of points on the splitting planes. We charge $O(1)$ time to each of these points. Because points on a splitting plane are not considered in the recursive calls, a point gets charged only once this way. Hence, the total time needed for all kd-splits is $O(n)$.

Now consider an octree split. Let n_C be the number of points in V_C. The time we need to perform the split with the plane orthogonal to the x-axis is $O(l \log l)$, where $l \leqslant n_C/2$ is the number of points in the smaller subset, and k is the number of points on the splitting plane. The $O(k)$ time can again be charged to the points on the splitting plane. The $O(l \log l)$ time is accounted for by charging $O(\log l)$ time to the l points in the smaller subset. (We assume for simplicity that $l \neq 0$. This need not be the case. However, l cannot be zero in all three splits in an octree split, otherwise we would have performed a kd-split. Hence, the simplification does not influence the bounds.) Because any point that gets charged goes to a set that has been halved in size, a point is charged $O(\log n)$ times. The lemma follows. □

Remark: The first stage can also be implemented to run in $O(n \log n \log \log n)$ time, by sorting the lists for the smaller subset in an octree split in $O(l \log \log l)$ time. This is possible by using Van Emde-Boas trees [20] after normalizing the coordinates of the points in $O(n \log n)$ time at the start of the algorithm. Details are given in the full paper. We do not expect this more complicated approach to be faster in practice, however.

The second stage. The second stage of the decomposition partitions the cells of the intermediate decomposition further, so that each cell in the final BSP is intersected by only one object. This is done in the standard way: For a cell C of the intermediate decomposition, let S_C be the set of object fragments inside C; the cell C is partitioned by taking planes through the facets of the objects in S_C in an arbitrary order, until there is only one object left in each cell.

To carry out the second stage of the algorithm, we need to know for each cell C in the intermediate decomposition the set S_C of object fragments lying in it. In the plane a simple plane sweep algorithm can compute the sets S_C in $O(n \log n)$ time in total, but in three and higher dimensions we need a different approach. Here the idea is to maintain the set of object fragments intersecting a subspace while we compute the intermediate decomposition. To do this efficiently, we maintain the extreme points in the x-, y-, and z-direction of the object fragments and we applying the tandem search technique explained above.

3 The analysis

We now analyze the method of the previous section under the assumption that the set S of objects in the scene satisfies the following condition:

bounding-box-fitness condition: There is a constant C such that any cube not containing a vertex of one of the bounding boxes of the objects in S is intersected by at most C objects in S.

It is essential that this condition only speaks of *cubes* without bounding box vertices in their interior. If the condition would speak of arbitrarily shaped boxes it would be a lot stronger, but it would also be highly unrealistic. We believe that the bounding-box-fitness condition is rather realistic, despite its rather technical flavor. Under the bounding-box-fitness condition the following lemma holds.

Lemma 3. *Any cell in the intermediate decomposition is intersected by $O(1)$ objects from S.*

Proof. Let C be a cell of the intermediate decomposition. By construction, C does not contain a vertex of any bounding box in its interior. Hence, if C is a cube then the lemma follows from the bounding-box-fitness condition. Now suppose that C is not a cube. Then C was created as one of the empty cells when a kd-split was performed. In this case C can be covered by a constant number of cubes that are contained in the union of all empty cells created at this step. This

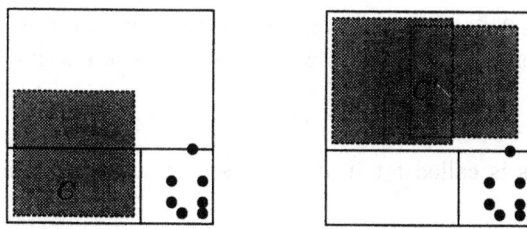

Fig. 4. Covering cells of the intermediate decomposition by cubes.

is illustrated for the planar case in Figure 4; for clarity, the squares that cover the cells C and C' are shown slightly smaller than they actually are. Because the cubes that cover C are inside the union of the empty cell they contain no vertices in their interior, so they are intersected by $O(1)$ objects. Hence, C is intersected by $O(1)$ objects as well. $\qquad\square$

So after the first stage of the algorithm we have an intermediate BSP of linear size such that each cell is intersected by $O(1)$ objects. Because the second stage is performed on a constant number of objects inside each cell, it does not increase the asymptotic size of the BSP or its construction time (provided that the objects are polygonal and of constant complexity, or are convex.) Hence, we can construct a BSP of linear size in $O(n \log^2 n)$ time.

Theorem 4. *Let S be a polygonal scene in $\mathbb{R}^d$ consisting of non-intersecting polygonal objects of constant complexity for which the bounding-box-fitness condition holds. Then there exists a linear size binary space partition for the objects in S. This binary space partition can be constructed in $O(n \log^2 n)$ time.*

Remark: One may think that the bounding-box-fitness condition is strong enough so that we could use a standard octree or kd-tree decomposition in the first stage of our algorithm. This is not true. If the objects are small compared to the bounding box of the whole scene, then an octree decomposition would generate too many useless splits. A standard kd-tree (where we use planes orthogonal to the coordinate axes in a round-robin fashion, always splitting the set of bounding box vertices into two sets of equal size) also does not work: in the full paper we give an example of a planar scene that satisfies the bounding-box-fitness condition for which a kd-tree results in $\Omega(n \log n)$ fragments. Thus our scheme is superior in theory. This does not mean that it is necessarily better in practice; experimental research is needed to see which decomposition is best.

We now prove that fatness implies the bounding-box-fitness condition. Intuitively, an object is called fat if it does not contain any long and skinny parts. Van der Stappen [16] gives an extensive treatment of fatness in the context of motion planning. Fatness can be defined formally in various ways, which are basically all equivalent. We use the following definition.

Definition 5. Let $0 \leqslant \alpha \leqslant 1$ be a constant. An object o in $\mathbb{R}^d$ is α-fat if, for any sphere σ whose center lies in o and whose boundary intersects o, the following holds
$$\text{Volume}(o \cap \sigma) \geqslant \alpha \text{Volume}(\sigma).$$

A set S of objects is called fat if all objects in S are α-fat for some constant $0 \leqslant \alpha \leqslant 1$.

(If the objects in S are convex, then the definition is equivalent to the following, simpler definition: an object is fat if its volume is at least β times the volume of its minimal enclosing d-dimensional sphere for some constant β.) Van der Stappen [16] proves that fat objects have the following property: Let $\rho(o)$ denote the radius of the minimal enclosing hypersphere of an object o.

Theorem 6. (van der Stappen [16]) *Let A be a region in $\mathbb{R}^d$ and let F be a set of non-intersecting fat objects. Suppose that there is a constant c such that $\rho(o) \geqslant c\rho(A)$ for all $o \in F$. Then the number of objects from F intersecting A is $O(1)$.*

Intuitively, this theorem states that the region A cannot be intersected by many fat objects that are relatively large compared to A. Using this result we can prove that a set of fat objects satisfies the bounding-box-fitness condition.

Lemma 7. *Any set S of non-intersecting fat objects in $\mathbb{R}^d$ satisfies the bounding-box-fitness condition.*

Proof. Let C be a cube that does not have bounding box vertices of objects in S in its interior. Consider an object $o \in S$ that intersects C. Because the bounding box vertices of o do not lie in the interior of C, the diameter of o must be at least the edge length of C. Hence, there is a constant c such that $\rho(o) \geqslant c\rho(C)$. The lemma now follows from Theorem 6. □

Although fatness implies the bounding-box-fitness condition, the reverse is certainly not true. Consider as an example an architectural model. The ceilings in such a model are very thin in the z-direction, so they are not fat. A similar observation can be made for the walls. If, however, the rooms have a reasonable shape (not extremely long and narrow) then an architectural model can be shown to satisfy the bounding-box-fitness condition.

The following result immediately follows from Theorem 4 and Lemma 7.

Corollary 8. *Let S be a polygonal scene in $\mathbb{R}^d$ consisting of n non-intersecting fat objects of constant complexity. Then there exists a linear size binary space partition for the objects in S. This binary space partition can be constructed in $O(n \log^2 n)$ time.*

Curved objects. The only place where we need that the objects are polygonal is in the second stage of the algorithm, where we separate the objects using planes through the facets of the objects. If the objects are curved we need a different strategy for this. If there is such a strategy—more precisely, if any pair of objects can be separated with a constant number of planes—then our method still works. For example, a linear size BSP exists for any set of convex curved objects that satisfies the bounding-box-fitness condition.

4 Concluding remarks and directions for future research

We presented a new method for constructing BSPs. For scenes with the bounding-box-fitness condition—a property we suspect many realistic scenes have—the method produces a linear size BSP in $O(n \log^2 n)$ time. We also proved that any set of disjoint fat objects in $\mathbb{R}^d$ satisfies the bounding-box-fitness condition, which thus implies that any set of disjoint fat objects admits a linear size BSP. The method is quite simple and it should perform well in practice. Of course, the validity of this claim should be verified experimentally. In general, it is important that experimental research be done to compare decomposition schemes, such as octrees and BSPs. It would also be nice to test how realistic the bounding-box-fitness condition is in practice.

For many geometric problems in computer graphics there is a large discrepancy between practice and theoretical results from computational geometry. In particular, theoretical results are often complicated, which can make them inefficient in practice. The reason for this is that they are geared to the worst-case scenario. But the worst-case examples are usually quite artificial and do not occur in practice. It is therefore important to understand the properties of realistic scenes and use these properties to develop algorithm that are provably efficient in practical situations. The goal of the present paper is to do this for BSP trees, but we think this should be done for other problems as well. If our experiments will show that most realistic scenes satisfy the bounding-box-fitness condition—and we expect this to be the case—then this will give a possible handle to tackle these other problems.

References

1. P. K. Agarwal, M. J. Katz, and M. Sharir. Computing depth orders and related problems. In *Proc. 4th Scand. Workshop Algorithm Theory*, volume 824 of *Lecture Notes in Computer Science*, pages 1–12, 1994.
2. C. Ballieux. *Motion planning using binary space partitions*. Report Inf/src/93-25, Utrecht University, 1993.
3. M. de Berg, M. de Groot, and M. Overmars. New results on binary space partitions in the plane. In *Proc. 4th Scand. Workshop Algorithm Theory*, volume 824 of *Lecture Notes in Computer Science*, 1994.
4. M. de Berg, M. Overmars, and O. Schwarzkopf. Computing and verifying depth orders. *SIAM J. Comput.*, 23(2):437-446, 1994.
5. N. Chin and S. Feiner. Near real time shadow generation using bsp trees. In *Proc. SIGGRAPH'89*, pages 99–106, 1989.
6. A. Efrat, M. Sharir, and G. Rote. On the union of fat wedges and separating a collection of segments by a line. *Comput. Geom. Theory Appl.*, 3:277-288, 1994.
7. H. Fuchs, Z. M. Kedem, and B. Naylor. On visible surface generation by a priori tree structures. *Comput. Graph.*, 14(3):124-133, 1980.
8. Marc van Kreveld. On fat partitioning, fat covering, and the union size of polygons. In *Proc. 3rd Workshop Algorithms Data Struct.*, volume 709 of *Lecture Notes in Computer Science*, pages 452–463, 1993.
9. J. Matousek, J. Pach, M. Sharir, S. Sifrony, and E. Welzl. Fat triangles determine linearly many holes. *SIAM J. Comput.*, 23:??–??, 1994.
10. B. Naylor, J. A. Amatodes, and W. Thibault. Merging BSP trees yields polyhedral set operations. *Comput. Graph.*, 24(4):115-124, August 1990.
11. M. H. Overmars. Point location in fat subdivisions. *Inform. Process. Lett.*, 44:261-265, 1992.
12. M. S. Paterson and F. F. Yao. Efficient binary space partitions for hidden-surface removal and solid modeling. *Discrete Comput. Geom.*, 5:485-503, 1990.
13. M. S. Paterson and F. F. Yao. Optimal binary space partitions for orthogonal objects. *J. Algorithms*, 13:99-113, 1992.
14. H. Samet. *Applications of Spatial Data Structures*. Addison Wesley, Reading, MA, 1990.
15. H. Samet. *The Design and Analysis of Spatial Data Structures*. Addison-Wesley, Reading, MA, 1990.
16. F. van der Stappen. *Motion Planning Amidsts Fat Obstacles*. Ph. D. Thesis, Utrecht University, 1994.
17. S. J. Teller and C. H. Séquin. Visibility preprocessing for interactive walkthroughs. *Comput. Graph.*, 25(4):61-69, July 1991.
18. W. C. Thibault and B. F. Naylor. Set operations on polyhedra using binary space partitioning trees. In *Proc. SIGGRAPH'87*, pages 153-162, 1987.
19. A. F. van der Stappen and M. H. Overmars. Motion planning amidst fat obstacles. In *Proc. 10th Annu. ACM Sympos. Comput. Geom.*, pages 31-40, 1994.
20. P. van Emde Boas, R. Kaas, and E. Zijlstra. Design and implementation of an efficient priority queue. *Math. Syst. Theory*, 10:99-127, 1977.

Geometric Pattern Matching in d-Dimensional Space*

L. Paul Chew[1]   Dorit Dor[2]   Alon Efrat[2]   Klara Kedem[3]

[1]Department of Computer Science, Cornell University, Ithaca, NY 14853, USA
[2]School of Mathematical Sciences, Tel Aviv University, Tel Aviv 69978, Israel
[3]Department of Math and CS, Ben Gurion University, Beer-Sheva 84105, Israel

Abstract. We show that, using the L_∞ metric, the minimum Hausdorff distance under translation between two point sets of cardinality n in d-dimensional space can be computed in time $O(n^{(4d-2)/3} \log^2 n)$ for $d > 3$. Thus we improve the previous time bound of $O(n^{2d-2} \log^2 n)$ due to Chew and Kedem. For $d = 3$ we obtain a better result of $O(n^3 \log^2 n)$ time by exploiting the fact that the union of n axis-parallel unit cubes can be decomposed into $O(n)$ disjoint axis-parallel boxes. We prove that the number of different translations that achieve the minimum Hausdorff distance in d-space is $\Theta(n^{\lfloor 3d/2 \rfloor})$. Furthermore, we present an algorithm which computes the minimum Hausdorff distance under the L_2 metric in d-space in time $O(n^{\lceil 3d/2 \rceil + 1} \log^3 n)$.

1 Introduction

We consider the problem of finding the resemblance, under translation, of two point sets in d-dimensional space for $d \geq 3$. In many matching applications, objects are described by d parameters; thus a single object corresponds to a point in d-dimensional space. One would like the ability to determine whether two sets of such objects resemble each other. A 3D example comes from molecular matching, where a molecule can be described by its atoms, represented as points in 3-space.

The tool that we suggest here for measuring resemblance is the well-researched minimum Hausdorff distance under translation. The distance function we use is the L_∞ metric. One advantage of using the Hausdorff distance is that it does not assume equal cardinality of the point sets. It measures the maximal *mismatch* between the sets when one point set is allowed to translate in order to minimize this mismatch. Two point sets are considered to be similar if this mismatch is small. We assume that the cardinalities of the sets are n and $m = o(n)$ and we express our results in terms of n.

There have been several papers on the subject of point set resemblance using the minimum Hausdorff distance under translation. Huttenlocher et al. [11, 12] find the

---

* The work of the first and fourth authors was partly supported by AFOSR Grant AFOSR-91-0328. The first author was also supported by ONR Grant N00014-89-J-1946, by ARPA under ONR contract N00014-88-K-0591, and by the Cornell Theory Center which receives funding from its Corporate Research Institute, NSF, New York State, ARPA, NIH, and IBM Corporation.

minimum Hausdorff distance for point sets in the plane in time $O(n^3 \log n)$ under the L_1, L_2, or L_∞ metrics. For point sets in 3-dimensional space their algorithm, using the L_2 metric, runs in time $O(n^{5+\epsilon})$. The method used in [12] cannot be extended to work under L_∞.

Chew and Kedem [7] show that, when using the L_∞ metric in the plane, the minimum Hausdorff distance can be computed in time $O(n^2 \log^2 n)$. This is a somewhat surprising result, since there can be $\Omega(n^3)$ different translations that achieve the (same) minimum [7, 18]. They [7] further extend their technique to compute the minimum Hausdorff distance between two point sets in d-dimensional space, using the L_∞ metric, achieving a time bound of $O(n^{2d-2} \log^2 n)$ for a fixed dimension d.

We show in this paper that, using the L_∞ metric, the minimum Hausdorff distance between two point sets can be found in time $O(n^3 \log^2 n)$ for $d = 3$, and in time $O(n^{(4d-2)/3} \log^2 n)$ for any fixed $d > 3$.

To estimate the quality of the time complexity of our algorithms, it is natural to seek the number of different translations that achieve the minimum Hausdorff distance between two point sets in d-dimensional space. More precisely, the number of connected components of feasible translations in the d-dimensional translation space. We show that this number is $\Theta(n^{\lfloor 3d/2 \rfloor})$ in the worst case. Note that, as for the planar case solved in [7], the runtime of the algorithms which we present for a fixed $d \geq 3$ is significantly lower than the number of the connected components in the d-dimensional translation space.

Our algorithms can be easily extended within the same time bounds to tackle the problem of *pattern matching in the presence of spurious points*, sometimes called *outliers*. In this problem we seek a small subset $X \subseteq A \cup B$ of points (whose existence is a result of noise) containing at most a pre-determined number k of points, such that $A \setminus X$ can be optimally matched, under translation, to $B \setminus X$.

Many optimization problems are solved parametrically by finding an oracle for a *decision problem* and then using this oracle in some parametric optimization scheme. In this paper we follow this line by developing an algorithm for the *Hausdorff decision problem* (see definition in the next section) and then using it as an oracle in the Frederickson and Johnson [10] optimization scheme. For the oracle in 3-space we prove that a set of n unit cubes can be decomposed into $O(n)$ disjoint axis-parallel boxes. We then apply the orthogonal partition trees (OPTs) described by Overmars and Yap [16] to find the maximal depth of disjoint axis-parallel boxes. We show that this suffices to answer the Hausdorff decision problem in 3-space. For $d > 3$ there is a super-linear lower bound on the number of boxes obtained by disjoint decomposition of a union of boxes (see [4]); thus we cannot use a disjoint decomposition of unit cubes. Instead, we build a decision-problem oracle by developing and using a modified, nested version of the OPT.

When using L_2 as the underlying metric we show that there can be $\Omega(n^{\lfloor 3d/2 \rfloor})$ connected componnents in the translation space, and that the complexity of the space of feasible translations is $O(n^{\lceil 3d/2 \rceil})$. We present an algorithm which computes the minimum Hausdorff distance under the L_2 metric in d-space in time $O(n^{\lceil 3d/2 \rceil + 1 + \delta} \log^3 n)$ for any $\delta > 0$.

The paper is organized as follows: In Section 2 we define the *minimum Hausdorff distance problem*, and describe the *Hausdorff distance decision problem*. In Section 3 we show that the union of n axis-parallel unit cubes in 3-space can be decomposed into $O(n)$ disjoint axis-parallel boxes, and use the orthogonal partition trees of [16] to solve the Hausdorff distance decision problem in 3-space. For $d > 3$, our algorithm is more involved and hence its description is separated into two sections; Section 4 contains a relaxed version of our data structures and an oracle which runs in time $O(n^{3d/2-1} \log n)$. In Section 5 we modify the data structures of the relaxed version and obtain an $O(n^{(4d-2)/3} \log n)$ runtime oracle. In Section 6 we show briefly how we plug the decision algorithm into the Frederickson and Johnson optimization scheme. Bounds on the number of translations that minimize the Hausdorff distance are presented in Section 7. The algorithm for the minimum Hausdorff distance under the L_2 metric is discussed in Section 8. Open questions appear in Section 9.

Since most of the spatial objects we deal with in this paper are axis-parallel cubes, axis-parallel boxes and axis-parallel cells, we will omit from now on the words 'axis-parallel' and talk about cubes, boxes and cells.

2 The Hausdorff Distance Decision Problem

The well-known *Hausdorff distance* between two point sets A and B is defined as

$$H(A, B) = \max(h(A, B), h(B, A))$$

where the *one-way* Hausdorff distance from A to B is

$$h(A, B) = \max_{a \in A} \min_{b \in B} \rho(a, b).$$

Here, $\rho(\cdot, \cdot)$ represents a familiar metric on points: for instance, the standard Euclidean metric (the L_2 metric) or the L_1 or L_∞ metrics. In this paper, unless otherwise noted, we use the L_∞ metric. In dimension d, an L_∞ "sphere" (i.e., a set of points equidistant from a given center point) is a d-cube.

The *minimum Hausdorff distance* between two point sets is the Hausdorff distance minimized with respect to all possible translations of the point sets. Huttenlocher and Kedem [11] observe that the minimum Hausdorff distance is a metric on point sets (and more general shapes) that is independent of translation. Intuitively, it measures the maximum mismatch between two point sets after the sets have been translated to minimize this mismatch. For the minimum Hausdorff distance the sets do not have to have the same cardinality, although, to simplify our presentation, we assume that both point sets are of size $\Theta(n)$.

As in [7] we approach this optimization problem by using the *Hausdorff decision problem* with parameter $\varepsilon > 0$ as a subroutine for a search in a sorted matrix of ε values. We define the *Hausdorff decision problem* for a given ε to be the question of whether the minimum Hausdorff distance under translation is bounded by ε. We say that the Hausdorff decision problem for sets A and B and for ε is *true* if there exists a translation t such that the Hausdorff distance between A and B shifted by t is less than or equal to ε.

We follow the approach taken in [7], solving the Hausdorff decision problem by solving a problem of the intersection of unions of cubes in the (d-dimensional) *translation space*. Let A and B be two sets of points as above, let ε be a positive real value, and let C_ε be a d-dimensional cube, with side size 2ε and with the origin at its center (the L_∞ "sphere" of radius ε). We define the set A_ε to be $A \oplus C_\varepsilon$, where $\oplus$ represents the Minkowski sum. Consider the set $A_\varepsilon \oplus -b$ where $-b$ represents the reflection of the point b through the origin. This set is the set of translations that map b into A_ε. The set of translations that map all points $b \in B$ into A_ε is then $\cap_{b \in B}(A_\varepsilon \oplus -b)$; we denote this set by $S(A, \varepsilon, B)$. It can be shown [7] that the Hausdorff decision problem for point sets A and B and for ε is true iff $S(A, \varepsilon, B) \cap -S(B, \varepsilon, A) \neq \phi$. We restrict our attention to the problem of determining whether $S(A, \varepsilon, B)$ is empty; extending our method to determining whether the intersection of this set with $-S(B, \varepsilon, A)$ is empty is reasonably straightforward.

Another way to look at the Hausdorff decision problem is to assign a different color to each $b \in B$ and to paint all the cubes of $A_\varepsilon \oplus -b$ for a particular b, in its assigned color. Now we can look at $A_\varepsilon \oplus -b$ as a union of cubes of one color which we call a *layer*. We thus have n layers in n different colors, one layer for each point $b \in B$. A point $p \in \mathbb{R}^d$ is *covered* by a color i if there exists a cube in the layer of that color containing p (i.e., if $p \in A_\varepsilon \oplus -b_i$). The *color-depth* of p is the number of layers containing p. Our aim is thus to determine if there is a point p of color-depth n.

3 The Decision Problem in 3 Dimensions

Overmars and Yap [16] address the question of determining the volume of a union of N d-boxes. Using a data structure they call an *orthogonal partition tree*, which we abbreviate as OPT, they achieve a runtime of $O(N^{d/2} \log N)$. They also observe that their data structure can be used to report other measures within the same time bound. One problem that can be easily solved using their data structure is the *maximum coverage* for a set S of N d-boxes. Defining the *coverage* of a point $p \in \mathbb{R}^d$ to be the number of (closed) d-boxes that contain p, the *maximum coverage* is $\max\{coverage(p) \mid p \in \mathbb{R}^d\}$.

Maximum coverage is almost what we need for the Hausdorff decision problem, but instead we need the maximum color-depth. The difference is that maximum coverage counts the number of different boxes while we need to count the number of different colors (layers). These two concepts are the same if, for each color, all the boxes of that color are disjoint. To achieve this, we first decompose each layer into $O(n)$ disjoint boxes; then we apply the OPT method to compute the maximum coverage (which will now equal the maximum color-depth).

Theorem 1. *The union of n unit cubes in $\mathbb{R}^3$ can be decomposed, in time $O(n \log n)$, into $O(n)$ boxes whose interiors are disjoint.*

Proof: For space limitations we just outline the proof. We slice the 3-dimensional space by planes parallel to the z axis at $z = 0, 1, 2, \ldots$ (wlog all cubes have nonnegative coordinates). For each integer i, n_i denotes the number of of cubes intersected by the plane $z = i$, $n = \sum n_i$, and E is the portion of the union of cubes that lies within the slab bounded by $z = i$ and $z = i + 1$. Since the complexity of the boundary of the union of n unit 3-cubes is linear in the number of cubes (e.g., [4]), the complexity of the boundary of E is $O(n_i + n_{i+1})$. To end the proof, we show that E can be decomposed into $O(n_i + n_{i+1})$ boxes with disjoint interiors. $\quad\square$

Applying this theorem to each of the n colors, we decompose each layer (recall that a layer is the union of all cubes of a single color) into $O(n)$ disjoint boxes, getting a total of $N = O(n^2)$ boxes, where boxes of the same color do not overlap. We can now apply the Overmars and Yap algorithm on these boxes, getting an answer to the Hausdorff decision problem in time $O(n^3 \log n)$. This gives us the following theorem.

Theorem 2. *For point sets in 3-space the Hausdorff decision problem can be answered in time $O(n^3 \log n)$.*

4 Higher Dimensions: the Relaxed Version

The decomposition method used for 3-space cannot be extended efficiently to work for $d > 3$, since as Boissonnat et al. [4] have recently shown, the complexity of the union of n d-dimensional unit cubes is $\Theta(n^{\lfloor d/2 \rfloor})$; thus a single layer (the union of cubes for a single color) cannot be decomposed into $O(n)$ disjoint boxes. Note that we cannot use the Overmars and Yap data structure and algorithm directly for the set of n^2 colored cubes because of the possible overlapping of cubes of the same color. Our method is therefore to augment the OPT [16], adding capabilities that efficiently handle the overlapping of same-color cubes.

We describe very briefly the OPT of Overmars and Yap [16]. To use the OPT for computing the measure of N d-boxes, Overmars and Yap [16] treat this *static d-dimensional* problem as a *dynamic $(d-1)$-dimensional* problem: they sweep d-space using a $(d-1)$-dimensional hyperplane h. During this sweep, cubes enter into h (start intersecting h) and leave it afterwards. Each entering cube causes an insertion update to the OPT and each leaving cubes causes a deletion update. Both insertions and deletions involve some computation concerning the required measure. Let $Q = \{q_1, \ldots, q_N\}$ be a set of N boxes contained in $(d-1)$-space. An OPT $\mathcal{T}$ defined for Q is a binary tree such that each node δ is associated with a box-like $(d-1)$-cell C_δ that contains some part of the $(d-1)$-space. For each node δ, the cell C_δ is the disjoint union of the cells associated with its children, $C_{\delta_{left}}$ and $C_{\delta_{right}}$. The cell C_{root} is the whole $(d-1)$-space. We say that a cell C_δ is an ancestor (resp. descendent) of a cell $C_{\delta'}$ if δ is an ancestor (resp. descendent) of δ' in $\mathcal{T}$. Note that this is just the containment relation between cells. We say that a box q is a *pile* with respect to a cell C_δ if (1) $C_\delta \not\subseteq q$ and (2) for at least $d-2$ of the axes, the projection of q on the axes contains the projection of C_δ. If $C_\delta \subseteq q$ we say that q *covers* C_δ.

The tree structure of the OPT is fixed though the data at the nodes are changed with the insertion and deletion of cubes throughout the algorithm. Some attributes of the OPT structure and the measure algorithm are ([16]):

(A1) Each cell C_δ stores those boxes of Q that cover C_δ, but don't cover the parent of C. (In this way, the OPT is an extension of the well known segment tree.)

(A2) Every leaf-cell also stores all boxes partially covering it (as piles), and there are $O(\sqrt{N})$ such boxes.

(A3) Each box q partially covers $O(N^{d/2-1})$ leaf-cells (as piles with respect to the corresponding leaf cells).

(A4) The height of the tree is $O(\log N)$.

(A5) The number of nodes in the tree is $O(N^{(d-1)/2})$.

(A6) The measure is computed in time $O(N^{d/2} \log N)$.

As mentioned above we would like to implement these trees for $N = O(n^2)$ *colored* cubes (n cubes in each of the n colors), and find whether there is a point covered by n colors. Two problems arise due to the possible overlap of cubes of the same color: (i) cubes of the same color might appear several times on a path from the root to a leaf, and (ii) there might be an overlap of piles or cubes of the same color in a leaf cell.

Let us first explain how we handle efficiently problem (ii). For a given leaf-cell C_δ denote the set of piles with respect to C_δ by P. We define $colors(P)$ to be the set of colors of all the piles of P. At each deletion or insertion of a pile to C_δ (and to P), we want to determine quickly whether there exists a point in C_δ covered by all colors of $colors(P)$. Note that for every color $i \in colors(P)$ the region in C_δ consisting of points *not* covered by color i is a set of disjoint boxes $G_{i\delta}$. Consider an addition of a pile of color i to C_δ. If the pile is contained in another pile of that color, then the addition does not contribute anything to $G_{i\delta}$. Otherwise, the addition may decrease the size of an existing box in $G_{i\delta}$ or add one box to $G_{i\delta}$ (because a pile is partial to an input unit cube and therefore slicing a cell into two uncovered boxes can be done in at most one axis). Thus at every instance the number of boxes in $G_{i\delta}$ for a given color i is linear in the number of piles of that color at that instance.

Thus determining whether there is a point in C_δ which is covered by all colors of P is equivalent to determining wheteher there is a point in C_δ which is out of all boxes of the sets $G_{i\delta}$, $i \in colors(P)$. This question, in its turn, is equivalent to determining whether the volume of the union $\cup_i G_{i\delta}$ is equal to the volume of C_δ. Notice that by posing the question in this way we handle the overlapping of piles of the same color and convert this question into a volume problem in $(d-1)$-space which can be answered by applying the volume algorithm in [16] on each leaf cell separately, as we explain below. This means that we have to define and maintain an OPT T_δ for each leaf cell C_δ. We call the OPT's constructed on leaves of T *leaf trees*. These trees maintain the boxes $G_{i\delta}$ which arise in the course of the algorithm. In the next section we will describe in detail the implementation and analysis of this step. The intuition however is that with each insertion of a pile q of color i to a leaf-cell δ, we first delete the set of boxes $G_{i\delta}$ from T_δ, then we recompute $G_{i\delta}$, insert the new boxes of $G_{i\delta}$ into T_δ computing the volume on the fly. The operations for deleting a pile are symmetric.

Regarding (i), we do not want to count a color twice (for the depth) when two cells along a path from the root of T to a leaf have two cubes of that color. Therefore, for each node δ, we maintain the "Active" colors in this node — that is, colors that cover this node and do not cover any of its ancestors along the path from δ to the root of T. The active colors will be the ones taken into account in the computation of the depth as we show below.

We now turn to the description of the fields of each node δ in our augmented OPT T. Notice that for some fields we make a distinction between leaf nodes and internal nodes.

- L_TOT_δ stores for a leaf δ, all cubes covering C_δ (either partially or completely) but not covering its parent. If δ is an internal node of T, then L_TOT_δ stores all cubes covering C_δ, but not covering its parent.
  The field L_TOT_δ is organized as a balanced search tree, sorted by colors and

lexicographicállly by cubes, so that determining whether a color i appears in L_TOT_δ, as well as inserting a cube into or deleting a cube from L_TOT_δ, is done in time $O(\log n)$.

- $Active_\delta$ stores for a leaf δ, all cubes covering C_δ (either partially or completely), whose color does not cover completely an ancestor cell of C_δ. For an internal node δ, $Active_\delta$ stores all colors i which cover C_δ, and no ancestor cell of C_δ is covered by color i. The field $Active_\delta$ is a balanced search tree sorted by colors. For leaf nodes the search tree is also sorted lexicographicállly by cubes, so that determining whether any cube of color i appears in $Active_\delta$, as well as inserting or deleting a color (or a cube, if δ is a leaf) is done in time $O(\log n)$.

- N_Active_δ is a counter which always contains zero if δ is a leaf. If δ is an internal node, then N_Active_δ contains the number of different colors in $Active_\delta$. This counter is updated whenever an element is inserted to or deleted from $Active_\delta$.

- $Depth_\delta$ is the maximal depth "achieved" by cubes stored at the $Active_\delta$ fields of δ and its descendent nodes. Formally, if δ is a leaf, then if there is a point in C_δ covered by all colors in $Active_\delta$ then $Depth_\delta$ is the number of different colors in $Active_\delta$, otherwise $Depth_\delta = 0$. If δ is an internal node, then $Depth_\delta$ is defined recursively by:

$$Depth_\delta = \max\{\ Depth_{\delta_{left}} + N\_Active_{\delta_{left}},\ \ Depth_{\delta_{right}} + N\_Active_{\delta_{right}}\ \}$$

When a (colored) cube q is inserted into T or deleted from T, we first update recursively the relevant fields L_TOT_δ. This task is simple, and details are omitted. Next, we update the $Depth_\delta$ fields of the leaves affected by q (see below). Updating the $Active$ fields of all nodes of T affected by q, is done in the following way: when a new cube q of color i is inserted, and we note that it completely covers a cell C_δ and that no cell on the path from the root of T to δ is coverd by color i, then we insert color i to $Active_\delta$ and delete all appearances of color i from the $Active$ fields of all nodes in the subtree rooted at δ (by scanning all the nodes in the sub-tree rooted at δ). When a cube q of color i is deleted from T, we find the nodes that have q in their $Active$ field, and for each such node δ we we search the sub-tree rooted at δ for its highest nodes that contain color i in their L_TOT field. The cubes found in L_TOT are then inserted to the respective $Active$ fields of these nodes. Details of these operations are presented in procedures Insert_To_Active and Delete_From_Active in the Appendix. The $Depth$ fields for each internal node are updated on the fly (see Appendix) by the recursive definition of $Depth$.

Recall that we find the depth $Depth_\delta$ in a leaf cell δ using the method of [16] for volume computation of the union of boxes $\cup_i G_{i\delta}$. This requires maintaining $d-1$-dimensional OPT trees T_δ for each leaf δ. When a pile is added (or deleted) from T_δ we update the nodes of T_δ that are affected by the box, and tracking the path from these nodes to the root of T_δ we thus update the volume of C_δ.

An underlying assumption in constructing an OPT is that all the input boxes have to be known in advance. Clearly this is not the case with the boxes $G_{i\delta}$ since these boxes are constructed in the course of the algorithm. In order to construct all the boxes $G_{i\delta}$ in advance we first perform a "dummy" execution of the algorithm, in which we maintain the list of all boxes $G_{i\delta}$ stored at each instance in each of the leaves. For the actual running of the algorithm we can assume that the boxes for the trees T_δ for all leaves δ are known in advance.

The correctness of the algorithm follows from the next claim:

Claim 3 *After the Depth fields are updated, there is a point of depth n on the hyperplane h if and only if $n = Depth_{root} + N\_Active_{root}$*

Proof omitted in this version for lack of space.

It follows that in order to answer the Hausdorff decision problem we only have to compare $Depth_{root} + N\_Active_{root}$ to n after each insertion or deletion of a cube.

Time and space analysis

We now bound the time required by the algorithm. Since operations on leaf-cells consume the larger chunk of time of our algorithm, we can afford, when inserting or deleteing a cube q of color i, to scan all nodes in T, and to update all the internal nodes of T affected by q. There are $N = n^2$ input cubes and $O(N^{(d-1)/2}) = O(n^{d-1})$ nodes in T (Attr. A5), therefore the total time spent on inner nodes of T is $n^2 \times O(n^{d-1}) \times O(\log n) = O(n^{d+1} \log n)$, where the log factor is caused by searching the trees associated with the fields in each node.

In bounding the leaf-cells operations, we first analyze the time required to insert or delete a cube q of color i to one of the leaf trees T_δ, and maintaining $Depth_\delta$. As we said above, each insertion of a pile into δ might create at most two new boxes in $G_{i\delta}$ and cause the removal of at most two old such boxes. So, at insertion (or deletion) we recompute $G_{i\delta}$ by deleting up to two boxes from T_δ and inserting up to two boxes to T_δ, while maintaining the volume of $\cup_i G_{i\delta}$.

By Attribute A2, the number of cubes stored at a leaf δ of T is $O(\sqrt{N}) = O(n)$, this implies that the number of boxes of $G_{i\delta}$ for all i is $O(n)$. Combining this fact with Theorem 4.4 of [16], we get that every leaf tree T_δ has $O(n^{(d-1)/2})$ nodes (Attr. A5) and that the time required for one insertion or deletion of a box to T_δ, including updating the volume information at the root of T_δ, is only (Attr. A3)

$$O(n^{d/2-1} \log n). \tag{1}$$

Let us bound now the time required to perform all insertion and deletion operations on all the trees T_δ. We apply an amortized-type argument to get a good bound for the total running time of the algorithm. Let q be a cube, and δ a leaf of T. We claim that during the course of the algorithm, q is inserted into $Active_\delta$ at most once, and deleted at most once. This is true since q is deleted when either the sweeping hyperplane h stops intersecting it or when h intersects a new cube q' of the same color that completely covers C_δ. In both cases q will not reappear.

By Attribute A3, a cube partially covers $O(n^{d-2})$ leaf-cells of T. Hence the total number of leaf trees affected by the n^2 input cubes, is

$$n^2 \times O(n^{d-2}) \tag{2}$$

Multiplying Equations 1 and 2 we get:

Theorem 4. *For $d > 3$ we can answer the Hausdorff decision problem in time $O(n^{3d/2-1} \log n)$.*

Turning now to space requirements, by Theorem 4.1 of [16], the space required for each of the trees T_δ is $O(n^{(d-1)/2})$. Since in T there are $O(n^{d-1})$ leaves, the total space required for the trees is $O(n^{3(d-1)/2})$. The space requirement can be reduced to $O(n^2)$, using ideas similar to the ones in Section 5 of Overmars and Yap [16]. Analysis of this claim will appear in the full version.

5 Higher Dimensions: the Improved Version

In this section we improve the relaxed algorithm, where we had two phases of OPT's. The first phase OPT, T, stored the decomposition of the $(d-1)$-space. The second phase OPT's T_δ described the leaf-cells δ of T. In order to distinguish between the cells in the trees of the two phases we call the cells of the second phase *sub-cells*.

The main idea for the improved oracle is to try to decrease the number of operations spent on the leaf trees by decreasing the number of these trees and storing more information in each of them. In order for the display of the improved oracle to be self-contained we first review briefly the decomposition of the $(d-1)$-space as described in [16], using similar notations (for more details see [16]).

Given N $(d-1)$-boxes $Q = \{q_1, \ldots, q_N\}$ we wish to decompose the $(d-1)$-space into $N^{(d-1)/2}$ cells such that attributes A1–A5 from Section 4 hold. We define an *i-boundary* of a box q to be a $(d-2)$-face of q orthogonal to the i-th axis. For axis 1 we split the $(d-1)$-space into $2\sqrt{N}$ slabs, such that each slab contains at most $\sqrt{N}$ 1-boundaries of the input boxes. We reapeat the process on axes j, $j = 2, \ldots, d-1$. For the j-th axis we consider each slab s of level $j-1$ in the following way. We define V_s^1 to be the boxes that have a j'-boundary intersecting s for some $j' < j$, and V_s^2 to be the boxes that intersect s but do not have any j'-boundary intersecting s for $j' < j$; we now split the slab s by a $(d-2)$-flat at every j-boundary of a box in V_s^1 and at every $\sqrt{N}$-th j-boundary of a box in V_s^2. As a result, each slab of level $j-1$ is split into up to $2j\sqrt{N}$ slabs of level j. This is, in essence, the cell decomposition and the OPT of Overmars and Yap, which we have used in the relaxed version.

We now show how we construct our OPT T so that it will have less nodes. This implies that we load more information on the leaf cells. The invariant that each leaf-cell in T is intersected just by piles is not maintained any more. We define *non-piles* as follows: If a box q intersects C_δ and is not a pile with respect to C_δ, nor does it cover C_δ completely then we call q a *non-pile* with respect to δ. The tree T of phase one is generated by the following procedure. Recall that we have $N = n^2$ cubes.

Cell_Decomposition

Split the $(d-1)$-space into $N^{1/3} = 2n^{2/3}$ slabs in the first axis, each of whose interiors contains at most $2N^{2/3} = n^{4/3}$ 1-boundaries of boxes.

Repeat for $j = 2, \ldots, d-1$. Define V_s^1 and V_s^2 as before. For the j-th axis, we consider a slab s of level $j-1$. Split s with a $(d-2)$-flat at every $N^{1/3}$-th j-boundary of a box in V_s^1 and at every $N^{2/3}$-th j-boundary of a box in V_s^2.

Notice that in the process of *Cell_Decomposition*, we do not put $(d-2)$-flats at every j-boundary of boxes from V_s^1, but at every $n^{2/3}$ of them. We denote by $F_{i\delta}$ the set of non-pile boxes of color i in C_δ, for $i = 1, \ldots, n$.

Lemma 5. *In the modified orthogonal partition tree generated by the Cell_Decomposition procedure on $N = n^2$ boxes, each leaf-cell contains at most $O(n^{2/3})$ non-pile boxes $F_{i\delta}$ and at most $O(n^{4/3})$ pile boxes.*

Proof: A non-pile box $F \in F_{i\delta}$, with respect to the leaf-cell C_δ, contains at least two boundaries of F in C_δ (say b_1-boundary and b_2-boundary where $b_1 < b_2$). Consider level b_2 in the Cell_Decomposition and let s be the slab of level $b_2 - 1$ that contains C_δ. As $b_1 < b_2$, the b_1-boundary is contained in the slab s and the box F is in V_s^1. At level b_2 we split the slab s at each $n^{2/3}$-th b_2-boundary of boxes from V_s^1. Therefore, C_δ contains at most $n^{2/3}$ non-pile boxes such that their largest boundary in C is b_2 and the number of non-pile boxes in C_δ does not exceed $O(n^{2/3})$. $\qquad\square$

Using the Cell_Decomposition procedure we obtain the following properties:

(i)   The number of cells in T is $O(n^{2(d-1)/3})$.
(ii) Each $(d-1)$-box partially covers $O(n^{2(d-2)/3})$ cells of T.
(iii) Each leaf cell C_δ contains at most $n^{4/3}$ boxes partially covering C_δ.
(iv) Each leaf cell intersects at most $n^{2/3}$ non-pile boxes with respect to that leaf.

To verify these invariants, observe that each slab was split into $2n^{2/3}$ and, as we are dealing with $(d-1)$-space, the number of cells is $O(n^{2(d-1)/3})$. At the moment we split the slabs generated for axis $j-1$ with respect to j-th axis, there are $O(n^{2(j-1)/3})$ slabs. A box that partially covers the cell, has at least one axis in which it does not cut through the slabs. Therefore, each box cuts through at most $O(n^{2(j-1)/3} \times n^{2(d-1-j)/3}) = O(n^{2(d-2)/3})$ cells. The last two properties follow immediately from the way that the slabs are split.

We now modify the second-phase trees. For each leaf-cell δ, we calculate as before the boxes $G_{i\delta}$ consisting of points that are not covered by any *pile* of color i at a certain instance (ignoring the non-pile boxes). If we now construct the standard OPT for the $N' = n^{4/3}$ boxes ($G_{i\delta}$), each slab at level j is split into $\sqrt{N'} = n^{2/3}$ slabs. As the total number of non-piles in each leaf-cell is only $O(n^{2/3})$, we may also split each slab at each non-pile boundary without increasing the complexity of the leaf-cells. Note that the value $n^{2/3}$ was selected as the smallest number such that each leaf-cell will contain no more than $O(\sqrt{n^{4/3}})$ non-pile boxes.

The following procedure describes the construction of the augmented OPT T_δ for a leaf-cell δ of T.

Sub_Cell_Decomposition

Calculate the piles $\mathcal{G} = \{G_{i\delta}\}$ and the non-piles $\mathcal{F} = \{F_{i\delta}\}$, of the leaf-cells, for $i = 1, \ldots, n$. Let $N' = |\mathcal{G}| = O(n^{4/3})$. Split the first axis at all $\sqrt{N'}$ 1-boundaries of boxes in $\mathcal{G}$ and at each 1-boundary of a non-pile box.

For $j = 2, \ldots, d-1$ consider the slab s generated at axis $j-1$ and let V_s^1 be the boxes of $\mathcal{G}$ that have a j'-boundary in s for some $j' < j$. Let V_s^2 be the boxes of $\mathcal{G}$ that intersect s but do not have any j'-boundary in s for $j' < j$. Split s with respect to the j-th axis at each j-boundary of boxes in V_s^1, at each $\sqrt{N'}$-th j-boundary of boxes in V_s^2 and at each j-boundary of a non-pile in s.

Sub_Cell_Decomposition procedure obtains the following properties for leaf tree T_δ:

(i) The number of sub-cells in T_δ is $O(n^{2(d-1)/3})$.
(ii) Each input box from $\mathcal{G} \cup \mathcal{F}$ partially covers $O(n^{2(d-2)/3})$ leaf sub-cells.

(iii) Each leaf sub-cell intersects at most $O(n^{2/3})$ boxes of $\mathcal{G} \cup \mathcal{F}$.

(vi) The boxes stored in a leaf sub-cell either totally cover this leaf sub-cell or are piles with respect to the leaf cell.

The proof is as for the trees of phase 1. By the $Sub\_Cell\_Decomposition$ procedure, each non-pile box is either disjoint from a leaf-sub-cell or completely contains it. Therefore, we may use the modified OPT's $\mathcal{T}_\delta$ to store $N' = O(n^{4/3})$ disjoint boxes that appear as piles in some leaf-sub-cells and $O(n^{2/3})$ other boxes that are either disjoint or contain each leaf-sub-cells.

We next describe how to use no extra time but an extra $\log n$ factor of space in order to update the $Active_\delta$ field correctly. For simplicity we first show how to use extra n factor of space. We store at each node two arrays: Non_Active_δ and $color_\delta$ each of length n (the number of different colors). The field $color_\delta(i)$ indicates if there exists a box of color i which is active in a descendant of δ but does not contain C_δ itself. The field $Non\_Active_\delta(i)$ will indicate if there exists a box of color i (not containing C_δ) which intersects a descendant δ' of δ but does not appear in $Active_{\delta'}$. These fields are updated easily in constant time at each node on the path from δ to the root of $\mathcal{T}$ (since $color_\delta(i) = color_{\delta_{left}}(i)$ or $color_{\delta_{right}}(i)$ and $Non\_Active_\delta(i) = Non\_Active_{\delta_{left}}(i)$ or $Non\_Active_{\delta_{right}}(i)$). In order to use only $O(\log n)$ (amortized) space at each node, we replace the arrays by linked lists at each node except for the root. The lists contain only those colors for which the value is $True$. This implies a $\log n$ factor of space as each box may cause a $color_\delta(i)$ (or $Non\_Active_\delta(i)$) to become $True$ on the $\log n$ nodes on the path to the root. In order to update these lists in time $O(1)$, we keep at each $color_\delta(i)$ pointers to $color_{\delta_{left}}(i)$ and $color_{\delta_{right}}(i)$ (the same applies for Non_Active). At each update we start at the root of $\mathcal{T}$, finding the pointers to its children and so on.

Finally, the difference between the relaxed version and the improved version is just the decomposition procedures. The overall complexity is now $O(n^2 \times n^{2(d-2)/3} \times n^{2(d-2)/3} \times \log n) = O(n^{(4d-2)/3} \log n)$ as we sweep n^2 boxes and each box updates $O(n^{2(d-2)/3})$ cells and $O(n^{4(d-2)/3})$ sub-cells.

6 Finding the Minimum Hausdorff Distance

Now we want to determine the minimum ε for which the intersection is still non-empty. It is easy to see that the desired minimum value is achieved at some ε_0 for which two cubes touch each other. There are $O(n^4)$ such ε values. We can afford to use a simple binary search on these values for dimension $d \geq 4$. For $d = 3$ this is too costly, hence we apply here the method of Frederickson and Johnson [10] for solving an optimization problem (see [7] for details). We get

Theorem 6. *Given point sets A and B in d-space and using the L_∞ metric, the minimum Hausdorff distance between A and B under translation can be determined in time $O(n^3 \log^2 n)$ if $d = 3$ and in time $O(n^{(4d-2)/3} \log^2 n)$ for $d > 3$, where $n = \max\{|A|, |B|\}$.*

7 Combinatorial Bounds

In this section we show combinatorial results on the region (in translation space) of translations that minimize the Hausdorff distance between two point sets.

Theorem 7. *Let $\mathcal{L} = \{\mathcal{L}_1, \ldots, \mathcal{L}_m\}$ be a collection of layers in d-space where each layer $\mathcal{L}_i$ is the union of n unit cubes and let T denote the intersection of the layers. Then the number of vertices of T is $O(m^d n^{\lfloor d/2 \rfloor})$.*

Proof: Consider a vertex v of T. Vertex v must be the result of an interaction between $d' \leq d$ layers; this is because, in dimension d, no vertex can involve more than d layers. Consider just the layers that interact to create v and call this set of layers $\mathcal{L}'$.

We claim that v, a vertex of the intersection of the layers of $\mathcal{L}'$, is also a vertex of the union of the layers of $\mathcal{L}'$. For v not to be a vertex of the union it would have to be hidden by a cube in some layer, say $\mathcal{L}_i \in \mathcal{L}'$. But if this were the case then, by the way $\mathcal{L}'$ was defined, layer $\mathcal{L}_i$ could not possibly have interacted with the other layers of $\mathcal{L}'$ to create v in the first place. Thus, each vertex of T is also a vertex of a union of ($\leq d$) layers.

The bound now follows from some simple counting. There are $O(m^d)$ possible sets of layers where the cardinality of each set is $\leq d$. Each such set of layers contains $O(dn)$ cubes. As mentioned above, a recent result of [4] states that the union of N unit boxes in d-space has complexity $O(N^{\lfloor d/2 \rfloor})$. We multiply $O(m^d)$ by $O((dn)^{\lfloor d/2 \rfloor})$ (and note that factors involving only d can be treated as constant and absorbed into the big-O notation) to get the bound in the theorem. $\square$

Note that the bound of $O(m^d n^{\lfloor d/2 \rfloor})$ on the number of vertices of T also bounds, under general position assumption, the complexity T, and the number of connected regions of T. For our problem $m = n$, thus the upper bound on the number of translations is $O(n^{\lfloor 3d/2 \rfloor})$. We now turn to present a construction that shows a lower bound on the number of translations.

Theorem 8. *Let $\mathcal{L}$ be defined as in Theorem 7. Then the number of connected components in $T = \cap_i \mathcal{L}_i$ is $\Omega(n^{\lfloor 3d/2 \rfloor})$.*

Proof: Our construction extends an earlier construction due to Rucklidge [18] that he developed for the two-dimensional case. We first show how we can construct $\Omega(n^3)$ connected components in T for $d = 2$. We define an *i-corridor* as a region enclosed between two stairs of (2-dimensional) cubes as shown in Figure 7(a), when the cubes are all of the color i, and each step of the "stairs" is of size δ, when δ is a very small positive fixed parameter (depends on n), to be determined later. Each i-corridor is generated by $2n$ cubes, and hence has n stairs.

We generate n corridors, each of a different color, using two batches of $n/2$ corridors. The i-th corridor in the first batch is generated by a shift of $i\delta'$ of the base corridor (in both x and y directions). The second batch, whose colors are $i = n/2 + 1, \ldots, n$, is generated by translating the first batch by $\delta/2$ in both x and y directions (see Figure 7(b)).

Note that each corridor of one of the batches intersects each corridor of the other batch in $\Omega(n)$ intersection points, yielding $\Omega(n)$ connected regions in the

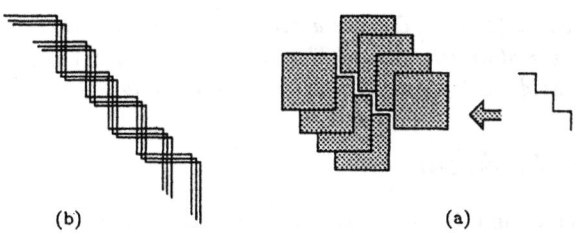

Fig. 1. Lower bound construction

interior of these two corridors. Let S denote the set of (colored) cubes that define these corridors. Note also that every point in the neighborhood of the corridors, but outside all corridors in covered by all n colors in S. Hence there are $\Omega(n^3)$ connected regions of the intersection of all cubes generated these corridors.

We now show how to generate $O((n^3)^k)$ connected regions in $2k$-space. Let S be the set of cubes as above. Given kn colors we denote by $S|_i$ a set of cubes which is identical to the original set S but the colors of the cubes are $(i-1)n+1, \ldots, in$ instead of $1, \ldots, n$. In $2k$-space we define k sets $S_1, \ldots, S_k$ of $|S|$ cubes each such that the projection of S_i on the $2i-1$ and $2i$ axes is identical to $S|_i$. The other $2(k-1)$ coordinates of the centers of the boxes in S_i, are zero. The colors of the cubes in S_i and $S_{j \neq i}$, are disjoint.

Our claim is that the corridors defined by $S_1, \ldots, S_k$ introduce $\Omega((n^3)^k)$ connected components in the region covered all kn colors. To verify this claim, let $X_{j_1, \ldots, j_k} = (x_1, x_2, \ldots, x_{2k-1}, x_{2k})$ be a $2k$-dimensional point where (x_{2i-1}, x_{2i}) is a (two-dimensional) point in the j_i-th connected regions of S and where $1 \leq j_i \leq \Omega(n^3)$. Let $X' = X_{j_1', \ldots, j_k'}$ and $X'' = X_{j_1'', \ldots, j_k''}$ be two different points and let i denote an index such that $j_i' \neq j_i''$ (this is well defined as X' and X'' are different). Let $C_i = (i-1)n+1, \ldots, in$ be the set of colors in S_i. Consider a line l (in the 2-space) that connects the projection of X' and X'' on coordinates $2i-1$ and $2i$. The endpoints of l (namely the projections of X' and X'') are not covered be exactly two colors from C_i. By definition of S, there is a point on l which is covered by all colors of C_i. Hence, X' and X'' could not possibly be on the same connected region and this implies $\Omega((n^3)^k)$ regions (each corresponds to a selection of j_i's). $\qquad \square$

The bound we presented above showed that there can be $O(n^{\lfloor 3d/2 \rfloor})$ translations t that minimize the *one-way* Hausdorff distance, but can be easily generalized to bound the two-way Hausdorff distance as well.

Remark. The proof above generalizes with small changes to non-congruent cubes and to balls (the L_2 norm). If the cubes defineding $\mathcal{L}$ are not congruent then the bound we get in Theorem 8 is larger - $\Theta(n^{\lceil 3d/2 \rceil})$. This is because for $d = 1$ the number of translations is $\Theta(n^2)$. Similarly, if each $\mathcal{L}_i$ above is the union of d-dimensional balls, then using the standard lifting transformation $\lambda : \mathbb{R}^d \to \mathbb{R}^{d+1}$ defined by $\lambda((x_1, \ldots, x_d)) \equiv (x_1 \ldots, x_d, x_1^2 + \ldots + x_d^2)$ their union can be expressed as the lower envelope of n hyperplanes in the $(d+1)$-space, which by [8] has complexity $O(n^{\lceil (d+1)/2 \rceil})$. Plugging this into the above proof shows:

Theorem 9. *Let $\mathcal{L} = \{\mathcal{L}_1, \ldots, \mathcal{L}_m\}$ be a collection of layers, where each layer $\mathcal{L}_i$ is the union of n cubes of arbitrary size. Then the complexity of $\cap_i \mathcal{L}_i$ is $O(m^d n^{\lceil d/2 \rceil})$. The same bound holds in the case that each $\mathcal{L}_i$ is the union of n balls.*

8 Using the L_2 Norm

In this section let A and B be two sets of n points in d-space, for $d \geq 4$ and let $H_2(A, B)$ denote the Hausdorff Distance between them when L_2 is the underlying metric. We have the following algorithmic result. (Note that [11, 12] presented an algorithm for this problem in 3-space with time $O(n^{5+\epsilon})$ for any $\epsilon > 0$.)

Theorem 10. *Let A and B be as above. A translation t that minimizes $H_2(A, t+B)$ can be found in expected time $O(n^{\lceil 3d/2 \rceil + 1} \log^3 n)$.*

Proof: Due to space limitations we present only a brief sketch of the proof here. As in the L_∞ case, we develop an algorithm for the Hausdorff decision problem and then combine it with the parametric search paradigm of Megiddo [15] to create an algorithm for finding the minimum Hausdorff distance.

To solve the decision problem, we show that there are just $O(n^{\lceil 3d/2 \rceil})$ *critical points* that are potential candidates for vertices of regions with color-depth n. Our counting method is analogous to that used in the proof of Theorem 7.

Each such *critical point* must be tested to see if it is covered by all the colors. To do this we use a result of [14] to build in time $O(n^{\lceil d/2 \rceil} \log^{O(1)} n)$ a data structure of size $O(n^{\lceil d/2 \rceil} \log^{O(1)} n)$ which can answer a coverage query in time $O(\log n)$. Initially, it looks as though such a data structure is needed for each color, but since the set of spheres for one color is just a translation of the set of spheres for each other color, we really need just one copy of this data structure. A single critical point must be tested for every color taking time $O(n \log n)$. Thus the total time for solving the Hausdorff decision problem is $O(n^{\lceil 3d/2 \rceil + 1} \log n)$. The extra log factors in the theorem are due to the overhead of using parametric search to solve the corresponding optimization problem.

$\square$

9 Open Problems

- Are L_∞ Hausdorff decision problems inherently easier than the ones for L_2? Is there a single technique that solves both types of problems efficiently.
- Given a d-dimensional box C and n boxes $G_1, \ldots, G_n$ lying inside C, is it possible to determine if $\cup_1^n G_i = C$ in time $o(n^{d/2})$? Finding such a technique would immediately improve the running time of our L_∞ algorithm.
- Finally, our techniques do not take advantage of the fact that the set of cubes for one color is really just a translation of the set of cubes for each other color. Is there some way to use this information to design a better algorithm?

Acknowledgments

We would like to thank Micha Sharir for useful discussions and help in the proof of Theorem 7. Thanks also to Reuven Bar-Yehuda for contributing to the algorithm of Section 4.

278

References

[1] P. K. Agarwal, B. Aronov and M. Sharir, Computing Envelopes in Four Dimensions with Applications, *Proc. 10th Annual ACM Symposium on Computational Geometry* 1994, 348–358.

[2] P. K. Agarwal, M. Sharir and S. Toledo, Applications of Parametric Searching in Geometric Optimization, *Proc. 3rd ACM-SIAM Symposium on Discrete Algorithms*, 1992, 72–82.

[3] N. M. Amato, M. T. Goodrich and E. R. Ramos, Parallel Algorithms for Higher-Dimensional Convex Hulls, *Proceedings 6 Annual ACM Symposium on Theory of Computing* 1994, 683–694.

[4] J. Boissonnat, M. Sharir, B. Tagansky and M. Yvinec, Voronoi Diagrams in Higher Dimensions under Certain Polyhedral Distance Functions, *Proc. 11th Annual ACM Symposium on Computational Geometry* 1995, 79–88.

[5] H. Brönnimann, B. Chazelle and J. Matoušek, Product Range Spaces, Sensitive Sampling, and Derandomization, *Proc. 34th Annual IEEE Symposium on the Foundations of Computer Science* 1993, 400–409.

[6] B. Chazelle, H. Edelsbrunner, L. Guibas and M. Sharir, A Singly Exponential Stratification Scheme for Real Semi-algebraic Varieties and Its Applications, *Theoretical Computer Science* 84 (1991), 77–105.

[7] L.P. Chew and K. Kedem, Improvements on Geometric Pattern Matching Problems, *Algorithm Theory – SWAT '92*, edited by O. Nurmi and E. Ukkonen, Lecture Notes in Computer Science #621, Springer-Verlag, July 1992, 318–325.

[8] H. Edelsbrunner, *Algorithms in Combinatorial Geometry*, Springer-Verlag, Heidelberg, Germany, 1987.

[9] A. Efrat and M. Sharir, A Near-linear Algorithm for the Planar Segment Center Problem, *Discrete and Computational Geometry*, to appear.

[10] G.N. Frederickson and D.B. Johnson, Finding kth Paths and p-Centers by Generating and Searching Good Data Structures, *Journal of Algorithms*, 4 (1983), 61–80.

[11] D.P. Huttenlocher and K. Kedem, Efficiently Computing the Hausdorff Distance for Point Sets under Translation, *Proceedings of the Sixth ACM Symposium on Computational Geometry* 1990, 340–349.

[12] D.P. Huttenlocher, K. Kedem and M. Sharir, The Upper Envelope of Voronoi Surfaces and its Applications, *Discrete and Computational Geometry*, 9 (1993), 267–291.

[13] J. Matoušek, Reporting Points in Halfspaces, *Comput. Geom. Theory Appls.* 2 (1992), 169–186.

[14] J. Matoušek and O. Schwarzkopf, Linear Optimization Queries, *Proc. 8th Annual ACM Symposium on Computational Geometry* 1992, 16–25.

[15] N. Megiddo, Applying Parallel Computation Algorithms in the Design of Serial Algorithms, *J. ACM* 30 (1983), 852–865.

[16] M. Overmars and C.K. Yap, New Upper Bounds in Klee's Measure Problem, *SIAM Journal of Computing*, 20:6 (1991), 1034–1045.

[17] F.P. Preparata and M.I. Shamos, *Computational Geometry*, Springer-Verlag, New York, Berlin, Heidelberg, Tokyo, 1985.

[18] W. Rucklidge, Lower Bounds for the Complexity of the Hausdorff Distance, *Proc. 5th Canadian Conference on Computational Geometry* 1993, 145–150.

[19] W. Rucklidge, Private Communication.

[20] R. Seidel, Small-dimensional Linear Programming and Convex Hulls Made Easy, *Discrete and Computational Geometry*, 6 (1991), 423–434.

Appendix

The operations for inserting a cube q into, or deleting q from a leaf-cell C_δ are done by the procedures Insert_to_Leaf and Delete_from_Leaf. They have the side effect of updating the field $Active_\delta$. In addition, they return $|colors(Active_\delta)|$ if there is a point in C_δ covered by all colors of $Active_\delta$, or 0 otherwise. The following procedures update the $Active$ field in each node of $\mathcal{T}$, and the $depth$ fields of the leaves of $\mathcal{T}$.

Insert_To_Active(q, δ)
if q is disjoint to C_δ then return ;
if δ is a leaf then
      $depth_\delta \leftarrow$ Insert_to_Leaf(q, δ);
else   /* δ is an internal node */
      if q does not cover C_δ completely then
            Insert_To_Active(q, δ_{left}) ; Insert_To_Active(q, δ_{right})
      else   /* q covers C_δ completely */
            if $color(q)$ in $Active_\delta$ or in $Active_\lambda$, λ an ancestor of δ then return
            else   /* No ancestor of δ is covered by $color(q)$ */
                insert $color(q)$ into $Active_\delta$
                for every descendant λ of δ
                      if λ is an internal node and $color(q) \in Active_\lambda$ then
                            delete $color(q)$ from $Active_\lambda$
                      if λ is a leaf then
                            for every q' of color $color(q)$ in $Active_\lambda$
                                set $Depth_\lambda \leftarrow$ Delete_from_Leaf(q', λ)

Delete_From_Active(q, δ)
if q is disjoint to C_δ then return ;
if q partially covers C_δ and δ is an internal node then
      Delete_From_Active(q, δ_{left}) ; Delete_From_Active(q, δ_{righ})
else
      if δ is a leaf then
            $Depth_\delta \leftarrow$ Delete_from_Leaf(q, δ)
      else   /* δ is an internal node, covered completely by q */
      if no ancestor of δ contains $color(q)$ in its $Active$ list then
            recover$(color(q), \delta)$
      else   /*No such ancestor exists */
            return

Recover $(color(q), \lambda)$
if $color(q) \in L\_TOT_\lambda$ then
      if λ is an internal node then
            Insert $color(q)$ into $Active_\lambda$
            return
      else   /* λ is a leaf containing cubes of $color(q)$ in L_TOT_λ */
            for each box q' of color $color(q)$ in L_TOT_λ
                $Depth_\lambda \leftarrow$ Insert_to_Leaf(q', λ)
else   /* no box of color $color(q)$ in $Active_\lambda$ */
      if λ is an internal node then
            Recover$(color(q), \lambda_{left})$; Recover$(color(q), \lambda_{right})$

Finding the Constrained Delaunay Triangulation and Constrained Voronoi Diagram of a Simple Polygon in Linear-Time [1]

(Extended Abstract)

Cao An Wang
Department of Computer Science
Memorial University of Newfoundland
St. John's, NFLD, Canada A1C 5S7
email: wang@garfield.cs.mun.ca

Francis Chin
Department of Computer Science
The University of Hong Kong, Hong Kong
email: chin@csd.hku.hk

Abstract

In this paper, we present a $\Theta(n)$ time worst-case deterministic algorithm for finding the constrained Delaunay triangulation and constrained Voronoi diagram of a simple n-sided polygon in the plane. Up to now, only an $O(n \log n)$ worst-case deterministic and an $O(n)$ expected time bound have been shown, leaving an $O(n)$ deterministic solution open to conjecture.

1 Introduction

Delaunay triangulation and *Voronoi diagram*, duals of one another, are two fundamental geometric constructs in computational geometry. These two geometric constructs for a set of points as well as their variations have been extensively studied [PrSh85, Aure90, BeEp92]. Among these variations, Lee and Lin [LeLi86] considered two problems related to *constrained Delaunay triangulation*[2]: (1) the Delaunay triangulation of a set of points constrained by a set of non-crossing line segments and (2) the Delaunay triangulation of the vertices of a simple polygon constrained by its edges. They proposed an $O(n^2)$ algorithm for the first problem and an $O(n \log n)$ algorithm for the second one. While the

---

[1] This work is supported by NSERC grant OPG0041629.
[2] Same as *generalized Delaunay triangulation* as defined in [LeLi86]

$O(n^2)$ upper bound for the first problem was later improved to $\Theta(n\ log\ n)$ by several researchers [Chew87, WaSc87, Seid88], the upper bound for second has remained unchanged and the quest for an improvement has become a recognized open problem [Aggr88, Aure91, BeEp92].

Recently, there have been some results related to this open problem on the Delaunay triangulation of simple polygons. Aggarwal, Guibas, Saxe, and Shor [AGSS89] showed that the constrained Delaunay triangulation of a convex polygon can be constructed in linear time. Chazelle [Chaz90] presented a linear time algorithm for finding an 'arbitrary' triangulation of a simple polygon. Klein and Lingas showed that this problem for L_1 metrics can be solved in linear time [KlLi92], and this problem for the Euclidean metrics can be solved in expected linear time by a randomized algorithm [KlLi93]. These efforts all seem to point toward a linear solution to the Delaunay triangulation of simple polygons and supports the intuition that the simple polygon problem is easier than the non-crossing line segment problem.

In this paper, we settle this open problem by presenting a deterministic linear time worst-case algorithm. Our approach follows that of [KlLi93]: (i) first decomposing the given simple polygon into a set of simpler polygons, called *pseudo-normal histograms*, then (ii) constructing the constrained Delaunay triangulation of each normal histogram, and finally (iii) merging the constrained Delaunay triangulations of all these normal histograms to get the result. In this 3-step progress, the first and third were shown to be possible in linear time, but the second step was done in expected linear time by a randomized algorithm. Our contribution is to show how this second step can be done in linear worst case time deterministically.

The organization of the paper is as follows. In Section 2, we review some definitions and known facts, which are related to our method. In Section 3, we concentrate on how to construct the constrained Delaunay triangulation or constrained Voronoi diagram of a normal histogram in linear time.

2 Preliminaries

The Constrained Delaunay Triangulation [LeLi86, Chew87, WaSc87, Seid88] of a set of non-crossing line segments L, denoted by $CDT(L)$, is a triangulation of the endpoints S of L satisfying the following two conditions: (a) the edge set of $CDT(L)$ contains L, and (b) when the line segments in L are treated as obstacles, the interior of the circumcircle of any triangle of $CDT(L)$, say $\Delta ss's''$, does not contain any endpoint in S visible to all vertices s, s', and s''.

In [Seid88, Ling89, JoWa93], the proper dual for the constrained Delaunay triangulation problem has been defined as the **Constrained (or called Bounded) Voronoi diagram** of L, denoted by $V_c(L)$. It extends the standard Voronoi diagram by: (i) imagining two sheets or half planes attached to each side of the obstacle line segments; (ii) for each sheet, there is a well-defined Voronoi diagram that is induced by only the endpoints on the other side of the

282

sheet excluding the obstacle line segment attached to the sheet; (iii) the standard Voronoi diagram is augmented by the Voronoi diagrams induced by the sheets. For simplicity, we omit the word 'constrained' over Voronoi diagrams in this paper as all the Voronoi diagrams are deemed to be constrained unless they are explicitly stated to be standard Voronoi diagrams.

2.1 Pseudo Normal Histograms (PNH)

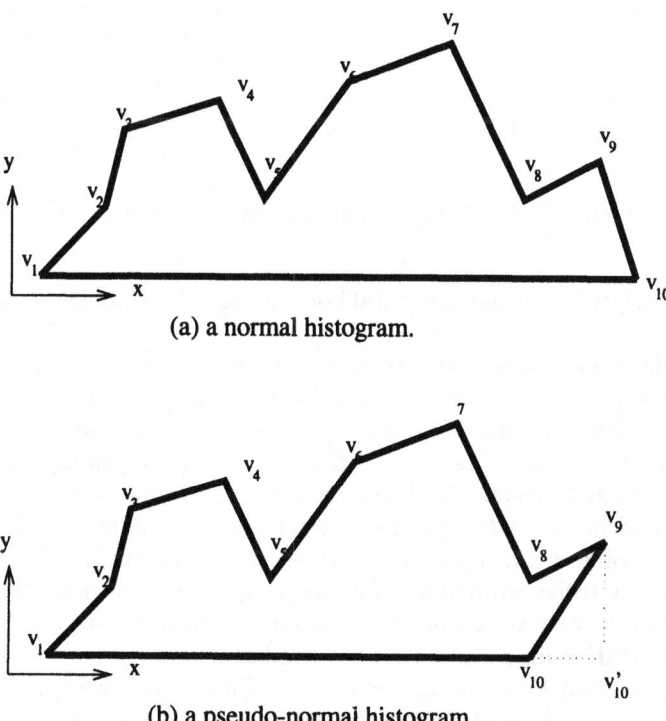

(a) a normal histogram.

(b) a pseudo-normal histogram.

Figure 1: Normal histograms and pseudo-normal histogram

A **normal histogram (NH)** [DjLi89] is a monotone polygon w.r.t. one of its edges, called *bottom edge*, such that all the vertices of the polygon lie on the same side of the line extending the bottom edge (Figure 1(a) gives an example). A **pseudo-normal histogram (PNH)** [KlLi93] can be viewed as a NH missing one of its bottom corners, i.e., a PNH can be transformed into an NH by adding a right-angle triangle at its bottom (Figure 1(b)).

2.2 Decomposition of a simple polygon into PNH's

Figure 2 illustrates how a polygon P is decomposed into 13 PNH's. PNH_1 is associated with the vertical bottom edge e missing its upper bottom corner;

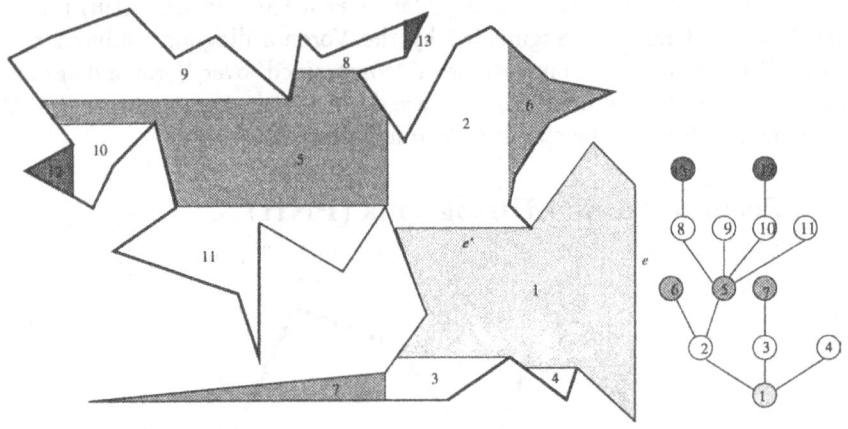

Figure 2: A decomposition of P into a tree of PNH's

PNH_2, associated with the horizontal bottom edge e', is missing its left bottom corner, etc.

A simple polygon P with n vertices can be decomposed into PNH's in $O(n)$ time according to [KlLi93] when provided with what is known as the *horizontal* and *vertical visibility maps* of P, which in turn can be obtained in linear time according to [Chaz90]. A **diagonal** of P is a line segment joining two vertices of P and lying entirely inside P, while a **chord** of P is a line segment, parallel to the designated bottom edge, joining a vertex and a boundary point of P (such a boundary point is called **pseudo-vertex**), and lying entirely inside P. The **horizontal visibility map** of a simple polygon P is a set of chords trapozoidalizing P so that every vertex of P is associated with at most two chords. The **vertical visibility map** can be defined similarly.

In order to find $V_c(P)$ in deterministic in linear time, what remains to be solved is the construction of the constrained Voronoi diagram of a pseudo-normal histogram efficiently. In [KlLi93], a randomized algorithm is introduced to find the Voronoi diagram of an NH in expected linear time. The Voronoi diagram of the corresponding PNH can then be obtained by removing the bottom vertex from the Voronoi diagram of this NH and this can be done in time linearly proportion to the size of the NH. In the next section, we shall concentrate our effort to design a linear time deterministic algorithm for constructing the Voronoi diagram of an NH and thus can find $V_c(P)$ in deterministic in linear time.

3 Finding the Constrained Voronoi Diagram of an NH

Given a normal histogram H, H is decomposed recursively into a tree, say T_I, of smaller normal histograms called **influence normal histograms (INH)**, where a node of T_I corresponds to an INH and an edge of T_I indicates an adjacency between two INH's (Figure 3).

The decomposition ensures that the portion of Voronoi diagram $V_c(H)$ in each INH can only be affected by its own vertices and the vertices of its sons and nothing beyond. In general, the Voronoi cells of $V_c(H)$ associated with vertices of an INH might cross its bottom edge and share edges with Voronoi cells associated with vertices of its parent, but not with those of its brothers nor its grandparents. Similarly, the Voronoi cells of an INH would not share any boundary with those of its grandsons. This property implies that, should the Voronoi diagrams of the INH's ($V_c(INH)$) be given, the repeatedly merging of the Voronoi diagrams of the adjacent INH's can be done in time linearly proportional to the sum of their sizes.

Let $V(p)$ denote the Voronoi cell associated with vertex p in a Voronoi diagram. A point p in a normal histogram H is called an **influence point** if the Voronoi cell $V(p)$ in $V_c(H \cup \{p\})$ will cross H's bottom edge e. The set of influence points is called the **influence region** IR w.r.t. bottom edge e. Consider Figure 3, the IR of H is the region enclosed by: $\widehat{v_1 v_5}$, $\overline{v_5 v_6}$, $\overline{v_6 v_7}$, $\overline{v_7 v_8}$, $\overline{v_8 v_9}$, $\overline{v_9 v_{10}}$, $\widehat{v_{10} v_{25}}$, $\overline{v_{25} v_{28}}$, $\widehat{v_{28} v_{34}}$, $\overline{v_{34} v_{35}}$, and $\overline{v_{35} v_1}$ (the bottom edge e), where $\overline{xy}$ and $\widehat{xy}$ represent respectively the straight line and the arc joining vertices x and y. The **root** (or **root INH**) of T_I is defined as the NH enclosing all influence points of H and containing only edges (or parts of edges) and chords of H such that all its horizontal chords would intersect the IR of H (i.e., the *smallest NH* containing IR). As an example, the root of INH is indicated by the white region in Figure 3. Let us now consider the part of H excluding the root INH, which consists of zero or more disjoint polygons. Each polygon is also an NH with a chord as its bottom edge. As given in Figure 3, H is decomposed into a root INH and 6 other NH's, i.e., the NH's above chords $v_3' v_3$, $v_3 v_3''$, $v_{12}' v_{12}$, $v_{13} v_{13}'$, $v_{25} v_{25}'$, and $v_{28} v_{28}'$. For example, the NH above chord $v_{28} v_{28}'$ is $(v_{28}, v_{29}, v_{30}, v_{31}, v_{32}, v_{28}')$. The decomposition can be recursively applied to each of these NH's.

Since any node of T_I does not contain the influence points of its parent by the definition of INH, the Voronoi cell associated with a vertex of any node in T_I could not cross the bottom edge of its parent. As the Voronoi cells associated with the internal vertices of an INH never share any edges with the Voronoi cells of its brother INH (to be shown later), the part of $V_c(H)$ within the root can be merged by the Voronoi diagram of the root INH with those of its sons in $O(m_0 + \sum_{i=1}^{s} m_i)$ time, where m_0 is the number of vertices of the root, s is the number of its sons and m_i is the number of vertices of its ith son. Let $M(n)$ denote the merging time for constructing $V_c(H)$ with $|H| = n$ when provided with the Voronoi diagram of every INH in T_I. Then, we have

$M(n) = C * (m_0 + \sum_{i=1}^{s} m_i) + \sum_{i=1}^{s} M(n_i)$ where C is a constant and n_i is the number of vertices of the ith subtree. As $n = m_0 + \sum_{i=1}^{s} n_i$, we can show that $M(n) = C(2n - m_0)$ by induction. Thus, the total merging time is $O(n)$.

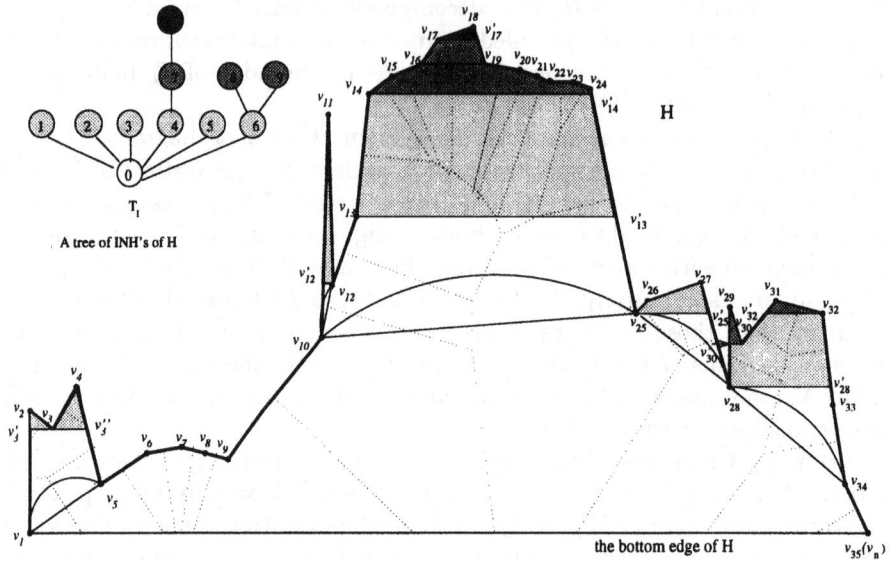

Figure 3: Decomposition of H into INH's

In the following sections, we shall prove the properties of influence region (IR) and influence normal histogram (INH) which allows us to do efficient merging and identification.

3.1   Influence Region

Let H_V be the subset of vertices of a normal histogram H such that the associated Voronoi cells of H_V cross H's bottom edge. Assume that H_V is ordered according to H. As H is an NH, H_V will also be an NH sharing the same bottom edge as H. Let us consider the example given in Figure 3 again, H_V is indicated by the sequence of vertices $(v_1, v_5, v_6, v_7, v_8, v_9, v_{10}, v_{25}, v_{28}, v_{34}, v_{35})$.

Lemma 1 *All points in H_V are influence points.* □

In general, IR includes some regions not belonging to H_V. Let e be a boundary edge of H_V. If e is also an edge of H, then e must be an edge of IR, e.g., $\overline{v_5 v_6}$, $\overline{v_6 v_7}$, $\overline{v_7 v_8}$, etc. in Figure 3. However, if e is a diagonal of H, then IR could include some region of H above e. Each of these regions is defined as follows. Assume $e = \overline{uw}$ is a diagonal. O_{uw} denotes the region in H above e (i.e., outside H_V) and below the circular arc $\widehat{uw}$ where its center is on the bottom edge. $O_{v_1 v_5}$, $O_{v_{10} v_{25}}$, $O_{v_{25} v_{28}}$ and $O_{v_{28} v_{34}}$ are such examples in Figure 3.

Theorem 1 $IR = (\cup_{\overline{uw} \in D} O_{uw}) \cup H_V$, where D is the set of edges of H_V which are diagonals of H. □

Collorary Let $H_V = (v_0, v_1, ..., v_n)$ for a normal histogram H, then IR w.r.t. H can be defined by the same sequence of vertices of H_V by replacing all diagonals $\overline{v_i v_{i+1}}$ of H in the sequence of H_V by an arc $\widehat{v_i v_{i+1}}$.

3.2 Influence Normal Histogram (INH)

By definition, an INH would contain all the edges in IR or H_V (Theorem 1) which are also edges in H (e.g., $\overline{v_5 v_6}$, $\overline{v_6 v_7}$, $\overline{v_7 v_8}$ etc. in Figure 3). The remaining edges of an INH would be those chords and edges of H enclosing O_{uw} for every $\overline{uw} \in D$ as O_{uw} is part of IR (Theorem 1). Let H_B be the part of the INH above each diagonal $\overline{uw}$ and enclosing O_{uw}. For example, as in Figure 3, the H_B's are (v_1, v_3, v_3, v_5), (v_{10}, v_{12}, v_{12}), $(v_{10}, v_{12}, v_{13}, v'_{13}, v_{25})$, $(v_{25}, v'_{25}, v_{28})$ and $(v_{28}, v'_{28}, v_{33}, v_{34})$.

Now, we can have a precise description of INH. There are two types of vertices in INH, the vertices of H_V and the vertices of H_B's, with one H_B for each edge in D. Thus, any vertex in INH which is not in H_V will be in H_B, and the endpoints of any edge in D will be vertices in both H_V and H_B. In the following we shall describe the properties of H_B and H_V and show that the Voronoi diagram of an INH can be constructed in linear time.

A **monotonic histogram** is an NH such that the bottom edge is on the x-axis and both the $x-$ and $y-$coordinates of the vertices along the boundary are monotonically non-decreasing. A **bitonic histogram** is an NH such that both the $x-$ and $y-$coordinates of the vertices along the boundary are first monotonically non-decreasing and then monotonically non-increasing.

Lemma 2 H_B is bitonic. □

Lemma 3 The Voronoi diagrams of H_B and H_V can be constructed in linear time.

Proof It is shown in [DjLi89] that the Voronoi diagram of a monotonic histogram can be constructed in linear time. By the fact that H_B can be partitioned into two monotonic polygons by the vertical straight line through its highest vertex or edge (Lemma 2), the Voronoi diagrams of two such monotonic polygons can be merged in linear time [JoWa93,KlLi93]. Thus, $V_c(H_B)$ can be found in linear time. The Voronoi diagram of a H_V can be constructed in linear time due to [AGSS89]. □

The following lemma (proof ommitted) shows that the Voronoi diagrams of two H_B's cannot affect each other.

Lemma 4 Given an INH with its attached H_B's, and let x and y be two non-H_V vertices in two different H_B's, then the Voronoi cells, $V(x)$ and $V(y)$, cannot share any point in H_V. □

Theorem 2 *The Voronoi diagram of an INH can be constructed in time linearly proportional to its size.*

Proof By Lemma 3, all the Voronoi diagrams of H_V and H_B's can be constructed in linear time, since each H_B shares an edge with H_V, by [JoWa93, KlLi93] and Lemma 4, the Voronoi diagrams of each H_B and H_V can be merged in time proportional to the number of Voronoi edges shared by them. As different H_B's do not interfere each other by Lemma 4, the merging time is linearly proportional to the number of Voronoi edges shared by H_B's and H_V, i.e. the size of INH. □

3.3 Region Identification

In this section, we shall present an algorithm which can identify the INH in an NH in time linearly proportional to the size of the INH. Chazelle's linear time algorithm [Chaz90] is first applied to an NH to obtain its horizontal visibility map (Figure 4). Because of the property of a normal histogram, H can be represented by a **partition tree** T_P, in which each tree node represents a chord in the map and each tree edge represents the adjacency of two chords. Let $n(v)$ denote the chord(s) associated with vertex v in H. If there are two chords in $n(v)$, $n^L(v)$ and $n^R(v)$ denote the left and right chord respectively (Figure 4). With the partition tree, the INH to be identified can be represented as a rooted subtree of T_P [3]. For example, the INH indicated by the shaded area can be represented by the rooted subtree as marked in Figure 4. The algorithm to identify the INH is based on tree traversal. In order to achieve the linear time complexity, only those tree nodes relevant to the INH will be traversed. Thus, one of the key steps in the tree traversal is the pruning condition, i.e., when the traversal of a subtree can be terminated. The other key step is the identification of the vertices of H_V so that we can partition the INH into H_B's and H_V for the construction of Voronoi diagrams as described in the previous section.

Let b_{uw} denote the perpendicular bisector of vertices u and w, and l_v be the perpendicular line segment from v to the bottom edge. The following two lemmas give necessary conditions for a vertex v of H to be and not to be a vertex of H_V. Based on Theorem 1 and the definition of H_B, we can ensure that a visited vertex, which is not a vertex of H_V, shall be a vertex of H_B.

Lemma 5 *For any vertex v of H, if l_v does not intersect with any b_{uv}, where $u \in (H - \{v\})$, then v is a vertex of H_V.* □

Lemma 6 *Let $u, v,$ and w be vertices of H in this ordering. If v lies outside O_{uw}, then v cannot be a vertex of H_V.* □

Vertex v can lie outside O_{uw} in two different ways, with $n(v)$ intersecting and not intersecting O_{uw}. As $O_{uw} \subseteq IR$ (Theorem 1), if $n(v)$ intersects O_{uw}, v will become one of the boundary vertices of INH. Since v is not in H_V, it must

---

[3] A rooted subtree of t has the property that the root of t is also the root of the subtree.

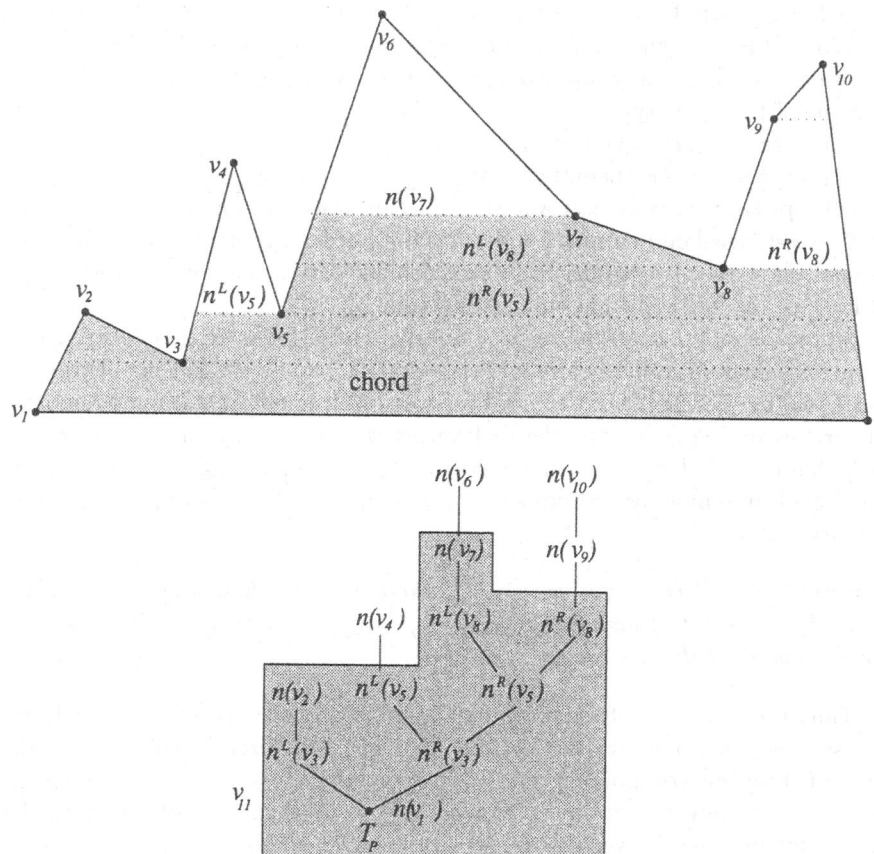

Figure 4: A demonstration for INH and its tree

be a vertex of H_B. If $n(v)$ does not intersect O_{uw}, any vertex above v cannot be a vertex of INH as its associated chord will not intersect O_{uw}. Thus the tree traversal can be terminated at v. In fact, only the first (lowest) vertex during the tree traversal, whose chord does not intersect O_{uw}, can be a vertex of INH or H_B. Based on this pruning condition, all the visited vertices will be either in H_V or in H_B.

When $n(v)$ is visited, if $n(v)$ intersects IR then we shall identify v as a vertex of the INH and continue the tree traversal to visit v's son(s). If v is the left(right)-endpoint of the chord, i.e., the IR is on its right[4], then the INH to be identified must be on the right(left)-hand side of v and v belongs to the left (right) boundary of the INH. Thus, v is put to the left (right) list $L_L(L_R)$ of vertices. As the chords of these two lists of vertices always intersect IR, L_L and

---

[4]Note that if there are two chords associated with v, v can be the right endpoint of one and the left endpoint of the other. The following discussion will still hold if we consider the chord one at a time.

L_R will eventually form the boundary of the INH.

Without loss of generality, assume the parent of $n(v)$ intersects IR and $n(v)$ is being visited. Based on Lemma 5 and Lemma 6, vertex v is tested and classified into one of the three types:

(a) a vertex in H_V (Lemma 5),

(b) a vertex in H_B (Lemma 6), or

(c) a **potential vertex** if we cannot decide whether it is in H_V or H_B.

The test based on Lemmas 5 and 6 has to consider all vertices of H. However, we shall show that it is sufficient to consider the test on the two last vertices, u_* and w_*, of the left and right sublists, L_L^B and L_R^B.

Since L_L and L_R are handled similarly, without loss of generality, let us simply consider L_L. Just before v is visited, let $L_L(v)$ be the left list of vertices, and $L_L'(v)$ be the left list of potential vertices. We can obtain $L_L'(v)$ by removing all vertices in $L_L(v)$ (except the first vertex which is always in H_V) known to be in H_B or H_V. Let $L_L^B(v)$ and $L_R^B(v)$ be the corresponding sublists of $L_L'(v)$ and $L_R'(v)$ by removing vertices known to be in H_B after v is visited and taken into account.

Lemma 7 *Let $L_L'(v) = (u_0, u_1, ..., u_k')$ where u_0 is the first vertex visited, and is in H_V. The set of bisectors $b_{u_0 u_1}, b_{u_1 u_2}, ... b_{u_{k'-1} u_k'}$ would not intersect each other above the bottom edge.* □

Thus the bisectors of every pair of neighboring vertices in $L_L'(v)$ will not intersect each other above the bottom edge and the Voronoi cell of x, for any $x \in L_L'(v)$, in the Voronoi diagram $V_c(L_L'(v) \cup \{v\})$, will cross the bottom edge. We shall show how to obtain $L_L^B(v)$ from $L_L'(v)$. Without loss of generality, let the left list of potential vertices $L_L'(v)$ be $(u_0, u_1, ..., u_{k'})$ just before v is visited. Vertex $u_{k'}$ is then first tested against v to see whether it is in H_B or remains in $L_L'(v)$. Vertex $u_{k'}$ is in H_B if $b_{u_{k'-1} u_{k'}}$ and $b_{u_{k'} v}$ intersect each other above the bottom edge. If $u_{k'}$ is found to be in H_B, the test will be carried on $u_{k'-1}$ and so on until we find the first u_k, $0 \le k \le k'$, that remains in $L_L'(v)$, i.e., $b_{u_{k-1} u_k}$ does not intersect with $b_{u_k v}$ above the bottom edge.

Backtrack()
$k \leftarrow k'$
while $((k \ge 1)$ **and** $(b_{u_{k-1} u_k}$ intersects $b_{u_k v}$ above the bottom edge$)$ **do**
    begin remove u_k from the list as u_k is a vertex of H_B; $k \leftarrow k - 1$ **end**

Effectively, $L_L'(v)$ can be recognized as a stack with $u_* = u_k'$ as its top element. Lemma 7 guarantees that because $b_{u_{k'-1} u_{k'}}$ does not intersect $b_{u_{k'} v}$ above the bottom edge, $b_{u_{p-1} u_p}$ would not intersect with $b_{u_p v}$ above the bottom edge for all p, $0 \le p \le k'$. Thus the test can be stopped as soon as the first element in $L_L'(v)$ that cannot be determined to be in H_B is identified. The resulting list is denoted by $L_L^B(v) = (u_o, ..., u_k)$.

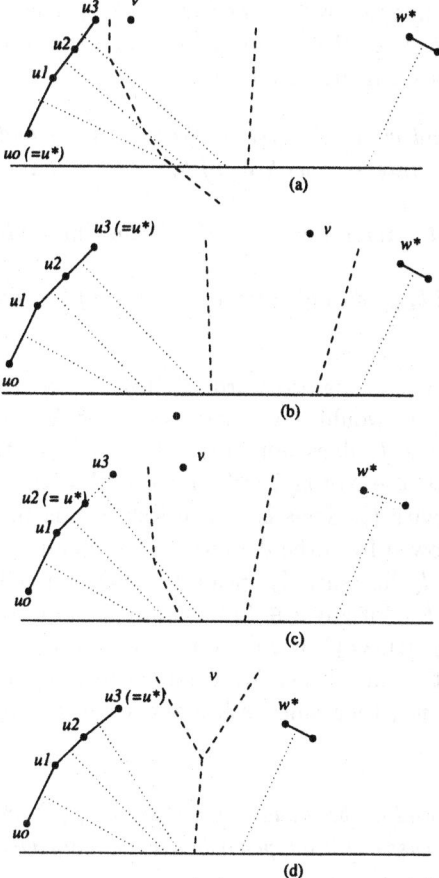

Figure 5: A demonstration for $L'_L(v)$ and $L^B_L(v)$

Note that some vertices in $L^B_L(v)$ might know to be in H_V, but for the simplicity of the algorithm they are not tested at this moment and remain in $L^B_L(v)$ as potential vertices. They will be recognized as vertices in H_V eventually.

Consider the instances in Figure 5 to illustrate the relationship between $L^B_L(v)$ and $L'_L(v)$, starting from u_0, the only vertex known to be in H_V, and after visited $n(u_1)$, $n(u_2)$ and $n(u_3)$, we still cannot decide whether or not u_1, u_2 and u_3 are vertices of H_V. Thus, vertices u_1, u_2 and u_3 are potential vertices and u_0, u_1, u_2, u_3 are vertices in in the left list, $L'_L(v)$. When v is being visited, (Figure 5) shows the four different cases: (a) u_1, u_2, and u_3 are vertices in H_B, and only u_0 is in $L^B_L(v)$ with $u_* = u_0$. (b) all u_1, u_2, and u_3 remain as potential vertices and are in $L^B_L(v)$ with $u_* = u_3$. We can show later that v belongs to H_V, when u_* is also considered, (c) u_1, u_2 are in $L^B_L(v)$ with $u_* = u_2$. Vertex u_3 should be known to be in H_B when considered with w before v is visited. We can show later that v is in H_B too. (d) same as (b) except that we can show that v is in H_B.

The following lemmas show that we have to keep track only the lists of left and right potential vertices, $L_L^B(v)$ and $L_R^B(v)$, and the test on v can be carried out only on their last elements u_* and w_*.

Lemma 8 *Let u_* and w_* be the last vertices in $L_L^B(v)$ and $L_R^B(v)$ respectively.*

(a) b_{u_*v} does not intersect with l_v iff b_{xv} does not intersect with l_v for all $x \epsilon L_L(v)$,

(b) b_{vw_*} does not intersect with l_v iff b_{xv} does not intersect with l_v for all $x \epsilon L_L(v)$, and

(c) both b_{u_*v} and b_{vw_*} do not intersect with l_v iff b_{xv} does not intersect with l_v for all $x \epsilon H$.

Proof (a) "If" part is straightforward. "Only if" part: all those vertices in $L_L(v)$ but not in $L_L^B(v)$ would have their associated Voronoi cells not crossing the bottom line. Since l_v does not intersect with b_{u_*v}, this would guarantee by Lemma 7 that all the Voronoi cells associated with the vertices of $L_L(v)$ would not intersect with l_v. Thus l_v lies entirely inside the Voronoi cell of v in $V_c(L_L(v) \cup \{v\})$ followed from the definition of Voronoi cell. (b) Similar to (a). (c) By the premise, l_v lies entirely inside the Voronoi cell of v in $V_c(L_L(v) \cup L_R(v) \cup \{v\})$. Thus, b_{vx} for any $x \in L_L(v) \cup L_R(v)$ cannot intersect with l_v. For $x \in (H - L_L(v) \cup L_R(v) \cup \{v\})$, if x lies on the left of u_* or the right of w_*, then x cannot be visible to v and hence b_{xv} cannot affect the Voronoi cell of v. If x lies between u_* and w_*, then since x lies above the chord $n(v)$ and hence b_{xv} cannot intersect l_v. $\square$

Theorem 3 *Let u_* and w_* be the last elements in $L_L^B(v)$ and $L_R^B(v)$ respectively. Then, the following cases will happen when v is being visited (note that $n^L(v)$ and $n^R(v)$ might be null)*

(a) if none of b_{u_*v} and b_{vw_*} intersect with l_v, then $v \in H_V$,

(b) if both b_{u_*v} and b_{vw_*} intersect each other above bottom line then $v \epsilon H_B$, and

(c) if neither (a) nor (b) is satisfied, then v is a potential vertex.

Proof (a) By Lemma 8 (c), b_{xv} would not intersect with l_v for all $x \in H$ and v would be a vertex in H_V (Lemma 5). (b) The intersection of b_{u_*v} and b_{uw_*} above the bottom line implies that v would lie outside $O_{u_*w_*}$. By Lemma 6, $v \notin H_V$. Since it is assumed that the parent of $n(v)$ intersects IR, either $n(v)$ intersects IR or $n(v)$ is the first (lowest) chord not intersecting IR. Thus $v \epsilon H_B$ by the definition of INH. (c) By definition of potential vertex. $\square$

Before describing the whole algorithm in detail, let us consider how and why the tree traversal works. The pruning only applies to those chords above those edges in D, i.e., those edges in H_V and are the diagonals of H. In particular, the pruning will apply to those chords that do not go through the IR. For example, as shown in Figure 3, $n(v_3)$ above diagonal $\overline{v_1 v_5}$, $n(v_{12})$ and $n(v_{13})$ above diagonal $\overline{v_{10} v_{25}}$, $n(v_{25})$ above diagonal $\overline{v_{25} v_{28}}$ and $n(v_{28})$ above diagonal

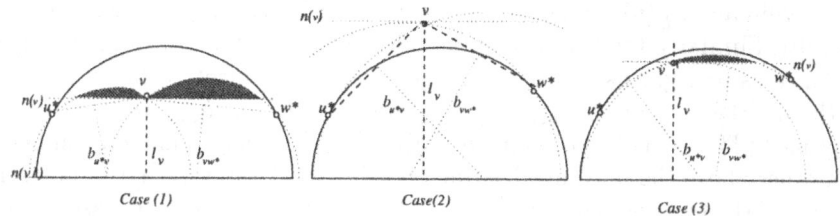

Figure 6: For the proof of Lemma 11

$\overline{v_{28}v_{34}}$. By the time the nodes in T_p corresponding to these chords are visited, the nodes in T_p corresponding to the endpoints of their associated diagonals should have been traversed and identified. For example, when $n(v_{12})$ and $n(v_{13})$ are visited, $n(v_{10})$ and $n(v_{25})$ should have been visited. Then we can determine the IR extended above the diagonal $\overline{v_{10}v_{25}}$, i.e., $O_{v_{10}v_{25}}$, and further determine whether a chord above the diagonal can be pruned or not. When v is visited, the bisectors, b_{u_*v} and b_{vw_*} for $u_*, w_* \in L_L^B(v) \cup L_R^B(v)$, have a lot to do with pruning. If b_{u_*v} intersects l_v above the bottom line, O_{u_*v} would lie totally below $n^L(v)$ and thus all descendants of (all chords above) $n^L(v)$ can be pruned. Alternatively, if b_{u_*v} intersects the bottom line and not l_v, some portion of O_{u_*v} would lie above $n^L(v)$ and descendants of $n^L(v)$ may not be pruned. Similarly b_{vw_*} and l_v have the same to do with pruning.

Theorem 4 *Let u_* and w_* be the last elements in $L_L^B(v)$ and $L_R^B(v)$ respectively. Then, the following cases will happen when v is being visited (note that $n^L(v)$ and $n^R(v)$ might be null) (refer to Figure 6)*

(a) The subtree rooted at $n^L(v)$ is pruned during tree traversal iff b_{u_*v} crosses l_v,

(b) The subtree rooted at $n^R(v)$ is pruned during tree traversal iff b_{vw_*} crosses l_v,

Proof The proofs for (a) and (b) are similar and only the proof for (a) is shown. Bisector b_{u_*v} intersect with l_v iff O_{u_*v} lies totally below v iff all chords above $n^L(v)$ do not intersect IR iff $n^L(v)$ can be pruned. $\square$

The above theorem describes the conditions when a subtree can be pruned. Subtrees are pruned as long as the representing INH's do not contain any IR region. When a subtree is pruned, the corresponding pruned portion of the normal histogram will also be a smaller normal histogram with the pruned chord, i.e., the root of the subtree, as the bottom edge. For example, the pruned position $(v_{28}, v_{29}, v_{30}, v_{31}, v_{32}, v_{28'})$ is an NH with $\overline{v_{28}v_{28'}}$ (the pruned chord) as bottom edge. The Voronoi diagrams of pruned portions can then be constructed recursively.

We shall describe the actions taken when v is visited after obtaining $L_L^B(v)$ and $L_R^B(v)$ for $L_L'(v)$ and $L_R'(v)$.

(a) If b_{u_*v} and b_{vw_*} do not cross l_v then we have $v \in H_V$ and both $n^L(v)$ and $n^R(v)$ are not pruned (Theorem 4). The traversal of the subtree rooted at

$n^L(v)$ will have $L^B_L(v)$ remained as its left list and v as the only element in its right list. Similarly, the traversal of the subtree rooted at $n^R(v)$ will have $L^B_R(v)$ remained as its right list and v as the only and first element in its left list.

(b) If either one of the b_{u_*v} and b_{vw_*} intersects l_v, say b_{u_*v}, then $n^L(v)$ will be pruned (Theorem 4). According to Theorem 3, if both b_{u_*v} and b_{vw_*} intersect each other above the bottom edge, then v is in H_B. Otherwise, v is a potential vertex and the tree traversal on $n^R(v)$ will be continued with v as the last element in the left list, i.e., vertex v is appended to $L^B_L(v)$ to form $L'_L(v')$, where $n(v')$ is the son of $n^R(v)$. Similar action will be carried out if b_{vw_*} intersect l_v.

(c) If both b_{u_*v} and b_{vw_*} intersect l_v, then both $n^L(v)$ and $n^R(v)$ are pruned (Theorem 4). According to Theorem 3, v will be in H_B. The tree traversal will be terminated at $n(v)$ and all the vertices in $L^B_L(v)$ and $L^B_R(v)$ will become vertices in H_V.

3.4 Complexity analysis

Our method for constructing the Constrained Voronoi diagram of a simple polygon P mainly relies on the efficiency of the identification of the INHs from an NH. Since the identification for different INHs is executed recursively, we shall only consider the root INH of an NH.

As described previously, when we traverse tree T_I of an NH to identify an INH, we visit each vertex of the INH exactly once. Those vertices have not been visited in the traversal of T_I cannot belong to the root INH. Therefore, we only need to show that each visited vertex is tested in constant times to classify it in H_V or in H_B.

Let us consider a vertex v. After a test, v can be classified into one of the three cases: (i) $v \epsilon H_V$, (ii) $v \epsilon H_B$, and (iii) v is a potential vertex. In cases (i) and (ii), v is stored in a list representing H_V or H_B and never to be tested again. In case (iii), v is stored in the left or right list and could be repeatedly tested when the descendants of v are visited. However, once vertex v is identified to be a vertex in H_B, v will never be tested again. Thus, we can argue that the time for visiting a vertex is constant when amortized over a sequence of tests. To see this, our analysis assumes that one unit credit should have been assigned to each potential vertex in L'_L and L'_R. Two unit credits are allocated for each test, one for the cost in carrying the test itself and the other is for assigning to the vertex should it be identified as a potential vertex. The test on classifying a vertex in L'_L or L'_R to be in H_B will be spent on the unit credit associated with it.

It is not difficult to see that linked lists and stack can be used to keep track the vertices in H_V and H_B's. In particular, insertion and concatenation operations on H_V and H_B's can be executed in constant time.

The time complexity analysis for constructing of Voronoi diagrams of INH and NH as well for P is obvious in the previous sections. We shall conclude the above analysis by a theorem without formal proof.

Theorem 5 $CDT(P)$ *can be found in* $\Theta(|P|)$ *time for simple polygon* P.

4 References

[Aure91] Aurenhammer A., (1991), 'Voronoi diagrams: a Survey', *ACM Computing Surveys* 23. pp.345-405.

[Aggr88] Aggarwal A., (1988), 'Computational Geometry', *MIT Lecture Notes* 18.409 (1988).

[AGSS89] Aggarwal A., Guibas L., Saxe J., and Shor P., (1989), 'A linear time algorithm for computing the Voronoi diagram of a convex polygon', *Disc. and Comp. Geometry* 4, pp.591-604.

[BeEp92] Bern M. and Eppstein D., (1992), 'Mesh generation and optimal triangulation', *Technical Report, Xero Palo Research Center.*

[DjLi89] Djidjev H., and Lingas A., (1989) 'On computing the Voronoi diagram for restricted planar figures', *Lecture Notes on Computer Science*, pp.54-64.

[Chaz90] Chazelle B., (1991), 'Triangulating a simple polygon in linear time', *Disc. and Comp. Geometry*, Vol.6 No.5, pp.485-524.

[Chew87] Chew P., (1987), 'Constrained Delaunay Triangulation', *Proc. of the 3rd ACM Symp. on Comp. Geometry*, pp.213-222.

[KlLi92] Klein R., and Lingas A., (1992), 'A linear time algorithm for the bounded Voronoi diagram of a simple polygon in L_1 metrics', *Proc. the 8th ACM Symp. on Comp. Geometry*, pp124-133.

[KlLi93] Klein R., and Lingas A., (1993), 'A linear time randomized algorithm for the bounded Voronoi diagram of a simple polygon', *Proc. the 9th ACM Symp. on Comp. Geometry*, pp124-133.

[LeLi86] Lee D. and Lin A., (1986), 'Generalized Delaunay triangulations for planar graphs', *Disc. and Comp. Geometry* 1, pp.201-217.

[Ling87] Lingas, A., (1987), 'A space efficient algorithm for the Constrained Delaunay triangulation', *Lecture Notes in Control and Information Sciences*, Vol. 113, pp. 359-364.

[LeLi90] Levcopoulos C. and Lingas, A., (1990), 'Fast algorithms for Constrained Delaunay triangulation', *Lecture Notes on Computer Science* Vol. 447, pp.238-250.

[PrSh85] Preparata F. and Shamos M., (1985), *Computational Geometry*, Springer-Verlag.

[Seid88] Seidel R., (1988), 'Constrained Delaunay triangulations and Voronoi diagrams with obstacles', *Rep. 260, IIG-TU Graz*, Austria, pp. 178-191.

[WaSc87] Wang C. and Schubert L., (1987), 'An optimal algorithm for constructing the Delaunay triangulation of a set of line segments', *Proc. of the 3rd ACM Symp. on Comp. Geometry*, pp.223-232.

[JoWa93] Joe B. and Wang C., (1993), 'Duality of Constrained Delaunay triangulation and Voronoi diagram', *Algorithmica*, pp.

[Wang93] Wang C., (1993), 'Efficiently updating the constrained Delaunay triangulations', *BIT*, 33 (1993), pp. 176-181.

External-Memory Algorithms
for Processing Line Segments
in Geographic Information Systems

(extended abstract)

Lars Arge[1*], Darren Erik Vengroff[2**], and Jeffrey Scott Vitter[3***]

[1] BRICS[†], Department of Computer Science, University of Aarhus, Aarhus, Denmark
[2] Department of Computer Science, Brown University, Providence, RI 02912, USA
[3] Department of Computer Science, Duke University, Durham, NC 27708, USA

Abstract. In the design of algorithms for large-scale applications it is essential to consider the problem of minimizing I/O communication. Geographical information systems (GIS) are good examples of such large-scale applications as they frequently handle huge amounts of spatial data. In this paper we develop efficient new external-memory algorithms for a number of important problems involving line segments in the plane, including trapezoid decomposition, batched planar point location, triangulation, red-blue line segment intersection reporting, and general line segment intersection reporting. In GIS systems, the first three problems are useful for rendering and modeling, and the latter two are frequently used for overlaying maps and extracting information from them.

1 Introduction

The Input/Output communication between fast internal memory and slower external storage is the bottleneck in many large-scale applications. The significance of this bottleneck is increasing as internal computation gets faster, and especially as parallel computing gains popularity [23]. Currently, technological advances are increasing CPU speeds at an annual rate of 40–60% while disk transfer rates are only increasing by 7–10% annually [24]. Internal memory sizes are also increasing, but not nearly fast enough to meet the needs of important

* Supported in part by the ESPRIT II Basic Research Actions Program of the EC under contract No. 7141 (Project ALCOM II). This work was done while a Visiting Scholar at Duke University. Email: `large@daimi.aau.dk`.
** Supported in part by the U.S. Army Research Office under grant DAAH04–93–G–0076 and by the National Science Foundation under grant DMR–9217290. This work was done while a Visiting Scholar at Duke University. Email: `dev@cs.brown.edu`.
*** Supported in part by the National Science Foundation under grant CCR–9007851 and by the U.S. Army Research Office under grants DAAL03–91–G–0035 and DAAH04–93–G–0076. Email: `jsv@cs.duke.edu`.
† Acronym for Basic Research in Computer Science, a Center of the Danish National Research Foundation.

large-scale applications, and thus it is essential to consider the problem of minimizing I/O communication.

Geographical information systems (GIS) are a rich source of important problems that require good use of external-memory techniques. GIS systems are used for scientific applications such as environmental impact, wildlife repopulation, epidemiologic analysis, and earthquake studies and for commercial applications such as market analysis, utility facilities distribution, and mineral exploration [17]. In support of these applications, GIS systems store, manipulate, and search through enormous amounts of spatial data [13, 18, 25, 27]. NASA's EOS project GIS system [13], for example, is expected to manipulate petabytes (thousands of terabytes, or millions of gigabytes) of data!

Typical subproblems that need to be solved in GIS systems include point location, triangulating maps, generating contours from triangulated elevation data, and producing map overlays, all of which require manipulation of line segments. As an illustration, the computation of new scenes or maps from existing information—also called map overlaying—is an important GIS operation. Some existing software packages are completely based on this operation [27]. Given two thematic maps (piecewise linear maps with, e.g., indications of lakes, roads, pollution level), the problem is to compute a new map in which the thematic attributes of each location is a function of the thematic attributes of the corresponding locations in the two input maps. For example, the input maps could be a map of land utilization (farmland, forest, residential, lake), and a map of pollution levels. The map overlay operation could then be used to produce a new map of agricultural land where the degree of pollution is above a certain level. One of the main problems in map overlaying is "line-breaking," which can be abstracted as the red-blue line segment intersection problem.

In this paper, we present efficient external-memory algorithms for large-scale geometric problems involving collections of line segments in the plane, with applications to GIS systems. In particular, we address region decomposition problems such as trapezoid decomposition and triangulation, and line segment intersection problems such as the red-blue segment intersection problem and more general formulations.

1.1 The I/O Model of Computation

The primary feature of disks that we model is their extremely long access time relative to that of solid state random-access memory. In order to amortize this access time over a large amount of data, typical disks read or write large blocks of contiguous data at once. To increase bandwidth parallel disks can be used. In our model, in a single I/O operation, each of D disks can transmit a block of B contiguous data items. Our problems are modeled by the following parameters:

$$N = \# \text{ of items in the problem instance;}$$
$$M = \# \text{ of items that can fit into internal memory;}$$
$$B = \# \text{ of items per disk block;}$$
$$D = \# \text{ of parallel disks,}$$

where $M < N$ and $1 \leq DB \leq M/2$. Depending on the size of the data items, typical values for workstations and file servers in production today are on the order of $M = 10^6$ or 10^7 and $B = 10^3$. Values of D range up to 10^2 in current disk arrays. Large-scale problem instances can be in the range $N = 10^{10}$ to $N = 10^{12}$.

In order to study the performance of external-memory algorithms, we use the standard notion of I/O complexity [28]. We define an *input/output operation* (or simply *I/O* for short) to be the process of simultaneously reading or writing D blocks of data, one block to or from each of the D disks. The I/O complexity of an algorithm is simply the number of I/Os it performs. For example, reading all of the input data requires N/DB I/Os, since we can read DB items in a single I/O. We will use the term *scanning* to describe the fundamental primitive of reading (or writing) all items in a set stored contiguously on external storage by reading (or writing) the blocks of the set in a sequential manner.

For the problems we consider we define two additional parameters:

$$K = \# \text{ of queries in the problem instance;}$$
$$T = \# \text{ of items in the problem solution.}$$

Since each I/O can transmit DB items simultaneously, it is convenient to introduce the following notation:

$$n = \frac{N}{DB}, \qquad k = \frac{K}{DB}, \qquad t = \frac{T}{DB}.$$

We will say that an algorithm uses a linear number of I/O operations if it uses at most $O(n)$ I/Os to solve a problem of size N. For convenience we also define

$$m = \frac{M}{B},$$

which is the optimal degree of recursion or branching used in efficient external-memory algorithms. Note that there is no D term in the denominator of the definition of the branching factor m. Use of "striping" [28] corresponds to treating the D disks conceptually as one disk with a larger block size of DB; the resulting branching factor from striping is thus M/DB rather than m, which works well when D is moderately sized as in current systems, but can cause loss of efficiency when D is very large.

1.2 Previous Results in I/O-Efficient Computation

Early work on I/O algorithms concentrated on algorithms for sorting and permutation-related problems [1, 20, 21, 28]. External sorting requires $\Theta(n \log_m n)$ I/Os,[1] which is the external-memory equivalent of the well-known $\Theta(N \log N)$ time bound for sorting in internal memory. More recently researchers have designed external-memory algorithms for a number of problems in different areas.

---

[1] We define for convenience $\log_m n = \max\{1, (\log n)/\log m\}$.

Most notably I/O-efficient algorithms have been developed for a large number of computational geometry [16] and graph problems [12]. In [5] a general connection between the comparison-complexity and the I/O complexity of a given problem is shown, and in [4] alternative solutions for some of the problems in [12] and [16] are derived by developing and using dynamic external-memory data structures.

1.3 Our Results

In this paper, we combine and modify in novel ways several of the previously known techniques for designing efficient algorithms for external memory. In particular we use the *distribution sweeping* and *batched filtering* paradigms of [16], the *buffer tree* data structure of [4], and the deterministic distribution methods for parallel disks in [20].[2] In addition we also develop a powerful new technique that can be regarded as a practical external-memory version of batched fractional cascading on an external-memory version of a segment tree. This enables us to improve on existing external-memory algorithms as well as to develop new algorithms and thus partially answer some open problems posed in [16].

In Section 2 we introduce the *endpoint dominance problem*, which is a subproblem of *trapezoid decomposition*. We introduce an $O(n \log_m n)$-I/O algorithm to solve the endpoint dominance problem, and we use it to develop an algorithm with the same asymptotic I/O complexity for trapezoid decomposition. In Section 3 we show how trapezoid decomposition can be used to get $O(n \log_m n)$-I/O algorithms for *planar point location* and *triangulation of simple polygons*. In Section 4 we give external-memory algorithms for line segment intersection problems. First we introduce an $O(n \log_m n)$-I/O *segment sorting* algorithm based on endpoint dominance, and we then show how this result can be used to develop an $O(n \log_m n + t)$-I/O algorithm for *red-blue line segment intersection*. Finally, we discuss an $O((n + t) \log_m n)$-I/O algorithm for the *general segment intersection* problem.

Our results are summarized in Table 1. For all but the batched planar point location problem, no algorithms specifically designed for external memory were previously known. The batched planar point location algorithm that was previously known [16] only works when the planar subdivision is monotone, and the problems of triangulating a simple polygon and reporting intersections between other than orthogonal line segments are stated as open problems in [16].

For the sake of contrast, our results are also compared with modified internal-memory algorithms for the same problems. In most cases, these modified algorithms are plane-sweep algorithms modified to use B-tree-based dynamic data structures rather than binary tree-based dynamic data structures, following the example of a class of algorithms studied experimentally in [11]. Such modifications lead to algorithms using $O(N \log_B n)$ I/Os. For two of the algorithms the known optimal internal-memory algorithms [7, 8] are not plane-sweep algorithms

---

[2] For brevity in this extended abstract, we restrict discussion to the one-disk model where $D = 1$. In the full version of this paper, we show how to use techniques from [20] to extend our results to the general $D > 1$ model.

Problem	I/O bound of new result	Result using modified internal memory algorithm
Endpoint dominance.	$O(n \log_m n)$	$O(N \log_B n)$
Trapezoid decomposition.	$O(n \log_m n)$	$O(N \log_B n)$
Batched planar point location.	$O((n + k) \log_m n)$	
Triangulation.	$O(n \log_m n)$	$\Omega(N)$
Segment sorting.	$O(n \log_m n)$	$O(N \log_B n)$
Red-blue line segment intersection.	$O(n \log_m n + t)$	$O(N \log_B n + t)$
Line segment intersection.	$O((n + t) \log_m n)$	$\Omega(N)$

Fig. 1. Summary of results.

and can therefore not be modified in this manner. It is difficult to analyze precisely how those algorithms perform in an I/O environment; however it is easy to realize that they use at least $\Omega(N)$ I/Os. The I/O bounds for algorithms based on B-trees have a logarithm of base B in the denominator rather than a logarithm of base m. But the most important difference between such algorithms and our results is the fact that the updates to the dynamic data structures are handled on an individual basis, which leads to an extra multiplicative factor of DB in the I/O bound, which is very significant in practice.

As mentioned, the red-blue line segment intersection problem is of special interest because it is an abstraction of the important map-overlay problem, which is the core of several vector-based GISs [2, 3, 22]. Our red-blue line segment intersection algorithm is optimal because the external-memory lower bound technique of [5] can be applied to the internal-memory lower bound of $\Omega(N \log N + T)$ to get an $\Omega(n \log_m n + t)$ I/O lower bound on the problem. Although a time-optimal internal-memory algorithm for the general intersection problem exists [8], a number of simpler solutions have been presented for the red-blue problem [6, 9, 19, 22]. Two of these algorithms [9, 22] are not plane-sweep algorithms, but both sort segments of the same color in a preprocessing step with a plane-sweep algorithm. The authors of [22] claim that their algorithm will perform well with inadequate internal memory owing to the fact that data are mostly referenced sequentially. A closer look at the main algorithm reveals that it can be modified to use $O(n \log_2 n)$ I/Os in the I/O model, which is only a factor of $\log m$ from optimal; unfortunately, the modified algorithm still needs $O(N \log_B n)$ I/Os to sort the segments, which is a factor of DB times the optimal I/O bound for the intersection problem.

2 Trapezoidal Decomposition

In order to solve the trapezoidal decomposition problem, we will rely on a solution to a subproblem that we call the *endpoint dominance problem* (EPD): Given N non-intersecting line segments in the plane, find the segment directly above each endpoint of each segment.

With an I/O-efficient algorithm for EPD, we can construct an I/O-efficient algorithm for trapezoidal decomposition, as described in the following lemma:

Lemma 1. *If EPD can be solved in $O(n \log_m n)$ I/Os, then so can trapezoid decomposition.*

Proof. We solve two instances of EPD, one to find the segments directly above each segment endpoint and one (with all y coordinates negated) to find the segment directly below each endpoint. We then compute the locations of all $O(N)$ vertical trapezoid edges. This is done by scanning the output of the two EPD instances in $O(n)$ I/Os. To explicitly construct the trapezoids, we sort all trapezoid vertical segments by the IDs of the input segments they lie on, breaking ties by x coordinate. This takes $O(n \log_m n)$ I/Os [4, 21]. Finally, we scan this sorted list, in which we find the two vertical edges of each trapezoid in adjacent positions. The total amount of I/O used is thus $O(n \log_m n)$. □

As described below, we solve EPD by building a data structure inspired by the buffer tree data structure of [4]. We then perform $O(N)$ queries, one for each line segment endpoint, to solve the EPD problem.

2.1 Buffer Trees and External-Memory Segment Trees

Buffer trees are data structures that can support the processing of a batch of N updates and K queries on an initially empty dynamic data structure of elements from a totally ordered set in $O((n + k) \log_m n + t)$ I/Os [4]. They can be used to implement sweepline algorithms in which the entire sequence of updates and queries is known in advance. The queries that such sweepline algorithms ask of their dynamic data structures need not be answered in any particular order; the only requirement on the queries is that they must all eventually be answered. Such problems are known as *batch dynamic problems* [14]. For the problems we are considering in this paper, the known internal-memory solutions cannot be stated as batched dynamic algorithms (since the updates depend on the queries) or else the elements involved are not totally ordered. We are led instead to other approaches.

The *segment tree* is a well-known dynamic data structure that is amenable to implementation as a buffer tree. A segment tree stores a set of segments in one dimension. Given a query point, it returns the segments that contain the point. Such queries are called *stabbing queries*. An external-memory segment tree based on the approach in [4] is shown in Figure 2. The tree is perfectly balanced over the endpoints of the segments it represents and has branching factor $\sqrt{m/4}$. Each leaf represents $M/2$ consecutive segment endpoints. The first level of the tree partitions the data into $\sqrt{m/4}$ slabs σ_i, separated by dotted lines on Figure 2. Multi-slabs are defined as contiguous ranges of slabs, such as for example $[\sigma_1, \sigma_4]$. There are $m/8 + \sqrt{m}/4$ multi-slabs. The key point is that the number of multi-slabs is a quadratic function of the branching factor. The reason why we choose the branching factor to be $\Theta(\sqrt{m})$ rather than $\Theta(m)$ is so that we have room in internal memory for two blocks for each of the $\Theta(m)$ multi-slabs. The smaller branching factor at most about doubles the height of the tree.

Segments such as $\overline{CD}$ that completely span one or more slabs are called *long segments*. A copy of each long segment is stored in the largest multi-slab it

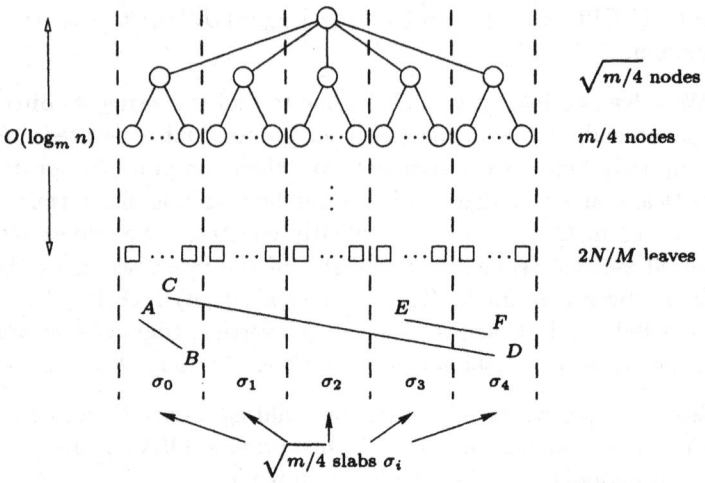

Fig. 2. An external-memory segment tree based on a buffer tree over a set of N segments, three of which, $\overline{AB}$, $\overline{CD}$, and $\overline{EF}$, are shown.

spans. Thus, $\overline{CD}$ is stored in $[\sigma_1, \sigma_3]$. All segments that are not long are called *short segments* and are not stored in any multi-slab. Instead, they are passed down to lower levels of the tree where they may span recursively defined slabs and be stored. $\overline{AB}$ and $\overline{EF}$ are examples of short segments. The portions of long segments that do not completely span slabs are treated as small segments. There are at most two such synthetically generated short segments for each long segment. Total space utilization is $O(n \log_m n)$ blocks.

To answer a stabbing query, we simply proceed down a path in the tree searching for the query value. At each node we encounter, we report all the long segments associated with each of the multi-slabs that span the query value.

Because of the size of the nodes and auxiliary multi-slab data, the buffer tree approach is inefficient for answering single queries. In batch dynamic environments, however, it can be used to develop optimal algorithms. In [4], techniques are developed for using external-memory segment trees in a batch dynamic environment such that inserting N segments in the tree and performing K queries requires $O((n+k) \log_m n + t)$ I/Os. In applications like EPD, where it is possible to process all N updates and then process all K queries, the technique reduces to batch filtering [16], in which we push all queries through a given level of the tree before moving on to the next level.

In order to solve the EPD problem, we modify the external segment tree described above so that the segments fully spanning a given multi-slab are stored in y-order. This requires two significant improvements over existing techniques. First, as discussed in Section 2.2, the tree construction techniques of [4] must be modified in order to guarantee optimal performance when the tree is built. Second, as discussed in Section 2.3 the batch filtering procedure must be augmented using techniques similar to fractional cascading [10].

2.2 Constructing Extended External Segment Trees

We will construct what we call an *extended external segment tree* using an approach based on distribution sweeping [16]. When we are building an external segment tree on non-intersecting segments in the plane we can talk about the order of segments in the same multi-slab just by comparing the order of their endpoints on one of the boundaries. An extended external segment tree is just an external segment tree as described in the last section built on non-intersecting segments, where the segments in each of the multi-slabs are sorted. In order to construct such a structure, we first use an optimal sorting algorithm [4, 21] to create a list of all the endpoints of the segments sorted by x-coordinate. This list is used during the whole algorithm to find the medians we use to split the interval associated with a given node into $\sqrt{m/4}$ vertical slabs. We now construct the $O(m)$ sorted segment lists associated with the root in the following way: First we scan through the segments and distribute the long segments to the appropriate multi-slab list. This can be done I/O-efficiently because we have enough internal memory to hold a block of segments for each multi-slab list. Then we sort each of these lists individually with an optimal sorting algorithm. Finally, we recursively construct an extended external segment tree for each of the slabs. The process continues until the number of endpoints in the subproblems falls below $M/2$.

Unfortunately this simple algorithm would use $O(n \log_m^2 n)$ I/Os, as we use $O(n \log_m n)$ I/Os to sort the multi-slab lists on each level of the recursion. To avoid this problem, we modify our algorithm such that each segment is contained only once in a list being sorted. This is done by talking advantage of the fact that, in a given node v, we already know the order of the segments that were treated as long segments in v's parent. Details will appear in the full version of this paper.

Lemma 2. *An extended external segment tree on N segments in the plane can be constructed in $O(n \log_m n)$ I/O operations.*

2.3 Filtering Queries Through an Extended External Segment Tree

Having constructed an extended external segment tree, we can now use it to find the segments directly above each of a series of K query points. In solving EPD, we have $K = 2N$, and the query points are the endpoints of the original segments. To find the segment directly above a query point p, we examine each node on the path from the root of the tree to the leaf containing p's x coordinate. At each such node, we find the segment directly above p by examining the sorted segment list associated with each multi-slab containing p. This segment can then be compared to the segment that is closest to the query point p so far, based on segments seen further up the tree, to see if it is the new globally closest segment. All K queries can be processed through the tree at once using batch filtering [16].

Unfortunately, the simple approach outlined in the preceding paragraph is not efficient. There are two problems that have to be dealt with. First, we must be able to look for a query point in many of the multi-slabs lists corresponding to a given node simultaneously. Second, searching for the position of a point in the sorted list associated with a particular multi-slab may require many I/Os. To solve the first problem, we will take advantage of the internal memory that is available to us. The second problem is solved with a notion similar to fractional cascading [9, 10, 26]. The idea behind fractional cascading on internal-memory segment trees is that instead of searching for the same element in a number of sorted lists, we augment the list at a node with sample elements from lists at the node's children. We then build bridges between the augmented list and corresponding elements in the augments lists of the node's children. These bridges obviate the need for full searches in the lists at the children. We take a similar approach for our external-memory problem, except that we send sample elements from parents to children. Furthermore, we do not use explicit bridges.

As a first step towards a solution based on fractional cascading, we preprocess the extended external segment tree in the following way: For each internal node, starting with the root, we produce a set of sample segments. We begin by scanning through all $O(m)$ of the multi-slab lists. For each of the $\sqrt{m/4}$ slabs (*not* multi-slabs) we produce a list of samples of the segments in the multi-slab lists that span it. The sample list for a slab consists of every $(2\sqrt{m/4})$th segment that spans it. For every slab we then augment the multi-slab lists of the corresponding child by merging the sampled list with the multi-slab list of the child that contains segments spanning the whole x-interval. This merging happens before we proceed to preprocessing the next level of the tree. At the lowest level of internal nodes, the sampled segments are passed down to the leaves.

We now prove a crucial lemma about the I/O complexity of the preprocessing steps and the space of the resulting data structure:

Lemma 3. *The preprocessing described above uses $O(n \log_m n)$ I/Os. After the preprocessing there are still $O(N)$ segments stored in the multi-lists on each level of the structure. Furthermore, each leaf contain less than M segments.*

Proof. Before any samples are passed down the tree, we have at most $2N$ segments represented at each level of the tree. Let N_i be the number of long segments, both original segments and segments sent down from the previous level, among all the nodes at level i of the tree after the preprocessing step. At the root, we have $N_0 \le 2N$. We send at most $N_i/(2\sqrt{m/4}) \cdot \sqrt{m/4} = N_i/2$ segments down from level i to level $i + 1$. Thus, $N_{i+1} \le 2N + N_i/2$. By induction on i, we can show for all i that $N_i = \left(4 - (1/2)^{i-1}\right) N = O(N)$. As enough internal memory is available to merge all the multi-slab lists at a node in a single pass, each segment on a given level is read and written only once. The number of I/Os used at level i of the tree is thus $O(n_i)$, where $n_i = N_i/B$. Since there are $O(\log_m n)$ levels, we in total use $O(n \log_m n)$ I/Os.

Before preprocessing, there were at most $M/2$ endpoints in each leaf. By an inductive argument similar to that given above, we show that the number of additional segments passed down the leaf is at most $M/2$, and thus the preprocessed leaves are of size at most M. $\qquad\square$

Having preprocessed the tree, we are now ready to filter the K query points through it. Since our fractional cascading construction is done "backwards", we filter queries from the leaves to the root rather than from the root to the leaves. To start off, we sort the query points by their x coordinates in $O(k \log_m k)$ I/Os. For every query point, we determine which leaf it belongs to and which of the segments stored in that leaf is directly above it (called *the dominating segment*). This can be done in $O(k + n)$ I/Os, because all the segments in a leaf fit in internal memory, and thus we can use an internal-memory algorithm to find the dominating segments. In order to prepare for the general step of moving queries up the tree, we sort the queries that belong to each leaf, based on the order of their dominating segments. This takes $O(k \log_m k)$ I/Os.

Now we go through $O(\log_m n)$ filtering steps, one for each level of the tree. Each filtering step begins with a set of queries at a given level, partitioned by the nodes at that level and ordered within the nodes by the order of the segments found so far to be directly above the query points. This corresponds to the output of the leaf processing. The filtering step should produce a similar input for the next level up the tree.

To produce the output of a given node for use at the next level up the tree, we "merge" the queries associated with its children and the node's multi-slab lists. The precise details of how this merge is done will appear in the full version of this paper. The general idea is as follows: Consider a slab s of the $\sqrt{m/4}$ slabs and the Q queries corresponding to it. Because we sampled every $(2\sqrt{m/4})$th segment spanning s and sent it down the tree, and because the queries are sorted according to the segment found to be directly above them so far, we know that the list of queries cannot be "too unsorted" relative to their positions in the sorted order of segments spanning s. As a matter of fact, we know that all queries lying between two consecutive sampled segments l_i and l_j will appear together in the list of queries. Since the number of segments between l_i and l_j is at most $2\sqrt{m/4}$, we can distribute the Q queries (and thus find the dominating segments on this level) in Q/B I/Os using $2\sqrt{m/4}$ blocks of internal memory. We simply reserve a block for each segment between l_i and l_j and scan through the Q queries distributing them according to the segment immediately above them. The key property is that we can process all the slabs simultaneously using only $(2\sqrt{m/4}) \cdot (\sqrt{m/4}) + (m/8 + \sqrt{m/4}) < m$ blocks of internal memory.

In the full paper all the details in the merge are given and it is shown that we can do one filtering step in $O(n+k)$ I/Os. When the filtering process reaches the root, we have the correct answers to all queries. The total I/O complexity of the algorithm is given by the following theorem:

Theorem 4. *An extended external segment tree on N non-intersecting segments in the plane can be constructed, and K query points can be filtered through the structure in order to find the dominating segments for all these points, in $O((n + k) \log_m n)$ I/O operations.*

Proof. According to Lemmas 2 and 3, construction and preprocessing together require $O(n \log_m n)$ I/Os.

Assuming that $K \leq N$, sorting the K queries takes $O(n \log_m n)$. Filtering the queries up one level in the tree takes $O(n)$ I/Os, resulting in an overall I/O complexity of $O(n \log_m n)$. When $K > N$, we can break the problem into $K/N = k/n$ sets of N queries. Each set of queries can be answered as shown above in $O(n \log_m n)$ I/Os, giving a total I/O complexity of $O(k \log_m n)$. □

This immediately gives us the following bound for EPD, for which $K = 2N$.

Corollary 5. *The endpoint dominance problem can be solved in* $O(n \log_m n)$ *I/O operations.*

3 Direct Applications

We now consider two direct applications of our external trapezoidal decomposition algorithm.

The *multi-point planar point location problem* is the problem of reporting the location of K query points in a planar subdivision defined by N line segments. In [16] an $O((n+k) \log_m n)$-I/O algorithm for this problem is given for *monotone* subdivisions of the plane. This result can be extended to arbitrary subdivisions using the algorithm developed in the previous section. Theorem 4 immediately implies the following:

Lemma 6. *The multi-point planar point location problem can be solved using* $O((n + k) \log_m n)$ *I/O operations.*

After computing the trapezoid decomposition of a simple polygon, the polygon can be *triangulated* in $O(n)$ I/Os using a slightly modified version of an algorithm from [15]:

Lemma 7. *A simple polygon with N vertices can be triangulated in* $O(n \log_m n)$ *I/O operations.*

4 Line Segment Intersection

In order to construct the optimal algorithm for the red-blue line segment intersection problem, we will first consider the problem of sorting a set of non-intersecting segments.

To sort N non-intersecting segments, we first solve EPD on the input segments augmented with the segment S_∞ with endpoints $(-\infty, \infty)$ and (∞, ∞). The existence of S_∞ ensures that all input segment endpoints are dominated by some segment. We define an aboveness relation $\searrow$ on elements of a non-intersecting set of segments S such that $\overline{AB} \searrow \overline{CD}$ *if and only if* either $(C, \overline{AB})$ or $(D, \overline{AB})$ is in the solution to EPD on S.[3] Similarly, we solve EPD with negated y coordinates and a special segment $S_{-\infty}$ to establish a belowness relation $\nearrow$.

---

[3] $(A, \overline{BC})$ denotes that $\overline{BC}$ is the segment immediately above A.

Sorting the segments then corresponds to extending the partial order defined by $\searrow$ and $\nearrow$ to a total order. If segment $\overline{AB}$ precedes segment $\overline{CD}$ in such a total order, and if we can intersect both $\overline{AB}$ and $\overline{CD}$ with the same vertical line l, then the intersection between l and $\overline{CD}$ is above the intersection between l and $\overline{AB}$. This property is important in the red-blue line segment intersection algorithm.

In order to obtain a total order we define a directed graph $G = (V, E)$ whose nodes consist of the input segments and the two extra segments S_∞ and $S_{-\infty}$. The edges correspond to elements of the relations $\searrow$ and $\nearrow$. For each pair of segments $\overline{AB}$ and $\overline{CD}$, there is an edge from $\overline{AB}$ to $\overline{CD}$ iff $\overline{CD} \searrow \overline{AB}$ or $\overline{AB} \nearrow \overline{CD}$. To sort the segments we simply have to topologically sort G. As G is a planar s,t-graph of size $O(N)$ this can be done in $O(n \log_m n)$ I/Os using an algorithm of [12].

Lemma 8. *A total ordering of the N non-intersecting segments can be found in $O(n \log_m n)$ I/Os.*

4.1 Red-Blue Line Segment Intersection

Using our ability to sort segments as described in the previous section, we can solve the red-blue line segment intersection in the optimal number of I/Os using a technique based on distribution sweeping [16].

Given input sets S_r of red segments and S_b of blue segments, we construct two intermediate sets T_r and T_b consisting of the red segments and the blue endpoints, and the blue segments and the red endpoints, respectively. Both T_r and T_b are of size $O(|S_r| + |S_b|) = O(N)$. We sort both T_r and T_b in $O(n \log_m n)$ I/Os using the algorithm from the previous section, and from now on we assume they are sorted.

We now locate segment intersections by distribution sweeping [16] with a branching factor of $\sqrt{m}$. The structure of distribution sweeping is that we divide the plane into $\sqrt{m}$ slabs, not unlike the way the plane was divided into slabs to build an external segments tree in Section 2.1. We define *long segments* as those crossing one or more slabs and *short segments* as those completely contained in a slab. Furthermore, we shorten the long segments by "cutting" them at the right boundary of the slab that contain their left endpoint, and at the left boundary of the slab containing their right endpoint. This may produce up to two new short segments for each long segment. Below we sketch how to report all T_i intersections between the long segments of one color and the long and short segments of the other color in $O(n + t_i)$ I/Os. We then partition sets T_r and T_b (and the new short segments) into $\sqrt{m}$ parts, one for each slab, and we recursively solve the problem on the short segments contained in each slab to locate their intersections. Each original segment is represented at most twice at each level of recursion, thus the total problem size at each level of recursion remains $O(N)$ segments. Recursion continues through $O(\log_m n)$ levels until the subproblems are of size $O(M)$ and thus can be solved in internal memory. This gives us the following result:

Theorem 9. *The red-blue line segment intersection problem on N segments can be solved in $O(n \log_m n + t)$ I/O operations.*

Now, we simply have to fill in the details of how intersections involving long segments are found in $O(n + t_i)$ I/Os.

Because T_r and T_b are sorted, we can locate interactions between long and short segments (both original and new short segments produced by cutting long segments) using a slightly modified version of the distribution-sweeping algorithm for solving orthogonal segment intersection [16]. We use the modified algorithm twice and treat long segments of one color as horizontal segments and short segments of the other color as vertical segments. Just as in the orthogonal case, all intersections are located and reported in $O(n + t_i)$ I/Os.

In order to report intersections between long segments of different color we again use the notion of multi-slabs as in Section 2. First we scan through T_r and distribute the long red segments into the $O(m)$ multi-slab lists. Next we scan through the blue segments in T_b, and for every long blue segment we look for and report intersections with the red segments in each of the appropriate multi-slab lists. Because T_b and T_r are sorted (which also means that all the multi-slab lists are sorted), we can report all intersections in $O(n + t_i)$ I/Os. Details will appear in the full paper.

4.2 General Line Segment Intersection

In this section we sketch the algorithm for the general line segment intersection problem. Details will appear in the full paper.

The general line segment intersection problem cannot be solved by distribution sweeping as in the red-blue case, because the $\nearrow$ and $\searrow$ relations for sets of intersecting segments are not acyclic, and thus the preprocessing phase to topologically sort the segments cannot be used to establish an ordering for distribution sweeping. However, as we show below, extended external segment trees can be used to establish enough order on the segments to make distribution sweeping possible. The general idea in our algorithm is to build an extended external segment tree on all the segments, and during this process to eliminate *on the fly* any inconsistencies that arise because of intersecting segments. This leads to a solution for the general problem that integrates all the elements of the red-blue algorithm into one algorithm. In this algorithm, intersections are reported both during the construction of an extended external segment tree and during the filtering of endpoints through the structure.

Constructing an extended external segment tree on intersecting segments. In the construction of an extended external segment tree in Section 2.2 we relied on the fact that the segments did not intersect in order to establish an ordering on them. The main contribution of our algorithm is a mechanism for breaking *long segments* into smaller pieces every time we discover an intersection during the construction of the multi-slab lists of a given node. In doing so we manage to construct an extended segment tree with no intersections between long segments stored in the multi-slab lists of the same node.

The main idea in the construction of the extended external segment tree in Section 2.2 was to use an approach based on distribution sweeping. The root node was built by distributing the long segments into the multi-slab lists, which were then sorted individually. The rest of the structure was then built recursively. Now as mentioned we cannot just sort the multi-slab lists in the case of intersecting segments. Instead we sort the lists according to left (or right) segment endpoint. The basic idea in our algorithm is now the following: We initialize one of the external-memory priority queues developed in [4] for each multi-slab list. Segments in these queues are sorted according to the order of the their endpoint on one of the boundaries the queue corresponds to, and the queues are structured so that a delete-min operation returns the topmost segment. First we remove intersection between segments stored in the same multi-slab list by scanning through the lists individually looking for intersecting segments. Every time we detect an intersection we remove one of the segments from the list and break it at the intersection point. This creates two new segments. If either one of these segments are long, we insert it in the priority queue corresponding to the appropriate multi-slab list. We also report the detected intersection. Next we remove intersections between segments in different multi-slab lists, and between the newly produced segments in the priority queues and segments in the multi-slab lists as well as other segments in the queues. To do so we repeatedly look at the top segment in each of the priority queues and the multi-slab lists. If any of these segments intersect, we report the intersection and break one of the segments as before. In the full paper we prove that if none of the top segments intersect, the topmost of them cannot be intersected at all. Therefore we can remove it and store it in a list that eventually becomes the final multi-slab list for the multi-slab in question. When we have processed all segments in this way, we end up with $O(m)$ sorted multi-slab list of non-intersecting segments.

The I/O usage of the construction algorithm can be analyzed similarly to the analyses in Section 2.2. The main difference between the two algorithms is the extra I/Os used to manipulate the priority queues. As the number of operations done on the queues is $O(T)$, the following lemma follows from the $O((\log_m n)/B)$ I/O bound on the insert and delete-min operations proven in [4].

Lemma 10. *An extended external segment tree can be constructed on intersecting segments in $O((n+t)\log_m n)$ I/O operations, where $T = B \cdot t$ is the number of inconsistencies (intersections) removed (reported).*

Filtering queries through the structure. We have constructed an external extended segment tree on the N segments, and in this process we have reported some of the intersections between them. The intersections that we still have to report must be between segments stored in different nodes. Now segments stored as long segments in a node v are small in all nodes on the path from v to the root. Thus if we have in each node a list of the endpoints of the segments stored in nodes on lower levels of the structure, sorted according to the long segment immediately above them, we can report the remaining intersections with algorithms similar to those used in Section 4.1. So in order to report the remaining intersections we preprocess the structure and filter the $O(N+T)$ endpoints of the segments stored in the tree through the structure as we did in Section 2.3. At the same time we report intersections with algorithms similar to those used in Section 4.1. In the full paper we show how this can all be done in $O((n+t)\log_m n)$ I/Os, which combined with Lemma 10 leads to the following:

Theorem 11. *All T intersections between N line segments in the plane can be reported in $O((n+t)\log_m n)$ I/O operations.*

5 Conclusions

We have developed a new technique which is a variant of fractional cascading designed for external memory. We have combined this technique with previously known techniques in order to obtain new efficient external-memory algorithms for a number of problems involving line segments in the plane.

The following two important problems, which are related to those we have discussed in this paper, remain open:

- If given the vertices of a polygon in the order they appear around its perimeter, can we triangulate the polygon in $O(n)$ I/O operations?
- Can we solve the general line segment intersection reporting problem in the optimal $O(n\log_m n + t)$ I/O operations?

References

1. A. Aggarwal and J. S. Vitter. The input/output complexity of sorting and related problems. *Communications of the ACM*, 31(9):1116–1127, Sept. 1988.
2. D. S. Andrews, J. Snoeyink, J. Boritz, T. Chan, G. Denham, J. Harrison, and C. Zhu. Further comparisons of algorithms for geometric intersection problems. In *Proc. 6th Int'l. Symp. on Spatial Data Handling*, 1994.
3. ARC/INFO. *Understanding GIS—the ARC/INFO method*. ARC/INFO, 1993. Rev. 6 for workstations.
4. L. Arge. The buffer tree: A new technique for optimal I/O-algorithms. In *Proc. of 4th Workshop on Algorithms and Data Structures*, 1995.
5. L. Arge, M. Knudsen, and K. Larsen. A general lower bound on the I/O-complexity of comparison-based algorithms. In *Proc. of 3rd Workshop on Algorithms and Data Structures, LNCS 709*, pages 83–94, 1993.
6. T. M. Chan. A simple trapezoid sweep algorithm for reporting red/blue segment intersections. In *Proc. 6th Can. Conf. Comp. Geom.*, 1994.
7. B. Chazelle. Triangulating a simple polygon in linear time. In *Proc. IEEE Foundation of Comp. Sci.*, 1990.
8. B. Chazelle and H. Edelsbrunner. An optimal algorithm for intersecting line segments in the plane. *JACM*, 39:1–54, 1992.

9. B. Chazelle, H. Edelsbrunner, L. J. Guibas, and M. Sharir. Algorithms for bichromatic line-segment problems and polyhedral terrains. *Algorithmica*, 11:116–132, 1994.

10. B. Chazelle and L. J. Guibas. Fractional cascading: I. a data structuring technique. *Algorithmica*, 1:133–162, 1986.

11. Y.-J. Chiang. Experiments on the practical I/O efficiency of geometric algorithms: Distribution sweep vs. plane sweep. In *Proc of 4th Workshop on Algorithms and Data Structures*, 1995.

12. Y.-J. Chiang, M. T. Goodrich, E. F. Grove, R. Tamassia, D. E. Vengroff, and J. S. Vitter. External-memory graph algorithms. In *Proc. ACM-SIAM Symp. on Discrete Alg.*, pages 139–149, Jan. 1995.

13. R. F. Cromp. An intellegent information fusion system for handling the archiving and querying of terabyte-sized spatial databases. In *S. R. Tate ed., Report on the Workshop on Data and Image Compression Needs and Uses in the Scientific Community, CESDIS Technical Report Series, TR-93-99*, pages 75–84, 1993.

14. H. Edelsbrunner and M. H. Overmars. Batched dynamic solutions to decomposable searching problems. *Journal of Algorithms*, 6:515–542, 1985.

15. A. Fournier and D. Y. Montuno. Triangulating simple polygons and equivalent problems. *ACM Trans. on Graphics*, 3(2):153–174, 1984.

16. M. T. Goodrich, J.-J. Tsay, D. E. Vengroff, and J. S. Vitter. External-memory computational geometry. In *Proc. of IEEE Foundations of Comp. Sci.*, pages 714–723, Nov. 1993.

17. L. M. Haas and W. F. Cody. Exploiting extensible dbms in integrated geographic information systems. In *Proc. of Advances in Spatial Databases, LNCS 525*, 1991.

18. R. Laurini and A. D. Thompson. *Fundamentals of Spatial Information Systems*. A.P.I.C. Series, Academic Press, New York, NY, 1992.

19. H. G. Mairson and J. Stolfi. Reporting and counting intersections between two sets of line segments. In *R. Earnshaw (ed.), Theoretical Foundation of Computer Graphics and CAD, NATO ASI Series, Vol. F40*, pages 307–326, 1988.

20. M. H. Nodine and J. S. Vitter. Large-scale sorting in parallel memories. In *Proc. of 3rd Annual ACM Symp. on Parallel Algorithms and Architectures*, pages 29–39, July 1991.

21. M. H. Nodine and J. S. Vitter. Deterministic distribution sort in shared and distributed memory multiprocessors. In *Proc. 5th ACM Symp. on Parallel Algorithms and Architectures*, pages 120–129, June–July 1993.

22. L. Palazzi and J. Snoeyink. Counting and reporting red/blue segment intersections. In *Proc. of 3th Workshop on Algorithms and Data Structures, LNCS 709*, pages 530–540, 1993.

23. Y. N. Patt. The I/O subsystem—a candidate for improvement. *Guest Editor's Introduction in IEEE Comp.*, 27(3):15–16, 1994.

24. C. Ruemmler and J. Wilkes. An introduction to disk drive modeling. *IEEE Comp.*, 27(3):17–28, Mar. 1994.

25. H. Samet. *Applications of Spatial Data Structures: Computer Graphics, Image Processing, and GIS*. Addison Wesley, MA, 1989.

26. V. K. Vaishnavi and D. Wood. Rectilinear line segment intersection, layered segment trees, and dynamization. *Journal of Algorithms*, 3:160–176, 1982.

27. M. J. van Kreveld. Geographic information systems. Technical Report INF/DOC-95-01, Utrecht University, 1995.

28. J. S. Vitter and E. A. M. Shriver. Algorithms for parallel memory, I: Two-level memories. *Algorithmica*, 12(2–3):110–147, 1994.

Optimized Binary Search and Text Retrieval[1]

Eduardo Fernandes Barbosa[2]   Gonzalo Navarro[3]

Ricardo Baeza-Yates[3]   Chris Perleberg[3]   Nivio Ziviani[2]

Abstract. We present an algorithm that minimizes the expected cost of indirect binary search for data with non-constant access costs, such as disk data. Indirect binary search means that sorted access to the data is obtained through an array of pointers to the raw data. One immediate application of this algorithm is to improve the retrieval performance of disk databases that are indexed using the suffix array model (also called PAT array). We consider the cost model of magnetic and optical disks and the anticipated knowledge of the expected size of the subproblem produced by reading each disk track. This information is used to devise a modified binary searching algorithm to decrease overall retrieval costs. Both an optimal and a practical algorithm are presented, together with analytical and experimental results. For 100 megabytes of text the practical algorithm costs 60% of the standard binary search cost for the magnetic disk and 65% for the optical disk.

KEY-WORDS: Optimized binary search, text retrieval, PAT arrays, suffix arrays, magnetic disks, read-only optical disks, CD-ROM disks.

1 Introduction

To provide efficient information retrieval in large textual databases it is worthwhile to preprocess the database and build an index to decrease search time. One important type of index is the PAT array [Gon87, GBY91] or suffix array [MM90], which is a compact representation of a digital tree called a PAT tree, reducing space requirements by only storing the external nodes of the tree. A PAT tree or suffix tree [Knu73] is a Patricia tree [Mor68] built on the positions of interest in the text database. Each position of interest (or index point) is called a semi-infinite string or suffix, defined by its starting position and extending to the right as far as needed to guarantee uniqueness. In a PAT array the data

[1] The authors wish to acknowledge the financial support from the Brazilian CNPq - Conselho Nacional de Desenvolvimento Científico e Tecnológico, Fondecyt Grant No. 1930765, IBM do Brasil, Programa de Cooperación Científica Chile-Brasil de Fundación Andes, and Project RITOS/CYTED. We also wish to acknowledge the fruitful suggestions from an anonymous referee.
[2] Departamento de Ciência da Computação, Universidade Federal de Minas Gerais, Belo Horizonte, Brazil
[3] Departamento de Ciencias de la Computación, Universidad de Chile, Santiago, Chile

access is provided through an indirect sorted array of pointers to the data. This array allows fast retrieval using an indirect binary search on the text.

When all data elements have constant access time, then binary search minimizes the number of accesses needed to search sorted data and also minimizes the total search time. However, some applications do not have data with constant access costs, including applications that have data distributed across a disk, where data near to the current disk head position costs less to access than data further away. For these applications we show that it is possible to improve the total search cost, which is equivalent to the number of data accesses multiplied by the average access cost.

The full inverted file is the most common type of index used in information retrieval systems. It is composed of a table and a list of occurrences, where an entry in the table consists of a word and a list of addresses in the text corresponding to that word. In general, inverted files need a storage overhead between 30% and 100%, depending on the data structure and the use of stopwords, and the search time is logarithmic. Similar space and time complexity can be achieved by PAT arrays. The great advantage of PAT arrays is its potential use in other kind of searches that are difficult or inefficient over inverted files. That is the case when searching for a long sequence of words, some types of boolean queries, regular expression searching, longest repetitions and most frequent searching [GBY91]. Consequently, PAT arrays should be considered seriously when designing a text searching method for text databases that are not updated frequently.

Due to the high non-constant retrieval costs inherent to the disk technology, a naive implementation of PAT arrays for large textual databases may result in a poor performance. A solution to improve the performance of text retrieval from disk was proposed in [BYBZ94], using two models of hierarchy of indices. They consist of a two-level hierarchy model that uses the main memory and one level of external storage (magnetic or optical devices) and a three-level hierarchy model that uses the main memory and two levels of external storage (magnetic and optical devices). Performance improvement is achieved in both models by storing most of higher index levels in faster memories, so that only a reduced interval of the PAT array is left on the disk, thus decreasing the number of accesses in the slowest device of the hierarchy.

The main goal of this paper is to present an algorithm that improves the expected retrieval time of the indirect binary search in an interval of a PAT array. For both magnetic and optical disks, we are interested in reducing the time complexity of the search in the last level, because optical disks have poor random access performance, more than a magnitude slower than magnetic disks, and magnetic disks are more than a magnitude slower than main memory. As magnetic and optical disks have non-uniform data access times depending on the current head position, our algorithm optimizes the total retrieval time, not the total number of disk accesses. The algorithm takes into account both the expected partition produced by reading each track and the cost of accessing that track.

2 Searching in a PAT Array on Disks

A PAT array is an array of pointers to suffixes, providing sorted accesses to all suffixes of interest in the text. With a PAT array it is possible to obtain all the occurrences of a string prefix in a text in logarithmic time using binary search. The binary search is indirect since it is necessary to access both the index and the text to compare a search key with a suffix. Figure 1 illustrates an example of a text database with nine index points and the corresponding PAT array.

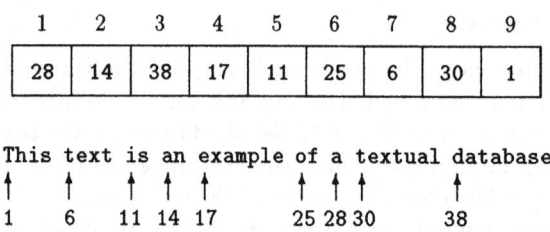

Fig. 1. PAT array or suffix array

To search for the prefix *tex* in the PAT array presented in Figure 1 we perform an indirect binary search over the array and obtain [7, 8] as the result (position 7 of the PAT array points to the suffix beginning at the position 6 in the text, and position 8 of the PAT array points to the suffix beginning at position 30 in the text). The size of the answer in this case is 2, which is the size of the array interval. Searching in a PAT array takes at most $2\,m \log n$ character comparisons and at most $4 \log n$ disk accesses, where n is the number of indexing points and m is the length of a given query. Building a PAT array is similar to sorting variable length records, at a cost of $O(n \log n)$ accesses on average. The extra space used by a PAT array is only one pointer per index point.

When the hierarchy model is used, the last phase of the search is performed by an indirect binary search in which a reduced interval of the PAT array is stored in main memory, as a PAT block with b elements. The hierarchy model [BYBZ94] divides the PAT array into equal-sized blocks with b elements and moves one element of each block to main memory together with a fixed amount of characters from the text related to the element selected from each block. Using this information, a preliminary binary search can be performed directly in the upper level. At the end of this search we know that the desired answer is inside the b elements of the PAT block. This small PAT block is transferred to main memory and the exact answer is found through an indirect binary search between the PAT block in main memory and text suffixes in disk. This way, the cost is no more than $2 \log_2 b$ disk accesses. This is the part of the search that we intend to improve.

The entries of the PAT block are pointers to random positions in the text file stored on disk. Consequently, the sequence of disk positions to be visited during a binary search produces random displacements of the disk access mechanism.

314

Our strategy to reduce the overall binary search time looks for pivots that need little disk head movement and that bisect the PAT block as closely as possible. Each disk head movement produced by accessing a pivot changes the costs of accessing the remaining potential pivots, changing the problem with each search iteration. This problem seems closely related to a binary search over nodes with different access probabilities, for which an optimal binary search tree can be constructed by moving the nodes with higher access probability to the root[Knu73]. A solution for searching in a non-uniform access cost memory is also closely related[AACS87, Kni88]. However, these solutions do not directly apply, since in our problem the costs vary dynamically, depending on the current position of the disk reading mechanism.

The following definitions are used in the presentation of the optimal and the practical algorithms:

1. Let b be the size of the current PAT block (a reduced interval of the PAT array initially transferred from disk to main memory). Note that b is reduced at each iteration in the optimal and the practical algorithms.

2. Let $Cost(h, t, n_s)$ (*cost function*) be the time needed to read n_s sectors from track t, with the reading mechanism being currently on track h. Thus

$$Cost(h, t, n_s) = Seek(|h - t|) + Latency + n_s \times Transfer \qquad (1)$$

Although the costs vary between magnetic and optical disks, in both cases there are three components: *seek time*, *latency time*, and *transfer time*. Seek time is the time needed to move the disk head to the desired disk track and therefore depends on the current head position. Latency time (or rotation time) is the time needed to wait for the desired sector to pass under the disk head. The average latency time is constant for magnetic disks and variable for CD-ROMs, which rotate faster reading inner tracks than when reading outer tracks. In our analysis, we use an average CD-ROM latency. Transfer time is the time needed to transfer the desired data from the disk head to main memory.

We ignore some details in this definition. For example, because we assume a constant latency for each access, reading two sectors in the same cylinder and rotational position (but different surfaces) of a multidisk drive will cost two constant latency times. It is clear that defining the access cost as function of sector is more accurate, however we have chosen to define it as a function of track (seek time plus a constant latency time) for simplicity.

3. Let an *useful sector* be a sector that contains a piece of text with at least one related pointer present in the current PAT block.

4. Let $size(t)$ be the number of useful sectors in a track t.

5. Let $newsize(t)$ (*reduction function*) be the expected number of useful sectors of the next iteration after reading track t.

Assuming that the search for a pointer can occur anywhere in the PAT block with equal probability, then the probability that a given segment is selected for the next iteration is proportional to its size. More formally, suppose the positions of a PAT block of size b are numbered $1..b$, and that track t "owns" positions $p_1, p_2, ...p_k$ of the PAT block. Figure 2 shows an instantiation of a partition of the PAT block containing 8 segments generated by the text keys of a track t. After reading track t we can compare the search key with the text keys and only one of the segments becomes the next subproblem.

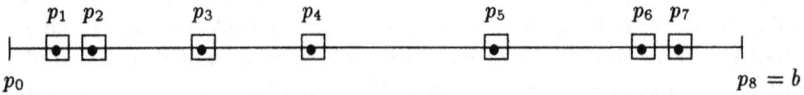

Fig. 2. A PAT block partition generated by all text keys of a track

Thus

$$newsize(t) = \sum_{i=1}^{k+1} (Length\ of\ segment\ i) \times (Prob\ of\ i\ being\ selected)$$

that is

$$newsize(t) = \sum_{i=1}^{k+1} \frac{(p_i - p_{i-1} - 1)^2}{b} \qquad (2)$$

where $p_0 = 0$ and $p_{k+1} = b$.

Assuming that all positions have the same probability, if one only knows that a track owns k positions without knowing which positions, then, the expected size of the next iteration (by counting over all possible values for p_i) is

$$newsize(k) = \frac{(b-k)(2b-k)}{b(k+2)} \approx \frac{2}{k+2}b \qquad (3)$$

where the approximation holds for $k << b$. An important improvement that an optimal algorithm can make is to search for tracks owning two or more PAT block positions, since it would drastically reduce the new size at a negligible additional cost. The optimal and practical algorithms that we present take full advantage of this idea.

6. Let $C(h, 1..b)$ be the cost of the optimal algorithm when the reading mechanism is currently on track h and $1..b$ is the current portion of the PAT block where we are searching.

3 An Optimal Algorithm

First, we present an optimal algorithm, using the definitions above, and show its impracticability. Second, using some simplifying assumptions, a feasible optimal algorithm is presented, still too costly for most uses, but useful for comparative purposes.

From Eq. (1) the track t to be read next is the one that minimizes

$$Cost(h, t, size(t)) + C(t, from..to)$$

where $from..to$ is the selected segment from the comparisons made possible by reading track t.

Thus, the optimal algorithm satisfies the recurrence equation

$$C(h, 1..b) = min_t(Cost(h, t, size(t)) + C(t, from..to)) \qquad (4)$$

Unfortunately, it is not possible to know in advance which segment $from..to$ will be selected after reading t, so it is not possible to recursively compute the C function of the right side of Eq. (4).

However, we can develop an algorithm that optimizes the *expected* cost by replacing C with the *expected* C. In this case, our recurrence is converted to

$$C(h, 1..b) = min_t(Cost(h, t, size(t)) + \sum_{i=1}^{k+1} \frac{(p_i - p_{i-1} - 1)}{b} C(t, p_{i-1} + 1, p_i - 1))$$

It is then possible to take each candidate segment and recursively compute the cost of the algorithm provided this segment is selected and the search starts from t, and then sum up the costs for all segments weighted by the probability for each segment to be selected. With a naive implementation, this strategy requires $O(3^b)$ time. The time can be reduced to $O(b^4)$ (with $O(b^3)$ space) by using dynamic programming. The time cost makes it impractical for use in situations where b is large, since calculations could demand more time than the savings produced by the smart search strategy. We do not include the pseudocode of this algorithm for lack of space.

We develop a simpler heuristic strategy by weakening the definition of *expected* C, interpreting it as if no information about the contents of the PAT block is available (of course we use it for the *Cost* function in Eq. (1), but not for the C of the right side). That means to use a C averaged over all possible PAT array block contents, and to use some weighting strategy to favor those tracks whose neighborhood is "good", in the sense of owning a large number of positions of the current PAT block. More formally,

$$C(h, b) = min_t(Cost(h, t, size(t)) + weight(t) \times C'(newsize(t)))$$

where $C'(x)$ is an average estimate of the cost of the algorithm with a PAT block of size x (see the next section).

In the next section we present a practical algorithm which follows this idea.

4 A Practical Algorithm

To design a practical algorithm from the general heuristic principle stated in the previous section, we need the following: a suitable weighting function, and

317

an estimation of the average cost of the algorithm, which is part of the of the definition of the very same algorithm.

The simplest weighting function is a uniform one, that is $weight(t) = 1$ for all tracks. This is equivalent to not taking into account the neighborhood of tracks, but only their contents and distance from current position. We show that this simple strategy is quite close to optimal, so the effort of making a more complex analysis at each iteration is not worth doing. Figure 3 presents the practical algorithm.

```
Search (bPAT, head)
  while (size of bPAT > 0)
  { compute S = set of useful tracks (which own a position of bPAT)
    compute newsize(s), for each s in S (recall Eq. (2))
    t = s in S which minimizes Cost(head,s,size(s)) + C'(newsize(s))
    move to t and read all useful sectors
    bPAT = appropriate new partition (after search key comparison
                                      with keys read)
    head = t
  }
```

Fig. 3. Practical algorithm

Note that the better candidates for selection are those tracks that either are near to the current head position or generate a good partition of the PAT block. A track near to the current position may avoid an expensive seek cost and, on the other hand, a candidate track that owns many positions increases the probability of making the next partition (*newsize*) much smaller than $b/2$.

The next section presents an approximate analysis of this algorithm for magnetic and optical disks. By using these formulas to estimate the cost of smaller instances of the problem, we are able to complete the definition of the algorithm, thus eliminating its self-reference.

Note that it is possible to apply the practical algorithm until obtaining a PAT block size small enough to be tractable with the optimal strategy that uses the $O(b^4)$ time dynamic programming mentioned in the previous section. It has to be experimentally determined whether this improvement is worth doing for small sizes.

Observe that we can traverse the PAT block from left to right, and keep the set of useful tracks. At the same time we can compute the sum of squares of the segments of the partition that each track produces in the PAT block (recall Eq. (2)), since it determines the average size of the subproblem that that track generates (*newsize*). Note that if the PAT block is traversed from one side to another, it is easy to accumulate the sum of squares, by recording the previous node owned by each track, together with the current sum of squares. This way, both S and *newsize* can be computed in one pass, that is, $O(b)$ time.

In the average case, this algorithm is $O(b)$ in each iteration (note that b

decreases at each step), since at most b tracks may be useful and they may be stored in a hash table to achieve constant search cost (when searching for a track in S). Of course it is $O(b \log b)$ in the worst case. The space requirement is $O(b)$.

In the next section we show that this algorithm makes, on average, less than $\log_{\frac{1}{\omega}}(b+1)$ iterations, where $\frac{1}{2} \le \omega < 1$ is the expected reduction in the size of the PAT block (i.e. the size of the PAT block at iteration i is bw^i). Comparing this with classical binary search, we note that more iterations are required. The total average CPU cost is

$$b \left(\sum_{i=0}^{\log_{\frac{1}{\omega}}(b+1)-1} \omega^i \right) = \frac{b^2}{(b+1)(1-\omega)} \approx \frac{b}{1-\omega}$$

which is linear. The worst case occurs when the search for each candidate takes $\log b$, making each iteration $O(b \log b)$:

$$\sum_{i=0}^{\log_{\frac{1}{\omega}}(b+1)-1} b\omega^i \log_2(b\omega^i) \approx \frac{b}{1-\omega} \log_{\frac{1}{\omega}} b$$

5 Analytical Results

In this section we evaluate the practical algorithm, both for magnetic and optical disks. In each case, an analytical cost model is presented and analytical bounds for the algorithm under this cost model are obtained.

5.1 Magnetic Disks

The following definitions are used in the analysis of the practical algorithm for magnetic disks:

1. Let $C(b)$ be the the retrieval cost for magnetic disks when retrieving from a PAT block with b elements.
2. Let σ be the sum of the latency and transfer time (which really depends on the number of sectors to read) and let θ be the seek time per track.
3. Let T be the number of tracks occupied by the text file and let ΔT be the distance that separates any two disk tracks.
4. Let δ be a small number of central positions in the current PAT block.

A Seek Cost Model for Magnetic Disks The cost function of magnetic disks may be modeled by a function of the form

$$f(\Delta T) = \sigma + \theta \Delta T$$

According to [HP90], typical disks have 500–2000 tracks by surface, each of them divided in 32 sectors. Sectors hold 512–2048 bytes. The typical value for latency is 8.3 ms, while transfer rates vary from 1 to 4MB per second. Average seek times range from 12 ms to 20 ms. Disks have from 1 to 20 plates, that is, 2 to 40 surfaces. The set of tracks from all surfaces which are at the same distance to the center is called a *cylinder*. For practical purposes, one can treat a disk with k surfaces and 32 sectors as if it had only one surface, but whose tracks held $32 \times k$ sectors, with the same latency (8.3 ms). So the following discussion assumes only one surface, although the number of surfaces must be taken into account when calculating the number of tracks (cylinders) required by a file of a specified size.

Average seek time means the sum of all possible head displacements, divided by the number of possible displacements. This is

$$\frac{2}{T^2} \sum_{i=1}^{T} \sum_{j=1}^{i} (i-j) = \frac{T^2 - 1}{3T} \approx \frac{T}{3} \tag{5}$$

From the above discussion, we get the following values (in milliseconds):

$$\sigma = 8.3 + n_s \times (0.125..2.0)$$
$$\theta = 0.018..0.12$$

For our purposes it is better for the file to be contiguously allocated on the disk, to reduce seek time. That also means that it should use as least cylinders as possible, so it should fill cylinders as completely as possible.

In many environments the sectors composing a file may be scattered on the disk. This obviously degrades the performance of any algorithm, although our algorithms are also optimal (in their own sense) under this situation. Another problem is that under different operating system policies, the cost model may vary. For example, some disk administrators do not serve requests that would make the disk head switch to the opposite direction of movement until the last request in the current direction is served. Under this scheme, those tracks that are following the current direction are much cheaper than the others. Both algorithms are able to handle all of these complications provided the cost function is appropriately defined. However, in the analysis we assume contiguous allocation and the simple cost model, which is optimistic if the file is scattered on the disk.

Analysis of the Algorithm It is useful to compare the performance of our practical algorithm with that of the standard binary search. The cost of each binary search step includes one seek, one rotational latency, and one transfer. Since the seek is random, we may use Eq. (5) to show that on average, 1/3 of the disk surface is traversed. The number of steps needed to complete the search is $\log_2(b+1)$, where b is the initial PAT block size. Thus we have

$$Binary\ Search\ Magnetic\ Cost\ (b) = \left(\sigma + \frac{T}{3} \theta \right) \log_2(b+1) \tag{6}$$

Now we turn our attention to our algorithm. Since we are not able to analyze the real algorithm, we use a simplified model, whose predictions are to be experimentally tested against the real algorithm, to show its precision. It is important to note that this model is an *upper bound* for the expected case of the algorithm, so its predictions are always pessimistic.

The model is as follows. Suppose there are no tracks with more than one useful sector (this is worse than reality). At each step, we select the δ central positions of our PAT block, and read the nearest track which owns some of those central positions. The process continues until the PAT block size is $\leq \delta$. At this point, we traverse the disk from one end to the other, in one pass, reading any useful track, until the PAT block becomes empty. Since the real algorithm considers all (useful) tracks and selects the best one taking into account just seek cost and the generated partition, this model can never make a better decision than the real algorithm.

We first obtain the expected size of the new PAT block. This is (by using the same idea of Eq. (2))

$$\frac{1}{\delta b} \sum_{i=\frac{b-\delta}{2}}^{\frac{b+\delta}{2}-1} (i-1)^2 + (b-i)^2 = \frac{b}{2} - 1 + \frac{\delta^2 + 14}{6b} \leq \frac{b}{2} + \frac{\delta}{6} + \frac{4}{3}$$

and the bound is obtained by considering $b \geq \delta$. Note that the bound is of the form $Xb + Y$, with $X = 1/2$ and $Y = \delta/6 + 4/3$.

The next step is to obtain the average seek cost needed to access the nearest of the δ tracks, from a total of T. By summing over all possible positions at this track we can prove that the seek cost is approximately $T/2\delta$.

Thus, a bound of the cost for size b (until obtaining a block of size $\leq \delta$) is

$$C(b) = \sigma + \theta\frac{T}{2\delta} + C(Xb + Y) \qquad (b > \delta)$$

By unfolding the right side of this recurrence, we get its closed expression

$$C(b) = \left(\sigma + \theta\frac{T}{2\delta}\right) \log_2\left(\frac{6}{3\delta - 8}b - \frac{3\delta + 16}{3\delta - 8}\right) + C(\delta)$$

$$\leq \left(\sigma + \theta\frac{T}{2\delta}\right) \log_2\left(\frac{6}{3\delta - 8}b\right) + C(\delta) \qquad (7)$$

The value of $C(\delta)$ corresponds to solve the PAT block of size δ, which consists of linearly traversing the disk surface and reading any useful track. On average, half of the δ tracks are read, and half of the surface is traversed. Thus,

$$C(\delta) = \sigma\frac{\delta}{2} + \theta\frac{T}{2}$$

which gives us the final cost expression

$$C(b) \leq \left(\sigma + \theta\frac{T}{2\delta}\right) \log_2\left(\frac{6}{3\delta - 8}b\right) + \sigma\frac{\delta}{2} + \theta\frac{T}{2}$$

It is possible to prove that the optimal value for δ is:

$$\delta = \sqrt{\frac{\theta T}{\sigma} \log_2 b} + O(1)$$

For example, for a 160 megabytes file, $b = 1000$, $T = 5000$, $\sigma = 10.3$, $\theta = 0.045$, we have $\delta = 15$, and $C(b) = 320$ milliseconds, a 37% of the cost of the standard binary search (Eq. (6) gives 850 milliseconds). The result obtained by numerically finding the optimal δ differs by less than one. For an actual disk with a track capacity of 32 sectors and 4 recording surfaces, simulations for the practical algorithm yield 34% of the cost of the standard algorithm.

5.2 CD-ROM Disks

The following definitions are used in the analysis of the practical algorithm for CD-ROM disks:

1. Let $C(b)$ be the the retrieval cost for CD-ROM disks when retrieving from a PAT block with b elements.
2. Let c be the sum of the latency and transfer time by sector read and let t_s be the seek time.
3. Let T be the number of tracks occupied by the text file. Let ΔT be the distance that separates any two disk tracks.
4. Let Q be the span size, the capability of accessing nearby tracks from the current position with no displacement of the reading mechanism.
5. Let α be the growing rate of the seek time as a function of the displacement of the access mechanism (in tracks) for short seeks.
6. Let β be the growing rate of the seek time as a function of the displacement of the access mechanism (in tracks) for long seeks.

A Seek Cost Model for CD-ROM Disks The cost function of the CD-ROM drive is highly dependent on disk position and the amount of the displacement of the access mechanism. An important feature to be considered is the *span size* capability Q. In actual CD-ROM drives the span size is up to 60 tracks, depending on the type of the drive. The data access located within span boundaries requires a seek time of only 1 millisecond per additional track, while the access of tracks outside the span size may require 200 to 600 milliseconds.

The set of tracks covered by a span in a CD-ROM might be compared to the set of tracks belonging to a cylinder in a magnetic disk. In [BZ92] the set of tracks inside a span is considered as an *optical cylinder*. Thus, the data access in CD-ROM disks has two modes: (i) *proximal access*, for tracks inside the optical cylinder, and (ii) *non-proximal access*, for tracks outside the optical cylinder boundaries. These two modes are also known as *short seeks* and *long seeks*, respectively.

Other components of the access time to a given sector are the rotational latency and the transfer time from disk to main memory. The rotational latency is directly proportional to the position of the data on the disk, due to the constant linear velocity (CLV) physical format, costing from 65 milliseconds (inner track) to 153 milliseconds (outer track) to locate a sector. The transfer time is directly proportional to the amount of data transferred from disk to main memory, at the constant rate of 150 kilobytes per second (300 or 600 kilobytes per second in some drives). Any data in the CD-ROM is accessed by giving the physical address of the corresponding sector, and the sector size is always 2048 bytes. So, the sum of latency and transfer time by sector read is

$$c = (65..153) + 13$$

The seek time may be linearized, considering a slope between 0.02 and 0.04 milliseconds per track for non-proximal accesses and 1 millisecond per track for proximal accesses. The expression for t_s is:

$$t_s = f(\Delta T) = \alpha \times \Delta T \quad for \ \Delta T \leq Q \ (proximal \ access)$$

$$t_s = f(\Delta T) = t_0 + \beta \Delta T \quad for \ \Delta T > Q \ (non - proximal \ access)$$

where t_0 represents the seek time for the first track outside the boundaries of the current optical cylinder, therefore requiring a seek. Some typical values for α, β and Q are: $\alpha \approx 1$ millisecond/track, $0.02 \leq \beta \leq 0.04$ milliseconds/track, $200 \leq t_0 \leq 600$ milliseconds and $1 < Q \leq 60$ tracks.

We have also to consider that the optical head adjusts itself every time a new access is done, centering the anchor point on the track it has just moved to.

Analysis of the Algorithm We begin this section with the analysis of binary search on this cost model. Since the probability for a random track to be within the span size is negligible, we have

$$Binary \ Search \ Optical \ Cost \ (b) = \left(c + t_0 + \frac{T}{3}\beta \right) \ \log_2(b+1) \qquad (8)$$

We use a different model to approximately analyze the behavior of our algorithm on the optical cost model, since the one used for the magnetic case is far from optimal here. The idea is as follows: at any time, if there is a text key within the span size, we read it; else we read the track owning the middle position of the PAT block. Again, this model cannot perform better than our algorithm, on average, since we include both situations in the practical algorithm.

Assuming that we read any sector within the span size, this sector is at random, so using Eq. (3) with $k = 1$, the expected size of the new PAT block after reading that sector is bounded by 2/3 b, while of course the non-proximal access cuts the PAT block by half.

Since the disk head is in the middle of the span size, the expected cost for the proximal access is

$$A = c + \alpha\frac{Q}{4}$$

while for the non-proximal access, since the track to read is at random but surely outside the span, we have the expected cost

$$B \;=\; c + t_0 + \beta \left(\frac{Q}{2} + \frac{T - Q}{3} \right)$$

Finally, the probability for the nearest track to be outside of the span size is

$$\left(1 - \frac{Q}{T} \right)^b \;=\; \rho^b$$

where A, B and ρ are used as shorthands.

Then, the cost expression satisfies the following recurrence

$$C(b) \;=\; (1 - \rho^b) \left(A + C \left(\frac{2}{3} b \right) \right) + \rho^b \left(B + C \left(\frac{1}{2} b \right) \right) \tag{9}$$

with the border condition $C(0) = 0$. Note that $A < B$.

Although this recurrence is hard to solve, it is possible to numerically compute any desired value. In order to provide a deeper insight on the complexity of this algorithm, we first prove a bound for $C(b)$ and then present an approximation, useful to compare the algorithm against the standard binary search. By using induction and bounding summations with integrals, we can prove that:

$$C(b) \;\leq\; A \log_{\frac{3}{2}} b \;+\; B + \left(B - A \log_{\frac{3}{2}} 2 \right) \left(\rho + \log_{\frac{3}{2}} \frac{1}{1 - \rho} \right)$$

In order to be able to compare with binary search, it is mandatory to find a tight value for the constant term, although it may not be a formal bound. The idea is to replace the sum of the costs for all traversed values of b by an integral, thus we have to use a logarithmic scale. By using this technique, we can prove that an approximate solution for this recurrence is:

$$C(b) \;\approx\; \frac{c + \alpha \frac{Q}{4}}{\log_2 3 - 1} \log_2 (b + 1) \;+\; \left(c + t_0 + \beta \frac{T}{3} - \frac{c + \alpha \frac{Q}{4}}{\log_2 3 - 1} \right) \log_2 \frac{1}{1 - \rho}$$

This final formula does give us a good understanding of the performance of the algorithm, and although it is not formally an upper bound, it is tight enough to extract percentages to compare it against the standard binary search. For example, for $b = 1000$, $T = 5000$ (equivalent to approximately 120 megabytes), $c = 125$, $\alpha = 1$, $\beta = 0.03$, $t_0 = 400$ and $Q = 50$, this final approximation gives us $C(b) = 4601$, 80.2% of the cost of the binary search (Eq. (8) gives 5726 milliseconds). By computing directly from the recurrence (9), we get $C(b) = 4491$ (78.4%). Simulations for the practical algorithm yield 70% of the cost of the standard binary search.

6   Experimental Results

We developed a simulation program to perform the actions of the practical algorithm. The simulator maps the text file on the disk sectors and tracks, either magnetic or optical, and computes the exact time needed to access and read any disk position. For a text with n index points and a PAT block with b elements, the simulator generates b random pointers in the range $1..n$. These pointers represent a set of random disk text positions which are stored in a table with b entries. The track number corresponding to each entry is also computed and stored in the table. By definition, all text index points associated to PAT array entries are in lexicographic order. We use this property to associate an integer (in ascending order from left to right) to each PAT block entry as a text representation.

The parameters of interest in the simulation are: the text size (in our experiments we used texts ranging from 1 to 245 Megabytes), the PAT block size, b (usually ranging from 256 to 2048 elements), and the access time function for the disk and reading device, either magnetic or optical. We consider an average word length of $\overline{W} = 6$ characters. Thus, given a text size with M bytes, the number of index points of the corresponding PAT array is given by $n = M/\overline{W}$. We assume that all files are contiguously stored in the disk starting at track 1.

For each iteration of the practical algorithm the simulator scans the current PAT block and sort the tracks in ascending order, so that we can compute the sum of squares as described in Eq. (2). Then, a next track is selected according to the cost minimization criteria of the practical algorithm and a new partition in the current PAT block is obtained, until the search key is found. We run a set of 400 successful random searches for each text and PAT block size, both for optical and magnetic disks. For comparison purposes, the same set of random pointers and search key for each simulation run is used by the practical algorithm and by the standard binary search algorithm.

Table 1 presents the results for magnetic disks and Table 2 presents the results for CD-ROM disks. The values in both tables represent the gain (*Practical-Cost / StandardBinaryCost*) in percentage terms (*StardardBinaryCost = 1*). All values are within 95% confidence interval.

$SectorLength = 512$ bytes; $TrackCapacity = 64$ sectors;
$Surfaces = 8$; $\sigma = 8.3ms$; $\theta = 0.045$; $(BinaryCost = 1.00)$;

PAT Block Size (b)	Text 1.0MB	Text 15.4MB	Text 30.7MB	Text 61.4MB	Text 122.9MB	Text 245.8MB
256	0.22±0.08	0.65±0.14	0.71±0.2	0.67±0.1	0.60±0.14	0.52±0.12
512	0.19±0.06	0.62±0.1	0.68±0.16	0.72±0.1	0.68±0.18	0.56±0.1
1024	0.18±0.04	0.56±0.16	0.65±0.1	0.70±0.18	0.65±0.12	0.55±0.08
2048	0.15±0.04	0.50±0.08	0.59±0.1	0.62±0.1	0.60±0.1	0.54±0.1

Table 1. Performance gain (*PracticalCost/BinaryCost*) for magnetic disks

$$Q = 30; \quad \alpha = 1; \quad t_0 = 300; \quad \beta = 0.03; \quad (BinaryCost = 1.00);$$

PAT Block Size (b)	Text 1.0MB	Text 15.4MB	Text 30.7MB	Text 61.4MB	Text 122.9MB	Text 245.8MB
256	0.48±0.1	0.51±0.1	0.60±0.16	0.72±0.15	0.78±0.14	0.80±0.2
512	0.41±0.14	0.50±0.12	0.58±0.12	0.70±0.2	0.78±0.16	0.78±0.14
1024	0.35±0.1	0.44±0.1	0.55±0.12	0.65±0.18	0.70±0.1	0.76±0.14
2048	0.32±0.08	0.42±0.06	0.50±0.08	0.61±0.12	0.63±0.08	0.70±0.14

Table 2. Performance gain ($PracticalCost/BinaryCost$) for CD-ROM disks

Some comments can be derived from the experimental results, as follows:

1. We observe that this process presents an intrinsically large variance. However, this is not a restriction since the probability of the standard binary search to perform better than the practical algorithm is very small. For example, if the desired answer is exactly at the root of the binary search process, then the standard algorithm is faster than ours. However, considering the values of b we are using, the probability of the answer to be at the root level of a subtree corresponding to a PAT block is very small. We observed that the practical algorithm performs better than the standard binary search in more than 95% of the cases, for all the parameters used in our experiments.

2. The results for magnetic disks have shown a non-monotonic variation of the gain. We found that the disk parameters in the cost function have different weights in the overall retrieval cost, depending on how much the file is spread on the disk tracks. Small files occupy few tracks in the disk and each track owns many positions of the PAT block, which makes the savings on latency larger than the savings on seek costs. Large files, distributed in many tracks on the disk, gives more margin for savings on seek costs. We verified experimentally this conclusion, by cancelling separately the influence of the seek costs and latency costs in the simulator. The gain is monotonic on both file size and b when we consider only the latency cost (with seek cost null), and non-monotonic on file size and constant on b when we consider only seek cost (with latency cost null).

3. The experimental average head displacement, in tracks, using the standard binary search is $0.31T \leq AverageHeadDisp \leq 0.37T$, which matches quite closely the assumption of $T/3$ used in our analysis. The same measure for the practical algorithm presents no significant difference for small files: for instance, a 1 megabyte file has an experimental average head displacement of $0.35T$. However, for large files the practical algorithm beats the standard binary search: for instance, for files ranging from 100 to 245 megabytes we obtained an average head displacement of only $0.1T$. This result confirms that the savings on seek costs have more weight for larger files.

4. Finally, we compared the cost reduction in terms of the time needed to search for a given key. For example, a text file of 30.7 megabytes, stored in a magnetic disk, with a PAT block of 1024 elements, has an average cost of 100 milliseconds using the standard binary search and 60 milliseconds using the practical algorithm. The same file stored in a CD-ROM disk has an average cost of 3.5 seconds using the stardard binary search and 1.8 seconds using the practical algorithm.

References

[AACS87] A. Aggarwal, B. Alpern, K. Chandra and M. Snir. "A Model for Hierarchical Memory", *Proc. of the 19th Annual ACM Symp. of the Theory of Computing*, New York, 1987, 305-314.

[BYBZ94] R. Baeza-Yates, E.F. Barbosa and N. Ziviani. Hierarchies of indices for text searching. In *Proceedings RIAO'94 Intelligent Multimedia Information Retrieval Systems and Management*, pages 11–13. Rockefeller University, New York, Oct. 1994.

[BZ92] E. F. Barbosa and N. Ziviani. Data structures and access methods for read-only optical disks. In R. Baeza-Yates and U. Manber, editors, *Computer Science: Research and Applications*, pages 189–207. Plenum Publishing Corp., 1992.

[GBY91] G. H. Gonnet and R. Baeza-Yates. *Handbook of Algorithms and Data Structures.* Addison-Wesley, 1991.

[Gon87] G. H. Gonnet. PAT *3.1: An Efficient Text Searching System.* User's Manual. Center for the New Oxford English Dictionary. University of Waterloo, Waterloo, Canada, 1987.

[HP90] J. L. Hennesy and D. A. Patterson. *Computer Architecture. A Quantitative Approach.* Morgan Kaufmann Publishers, Inc., 1990.

[Kni88] W.J. Knight. Search in an Ordered Array having Variable Probe Cost. *SIAM J. of Computing* 17 (6), Dec. 1988, 1203-1214.

[Knu73] D.E. Knuth. *The Art of Computer Programming: Sorting and Searching*, volume 3. Addison-Wesley, Reading, Massachusetts, 1973.

[MM90] U. Manber and G. Myers. Suffix Arrays: A new method for on-line string searches. *ACM-SIAM Symposium on Discrete Algorithms*, pages 319–327, Jan. 1990.

[Mor68] D. R. Morrison. PATRICIA - Practical Algorithm To Retrieve Information Coded in Alphanumeric. *Journal of the ACM*, 15(4):514–534, 1968.

On Using q-Gram Locations
in Approximate String Matching*

Erkki Sutinen and Jorma Tarhio

Department of Computer Science
P.O. Box 26 (Teollisuuskatu 23)
FIN-00014 University of Helsinki, Finland
E-mail: {sutinen,tarhio}@cs.Helsinki.FI

Abstract. Approximate string matching with k differences is considered. Filtration of the text is a widely adopted technique to reduce the text area processed by dynamic programming. A sublinear filtration algorithm is presented. The method is based on the locations of the q-grams in the pattern. Samples of q-grams are drawn from the text at fixed periods, and only if consecutive samples appear in the pattern approximately in the same configuration, the text area is examined with dynamic programming. Practical experiments show that this approach gives better filtration efficiency than an earlier method.

1 Introduction

Background. Efficient solutions for approximate string matching are useful in many areas, such as molecular biology, text databases, and data communications. We will consider the k *differences problem*, a version of the approximate string matching problem. Given integer k and two strings, *text* $T = T[1 \ldots n]$ and *pattern* $P = P[1 \ldots m]$ over some alphabet Σ of size c, the task is to find (the end points of) all approximate occurrences of P in T. An approximate occurrence signifies substring p of T such that at most k editing operations (insertions, deletions, changes) are needed to convert p to P, in other words *edit distance* $d(P, p)$ is at most k.

There are numerous algorithms proposed for this problem. A natural solution is a modification of dynamic programming. This approach leads to $O(nk)$ algorithms [Ukk85, GaP89]. Because processing of all the text positions with dynamic programming is rather slow, many filtering techniques [Ukk92, TaU93, PeW93, ChL94, ChM94] have been proposed to reduce the text area necessary to examine using dynamic programming. Some of these approaches lead even to algorithms which are sublinear on the average.

It is typical for the k differences problem that none of the solutions is the best for every combination of problem parameters m, k, and c [JTU91]. Especially large k is troublesome for small alphabets. The key factor is $f = \frac{n-n_p}{n}$, the efficiency of the filtration phase, where n_p is the number of text positions left

---

* The work was supported by the Academy of Finland.

for the dynamic programming phase. Good filtration efficiency is crucial for the practical speed of an algorithm. Besides saving checking time, filtration also consumes resources, which should be taken into account in designing efficient filtration techniques.

One way to reduce n_p is to develop necessary conditions for a text area to include an approximate match of the pattern [GrL89, Ukk92, Tak94]. These conditions often deal with q-grams of the pattern, i.e. continuous substrings of length q. The idea is that whenever an approximate match occurs, it has to resemble the original pattern. This resemblance is reflected by the existence of the *same* q-grams both in the pattern and in its approximate match.

Takaoka [Tak94] presents an efficient filtration technique based on sampling. In his method every hth q-gram of the text is drawn as a sample. If a sample appears in the pattern, a neighborhood of the sample is examined using dynamic programming. Takaoka's method is a simplification of the Chang-Lawler algorithm [ChL94].

Sketch of the solution. Besides the condition for the number of common q-grams in the pattern and its approximate match, one may also utilize the fact that the preserved q-grams have to be *approximately at the same locations* both in the pattern and in its approximate match. We will present a new sublinear filtration technique based on a sampling scheme similar to Takaoka's approach and on approximate locations of the q-grams in the pattern.

An approximate location of a q-gram in the pattern is defined by dividing the pattern into blocks using sampling step $h \geq q$. Let P_0 be $(k+2)h$ characters long prefix of the pattern. We cut P_0 into $k+2$ blocks of h positions and extend each block with $k+q-1$ positions to the right. Then two consecutive blocks have an overlap of $k+q-1$ positions.

In the text, we examine every hth q-gram as a sample. Let $d_1, d_2, \ldots$ be the samples. Because $h \geq q$, the samples do not overlap. We will show that a necessary condition for an approximate match is that at least two of the $k+2$ consecutive samples $d_{j-k-1}, \ldots, d_j$ match. In other words, $d_{(j-k-2)+i} \in Q_i$ holds for at least two indices i, $1 \leq i \leq k+2$, where Q_i is the set of the q-grams of the ith block of the pattern. We will also consider a more general case, where we require that at least s of the $k+s$ consecutive samples match.

In Fig. 1 there is an example, where $m = 40, k = 2, q = 3$, and $h = 9$. Samples have been boxed. We have $d_2 \in Q_1, d_3 \in Q_2$, and $d_4 \in Q_3$ and the count of positive samples is three at $d_5 = $ PQS so that there is a potential approximate occurrence of the pattern.

We use the shift-add approach [BaG92] to compute the sum of matches for the $k+2$ consecutive samples. By doing this, we actually reduce the k differences problem to a variation of the k mismatches problem, where each position of the pattern has a set of accepted characters of its own. In our approach, the "transformed" pattern contains $k+2$ positions (i.e. samples) and each of them has a set of accepted q-grams. A similar transformation is applied to single characters in [TaU93].

```
PATTERN = abcdefghijklmnopqrstABCDEFGHIJKLMNOPQRST

BLOCKS =   abcdefghijklm = Q₁
              jklmnopqrstAB = Q₂
                  stABCDEFGHIJK = Q₃
                      HIJKLMNOPQRST = Q₄



SAMPLES


              d₁       d₂       d₃       d₄       d₅
...xxx...yyy...zzz...abcdefghxijklmnopqrstABCDEFGHIJKLMNOPOST ...
       count:     0        0        0        1        3
```

Fig. 1. *Example of sampling.*

In earlier studies, only Holsti and Sutinen [HoS94] use the locations of the q-grams, but they consider only static texts and the details of their method are different.

Results. Let us assume that individual characters in P and T are chosen randomly. We will show that the asymptotic bound for the filtration efficiency of our method is $\Omega(1 - \frac{m+k^2}{c^q})$. The average time complexity is $O(\frac{kn}{m} \log_c m)$ for small values of k when $q = \log_c m$.

We carried out some experiments and compared our method with Takaoka's method which is among the best in practice. The filtration efficiency of our method was much better for a large range of problem parameters. For example, the number of text positions our algorithm processes with dynamic programming is less than $\frac{1}{50}$ of the corresponding number for Takaoka's method in the case of $c = 40$, $m = 40$, and $k = 8$.

Outline. The rest of this paper is organized as follows. In Section 2, we start with our sampling theorems, stating the necessary conditions for an approximate match of the pattern in a text area in terms of occurrences of q-grams. We present our algorithm in Section 3, and analyze it in Section 4. In Section 5 we review our preliminary experiments before giving concluding remarks in Section 6.

2 Sampling Based on q-Grams

Let $k \geq 0$, $q \geq 1$, and $s \geq 1$ be integers. In the text, every hth q-gram is examined as a sample. We call these samples q-*samples*. Distance h between the endpoints of two consecutive q-samples is the *sampling step*. Let $d_1, \ldots, d_{\lfloor n/h \rfloor}$ be the q-samples of the text. Let us assume that d_1 ends at position h.

Let us consider what the maximal value of h could be for $s = 2$. Let $p = T[j_1 \ldots j_2]$ be an approximate match of pattern P, i.e. $d(P,p) \leq k$. Since k

deletions produce the narrowest approximate match such that p is $m-k$ positions wide, p must include at least $m - k - q + 1$ q-grams. We require that p includes $k + 2$ non-overlapping substrings of equal size h. These conditions lead to the following bound:

$$h \leq \frac{m - k - q + 1}{k + 2}.$$

The basis for sampling is in Theorem 1.

Theorem 1. *Let p be a substring of T such that $d(P, p) \leq k$. If*

$$h = \lfloor \frac{m - k - q + 1}{k + s} \rfloor$$

is the sampling step, $h \geq q$, then at least s of the q-samples in p occur in P.

Proof. Let p be $T[i \ldots j]$. Let r be the number of q-samples in p. To estimate the lower bound for r, let us consider the situation where the leftmost q-sample of p starts as right as possible; in this case the leftmost q-sample in p starts at $(i-1) + h$. Since the rightmost possible q-sample in p starts at position $j - q + 1$ and the sampling step is h (see Fig. 2), we get the inequality for the starting position of the rth q-sample in p:

$$(i - 1) + rh > j - q + 1 - h.$$

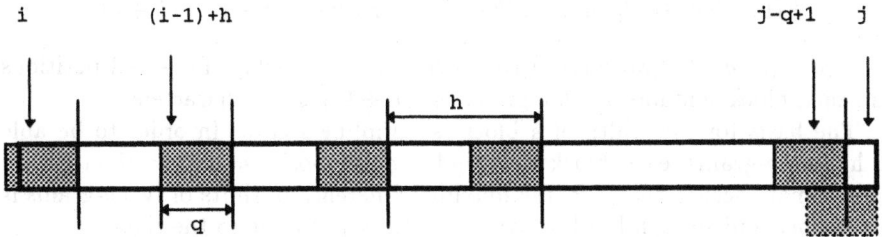

Fig. 2. *Locations of q-samples in $p = T[i \ldots j]$.*

Since $|p| = j - (i - 1)$, this equals

$$(r + 1)h > |p| - q + 1,$$

from which we get

$$r > \frac{|p| - q + 1}{h} - 1.$$

Since $h \leq \frac{m-k-q+1}{k+s}$ and $|p| \geq m - k$, we get

$$r > \frac{|p| - q + 1}{m - k - q + 1}(k + s) - 1 \geq k + s - 1. \tag{1}$$

Let t be the number of the q-samples in p which are q-grams of P. We make an antithesis: $t \leq s - 1$.

According to inequality (1), p includes $r > k + s - 1$ q-samples. Now $r - t$ q-samples in p do not occur in P. Because

$$r - t > k + s - 1 - t \geq k + s - 1 - (s - 1) = k$$

and because the q-samples do not overlap $(h \geq q)$, p includes more than k differences with P. Thus $d(P, p) > k$ holds, which is a contradiction. □

Also Takaoka's method [Tak94] is based on Theorem 1, and a similar idea is also presented by Wu and Manber [WuM92]. In Takaoka's method, if a q-sample occurs anywhere in the pattern, the neighborhood of this sample is checked with dynamic programming. Thus Takaoka considers only the case $s = 1$. Takaoka presents a similar theorem without a proof for a *fixed* position of P.

Our approach is different, because our algorithm utilizes the locations of the q-grams in P. The algorithm examines $k + s$ consecutive q-samples together. Let p be an approximate match of P in T. Let us assume that h has been selected according to Theorem 1 so that there are at least $k + s$ consecutive q-samples in p. It turns out that then at least s of those samples must exist in the pattern, and these samples must have the same relative locations in both the pattern and the text. This requirement is stronger than the condition of Theorem 1.

We select $r = k + s$ fixed blocks from the pattern:

$$P[1 \ldots h + d], P[h + 1 \ldots 2h + d], \ldots, P[(r - 1)h + 1 \ldots rh + d],$$

where $d = k + q - 1$. Two consecutive blocks have an overlap of $k + q - 1$ positions and each block contains $h + k$ q-grams and $h + k + q - 1$ characters.

The basis for the width of a block is sampling step h. In order to be able to handle q-grams, each block is extended $q - 1$ positions to the right. In an approximate occurrence of P, the maximal difference of shifts of two q-grams is k positions and so each block is extended still k positions to the right.

Note that the last $m - rh - d$ q-grams of P do not necessarily occur in any block, when $rh + d < m$. This is an advantage in filtration.

Let Q_i denote the set of the q-grams of the ith block. Our approach is based on the following theorem.

Theorem 2. *Let h be as in Theorem 1. Let $p = T[i \ldots j]$ be an approximate match of P, i.e. $d(P, p) \leq k$. Then for any sequence of $k + s$ consecutive q-samples $d_{b+1}, \ldots, d_{b+k+s}$ included by p, there is integer t such that $d_{b+i+t} \in Q_i$ holds for at least s of the samples.*

Sketch of the proof. Let p include r q-samples. Theorem 1 implies that $r \geq k + s$. Let us consider an arbitrary sequence of $k + s$ consecutive q-samples in p. We know that at least s of the samples occur in P, because otherwise $d(P, p)$ would be greater than k. Let these s samples be $R = \{d_{b_1}, \ldots, d_{b_s}\}$ and let $e_1, \ldots, e_s$ be the end positions of their occurrences in P.

Let us align the pattern with the text according to d_{b_1}. Let $S(i) = e_i - e_1 - (b_i - b_1) * h$ be the shift of d_{b_i} in P. Let i_{min} and i_{max} be the indices of the samples in R with which $S(i)$ get its minimum and maximum, respectively. The definitions imply $S(1) = 0$ and $S(i_{min}) \leq 0 \leq S(i_{max})$.

Clearly it is possible to select R and the corresponding occurrences in P in such a way that $S(i_{max}) - S(i_{min}) \leq k$ holds, because otherwise $d(P, p)$ would be greater than k.

Let us denote the start and end positions of block Q_j by c_j and g_j, respectively. Let d_{b_1} occur in $Q_a = P[c_a \ldots g_a]$. To complete the proof, it is sufficient to show that $c_x \leq e_1 + S(i_{min}) - q + 1$ and $e_1 + S(i_{max}) \leq g_x$ hold for $x = a - 1$, a, or $a + 1$. Then $d_{b_i} \in Q_{x+(b_i-b_1)}$ is clearly satisfied for every $d_{b_i} \in R$, which means that the value of t is $b_1 - b - x$.

There are three cases to consider.

(i) Let us assume that $g_a - e_1 \geq k$ and $e_1 - (c_a + q - 1) \geq k$. Now both $c_a \leq e_1 + S(i_{min}) - q + 1$ and $e_1 + S(i_{max}) \leq g_a$ are clearly satisfied.

(ii) Let us then consider the case $g_a - e_1 < k$. If $g_a - (e_1 + S(i_{min})) \geq k$ holds, both $c_a \leq e_1 + S(i_{min}) - q + 1$ and $e_1 + S(i_{max}) \leq g_a$ are satisfied. If $g_a - (e_1 + S(i_{min})) < k$, then $c_{a+1} \leq e_1 + S(i_{min}) - q + 1$ and $e_1 + S(i_{max}) \leq g_{a+1}$ hold.

(iii) The case $e_1 - (c_a + q - 1) < k$ is symmetric with case (ii). $\square$

The bounds for the location of an approximate match are determined by the following theorem, when we have found enough matching q-samples.

Theorem 3. *Let us assume that s of $k+s$ consecutive q-samples $d_{b+1}, \ldots, d_{b+k+s}$ satisfy $d_{b+i} \in Q_i$ where q-sample d_{b+k+s} ends at text position j. Then an approximate occurrence of the pattern is located in text area*

$$T[j - (k + s)h - 2k - q + 2 \ldots j + m - (k + s - 1)h + k - q].$$

The width of the text area is $m + 3k + h - 1$.

Proof. Let $d_{b+t} = u$ be one of the q-samples satisfying $d_{b+t} \in Q_t$, $1 \leq t \leq k + s$. Let d_{b+t} end at position j'. We set $\Delta = j - j' = (k + s - t)h$.

Let us first study end position j_R of an approximate occurrence of P. We consider the case when j_R reaches its maximum value. This happens when q-gram u occurs at the leftmost possible position in block Q_t, i.e. the end position of u in P is $i_L = (t - 1)h + q$ (see Fig. 3). The length of the suffix of P to the right of i_L is trivially $m_R = m - i_L$.

Since we allow k differences between P and its approximate match in T, the approximate match cannot reach more than $m_R + k$ positions over j'. This means that

$$
\begin{aligned}
j_R &= j' + m_R + k \\
&= j - \Delta + m - i_L + k \\
&= j - (k + s - t)h + m - (t - 1)h - q + k \\
&= j + m - (k + s - 1)h + k - q.
\end{aligned}
$$

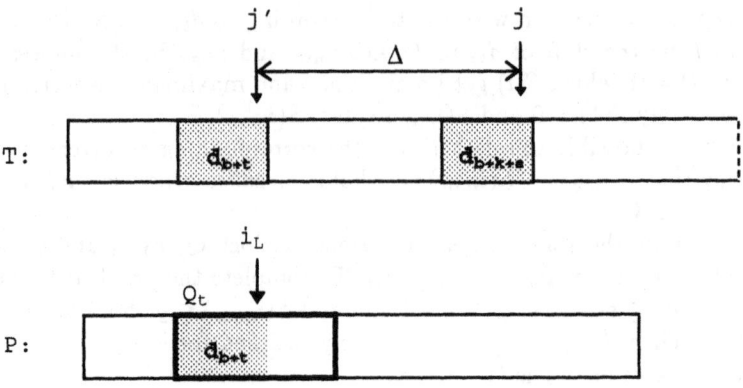

Fig. 3. *Locations of d_{b+t} in T and P.*

To find out the leftmost possible starting position j_L for the approximate match, we examine the case when j_L reaches its minimum. Now, u occurs at the rightmost position inside block Q_t, that is, it ends at position $i_R = th + q - 1 + k$.

For the same reason as in the case above, the approximate match in text T cannot start before

$$
\begin{aligned}
j_L &= j' - (i_R - 1) - k \\
&= j - \Delta - th - q + 1 - k + 1 - k \\
&= j - (k + s - t)h - th - 2k - q + 2 \\
&= j - (k + s)h - 2k - q + 2.
\end{aligned}
$$

The width of the text area is $m + 3k + h - 1$. □

3 Algorithm

We will reduce the k differences problem to a generalized k mismatches problem, where each position of the pattern has a set of accepted characters of its own. We consider q-grams as the alphabet, q-samples as text $T' = d_1 \ldots d_{n'}$, and blocks of the original pattern as pattern $P' = Q_1 \ldots Q_{m'}$, where $n' = \lfloor n/h \rfloor$ and $m' = k + s$. A approximate match of P' with at most k mismatches ends at j, if $T'[j - m' + i] \in P[i]$, that is $d_{j-m'+i} \in Q_i$, holds for at least $m' - k$ indices i, $1 \le i \le m'$.

The transformed problem can be efficiently solved using the shift-add technique [BaG92]. We define bit matrix B as follows: $B[d, j] = 1$, if q-gram d belongs to Q_j, otherwise $B[d, j] = 0$. For each q-gram d, $B[d, *]$ gives the block profile of d.

Array $M[1 \ldots m']$ is used to compute the number of matching q-samples in an alignment of the pattern $T'[i \ldots i + m' - 1]$. An approximate match with at most k mismatches is found when $M[m'] \ge m' - k = s$. Initially, M consists of 0's. Array M is updated at each text position as follows (see also Fig. 4):

$$\textbf{for } j := m' \textbf{ downto } 2 \textbf{ do } M[j] := M[j-1];$$
$$\textbf{for } j := 1 \textbf{ to } m' \textbf{ do } M[j] := M[j] + B[d,j];$$

In practice, the next value of M is evaluated using bit parallel operations. Implementation details are discussed in the end of this section.

SAMPLES

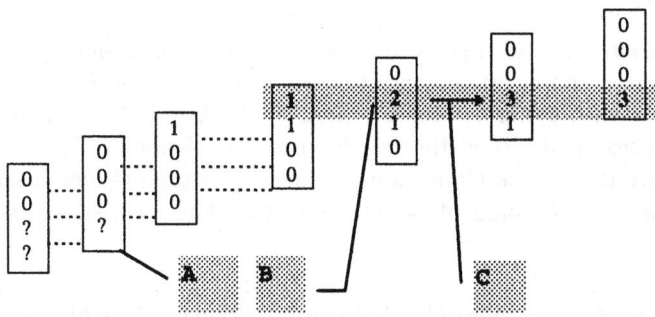

Fig. 4. *Computing the number of matching q-samples: (A) The three first elements of array M are zero, because none of the samples 'xxx', 'yyy', and 'zzz' starts an approximate match. (B) $M[2] = 2$, because samples 'ijk' and 'rst' belong to blocks Q_1 and Q_2. (C) The value of $M[2]$ is shifted to $M[3]$, corresponding to the new phase. Since sample 'GHI' belongs to block Q_3, $M[3] := M[2] + 1$.*

We are now ready to present our algorithm for approximate string matching.

Algorithm A.
1. preprocess P;
2. **for** $i := 1$ **to** m' **do** $M[i] := 0$;
3. **for** $j := h$ **to** n **step** h **do**
4. **begin**
5. $d := T[j - q + 1 \ldots j];$
6.   Shift_add($M, B[d, *]$);
7.   **if** $M[m'] \geq m' - k$ **then**
8. $DP(T[j - m'h - 2k - q + 2 \ldots j + m - (m' - 1)h + k - q], P);$
9. **end**

In Algorithm A, procedure DP searches for approximate matches of pattern P in text area $T[i_1 \ldots i_2]$. This procedure evaluates edit distance matrix

$d[0\ldots(i_2 - i_1 + 1), 0\ldots m]$ using dynamic programming, with initial values $d[i,j] = 0$ for $j = 0$ and $d[i,j] = j$ for $i = 0$:

$$d[i,j] = \min \begin{cases} d[i-1,j-1] + \delta_{T[i_1+i-1]=P[j]} \\ d[i-1,j]+1 \\ d[i,j-1]+1. \end{cases}$$

Above,

$$\delta_{a=b} = \begin{cases} 0 \text{ if a=b} \\ 1 \text{ otherwise.} \end{cases}$$

The necessary and sufficient condition for an approximate match of P ending at text position i is $d[i - i_1 + 1, m] \le k$.

A useful heuristic. Let us assume that Algorithm A has found a potential approximate match ending at text position j. Instead of checking this potential match directly with dynamic programming, we backtrack $(m' - 1)h + \lfloor\frac{h}{2}\rfloor$ positions in the text and restart the search with new q-samples. The restarting is permitted only if $j - j_p$ is large enough, where j_p is the previous backtracking position, otherwise checking phase DP is called. This heuristic works well in practice.

Implementation details. Index Q tells for each q-gram of P the blocks containing that q-gram. The index is constructed during the preprocessing of the pattern.

Let us consider the case $s = 2$. Because the sufficient number of positive q-samples in a text area is then two, only two bits are needed for an element of M. To calculate efficiently the next value for M, we use the shift-add technique. Thus, two bits are also reserved for each block in an element of index Q.

4 Analysis

To analyze the efficiency of Algorithm A, it is essential to estimate its filtration efficiency f_A. The filtration efficiency equals to the matching probability of the mismatch problem.

In the following, we concentrate on the case $s = 2$. We assume that the texts and patterns are generated according to the i.i.d. model, i.e., characters are independently and identically distributed.

We denote by P_c the probability that P' matches, i.e.

$$P_c = Pr(\text{at least two samples match in } T'[j\ldots j + m' - 1]).$$

We define:

$$P_1(i) = Pr(d \in Q_i),$$
$$P_0(i) = Pr(d \notin Q_i) = 1 - P_1(i),$$

where d is a q-gram. Since $P_1(i) = P_1(1)$ and $P_0(i) = P_0(1)$ hold for each $i, 1 \le i \le k+2$, we define

$$P_1 = P_1(1), P_0 = P_0(1).$$

Using these definitions, we can reformulate P_c:

$$P_c = 1 - P_0^{k+2} - (k+2)P_1 P_0^{k+1}, \qquad (2)$$

because the number of configurations with exactly one match is $k+2$. Because a block includes at most $h+k$ different q-grams, we get an upper bound for P_1:

$$P_1 \le \frac{h+k}{c^q}.$$

By applying the formula of h we get:

$$P_1 \le \frac{\frac{m-k-q+1}{k+2} + k}{c^q}$$
$$\le \frac{m-k-q+1+k^2+2k}{kc^q}$$
$$\le \frac{m+k^2+k}{kc^q}.$$

By setting

$$p = 1 - \frac{m+k^2+k}{kc^q}$$

and noticing that $P_0 = 1 - P_1 \ge p$, we get a lower bound for $1 - P_c$:

$$1 - P_c \ge p^{k+2} + (k+2)P_1 p^{k+1} \ge p^{k+2}.$$

Since $1 - P_c$ is the same as filtration efficiency f_A, we have obtained the following estimate for f_A:

Theorem 4. *Filtration efficiency f_A of the Algorithm A for $s = 2$ is*

$$f_A \ge (1 - \frac{m+k^2+k}{kc^q})^{k+2}.$$

The bound of Theorem 4 is rough, and better estimates should be based on formula (2).

Next we estimate the time complexity of our algorithm. Let us consider separately the four major phases of the algorithm:

1. Preprocessing of the pattern creates index Q, implemented as a hash table of size $m - q + 1$. The natural mapping of q-gram d to integer $v(d)$ of base c needs $O(q)$ time using Horner's rule. Hashing $v(d)$ and handling possible collisions can be made in constant time on the average. Since each subsequent q-gram can be processed in $O(1)$ time using its predecessor (cf. the Rabin-Karp algorithm [KaR87]), the total time for hashing all the q-grams of the pattern is $O(m)$. Storing the locations of a q-gram of the pattern involves evaluating $2k/h \le 2k/q$ different blocks, and so the preprocessing time of the whole pattern is $O(mk/q)$.

2. Applying the shift-add approach needs $O(k/w)$ time for a q-sample, where w is the word size. A shift-add operation works in constant time for small values of k.

3. The processing of q-samples consists of two parts. Evaluating all the $\lceil n/h \rceil$ q-samples needs $O(nq/h)$ time, assuming that efficient hashing is applied. Since block profiles of q-samples are shift-added to M, the whole phase consumes $O(\frac{nqk^2}{mw})$ time, where we have used the approximation $h = \Omega(m/k)$, which holds for $k + q \leq \frac{1}{2}m + 1$ and $s = O(k)$. For small k, the time complexity is $O(nqk/m)$.

4. Dynamic programming is applied only to the filtered q-sample locations. Dynamic programming for a text of length n_0 requires $O(n_0 k)$ time. Based on Theorem 3 and Theorem 4, we conclude that the time complexity of this phase is $O(\frac{n}{h}(1 - f_A)(m+h)k) = O(\frac{n}{h}\frac{m+k^2+k}{c^q}(m+h)k)$, where we have used the approximation

$$(1 - \frac{m + k^2 + k}{kc^q})^{k+2} \leq 1 - (k+2)\frac{m + k^2 + k}{kc^q}.$$

The time for the processing of the q-samples dominates over the other phases; therefore, we can collect our results to the following theorem:

Theorem 5. *Let w be the word size in bits. The average time complexity of Algorithm A for $s = 2$ is $O(\frac{k^2 qn}{mw})$ in a general case and $O(\frac{n}{m}qk)$ for $k \leq w/2$.*

When $q = \log_c m$, the time complexity is $O(\frac{k^2 n}{mw} \log_c m)$ or $O(\frac{kn}{m} \log_c m)$. The latter is the same as the average time complexity of the Chang-Lawler algorithm and Takaoka's algorithm.

5   Experimental Results

We have compared the filtration efficiency of our algorithm for $s = 2$ with that of Takaoka's algorithm.

The texts and patterns in the first test were generated according to the i.i.d. model, i.e., the characters are independently and identically distributed. Table 1 shows the results for alphabet size $c = 40$. Alg. A is the basic version of our method and Alg A' is augmented with the backtracking heuristic. The text is 500,000 characters long, pattern length m is 40, and error level k varies from 0 to 12, i.e., from 0% to 30% (where the relative error is defined as k/m). We have counted the number of columns (i.e. the total width of the area) processed in the dynamic programming phase to evaluate the filtration efficiency of the algorithms.

As the results show, all the three algorithms lose their filtration power at error levels over 30%. This is because a higher error level k means a lower sampling step h and, therefore, also smaller q. In particular, our algorithms reach $q = 1$ at error level 30% corresponding to $k = 12$ in Table 1 (Takaoka's algorithm reaches

Table 1. *Processed columns for c = 40, m = 40, and n = 500,000.*

k	Takaoka	Alg. A	Alg. A'
0	440	58	58
2	1,496	65	54
4	2,448	65	56
6	4,056	71	65
8	5,996	83	69
9	191,737	21,356	440
10	261,605	29,729	1,362
11	251,332	59,746	5,052
12	272,392	500,000	500,000

$q = 1$ at $k = 13$). A sharp increase in the number of columns is characteristic for these algorithms, when h and q approach to one.

The optimal error level for these algorithms seems to be at about 0–20%. For these error levels, Takaoka's method evaluates under 1.2% of the columns; our algorithms do not examine more than 0.02% of the text columns. For the error levels 22.5–27.5%, Algorithm A' still preserves its efficiency while evaluating at most 1% of the columns, while Takaoka's method evaluates 38–50% of the columns.

Our preliminary implementations show that the number of evaluated columns is reflected also in the execution time of the algorithms. The difference between Takaoka's algorithm and Algorithm A' is very clear at the error levels of 22.5–27.5%: our algorithm is about four times faster.

Table 2. *Processed columns for an English text, n = 492,459.*

m	k	Takaoka	Alg. A'
4	1	441,591	193,623
8	1	71,650	850
8	2	440,606	367,015
8	3	492,366	488,726
16	1	153,708	716
16	2	120,713	1,455
16	3	174,011	7,751
16	4	333,872	491,958
16	5	492,438	492,427

Tests with other alphabets demonstrate similar behavior. Results from another test with an English text are shown in Table 2.

6 Concluding Remarks

We have presented a sublinear filtration algorithm for approximate string matching with k differences. Our experiments show that the new approach gives considerably better filtration efficiency than Takaoka's algorithm. It is possible to
apply the method also to static texts [SuT95].

The number of positive samples is a parameter of our method. Besides the
experiments reported for $s = 2$, we simulated the behavior of our algorithm for
values $s > 2$. The efficiency of our algorithm grows clearly when s increases. On
the other hand, an increase in s decreases h and q, and the shift-add operation
gets slower. The relationships between these parameters determine the limits of
the applicability of our method.

One might expect that our algorithm for $s = 1$ would be the same as
Takaoka's algorithm, but that is not the case. The speed and the filtration efficiency of these algorithms are almost the same, but their operations are different.

The idea of s positive samples can easily be incorporated with Takaoka's
method. The filtration efficiency of the resulting algorithm lies between that of
Takaoka's algorithm and our algorithm. However, this variation cannot outper-
form our approach, because the order of the q-grams is not taken into account.

Limited backtracking with a phase shift clearly improves the efficiency of our
method. It would be possible even to make several subsequent phase shifts after
a potential match. Every additional shift might improve the filtration, but on
the other hand, consumes more time. Finding an optimal value for the number of
shifts is left for further study. The idea of phases may also be applied in parallel
processing: h processors may be used in such a way that the ith processor starts
at text position i.

References

[BaG92]   R. Baeza-Yates and G. Gonnet: A new approach to text searching. *Communications of ACM* **35**, 10 (1992), 74–82.
[ChL94]   W. Chang and E. Lawler: Sublinear approximate string matching and biological applications. *Algorithmica* **12** (1994), 327–344.
[ChM94] W. Chang and T. Marr: approximate string matching and local similarity.
Combinatorial Pattern Matching, Proceedings of 5th Annual Symposium
(ed. M. Crochemore and D. Gusfield), *Lecture Notes in Computer Science*
807, Springer-Verlag, Berlin, 1994, 259–273.
[GaP89]   Z. Galil and K. Park: An improved algorithm for approximate string matching. *Proceedings of 16th International Colloquium on Automata, Languages
and Programming* (ed. M. Chytil et al.), *Lecture Notes in Computer Science*
372, Springer-Verlag, Berlin, 1989, 394–404.
[HoS94]   N. Holsti and E. Sutinen: Approximate string matching using q-gram
places. *Proc. Seventh Finnish Symposium on Computer Science* (ed. M.
Penttonen), University of Joensuu, 1994, 23–32.
[GrL89]   R. Grossi and F. Luccio: Simple and efficient string matching with k mismatches. *Information Processing Letters* **33** (1989), 113–120.

[JTU91]   P. Jokinen, J. Tarhio, and E. Ukkonen: A comparison of approximate string matching algorithms. Report A-1991-7, Department of Computer Science, University of Helsinki, 1991.

[KaR87]   R. Karp and M. Rabin: Efficient randomized pattern-matching algorithms. *IBM Journal of Research and Development* **31** (1987), 249–260.

[PeW93]   P. Pevzner and M. Waterman: A fast filtration algorithm for substring matching problem. *Combinatorial Pattern Matching, Proceedings of 4th Annual Symposium* (ed. A. Apostolico et al.), *Lecture Notes in Computer Science* **684**, Springer-Verlag, Berlin, 1993, 197–214.

[SuT95]   E. Sutinen and J. Tarhio: Information retrieval based on q-gram locations. In preparation.

[Tak94]   T. Takaoka: Approximate pattern matching with samples. *Proceedings of ISAAC '94, Lecture Notes in Computer Science* **834**, Springer-Verlag, Berlin, 1994, 234–242.

[TaU93]   J. Tarhio and E. Ukkonen: Approximate Boyer-Moore string matching. *SIAM Journal on Computing* **22**, 2 (1993), 243–260.

[Ukk92]   E. Ukkonen: Approximate string matching with q-grams and maximal matches. *Theoretical Computer Science* **92**, 1 (1992), 191–211.

[Ukk85]   E. Ukkonen: Finding approximate patterns in strings. *Journal of Algorithms* **6** (1985), 132–137.

[WuM92]   S. Wu and U. Manber: Fast text searching allowing errors. *Communications of ACM* **35**, 10 (1992), 83–91.

Routing with Bounded Buffers and Hot-Potato Routing in Vertex-Symmetric Networks*

Friedhelm Meyer auf der Heide and Christian Scheideler

Department of Mathematics and Computer Science
and Heinz Nixdorf Institute
University of Paderborn
33095 Paderborn, Germany

Abstract. In this paper we present and analyze on-line routing schemes with contant buffer size and hot-potato routing schemes for vertex-symmetric networks. In particular, we prove that for *any* vertex-symmetric network with n vertices, degree d, and diameter $D = \Omega(\log n)$, a randomly chosen function and any permutation can be routed in time

- $O(\log n \cdot D)$, with high probability (w.h.p.), if constant size buffers are available for each edge,
- $O(\log n \cdot D \log^{1+\epsilon} D)$ for any $\epsilon > 0$, w.h.p., if for each vertex buffers of size 3, independent of the degree of the network, are available.

The schedule for the second result can be converted into a hot-potato routing schedule, if a self-loop is added to each vertex.

E.g., for any bounded degree vertex-symmetric network with self-loops and diameter $O(\log n)$ (among them expanders) we obtain a hot-potato routing protocol that needs time $O(\log^2 n(\log \log n)^{1+\epsilon})$ for any $\epsilon > 0$ to route a randomly chosen function and any permutation, w.h.p..

Our protocols also allow bounds on the space requirements for vertices and packets in the network: we show that $O(D(\log \log D + \log d))$ space suffices for storing routing information in the vertices and $O(\log D)$ space suffices for storing routing information in the packets.

This is the first result about space-efficient routing where both the buffer size and the space for storing routing information is strongly bounded. Previous results are only known about routing protocols that either can reduce the buffer size or the space for storing routing information. For space-efficient hot-potato routing no general results are known.

In order to prove the results above we introduce a new off-line routing protocol for arbitrary networks which is fast even for vertex buffers of size 1. This bound can not be reached by any other non-trivial off-line routing protocol yet.

---

* email: {fmadh,chrsch}@uni-paderborn.de, Fax: +49-5251-603514. Supported in part by DFG-Sonderforschungsbereich 1511 "Massive Parallelität: Algorithmen, Entwurfsmethoden, Anwendungen", by DFG Leipniz Grant Me872/6-1, and by the Esprit Basic Research Action Nr 7141 (ALCOM II)

1   Introduction

Packet routing schedules have extensively been studied for a wide range of networks and routing strategies. Whereas much is known about the runtime under the condition that enough space for storing routing tables and an unbounded buffer size is available (see, e.g., [MV95]), little is known about how space restrictions such as bounded buffers or space bounds for storing routing tables influences the runtime. For hot-potato routing, even without space restrictions little is known so far about the routing time in networks.

Hot-potato routing or deflection routing is a variant of packet routing where the packets are always moving, i.e., they are treated as hot potatos. In each step a processor must send out all packets it received at the beginning of the step. The advantage of hot-potato routing is that packets are not stored between time steps as in the store-and-forward routing model. Thus buffers are not required except for input and output buffers in each vertex. Input buffers contain the packets before they start to move, output buffers contain them after they have reached their destinations. Packets in transit are never stored in buffers. This, together with a space-efficient design of rules how to move the packets forward, keeps the hardware cheap and routing cycles very fast. Because of these reasons hot-potato routing is especially useful for optical networks [AS92, M89, S90], where buffering would involve the packets to be stored in electronic media.

In this paper we present on-line routing schemes with bounded buffer size and hot-potato routing schemes, and analyze the time and space they need on vertex-symmetric networks. These schemes use a new space-efficient routing strategy based on simulations of networks (see [MS95]), and a new off-line routing protocol. This off-line protocol seems to be of independent interest.

1.1   Routing with Bounded Buffer Size

The *routing network* is represented by a connected graph $H = (V, E)$, where $V = [n]$ is the set of all vertices (or processors) and $E \subset V \times V$ is the set of all edges (or links) in H. Each $\{v, w\} \in E$ consists of two links, one in each direction.

Each vertex has one *input buffer* and one *output buffer*. Initially, the packets are stored in the input buffers. If a packet reaches its destination, it is stored in the output buffer. Packets in transit are stored in so-called *transit* buffers. We call a transit buffer *edge buffer* if it is attached to an edge and *vertex buffer* if it is attached to a vertex.

We only consider *oblivious* routing strategies, i.e., a packet with origin u and destination v has to travel along a prescribed *routing path* $p(u, v)$ in H. The set of these paths for all $\binom{n}{2}$ pairs $\{u, v\}$ of vertices in H is called a *path system* and denoted by $\mathcal{P}$. A *shortest path system* contains only paths $p(u, v)$ that are shortest paths from u to v in H. Clearly, any shortest path has to be *vertex-simple*, that is, it uses no vertex more than once.

A *packet* consists of a *source* $u \in V$, a *destination* $v \in V$, *routing information*, and a *message*. The source and destination need $\log n$ bits. Throughout this paper we restrict the routing information to be very small, namely of length at most $O(\log n)$. We assume the messages to have uniform length.

Given a path system $\mathcal{P}$ in H, a *routing protocol* consists of a *contention resolution protocol* and a *routing structure* for all vertices in H.

The contention resolution protocol controls how long incoming packets have to wait in a vertex before they are sent to the next vertex on their paths. It has to ensure that no two packets *collide* during the routing, that is, try to use the same edge at the same time.

The edge along which a packet has to be sent is determined with the help of a *routing structure* stored in v. This is a (static) data structure that, given the destination and the routing information of a packet, enables v to compute the next edge the packet has to use w.r.t. its path prescribed in $\mathcal{P}$, and (maybe) update the packet's routing information. We demand that this access needs constant time, i.e. a constant number of operations on $\log n$-bit words.

Routing is performed in synchronous rounds. In a *round*, each vertex uses its contention resolution protocol to choose one packet to forward. Then it uses its routing structure to compute the outgoing edge the packet has to go. If the buffer on the edge to be used is full, the packet is returned to the buffer it came from, otherwise its routing information is updated and the packet is forwarded.

Clearly, the following parameters greatly influence the time needed to route a function $f : V \rightarrow V$ in H:

- the *dilation* D of $\mathcal{P}$, that is, the length of the longest path in $\mathcal{P}$,
- the *congestion* C, i.e. the maximum number of routing paths $p(u, f(u))$ in $\mathcal{P}$ that pass through the same vertex in H, and
- the *buffer size* B available at edges or vertices to store packets.

In routing schemes with bounded buffers two events may occur that prevent the scheme from terminating: *deadlocks* and *livelocks*. A deadlock appears if a set of packets is not able to move forward any more. Livelocks occur if a set of packets is routed in such a way that they mutually deflect each other in an infinite loop from which they never recover. As we will see, our routing protocols are constructed in a way that deadlocks and livelocks can not appear.

1.2 Hot-Potato Routing

The hot-potato routing model differs from the routing model for bounded buffers only in three aspects.

- Hot-potato routing strategies do not need any transit buffers.
- Each time step, a vertex may receive and send out several packets, at most one packet per edge.
- Since packets can not be buffered, the contention resolution protocol now decides which packets to forward on their routing path, and which to route to other directions. Our protocol will make sure that a packet is never more than one edge away from its routing path.

1.3 Vertex-Symmetric Networks

In this paper we only deal with routing schemes for vertex-symmetric networks. This class is defined as follows.

Definition 1. A graph $H = (V, E)$ is called *vertex-symmetric* if for any pair u, v of vertices in H there exists an automorphism $\varphi : V \rightarrow V$ mapping u to v such that for the graph $H_\varphi = (V, E_\varphi)$ with $E_\varphi = \{\{\varphi(x), \varphi(y)\} \mid \{x, y\} \in E\}$ it holds $H_\varphi = H$.

Vertex-symmetric networks form a very general class and include most of the standard networks such as the d-dimensional torus, the butterfly, the hypercube, etc.. Furthermore, the best expanders that have an explicit construction are all Cayley graphs and therefore vertex-symmetric (see, e.g., [M94]).

The goal of this paper is to find a path system and a routing protocol for any vertex-symmetric network such that a small buffer size and little space suffices for storing routing structures in the vertices and routing information in the packets while still maintaining a fast routing time.

1.4 Previous Results

In the following the term *"with high probability"* (w.h.p.) means "with probability of at least $1 - \frac{1}{n^{-\alpha}}$", where n is the number of vertices and α is any constant.

We start with results on off-line routing. It is not difficult to see that $(C \cdot D + 1)D$ rounds suffice to route f using a hot-potato protocol. This upper bound follows from the fact that a packet may collide along its prescribed path with at most $C \cdot D$ other packets. A simple graph coloring argument yields that $C \cdot D + 1$ routing phases, each taking time D, suffice to guarantee that no two packets collide. In [LMR88] an off-line routing protocol is presented that works for networks with constant size edge buffers. (The required buffer sizes are not explicitly computed, but it seems they have to be fairly large.) They show that $O(C + D)$ time suffices to route all packets. On the other hand, if there is at least one edge that transmits C packets and one packet that traverses $O(D)$ vertices then packet routing takes $\Omega(C + D)$ time, even if arbitrarily large buffers are allowed.

Let us now turn to on-line routing. If no restrictions are imposed on space requirements then, according to [MV95], it holds for arbitrary networks with diameter D that any function f with congestion C can be routed on-line in time $O(D + C + \log n)$, w.h.p.. Their results can be used to prove that, for all vertex-symmetric networks with diameter $D = \Omega(\log n)$ and degree d, a randomly chosen function can be routed in time $O(D)$, w.h.p., if $O(n \cdot D \cdot \log d)$ space is available in each vertex and routing information of length $O(\log n)$ is available in each packet. This result was generalized by [MS95] where it is shown, e.g., that for every $s \in [2, n]$, a random function can be routed in time $O(\log_s n \cdot D)$ if $O(s \cdot D \cdot \log d)$ space is available in each vertex and $O(\log(s \cdot D))$ space is available for storing routing information in each packet. Unfortunately, all on-line protocols mentioned above need edge buffers of size $O(\log n)$, w.h.p..

In [LMRR94] it is shown that for any bounded-degree leveled network with depth L, any set of n packets whose paths have congestion C can be routed in time $O(C + L + \log n)$, using 1 buffer per edge. Leighton and Rao furthermore prove in [LR88] that for any network with flux α, any CRCW PRAM algorithm can be simulated with delay $O(\frac{\log^2 n}{\alpha})$, w.h.p., using a sufficiently large constant number of buffers per edge. This very general result has the drawback that the diameter of a network can be by a factor of $\log n$ lower than $\frac{\log n}{\alpha}$. In this paper, instead, we directly consider the diameter of a network.

Experimental results on simulations of hot-potato routing on various networks are documented in [AS92, GH92, M89]. In several of these papers a probabilistic analysis of simple protocols is presented, but various independence assumptions are made to make the analysis tractable.

Feige and Raghavan [FR92] present a simple deterministic hot-potato routing protocol that routes a random function on the $(n, 2)$-torus in $2n + O(\log n)$ steps, w.h.p..

Furthermore they present a simple deterministic routing protocol that routes a random function on the d-dimensional hypercube in $O(d)$ steps, w.h.p..

In [MW95] a probabilistic hot-potato routing protocol is presented that routes a random function on the (n, d)-torus in $dn + O(d^3 \log n)$ steps, w.h.p., if $d = O(n^\epsilon)$ with $0 < \epsilon < \frac{1}{2}$.

Apart from the Butterfly network, the torus, and the hypercube no other non-trivial results for hot-potato routing on vertex-symmetric networks are known so far.

1.5 New Results

Our main result is an upper bound for the routing time and space requirement for various space-efficient routing schemes on vertex-symmetric networks. In particular, we prove:

Main Theorem: *Let $H = (V, E)$ be any connected vertex-symmetric network with n vertices, degree d, and diameter $D = \Omega(\log n)$. Then a randomly chosen function and any permutation can be routed in time*

- *$O(\log n \cdot D)$, w.h.p., if sufficiently large, constant size vertex buffers and $O(D \log d)$ space per vertex are available,*
- *$O(\log n \cdot D \log^{1+\epsilon} D)$ for any $\epsilon > 0$, w.h.p., if vertex buffers of size 3 and $O(D(\log \log D + \log d))$ space per vertex are available.*

The schedule for the second result can be converted into a hot-potato routing schedule if a self-loop is added to each vertex. For all these schedules, $O(\log D)$ space is sufficient for storing routing information in each packet, w.h.p..

If H has degree two then an efficient hot-potato routing scheme is straightforward, since H must be a ring. So in the following we will only deal with vertex-symmetric networks with degree at least three.

We now give a short summary of the techniques used to derive the Main Theorem. Our approach is 'Routing via Simulation' introduced in [MS95]. For this purpose we will embed a network $G = (V, R)$ in H that is based on a Butterfly network. We will describe it precisely in Section 3. For this network we can prove the following result with a variant of Ranade's protocol (see [R91] or [L92]).

Vertex buffers of size 2 suffice to route a randomly chosen function in G using time $O(\log n)$, w.h.p..

Consider a network $H = (V, E)$. Fix a shortest path system $\mathcal{P}_R^H$ in H which contains shortest paths $p_H(u, v)$ in H only for pairs $(u, v) \in R$. Our routing strategy will simulate routing in G by routing in H:

Suppose, a packet with origin u and destination v travels along the path $p_G(u, v)$ in G. In order to simulate the traversal of an edge $(x, y) \in R$, it chooses the path $p_H(x, y)$.

Our way to implement routing with bounded buffer size in H will be to use off-line routing schemes for simulating one routing step in G. If we are satisfied with larger (still constant) bounds on vertex buffers, we may use the off-line protocol from [LMR88] mentioned in Section 2.

For vertex buffers of very small size or hot-potato protocols, we need a new off-line protocol. We prove the following result which may be of independent interest.

For any set of packets with vertex-simple paths having congestion C and dilation D, there is an off-line protocol for vertex buffers of size one that needs time

$$O((D + C\log(C + D))\log\log(C + D))(\log\log\log(C + D))^{1+\epsilon})$$

for any $\epsilon > 0$ to route all packets.

Let S_n be the set of all permutations on V. For a permutation $\pi \in S_n$ let $\pi \circ R = \{\{\pi(u), \pi(v)\} \mid \{u, v\} \in R\}$. We change the definition of the congestion C in a way that C denotes the maximum number of paths in a shortest path system $\mathcal{P}_{\pi \circ R}^H$ that cross each other at a vertex. Then, according to [MS95], it holds:

Let $H = (V, E)$ be defined as in the Main Theorem and $G = (V, R)$ be defined as in Section 3. Then there is an embedding $\pi \in S_n$ of G into H and a shortest path system $\mathcal{P}_{\pi \circ R}^H$ such that the congestion C is $O(D)$.

Therefore routing a random function f in H needs, w.h.p., at most time

- $O(\log n \cdot D)$ with vertex buffers of sufficiently large, constant size,
- $O(\log n \cdot D \log^{1+\epsilon} D)$ for any $\epsilon > 0$ with vertex buffers of size 3,
- $O(\log n \cdot D \log^{1+\epsilon} D)$ for any $\epsilon > 0$ in hot-potato mode if each vertex has a self-loop.

Using the Valiant-Brebner paradigm, it is not difficult to show that this result leads to protocols for routing arbitrary permutations in H within the same time bounds by sending the packets first to randomly chosen destinations before they are sent to those destinations prescribed by the permutation. Finally, using techniques similar to those in [MS95], we obtain the space bounds mentioned in the Main Theorem.

1.6 Organization of the Paper

The following section contains our new off-line routing protocol. Section 3 uses this protocol to establish a routing scheme with buffer size 3 per vertex and a hot-potato routing scheme for arbitrary networks. Finally, Section 4 contains space bounds for the contention resolution protocol, routing structures, and routing information that hold for all vertex-symmetric networks.

2 Off-line Routing Schemes

In this section we will present two off-line routing schemes. The following theorem follows directly from a result in [LMR88] (see also [LMR94]) by changing their model in a way that we consider vertices instead of edges.

Theorem 2. *For any set of packets with vertex-simple paths having congestion C and dilation D, there is an off-line schedule for vertex buffers of constant size that needs $O(C + D)$ time to route all packets.*

In [LMR88] and [LMR94] the buffer size is not explicitly computed, but it seems that it has to be fairly large, caused by the last of a sequence of refinements of schedules to develop this off-line schedule. In the following we present an off-line routing scheme which only needs one buffer for each vertex.

Theorem 3. *For any set of packets with vertex-simple paths having congestion C and dilation D, there is an off-line schedule for vertex buffers of size 1 that needs time*

$$O((D + C\log(C + D))\log\log(C + D))(\log\log\log(C + D))^{1+\epsilon})$$

for any $\epsilon > 0$ to route all packets.

Proof: Before proceeding, we need to introduce some notation. Let V be the set of all vertices in the network and $\mathcal{P}$ be a path system consisting of all paths the packets use to reach their destinations. Consider a schedule S for routing the packets. Let a path $P \in \mathcal{P}$ of length ℓ be represented as a sequence $((v_1, t_1), \ldots, (v_\ell, t_\ell)) \in (V \times \mathbf{N})^\ell$ of vertices and time steps the vertices of P are reached by the packet using P in S. We define $\mathcal{A}_k^S$ to be the collection of all paths $((v_1', t_1'), \ldots, (v_k', t_k'))$ for which there is a path $((v_1, t_1), \ldots, (v_\ell, t_\ell)) \in \mathcal{P}$ and an i such that $((v_1', t_1'), \ldots, (v_k', t_k')) = ((v_i, t_i), \ldots, (v_{k+i-1}, t_{k+i-1}))$ and $t_i, \ldots, t_{k+i-1}$ are consecutive time steps. Let $\mathcal{F}_k^S$ be the collection of paths $((v_1', t_1'), \ldots, (v_k', t_k'))$ in $\mathcal{A}_k^S$ that furthermore fulfill $t_1' = j \cdot k$ for an integer j. A T-*frame* consists of T consecutive time steps. For a fixed $A \in \mathcal{A}_k^S$ and T-frame F, the *frame congestion* $C_{A,F}^S$ is defined as

$$C_{A,F}^S = \text{number of packets that traverse a vertex in } A \text{ within frame } F \text{ in } S \;.$$

Our strategy for constructing an efficient schedule is to make a succession of refinements to an initial schedule S_0, in which each packet moves at every step until it reaches its destination. The proof uses the Lovász Local Lemma [AES92, p.55] at each refinement step.

Lemma 4 (Lovász Local Lemma). *Let $A_1, \ldots, A_n$ be a set of "bad" events in an arbitrary probability space. Suppose that each event A_i is mutually independent of a set of all the other events A_j but at most b, and that $Pr(A_i) \leq p$. If $ep(b+1) < 1$ then, with probability greater than zero, no bad event occurs.*

During these refinements, we choose suitable vertices for the packets to wait at. These vertices are called *secure* vertices. The secure vertices are selected in such a way that no two secure vertices are direct neighbors on a path of a packet. So if a packet decides to move forward along its path it can do so without violating the restriction of one buffer per vertex or moving waiting packets too far away from their secure vertices, even if all secure vertices are occupied. This can be done by simply exchanging packets if a packet wants to enter a secure vertex with a packet waiting at it and, as soon as the packet moves on, moving the waiting packet back to its secure vertex.

The first step of our refinement is to assign an initial delay to each packet. The delays are chosen from the range $[1, \alpha_1 C]$, where α_1 will be determined later. In the resulting schedule, S_1, a packet that is assigned a delay t waits in its starting vertex for t steps, then moves along its prescribed path without waiting until it enters its destination vertex. The time S_1 needs is at most $D + \alpha_1 C$. Let $\gamma_i = 4^{\log^*(C+D)-(i-1)}$ for any $i \geq 1$, $T_1 = 32\gamma_1 \log(C + D)$, and $T_i = 32\gamma_i \log T_{i-1}$ for any $i > 1$. We use the Lovász Local Lemma to show that if the delays are chosen randomly, independently, and uniformly, then with nonzero probability the frame congestion for any $A \in \mathcal{A}_{T_1}^{S_0}$ for some fixed frame of size $3T_1T_2$ is less than $\frac{T_1}{8\gamma_1}$. Thus, such a set of delays must exist.

To apply the Lovász Local Lemma, we associate a bad event with each path $A \in \mathcal{A}_{T_1}^{S_0}$. The bad event for A is that at least $\frac{T_1}{8\gamma_1}$ packets traverse the vertices in A in some fixed $3T_1T_2$-frame defined later. To show that there is a way of choosing the delays so that no bad event occurs, we need to bound the dependence b among the bad events and the probability p that a bad event occurs.

Calculating the dependence is straightforward. Whether or not a bad event occurs solely depends on the delays assigned to the packets that pass through a path A. Thus, the bad events for paths A and A' are independent unless some packet passes through a vertex in A and a vertex in A'. Clearly, at most $C \cdot T_1$ packets pass through A, and each of these packets passes through at most D other vertices. Since at most C packets pass through each of these vertices, there are at most $C \cdot T_1 \cdot D \cdot C \cdot T_1$ sets $A' \in \mathcal{A}_{T_1}^{S_0}$ that depend on A. Thus the dependence b of the bad events is at most $D(C \cdot T_1)^2$.

It remains to compute the probability that a bad event occurs. Let p be the probability of the bad event corresponding to path A for a fixed $3T_1T_2$-frame defined later. Then

$$p \leq \binom{T_1 \cdot C}{\frac{T_1}{8\gamma_1}} \left(\frac{3T_1T_2}{\alpha_1 C} \right)^{\frac{T_1}{8\gamma}} .$$

This expression is derived as follows for the case that there is no packet that moves through more than one vertex in A.

- There are $\binom{T_1 \cdot C}{\frac{T_1}{8\gamma_1}}$ ways to select $\frac{T_1}{8\gamma_1}$ packets out of at most $T_1 \cdot C$ that move through vertices in A.
- The probability that a packet crosses a vertex within the $3T_1T_2$-frame is at most $\frac{3T_1T_2}{\alpha C}$.

It is not difficult to show that this upper bound for p is also an upper bound for the situation that there are packets that move through several vertices in A. Clearly, by choosing $\alpha_1 \geq 24e\gamma_1 T_1 T_2$ it holds that $ep(b+1) < 1$. So, according to the Lovász Local Lemma, the packets can be given delays in such a way that no bad event occurs. Using these delays we obtain a schedule S_1.

We now want to assign a secure vertex to each path $A \in \mathcal{F}_{T_1 T_2}^{S_1}$ in such a way that there is no $A \in \mathcal{F}_{T_1 T_2}^{S_1}$ for which the distance between secure vertices is smaller than γ_1. With the help of the Lovász Local Lemma we will show that such an assignment of secure vertices indeed exists.

For each $A \in \mathcal{F}_{T_1 T_2}^{S_1}$, decompose A into T_2 disjoint paths $A_1, \ldots, A_{T_2} \in \mathcal{F}_{T_1}^{S_1}$ of length T_1. Each of these paths A_i chooses randomly and independently a candidate for the secure vertex of A. To apply the Lovász Local Lemma, we associate a bad event with each path $A \in \mathcal{F}_{T_1 T_2}^{S_1}$. The bad event for A is that none of the candidates chosen by $A_1, \ldots, A_{T_2}$ fulfills the requirement that all other candidates chosen by paths $A' \in \mathcal{F}_{T_1}^{S_1}$ that intersect A within the $3T_1T_2$-frame chosen for the A_i in A have a distance of at least γ_1 from this candidate.

Whether or not a bad event occurs solely depends on the choices of those paths $A' \in \mathcal{F}_{T_1}^{S_1}$ that intersect A. Thus the dependence b of the bad events is at most $\frac{T_1}{8\gamma_1} \cdot T_1 T_2$.

It remains to calculate the probability that a bad event occurs. Let p be the probability of the bad event corresponding to path A. Then

$$p \leq \left((2\gamma_1 + 1) \cdot \frac{T_1}{8\gamma_1} \cdot \frac{1}{T_1} \right)^{T_2} .$$

This expression is derived as follows. On a fixed path there are at most $2\gamma_1 + 1$ vertices at distance γ_1 from the chosen candidate. For each of these vertices within distance γ_1, at most $\frac{T_1}{8\gamma_1}$ other paths run through them for the fixed $3T_1T_2$-frame. The probability that one of these paths decides to choose this vertex as its candidate is $\frac{1}{T_1}$. Clearly, it holds that $ep(b+1) < 1$. So, according to the Lovász Local Lemma, there exists a secure vertex for every $A \in \mathcal{F}_{T_1 T_2}^{S_1}$. This completes the design of schedule S_1.

The idea behind refining schedule S_1 is to cut the paths the packets use in S_1 in pieces at the secure vertices such that each piece lies within two consecutive $T_1 T_2$-frames starting with time step $iT_1 T_2$ for an integer $i \geq 0$. Let schedule S_1^1 be a part of schedule S_1 consisting of the first two consecutive $T_1 T_2$-frames and all pieces of paths lying within them, as shown in the picture.

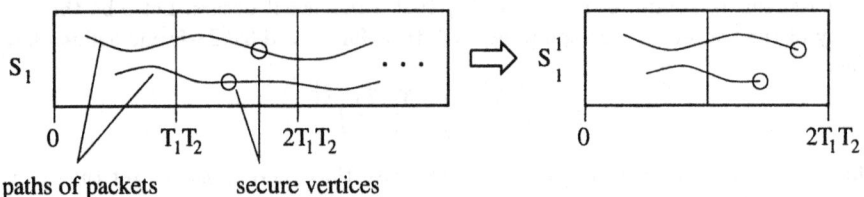

paths of packets   secure vertices

Figure 1: Construction of schedule S_1^1

We refine schedule S_1^1 in the following way. Those packets that have a piece of their path in S_1^1 choose additional delays from the range $[1, \alpha_2 T_1 T_2]$, where α_2 will be determined later. In the resulting schedule S_2^1 a packet that is assigned a delay t_1 in S_1^1 and t_2 in S_2^1 waits in its starting vertex for $t_1 + t_2$ steps, then moves along its prescribed path without waiting until it enters its secure vertex. Thus the time S_2^1 needs is at most $(2 + \alpha_2) T_1 T_2$. We again use the Lovász Local Lemma to show that if the new delays are chosen randomly, independently, and uniformly then, with nonzero probability, the frame congestion for any path $A \in \mathcal{A}_{T_2}^{S_1^1}$ for some fixed frame of size $3T_2 T_3$ is less than $\frac{T_2}{8\gamma_2}$. Thus, such a set of delays must exist.

To apply the Lovász Local Lemma, we associate a bad event with each path $A \in \mathcal{A}_{T_2}^{S_1^1}$. The bad event for A is that at least $\frac{T_2}{8\gamma_2}$ packets traverse the vertices in A in some $3T_2 T_3$-frame. To show that there is a way of choosing the delays so that no bad event occurs, we again need to bound the dependence b among the bad events and the probability p that a bad event occurs.

Whether or not a bad event occurs depends solely on the delays assigned to the packets that pass through a vertex set A. Thus, the bad events for paths A and A' are independent unless some packet passes through a vertex in A and a vertex in A'. Clearly, at most $\frac{T_1}{8\gamma_1}$ packets pass through A, and each of these packets passes through at most $2T_1 T_2$ other vertices within the first two $T_1 T_2$-frames. Since at most $\frac{T_1}{8\gamma_1}$ packets pass through each of these vertices, there are at most $\frac{T_1}{8\gamma_1} \cdot 2T_1 T_2 \cdot \frac{T_1}{8\gamma_1} \cdot T_2$ sets $A' \in \mathcal{A}_{T_2}^{S_1^1}$ that depend on A. Thus the dependence b of the bad events is at most $T_1^3 \cdot T_2^2$.

It remains to compute the probability that a bad event occurs. Let p be the probability of the bad event corresponding to path A for a fixed $3T_2 T_3$-frame. Then

$$p \leq \left(\frac{\frac{T_1}{8\gamma_1}}{\frac{T_2}{8\gamma_2}} \right) \left(\frac{T_2 \cdot 3T_2 T_3}{\alpha_2 T_1 T_2} \right)^{\frac{T_2}{8\gamma_2}} .$$

This expression is derived in a similar way as above. Note that, since each of the $\frac{T_1}{4\gamma_1}$ packets may move through all vertices in A, we get a probability of at most $\frac{T_2 \cdot 3T_2 T_3}{\alpha_2 T_1 T_2}$.

Clearly, by choosing $\alpha_2 \geq \frac{3e}{2}T_3$ it holds that $ep(b+1) < 1$. Thus according to the Lovász Local Lemma, the packets can be given delays in such a way that no bad event occurs. Using these delays we obtain a schedule S_2^1.

We now want to assign a secure vertex to each path $A \in \mathcal{F}_{T_2 T_3}^{S_2^1}$ in such a way that there is no $A \in \mathcal{F}_{T_2 T_3}^{S_2^1}$ for which the distance between secure vertices is smaller than γ_2. With the help of the Lovász Local Lemma we will show that such an assignment of secure vertices indeed exists.

For each $A \in \mathcal{F}_{T_2 T_3}^{S_2^1}$, decompose A into T_3 disjoint paths $A_1, \ldots, A_{T_3} \in \mathcal{F}_{T_2}^{S_2^1}$. Each of these paths A_i chooses randomly and independently a candidate for the secure vertex of A. To apply the Lovász Local Lemma, we associate a bad event with each path $A \in \mathcal{F}_{T_2 T_3}^{S_2^1}$. The bad event for A is that none of the candidates chosen by $A_1, \ldots, A_{T_3}$ fulfills the requirement that all other candidates chosen by paths $A' \in \mathcal{F}_{T_2}^{S_2^1}$ that intersect A within the $3T_2 T_3$ frame chosen for the A_i in A have a distance of at least γ_2 from this candidate.

Whether or not a bad event occurs solely depends on the choices of those paths $A' \in \mathcal{F}_{T_2}^{S_2^1}$ that intersect A. Thus the dependence b of the bad events is at most $\frac{T_2}{8\gamma_2} \cdot T_2 T_3$.

It remains to calculate the probability that a bad event occurs. Let p be the probability of the bad event corresponding to path A. Then

$$ p \leq \left((2\gamma_2 + 1) \cdot \frac{T_2}{8\gamma_2} \cdot \frac{1}{T_2} + \frac{2\gamma_2 + 1}{\gamma_1} \right)^{T_2} . $$

This expression is derived as follows. On a fixed path there are at most $2\gamma_2 + 1$ vertices at distance γ_2 from the chosen candidate. For each of these vertices within distance γ_2, at most $\frac{T_2}{8\gamma_2}$ other paths run through them for the fixed $3T_2 T_3$-frame. The probability that one of these paths decides to choose this vertex as its candidate is $\frac{1}{T_2}$. Furthermore, the probability that a new secure vertex come to close to an old secure vertex is at most $\frac{2\gamma_2 + 1}{\gamma_1}$. Clearly, it holds that $ep(b+1) < 1$. So, according to the Lovász Local Lemma, there exists a secure vertex for every $A \in \mathcal{F}_{T_2 T_3}^{S_2^1}$. This completes the design of schedule S_2 for the first two $T_1 T_2$-frames in S_1.

Applying this calculation to all other consecutive pairs of $T_1 T_2$-frames in schedule S_1 we get a schedule S_2^i for every $2T_1 T_2$-frame starting at time $iT_1 T_2$ for an integer i. In order to build a schedule S_2 for our routing problem out of these schedules S_2^i, we paste S_2^i and S_2^{i+1} together by letting all packets that are active in both schedules wait at their secure vertices w.r.t. schedule S_1. Note that, if the $3T_1 T_2$-frames used for developing schedule S_1 are chosen appropriately, during schedule S_2^i and S_2^{i+1} no vertex is used twice as secure vertex since the two consecutive pairs of $T_1 T_2$-frames, S_2^i and S_2^{i+1} are built of, together cover $3T_1 T_2$ time steps in schedule S_1.

The refinements are continued recursively using a stretch factor of $\alpha_i \geq \frac{3e}{2}T_i$, until $T_i \geq T_{i-1}$, that is, T_i is a constant. At that stage a simple graph coloring argument yields a schedule S_i that needs time

$$ O((D + \alpha_1 C) \cdot (2 + \alpha_2) \cdot \ldots \cdot (2 + \alpha_{\log^*(C+D)})) $$
$$ = O((D + C \log(C+D) \log\log(C+D))(\log\log\log(C+D))^{1+\epsilon}) $$

for any $\epsilon > 0$. □

3 Routing with Bounded Buffers and Hot-Potato Routing in Arbitrary Networks

In this section we will show how efficient routing with bounded buffers and hot-potato routing can be done in arbitrary networks H with n vertices. Our strategy will be to find a suitable 1-1 embedding of the following graph G into H.

Definition 5. For $n \geq 4$ let $d = \max\{i \in \mathbb{N} \mid i \cdot 2^i \leq n\}$. Let $G = (V, R)$ be a network with vertex set

$$V = \{(\ell, x) \mid \ell \in \{0, \ldots, \lfloor n/2^d \rfloor\} \wedge x = (x_{d-1}, \ldots, x_0) \in \{0, 1\}^d \wedge 2^d \cdot \ell + x < n\},$$

and edge set R defined in the following way.

- For each $\ell \in [d]$, every vertex $(\ell, x) \in V$ is connected with (ℓ', x) and $(\ell', (x_{d-1}, \ldots, x_{\ell+1}, 1, x_{\ell-1}, \ldots, x_0))$ where $\ell' = \ell + 1$ if $2^d(\ell+1) + x < n$ and 0 otherwise.
- For each ℓ, $d \leq \ell \leq \lfloor n/2^d \rfloor$, each vertex $(\ell, x) \in V$ is connected with $(\ell+1, x)$ if $2^d(\ell+1) + x < n$ and $(0, x)$ otherwise.

G is chosen in such a way that it consists of a Butterfly network connected back-to-back by linear arrays of length at most $\log n + 2$. A picture will clarify this for $n = 22$.

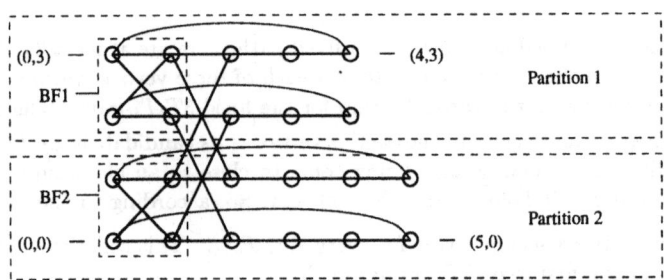

Figure 2: Network G with 22 vertices

Suppose each vertex in G wants to send a packet stored in its input buffer to a random destination in G. We will route these packets in two phases.

- **Phase 1:** Route all packets in Partition 1 to their destinations in two subphases, each using the Ranade protocol. During the first subphase only those packets become active that have their random destinations in Partition 1. The picture below clarifies how the packets have to be routed in this phase.

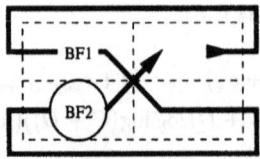

Figure 3: The first subphase

The first time the packets move through BF1 only those edges connecting vertices of different rows are used. In BF2 the packets are routed to the rows of their random destinations. When packets move back from BF2 to Partition 1, only edges connecting vertices of the same row are used afterwards.

During Subphase 2 only those packets become active that have their random destinations in Partition 2. In this case the packets in Partition 1 are first moved to Partition 2 and then routed to their destinations analogous to Subphase 1.

The analysis of Ranade's protocol in [R91] can easily be modified such that each edge in G only needs one buffer to ensure that, w.h.p., after $O(\log n)$ rounds Phase 1 is completed. (Note that the input and output buffers do not count as buffers here.)

– **Phase 2:** Route all packets stored in Partition 2 in a similar way as done for Partition 1 in Phase 1.

So altogether two buffers at each vertex in G suffice to route the packets in G according to a randomly chosen function f with the above scheme in $O(\log n)$ rounds, w.h.p..

Let $G = (V, R)$ be embedded 1-1 into H via a permutation $\pi \in S_n$. Fix a shortest path system $\mathcal{P}^H_{\pi \circ R}$ in H. Suppose now we want to simulate the Ranade protocol on G by H. Then we can simulate each routing step of G on-line by using the off-line protocol for routing packets along all paths in $\mathcal{P}^H_{\pi \circ R}$. So H needs two buffers per vertex for simulating the protocol described above, and one buffer per vertex to be able to use our off-line protocol presented in Theorem 3 for simulating one routing step in G. From this we can conclude the following theorem.

Theorem 6. *Fix a 1-1 embedding of G into H. Let C denote its congestion and D its dilation. Then a random function can be routed in time*

– $O(\log n \cdot (C + D))$, *w.h.p., if vertex buffers of sufficiently large constant size are available;*

– $O(\log n \cdot (D + C) \log^{1+\epsilon}(C + D))$ *for any $\epsilon > 0$, w.h.p., if vertex buffers of size 3 are available.*

If we add a self-loop to each vertex we can easily coordinate the second routing protocol in Theorem 6 in such a way that the $3n$ buffers can be simulated by the edges of H. (Note that each edge consists of two links capable of transporting one packet per time step.) Thus the following result holds.

Corollary 7. *The schedule for the second result in Theorem 6 can be converted into a hot-potato routing schedule, if every vertex in H has degree at least 3 and a self-loop.*

In order to route an arbitrary permutation we subdivide the n packets into α sets of size $\frac{n}{\alpha}$ for sufficiently large α. For each of these sets, the packets are first sent to randomly chosen rows before they are sent to their destinations. Note that the packets can be evenly distributed among the vertices of their randomly chosen rows using the fact that for sufficiently large α it holds w.h.p. that for any row the number of packets that randomly chose it is at most the number of vertices in the row. Thus, using the Valiant-Brebner paradigm, the following corollary holds.

Corollary 8. *The schedules presented in Theorem 6 can be used to obtain a routing protocol that routes any permutation in H in time*

– $O(\log n \cdot (C + D))$, *w.h.p., if vertex buffers of sufficiently large constant size are available;*

– $O(\log n \cdot (D + C) \log^{1+\epsilon}(C + D))$ *for any $\epsilon > 0$, w.h.p., if vertex buffers of size 3 are available.*

4 Space-Efficient Routing Structures and Contention Resolution Protocols

In this section we will present a space-efficient routing structure and contention resolution protocol for our routing schemes. In particular, we prove the following theorem.

Theorem 9. *Let H be defined as in the Main Theorem. Then the routing protocols for Theorem 6 can be implemented in such a way that*

(1) space $O(D \log d)$ for storing routing structures in a vertex,
(2) space $O(\log D)$ for storing routing information in a packet,
(3) space $O(D)$ for the contention resolution protocol if the off-line protocol presented in [LMR88] is used,
(4) space $O(D \log \log D)$ for the contention resolution protocol if the off-line protocol in Theorem 3 is used,

suffice to route a random function f in H, w.h.p..

Proof: With a proof similar to the proof of Theorem 4.1 in [MS95] it can be shown that, for a suitable embedding π of G into H and a suitable path system $\mathcal{P}^H_{\pi \circ R}$ simulating the edges of G, a path system can be constructed for H that needs space $O(D \log d)$ in each vertex, and space $O(\log D)$ for storing routing information in a packet.

In remains to prove the space bounds for the contention resolution protocol. Let T_v denote a table that associates to each packet traversing v a number of steps it has to wait there before moving on.

The off-line routing protocol presented in [LMR88] is constructed in such a way that all packets that traverse v wait there for at most a constant number of time steps. The packet that starts at v may wait for at most $O(D)$ time steps. Thus T_v needs space $O(D)$.

According to the proof of the off-line schedule presented in Theorem 3, a packet may wait at most $O(D \log^{1+\epsilon} D)$ time steps for any $\epsilon > 0$ before it is started. Once a packet is started it may wait at a vertex for at most $O(\log^{1+\epsilon} D)$ time steps for any $\epsilon > 0$. Thus T_v needs space $O(D \log \log D)$. This proves the theorem. □

5 Conclusions

Very recently, we used the techniques described above to obtain a *deterministic* scheme that routes arbitrary permutations within the same time and space bounds as stated in the Main Theorem. This is achieved by an embedding of a variant of the Multibutterfly network (see [U92]) instead of the variant of the Butterfly network described in Definition 5.

6 Acknowledgements

We would like to thank Berthold Vöcking for several helpful comments on an early draft of the paper.

References

[AES92] N. Alon, P. Erdős, J. Spencer. *The Probabilistic Method.* Wiley Interscience Series in Discrete Mathematics and Optimization, John Wiley & Sons, 1992.

[AS92] A. Acampora, S. Shah. Multihop Lightwave Networks: a Comparison of Store-and-Forward and Hot-Potato Routing. *IEEE Transaction on Communications,* pp. 1082-1090, 1992.

[FR92] U. Feige, P. Raghavan. Exact Analysis of Hot-Potato Routing. In *Proc. of the 33rd Symp. on Foundations of Computer Science,* pp. 553-562, 1992.

[GH92] A. Greenberg, B. Hajek. Deflection Routing in Hypercube Networks. *IEEE Transactions on Communications,* pp. 1070-1081, 1992.

[L92] F.T. Leighton. *Introduction to Parallel Algorithms and Architectures: Arrays, Trees, Hypercubes.* Morgan Kaufmann, San Mateo, CA, 1992.

[LMR88] F.T. Leighton, B.M. Maggs, S.B. Rao. Universal Packet Routing Algorithms. In *Proc. of the 29th Ann. Symp. on Foudations of Computer Science,* pp. 256-271, 1988.

[LMRR94] F.T. Leighton, B.M. Maggs, A.G. Ranade, S.B. Rao. Randomized Routing and Sorting on Fixed-Connection Networks. *Journal of Algorithms* 17, pp. 157-205, 1994.

[LMR94] F.T. Leighton, B.M. Maggs, S.B. Rao. Packet Routing and Job-Shop Scheduling in O(Congestion + Dilation) Steps. *Combinatorica* 14, pp. 167-186, 1994.

[LR88] F.T. Leighton, S.B. Rao. An Approximate Max-Flow Min-Cut Theorem for Uniform Multicommodity Flow Problems with Applications to Approximation Algorithms. In *Proc. of the 29th Ann. Symp. on Foudations of Computer Science,* pp. 422-431, 1988.

[M89] N. Maxemchuk. Comparison of Deflection and Store-and-Forward Techniques in the Manhattan Street and Shuffle-Exchange Networks. In *Proc. of the IEEE INFOCOM,* pp. 800-809, 1989.

[M94] M. Morgenstern. Existence and Explicit Constructions of $q + 1$ Regular Ramanujan Graphs for Every Prime Power q. *Journal of Comb. Theory, Series B* 62, pp. 44-62, 1994.

[MS95] F. Meyer auf der Heide, C. Scheideler. Space-Efficient Routing in Vertex-Symmetric Networks. To appear at *SPAA 95.*

[MV95] F. Meyer auf der Heide, B. Vöcking. A Packet Routing Protocol for Arbitrary Networks. In *Proc. of the 12th Symp. on Theoretical Aspects of Computer Science,* pp. 291-302, 1995.

[MW95] F. Meyer auf der Heide, M. Westermann. Hot-Potato Routing on Multi-Dimensional Tori. To appear at WG95 (21st Workshop on Graph-Theoretical Concepts in Computer Science), June 1995.

[R91] A.G. Ranade. How to Emulate Shared Memory. *Journal of Computer and System Sciences* 42, pp. 307-326, 1991.

[S90] T. Szymanski. An Analysis of Hot-Potato Routing in a Fiber Optic Packet Switched Hypercube. In *Proc. of the IEEE INFOCOM,* pp. 918-925, 1990.

[U92] E. Upfal. An O(log N) Deterministic Packet Routing Scheme. *Journal of the ACM* 39, 1992.

Load Balancing for Response Time

Jeffery Westbrook

Department of Computer Science, Yale University, New Haven, CT 06520.

Abstract. A centralized scheduler must assign tasks to servers, processing on-line a sequence of task arrivals and departures. Each task runs for
an unknown length of time, but comes with a weight that measures resource utilization per unit time. The *response time* of a server is the sum
of the weights of the tasks assigned to it. The goal is to minimize the maximum response time, *i.e.*, load, of any server. Previous papers on on-line
load balancing have generally concentrated only on keeping the current
maximum load bounded by some function of the maximum off-line load
ever seen. Our goal is to keep the current maximum load on an on-line
server bounded by a function of the current off-line load. Thus the loads
are not permanently skewed by transient peaks, and the algorithm takes
advantage of reductions in total weight. To achieve this, the scheduler
must occasionally reassign tasks, in an attempt to decrease the maximum
load. We study several variants of load balancing, including identical machines, related machines, restricted assignment tasks, and virtual circuit
routing. In each case, only a limited amount of reassignment is used but
the load is kept substantially lower than possible without reassignment.

1   Introduction

This paper examines on-line or dynamic load balancing problems that arise
in heterogeneous distributed systems containing workstations, I/O devices, and
variable-bandwidth communication channels. A typical instance consists of a
fixed set V of n servers and a set of tasks U to be processed by those servers.
Each task $u \in U$ is made up of a sequence of subtasks, or basic steps, and has a
weight, w_u, which is in some way a measure of the amount of service needed to
perform a basic step. The tasks arrive and depart on-line, that is, arrival time
and the number of subtasks are unknown in advance.

For example, the servers may be communication channels between two sites,
the tasks a video transmission made up of a stream of packets, and the weight
of a task will be the time required to transmit one packet, generally a linear
function of the packet size. The total number of packets to be sent is unknown,
however, as is the start time of each task. As another example, the servers may
be distributed database platforms, the tasks application programs performing a
sequence of accesses to the database, and the weights the time for an individual
access. Again, the start time of each application is unknown, as is the total
number of queries per task.

The *load* on a server is the sum of the weights of the tasks assigned to it. The
response time for a task is the time required to perform a basic unit of the task,

such as transmit a single packet or answer a single database query. Assuming that a server timeshares fairly among the tasks assigned to it, the response time is bounded by the load. Response time is important when a task involves an interaction between a human user and the servers, such as audio or video transmission or database queries. Response time is also important when each task is a thread of a parallel computation, comprised of subtasks separated by synchronization barriers. Each task to should get to the synchronization barriers at roughly the same rate. The *load balancing problem* is the problem of keeping the tasks distributed over the servers so that no server has too high a load, *i.e.*, the maximum load is minimized.

More formally, a load-balancing problem consists of a game between an algorithm and an adversary. At the start of round t, there is a set of *active tasks*, $U(t)$. In a round, the adversary may add a new task to $U(t)$, or it may remove a task from $U(t)$. In response, the algorithm must assign each new task to some server or set of servers. It may also reassign tasks among servers. We refer to t as the *time*. An optimal assignment of the active tasks minimizes the maximum load. Let λ_t^* be the maximum load in an optimal assignment of the tasks active at time t. For an on-line load-balancing algorithm A, let $A(t)$ denote the maximum load on any server at time t if A is used to determine assignment of tasks to servers. Note that λ_t^* depends only $U(t)$, whereas $A(t)$ in general depends on the entire history of task arrivals and departures.

We say A is c-competitive against *peak load* if at for all t, $A(t) \leq \max_{t' \leq t} c\lambda_{t'}^*$. That is, the maximum load on an on-line server is bounded by c times the maximum optimal load ever seen. We say A is c-competitive against *current load* if for all t, $A(t) \leq c\lambda_t^*$. That is, the maximum load on an on-line server is no more than c times the maximum load in an optimal assignment of the currently active tasks.

Almost all previous papers on competitive on-line load balancing have dealt only with peak load. In [1, 3, 5], tasks never depart, in which case current and peak load are equivalent. In [2, 4], tasks may depart but the algorithms are competitive against peak load only. In this paper, we examine the design of algorithms that are competitive against current load.

The distinction between peak load and current load is quite significant. The peak load measurement is useful in network design, where it can be used, in conjunction with an estimate of the peak optimum load, to determine how fast a server must be to ensure a given upper bound on response time. The current load measurement is important when the system is actually running. Suppose a large number of video transmissions across a network start simultaneously. Viewers will see jerky images, since each individual frame in the image is slow to arrive through the network. Nothing can be done, since the network is heavily loaded. But suppose that most of transmissions are short and end quickly, leaving only a few long-term transmissions. The algorithm that assigns tasks to channels may have assigned all the long transmissions to the same channel, not knowing in advance which transmissions were long and which were short. Even if the load on this channel is competitive with the prior peak load, it is of little consolation to the viewers, who must continue to watch a jerky transmission even though the short-term jobs have left and there is enough capacity to send the long-term transmissions at a fast rate. The behavior of load-balancing algorithms that are competitive against current load cannot be skewed by an initial peak. The current load measure is also used in the application of load balancing to network flow described in [10]. Naturally, an algorithm that is c-competitive against current load is c-competitive against peak load.

The video transmission example suggests that to be competitive against current load, the on-line scheduler must sometimes reassign tasks to different servers. This should not be done too often, however. Usually, a certain amount of preprocessing is done in assigning a task, such as setting the switches in a virtual network circuit, or building search data structures for database queries. The cost of doing this work is called the *restart cost*. By constantly reassigning tasks, the algorithm can achieve the optimal maximum load, but all the server time will be taken up by restarts. Our results explore the tradeoff between restart cost and load balance.

We consider the following specific load balancing problems. The survey paper by Lawler *et al.* [9] lists many others.

- Identical Machines: Each task u has an associated weight w_u and can be served by any one of the servers. All servers run at the same speed. The load on server v is the sum of the weights of tasks assigned to it.
- Related Machines: The same as the previous problem, except that each server v runs at a different speed, or has a capacity, cap_v. The load on v is the sum of the assigned weights divided by the capacity.
- Restricted Assignment: Each task has a unit weight but can only be served by one of some subset of the machines. The load on a machine is the number of assigned tasks.
- Virtual Circuit Routing: The servers are the edges of an undirected graph. Each edge has an associated capacity cap_e. A task u is a request for a connection between two nodes s_u, t_u. The set of servers assigned to the task must form a path between s_u and t_u. Each task has an associated weight w_u. The load on an edge is the sum of the weights of the connections using that edge divided by its capacity.

For all these problems, an algorithm that never reassigns tasks cannot be better than n-competitive against current load. Each problem contains the identical machines problem as a restriction. In the identical machines problem, an adversary may generate n^2 unit cost tasks, after which some server v must have load at least n. The adversary then deletes all tasks except for those on v. On the other hand, as noted in [4], Graham's [7] list processing heuristic for identical machines, which simply assigns each new task to the least loaded server, and never reassigns a task, is 2-competitive against peak load.

To measure the cost of reassignment, we assume that each task u has an associated restart cost r_u. The restart cost is incurred every time u is assigned or reassigned to some server or connection path. We assume that r_u depends only on u. That is, r_u does not vary with the time of assignment or the servers to which u is assigned. In addition, in all but the last section of this paper we assume that $r_u = c \cdot w_u$ for a fixed constant c. In the final section we give a general method to remove this last restriction, at the price of an increase in the competitive ratio. Dealing with restart costs that depend upon time and server is an intriguing open problem. The algorithms in this paper keep the total cost of reassignments bounded by some small function, usually linear, of the sum $S = \sum_{u \in U} r_u$. Since each task incurs its restart cost at least once, when it is first assigned, any algorithm must incur a total restart cost of at least S.

We first present a simple algorithm for the case of identical machines that is 6-competitive against current load, with total restart cost either $2S$ or $3S$, depending on whether the restart costs are unit or equal to a constant times the weight of a task, respectively. Section 3 gives a general *witness-based* strategy for constructing algorithms competitive against current load, and uses it to derive an $(8 + \epsilon)$-competitive algorithm for related machines, ϵ an arbitrarily small constant, with total restart cost $O(S)$. Previously, Azar *et al.* gave an algorithm for related machines that is 20-competitive against peak load [3], but not competitive against current load. Their algorithm never reassigns a task.

In Section 4 we study the restricted assignment problem, and give a rebalancing scheme that is parameterizable to trade off competitive ratio against restart cost; its best competitive ratio is $O(1)$ at a restart cost of $O(S \log n)$. Previously, Phillips and Westbrook [10] give a preemptive algorithm that is $O((\log n)/\rho)$-competitive against current load while incurring restart cost ρS, where $0 < \rho \leq 1$ is a user-specified parameter. Their algorithm works for arbitrary weights. Azar *et al.*[2] give an eager algorithm for the case of unit weights that is $O(1)$ competitive against peak load, but only if the optimum peak load is $\Omega(\log n)$. This algorithm reassigns each task $O(\log n)$ times, but is no better than n-competitive against current load.

In Section 5 we give an algorithm for virtual circuit routing that is $O(\log n)$-competitive against current load, with restart cost $O(S \log n \log(C/\text{cap}_{min}))$, where C is the sum of edge capacities and cap_{min} is the minimum edge capacity. This algorithm is based on an algorithm of Azar *et al.*[2] that is $O(\log n)$ competitive against peak load, not competitive against current load, and reassigns each task $O(\log n)$ times. Finally, in Section 6 we give a general method to handle restart cost functions of the form $r : U \to R^+$.

2 Identical Machines

In this section we give an algorithm that is 6-competitive against current load. It performs $2S$ reassignments in the case that the restart cost $r_u = 1$ for all $u \in U$, and $3S$ reassignments in the case that $r_u = cw_u$, c a constant, for all $u \in U$. There are several previous algorithms that are competitive against peak load with a ratio $2 - \epsilon$ for a small constant ϵ [6, 7, 8]. None of these reassign tasks, and none are better than n-competitive against current load.

First, consider a particularly simple case: $w_u = \omega$ for all $u \in U$, ω a constant. It is easy to show that $\lambda_t^* = \omega \lceil |U(t)|/n \rceil$. Furthermore, it is easy to maintain on-line an assignment of the active tasks that achieves this optimum load, and incurs total reassignment cost $2S$. When a task arrives, it is placed on a server with minimum load. When a task departs from server v_j, a task is moved from some server with maximum load to server v_j, unless v_j itself has maximum load. The restart cost of the reassignment is charged to the departing task.

Now suppose that the tasks weights are not all equal. At time t, let $w_{\min} = \min_{u \in U(t)} w_u$ and let $w_{\max} = \max_{u \in U(t)} w_u$. Let $a = \lfloor \log w_{\min} \rfloor$ and $b = \lfloor \log w_{\max} \rfloor$. Partition the jobs by size into $b - a + 1$ classes, so that class i contains all jobs $u \in U(t)$ such that $2^i \le w_u < 2^{i+1}$, $a \le i \le b$.

We treat each class independently. Within class i, we treat each job as if it had size 2^{i+1}, and run the unit-weight algorithm. If a new task arrives, it is assigned into the appropriate class, adjusting the partition upwards or downwards as necessary if the new task changes the bound a or b. When a task departs, the appropriate reassignment is done within its class, and the partition is again adjusted if the task departure changes a or b. Let U_t^i denote the set of tasks in class i.

Since class b contains at least one job, $\lambda_t^* \ge 2^b$. Also, $\lambda_t^* \ge \sum_{i=a}^{b} 2^i |U_t^i|/n$; this sum is a lower bound on the average total weight per processor. Conversely, the maximum load on any on-line server is at most the sum of the maximum loads in each class. Hence

$$A(t) \le \sum_{i=a}^{b} 2^{i+1} \lceil |U(t)^i|/n \rceil \le \sum_{i=a}^{b} 2^{i+1}(|U(t)^i|/n+1) \le 2 \left(\sum_{i=a}^{b} 2^i |U(t)^i|/n \right) + 4 \cdot 2^b.$$

These observations yield the following theorem.

Theorem 1. *The algorithm maintains load within 6 times the current optimum. Over all assignments and reassignments, the total cost of reassignments is order $2S$, where $S = \sum_{u \in U} r_u$, if $r_u = 1 \ \forall u \in U$, or $3S$, if $r_u = cw_u \ \forall u \in U$, c a constant.*

3 Related Machines

In the related machines problem, each machine v_j, $1 \le j \le n$, has a particular capacity or throughput, cap_j. Without loss of generality, let $\text{cap}_j \ge \text{cap}_{j+1}$ for all $1 \le j < n$. Let $W_j(t)$ denote the sum of the weights of tasks assigned to server v_j at time t. The load on server j at time t is given by $W_j(t)/\text{cap}_j$. An algorithm for the special case that tasks never depart is given in [1].

3.1 Witness-based partitions.

The paradigm used here and in the remaining sections of this paper is as follows. Assume the existence of a load-balancing algorithm, A, parameterized by λ, with the following properties. When a new task u is added to the active set U_t, A either assigns u so that the maximum load is at most $c(n)\lambda$, or reports (correctly) that in any assignment of $U_t \cup \{u\}$, the maximum load is greater than λ. In the first case A *accepts* u, and in the second case A *rejects* u. Algorithm A may perform rebalancing after task arrivals or departures, but the restart cost incurred will be limited. Algorithm A is called the *one-level* algorithm and is used as a subroutine.

We maintain a partition of the active tasks, $U(t)$, into "levels," $U_i(t)$, $a \le i \le q$, where $q \le \lceil \log \lambda_t^* \rceil$. The lower bound a depends on the particular application, but in general is chosen so that the restart cost can be bounded.

The partition changes as jobs arrive and depart, but will always satisfy the following condition: the load on a server due to jobs in $U_i(t)$ is at most $c(n)\lambda_i$, where $c(n)$ is some function of n and $\lambda_i = 2^i$. A partition that satisfies this conditions is called a $c(n)$-*witness partition*.

Lemma 2. *An on-line algorithm that maintains a $c(n)$-witness partition is $4c(n)$-competitive against current load.*

Proof. The on-line load is at most $\sum_{i=a}^{q} 2^i c(n) \le \sum_{i=-\infty}^{q} 2^i c(n) \le 2^{q+1} c(n) \le c(n) 2^{2+\log \lambda_t^*}$.

Each level in the partition is regarded as a separate instance of the load-balancing problem, and managed using the one-level algorithm. To assign a new task u, we proceed as follows. The value of a is adjusted as necessary. In the related machines problem, we will define a so a can only decrease. Then, starting at level $i = a$, we attempt to put u into level i, using the one-level algorithm. If the one-level algorithm accepts u, we are done. If it rejects u, we increment i by 1 and repeat. Once $i = \lceil \log \lambda_t^* \rceil$, the insertion must be accepted by definition of the one-level algorithm. Finally, we set $q = \max\{q, i\}$.

To delete task u, we remove it from whatever level it belongs to, using the one-level algorithm. This may cause local rebalancing within that level. The value of a is adjusted upward, if necessary. Then a global rebalancing phase occurs. Each task u in level-q is selected in turn, and an attempt is made to insert u into any level $i < q$. If u is accepted by a lower level, then the next task in level q selected and rebalancing continues. If level q is emptied of tasks, then q is decremented and global rebalancing resumes on the new maximum level. If u is rejected by all lower levels, and by level $q - 1$ in particular, then u is left in level q and global rebalancing terminates. By the definition of the one-level algorithm, a rejection implies $\lambda_t^* > 2^{q-1}$, which implies $q \le \lceil \log \lambda_t^* \rceil\}$. Task u is called a *witness for load* 2^q. By using witnesses, the algorithm can avoid compute the exact value of λ_t^* (although it could be computed by brute force simulation, if necessary).

The partition limits the restart cost incurred during global rebalancing in the following way. Suppose task u is reassigned from q down to q'. When u first arrived, it must have been rejected by level q' to end up in q. Since q' now accepts u, it must be that some number of tasks have departed from level q' in the meantime. The restart cost of moving u can be amortized against the restart costs of the departed jobs. The precise details of the amortization argument depend on the application.

3.2 Algorithm ONE-LEVEL.

In this section we give a one-level algorithm for related machines, called ONE-LEVEL. Given a parameter λ, ONE-LEVEL keeps the load bounded by $(1+\beta)\lambda$ where $\beta > 1$ is any constant. If the total weight of all tasks ever accepted is W, then ONE-LEVEL reassigns a total weight of $W(1 + 1/(\beta - 1))$. This scheme is an adaptation of the algorithm of [1].

Recall $W_j(t)$ is the total weight assigned to processor v_j at time t. We also define a quantity $M_j(t)$, which will roughly be the maximum weight ever on processor v_j since it was last rebalanced. For convenience, we will often drop the time parameter, when the time is fixed, and simply refer to W_j and M_j.

Insertion. Let u be a new task, with weight w_u. Let j be the maximum index such that $(W_j + w_u)/\text{cap}_j \leq (1 + \beta)\lambda$. If there is no such j, then reject task u. Otherwise, assign task u to processor j, increase W_j by w_u, and set $M_j = \max\{W_j, M_j\}$.

Deletion. Let u be the job that is departing, say from processor j. Decrease the value of W_j by w_t. Then apply the following rebalancing procedure:

1. Let $\ell = \max\{k \mid \sum_{j=1}^{k} M_j - \beta W_j \geq 0\}$
2. If there is no such ℓ, then stop.
3. Otherwise, for all $j \leq \ell$, set $M_j = 0$, $W_j = 0$. Preempt all jobs assigned to servers $v_1, ..., v_\ell$ and reassign them by re-inserting them one by one, using the above insertion algorithm.

The insertion and deletion routines preserve the following two properties:

Prop. 1. $M_j(t) \geq W_j(t)$ $\forall t, j$.
Prop. 2. $\sum_{j=1}^{k} (M_i - 2W_i) < 0$, for all $1 \leq k \leq n$.

Lemma 3. *Let u be a task assigned to processor v_k. For all $j > k$, $(M_j + w_u)/\text{cap}_j > (1 + \beta)\lambda$.*

Proof. Let t be the current time and $t' < t$ be the time at which ONE-LEVEL assigned u to v_k. Then $M_j(t) \geq M_j(t')$ for all $j > k$. Otherwise, M_j must have been decreased at some time $t'' > t'$. But M_j can only be decreased in a rebalancing operation, at which time all jobs on servers numbered lower than j are reassigned, contradicting the assertion that u was assigned to v_k at time $t' < t''$. Since u was not placed on server v_j, it must be that $(M_j(t') + w_u)/\text{cap}_u \geq (W_j(t') + w_u)/\text{cap}_u > (1 + \beta)\lambda$.

Lemma 4. *Suppose task u is rejected upon attempted insertion at time t. Let λ^* be the maximum load in the optimum assignment of the active tasks, $U(t)$, including u. Then $\lambda^* > \lambda$.*

Proof. Let ℓ be minimal such that $M_\ell/\text{cap}_\ell < \beta\lambda$. If there is no such ℓ, define $\ell = n + 1$.

Suppose $\ell = 1$. Since u is rejected it must be the case that $w_u)/\text{cap}_1 > (1 + \beta)\lambda - (W_1/\text{cap}_1$. Since $\lambda^* \geq w_u/\text{cap}_1$ and $W_1/\text{cap}_1 \leq M_1/\text{cap}_1 < \beta\lambda$, it follows that $\lambda^* > \lambda$.

Suppose $\ell > 1$. For all servers $j < \ell$, $M_j/\text{cap}_j \geq \beta\lambda$. Let X be the set of tasks currently assigned by ONE-LEVEL to servers $v_1, \ldots, v_{\ell-1}$. Given some optimal assignment of tasks to servers, consider where the tasks in X are located.

Suppose that in the optimal assignment, all tasks in X are also assigned to servers $v_1, \ldots, v_{\ell-1}$. Let W_j^* be the weight on v_j in the optimal assignment. We have $\lambda^* \geq W_j^*/\text{cap}_j$ for all $j < n$. Hence $\lambda^* \sum_{i=1}^{\ell-1} \text{cap}_i \geq \sum_{i=1}^{\ell-1} W_i^* \geq \sum_{i=1}^{\ell-1} W_i$. By Prop. 3.2, $\sum_{i=1}^{\ell-1} W_i > \sum_{i=1}^{\ell-1} M_i/\beta \geq \lambda \sum_{i=1}^{\ell-1} \text{cap}_i$. Thus $\lambda^* > \lambda$.

Now suppose that in the optimal assignment there is some task $x \in X$ that is assigned to some server v_k, $k \geq \ell$. Suppose ONE-LEVEL has assigned x to $v_{k'}$, $k' < \ell$. By Lemma 3, $(M_\ell + w_x)/\text{cap}_\ell > (1 + \beta)\lambda$. Since $M_\ell/\text{cap}_\ell < \beta\lambda$, it follows that $\lambda^* \geq w_x/\text{cap}_k \geq w_x/\text{cap}_\ell > \lambda$.

Lemma 5. *Over a series of m task insertions and departures, if the total weight of all tasks is W, the total weight of assigned and reassigned tasks is $(1 + \frac{1}{\beta-1})W$.*

Proof. We perform an amortized analysis. Let $M(t) = \sum_{j=1}^{n} M_j(t)$ and let $\Psi = M(t)/(\beta - 1)$ When task u is assigned to processor v_j, the amortized assigned weight is $w_u(1 + 1/(\beta - 1))$. . When u departs, Ψ does not change and the amortized weight reassigned is zero. Suppose a rebalancing step reassigns all tasks on servers $v_1, \ldots, v_k$. Let $\Delta W = \sum_{j=1}^{k} W_j$ and $\Delta M = \sum_{j=1}^{k} M_j$. The actual weight reassigned is ΔW. The potential initially decreases by $\Delta M/(\beta-1)$ as M_j's are set to zero, and then increases by at most $\Delta W/(\beta - 1)$ as the tasks are reassigned. By definition, $\Delta M - \beta \Delta W \geq 0$, which implies $\Delta W - \Delta M f \leq -(\beta - 1)\Delta W$. The net potential change is $(\Delta W - \Delta M)/(\beta - 1) \leq -\Delta W$. Thus the amortized reassigned weight is zero.

3.3 A witness-based algorithm for related machines.

Algorithm ONE-LEVEL is used to maintain a witness-based partition for related machines. To derive a good amortized bound on the restart cost, several modifications are made to the basic definitions of a witness based strategy.

First, call a task u *new* if it has never been assigned to a processor, otherwise *old*. Algorithm ONE-LEVEL is allowed to accept a new task into level i if it can be placed on some processor v_j without increasing the load beyond $(1 + \beta + \alpha)\lambda_i$, where $\alpha > 0$ is any constant. As before, ONE-LEVEL rejects an old task unless it can be placed without increasing the load beyond $(1 + \beta)\lambda_i$. Recall that the partition has levels $a \leq i \leq q$. The lower bound a is set to $\min_{u \in U(t)} \lfloor \log w_u/((1 + \beta + \alpha)\text{cap}_1) \rfloor$.

Second, in global rebalancing, determine q, the maximum occupied level, and v_j, the minimum numbered processor holding a task in level q. Select any task u'

assigned to j at level q. Attempt to insert u into level $q-1$ using algorithm ONE-LEVEL. If u is rejected, terminate rebalancing. If u is accepted, and reassigned from processor v_j in level q to processor v_k in level $q-1$, then W_{qj} and M_{qj} are decreased by w_u.

Theorem 6. *The witness-based algorithm is $4(1+\beta+\alpha)$-competitive against current load. If the total weight of all tasks is W, then the total weight that is reassigned is $W(1+(\beta^2+\beta+\alpha)/\alpha(\beta-1))$.*

Proof. It is not hard to verify Propositions 3.2 and 3.2 and Lemma 3 remain true. Hence Lemma 4 still holds for the modified algorithm. The competitive ratio follows from Lemma 2. To show the bound on total weight reassigned, we extend the amortized analysis of Lemma 5. Define a new value, $\hat{M}_{ij}$, which will roughly be the maximum total weight on processor v_j due to tasks in level i, since i was last the top level. This value is used only for purposes of analysis, and is not maintained by the algorithm. Upon any assignment of a task u to processor v_j in level i at time t, set $\hat{M}_{ij} = \max\{\hat{M}_{ij}, W_{ij}(t)\}$. During global rebalancing, if task u is reassigned from processor v_j in level q to some processor in level $q-1$, set $\hat{M}_{qj} = \hat{M}_{qj} - w_u$.

Let $M = \sum_{i=0}^{q}\sum_{j=1}^{n} M_{ij}$ and $\hat{M} = \sum_{i=0}^{q}\sum_{j=1}^{n}\hat{M}_{ij}$. Define

$$\Phi_{ij} = \frac{1+\beta}{\alpha}\hat{M} + \frac{1}{\beta-1}\left(1+\frac{1+\beta}{\alpha}\right)M.$$

The amortized weight assigned in inserting new task u is $w_u(1+(\beta^2+\beta+\alpha)/\alpha(\beta-1))$. The amortized weight assigned in a deletion is zero. Consider a ONE-LEVEL rebalancing, with reference to the proof of Lemma 5. First M decreases by some amount ΔM, and then M and $\hat{M}$ increase by at most ΔW, where ΔW is the total weight reassigned in the rebalancing. The change in Φ is therefore

$$\frac{-\Delta M}{\beta-1}\left(1+\frac{1+\beta}{\alpha}\right) + \frac{1+\beta}{\alpha}\Delta W + \frac{\Delta W}{\beta-1}\left(1+\frac{1+\beta}{\alpha}\right).$$

Since $\Delta W - \Delta M \le -(\beta-1)\Delta W$ (see Lemma 5) the net change in potential is $-\Delta W$, which pays for the actual weight reassigned.

Finally, consider a global rebalancing step, in which task u is moved from processor v_j in the top level, q, processor v_k in level $q-1$. The value of M is non-increasing, since M_{qj} decreases by w_u while $M_{(q-1)k}$ increases by at most w_u. The value of $\hat{M}_{qj}$ decreases by w_u, while $\hat{M}_{(q-1)k}$ increases by $w_u - s$, where s, the *slack*, is the difference between $\hat{M}_{(q-1)k}$ and $W_{(q-1)k}$ prior to reassigning u.

We claim that $s > \alpha\lambda_{q-1}\text{cap}_j$, for the following reason. If t is the current time, consider the time $t' < t$ at which u was new. The choice of a guarantees that level $q-1$ existed when u was new. Since u was rejected by level $q-1$, $(W_{(q-1)k}(t')+w_u)/\text{cap}_j > (1+\beta+\alpha)\lambda_{q-1}$. Since level $q-1$ has not been the top level since time t', $\hat{M}_{(q-1)k}$ has not decreased since time t'. Hence $\hat{M}_{(q-1)k} \ge$

$W_{(q-1)k}(t')$. On the other hand, since u can fit on v_j, $(W_{(q-1)k}(t) + w_u)/\text{cap}_j \leq (1 + \beta)\lambda_{q-1}$. Combining these observations yields $s > \alpha\lambda_{q-1}\text{cap}_j$.

We have that $w_u \leq (1+\beta)\lambda_{q-1}\text{cap}_j \leq s\frac{1+\beta}{\alpha}$. Therefore the potential decrease by at least w_u, and the amortized weight reassigned in a global rebalancing step is 0.

Corollary 7. *For any $\epsilon > 0$ there is a load-balancing algorithm for the related machines problem with competitive ratio $8 + \epsilon$ that incurs $O(S)$ startup cost.*

The corollary follows from Theorem 6 by choosing α and β sufficiently small, and using the restriction that $S = cW$ for some constant c.

4 Eager Load Balancing for Restricted Assignment

In the restricted assignment problem, each task u has an associated subset of servers on which it can be executed. It must be assigned to one server in that subset. If no tasks ever depart and tasks cannot be preempted, the best possible competitive ratio for both randomized and deterministic algorithms is $\Omega(\log n)$; this ratio is achievable with a simple greedy strategy [5]. Azar *et al.*[4] showed that when tasks both arrive and depart, no non-preemptive algorithm can be better than $O(\sqrt{n})$ competitive against peak load. An non-preemptive algorithm that is $O(\sqrt{n})$ competitive against peak load is given in [3].

In this section we give an *eager* algorithm for the case of unit weights. It is eager because tasks may be reassigned after both arrival and departure of other tasks. The previous *lazy* algorithms reassigned tasks only in response to other tasks being deleted from the system. In the restricted assignment problem, no lazy algorithm can improve on the $\Omega(\log n)$ bound for permanent tasks without reassignment. The competitive ratio of the eager algorithm is parameterized by a value $1 < q < \log n$, which determines both the ratio and amount of reassignments. It may be as good as $O(1)$-competitive against current load. The algorithm maintains a witness-based partition using the following one-level algorithm parameterized by λ.

Regard the problem as a game on a dynamic bipartite graph. On one side are the servers, V; on the other side are the tasks, U. An edge $\langle u, v \rangle$ indicates that u can be assigned to v. Edge $\langle u, v \rangle$ is *matching* if u is assigned to v. Given $X \subseteq U$, let $Y(X) = \{v \in V \mid \exists \langle u, v \rangle \in E, u \in X\}$. In other words, Y is the set of columns v such that some element in X has an edge to v. By the pigeonhole principle, $\lambda \geq \max_{X \subseteq U} |X|/|Y(X)|$.

Let $load(v)$ denote the load on $v \in V$, *i.e.*, the number of matching edges incident on v. A *balancing path* is an even-length sequence of alternating matched and unmatched edges $\{v_1, u_1\}, \{u_1, v_2\}, \{v_2, u_2\}, \ldots, \{u_{m-1}, v_m\}$ with the property that $load(v_i) < load(v_1)$ for $1 \leq i \leq m - 1$ and $load(v_m) < load(v_1) - \lambda$. A balancing path can be used to reduce the maximum load on $v_1, v_2, \ldots, v_m$ by reassigning u_i to v_{i+1} for $1 \leq i \leq m - 1$. The servers are *r-balanced* if there is no balancing path of length r or less.

Lemma 8. *If the n servers are 2q-balanced, $1 \leq q \leq \ln n$, then the maximum on-line load is $c\lambda$, where c satisfies the inequality*

$$\ln n/q \geq c(\ln c - 1) \qquad (1)$$

Proof. Let h be the maximum j such that there is an on-line server of height at least $j\lambda$. For all $j \geq 0$, let Y_i^j be the set of columns $v \in V$ that are reachable by an alternating path of length at most $2i$, $1 \leq i \leq q$, starting from a column of height at least $j\lambda$. Thus Y_0^j is the set of columns of height at least $j\lambda$. Let $y_i^j = |Y_i^j|$.

No column in Y_i^j has height less than $(j-1)\lambda$, since otherwise there would be a balancing path of length $2i \leq 2q$. Let X_i be the set of items on columns of height at least λi. We have $|X_i| \geq i\lambda y_0^i$.

For any j and $i \leq (q-1)$, the set Y_{i+1}^j contains all servers adjacent to some item that is currently placed on a server in Y_i^j. There are at least $\lambda(j-1)y_i^j$ such items, and hence by the pigeonhole principle

$$y_i^j (j-1)\lambda/y_{i+1}^j \leq \lambda. \qquad (2)$$

We also have (2) $y_0^j \geq y_q^{j+1}$. This follows from the observation that no server of load less than $j\lambda$ is reachable in $2q$ or fewer steps from a server of load $(j+1)\lambda$, or else there must be a $2q$-balancing path. Combining equations 2 and (2) we derive the following recurrence:

$$y_0^j \geq (j)^q y_0^{j+1} \qquad y_0^0 = n$$

Solving the recurrence, we find $n \geq y^h(h!)^q$. Since y^h is at least 1, this implies $n \geq (h!)^q$. Applying the natural logarithm and Stirling's approximation yields $\ln n/q \geq h(\ln h - 1)$.

Lemma 9. *For any q, all servers can be kept 2q-balanced using $O(qh)$ reassignments per insertion or deletion.*

Proof. To insert an item u, place it on any server v to which it is adjacent. This increases the height of v and may create a balancing path starting at v. It cannot, however, create a rebalancing path originating at any other node. Rebalance along any balancing path starting at v and terminating at v_1. At the conclusion of rebalancing, v is returned to its initial height, server v_1 has increased in height by 1, and all other servers have unchanged height. Recursively apply the same procedure starting at v_1. By the definition of a rebalancing path, $load(v_1) = load(v) - \lambda$ after rebalancing. Hence the procedure can only be applied $O(h)$ times before reaching a column of height 1. Each call performs $O(q)$ reassignments.

The case of deletion is similar, except that deleting an item from v may create a rebalancing path terminating at v, and recursive calls occur at servers that are increasing in height by λ.

Theorem 10. *The eager rebalancing witness-based algorithm is $O(h)$ competitive against current load and performs $O(qh)$ rebalances per item, where h is the value of c satisfying equation 1 for parameter q.*

Proof. Choose $a = 0$ in the partition and for global rebalancing take tasks off the most loaded processor in level q. For level i, define $\Phi_{ij} = (M_{ij} - W_{ij})qh$, where M_{ij} is the maximum load on column j in level i, and W_{ij} is the current load on column j. The proof is a straightforward case analysis.

Corollary 11. *For any $c < 1$, there is an algorithm that performs $O(\log n / \log\log n)$ reassignments and keeps the on-line load within a factor of $O((\log n)^c / \log\log n)$ of the current load.*

Corollary 12. *There is an algorithm that performs $O(\log n)$ reassignments and keeps the on-line load within a factor of $O(1)$ of the current load.*

Corollary 11 follows from setting $q = (\log n)^{1-c}$, and Corollary 12 from setting $q = \log n$. This improves on the results of [2] in two ways: it works for all values of the optimum load and it is competitive against current rather than peak load.

5 Virtual Circuit Routing

In the virtual circuit routing problem one is given a communication network modeled by an undirected graph. Each edge $e \in E$ has an associated capacity cap_e. A task u is a triple (w_u, s_u, t_u), indicating the need for a weight w_u connection between nodes s_u and t_u. An on-line algorithm must choose a path in the graph between s_u and t_u to serve as the virtual connection. The current weight on each edge, W_e, is increased by w_u. The load on an edge is W_e/cap_e. Recall we assume that the restart cost, $r_u = c \cdot w_u$. This restriction can be eased using the method of Section 6.

Azar *et al.* [2] give a one-level algorithm, parameterized by λ, that is $O(\log n)$ competitive and reassigns each task $O(\log n)$ times. We use this in a witness-based algorithm that is competitive against current load.

Let $C = \sum_{e \in E} \text{cap}_e$ and let $W(t) = \sum_{u \in U(t)} w_u$. Let λ_t^* be the optimum maximum load. We have $\lambda_t^* \geq W_e/\text{cap}_e$, or $\lambda_t^* \text{cap}_e \geq W_e$, $\forall e \in E$. Summing over E gives $\lambda_t^* \sum_{e \in E} \text{cap}_e \geq \sum_{e \in E} W_e \geq W(t)$, and hence $\lambda_t^* \geq W(t)/C$. On the other hand, $\lambda_t^* < W(t)/\text{cap}_{min}$, where $\text{cap}_{min} = \min_{e \in E}\{\text{cap}_e\}$.

In the witness algorithm, set the lowest level $a = \lfloor \log(W(t)/(2C \cdot c(n))) \rfloor$, where $c(n) = O(\log n)$ is the competitive ratio of the one-level algorithm; a is only adjusted, however, if the value of $W(t)$ has doubled or halved since a was last set. The highest level, q, is at most $\lceil \log W(t)/\text{cap}_{min} \rceil$. During global rebalancing, select tasks to move down from level q in any order. Deleting a task from q and inserting it into $j < q$ may cause rebalancing within levels q and j.

For task $u \in U(t)$, let $\ell(u)$ be the level containing u and $a(u, i)$ be the number of times u has been reassigned by the one-level algorithm for level i. Let $\Phi_u = [(c_1 \log n + 1)(\ell(u) - a + 1) - a(u, i)]w_u$, where $c_1 \log n$ is the maximum number of reassignments within a level. When u is newly arrived, $\ell(u) - a = O(\log W(t)/\text{cap}_{min} - \log(W(t)/(2C \cdot c(n))))$ and hence the amortized weight assigned on a new insertion is $O(w_u \log n \log(C \cdot c(n)/\text{cap}_{min}))$. The amortized weight reassigned when u is reassigned within a level or across levels is 0. If a decreases by 1, $W(t)$ has decreased by a factor of 2. The potential increases by $O(W(t) \log n)$. This is charged to the weight that has left, for an additional $O(\log n)$ per unit weight. Since $c(n) = O(\log n)$, the witness-based algorithm is $O(\log n)$ competitive against current load. Since $C/\text{cap}_{min} = \Omega(n)$, $\log(C \cdot c(n)/\text{cap}_{min}) = O(\log C/\text{cap}_{min})$. If the total weight of all tasks is W, the total reassignment cost is $O(W \log n \log C/\text{cap}_{min})$, which is $O(S \log n \log C/\text{cap}_{min})$.

6 Unrelated Weights and Startup Costs

In the previous sections we gave algorithms that provide competitive bounds while guaranteeing that the total weight that is assigned and reassigned is bounded by a small function f of the total weight of the input tasks. Under the assumption that $w_u = c \cdot r_u$, this implies that the total assignment cost is bounded by the same function f on the total startup cost. In this section we show how to extend these algorithms to handle any restart cost function of the form $r : u \to R^+$ (but still not independent of t and the servers involved in the restart.) Tasks are partitioned by startup cost per unit weight, r_u/w_u. Let a be the minimum value of this ratio over all input tasks and b the maximum value.

Choose a parameter $1 < \delta \leq (b/a)$. Partition the input tasks into $O(\log_\delta(b/a))$ classes such that task u is in class i if

$$a\delta^i \leq \frac{r_u}{w_u} < a\delta^{i+1}.$$

Within each class run the load-balancing algorithm appropriate to the problem being solved.

Lemma 13. *Let A be an algorithm that achieves a competitive ratio of c while the total weight of reassigned items is bounded by $g(W)$, where $W = \sum_{u \in U} w_u$ and $g(x) = \Omega(x)$. Then there exists an algorithm A_δ that achieves a competitive ratio of $c \log_\delta(b/a)$ and incurs a total reassignment cost $a\delta^{i+1}g(S/a\delta^i)$, where $S = \sum_{u \in U} r_u$.*

Proof. Let λ^* be the maximum over classes of the optimum load within that class. This is a lower bound on the optimum load for all tasks. Within each class no server has load greater than $c\lambda^*$, and hence maximum load due to all classes is $O(c \log_\delta(b/a)$. Within class i, the reassignment cost per unit weight is at most $a\delta^{i+1}$, hence the total cost of reassignments is $a\delta^{i+1}g(W)$. On the other hand, $W \geq S/(a\delta^i)$.

Using Lemma 13 we have the following:

- Algorithms for identical and related machines that are $O(\log_\delta(b/a))$ competitive against current load and incur total assignment cost $O(\delta S)$.

- An algorithm for the restricted machines problem that is $O(\log_\delta(b/a))$ competitive against current load and incurs reassignment cost $O(\delta \log n)$ when all weights are unit, and an algorithm for the restricted machines problem that is $O(\frac{1}{\rho} \log n \log_\delta(b/a))$ competitive against current load and incurs a reassignment cost $O(\rho \delta S)$ for arbitrary weights, where ρ is any parameter between 0 and 1.
- An algorithm for virtual circuit routing that is $O(\log n \log_\delta b/a)$ competitive against peak load and incurs a reassignment cost $O(\delta S \log n \log(C/\text{cap}_{min}))$.

All results follow by applying the lemma to algorithms presented herein, with the exception of the algorithm for general weights in the restricted subset case, which follows by applying the lemma to the algorithm in [10]

While the bounds of this section are not ideal, in that they depend on the value of a ratio between input costs, they are independent of the number of tasks. An interesting open problem is to find a scheme with competitive ratio solely a function of the number of machines, n.

References

1. J. Aspnes, Y. Azar, A. Fiat, S. Plotkin, and O. Waarts. On-line load balancing with applications to machine scheduling and virtual circuit routing. In *Proc. 25th ACM Symp. on Theory of Computing*, pages 623–631, 1993.
2. B. Awerbuch, Y. Azar, S. Plotkin, and O. Waarts. Competitive routing of virtual circuits with unknown duration. In *Proc. ACM/SIAM Symp. on Discrete Algorithms*, pages 321–330, 1994.
3. Y. Azar, B. Kalyanasundaram, S. Plotkin, K. Pruhs, and O. Waarts. Online load balancing of temporary tasks. In *Proc. 1993 Workshop on Algorithms and Data Structures (WADS 93), Lecture Notes in Computer Science 709*. Springer-Verlag, Aug. 1993.
4. Y. Azar, A. Karlin, and A. Broder. On-line load balancing. In *Proc. 33nd Symp. of Foundations of Computer Science*, pages 218–225, 1992.
5. Y. Azar, J. Naor, and R. Rom. The competitiveness of on-line assignments. In *Proc. 3rd ACM-SIAM Symp. on Discrete Algorithms*, pages 203–210, 1992.
6. Y. Bartal, A. Fiat, H. Karloff, and R. Vohra. New algorithms for an ancient scheduling problem. In *Proc. 24nd ACM Symp. on Theory of Computing*, 1992.
7. R. L. Graham. Bounds for certain multiprocessing anomalies. *Bell System Technical Journal*, 45:1563–1581, 1966.
8. D. R. Karger, S. J. Phillips, and E. Torng. A better algorithm for an ancient scheduling problem. In *Proc. 1994 ACM/SIAM Symp. on Discrete Algorithms*, 1994. To appear.
9. E. L. Lawler, J. K. Lenstra, A. H. Rinnooy Kan, and D. B. Shmoys. Sequencing and scheduling: Algorithms and complexity. In S. C. Graves, A. Rinnooy Kan, and P. Zipkin, editors, *Handbook of Operations Research and Management Science, Volume IV: Production Planning and Inventory*, pages 445–522. North-Holland, 1993.
10. S. Phillips and J. Westbrook. On-line load balancing and network flow. In *Proc. 1993 Symp. on Theory of Computing*, pages 402–411, Apr. 1993.

Self-Simulation for the
Passive Optical Star Model [*]

P. Berthomé [1]   Th. Duboux [1]   T. Hagerup [2]   I. Newman [3]   A. Schuster [4]

1- LIP, CNRS URA 1398, ENS-Lyon 69 364 LYON CEDEX 07, France.
E-mail: {berthome,duboux}@lip.ens-lyon.fr
2- Max-Planck-Institut für Informatik, D–66123 Saarbrücken, Germany.
E-mail: torben@mpi-sb.mpg.de
3- Mathematics and Computer Science, Haifa University, Haifa, Israel 31905.
E-mail: ilan@mathcs.haifa.ac.il
4- Department of Computer Science, Technion, Haifa, Israel 32000.
E-mail: assafs@cs.technion.ac.il

Abstract. In the context of parallel computing, optical technology offers simple interconnection schemes with straightforward layouts that support complex logical interconnection patterns. The Passive Optical Star (POS) is often suggested as a platform for implementing the optical network: logically it offers an all-to-all broadcast capability.

We investigate the *self-simulation* or *scalability* properties of the POS. A family of parallel machines is said to be self-simulating or scalable if reducing the number of processors by a factor of k (by going to a smaller member of the family) increases the computation time by a factor of (the optimal) $O(k)$.

We present a randomized algorithm for an n-processor POS that simulates a kn-processor POS with a slowdown of $O(k + \log^* n)$ using local control only, thus coming close to the self-simulation ideal of $O(k)$. We also analyze *direct* algorithms that send messages directly from their origin to their destination; for this case we prove that $\Theta(k^2)$ is the exact complexity.

1   Introduction

The interest in optical communication systems and optical computers has recently increased dramatically in the community dealing with the theory of algorithms. Indeed, the emerging high-bandwidth applications, such as voice/video services, huge distributed data bases, and network super-computing are driving the use of optics as the communication medium for the future [2,6]. This new

[*] This work was supported in part by ANM, C3, the French-Israeli grant for cooperation in Computer Science, by a grant from the Israeli Ministry of Science, and by the ESPRIT Basic Research Actions Program of the EU under contract No. 7141 (project ALCOM II).

technology suggests many new communication, routing and simulation problems. Here we deal with the technology of the *Passive Optical Star* (POS), where communication links consist of optical fibers with orders of magnitude more information bandwidth than what can be modulated electronically. The high bandwidth is utilized by the electronic interfaces by using the *Wavelength Division Multiplexing* technique (WDM): several transmissions may be carried by the same optical link, provided that the transmitters use different wavelengths [16]. A physical star topology is often suggested for implementing an optical network in which optical fibers are interconnected via an optical star coupler [11].

In POS-based networks each node consists of a processing element with a local memory (PE), a transmitter, and a receiver. At each step of the machine, each processor may tune its receiver to a given wavelength, its transmitter to another. Then it receives a message through its receiver, transmits a message through its transmitter, and performs local computations on the received data. If several receivers tune to the same wavelength, then all of them hear the message transmitted on that wavelength. Thus *broadcast* becomes an elementary operation, which may be carried out concurrently. If several transmitters tune to the same wavelength, all receivers on that wavelength receive a "noise" indication. The POS paradigm has been used for the design and implementation of several experimental networks [14, 15]. For example, Dowd [7] presented a network using a single optical passive coupler. Both transmitters and receivers are tunable and communicate using the slotted ALOHA protocol. Another important characteristic of the POS model is that the decisions of whether to transmit/receive, and using which wavelength, are taken locally by the PE depending on local data. Thus a transmitting processor may not know which, and how many (if any) receivers listen to it. Both this feature and the ease of broadcasting influence the nature of the algorithms that are designed for POS machines.

One may view the abstract POS model as a generalization of the Optical Communication Parallel Computer, OCPC. This model, introduced by Anderson and Miller [1], has recently attracted increased attention [8–10]. Similar to the POS, an n-processor OCPC is composed of n PEs, each having a tunable transmitter and a fixed receiver, which serve as its opto-electronic network interface. Each processor may send a message to any other processor, but no broadcasting is possible.

In this work we address one of the most fundamental aspects of the POS as a parallel computational model, namely, its ability of *self-simulation* (also known as *virtual parallelism* or *scalability*). The self-simulation capability of a parallel model is directly related to the efficiency and ease of algorithm design: it is desirable that the algorithm designer may assume that processors (and/or wavelengths) do not represent a critical resource when writing his or her programs. Once the algorithm that is designed for N processors is to be executed on a certain machine with only $n < N$ processors, a simulation of the N-processor program should be done by the actual n processors. Similarly, the communication carried out by the larger virtual machine must be "scaled down". This simulation is done essentially by mapping N/n program-processors to each real processor, which simulates the actions of these "virtual" processors. The main obstacle here is the simulation of any communication pattern employed by the simulated machine.

The remainder of the paper is organized as follows. After defining the model and the problem in Section 2, we show in Section 3 a tight bound of k^2 steps for direct simulations, even if full knowledge of the communication pattern is available. In Section 4, we provide a randomized algorithm to solve the general problem. We conclude the paper in Section 5.

2 The Model and Preliminaries

Let us briefly recall the important features of the *Passive Optical Star* network. Each processor in a POS network contains a receiver, a transmitter and a processing element with a memory unit (PE). All the receivers and transmitters are connected by optical fibers to a Passive Star Coupler, which is an all-to-all communication device with broadcast capability. The transmitters and receivers may tune to one of a range of wavelengths. Choosing a particular wavelength is done locally and independently of the choices made by the other processors in the system. Any message sent by a transmitter using a certain wavelength reaches all the receivers which tune to this wavelength. When two or more processors transmit on the same wavelength, all receivers tuned to this wavelength receive a "noise" indication, but no other useful information.

We assume a synchronous working mode, and that messages have a certain bounded length. The following operations are performed by every processor of the POS at every step. *Tuning:* the receiver and transmitter tune to their frequencies. *Communication:* each transmitter may send a message and every receiver stores the message which it receives on the wavelength to which it is tuned. *Computation:* some local computation is carried out.

Our goal is to simulate the communication of a large POS machine on a smaller one. We call this procedure *self-simulation*. First, we need to clearly define the state of the different transmitters and receivers during a single step of communication for a general POS having n PEs and whose optical devices can select one out of Λ wavelengths.

Definition 1. A *communication pattern* associated with a POS machine with n PEs and Λ available wavelengths is a bipartite graph $G = (T, R, E)$ such that: (i) $|T| = |R| = n$, $T = (t_1, \ldots, t_n)$, $R = (r_1, \ldots, r_n)$; (ii) $|E| \leq n$, the edges in E are labeled in $[1 \ldots \Lambda]$ and two edges have a vertex in common in T iff they have the same label; (iii) each element of R has at most one incident edge.

Thus, a communication pattern can be associated in a canonical fashion with an actual communication step in the POS machine: if there exists an edge labeled λ between t_i and r_j in G, then the transmitter of the i-th PE and the receiver of the j-th PE are tuned to the wavelength numbered λ. Furthermore, any communication without conflict in the POS is associated with a communication pattern in this way. However, in any real communication, two different transmitters can use the same wavelength, creating access conflicts. We extend the previous definition in order to reflect this property, as follows.

Definition 2. A *generalized communication pattern* associated with a POS machine with n PEs and Λ available wavelengths is a bipartite graph $G = (T, R, E)$ such that: (i) $|T| = |R| = n$, $T = (t_1, \ldots, t_n)$, $R = (r_1, \ldots, r_n)$; (ii) the edges in E are labeled in $[1 .. \Lambda]$ and two edges having a vertex in common in T have the same label; (iii) two edges having a vertex in common in R have the same label; (iv) each subgraph of G spanned by the edges with a fixed label is a complete bipartite graph.

In our simulation problem, we assume that the simulated machine is of size kn and has kn wavelengths, and that the simulating machine is of size n. We want to find an efficient way to simulate any communication step with the minimum number of wavelengths. For the sake of clarity, we shall refer to PEs in the simulated machine as *processors*, whereas they will be referred to as *clusters* in the simulating machine. Transmitting elements (resp. receiving elements) will be called *T-processors* (resp. *R-processors*) in the simulated machine and *T-clusters* (resp. *R-clusters*) in the simulating POS machine. Any cluster will simulate the actions of k processors. To be specific, the algorithm on processor j will be executed by cluster $\lceil j/k \rceil$.

Thus, given any (generalized) communication pattern $G = (T, R, E)$ associated with a POS of size kn, the simulation problem can be modeled by the induced bipartite multi-graph $G' = (T', R', E')$ by a "division" mapping. Both sets T' and R' are of size n, $T' = (t'_1, \ldots, t'_n)$ and $R' = (r'_1, \ldots, r'_n)$, and the set E' is the image of the set E under the following function: the edge $e = (t_i, r_j)$ labeled λ in E is mapped onto the edge $e' = (t'_{\lceil i/k \rceil}, r'_{\lceil j/k \rceil})$ labeled λ.

We assume that the global communication pattern is not known to the clusters. Each cluster gets only local information, namely, each T-cluster gets the k messages and the frequencies on which they are supposed to be sent, and each R-cluster gets the k frequencies to listen to. In particular, the T-processors do not know which are the R-processors that might be listening on their frequencies. Similarly, the R-processors do not know the identity of the T-processors that transmit on the wavelength to which they are listening. The only connection between them is given implicitly by the wavelength known to both.

An actual simulation of a communication step in ℓ steps is a protocol that schedules frequencies for each cluster in ℓ discrete time steps and the messages that each sending cluster should send. In such a protocol, some of the data transferred may be auxiliary data for internal use or unsuccessful transmission of real messages (due to collisions). However, every real message should eventually reach all its destinations in these ℓ steps.

In the sequel, we assume that both n and k are known to every cluster.

3   Direct Algorithms

In this section we analyze algorithms that are *direct*, i.e., a message is always sent directly from its origin to its final destination and is never stored in an intermediate location. We show that k^2 steps are necessary in this case even when the communication pattern is known in advance (the *offline* case), and that there is a matching upper bound.

3.1   Offline Direct Lower Bound

Theorem 3. *If only direct communication is possible, there is a communication pattern that requires k^2 communication steps, even if global knowledge is available to every processor in advance and the number of wavelengths available to the simulating machine is unbounded.*

Proof: Suppose that each of t T-clusters sends k distinct messages within fewer than k^2 time slots. Since a T-cluster can transmit at most one message per time slot, each T-cluster will have at least one message that is transmitted in at most $k - 1$ time slots. Choosing t sufficiently large, we can find k messages (from different T-clusters) that are transmitted only within *the same $k - 1$* time slots. But if each set of k messages (from different T-clusters) is to be received by some R-cluster, this is a contradiction.   ∎

3.2   Online Direct Upper Bound

We show here a deterministic direct online algorithm that completes the self-simulation in k^2 steps using the original set of frequencies (kn wavelengths in the worst case). It is convenient to identify the following restricted problem.

Definition 4. Assume that there is a globally known bound on the actual number of messages in a cluster. Namely, assume that every T-cluster contains at most a messages and that every R-cluster contains at most b messages; then we say that the problem is of (a, b)-*type*. The general problem is of (k, k)-type.

Lemma 5. *The simulation of an (a, b)-type pattern can be done in $a \cdot b$ steps by a deterministic direct algorithm that uses the same number of wavelengths as the simulated machine.*

Basically, the algorithm tries all possible pairs of transmitting-receiving wavelengths. The details are left to the reader.

Corollary 6. *The self-simulation of any communication pattern can be done deterministically in k^2 steps by an algorithm that uses only direct communication and kn wavelengths.*

By Theorem 3, the algorithm above is best possible among all direct algorithms. Note also that a very minimal global knowledge is assumed. In fact, even the size of the network need not be known.

Observation 7 *There is an (offline) schedule that self-simulates any d-bounded degree communication pattern, i.e., such that the corresponding induced graph G' on clusters has degree bounded by d, in d steps, and any permutation in k steps.*

Proof: This follows from the fact that the graph G' can be edge-colored by d colors so that adjacent edges are colored by different colors [12]. Then note that a permutation is a k-bounded degree communication pattern. ∎

4 Randomized Online Simulation

In this section, we present an improvement using random techniques. We use the notion of an event occurring *with high probability* (or w.h.p.) from [3]: the event occurs with probability at least $1 - 2^{-n^\epsilon}$, for some constant $\epsilon > 0$.

Theorem 8. *There is a step-by-step algorithm that simulates each step of a kn-processor, kn-wavelength POS by an n-processor n-wavelength POS in $O(k + \log^* n)$ steps w.h.p.*

We proceed in the remainder of this section to describe the simulation of a single communication step. The simulation of a full program is done by simulating each step at a time. The local information needed is kept by letting each physical cluster simulate a fixed set of logical processors. The high probability of success for a single step guarantees the high probability of the overall success, as long as the program is limited to a number of steps that is polynomial in n (for longer computations, a different but standard argument applies).

No-Collisions Assumption. We first deal with the communication patterns, as defined in Definition 1, where the communication is collision-free. Then, we show how to simulate generalized communication patterns.

The idea of the algorithm is essentially to redistribute the receivers and transmitters so that the resulting problem will be reduced to a collection of (a, b)-type problems with small $a \cdot b$, and then to use Lemma 5.

We divide the kn wavelengths of the simulated POS into k color classes, each of size n. This is done in advance in an arbitrary fixed way. Each receiver/transmitter can be classified according to its color class. Our aim is that the messages of each color class will be nearly equally split among the clusters (both for transmitters and receivers). Once this is achieved we will reduce the problem to k restricted problems that can be completed fast enough. In this balancing process, rather than moving real messages we move *tokens* that represent messages. Only after tokens arrive at their destinations, the corresponding messages are transmitted from the origins of the tokens to the sites where the tokens end or in the opposite direction. To this end the path taken by each token and the schedule when each "link" is used are saved and are used later on to schedule the tokens along the path or in the reverse direction. The full algorithm consists of four phases.

Algorithm Self-Simulation

Phase (1) Balance Receivers (i.e., balance their tokens).
Phase (2) Balance Transmitters (messages).
Phase (3) Transmit messages from intermediate locations found in Phase (2) to intermediate receiver locations found in Phase (1).
Phase (4) Transmit messages to final receiver locations.

4.1   PRAM Simulation and Semisorting

A major building block in our realization of the above outline is a reduction to the *semisorting* problem, in which elements with the same value are grouped together, possibly with some empty places between consecutive elements.

It will also be convenient to assume that the POS with n transmitters and n receivers contains, in addition, an extra $O(n)$ auxiliary processors for internal use. This causes only an additional constant slowdown factor in the simulation time.

Definition 9. The *semisorting* problem of size n is, given n integers $x_1, \cdots, x_n$ in the range $[1 .. n]$, to place them in an output array of $O(n)$ cells such that between two cells containing elements of the same value there is no cell containing an element of a different value (there may, however, be empty cells or cells containing other elements of the same value).

Theorem 10. *[3, Thm. 9.21] Semisorting problems of size n can be solved on a* COLLISION CRCW PRAM *using* $O(\log^* n)$ *time,* $\lceil n/\log^* n \rceil$ *processors and* $O(n)$ *space w.h.p.*

In the COLLISION model, in case of concurrent writing to a cell by several processors, a special collision symbol is stored in the cell. In order to use the result above as a building block in our self-simulation, we let the POS simulate the PRAM. The following observation is straightforward.

Lemma 11. *[4, Lem. 4.5] An n-processor* POS *with n wavelengths can simulate an n-processor* COLLISION CRCW PRAM *with shared memory of size $O(n)$ in a step-by-step fashion with only a constant-factor slowdown. In this simulation each processor simulates $O(1)$ cells.*

Corollary 12. *Semisorting problems of size n can be solved by an n-processor n-wavelength* POS *in $O(\log^* n)$ time w.h.p.*

4.2 Balance Receivers – Implementation

Our goal here is to balance the receivers so that for each color class i, every cluster eventually contains nearly the same number of receivers of color i. Each R-cluster creates a *token* for every message it should receive. A token contains the index of the creating cluster, the index of the corresponding receiver, and the color class of this receiver, i.e., of the wavelength to which it listens. During execution of different sub-phases, the scheduling, the paths and wavelengths that are used by the tokens are recorded in the transmitting and the receiving devices. Following this phase, the communication in the reverse direction can be reconstructed in a deterministic manner. Notice that such reconstruction is possible as long as the communication pattern *origins of tokens-to-destinations* defines a point-to-point communication.

The algorithm for token balancing is decomposed into several sub-phases. Recall that we start with up to kn tokens (one per message to be received), colored by k colors, and with n clusters that contain k tokens each.

Sorting in small groups. For the final algorithm to work fast enough we need to reduce the number of tokens to be linear in the number of clusters. For this we first sort in small groups and "compress" tokens, as follows. Let every k consecutive clusters form a group. Within each group, the tokens are sorted according to their color. Sorting k^2 keys in the range $[1..k]$ using k clusters can be done by several $O(k)$ steps algorithms. We simulate a sorting algorithm on a $k \times k$ mesh [13] that operates only within rows and within columns. Row operations can be performed locally in $O(k)$ steps, and column operations can be transformed into row operations by means of transposition, which is a fixed permutation that can be realized in k steps (Observation 7). Thus the whole procedure takes $O(k)$ steps and uses n wavelengths overall.

Packing into Super-Tokens. Within every cluster, all the tokens of the same color are "packed" into a *super-token* of that color. A super-token represents up to k tokens and contains the identity of the cluster in which it is created. Note that within a group of k clusters, at most $2k$ super-tokens are created.

Balance within groups. We distribute the super-tokens evenly within each group of k clusters, such that there are at most two super-tokens in each cluster. This can be done by moving the super-tokens one by one in $2k$ steps using n wavelengths overall.

Semisorting. We use the randomized semisorting algorithm (Corollary 12) to semisort the super-tokens by their colors. The total output array is of size bn for some constant b, namely we use bn auxiliary clusters. As a result we may assume that at the end of the semisorting sub-phase there is at most one super-token in every cluster. This sub-phase takes $O(\log^* n)$ steps, w.h.p., and uses n wavelengths. We remark that this is the *only* place where randomization is used.

Send back information. The current location of every super-token is sent back to its origin cluster. This operation involves reversing the paths taken by the super-tokens. Note that determining these paths involved semisorting, which in turn involved concurrent write. However, in the final result, the routing of the super-tokens is a point-to-point communication pattern, and thus can be reversed using the schedule and the routing data that was stored along the routing phases.

Unpacking. The actual tokens are sent to the place in which the super-token representing them ended. As each origin cluster contains at most k tokens and the destination contains just a single super-token, this communication can be realized in k steps: the origins transmit the tokens one by one while the destinations receive on fixed wavelengths. This operation uses n wavelengths.
As a result, there are at most k *simple* tokens in each cluster, all with the same color. As the super-tokens were semisorted (operation which creates "holes"), some of the clusters may contain no tokens. We assume from now on that each cluster contains exactly k tokens (of the same color), by introducing dummies if necessary. Empty clusters are considered to have the same color as the clusters in their intervals (although this color is not assumed to be actually known to the clusters). Thus we may assume that the i-th color class contains exactly $\lambda_i \cdot k$ tokens for some integral λ_i (this includes the dummy tokens).

Transpose. The tokens are redistributed among clusters so that each cluster gets approximately the same number of tokens from each color class (here the i-th color class is considered to be of size $\lambda_i \cdot k$). This is done by sending the token at the i-th place ($0 \le i \le kn - 1$) to the ($i \bmod n$)-th cluster. As this communication pattern is a fixed known permutation, it can be done in k steps using n wavelengths (by Observation 7).

At this point the tokens are sufficiently balanced for our purpose. Namely, as we have shown, the total number of tokens (including dummy tokens) is $O(kn)$, with k of them at a cluster. Furthermore, we ended up with $\lambda_i \cdot k$ tokens of color i. This latter number might be k times larger than the original number of tokens of color i (because of the "rounding up" to k). However $\sum_{i=1}^k \lambda_i \cdot k = O(kn)$. Let a_i be integral and such that $(a_i - 1)n \le \lambda_i \cdot k < a_i n$. The transpose phase ends with at most a_i elements of color i in each cluster. As we shall see, the complexity of the algorithm will be determined by the sum $\sum_{i=1}^k a_i$, for which $\sum_{i=1}^k a_i = \frac{1}{n} \sum_{i=1}^k n a_i = \frac{1}{n} \cdot O(kn) = O(k)$. We need global knowledge of the a_i's, which is obtained through an additional sub-phase as follows.

Determining the a_i's. The sub-phase proceeds in k rounds, one for each color. a_i is determined in the i-th round by letting each cluster transmit a signal in the first j steps if it stores j tokens of the i-th color (including dummies). The round is stopped on the first silent step, which is the $(a_i + 1)$-th step. The whole sub-phase takes time $k + \sum_{i=1}^{k} a_i = O(k)$.

Summarizing the above implementation, we get that the phase of Balancing Receivers takes $O(k + \log^* n)$ steps w.h.p. All the sub-phases are deterministic and take $O(k)$ steps, except for the semisorting, which takes $O(\log^* n)$ w.h.p. All the sub-phases use $O(n)$ wavelengths.

4.3 Implementing the Other Phases

(2) Balancing transmitters. The implementation of this phase is similar to the algorithm for balancing the receivers. Here, however, the tokens contain the actual messages and the corresponding wavelengths. Recording the communication involving tokens is not necessary here, since messages will not be moved back. It is still necessary, though, to (temporarily) record the communication involved in the movement of super-tokens.

By the assumption of no collisions, there are at most n transmitters of any given color. Therefore, as a result of this phase, we will have $O(1)$ messages of each color at each intermediate T-cluster. Similar to the algorithm for balancing the receivers, the algorithm here uses n wavelengths and takes $O(k + \log^* n)$ steps.

(3) Communication between intermediate clusters. As a result of balancing the receivers and balancing the transmitters (phases 1 and 2) each intermediate T-cluster contains $O(1)$ messages of each color and each R-cluster contains a_i messages of color i to receive. Thus the communication restricted to the i-th color is an $(O(1), a_i)$-type problem and can be carried out in $O(a_i)$ steps. This phase can be implemented in $O(k)$ steps, and only n wavelengths are used since the color classes are scheduled separately.

(4) Sending messages to final destination. Sending the messages from the intermediate R-clusters to the final destination R-clusters is done along the reverse paths (timing, wavelengths) created by the tokens in the balancing-receivers phase. This is done in $O(k)$ steps in a deterministic way. The data regarding the super-token movement is internal to the corresponding balancing-transmitters phase and is not used here.

4.4 The Generalized Communication Patterns

We now extend our self-simulation algorithm so that it also deals with generalized communication patterns, where the pattern may include collisions. Thus, in the general case, two or more processors may transmit on the same wavelength. This means that in the associated graph G, the in-degree of each R-processor may be greater than one. In such a case, an R-processor with degree greater than one should receive a "noise" signal.

Our approach to the general simulation problem is to split the communication problem into two parts, which will be treated separately. This reduction can be viewed entirely in terms of the simulated machine. The first part is the *many-to-one* communication, which treats the collisions: a message that is transmitted in the original computation using wavelength λ is moved to processor λ. In case that more than a single transmitter transmits on wavelength λ, processor λ should know that there is a collision (but need not know the messages). The second part is a *one-to-many* communication, where processor λ sends its message (or noise) to all receivers of the original computation that are using wavelength λ.

The second, one-to-many part is just an instance of the communication problem we have treated so far with the "no-collision" assumption, and thus can be solved in $O(k + \log^* n)$ steps w.h.p. For the part of many-to-one communication, processor λ needs to receive the message on wavelength λ, or a noise if there is more than one such message. Here we apply a variant of the algorithm treated so far. Phases 1, 2 and 4 can be implemented as before. In Phase 3, in contrast with what was the case before, there can be more than n transmitters of a given color, but no more than n receivers of a fixed color. In other words, the communication problem for color i is an $(a_i, O(1))$-type problem, where here a_i corresponds to the fact that there are at most $a_i n$ transmitters of color i. However, as a result of the balancing phase $\sum_{i=1}^{k} a_i = O(k)$. Thus, similarly to the one-to-many communication problem, the whole process for all i takes $O(k)$ steps. However, for each color i, the $(a_i, O(1))$-type problem is guaranteed to succeed (Lemma 5) only if there is no collision in the first place. Here, it either succeeds and the receiver λ can decide whether there is more then one message on its frequency, or it may fail, in which case the receiver λ can be sure that there is a collision.

5 Concluding Remarks

In this paper, we presented an online $O(k + \log^* n)$ randomized indirect self-simulation algorithm, matching the lower bound up to an $O(\log^* n)$ additive term. Our self-simulation algorithm uses an n-processor n-wavelength POS machine. This is optimal also with respect to the number of wavelengths.

A natural question is the tradeoff between time and the number of wavelengths. Our self-simulation algorithm is easily extended to give an $O((k + \log^* n)n/n')$ steps algorithm using $n' < n$ wavelengths. It is also not hard to see that this is optimal except for the $O(\log^* n)$ additive term.

Since the only step that requires randomness in the simulation is the semisorting one, we can show that the deterministic on-line self-simulation problem can be solved in $O(k + \log n)$ steps by using deterministic sorting on the CRCW PRAM [5]. We have also shown in [4] that the off-line problem can be solved in $5k$ steps. There are some related issues which were not treated here but were often raised, such as taking into account limitations on the number of wavelengths or the tuning time of the communicating elements.

Acknowledgment. The authors would like to thank Afonso Ferreira and Nashib Qadri for their helpful comments.

References

1. R.J. Anderson and G.L. Miller. Optical communication for pointer based algorithms. Technical Report CRI 88-14, Comp. Sci. Dept., USC, 1988.
2. R. Ballart and Y.C. Ching. SONET: Now It's the Standard Optical Network. *IEEE Comm. Magazine*, pages 8–15, March 1989.
3. H. Bast and T. Hagerup. Fast parallel space allocation, estimation and integer sorting (revised). Technical Report TR MPI-I-93-123, Max-Planck-Institut für Informatik, Saarbrücken, 1993.
4. P. Berthomé, Th. Duboux, T. Hagerup, I. Newman, and A. Schuster. Self-Simulation for the Passive Optical Star Model. Technical Report 95-13, LIP-ENS-Lyon, June 1995.
5. R. Cole. Parallel Merge Sort. *SIAM J. Comput.*, 17(4):770–785, 1988.
6. T. Cox, F. Dix, C. Hemrick, and J. McRoberts. SMDS: The Beginning of WAN Superhighways. *Data Communications*, April 1991.
7. P.W. Dowd. Random Access Protocols for High-Speed Interprocessor Communication Based on an Optical Passive Star Topology. *Journal of Ligthwave Technology*, 9(6):799–808, June 1991.
8. L.A. Goldberg, M. Jerrum, T. Leighton, and S. Rao. A Doubly Logarithmic Communication Algorithm for the Completely Connected Optical Communication Parallel Computer. In *Proc. 5th ACM SPAA*, pages 300–309, June 1993.
9. L.A. Goldberg, M. Jerrum, and P.D. MacKenzie. An $\Omega(\sqrt{\log \log n})$ Lower Bound for Routing in Optical Networks. In *Proc. 6th ACM SPAA*, pages 147–156, June 1994.
10. L.A. Goldberg, Y. Matias, and S. Rao. An Optical Simulation of Shared Memory. In *Proc. 6th ACM SPAA*, pages 257–267, June 1994.
11. G. Gravenstreter, R.G. Melhem, D.M. Chiarulli, S.P. Levitan, and J.P. Teza. The partitioned optical passive stars (pops) topology. In *IPPS*, 1995.
12. L. Lovász and M.D. Plummer. *Matching theory*. North Holland, Amsterdam, 1986.
13. J.M. Marberg and E. Gafni. Sorting in constant number of row and column phases on a mesh. *Algorithmica*, 3(4):561–572, 1988.
14. B. Mukherjee. WDM-Based Local Lightwave Networks Part I: Single-Hop Systems. *IEEE Networks*, pages 12–27, May 1992.
15. B. Mukherjee. WDM-Based Local Lightwave Networks Part II: Multi-Hop Systems. *IEEE Networks*, pages 20–32, July 1992.
16. R. Ramaswami. Multi-wavelength lightwave networks for computer communication. *IEEE Comm. Magazine*, 31:78–88, 1993.

Computing the Agreement of Trees with Bounded Degrees

Martin Farach[1], Teresa M. Przytycka[1*] and Mikkel Thorup[3]

[1] Department of Computer Science, Rutgers University, Piscataway, NJ 08855, USA.
E-mail: farach@cs.rutgers.edu, WWW: http://www.cs.rutgers.edu/~farach.
[2] Department of Mathematics and Computer Science, Odense University, Campusvej
55, 5230 Odense M, Denmark. E-mail: przytyck@imada.ou.dk.
[3] Department of Computer Science, University of Copenhagen, Universitetsparken 1,
2100 Kbh. Ø, Denmark. E-mail: mthorup@diku.dk, WWW:
http://www.diku.dk/~mthorup.

Abstract. The *Maximum Agreement Subtree* (MAST) is a well-studied
measure of similarity of leaf-labelled trees. There are several variants,
depending on the number of trees, their degrees, and whether or not
they are rooted. It turns out that the different variants display very
different computational behavior. We address the common situation in
biology, where the involved trees are rooted and of bounded degree, most
typically simply being binary.

- We give an algorithm which computes the MAST of k trees on n species
  where some tree has maximum degree d in time $O(kn^3 + n^d)$. This
  improves the Amir and Keselman FOCS '94 $O(kn^{d+1} + n^{2d})$ bound.
- We give an algorithm which computes the MAST of 2 trees with degree
  bound d in time $O(n\sqrt{d}\log^3 n)$. This should be contrasted with the
  Farach and Thorup FOCS '94 $O(nc^{\sqrt{\log n}} + n\sqrt{d}\log n)$ bound. Thus,
  for d a constant, we get an $O(n\log^3 n)$ bound, replacing the previous
  $O(nc^{\sqrt{\log n}})$ bound.

Both of our algorithms are quite simple, relying on the combinatorial
structure of the problem, rather than on advanced data structures.

1 Introduction

Let A be a set. Define a *leaf labeled tree*, T, on A to be a rooted tree with no
degree 1 nodes such that the leaves of T are uniquely labeled with the elements of
A. For convenience, we will identify the leaves with their labels when the tree is
understood. Leaf labeled trees are used in many settings to model clustering [3].
In particular, they are used in biology to model the evolution of species: the
leaves represent the species under consideration, and the internal nodes represent
posited ancestors.

Given a leaf labeled tree T on set A, and given $B \subseteq A$, the *topological
restriction of T to B*, written $T|B$, is the tree with vertex set $\{\mathrm{lca}^T(a, b) | (a, b) \in$

---

* Part of this work was done when this author was visiting DIMACS and University
of Warwick.

B^2}, where the arcs are defined such that for all $(a, b) \in B^2$, $\mathrm{lca}^{T|B}(a, b) = \mathrm{lca}^T(a, b)$. Here $\mathrm{lca}^T(x, y)$ is the least common ancestor of nodes x and y in T, for any tree T. Such a tree is uniquely defined by B. More operationally, we get $T|B$ from T by first removing all nodes without descendants in B, and then contracting any path of degree 1 nodes to a single edge. Given a set $B \subseteq A$ the tree $T|B$ can be computed in $O(n)$ time [4].

This restriction operator immediately implies a similarity measure on trees, as follows.

The Maximum Agreement Subtree Problem (MAST)
Input: A tuple $\mathbf{T} = (T_1, \ldots, T_k)$ of leaf labeled trees on some common set A.
Output: A maximum cardinality subset B of A such that $T_1|B, \ldots, T_k|B$ are isomorphic, where all isomorphisms are understood to map leaves with the same labels to each other.

Finden and Gordon [6] introduced the MAST problem in order to study the similarities of evolutionary trees, for which it turns out to be a useful tool. They gave a heuristic method for computing the maximum agreement subtree of two binary trees. Others have since given proper algorithms and fast running times for the two tree case, both with constant and unbounded degree, and for both rooted and unrooted trees (see [4, 5, 10]).

Since evolutionary trees are usually binary, the most studied case of MAST computation is that of finding the MAST of two binary trees, i.e. Finden and Gordon's original problem. We call this problem the BinMAST problem. The fastest algorithm known for BinMAST runs in $O(nc^{\sqrt{\log n}})$ time [5]. In this paper, we give an $O(n \log^3 n)$ time algorithm. Our algorithm can be generalized to any degree bound d, the resulting complexity being $O(n\sqrt{d} \log^3 n)$

In [1], Amir and Keselman argue that biologists often want to compute the MAST of many trees, but show that computing the MAST of just 3 trees with unbounded degree is $\mathcal{NP}$-hard. However, as noted above, in many applications, (some of) the involved trees have bounded degree, and Amir and Keselman considered the general problem of computing the MAST of k trees on n leaves where at least one tree has maximum degree d, a problem which we will refer to as the $k/n/d$-MAST. In [1, 2], they showed that the $k/n/d$-MAST can be computed in $O(kn^{d+1} + n^{2d})$ time. We present an algorithm for $k/n/d$-MAST which runs in time $O(kn^3 + n^d)$.

1.1 BinMAST

Steel and Warnow gave the first polynomial time algorithm for computing the MAST of two trees [10]. Henceforth we refer to their algorithm as the SW algorithm. The SW algorithm is a dynamic program that computes the local solution to the MAST problem for each pair of subtrees. This is done bottom-up, and for constant degree trees, each new local solution is found in constant time, the total time being $O(n^2)$ (for unbounded degree trees, each local computation involves computing a maximum bipartite matching, and the total time becomes $O(n^{2.5} \log n)$).

It is natural to try sparsifying the SW algorithm, restricting the local MAST computations to a significant $o(n^2)$ subset of the subtree pairs. In [5], such sparsification was done implicitly by a fairly intricate recursive scheme specifically designed to give an optimal algorithm for the case of unbounded degrees. As mentioned above, for binary trees their recursion takes $O(nc^{\sqrt{\log n}})$ time. Our algorithm is a simple dynamic program tailored to binary trees. A key feature of our algorithm is that we identify all significant subtree pairs initially before the dynamic program starts. We are thus able to radically simplify the method for actually computing the needed values, and therefore we have an algorithm which is both more efficient and much simpler. It solves the MAST problem for binary trees computing $O(n \log^2 n)$ local MAST values in $O(n \log^3 n)$ total time. The technique presented in this paper can quite easily be generalized to give an $O(n\sqrt{d} \log^3 n)$ time algorithm for degree d bounded trees. However, for unbounded degrees, the technique of [5] is still preferable.

The main idea is to combine a decomposition of the trees into paths with dynamic programming algorithms for two other optimization problems: a variant of the Maximum Weight Common Subsequence Problem (which, because of the interpretation in terms of bipartite graphs used in this paper, we call the Maximum Non-Crossing Matching problem) and a problem that we call the Maximum Crossing problem. These problems are formally presented in Sect. 2. In Sect. 3, we present our solution to the BinMAST problem.

1.2 $k/n/d$-MAST

The SW approach is naturally generalized to k trees, but there are $\Theta(n^k)$ k-tuples of subtrees, so the size of the dynamic program is exponential in k. As we shall discuss at the end of Sect. 3, it seems that none of our sparsification ideas for the SW algorithm for two trees can avoid exponentiallity in k.

The Amir-Keselman bound [1, 2] of $O(kn^{d+1} + n^{2d})$ is achieved by computing local MAST values for certain k-tuples of subtrees. Their sparsification is implicit in their recursive formulation of the problem. Our improvement to $O(kn^3 + n^d)$ is also based on a dynamic program, but a key is that we are **not** computing MAST values of k-tuples of subtrees. We are still considering such k-tuples, but, in contrast to all other efficient algorithms for MAST problems, we *compute a different function on the k-tuples*, allowing for our substantial savings. Our solution to the $k/n/d$-MAST problem is described in Sect. 4.

2 Definitions of the Maximum Non-Crossing Matching and the Maximum Crossing Pair Problems

Both problems are defined in the terms of bipartite graphs. Let $B = (I_0 \cup I_1, E)$ be a weighted bipartite graph, where $I_0 = \{1, \ldots, n_0\}$, $I_1 = \{1, \ldots, n_1\}$, and the weights of edges are non-negative. Let $|E| = m$ and let $n = \max(n_0, n_1)$. Assume, without loss of generality, that $n \le m$. The weight of edge (i, j) is denoted by $w(i, j)$. Extend the weight function to any pair (i, j), where $1 \le i \le n_0$ and

$1 \leq j \leq n_1$ by letting $w(i, j) = 0$ for all $(i, j) \notin E$. An edge (i, j) *dominates* (i', j') if $i \leq i'$ and $j \leq j'$. Two pairs $(i, j), (i', j')$ *cross* if $i \leq i'$ and $j \geq j'$ or $i \geq i'$ and $j \leq j'$. The crossing between such a pair is *proper* if $i \neq i'$ and $j \neq j'$.

Nested Maximum Non-Crossing Matching with Cuts. The *Maximum Non-Crossing Matching* (*MNCM*) problem is: given a weighted bipartite graph B, find a maximum total weight set of non-crossing edges. In this paper, we use a slightly more general version of the problem. The generalization goes in two ways. First, we are interesting in finding, for all edges (i, j), $MNCM(i, j)$, which is defined to be the *MNCM* of the graph restricted to the edges that are dominated by (i, j). We call this version of the problem the *Nested Maximum Non-Crossing Matching* problem (*MNCM**). The standard algorithms to solve the *MNCM* problem solve, in fact, the nested version of the problem. Our second extension is to assume that the graph also contains a special type of weighted edges, called *cut edges*. We are interested in the Non-Crossing Matching problem, subject to the restriction that if the matching contains a cut edge (i, j) then it cannot contain any edge dominated by (i, j). We call this problem the Maximum Non-Crossing Matching problem with Cuts Problem and denote it by $MNCM_c$. Finally, the Nested Maximum Non-Crossing Matching with Cuts Problem ($MNCM_c^*$) is that of computing, for all edges (i, j), the value $MNCM_c(i, j)$, which, as before, is defined to be the $MNCM_c$ of the graph restricted to the edges dominated by (i, j). Let $cut(i, j)$ denote the weight of the cut edge between i and j (if there is no cut edge between i and j, set $cut(i, j) = 0$).

The $MNCM_c^*$ problem can be solved in $O(n^2)$ time by dynamic programming using the following recurrence:

$$MNCM_c(i, j) = \begin{cases} 0 & \text{for } i < 0 \text{ or } j < 0 \\ \max \begin{cases} MNCM_c(i-1, j) \\ MNCM_c(i, j-1) \\ MNCM_c(i-1, j-1) + w(i, j) \\ cut(i, j) \end{cases} & \text{otherwise} \end{cases} \quad (1)$$

In the case of sparse graphs, the $O(m \log n)$ time (or $O(m \log \log n)$-time using integer priority queue operations of [9]) algorithm for *MNCM* [8] extends trivially to the $MNCM_c^*$ problem.

Nested Maximum Crossing problem. Consider a pair (i, j), where $0 < i \leq n_0$, $0 < j \leq n_1$. We say the crossing between e_1, e_2 *is dominated* by (i, j) if e_1 and e_2 cross properly and both of them are dominated by (i, j). The Maximum Crossing (*MC*) problem is that of finding the maximum total weight of a pair of crossing edges. As in the case of Maximum Non-Crossing Matching, we are interested in the "nested" version of the problem (*MC**). Namely, for any (i, j) where $i \in I_0$, $j \in I_1$ compute $MC(i, j)$, which is defined to be the *MC* of the graph restricted to the edges that are dominated by (i, j).

The *MC** problem can also be solved in $O(n^2)$ time by dynamic programming, based on the following recurrence:

$$MC(i, j) = \begin{cases} 0 & \text{if } i = 0 \text{ or } j = 0 \\ \max \begin{cases} MC(i-1, j) \\ MC(i, j-1) \\ \max\{w(i', j)|i' < i\} + \max\{w(i, j')|j' < j\} \end{cases} & \text{otherwise} \end{cases} \quad (2)$$

In fact, we need to solve a slightly more general problem:

The Red-Green Crossing problem. Let B be a bipartite whose edges are colored with two colors: red and green. Order the set of edges, E, by lexicographic order. (We represent the edges as pairs such that the endpoint that belongs to I_0 is the first element of a pair.) A *red-green* crossing is a crossing between a red edge and a green edge such that the green edge is lexicographically larger than the red edge. Our goal is to compute the solution to the nested version of the following problem: find the maximum weight red-green crossing. We refer to this problem as to the Maximum Red-Green Crossing problem and abbreviate it by $MRGC^*$. The algorithm for $MRGC^*$ will be used in our algorithm for the Maximum Agreement Subtree problem.

The MC^* problem can be reduced to the colored version by representing each edge as a pair of edges of the same weight and different colors.

In the full journal version of this paper will contain an $O(m \log n)$-time algorithm for the $MRGC^*$ problem.

3 The BinMAST Problem

Fix the two trees T_0, T_1 for which we want to solve the maximum agreement subtree problem. For any graph $G = (V, E)$, let $V(G) = V$. For any subtree T' of a rooted tree T, let $r(T')$ be the node in T' of minimum depth. For all $(v_0, v_1) \in V(T_0) \times V(T_1)$, let $\text{MAST}(v_0, v_1)$ denote the MAST of the subtrees rooted respectively at v_0 and v_1. Here the subtrees are understood to be topologically restricted to the intersection of their label sets. More formally, $\text{MAST}(v_0, v_1) = \text{MAST}(T_0|B, T_1|B)$ where $B = A(v_0) \cap A(v_1)$, and where $A(v)$ is the label set descending from v.

For each internal vertex v in T_i ($i = 0, 1$), let the *center child*, denoted $cntr(v)$ be the node with the maximum number of descending leaves, and let the *side child*, denoted $side(v)$ be the other child, with ties broken arbitrarily. Correspondingly, call the arc $(x, cntr(x))$ a *center arc*, and $(x, side(x))$ a *side arc*. By a *center path* we mean a maximum path using center arcs. Let $\prec$ order the center paths in each tree following the preordering of the roots of the center paths. Also, let $\prec$ be the corresponding lexicographic ordering of the center path pairs (P_0, P_1) where P_i is a center path in T_i. We will process the center path pairs, computing certain MAST-values in the order determined by $\prec$. We will use the recursive formula below, which is a simple reformulation of the one used in the SW algorithm [10]. For technical reasons, for any leaf v, we define

$side(v) = cntr(v) = \perp$, where $\perp$ is a special "vertex" not belonging to any tree.

$$
\text{MAST}(x_0, x_1) = \begin{cases} 0 \text{ if } x_0 = \perp \text{ or } x_1 = \perp. \text{ Otherwise:} \\ \max \begin{cases} a)1 \text{ if } (x_0, x_1) \text{ is an interesting leaf pair.} \\ b)\text{MAST}(x_0, cntr(x_1)), \text{MAST}(cntr(x_0), x_1) \\ c)\text{MAST}(cntr(x_0), cntr(x_1)) + \underline{\text{MAST}(side(x_0), side(x_1))} \\ d)\text{MAST}(side(x_0), x_1), \text{MAST}(x_0, side(x_1)) \\ e)\underline{\text{MAST}(side(x_0), cntr(x_1))} + \underline{\text{MAST}(cntr(x_0), side(x_1))} \end{cases} \end{cases} (3)
$$

The underlining indicate vertex pairs belonging center path pairs preceding that of (x_0, x_1) in $\prec$. To see the relationship with the previous recurrence formulas 1 and 2, think of $cntr(x_i)$ as $x_i - 1$. In the following, we shall see that we only need to compute relatively few MAST-values.

Interesting and significant vertex pairs. For each internal vertex v in T_i, $i = 0, 1$, let $A^s(v)$ denote the label set descending from $side(v)$. For leaves v, $A^s(v) = A(v)$. Note that if r is the root of a center path P, then $\{A^s(v)|v \in V(P)\}$ is a partitioning of $A(r)$.

A vertex pair $(v_0, v_1) \in V(T_0) \times V(T_1)$ is *interesting* if $A^s(v_0) \cap A^s(v_1) \neq \emptyset$.

Lemma 1. *The are at most $n \log^2 n$ interesting vertex pairs, and they can be identified in time $O(n \log^2 n)$.*

Proof: Any interesting vertex pair is witnessed by a label, and since the path from a leaf to the root pass at most $\log n$ side arcs, each of the n labels can witness at most $\log^2 n$ interesting pairs. ∎

A center path pair (P_0, P_1) is *interesting* if it contains an interesting vertex pair, or equivalently, if $A(r(P_0)) \cap A(r(P_1))) \neq \emptyset$. The size of (P_0, P_1), denoted $||P_0, P_1||$, is the number of interesting vertex pairs $(v_0, v_1) \in V(P_0) \times V(P_1)$. Thus, by Lemma 1, $\sum_{(P_0, P_1)} ||P_0, P_1|| \leq n \log^2 n$. A vertex pair (v_0, v_1) in an interesting center path pair (P_0, P_1) is said to be *significant* if either

(i) (v_0, v_1) is interesting,
(ii) $(v_0, v_1) = (r(P_0), r(P_1))$, or
(iii) v_0 or v_1 is the root of its center path and the other node belongs to an interesting vertex pair.

In relation to recurrence 3, note the following relationship between interesting and significant vertex pairs:

Lemma 2. *If $(v_0, v_1) \in V(P_0) \times V(P_1)$ is interesting then $(v_0, side(v_1))$, $(side(v_0), v_1)$, and $(side(v_0), side(v_1))$ are all significant, and belong to center path pairs preceding (P_0, P_1) in $\prec$.*

Proof: For $i = 0, 1$, let P_i be the path containing v_i, and let Q_i be the path with root $side(v_i)$. By definition (v_0, v_1) is interesting because $A^s(v_0) \cap A^s(v_1) \neq \emptyset$. Thus, v_0 is interesting with respect to some vertex in Q_1, so v_0 is interesting in (P_0, Q_1), so $(v_0, side(v_1))$ satisfies (iii) relative to $(P_0, Q_1) \prec (P_0, P_1)$. The case of $(side(v_0), v_1)$ is symmetric. Concerning $(side(v_0), side(v_1))$ we note that $(side(v_0), side(v_1))$ satisfies (ii) relative to $(Q_0, Q_1) \prec (P_0, P_1)$. ∎

In our dynamic program we want to compute the MAST-value for all significant vertex pairs. These MAST values are stored in a binary search tree, thus they can be accessed in $O(\log n)$ time. Note that there are at most $3 \cdot \|P_0, P_1\| + 1$ significant pairs in (P_0, P_1). Uninteresting center path pairs have no significant vertex pairs, so by Lemma 1, in total there are only $O(n \log^2 n)$ significant vertex pairs. Also note that in $O(n \log^2 n)$ time, we can identify all significant vertex pairs for all interesting center path pairs. We will compute the MAST-values of significant vertex pairs for one interesting path pair at a time, following the lexicographic ordering given by $\prec$. More specifically we wish show:

Proposition 3. *Let (P_0, P_1) be an interesting center path pair. Given the MAST-values for all significant pairs belonging to center path pairs $(Q_0, Q_1) \prec (P_0, P_1)$, we can compute the MAST-values for all significant pairs in (P_0, P_1) in $O(\|P_0, P_1\| \log n)$ time.*

Proof: The proof of the proposition essentially covers the rest of this section. Fix the interesting center path pair (P_0, P_1), and assume we have computed the MAST-values for all significant pairs belonging to center path pairs $(Q_0, Q_1) \prec (P_0, P_1)$. Thus, by Lemma 2, if $(v_0, v_1) \in V(P_0) \times V(P_1)$ is interesting, then $\text{MAST}(v_0, side(v_1))$, $\text{MAST}(side(v_0), v_1)$, and $\text{MAST}(side(v_0), side(v_1))$ are all known.

Modifying the recurrence formula. In order to relate to the *MNCM* and *MC* recurrences 1 and 2, we enumerate the vertices in each P_i from 0 to $|P_i| - 1$ starting from the leaf. Thus $x - 1$ replaces $cntr(x)$, and we set $\perp = -1$. Now, rewrite recurrence 3 as:

$$\text{MAST}(x_0, x_1) = \begin{cases} 0 \text{ if } x_0 = -1 \text{ or } x_1 = -1. \text{ Otherwise:} \\ \max \begin{cases} a)1 \text{ if } (x_0, x_1) \text{ is an interesting leaf pair.} \\ b)\text{MAST}(x_0, x_1 - 1), \text{MAST}(x_0 - 1, x_1) \\ c)\text{MAST}(x_0 - 1, x_1 - 1) + \text{MAST}(side(x_0), side(x_1)) \\ \quad \text{if } (x_0, x_1) \text{ is interesting.} \\ d)cut'(x_0, x_1) \text{ if } (x_0, x_1) \text{ is interesting.} \\ e)cut''(x_0, x_1) \text{ if } (x_0 + 1, x_1 + 1) \text{ is interesting} \\ \quad \text{or } (x_0, x_1) \text{ is significant.} \end{cases} \end{cases} \quad (4)$$

where cut' and cut'' replace the expressions in the last two lines of (3) and are explained below. Note that our condition in c) of (x_0, x_1) being interesting is valid in the sense that otherwise $\text{MAST}(side(x_0), side(x_1)) = 0$ and then c) follows by applying b) twice. Set $m = \|P_0, P_1\|$, that is, let m be the number of interesting vertex pairs in (P_0, P_1). Recall that the number of significant vertex pairs is bounded by $3m + 1 = O(m)$. Assuming appropriate definitions of cut' and cut'', we now have an $MNCM_c^*$ problem with $O(m)$ cut-edges corresponding to a), d), and e), $O(m)$ normal "matching" edges corresponding to c), and $O(m)$ "query" edges (x_0, x_1) for which we want to know $MNCM(x_0, x_1)$; namely the significant vertex pairs. At the moment, we may have some vertices that are not incident with any of these edges. However, from the significant edges, we can identify all the matching edges, and clearly, in $O(m \log n)$ time, we can sort the involved vertices so as to skip the superfluous vertices. Afterwards, we have an $O(m)$

sized $MNCM_c^*$ problem, which we know that we can solve in time $O(m \log n)$. This time bound also includes the $O(m)$ accesses of $O(\log n)$ time to look up needed MAST values.

The cut edges. We define

$$cut'(x_0, x_1) = \max\{\text{MAST}(side(x_0), x_1), \text{MAST}(x_0, side(x_1))\}. \qquad (5)$$

The reason why in 3d) we only need to consider the case where (x_0, x_1) are interesting is that otherwise, for $\{i, \bar{i}\} = \{0, 1\}$, $\text{MAST}(side(x_i), x_{\bar{i}}) = \text{MAST}(side(x_i), x_{\bar{i}} - 1) \le \text{MAST}(x_i, x_{\bar{i}} - 1)$. In other words, if (x_0, x_1) is not interesting then $b)$ overrules $d)$.

The value $cut''(x_0, x_1)$ replaces the last line in the recurrence (3). Unfortunately, there may be $\Omega(n^2)$ non-zero values of the sums $\text{MAST}(side(x_0), x_1 - 1) + \text{MAST}(x_0 - 1, side(x_1))$. We define cut'' as follows:

$$cut''(x_0, x_1) = \max \begin{cases} b) cut''(x_0 - 1, x_1), \ cut''(x_0, x_1 - 1) \\ e) \max\{\text{MAST}(side(x_0), x_1') \mid x_1' < x_1 \text{ and } (x_0, x_1') \text{ is} \\ \quad \text{interesting}\} + \max\{\text{MAST}(x_0', side(x_1)) \mid x_0' < x_0 \\ \quad \text{and } (x_0', x_1) \text{ is interesting}\} \end{cases} \qquad (6)$$

To see that this definition of cut'' is correct, first recall that if (x_0, x_1) is not interesting, for $\{i, \bar{i}\} = \{0, 1\}$, $\text{MAST}(side(x_i), x_{\bar{i}}') = \text{MAST}(side(x_i), x_{\bar{i}}' - 1)$. Hence $\text{MAST}(side(x_i), x_{\bar{i}} - 1)$ equals

$$\max\{\text{MAST}(side(x_0), x_{\bar{i}}') \mid x_{\bar{i}}' < x_{\bar{i}} \text{ and } (x_0, x_{\bar{i}}') \text{ is interesting}\}.$$

Thus 6e) is equal to 3e).

The point in 6b) is that it can substitute for 3b). More precisely, all calls to 3b) immediately preceding a call to 3e) are translated into calls to 6b)—any other calls to 3b) are replaced by calls to 4b). Thus, our entry (x_0, x_1) into 4e) is either a start query meaning that (x_0, x_1) is significant, or it is following a call to c) implying that $(x_0 + 1, x_1 + 1)$ is interesting. Thus we may conclude that recurrence formulas 4,5 and 6 are equivalent to recurrence formula 3.

Now, recurrence formula 4 is equivalent to the recurrence 2 of the Nested Red-Green Maximum Crossing Pair problem ($MRGC^*$). For each interesting pair (x_0, x_1), we have a green edge (g, x_0, x_1) with weight $\text{MAST}(side(x_0), x_1)$ and a red edge (r, x_0, x_1) with weight $\text{MAST}(x_0, side(x_1))$. As before, we note that in $O(m \log n)$ time, we can skip all vertices not incident with any red or green edges, and afterwards, we have an $MRGC^*$ problem is of size $O(m)$. In the journal version this problem will be solved in time $O(m \log n)$. This completes the computation of the MAST-values of all significant vertex pairs in (P_0, P_1), based on MAST-values of significant vertex pairs from center path pairs $(Q_0, Q_1) \prec (P_0, P_1)$. Thus follows Proposition 3. ∎

In conclusion, we have shown

Theorem 4. *The MAST problem on two binary trees can be solved in $O(n \log^3 n)$ time.*

Proof: By Lemma 1 there are at most $n\log^2 n$ interesting vertex pairs, and hence at most $n\log^2 n$ interesting center path pairs. These are sorted by the lexicographic ordering $\prec$ in time $O(n\log^3 n)$. Processing the path pairs (P_0, P_1) following this ordering, by Proposition 3, we can compute the MAST-values of all significant vertex pairs in (P_0, P_1) in $O(\|P_0, P_1\|\log n)$ time, hence in $O(n\log^3 n)$ total time. At the end, we can return $\text{MAST}(r(T_0), r(T_1))$ which by case (ii) is a significant pair in the center path pair containing both roots. $\blacksquare$

3.1 Comments

Consider, now, the case of trees with some small degree bound, d. Things become slightly more complicated in this case. The recurrence for $\text{MAST}(v_0, v_1)$ then involves a weighted bipartite matching, where the independent sets are the children of v_0 and v_1, and where the weight function is MAST. Without going into details, in [4, §2] it is shown that we can reduce these matchings so that in total they contain only $O(n\log^2 n)$ "interesting edges". Since the vertex sets in each matching are of size at most $2d$, using the best bound for weighted bipartite matching [7], we get that the matching work can be done in time $O(n\sqrt{d}\log^3 n)$, which is thus our bound for computing the MAST of two trees with degree bounded by d. This should be contrasted with the previous bound of $O(n^{1+o(1)} + n\sqrt{d}\log n)$ from [5]. The $n^{o(1)}$ term hides an expression which is $\Omega(c^{\sqrt{\log n}})$. Thus our bound of $O(n\sqrt{d}\log^3 n)$ is worse for large d, but better for small values of d, say, if $d = \text{polylog}\, n$. In particular, our result is superior for the common case of constant degree trees.

Our $n\log^2 n$ bound on the number of interesting vertex pairs (Lemma 1) generalize naturally to an $n\log^k n$ bound on the number of "interesting k-tuples" for k trees. Based on this observation, we would like to see our techniques for 2 binary trees generalized to an $O(n\log^{O(k)} n)$ time algorithm for k binary trees. It should be noted that for variable degrees, the problem fundamentally changes character for $k > 2$. In [1] Amir and Keselman showed MAST to be NP-hard for just 3 unbounded degree trees. Moreover, using an entirely different approach, they presented an $O(kn^{d+1} + n^{2d})$ bound for this problem [1, 2]. In the following section, we improved this bound to $O(kn^3 + n^d)$, which is $O(kn^3)$ for binary trees. Thus the suggested generalization of the above presented techniques would only be relevant for small values of k.

4 The $k/n/d$-MAST Problem

Fix a tuple $\mathbf{T} = (T_1, \ldots, T_k)$ of leaf labeled trees over a set A on size n, and let d be the degree bound of T_1. An *agreement set* is any set $B \subseteq A$ such that $T_1|B, \ldots, T_k|B$ are isomorphic. So MAST is the problem of finding a maximum cardinality agreement set. For simplicity, we will limit ourselves to finding the maximum cardinality of an agreement set. However, the algorithm below is easily modifiable to produce an actual maximum cardinality agreement set within the same time bounds.

For any pair $(a, b) \in A^2$, let $\mathrm{lca}^{\mathbf{T}}(a, b) = (\mathrm{lca}^{T_1}(a, b), \ldots, \mathrm{lca}^{T_k}(a, b))$. We will refer to the k-tuples generated in this way as $\mathrm{lca}^{\mathbf{T}}$-*tuples*. Any agreement set B defines a canonical representative for the isomorphic trees $T_1|B, \ldots, T_k|B$, as follows.

Definition 5. For any agreement set B of $\mathbf{T}$, the *agreement tree* $\mathbf{T}|B$, is the tree with vertex set $\{\mathrm{lca}^{\mathbf{T}}(a, b)| (a, b) \in B^2\}$, where the arcs are defined such that for all $(a, b) \in B^2$, $\mathrm{lca}^{\mathbf{T}}(a, b)$ is the least common ancestor of $\mathrm{lca}^{\mathbf{T}}(a, a)$ and $\mathrm{lca}^{\mathbf{T}}(b, b)$. The *size* of $\mathbf{T}|B$ is the cardinality of B.

Clearly $\mathbf{T}|B$ is uniquely defined and isomorphic to $T_1|B, \ldots, T_k|B$, as desired. All vertices in agreement trees are $\mathrm{lca}^{\mathbf{T}}$-tuples, of which there are only $O(n^2)$. Moreover, in time $O(kn^2)$, we can enumerate all the different $\mathrm{lca}^{\mathbf{T}}$-tuples, and tabulate $\mathrm{lca}^{\mathbf{T}}(\cdot, \cdot)$ accordingly by a simple tree traversal within each T_i. In the following, let $\mathbf{u}, \mathbf{v}, \mathbf{w}$ be $\mathrm{lca}^{\mathbf{T}}$-tuples.

Definition 6. For all $\mathbf{v}$, let $\mathrm{mast}(\mathbf{v})$ denote the maximum size of an agreement tree with root $\mathbf{v}$.

We will present a dynamic program that computes $\mathrm{mast}(\mathbf{v})$ for all $\mathbf{v}$ in time $O(n^3 + n^d)$. Note that $\mathrm{mast}((v_1, \ldots, v_k))$ is not the MAST-value of the subtrees descending from $(v_1, \ldots, v_k)$—this MAST-value should have been expressed as the maximum size of an agreement tree whose root $(w_1, \ldots, w_k)$ satisfies that w_i is an ancestor of v_i for all $i \leq k$. Computing mast-values instead of MAST-values in the dynamic program is one of the keys to our improvements to the Amir-Keselman $O(kn^{d+1} + n^{2d})$ bound. Having computed the n^2 mast-values, in time $O(n^2)$, we can find $\mathrm{MAST}(\mathbf{T})$ as $\max_{\mathbf{v}}\{\mathrm{mast}(\mathbf{v})\}$.

Let $\mathbf{v}$ *dominate* $\mathbf{w}$, written $\mathbf{v} \succ \mathbf{w}$, if $\mathbf{v}[i]$ is a strict ancestor of $\mathbf{w}[i]$, for all $i \leq k$. Thus if $\mathbf{w}$ descends from $\mathbf{v}$ in some agreement tree, then $\mathbf{v} \succ \mathbf{w}$. Unfortunately, it can be shown that $\succ$ is of size $\Theta(n^4)$, as is its transitive reduction. Hence $\succ$ cannot be computed within the desired time-bounds if $d = 2$. Let the domination $\mathbf{v} \succ \mathbf{w}$ have *direction* $\boldsymbol{\delta} = (\delta_1, \ldots, \delta_k)$ if for $i \leq k$, $\mathbf{w}[i]$ descends from, or is equal to, the δ_ith child of $\mathbf{v}$. In this case, denote $\boldsymbol{\delta}$ by $(\mathbf{v} \triangleright \mathbf{w})$, and call $\boldsymbol{\delta}$ an *active direction* from $\mathbf{v}$. If a is any *label descending* from $\mathbf{w}$, meaning that a is a label of a leaf descending from $\mathbf{w}[i]$ in $\mathbf{T}[i]$ for all $i \leq k$, then $\mathbf{v} \succ \mathrm{lca}^{\mathbf{T}}(a, a)$ and $(\mathbf{v} \triangleright \mathbf{w}) = (\mathbf{v} \triangleright \mathrm{lca}^{\mathbf{T}}(a, a))$. Thus, trivially, there are $O(n)$ active directions from $\mathbf{v}$. Two directions $\boldsymbol{\delta}_1, \boldsymbol{\delta}_2$ are *compatible*, denoted $\boldsymbol{\delta}_1 \perp \boldsymbol{\delta}_2$, if they differ in all coordinates.

Definition 7. Let $\mathbf{v}$ *properly dominate* $\mathbf{w}$, denoted $\mathbf{v} > \mathbf{w}$, if $\mathbf{v} \succ \mathbf{w}$ and there exists a $\mathbf{u} \prec \mathbf{v}$ such that $(\mathbf{v} \triangleright \mathbf{u}) \perp (\mathbf{v} \triangleright \mathbf{w})$. In this case, call $(\mathbf{v} \triangleright \mathbf{w})$ a *proper direction* from $\mathbf{v}$, and denote the set of such proper directions from $\mathbf{v}$ by $D(\mathbf{v})$.

Note that proper dominance is tight with respect to agreement trees in the following sense:

Observation 8. $\mathbf{v} > \mathbf{w}$ *if and only if* $\mathbf{w}$ *descends from* $\mathbf{v}$ *in some agreement tree.*

Restricting ourselves from $\prec$ to $<$ is worth a linear factor:

Lemma 9. *The ordering $<$ is of size $O(n^3)$, and can be computed in time $O(kn^3)$.*

Proof: We will prove the lemma by showing that for any $\mathbf{w}$ there are at most n $\mathtt{lca}^{\mathbf{T}}$-tuples $\mathbf{v} > \mathbf{w}$, and that all these $\mathbf{v} > \mathbf{w}$ can be found in time $O(kn)$.

For any $\mathbf{v} > \mathbf{w}$, there is a label a witnessing that the domination is proper, i.e. $\mathbf{v} \succ \mathtt{lca}^{\mathbf{T}}(a, a)$ and $(\mathbf{v} \triangleright \mathbf{w}) \perp (\mathbf{v} \triangleright \mathtt{lca}^{\mathbf{T}}(a, a))$. But a only witnesses one $\mathbf{v} > \mathbf{w}$; namely $\mathbf{v} = \mathtt{lca}^{\mathbf{T}}(a, b)$ where b is any label descending from $\mathbf{w}$. To compute the $\mathbf{v} > \mathbf{w}$ efficiently, we fix b, and let a range over all labels such that for no $i \leq k$, a is a label of a leaf descending from $\mathbf{w}[i]$ in $\mathbf{T}[i]$. The set of such a's is trivially computed in time $O(kn)$, and since there are fewer than n of these, the $\mathtt{lca}^{\mathbf{T}}(a, b)$ values are found in time $O(kn)$. ∎

Define the *compatibility graph* of $\mathbf{v}$ to be $G(\mathbf{v}) = (D(\mathbf{v}), E(\mathbf{v}))$, where $E(\mathbf{v}) = \{\{\delta_1, \delta_2\} | \delta_1 \perp \delta_2\}$. We are now ready to describe the basic recursion of our dynamic program:

Lemma 10. *For each internal $\mathtt{lca}^{\mathbf{T}}$-tuple $\mathbf{v}$, let $G(\mathbf{v}) = (D(\mathbf{v}), E(\mathbf{v}))$ be its compatibility graph. For each $\delta \in D(\mathbf{v})$, let $M[\mathbf{v}, \delta] := \max\{\mathtt{mast}(\mathbf{w}) | \mathbf{v} > \mathbf{w}$ and $(\mathbf{v} \triangleright \mathbf{w}) = \delta\}$ be the weight of δ in $G(\mathbf{v})$. Then*

$$\mathtt{mast}(\mathbf{v}) = \begin{cases} 1 & \text{if } v \text{ is a leaf;} \\ \mathtt{MWC}(G(\mathbf{v})) & \text{otherwise,} \end{cases}$$

where $\mathtt{MWC}(G(\mathbf{v}))$ is the weight of a maximum weight clique in $G(\mathbf{v})$.

Proof: By induction: If $\mathbf{v}$ is a leaf, by definition, $\exists a \in A$ such that $\mathbf{v} = \mathtt{lca}^{\mathbf{T}}(a, a)$, so $\mathtt{mast}(v) = 1$.

Fix an internal $\mathtt{lca}^{\mathbf{T}}$-tuple $\mathbf{v}$. Let $\delta^* = \{\delta_1, \ldots, \delta_l\}$, $l \geq 2$ be a maximum weighted clique in $G(\mathbf{v})$, and for $i \leq l$, choose $\mathbf{w}_i < \mathbf{v}$ such that $(\mathbf{v} \triangleright \mathbf{w}_i) = \delta_i$ and $M[\mathbf{v}, \delta_i] = \mathtt{mast}(\mathbf{w}_i)$. By induction, for each i, we have an agreement tree t_i rooted in $\mathbf{w}_i$ of size $\mathtt{mast}(\mathbf{w}_i)$. The tree with root $\mathbf{v}$ and the t_i as subtrees of $\mathbf{v}$ is an agreement tree of size $\sum_i \mathtt{mast}(\mathbf{w}_i) = \mathtt{MWC}(G)$. Thus $\mathtt{MWC}(G(\mathbf{v})) \leq \mathtt{mast}(\mathbf{v})$.

Conversely, consider an agreement tree t with root $\mathbf{v}$ and children $\mathbf{w}_1, \ldots, \mathbf{w}_l$. Thus t has size $\mathtt{mast}(\mathbf{v}) = \sum_i \mathtt{mast}(\mathbf{w}_i)$. Then $(\mathbf{v} \triangleright \mathbf{w}_i) \perp (\mathbf{v} \triangleright \mathbf{w}_j)$ $\forall i < j \leq l$. Since $l > 1$, this means that all directions $(\mathbf{v} \triangleright \mathbf{w}_i)$ are proper and that $\{(\mathbf{v} \triangleright \mathbf{w}_i)\}_i$ forms a clique in $G(\mathbf{v})$. Consequently $\mathtt{mast}(\mathbf{v}) = \sum_{i=1}^{l} \mathtt{mast}(\mathbf{w}_i) \leq \mathtt{MWC}(G(\mathbf{v}))$. ∎

In order to implement Lemma 10, first notice that each edge $\{\delta_1, \delta_2\}$ in $G(\mathbf{v})$ is witnessed by any pair of labels a, b, where $\delta_1 = (\mathbf{v} \triangleright \mathtt{lca}^{\mathbf{T}}(a, a))$ and $\delta_2 = (\mathbf{v} \triangleright \mathtt{lca}^{\mathbf{T}}(b, b))$. Then $\mathbf{v} = \mathtt{lca}^{\mathbf{T}}(a, b)$, so no label pair can witness more than one edge total in all the compatibility graphs. Summing up, we conclude:

Proposition 11. *In $O(kn^3)$ time we can tabulate:*

- $\forall a, b \in A : \mathtt{lca}^{\mathbf{T}}(a, b)$.

- *The partial order $<$. Enumerate all $\text{lca}^{\mathbf{T}}$-tuples by topological sort order in $<$.*
- *$\forall v \forall \text{lca}^{\mathbf{T}}(a, a) < v : (v \triangleright \text{lca}^{\mathbf{T}}(a, a))$.*
- *$\forall v : G(v) = (D(v), E(v))$. The directions in $D(v)$ are enumerated, so that each can be identified in constant time.*
- *$\forall v \forall \delta \in D(v) : W[v, \delta] := \{w | v > w \text{ and } (v \triangleright w) = \delta\}$.* ∎

In our implementation of Lemma 10, we will, in fact, consider all cliques in all $G(v)$. For worst-case considerations, these are generated efficiently as follows:

Proposition 12. *In $O(d^3(ne/d)^d)$, we can tabulate:*

- *$\forall v : C(v) = \{c | c \text{ is a clique in } G(v)\}$.*

Furthermore $\sum_v |C(v)| \in O(d(ne/d)^d)$

Proof: Consider a clique $\delta^* = \{\delta_1, \ldots, \delta_l\}$, $l \geq 2$, in $G(v)$. Then there is a label set $\mathbf{a}^* = \{a_1, \ldots, a_l\}$ witnessing this clique in the sense that $(v \triangleright \text{lca}^{\mathbf{T}}(a_i, a_i)) = \delta_i \ \forall i \leq l$. Furthermore, $v = \text{lca}^{\mathbf{T}}(a_i, a_j) \ \forall i < j \leq l$. Thus, $\mathbf{a}^*$ uniquely determines δ^* and v. Due to the degree bound on T_1 we know that $l \leq d$, so in conclusion, we can generate all cliques in all $G(v)$ by considering all label sets $\{a_1, \ldots, a_l\}$, $2 \leq l < d$, checking for each, in $O(l^2)$ time, whether for some v, $v = \text{lca}^{\mathbf{T}}(a_i, a_j) \ \forall i < j \leq l$. Thus all cliques can be generated in time $O(\sum_{l \leq d} l^2 \binom{n}{l}) = O(d^3(ne/d)^d)$, and there are $O(d(ne/d)^d)$ of them. ∎

We are now ready to present a concrete algorithm.

Algorithm A: Given the above tables, computes $\text{MAST}(\mathbf{T})$ by computation of $\text{mast}(v)$ for all $\text{lca}^{\mathbf{T}}$-tuples v.

A.1. For all labels a, set $\text{mast}(\text{lca}^{\mathbf{T}}(a, a)) = 1$.

A.2. For all internal $\text{lca}^{\mathbf{T}}$-tuples v in topological sort order in $<$ do:

A.2.1. For all $\delta \in D(v)$:

A.2.1.1. $M[v, \delta] := \max\{\text{mast}(w) | w \in W(v, \delta)\}$.

A.2.2. Set $\text{mast}(v) = \text{MWC}(G(v)) = \max_{c \in C(v)} \{\sum M[v, \delta_i] | c = (\delta_1, \ldots, \delta_l)\}$.

A.3. Return $\max_v \{\text{mast}(v)\}$.

Clearly, the steps up till step A.2.1.1, are computed in time $O(|<|) = O(n^3)$ total time. By Proposition 12, all calls to Step A.2.2 take time $O(d \cdot d(ne/d)^d) = O(n^d)$. Finally Step A.3 is done in $O(n^2)$ time. By Propositions 11 and 12, all preprocessing takes time $O(kn^3 + n^d)$, so in conclusion, we have shown

Theorem 13. *$k/n/d$-MAST can be computed in $O(kn^3 + n^d)$ time.*

4.1 Comments

As presented, the algorithm uses $O(kn^2 + n^3 + n^d)$ space. It is, however, not difficult to generate the cliques on the fly, thus reducing the space requirement to $O(kn^2 + n^3)$. In concrete implementations of our algorithm it would pay off to be careful in the clique generation. For example, in practice, we expect many of the label sets used in Proposition 11 to witness the same cliques, and reducing this kind of redundancy would likely give a major speed-up.

References

1. A. Amir and D. Keselman. Maximum agreement subtrees in multiple evolutionary trees. *Proc. of the 35th IEEE Annual Symp. on Foundation of Computer Science*, pages 758–769, 1994.
2. A. Amir and D. Keselman. Maximum agreement subtrees in multiple evolutionary trees – a correction. Personal Communication, 1995.
3. J-P. Barthélemy and A. Guénoche. *Trees and Proximity Representations*. Wiley, New York, 1991.
4. M. Farach and M. Thorup. Fast comparison of evolutionary trees (extended abstract). *Proc. of the 5th Annual ACM-SIAM Symposium on Discrete Algorithms*, pages 481–488, 1994.
5. M. Farach and M. Thorup. Sparse dynamic programming for evolutionary tree comparison. *Proc. of the 35th IEEE Annual Symp. on Foundation of Computer Science*, pages 770–779, 1994.
6. C. R. Finden and A. D. Gordon. Obtaining common pruned trees. *Journal of Classification*, 2:255–276, 1985.
7. H. Gabow and R. Tarjan. Faster scaling algorithms for network problems. *SIAM Journal on Computing*, 18(5):1013–1036, 1989.
8. G.Jacobson and K-P. Vo. Heaviest increasing/common subsequence problems. *Proc. of 3rd Combinatorial Pattern Matching Conference*, 1992.
9. D.B. Johnson. A priority queue in which initialization and queue operations take $O(\log \log D)$ time. *Math. Systems Theory*, 15:295–309, 1982.
10. M. Steel and T. Warnow. Kaikoura tree theorems: Computing the maximum agreement subtree. *Information Processing Letters*, 48:77–82, 1993.

Approximation algorithms for feasible cut and multicut problems

Bo Yu and Joseph Cheriyan *

Abstract. Let $G = (V, E)$ be an undirected graph with a capacity function $u : E \rightarrow \Re_+$ and let $S_1, S_2, \ldots, S_k$ be k commodities, where each S_i consists of a pair of nodes. A set S of nodes is called feasible if it contains no S_i, and a cut $(S, \overline{S})$ is called feasible if S is feasible. We show that several optimization problems on feasible cuts are NP-hard. We give a $(4 \ln 2)$-approximation algorithm for the minimum capacity feasible v^*-cut problem. The multicut problem is to find a set of edges $F \subseteq E$ of minimum capacity such that no connected component of $G \setminus F$ contains a commodity S_i. We show that an α-approximation algorithm for the minimum-ratio feasible cut problem gives a $2\alpha(1 + \ln T)$-approximation algorithm for the multicut problem, where T denotes the cardinality of $\bigcup_i S_i$. We give a new approximation guarantee of $O(t \log T)$ for the minimum capacity-to-demand ratio Steiner cut problem; here each S_i is a set of nodes and t denotes the maximum cardinality of a commodity S_i.

1 Introduction

Polynomial-time approximation algorithms for solving various NP-hard problems on graphs involving cuts and multicuts have recently attracted a great deal of research. For example, Dahlhaus et al introduce and study the multiway cut problem [DJPSY 94], Leighton and Rao study the sparsest cut problem [LR 88], and Garg et al study the multicut problem [GVY 93]; see also [KRAR 90, KPRT 94]. There are both practical and theoretical reasons behind this increased research interest. Practical applications include minimizing communication costs in parallel computers, partitioning files among the nodes of a network, deleting the minimum number of edges to get a bipartite graph, VLSI design, etc., see [DJPSY 94, LR 88, GVY 93]. Theoretical motivations include obtaining the best approximation guarantees possible, designing and analyzing simple general-purpose algorithms such as the greedy heuristic, and applying techniques from combinatorial optimization such as linear programming relaxations and duality theory.

* Department of Combinatorics & Optimization, University of Waterloo, Ontario, Canada N2L 3G1. Supported in part by NSERC grant no. OGP0138432 (NSERC code OGPIN 007).

We introduce and study a new class of NP-hard cut problems called *feasible cut* problems. The input consists of an undirected graph $G = (V, E)$ with a capacity function $u : E \rightarrow \Re_+$, as well as a set of pairs of nodes that we call *demand edges*. A cut $(S, \overline{S})$ is called *feasible* if S includes at most one end of any demand edge; possibly, $\overline{S}$ contains both ends of every demand edge. The minimum capacity v^*-cut problem is to find a feasible cut $(S, \overline{S})$ such that S contains a given node v^* and the cut has minimum capacity $u(S, \overline{S})$. One variant of the minimum-ratio feasible cut problem is to find a feasible cut $(S, \overline{S})$ with $\frac{u(S, \overline{S})}{|S|}$ minimum. In Section 2, we show that both problems are NP-hard. The problems without the feasibility condition are well known to be solvable in polynomial time. One of our motivations is to study the approximation guarantees achievable for familiar cut problems after incorporating the feasibility condition. Another motivation is that the minimum-ratio feasible cut problem may be used as a subroutine in approximately solving the multicut problem, a key problem that is being intensively studied [GVY 93, GVY 94, KPRT 94, BTV 95]. We give a 2.8-approximation algorithm for the minimum capacity feasible v^*-cut problem. For the minimum-ratio feasible cut problem, we give $O(1)$-approximation algorithms for some special cases. In Section 3, we present a variant of the greedy method for approximately solving NP-hard multicut problems. The first step is to formulate a variant of the multicut problem as a set covering problem (that may have a number of variables exponential in $|V|$), see also Bertsimas and Vohra [BV 94]; then, following the greedy method, we repeatedly find a cut that approximately minimizes a fractional objective function. In particular, we show that an α-approximation algorithm for the minimum-ratio feasible cut problem gives an $O(\alpha \log |V|)$-approximation algorithm for the multicut problem. Our work here gives a simple, general scheme for analyzing greedy heuristics for multicut problems; our scheme may be used to replace ad hoc arguments used previously, see e.g., [KRAR 90, Lemma 3.1] and [KPRT 94, Theorem 2.8].

In the Steiner cut problems introduced by Klein et al [KPRT 94], the input consists of $G = (V, E)$, $u : E \rightarrow \Re_+$, and k commodities $S_1, S_2, \ldots, S_k$ where each S_i is a set of nodes (possibly, $|S_i| > 2$). The Steiner multicut problem is to find a set of edges $F \subseteq E$ of minimum capacity such that no connected component of $G \backslash F$ contains a commodity S_i, i.e., F should separate each commodity $S_i (1 \leq i \leq k)$ in the sense that there exist two nodes v, w in S_i such that every $v - w$ path uses an edge of F. A node is called a *terminal* if it belongs to some commodity S_i. We use T to denote the number of terminals, $T = |\bigcup_i S_i|$, and t to denote the maximum cardinality of a commodity, $t = \max_i |S_i|$. In the *simple* multicut problem, each commodity has precisely two nodes; this problem was studied by Klein et al and Garg et al [KRAR 90], [GVY 93]; they proved approximation guarantees of $O(\log^3 T)$ and $4 \ln(T+1)$, respectively. The (Steiner) multicut problem was introduced by Klein et al [KPRT 94], who gave an $O(\log T \log k \log kt)$-approximation algorithm. In the minimum capacity-to-demand ratio Steiner cut problem, there is an additional input, namely, a nonnegative real-valued demand dem_i for each commodity S_i, $1 \le i \le k$, and the problem is to find a cut $(S, \overline{S})$ minimizing $\frac{u(S, \overline{S})}{dem(S, \overline{S})}$, where $dem(S, \overline{S})$ denotes the sum of the demands of the commodities separated by the cut, $\sum_{(i: \emptyset \ne S_i \cap S \ne S_i)} dem_i$. In Section 4, we give a new approximation guarantee for the latter problem that is independent of the number of commodities, k. Our approximation guarantee is $O(t \log T)$ compared to the $O(\log kt \log T)$ guarantee of [KPRT 94]. When t is fixed, then our guarantee is $O(\log T)$ versus $O(\log k \log T)$. Our analysis uses the method of Linial et al [LLR 94] and is simpler than the one in [KPRT 94].

1.1 Notation

We usually denote the graph under consideration by $G = (V, E)$, and use n for $|V(G)|$. For a set of nodes S, $\overline{S}$ denotes $V \setminus S$. We usually allow a nonnegative capacity $u : E \to \Re_+$ on the edges, and in some cases there is a nonnegative weight $w : V \to \Re_+$ on the nodes. For a set $S \subseteq V$ or $F \subseteq E$, $w(S)$ and $u(F)$ denote the sum of the weights of nodes in S and the sum of the capacities of the edges in F, respectively. For $S \subseteq V$, $\delta(S)$ denotes the set of edges that have exactly one end in S. If $\emptyset \ne S \ne V$, then $\delta(S)$ is called a *cut* and is also denoted by $(S, \overline{S})$; and $u(\delta(S)) = u(S, \overline{S})$ is called the capacity of the cut.

2 Feasible cut problems

In this section, we focus on the special case where each commodity has precisely two terminal nodes. We first show that three variants of the feasible cut problem are NP-hard. Then we give a 2.8-approximation algorithm for the minimum capacity v^*-cut problem; a $(2 - \epsilon)$-approximation algorithm (ϵ an absolute constant) would be a major result, since it would give a $(2 - \epsilon)$-approximation algorithm for the minimum node cover problem (Corollary 2). Finally, we study approximation algorithms for a variant that has a fractional objective function.

Given an undirected graph $G = (V, E)$ with a capacity function $u : E \to \Re_+$, and a set of commodities, where each commodity is a pair of nodes, we call every commodity a *demand edge*. A set of nodes is called *feasible* if it contains no commodity, i.e., it includes at most one end of any demand edge; possibly, no demand edge has an end in the set. A *feasible cut* $(S, \overline{S})$ is one such that S is feasible. Q denotes the set of end nodes of demand edges. We study three related problems on feasible cuts:

(P1): (*minimum-ratio feasible cut problem*) minimize $\frac{u(\delta(S))}{|S \cap Q|}$ such that S is a feasible set of nodes, $\emptyset \ne S \ne V$.

(P2): Given $w : V \to \Re_+$ (i.e., each node is given a nonnegative weight $w(v)$), minimize $u(\delta(S)) - w(S)$ such that S is feasible, $\emptyset \neq S \neq V$.

(P3): (*minimum capacity feasible v^*-cut problem*) Given a fixed node v^*, minimize $u(\delta(S))$ such that S is feasible and $v^* \in S \neq V$.

2.1 Hardness of feasible cut problems

Theorem 1. *Problems (P1), (P2) and (P3) are NP-hard.*

Proof.

Problem (P1): We reduce the minimum node cover problem to (P1). Let $\tilde{G} = (\tilde{V}, \tilde{E})$ be an instance of the minimum node cover problem. Clearly, $\tilde{G}$ has a node cover of size $\leq |\tilde{V}| - 1$. Assume that $\tilde{G}$ has no isolated nodes. We construct an instance of (P1) from $\tilde{G}$ as follows, see Figure 1:

- take two copies of $\tilde{V}$, say V and V', and an extra node z;
- form a perfect matching between V and V' with each matching edge having unit capacity;
- fix a node v' in V' and add an edge between v' and every other node in V' with capacity $+\infty$; also, add the edge $v'z$ with unit capacity;
- for each edge of $\tilde{G}$, we take the corresponding node pair of V to be a demand edge; also, we add the demand edges $zw, \forall w \in V'$. Note that $Q = \{z\} \cup V \cup V'$.

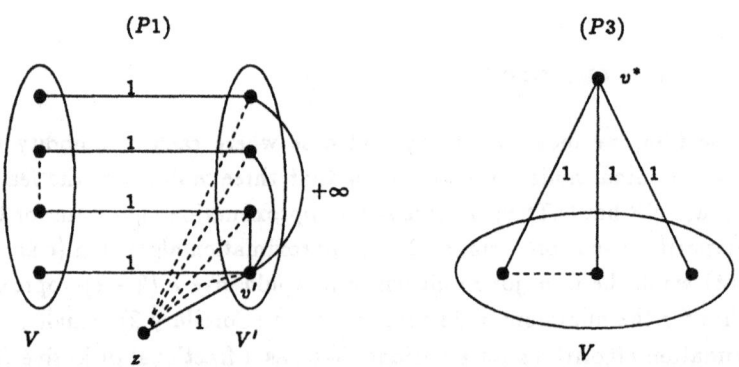

Fig. 1. Reducing the minimum node cover problem to problems (P1) (left) and (P3) (right). Dashed lines indicate demand edges.

We claim that if S is an optimal feasible set of (P1), then $V \backslash S$ gives a minimum node cover of the graph $\tilde{G}$, and that conversely, if N is a minimum node cover of $\tilde{G}$, then $(V \backslash N) \cup V'$ is an optimal feasible set of (P1). First, suppose that $S \cap V' = \emptyset$. Then $\frac{u(\delta(S))}{|S \cap Q|} = \frac{|S|}{|S \cap Q|} = 1$. Now, suppose that $S \cap V' \neq \emptyset$. Then $S \supseteq V'$, since there is a path of infinite capacity between any two nodes of

V', and so $\dfrac{u(\delta(S))}{|S \cap Q|} = \dfrac{u(\delta(S))}{|S|} = \dfrac{1 + |V \backslash S|}{2|V| - |V \backslash S|}$. Note that this ratio is minimum when $|V \backslash S|$ is minimum, and if $|V \backslash S| \le |V| - 1 = |\widetilde{V}| - 1$, then the ratio is less than 1. By the feasibility of S, every demand edge has one end in $\{z\} \cup (V \backslash S)$, so $V \backslash S$ corresponds to a node cover of $\widetilde{G}$. Hence the optimal feasible set S of (P1) must contain V' with $V \backslash S$ corresponding to a minimum node cover of $\widetilde{G}$. Our claim follows easily.

Problem (P2): We reduce the maximum independent set problem to (P2). Taking an instance $G = (V, E)$ of the maximum independent set problem, we assign each vertex in G with unit weight and each edge in G with zero capacity. Further we have a demand edge for each edge in G. Then for any feasible set S in the instance of (P2), the objective function $u(\delta(S)) - w(S)$ equals $-|S|$. Therefore by the feasibility condition, S is a maximum independent set of G if and only if S is an optimal feasible set of the instance of (P2).

Problem (P3): We reduce the minimum node cover problem to (P3). Given an instance $G = (V, E)$ of the minimum node cover problem, we construct an instance $G' = (V', E')$ of (P3) as follows, see Figure 1: (i) $V' = V \bigcup \{v^*\}$, $E' = \{v^* v | v \in V\}$ and each edge has unit capacity. (ii) Each edge in G gives a demand edge in G'. Then for any feasible set S with $v^* \in S$ we have $u(\delta(S)) = |V \backslash S|$, and $V \backslash S$ is a node cover in G by the feasibility of S. Hence $(S, \bar{S})$ is a minimum feasible v^*-cut in G' if and only if $\bar{S}$ is a minimum node cover in G. $\square$

From the above proof, we get two easy corollaries.

Corollary 2. *An α-approximation algorithm for problem (P3) (minimum capacity feasible v^*-cut) gives an α-approximation algorithm for the minimum node cover problem.*

Corollary 3. *The maximum capacity feasible cut problem (find a feasible cut $(S, \bar{S})$ of maximum capacity) is NP-hard; an α-approximation algorithm gives an α-approximation algorithm for the maximum independent set problem.*

The construction for Corollary 3 is a variant of the construction for problem (P1) in Theorem 1.

2.2 Algorithms for the minimum capacity feasible v^*-cut problem

We focus on problem (P3): given a fixed node v^*, find a feasible cut $(S, \bar{S})$ of minimum capacity with $v^* \in S$. In some special cases, the optimal solution can be found by network flow techniques (see Proposition 7). For the general problem, we present a $(4\ln 2)$-approximation algorithm.

Problem (P3) can be formulated as an integer program; relaxing the integrality constraints gives a linear program. There is a nonnegative length variable l_e for every edge e. Each node v is assigned with a potential d_v such that the potential difference across each edge is no more than the length of that edge. Furthermore, for each demand edge, the sum of the potentials of its two end nodes is at least one. The following LP expresses this. We constrain the special node v^* to have zero potential.

$$(LP1) \begin{cases} z^* = \text{minimize } \sum_e u_e l_e \\ \text{subject to} \\ d_w \le l_{vw} + d_v, \text{ for every edge } vw \\ d_v \le l_{vw} + d_w, \text{ for every edge } vw \\ d_s + d_t \ge 1, \text{ for every demand edge } st \\ d_{v^*} = 0 \\ l_e \ge 0. \end{cases}$$

Theorem 4. *Given an instance of (P3), there is a polynomial algorithm to find a feasible set S with $v^* \in S$ such that*

$$z^* \le u(\delta(S^{opt})) \le u(\delta(S)) \le (4 \ln 2) z^*,$$

where S^{opt} denotes the optimal set.

We will use the region growing technique and the following lemma due to Garg et al [GVY 93].

Lemma 5. *Let $l : E \to \Re_+$ be a length function on $G = (V, E)$, and let $u : E \to \Re_+$ be a capacity function. For any positive ϵ and p, and any node v^*, there exists a set S containing v^* such that $u(\delta(S)) \le \epsilon(1 + \dfrac{1}{p}) \sum_e u_e l_e$, and for every node v in S, the shortest $v^* - v$ path with respect to l has length less than $\epsilon^{-1} \ln(p+1)$.*

Lemma 6. *If we choose $\epsilon = 2 \ln(p+1)$, then the set S in Lemma 5 is feasible.*

Proof. Since $\epsilon = 2 \ln(p+1)$, every node in S is at a distance less than $1/2$ from v^*. From the constraint $d_s + d_t \ge 1$ of $(LP1)$, for each demand edge st, either $d_t \ge 1/2$ or $d_s \ge 1/2$, which implies S contains no demand edges. □

Proof. (Theorem 4) The proof follows from Lemmas 5 and 6 by choosing $p = 1$ and $\epsilon = 2 \ln 2$. □

The next result gives a polynomial algorithm for the optimal feasible v^*-cut for several special cases. For an instance of problem (P3), let F denote the set of demand edges; the graph (V, F) is called the *demand graph*.

Proposition 7. *If the number of maximal independent sets in the demand graph (V, F) is polynomial in n, and all the maximal independent sets can be found in polynomial time, then a minimum capacity feasible v^*-cut can be found in polynomial time.*

Proof. For each maximal independent set $I \subseteq Q$ of the demand graph, we find a cut of minimum capacity separating $\{v^*\}$ from $Q \setminus I$ by solving a single maximum flow problem (the source is v^* and the sinks are $Q \setminus I$). The best of these cuts is the required output. □

Corollary 8. *If the demand graph (V, F) is one of the following, then a minimum capacity feasible v^*-cut can be found in polynomial time:*

(a) a graph with a bounded number of edges, i.e., $|F| = O(1)$,
(b) a clique,
(c) a complete bipartite graph,
(d) the complement of a triangulated graph,
(e) the complement of a bipartite graph, or
(f) the complement of a line graph.

See p. 302 of [GLS 88] for similar results on the maximum independent set problem. Unfortunately, problems (P1), (P2) and (P3) remain NP-hard for the special case when the demand graph is bipartite. This follows because each of these problems can be transformed into an "equivalent" problem such that no two demand edges have a node in common, i.e., the demand edges form a matching. To see this, repeatedly split a node incident with two or more demand edges into two new nodes joined by a new edge with a huge capacity.

2.3 Approximating minimum-ratio feasible cuts

We focus on finding a minimum-ratio feasible cut. It is well known that a polynomial algorithm for the minimum capacity feasible v^*-cut problem gives a polynomial algorithm for the minimum-ratio feasible cut problem [GM 84, Appendix 5]. Consequently, we can efficiently solve the minimum-ratio feasible cut problem for the special cases in Corollary 8. In general, an approximation algorithm for the minimum capacity feasible v^*-cut problem does *not* give an approximation algorithm for the minimum-ratio feasible cut problem. The main result of this subsection is an $O(1)$-approximation algorithm for the minimum-ratio feasible cut problem for the special case when the feasible shore B of a minimum capacity feasible cut $(B, \overline{B})$ has ≥ 0.64 of the terminals. Recall that Q denotes the set of terminal nodes. For related work, see [HFKI 87].

Proposition 9. *Given an α-approximation algorithm for the minimum capacity feasible v^*-cut problem, and assuming that there is a minimum capacity feasible* cut $(B, \overline{B})$ such that $|B \cap Q| \geq (\epsilon + 1 - \frac{1}{\alpha})|Q|$, $1 > \epsilon > 0$, there is a $(1/\epsilon)$-approximation algorithm for the minimum-ratio feasible cut problem.

Proof. Assume every cut has positive capacity and $Q \neq \emptyset$ (i.e., there is a demand edge). Let the feasible set S^* achieve the optimal ratio λ^*, i.e.,

$$\lambda^* = \frac{u(\delta(S^*))}{|S^* \cap Q|} \leq \frac{u(\delta(S))}{|S \cap Q|}, \quad \forall S \text{ feasible}, \emptyset \neq S.$$

Here is the standard transformation to the minimum capacity feasible v^*-cut problem: We "linearize" the problem to an instance of problem (P2)

$$\text{minimize } \{u(\delta(S)) - \lambda|S \cap Q| : S \text{ feasible}, \emptyset \neq S \cap Q\},$$

where λ is a nonnegative parameter. The optimal value of the linearized problem is zero iff we fix $\lambda = \lambda^*$; also, λ is less (greater) than λ^* iff the optimal value of the linearized problem is positive (negative). To find λ^*, we execute a binary search over the interval $[\hat{u}/n, U]$, where $\hat{u} = \min_e u_e$ and $U = \sum_e u_e$. Each iteration of the binary search solves the linearized problem with the current value of λ. To solve the linearized problem, we transform it to a variant of the minimum capacity feasible v^*-cut problem. We add a new node v^* and all the edges v^*w, $w \in Q$, and fix the capacity of each new edge at λ. The goal of our variant is to find a cut $(S', \overline{S'})$ of minimum capacity such that $S' = \{v^*\} \cup S$, S is feasible and $\emptyset \neq S \cap Q$, i.e., the feasible shore S' should contain v^* as well as a terminal node of the original problem. This can be achieved as follows: for each terminal node $w \in Q$, construct an instance of the minimum capacity feasible v^*-cut problem by "contracting" nodes v^* and w into a new node v^*. The best of the $|Q|$ feasible v^*-cuts gives the desired cut $(S', \overline{S'})$. In terms of the v^*-cut problem, the capacity of $(S', \overline{S'})$ is $u(\delta(S)) + \lambda|\overline{S} \cap Q|$. Subtracting $\lambda|Q|$, we get the optimal value of the linearized problem.

The rest of the proof focuses on the solution of the linearized problem for a fixed value of the parameter λ, however, we keep the notation of the original minimum-ratio feasible cut problem, i.e., S, S^*, S_λ denote a feasible shore without the new node v^*. The objective function for the feasible v^*-cut problem as a function of λ is

$$g(\lambda) = \min_{S \text{ feasible}, \emptyset \neq S \cap Q} \{u(\delta(S)) + \lambda|\overline{S} \cap Q|\} = u(\delta(S_\lambda)) + \lambda|\overline{S_\lambda} \cap Q|,$$

where S_λ denotes the optimal set when the parameter is fixed at λ. The function $g(\lambda)$ is the "lower envelope" of the linear functions $u(\delta(S)) + \lambda|\overline{S} \cap Q|$ ($\forall S$ feasible, $\emptyset \neq S \cap Q$), each of which has positive slope, and so $g(\lambda)$ is a piece-wise linear, concave function with positive slope, see Figure 2. Note that $g(\lambda^*) = u(\delta(S^*)) + \lambda^*|\overline{S^*} \cap Q| = \lambda^*|Q|$. Let λ_1 and λ_2 denote the values of the parameter such that

$$\lambda_1|Q| = g(\lambda_1)/\alpha \quad \text{and} \quad \lambda_2|Q| = \alpha \cdot g(\lambda_2).$$

402

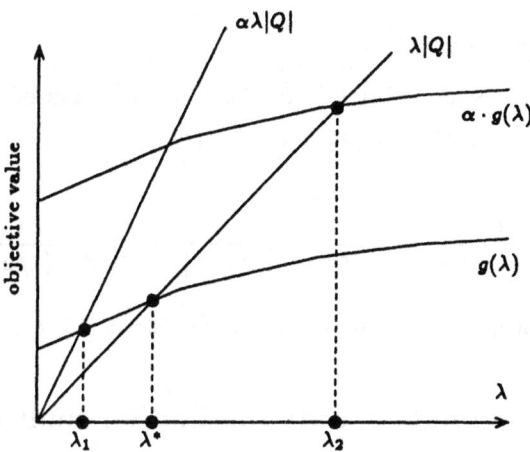

Fig. 2. An illustration of the parametrized objective function $g(\lambda)$, and of $\lambda_1, \lambda^*, \lambda_2$.

Observe that if the parameter λ does not lie in the interval $[\lambda_1, \lambda_2]$, then the current iteration of the binary search (in the standard method for the minimum-ratio problem) gives the correct decision: if $\lambda < \lambda_1$, then the smallest possible objective value found by the approximation algorithm for the feasible v^*-cut problem is $g(\lambda)$, and since $1/\alpha$ of this value is $> \lambda|Q|$, we decide correctly that $\lambda^* > \lambda$; the other case, $\lambda > \lambda_2$, is similar. On the other hand, if λ is in the interval $[\lambda_1, \lambda_2]$, then we cannot make the correct decision. In this case, our approximation algorithm for the minimum-ratio problem terminates, and returns the current value of λ and the associated feasible set S, $\emptyset \neq S \bigcap Q$, as the approximately optimal solution.

We claim that under the hypothesis of the proposition $\lambda_1 \geq \epsilon \lambda^*$ and $\lambda_2 \leq \lambda^*/\epsilon$. The proposition follows from this claim. Now, we prove the claim. For the feasible shore B of the minimum capacity feasible cut in the hypothesis, i.e., $B = S_\lambda$ for $\lambda = 0$, let β denote $|B \bigcap Q|/|Q|$, and note that for all $\lambda \geq 0$, $|S_\lambda \bigcap Q|/|Q| \geq \beta$.

Since $\alpha \lambda_1 |Q| = g(\lambda_1) = u(\delta(S_{\lambda_1})) + \lambda_1 |\overline{S_{\lambda_1}} \cap Q|$,

$$(\alpha - 1)\lambda_1 \frac{|Q|}{|S_{\lambda_1} \cap Q|} = \frac{u(\delta(S_{\lambda_1}))}{|S_{\lambda_1} \cap Q|} - \lambda_1 \geq \lambda^* - \lambda_1.$$

Hence, $\lambda_1 \geq \dfrac{1}{1 + (\alpha - 1)/\beta} \lambda^* \geq \epsilon \lambda^*$, since by the hypothesis $\beta \geq \epsilon + 1 - \frac{1}{\alpha} \geq \beta\epsilon + \frac{\alpha-1}{\alpha}$ (since $\beta \leq 1$) $\geq \beta\epsilon + (\alpha - 1)\epsilon$ (since $\epsilon \leq \frac{1}{\alpha}$), hence $\frac{\beta}{\beta+(\alpha-1)} \geq \epsilon$.

For λ_2 we have,

$$\frac{\lambda_2 |Q|}{\alpha} = g(\lambda_2)$$
$$\leq u(\delta(S^*)) + \lambda_2 |\overline{S^*} \cap Q|, \quad \text{by definition of } g(\lambda_2)$$
$$= \lambda^* |Q| + (\lambda_2 - \lambda^*)|\overline{S^*} \cap Q|.$$

Hence,
$$\lambda_2 \leq \frac{\alpha|S^* \cap Q|}{|Q| - \alpha|\overline{S^*} \cap Q|} \; \lambda^* \leq \frac{\alpha|Q|}{|Q| - \alpha|\overline{S^*} \cap Q|} \; \lambda^* \leq \frac{1}{\beta + (1/\alpha) - 1} \; \lambda^* \leq \frac{\lambda^*}{\epsilon},$$

where the last inequality follows from the hypothesis $\beta \geq \epsilon + 1 - \frac{1}{\alpha}$. □

3 An approximate greedy method for multicut problems

We present a greedy method for approximately solving NP-hard multicut problems; this method applies to Steiner multicut problems too. First, we formulate a variant of the multicut problem as a set covering problem with an exponential (in n) number of sets. This follows Bertsimas and Vohra [BV 94]; however, Bertsimas and Vohra never gave a polynomial-time approximation algorithm for this formulation. We directly apply an approximate greedy heuristic to the set covering problem: each iteration of this heuristic solves a minimum-ratio feasible cut subproblem. The subproblem turns out to be NP-hard too. However, if we can find an α-approximation for the subproblem, then our iterative method finds an $O(\alpha \log n)$-approximation to the multicut problem.

Recall that a set $S \subseteq V$ is *feasible* if it contains no commodity S_i, $1 \leq i \leq k$. Let $\mathcal{F}$ denote the family of all feasible sets. The goal is to "cover" all the terminals using feasible sets. For each feasible set S chosen in the covering, all edges in $\delta(S)$ are deleted from G, i.e., the multicut corresponding to this formulation is the union of $\delta(S)$ over all S in the covering. The following integer program (SC) for the set covering formulation has a 0–1 variable x_S for each feasible set S. The optimal value of (SC) is at least half and at most twice the capacity of an optimal multicut.

$$
\begin{array}{lll}
(SC) & \text{minimize} & \displaystyle\sum_{S \in \mathcal{F}} u(\delta(S)) x_S \\
& \text{subject to} & \displaystyle\sum_{S \in \mathcal{F}: v \in S} x_S \geq 1, \qquad \forall v \in Q = \bigcup_i S_i \\
& & x_S \in \{0, 1\}, \qquad\qquad \forall S \in \mathcal{F}.
\end{array}
$$

Recall that the greedy heuristic for solving the set covering problem repeatedly chooses a set S minimizing the ratio c_S/N'_S, where c_S is the coefficient of S in the objective function, and N'_S is the number of nonzero coefficients in S's column in the current constraints matrix [Ch 79]. The greedy heuristic is guaranteed to produce a set covering with objective value $\leq (1 + \ln N_{max})z$, where N_{max} is the maximum over all columns of the number of nonzero coefficients and z is the optimal value (actually, z is the optimal value of the LP relaxation). For the multicut problem, the greedy heuristic applied to (SC) gives an approximation guarantee of $2(1 + \ln T) = O(\log T)$. Unfortunately, each iteration of the greedy heuristic applied to (SC) has to solve the NP-hard minimum-ratio feasible cut problem, since we have to find a feasible set S minimizing $u(\delta(S))/|S \cap Q|$. Our next result shows that if we can find an α-approximation to the minimum-ratio feasible cut problem, then by iterating this we can find a $2\alpha(1 + \ln T)$-approximation to the multicut problem via (SC).

Theorem 10. *Consider an approximate greedy heuristic for the set covering problem that in each iteration finds a set $\tilde{S}$ such that*

$$\frac{c_{\tilde{S}}}{N'_{\tilde{S}}} \leq \alpha \cdot \min_{S} \frac{c_S}{N'_S}$$

where c_S is the coefficient of S in the objective function, and N'_S denotes the number of nonzero coefficients in S's column in the current constraints matrix (after deleting rows of points already covered). Then the final set covering found by the heuristic has objective value at most $\alpha(1 + \ln N_{max})$ times the optimal value.

Proof. The proof is similar to Chvatal's analysis of the greedy heuristic [Ch 79]. See also Lemma 3.2.1 in [RV 93].

Consider the LP relaxation of the set covering problem and its dual

$$(P) \begin{cases} \min \sum_S c_S x_S \\ \text{subject to} \\ \sum_{S:i \in S} x_S \geq 1, \text{ for each } i \\ x \geq 0 \end{cases} \quad (D) \begin{cases} \max \sum_i y_i \\ \text{subject to} \\ \sum_{i \in S} y_i \leq c_S, \text{ for each } S \\ y \geq 0. \end{cases}$$

We show that the heuristic constructs a feasible dual solution y such that its objective value is $\geq \dfrac{Z_H}{\alpha(1 + \ln N_{max})}$, where Z_H is the objective value of the set covering found by the heuristic. The theorem follows because for every feasible dual solution y, and an optimal solution x of (P),

$$\sum_i y_i \leq \sum_S c_S x_S \leq Z^*,$$

where Z^* is the optimal value of the set covering problem.

Let $\tilde{S}_1, \tilde{S}_2, \tilde{S}_3, \ldots$ be the sequence in which the heuristic chooses sets, and let $\tilde{S}_{f(i)}$ be the first set chosen by the heuristic that contains element i. For each element i, let $w_i = \dfrac{c(\tilde{S}_{f(i)})}{|\tilde{S}_{f(i)} \setminus (\tilde{S}_1 \cup \ldots \cup \tilde{S}_{f(i)-1})|}$. Clearly, $\sum_i w_i = Z_H$.

Now we claim that for each set S, $\sum_{i \in S} w_i \leq c_S \cdot \alpha(1 + \ln|S|)$. We prove the claim as follows. Order the elements in S in the reverse order in which they were first covered: $i_1, i_2, \ldots, i_{|S|}$. Consider i_l. When i_l is first covered, say by $\tilde{S}_p$, S has at least l uncovered elements, so $|S \backslash (\tilde{S}_1 \bigcup \ldots \bigcup \tilde{S}_{p-1})| \geq l$. Hence, by our assumption,

$$\frac{c_{\tilde{S}_p}}{|\tilde{S}_p \backslash (\tilde{S}_1 \bigcup \ldots \bigcup \tilde{S}_{p-1})|} \leq \alpha \cdot \frac{c_S}{|S \backslash (\tilde{S}_1 \bigcup \ldots \bigcup \tilde{S}_{p-1})|} \leq \frac{\alpha c_S}{l},$$

so $w_{i_l} \leq \frac{\alpha c_S}{l}$. The claim follows since $\sum_{i_l \in S} w_{i_l} \leq \alpha c_S \sum \frac{1}{l} \leq \alpha c_S (1 + \ln|S|)$. To get the dual solution y, let $y_i = \frac{w_i}{\alpha(1 + \ln|N_{max}|)}$. $\square$

The above theorem applies to another set covering formulation of a variant of the multicut problem. Instead of covering by feasible sets, we allow arbitrary node sets. We have a 0–1 variable x_S for every set $S \subseteq V$. The objective function is similar to the one in (SC), but the constraints are different. For each commodity $S_i \subseteq V$, $1 \leq i \leq k$, we require that at least one of the sets S chosen in the covering "separates" S_i, i.e., S includes some node of S_i and does not include some other node of S_i. The optimal value of (SC$'$) is at least half and at most twice the capacity of an optimal multicut.

$$
\begin{aligned}
(SC') \quad &\text{minimize} \quad \sum_{S \subseteq V} u(\delta(S)) x_S \\
&\text{subject to} \quad \sum_{S \subseteq V: \emptyset \neq S \cap S_i \neq S_i} x_S \geq 1, \qquad \forall i = 1, \ldots, k \\
&\qquad\qquad\qquad\qquad\quad x_S \in \{0, 1\}, \qquad \forall S \subseteq V.
\end{aligned}
$$

Now, the greedy heuristic iteratively finds a set S minimizing $\frac{u(S, \bar{S})}{dem(S, \bar{S})}$, where $dem(S, \bar{S})$ denotes the number of commodities separated by $(S, \bar{S})$. (This agrees with our notation in Sections 1 and 4 since each commodity here has unit demand.) The problem of finding a cut minimizing the capacity-to-demand ratio is NP-hard, however, extensive research has been devoted to designing approximation algorithms. For the case of $|S_i| = 2$, $1 \leq i \leq k$, there is an $O(\log T) = O(\log k)$ approximation algorithm due to Linial et al [LLR 94], see also [KRAR 90]. For the general case, approximation guarantees of $O(\log kt \log T)$ and $O(t \log T)$ can be achieved in polynomial time; these results are due to Klein et al [KPRT 94], and Section 4 of this paper, respectively. We obtain approximation guarantees of $O(\log^2 k)$ and $O(\min(t, \log kt) \log T \log k)$ for the simple and Steiner multicut problems, respectively, by directly applying Theorem 10. However, this does not improve on the approximation guarantees of [GVY 93] and [KPRT 94], respectively.

4 A new approximation guarantee for minimum-ratio Steiner cuts

Let S_i denote the set of terminals for commodity i, where $1 \leq i \leq k$. We say that a cut $(X, \overline{X})$ separates terminal set S_i if $\emptyset \neq X \cap S_i \neq S_i$. A demand dem_i is associated with every commodity i, $dem : \{1, \ldots, k\} \to \Re_+$. The demand $dem(X, \overline{X})$ across a cut $(X, \overline{X})$ is the sum of the demands of the separated commodities. Recall that the minimum capacity-to-demand ratio Steiner cut problem is to find a cut that minimizes the ratio of the capacity of the cut and the demand across the cut.

Theorem 11. *Given an instance of the minimum capacity-to-demand ratio Steiner cut problem, there is a (deterministic) polynomial algorithm to find a cut $(X, \overline{X})$ such that*
$$z^* \leq \min_{\emptyset \neq Y \neq V} \left\{ \frac{u(Y, \overline{Y})}{dem(Y, \overline{Y})} \right\} \leq \frac{u(X, \overline{X})}{dem(X, \overline{X})} \leq O(t \log T) z^*.$$

Here, z^* denotes the optimal value of the LP relaxation (LP2) (see below) of the problem, t denotes $\max_i |S_i|$, and T denotes $|\bigcup_i S_i|$.

We will use the following result due to Linial et al [LLR 94]. For a graph G and length function $l : E \to \Re_+$, let $dist_l(v, w)$ denote the length of a shortest $v - w$ path with respect to l.

Proposition 12. *Given a graph G, a length l_e on each edge e, and a set of nodes Q, there is a deterministic polynomial algorithm that constructs an l_1-metric $\rho : V \times V \to \Re_+$ such that*
1. *for every pair of nodes $\{v, w\}$ in Q, $\dfrac{dist_l(v, w)}{O(\log |Q|)} \leq \rho(v, w) \leq dist_l(v, w)$;*
2. *for every pair of nodes $\{v, w\}$ in V, $\rho(v, w) \leq dist_l(v, w)$.*

The next fact is well known [AD 91].

Fact 13 *Every l_1-metric on node set V can be written as a nonnegative linear combination of incidence vectors of cuts of the complete graph on V.*

Proof. (Theorem 11) Let $l : E \to \Re_+$ be an optimal solution for the following LP relaxation of our problem. This LP relaxation is due to Klein et al [KPRT 94].

$$(LP2) \begin{cases} z^* = \text{minimize } \sum_e u_e l_e \\ \text{subject to} \\ \sum_{i=1}^{k} dem_i \cdot l(S_i) = 1 \\ l_e \geq 0, \end{cases}$$

where $l(S_i)$ denotes the minimum length of a spanning tree of the distance network $D_G(S_i)$ [HRW 92]. In more detail, given G, $l : E \to \Re_+$ and $S_i \subseteq$

$V, D_G(S_i)$ consists of the complete graph on the node set S_i and edge lengths $dist_l$, i.e., the length of an edge vw, $v, w \in S_i$, equals the length $dist_l(v, w)$ of a shortest v, w path in G with respect to l. We use F_i to denote the set of edges vw of a minimum spanning tree of $D_G(S_i)$, so $l(S_i) = \sum_{vw \in F_i} dist_l(v, w)$.

Then,

$$z^* = \sum_e u_e l_e = \frac{\sum_e u_e l_e}{\sum_{i=1}^{k} dem_i \cdot l(S_i)}$$

$$\geq \frac{\sum_{vw \in E} u_{vw} dist_l(v, w)}{\sum_{i=1}^{k} dem_i(\sum_{vw \in F_i} dist_l(v, w))}$$

$$\geq \frac{1}{O(\log T)} \frac{\sum_{vw \in E} u_{vw} \rho(v, w)}{\sum_{i=1}^{k} dem_i(\sum_{vw \in F_i} \rho(v, w))},$$

where ρ is an l_1-metric satisfying the two properties in Proposition 12, and we take the set Q in the proposition to be $\bigcup_{i=1}^{k} S_i$. By Fact 13, there exists a cut $(X, \overline{X})$ such that this quantity is

$$\geq \frac{1}{O(\log T)} \frac{\sum_{vw \in (X, \overline{X})} u_{vw}}{\sum_{i=1}^{k} dem_i \cdot |(X, \overline{X}) \cap F_i|}$$

$$\geq \frac{1}{O(\log T)} \frac{u(X, \overline{X})}{\sum_{(i:\emptyset \neq S_i \cap X \neq S_i)} dem_i \cdot (|S_i| - 1)}$$

$$\geq \frac{1}{(t - 1)O(\log T)} \cdot \frac{u(X, \overline{X})}{dem(X, \overline{X})}.$$

Given the l_1 metric ρ in the form of an embedding of (V, ρ) into a real space with l_1 norm, the cut $(X, \overline{X})$ can be found in polynomial time, see [LLR 94]. □

5 Conclusions

We conclude with two open problems. Is the problem of finding a feasible cut of minimum capacity NP-hard? Note that this is a variant of problem (P3) where we drop the constraint on the node v^*. Is there an $O(1)$ approximation algorithm for the minimum-ratio feasible cut problem? Our algorithm in Section 2.3 needs a "strong" assumption to give an $O(1)$ approximation guarantee, namely, there exists a minimum capacity feasible cut $(B, \overline{B})$ such that B contains a large fraction of the terminal nodes. Note that such a cut $(B, \overline{B})$, if present, is an $O(1)$ approximation to the minimum-ratio feasible cut. The problem of finding such a cut $(B, \overline{B})$ is a bicriteria optimization problem, but unfortunately, we do not know how to solve it in polynomial time. Under the hypothesis of Proposition 9, our algorithm in Section 2.3 is an $O(1)$ approximation algorithm for this bicriteria optimization problem.

References

[AD 91] D. Avis and M. Deza, *The cut cone, L^1 embeddability, complexity, and multicommodity flows*, Networks, 21 (1991), pp. 595-617.

[BV 94] D. Bertsimas and R. Vohra, *Linear programming relaxations, approximation algorithms and randomization: a unified view of covering problems*, manuscript, 1994.

[BTV 95] D. Bertsimas, C. Teo and R. Vohra, *Nonlinear formulations and improved randomized approximation algorithms for multicut problems*, Proc. 4th I.P.C.O, pp. 29–39, LNCS 920, Springer-Verlag, Berlin, 1995.

[Ch 79] V. Chvátal, *A greedy heuristic for the set-covering problem*, Mathematics of Operations Research, 4 (1979), pp. 233-235.

[DJPSY 94] E. Dahlhaus, D. S. Johnson, C. H. Papadimitriou, P. D. Seymour and M. Yannakakis, *The complexity of multiterminal cuts*, SIAM J. Computing 23 (1994), pp. 864–894.

[GVY 93] N. Garg, V. Vazirani and M. Yannakakis, *Approximate max-flow min-(multi)cut theorems and their applications*, Proc. 25th ACM S.T.O.C., 1993, pp. 698–707.

[GVY 94] N. Garg, V. Vazirani and M. Yannakakis, *Multiway cuts in directed and node weighted graphs*, Proc. 21st I.C.A.L.P., LNCS 820, Springer-Verlag, Berlin, 1994.

[GM 84] M. Gondran and M. Minoux, *Graphs and algorithms*, Wiley, New York, 1984.

[GLS 88] M. Grötschel, L. Lovász and A. Schrijver, *Geometric algorithms and combinatorial optimization*, Springer-Verlag, Berlin, 1988.

[HFKI 87] S. Hashizume, M. Fukushima, N. Katoh and T. Ibaraki, *Approximation algorithms for combinatorial fractional programming problems*, Mathematical Programming 37 (1987), pp. 255–267.

[HRW 92] F. K. Hwang, D. S. Richards and P. Winter, *The Steiner tree problem*, North-Holland, Amsterdam, 1992.

[KRAR 90] P. Klein, S. Rao, A. Agrawal, and R. Ravi, Preliminary version in *Approximation through multicommodity flow*, Proc. 31st IEEE F.O.C.S., 1990, pp. 726–737.

[KPRT 94] P. Klein, S. Plotkin, S. Rao, and E. Tardos, *Approximation algorithms for Steiner and directed multicuts*, manuscript, 1994.

[LR 88] F. T. Leighton and S. Rao, *An approximate max-flow min-cut theorem for uniform multicommodity flow problems with applications to approximation algorithms*, manuscript, September 1993. Preliminary version in Proc. 29th IEEE F.O.C.S., 1988, pp. 422–431.

[LLR 94] N. Linial, E. London and Y. Rabinovich, *The geometry of graphs and some of its algorithmic applications*, manuscript, April 1995. Preliminary version in Proc. 35th IEEE F.O.C.S., 1994, pp. 577–591.

[RV 93] S. Rajagopalan and V. Vazirani, *Primal-dual RNC approximation algorithms for (multi)-set (multi)-cover and covering integer programs*, Proc. 34th IEEE F.O.C.S., 1993, pp. 322-331.

On Parallel versus Sequential Approximation. *

Maria Serna and Fatos Xhafa

Departament de Llenguatges i Sistemes
Universitat Politècnica de Catalunya
Pau Gargallo 5, 08028-Barcelona
email: mjserna,fatos@goliat.upc.es

Abstract. Here we deal with the class NCX of optimization problems that are approximable within constant ratio in NC. We first introduce a new kind of reduction that preserves the relative error of the approximate solutions and show that the class NCX has *complete* problems for this reducibility. An important subset of NCX is the class Max SNP, we show that Max SNP complete problems have a threshold on the parallel approximation ratio that is, there are constants c_1, c_2 such that although the problem can be approximated in P within c_1 it cannot be approximated in NC within c_2 unless P=NC. This result is attained by showing that the problem of approximating the value obtained through a non-oblivious local search algorithm is P-complete for some values of the approximation ratio. We finally show that approximating within ϵ using non-oblivious local search is in average NC.

1 Introduction.

It is already well known that there are no polynomial time algorithms for *NP-hard problems* unless P=NP, therefore for such problems the attention have been focused in finding (in polynomial time) approximate solutions, in this paper we consider NP optimization problems with polynomially bounded, in the input's length objective function. The class APX consists of those NPO problems whose solutions can be approximated in polynomial time, with relative error bounded by a constant. In order to give a precise characterization of such complexity classes, Papadimitriou and Yannakakis [PY91] used Fagin's syntactic definition of the class NP and introduced the classes Max NP and Max SNP. They proved that any problem in Max NP can be approximated (in polynomial time) with constant ratio and many problems were shown to be Max SNP-complete under *L-reductions* (linear reductions). Later [KMSV94] proved that the class APX coincides with the closure under *E-reductions* of Max NP and Max SNP.

In the parallel setting we have an analogous situation. We consider the class of problems that are approximable within constant ratio in NC that we denote as NCX. Many properties are common for the class NCX and APX. For example, in [DSST95] it was shown that Max SNP is contained in NCX, to do so they introduced L-reductions under the log space criterium and proved that all known Max SNP complete problems are complete under this reducibility.

We first consider the possibility, for the class NCX, of having complete problems. To do so we define some kind of reduction, called *NCAS-reduction*, that preserves the "relative error" of approximations. This reduction is a generalization of the *L+log space* [PY91] in the following sense: L-reductions relate the optima of both problems, while NCAS-reductions only has the property that good approximations to one problem correspond to good approximations to the other. We show that *Max Bounded Weighted Sat* is complete for NCX, notice that this problem is also complete for the class APX under PTAS reductions [ADP80].

One general approach when dealing with hard combinatorial problems is to use a *local search* algorithm. Starting from an initial solution, the algorithm moves to a better one among its neighbors,

* This research was supported by the ESPRIT BRA Program of the EC under contract no. 7141, project ALCOM II.

until a *locally optimal* solution is found. This kind of approach is crucial for Max SNP problems. In [KMSV94] is provided a characterization of Max SNP in terms of a class of problems called Max k-CSP, their results show that a simple non-oblivious local search algorithm provides a polynomial time approximation algorithm. Thus every Max SNP problem can be approximated within constant factor by such algorithms. Furthermore the ratios achieved using this algorithms are comparable to the best-known ones.

We analyze the parallel complexity of such approach. We first define what we call a *local problem* in which we ask for the value of the local optimal solution attained accordingly to a prespecified local search algorithm. We show that the problem corresponding to non-oblivious local search is P-complete, furthermore it cannot be approximated in NC for some ratios unless P=NC. We use this result to show that there exists a threshold on the parallel approximation ratio of Max SNP complete problems, that is, there are constants ϵ_0, ϵ_1 such that the problem can be approximated in NC within ϵ_0, but not within ϵ_1 unless P=NC. In particular we show that the problems Max Cut and Max 3SAT can be approximated in NC within 1 but not within $1 - \epsilon$ for any ϵ.

Although this results means that we cannot achieve the best ratios, for Max SNP complete problems, in NC we analyze the expected behaviour of a general non-oblivious local search algorithm. We show that the expected number of iterations is polylogarithmic, when the search starts from a random initial solution and using a quite general improvement model.

2   Basic definitions and problems.

Let Π be an NP Optimization (NPO) problem, whose objective function is polynomially bounded with respect to the input length. Let I_Π denote the set of instances and let $Sol_\Pi(x)$ denote the solution set to instance x. For any solution S, $S \in Sol_\Pi(x)$, let $V(x,S)$ be the value of the objective function on S and let $OPT(x)$ be the optimum value to instance x.

Definition 1. An algorithm A approximates the problem Π within a factor ϵ if it finds a solution S such that

$$\frac{1}{1+\epsilon} \leq \frac{V(x,S)}{OPT(x)} \leq 1 + \epsilon \qquad (1)$$

In this case we say that S has relative error within ϵ from the optimum solution and use the notation $E_\Pi(x,S) \leq \epsilon$

We note here that this definition applies uniformly for either maximization or minimization problems.

Definition 2. Given an instance x of any NPO problem Π and any $\epsilon > 0$, the $\epsilon - \Pi$ problem is: compute a value $m(x)$ such that

$$\frac{1}{1+\epsilon} \leq \frac{m(x)}{OPT(x)} \leq 1 + \epsilon \qquad (2)$$

Notice that the best approximations are achieved for $\epsilon \to 0$. Based on the definition of *PTAS reductions* ([ADP80], [CS81], [PM81]) we define the error preserving reductions in NC.

Definition 3. Let A and B be NPO problems. We say that the problem A is NCAS-reducible to B ($A \leq_{NCAS} B$) if three functions f,g,c exist such that the following conditions hold:

(a) For any $x \in I_A$, $f(x) \in I_B$ and the function f is computable in NC with respect to $|x|$.
(b) For any $x \in I_A$ and for any $y \in Sol_B(f(x))$, $g(x,y) \in Sol_A(x)$ and the function g is computable in NC with respect to both $|x|$ and $|y|$.

(c) $c : (0,1) \rightarrow (0,1)$, a rational (onto) valued function.

(d) For any $z \in I_A$ and for any $y \in Sol_B(f(x))$ and for any rational ϵ, $\epsilon \in (0,1)$, if $E_B(f(x),y) \leq c(\epsilon)$ then $E_A(x, g(x,y)) \leq \epsilon$.

From this definition we have that NCAS reduction preserves the relative error of approximation that is, when $A \leq_{NCAS} B$ then if we can find in NC approximate solutions for B implies that we can find in NC also approximate solutions for A. On the other hand this kind of reduction also "transmits" the non–approximability from A to B.

Recall the $L - reduction$ as defined in [PY91]. If we put the condition for a $L - reduction$ to be $log - space$ then clearly, we have that $NCAS - reduction$ is a generalization of $L - reduction$ (we choose $c(\epsilon) = \frac{\epsilon}{\alpha\beta}$ where α and β are the constants of $L - reduction$).

The following properties are immediates:

Proposition 4. *If $A \leq_{NCAS} B$ and $B \in NCX$ then $A \in NCX$.*

Proposition 5. *The reduction $\leq_{NCAS}$ is reflexive and transitive.*

Proposition 6. *If the problem A NCAS reduces to B and A cannot be approximated within some ϵ, $\epsilon \leq \epsilon_0$ then B cannot be approximated within some ϵ', $\epsilon' \leq c(\epsilon_0)$ where c is the function given by the NCAS reduction.*

Definition 7. Let A be a problem in NCX. We say that A is complete for NCX iff for any $B \in NCX$, $B \leq_{NCAS} A$.

Through the paper we will consider the following problems:

Weighted Max Cut.[GJ79]
Given a graph $G = (V, E)$ with positive weights on the edges, find a partition of V into two sets V_1, V_2, such that the sum of the weights of the edges between V_1 and V_2 is maximized. When all the weights are equal to one then we have an instance of unweighted Max Cut.

Weighted Max k-Sat.[SY91]
Given a Boolean formula in CNF where each clause contains at most k literals and has a positive integer weight, find an assignment of 0/1 to all variables that maximizes the sum of the weights of the satisfied clauses. In the particular case $k = 3$ we have the problem of Max 3Sat.

Weighted Not All Equal kSat.[PY91]
We are given a set of weighted clauses with at most k literals of the form $NAE(x_1, \cdots, x_k)$ where each x_i is a literal or a constant 0/1. Such a clause is satisfied if its constitutes do not have all the same value. Find an assignment to the variables such that the sum of the weights of the satisfied clauses is maximized. When the clauses do not contain negated literals the problem is called POS NAE kSAT.

Max Bounded Weighted SAT[ACP94].
Given a set of clauses C over a set of variables $\{x_1, x_2, \cdots, x_n\}$ and weights $\{w_1, w_2, \cdots, w_n\}$ to the variables such that

$$W \leq \sum_{i=1}^{n} w_i \leq 2W$$

find a truth assignment to the variables that maximizes the measure function:

$$V(C, \tau) = \begin{cases} max(W, \sum_{i=1}^{n} w_i\tau(x_i)) & \text{if } \tau \text{ satisfies all the clauses of C,} \\ W & \text{otherwise} \end{cases}$$

Circuit True Gates Problem (CTGP)[Ser91]

Given an encoding of a boolean circuit C together with an input assignment, compute the number of true gates, denoted by TG(C).

3 NCX-Completness.

In order to prove completeness result for NCX we will consider the *Max Bounded Weighted SAT* problem. Firstly, we observe that this problem is in NCX. For that, we note that the assignment $x_i = 1$, $1 \leq i \leq n$, has measure either W or $\sum_{i=1}^{n} w_i$ and therefore gives an approximation with error $\frac{1}{2}$, that is a 1-approximation according to our definition.

Theorem 8. *MAX BOUNDED WEIGHTED SAT is NCX-complete.*

Proof. Let Π be an optimization problem in NCX (suppose Π is a maximization one). In order to reduce Π to *Max Bounded Weighted SAT* we first reduce it, using a NCAS reduction, to another problem Δ and then reduce Δ to *Max Bounded Weighted SAT*. Our reduction is based in [ACP94] but extended to the parallel setting.

Let T be the NC δ-approximation algorithm for Π. The problem Δ is as follows: The instances of Δ are those of Π and its measure function, for instance x and solution y, $V_\Delta(x,y)$ is:

$$V_\Delta(x,y) = a(x,\delta) + max\{V_\Pi(x,y), t(x)\}$$

where

$$a(x,\delta) = \begin{cases} \frac{2\delta-1}{1-\delta} t(x) \text{ if } \delta > \frac{1}{2}, \\ 0 \qquad\qquad \text{otherwise} \end{cases} \quad \text{and} \quad t(x) = V_\Pi(x, T(x))$$

In other words, in the new problem Δ is included the approximation algorithm $T(x)$. The idea is to obtain a problem with bounded measure since we want to reduce it to a *weighted* problem. In fact, if we denote by $l(x) = a(x,\delta) + t(x)$ then the following holds [ACP94]:

$$l(x) \leq V_\Delta(x,y) \leq 2l(x)$$

which means that the measure function of Δ satisfies the same kind of constraint as *Max Bounded Weighted SAT*.

Now, the reduction from Δ to *Max Bounded Weighted SAT* goes like that. Given an instance x of Δ, we apply Cook's theorem (see [GJ79]). Then we will have a transformation (in polynomial time) from the problem Δ to *SAT*. But, we can achieve this transformation in *log − space*. Indeed, all the information we need each step of computation (in the work tape) for the variables in order to write the formula is: the step of the computation i, the index k of the actual state q_k of the machine, the index j of the tape square where read-write head is scanning and the index l of the bit of the input x, $x = s_{k_1} s_{k_2} \cdots s_l \cdots s_{k_n}$ that is scanning the machine. Therefore, the amount of the space we need in the work tape is logarithmic in the size of the input x since all these indices are bounded by a polynomial in the size of x. In other terms, the main need for the memory in such construction is for counting up to a polynomial in the length of the input and this can be done in logarithmic space.

Let ϕ_x be the boolean formula obtained. Let denote $y_1, y_2, \cdots, y_r$ the variables that describe the solution y and by $m_1, m_2, \cdots, m_s$ the boolean variables that give the solution $V_\Delta(x,y)$. Now, assign weights to the variables. The variables m_i's receive weights 2^{s-i} and all other variables are assigned the weight 0. So, we have an instance of Weighted SAT. Since the measure $V_\Delta(x,y)$ is bounded then we have an instance of *Bounded Weighted SAT* (the constant W depends on the bound of the problem Δ). Furthermore, for any truth assignment which satisfies the formula, to recover a solution y is straightforward (we look at the values of y_i's). On the other hand, this transformation guarantees the relative error because $V_\Delta(x, y)$ is equal to the sum of the weights of the *true* variables.

For the rest of the theorem we have to prove that Π is NCX-reduced to Δ. The transformation is the following:

- For any instances x, $f(x) = x$.
- For any instance x and any solution y that corresponds to instance $f(x)$,

$$g(x, y) = \begin{cases} y & \text{if } t(x) \leq V_\Delta(x, y), \\ T(x) & \text{otherwise} \end{cases}$$

- For any rational ϵ, $\epsilon \in (0, 1)$,

$$c(\epsilon) = \begin{cases} \frac{1-\delta}{\delta}\epsilon & \text{if } \delta > \frac{1}{2}, \\ \epsilon & \text{otherwise} \end{cases}$$

This transformation preserves the relative error when passing from solutions of Δ to solutions of Π (see details in [ACP94]). Since $NCAS$ reductions compose we have that $\Pi \leq_{NCAS} Max$ *Bounded Weighted SAT*. $\square$

4 The complexity of local search.

The complexity of local search algorithms has been extensively treated in [JPY88]. It results that the time needed by any local search algorithm to find a locally optimal solution depends on the neighborhood structure used. So, there are local search algorithm for which locally optimal solutions are not known to be computable in polynomial time.

Definition 9. (Local Search Algorithm)
Given a solution S of an NPO problem Π, the δ-neighborhood of S, denoted $N(S, \delta)$, is the set of all solutions S' that have distance at most δ from S,

$$N(S, \delta) = \{S' \mid \mathcal{D}(S, S') \leq \delta\}$$

where $\mathcal{D}(S, S')$ is the Hamming distance between S and S'. A solution S is locally optimal iff

$$\forall S' \in N(S, \delta), \ V(x, S) \geq V(x, S').$$

A local search algorithm starts from an initial solution and each iteration moves from the current solution S_i to some solution $S_{i+1} \in N(S, \delta)$ with better cost, until it arrives at a locally optimal solution.

This kind of local search is called *standard local search* or *oblivious local search*. A more generalized method for local search, *non–oblivious local search* is given in [KMSV94]. The non–oblivious local search was shown to be more powerful than the oblivious one since it permits to explore in both directions: that of the objective function and of the distance function. As a consequence, non–oblivious local search algorithms were used to approximate within a constant factor all Max SNP problems.

Definition 10. (Non–oblivious Local Search Algorithm)
A non–oblivious local search algorithm is a local search algorithm whose weight function is defined to be

$$\mathcal{F}(I, S) = \sum_{\mathbf{x}} \sum_{i=1}^{r} p_i \Phi(I, S, \mathbf{x})$$

where r is a constant, Φ_i's are quantifier–free first–order formulas and the profits p_i are real constants. The distance function D is any arbitrary polynomial time computable function.

We consider the problem of asking for the value of the local optimum achieved by a non-oblivious local search algorithm, in particular we introduce the problem for Max Cut:

Local Max Cut.

Given an instance of Max Cut and an initial solution S, find a locally optimum solution, achieved through non-oblivious local search. starting from S.

We show that the Local Max Cut problem is non approximable in NC, unless P=NC. We notice here that we do not refer to any method (*standard local search, non−oblivious local search, etc.*) used to find the locally optimal solution. The proof uses a reduction from the *True Gates* problem, recall that the ϵ approximation to this problem was shown to be P−complete for any $\epsilon \in [0, 1)$ [Ser91].

Theorem 11. *There is an $\epsilon_1 > 0$ such that the ϵ−Local Max Cut problem is P−complete for any $\epsilon < \epsilon_1$*

Proof. We make use of the reduction given in [SY91] from Circuit Value to the Local Max Cut. The reduction goes through three stages: in the first one the instance of CVP is reduced to an instance of POS NAE 3SAT, in the second POS NAE 3SAT is reduced to Weighted Max Cut and finally we transform the last problem into an instance of the unweighted Max Cut problem. Here we give only the variables and the weights for them, further details can be found in [SY91].

The variables:

For each gate g_i there is introduced a variable (denoted with the same symbol) g_i and the following groups of the variables:

control variables: $\{z_i\}$, $\{y_i\}$ (in fact $z_i = \overline{y_i}$)

and local variables:

$$\{\alpha_i^1\}, \{\alpha_i^2\}, \{\delta_i^1\}, \{\delta_i^2\}$$
$$\{\beta_i^1\}, \{\beta_i^2\}, \{\beta_i^3\}, \{\gamma_i^1\}, \{\gamma_i^2\}, \{\gamma_i^3\}$$
$$\{\omega_i\}$$

are also introduced.

Weights:

To each variable is associated a weight (see Table 1.) where the weight of the variable v is denoted by $|v|$:

$$|g_i| = 100(2n + 1 - i) + 60$$
$$|z_i| = 100(2n + 1 - i) = |g_i| - 60$$
$$|y_i| = 100(2n + 1 - i) + 10 = |g_i| + 10$$
$$|\alpha_i^k| = |g_i| + 10, \quad |\delta_i^k| = |g_i| + 10, \qquad k = 1, 2$$
$$|\beta_i^k| = |g_i|, \quad |\gamma_i^k| = |g_i|, \qquad k = 1, 2, 3$$
$$|\omega_i^k| = |\delta_i^k| - |g_i| = 10$$

Table 1.

Having the instance I of POS NAE 3−SAT, the instance of Max Cut is constructed as follows:

- There is one vertex for each variable and two vertices labelled by 0 and 1.
- For every clause NAE(x,y) with weight W in I, it is included an edge with the same weight and for each clause NAE(x,y,z), three edges (x,y), (x,z), (y,z) with weight $W/2$ each are included.
- The weights of the clauses in the instance I are defined as function of the variable's weights in such a way that the following two properties hold:
  1. an edge connecting two variable vertices u, v has weight equal to the product $|u| \cdot |v|$
  2. the weight of the edge connecting a variable vertex v to a constant vertex 0/1 is a multiple of the weight $|v|$.

Therefore, the instance of Weighted Max Cut constructed above has the property that any locally optimal solution (locally optimal cut) induces a truth assignment that is locally optimal for POS NAE 3−SAT. The final instance for unweighted Max Cut is obtained by replacing every vertex by and adequate subgraph, this is done in such a way that the new graph is unweighted and verifies the above property for locally optimal solutions. Going back to CVP that means that the input variables and gate variables in such assignment are consistent with the circuit. In other words, the values of the input variables coincide with the given input of the circuit and the gate variables have the value that is computed by the corresponding gates on that input.

Now, let v be a vertex. From properties 1. − 2., the total weight of its incident edges is a multiple of $|v|$, denoted $d(v) \cdot |v|$. Therefore the total weight of the cut (V_1, V_0) will be

$$W(V_1, V_0) = \sum_{v \in V_1} d(v) \cdot |v| \qquad (3)$$

where V_1 is the set of vertices corresponding to the true variables. Now, we express the weights of the variables (more precisely the weights of *control and local variables*) in terms of $|g_i|$ as in Table 1. Thus the weight of the cut given in (3) is written as

$$W(V_1, V_0) = \sum_{g \in TG} f(|g|) \qquad (4)$$

where TG denotes the set of true gates and f is a rational function. But we can always find constants m and M such that

$$m \cdot TG(C) \le \sum_{g \in TG} f(|g|) \le M \cdot TG(C) \qquad (5)$$

Therefore, from (4) and (5) it results that we cannot compute (or approximate) in NC the value of a locally optimal cut for any $\epsilon < \epsilon_1$ (ϵ_1 is a function of m and M). because it would imply that we can compute (approximate) the number of true gates (TG) of a circuit. □

We can define in the same way as *Local Max Cut*, the Non−oblivious Local Search Problem. Using arguments similars to those of Theorem 2 we can construct. instead of a Max Cut instance, an instance of Non−oblivious Local Search Problem. Therefore we have the following:

Corollary 12 *Approximating a Non−oblivious Local Search problem is P−complete for some values of ϵ.*

Suppose we have a problem Π and an algorithm A that approximates it for some ϵ_0 in polynomial time (for example, the standard local search algorithm). Furthermore, suppose that the value given by this algorithm cannot be approximated in NC for any $\epsilon < \epsilon_1$. In this situation, we naturally ask whether there is a threshold such that the problem Π itself cannot be approximated in NC.

Theorem 13. *Let x be an instance of an NPO problem Π and suppose that the algorithm A approximates Π within ϵ_0. Then, if the value $A(x) = V(x, S)$ cannot be approximated in NC for $\epsilon < \epsilon_1$, for some $\epsilon_1 > \epsilon_0$ then Π cannot be approximated in NC for $\epsilon < \epsilon_2$ for some ϵ_2 that depends on ϵ_0 and ϵ_1.*

Proof. Since A approximates Π within ϵ_0 we have that

$$\frac{1}{1 + \epsilon_0} \le \frac{A(x)}{OPT_\Pi(x)} \le 1 + \epsilon_0. \qquad (6)$$

Suppose the contrary, that there is an NC algorithm B that approximates Π within some ϵ, $0 \leq \epsilon < \epsilon_2$, that is

$$\frac{1}{1+\epsilon} \leq \frac{B(x)}{OPT_\Pi(x)} \leq 1 + \epsilon \qquad (7)$$

Now, we can write

$$\frac{B(x)}{A(x)} = \frac{B(x)}{OPT_\Pi(x)} \cdot \frac{OPT_\Pi(x)}{A(x)}$$

and therefore from (6) and (7) we have

$$\frac{1}{(1+\epsilon_0)(1+\epsilon)} \leq \frac{B(x)}{A(x)} \leq (1+\epsilon_0)(1+\epsilon) \qquad (8)$$

If we chose ϵ_2 such that $\epsilon_0 + \epsilon_0(1+\epsilon_2) = \epsilon_1$ then the inequalities (8) mean that we can approximate $A(x)$ within $\epsilon < \epsilon_1$ and that contradicts the supposition. $\square$

It was shown in [Lub88] that Max Cut can be approximated in NC within 1. Recall that Max Cut is Max SNP complete under $L-reductions$ [PY91]. Moreover, it was also shown to be complete for that class under *logspace $L-reductions$* [DSST95], therefore any Max SNP complete problem is approximable within a constant factor in NC. In [ACP94] it is given a 1-approximation for Max 3SAT in P, we can achieve the same ratio in NC:

Proposition 14. *Max 3SAT can be approximated in NC within 1.*

Proof. We consider the following algorithm:

- Let V_j be the set of clauses where the variable x_j or its negation appears.
  $V_j = \{C_i \mid x_j \in C_i \text{ or } \overline{x_j} \in C_i\}$ for $j \geq 1$. Denote by n_j its cardinality, $n_j = |V_j|$.
- Sort the sequence of the sets V_j in nonincreasing order of their cardinalities.
- Do a partition of the sets V_j that is, take $V_j := V_j - \cup_{i<j} V_i$
- For all V_j compute: n_j' =number of appearances of x_j in clauses of V_j;
  n_j'' =number of appearances of $\overline{x_j}$ in clauses of V_j;
- For all j, if $n_j' \geq n_j''$ then assign $x_j := True$ else assign $x_j := False$

The assignment found above satisfies at least $m/2$ clauses. For that, we note that x_j satisfies at least $n_j/2$ clauses, therefore at least $\sum_1^k \frac{n_j}{2} = \frac{m}{2}$ clauses are satisfied.
It is straightforward to see that this algorithm can be efficiently implemented in a PRAM using $O(n)$ processors and in $O(\log n)$ time. $\square$

Now, since Max Cut is *logspace* complete for Max SNP we have that:

Theorem 15. *For every Max SNP complete problem Π, exist ϵ_0, ϵ_1, $\epsilon_1 \leq \epsilon_0$, such that Π can be approximated in NC for any $\epsilon \geq \epsilon_0$ but cannot be approximated in NC for any $\epsilon < \epsilon_1$.*

5 Expected performance of local search algorithms

Recall from the definition of local search that, given an instance of the problem and a solution to it, we must be able to determine in polynomial time whether the solution is locally optimal and, if not, to generate a neighboring solution of improved cost. That, on turn, means that we are considering NPO problems whose domain of feasible solutions has cardinality polynomial in the input size. We are interested in the expected number of iterations of any local improvement algorithm for such problems under any reasonable probabilistic model.

Let's give first some notations and considerations. Given a problem $\Pi \in NPO$, we may consider the set of its feasible solutions as a $q \log n$–hypercube where n is the input size and q a constant that depends only on the instance. We can suppose that the values of the objective function f are distinct. Therefore, the vertices of the hypercube can be ordered from high to lower functional values and is called an *ordering*. Given a set S of vertices in the hypercube, by $\mathcal{B}(S)$ is denoted the boundary of S that consists of all vertices not in S that are adjacent to some vertex in S, that is

$$\mathcal{B}(S) = \{y \mid \exists x \in S, \; x \text{ and } y \text{ are adjacent}\}$$

A local improvement algorithm (see [Tov86], for example), in its standard form, is:

1. start at some random vertex x;
2. choose a vertex y adjacent to x such that $f(y) > f(x)$. If no such y exist, then stop.
3. set x equal y and iterate (2.);

If the local and global optima coincides, then this algorithm can be seen as a walk to the root of a tree whose vertices represent feasible solutions and the root represents the local optima. Otherwise, in the case of problems with multiple local optima, the algorithm generates a forest with as many trees as local optima there are. In the first case, the height of the tree gives us the maximum number of iterations done by the algorithm in order to find the optima. In the later, the number of iterations is given by the forest's depth.

In order to evaluate the expected number of iterations we must precise how do we choose at step 2. Many reasonable probability distributions exist [Tov86]. Here we will consider the *boundary distribution*. This is defined as follows. Let $\mathcal{B}(i)$ be the boundary of the vertices chosen until step i. Then, the $i + 1$st vertex is chosen uniformly in the boundary $\mathcal{B}(i)$. In fact, an even more simplified model will be considered. Instead of choosing randomly and uniformly from $\mathcal{B}(i)$, we consider the model where the $i + 1$st vertex is chosen uniformly from a subset of $\mathcal{B}(i)$, namely that of the deepest vertices generated so far. It results that this process *stocastically dominates* the first one, in the sense that the expected height in the second model is greater than or equal to that of the first one. So it suffices to find an upper bound for the expected height of the tree generated in the second model. A formal definition of *stochastic dominance* [Roh76] follows.

Definition 16. If X and Y are two random variables with distribution functions $F_x(t)$ and $F_y(t)$, then we say that X stochastically dominate Y $(X \succeq Y)$, if $F_x(t) \leq F_y(t)$, $\forall t$.

It is clear that if X stochastically dominates Y then $E(X) \geq E(Y)$. The definition given above is extended to sequences of random variables.

Definition 17. Let $X = X_1, X_2, \ldots$ and $Y = Y_1, Y_2, \ldots$ be two sequences of random variables. We say that X stochastically dominates Y if $\forall i, \; X_i \succeq Y_i$.

Let $r = q \log n$. Let $P_k = \sum_{j=0}^{k} \binom{r}{j}$. The following lemma gives a lower bound on the size of the boundary of a set of vertices in the r–hypercube.

Lemma 18. *Let $S(i)$ be the size of the smallest boundary of a set of size i. Then,*

- $S(i) = \binom{r}{k+1}$, *if $i = P_k$.*

otherwise:

- $\binom{r}{k+1} \leq S(i)$, *if $P_k < i < P_{k+1}$ and $k \leq (n-3)/2$.*
- $\binom{r}{k+2} \leq S(i)$, *if $P_k < i < P_{k+1}$ and $k \geq (n-3)/2$.*

Proof (Idea): We start from the intuitive idea that in the $q \log n$–hypercube, the boundary of a set of i vertices (as above) has at least $(q log n)^k$ vertices and then apply Kruskal–Katona theorem [Kle81] that shows how to find a set of i vertices in a hypercube with minimal boundary size. □

The stochastic process described below, will stochastically dominate the pathlengths of the vertices of the tree. This process is called the *largest k-process* and is denoted by L^k. Let $k = k_1, k_2, ...$ be a sequence of positive integers. The *largest k-process* is the the sequence of random variables $L^k = L_1^k, L_2^k, ...$ whose values are generated as follows: $L_1^k = 1$; given the values of $L_1^k, L_2^k, ..L_{i-1}^k$, choose randomly and uniformly one of the k_i largest values and set this value plus one to L_i^k.

Lemma 19. *Given a set of i vertices on the hypercube, let $B(i)$ denote a lower bound on its size. Let $k = k_1, k_2, ..., k_{n}$ be the sequence of integers where $k_i = \max(1, \lfloor B(i)/(q \log n - 1) \rfloor)$ and let $H = \{H_i\}$ be the sequence of random variables such that H_i denotes the height of the vertex i in the tree generated by the algorithm under boundary distribution. Then $L^k \succeq H$.*

Proof (Idea): Notice that L^k is generated by choosing among the largest values, that means choosing among the deepest vertices generated so far. This fact assures that L^k stocastically dominates the pathlength of the vertices. Furthermore, by choosing $k_i = max(1, \lfloor B(i)/(q \log n - 1) \rfloor)$ it is guaranteed that the vertices are chosen accordingly the boundary distribution. □

From this lemma, an upper bound for L^k is also an upper bound for the maximum pathlength on the tree. Let denote by μ_k, the *growth rate* of L^k, its average increase. The key fact is the following theorem [AP81].

Theorem 20. *[AP81] Let m be a positive integer and let M be a sequence of m's. Then the expected rate of growth, μ_m, of the sequence L^M is less than or equal to e/m for large m.*

Now, we are able to state the following result:

Theorem 21. *The expected number of iterations of any local improvement algorithm is less than:*

(a) $e\alpha \log^2 n \log \log n$, if the problem is local–global and the probability distribution used is the boundary distribution.

(b) $e\beta \log n$, if the problem has multiple optima and under any probability distribution.

where α and β are constants that depend only on the instances.

Proof. The idea is to see the *largest k-process* as formed of subsequences each of them simulated for a fixed m. The rate growth for each subsequence is then given by Theorem 6. Let $s = \lfloor (r-1)/2 \rfloor$ and divide the set of 2^r vertices of the r–hypercube into the segments:

$$1 \leq i \leq P_s, \; P_s < i \leq P_{s+1}, \cdots, P_{2s} < i \leq P_{2s+1}$$

The pathlengths of the vertices of the tree corresponding to each segment j, $1 \leq j \leq r$ are stocastically dominated by the subsequence of L^k with $m_j = k_j$, where k_j is given in Lemma 2. Thus, $L^k = L^{m_1}, L^{m_2}, \cdots, L^{m_r}$. Therefore, the total expected height is less than

$$\sum_{i=1}^{P_s} e(r-1)/B(i) + \sum_{P_s+1}^{2^r} e(r-1)/B(i)$$

$$1 + e(r-1) \sum_{k=1}^{s} \binom{r}{k} / \binom{r}{s} + e(r-1) \sum_{k=s}^{r-1} \binom{r}{k} / \binom{r}{s+1} + 1$$

$$\approx 2 + e(r-1)^2 + e(r-1)(r/2 + r \log r/2)$$

$$< eq^2 \log^2 n \log \log n$$

So, this is an upper bound for the expected number of iterations of the algorithm.

The proof for the case (b) uses similar arguments. We notice that in this case no matter how do we choose the vertex but, however, the way we choose must assure that all the *orderings* are equally likely. □

Notice that from this theorem we have that in particular oblivious and non-oblivious local search problems are in average NC, just use local improvement algorithms under any reasonable probabilistic model.

References

[ACP94] G. Ausiello, P. Crescenzi, and M. Protasi. Approximate Solution of NP Optimization Problems. Technical report, Dipartimento di Scienze dell'Informazione, Università degli Studi di Roma "La Sapienza", 1994.

[ADP80] G. Ausiello, A. D'Atri, and M. Protasi. Structure preserving reductions among convex optimization problems. *Journal of Computer and System Sciences*, 21:136–153, 1980.

[AP81] D. Aldous and J. Pitman. *The asymptotic speed and shape of a particle system.* Cambridge University Press, 1981.

[CS81] B. Corte and R. Schrader. On the existence of fast approximation schemes. In *Nonlinear Programming*, pages 415–437. Academic Press, 1981.

[DSST95] J. Díaz, M.J. Serna, P. Spirakis, and J. Torán. Paradigms for fast parallel approximations. *Manuscript in preparation*, 1995.

[GJ79] M.R. Garey and D.S. Johnson. *Computers and Intractability - A Guide to the Theory of NP-Completeness.* W. H. Freeman and Co., 1979.

[JPY88] D. Johnson, Ch. Papadimitriou, and M. Yannakakis. How easy is local search? *Journal of Computer and System Sciences*, 37:79–100, 1988.

[Kle81] J.K. Kleitman. Hypergraphic extremal properties. In *Proceedings of the 7th British Combinatorial Conference*, pages 59–78, 1981.

[KMSV94] S. Khanna, R. Motwani, M. Sudan, and U. Vazirani. On syntactic versus computational views of approximability. In *Proceedings 35th IEEE Symposium on Foundations of Computer Science*, pages 819–830, 1994.

[Lub88] M. Luby. Removing randomness in parallel computation without a processor penalty. In *Proceedings 29th IEEE Symposium on Foundations of Computer Science*, pages 162–173, 1988.

[PM81] A. Paz and S. Moran. Non deterministic polynomial optimization problems and their approximation. *Theoretical Computer Science*, 15:251–277, 1981.

[PY91] C.H Papadimitriou and M. Yannakakis. Optimization, Approximation, and Complexity Classes. *Computer and System Sciences*, 43:425–440, 1991.

[Roh76] V. Rohatgi. *An introduction to probability theory and mathematical statistics.* John Wiley, 1976.

[Ser91] M.J. Serna. Approximating linear programming is log-space complete for P. *Information Processing Letters*, 37:233–236, 1991.

[SY91] A.A. Schaffer and M. Yannakakis. Simple local search problems that are hard to solve. *SIAM Journal of Computing*, 20:56–87, 1991.

[Tov86] C. Tovey. Low order polynomial bounds on the expected performance of local improvement algorithms. *Journal of Mathematical Programming*, 35:193–224, 1986.

An Efficient and Effective Approximation Algorithm for the Map Labeling Problem *

Frank Wagner [†]   Alexander Wolff [‡]

Freie Universität Berlin

Abstract

The *Map Labeling Problem* is a classical problem of cartography. There is an approximation algorithm A which is theoretically optimal: A has optimal running time and guarantees a label size of 50 percent of the maximum. Unfortunately A is useless in practice as it typically produces results that are intolerably far off the optimal size. On the other hand there is a heuristic with good practical results, which is used in real applications.

Recently a hybrid algorithm was suggested that first runs A and then uses its result to control the heuristic.

In this paper we integrate the two parts of the hybrid method into an efficient and effective approximation algorithm. In addition we include a strategy to improve the empirical quality of the results significantly. The resulting algorithm B
- guarantees optimal approximation quality and runtime behaviour, and
- yields results closer to the optimum than the best heuristic known so far.

The sample data used in the experimental evaluation consists of three different classes of random problems and a selection of problems arising in the production of groundwater quality maps by the authorities of the City of München.

1 Introduction

Map lettering is one of the classical key problems that has to be solved in the process of map production. Usually the map producer does not only want to show the exact geographic positions of the features depicted but also explain properties of these features. She has to arrange this information on the map so that:
— for every piece of information it is intuitively clear which feature is described;
— the information is of legible size;
— different texts do not overlap.
These and in addition a lot of esthetic criteria are described by Imhof [I75] in an attempt to characterize good quality map lettering having mostly manual map making in mind. Nowadays there is an increasing need for large, especially technical maps, for which legibility is much more important than beauty.

---

*This work was done at the Institut für Informatik, Fachbereich Mathematik und Informatik, Freie Universität Berlin, Takustraße 9, 14195 Berlin-Dahlem, Germany. It was supported by the ESPRIT BRA Project ALCOM II

[†] wagner@math.fu-berlin.de
[‡] awolff@inf.fu-berlin.de

Figure 1: A valid labeling   Figure 2: An optimal labeling for the example of Figure 1

The application which brought the problem to our attention is the design of groundwater quality maps by the municipal authorities of the City of München. They have a net of drillholes spread over the city. The map has to contain the location of these holes and for every hole a block of measuring results such as the concentration of certain chemicals.

The growing importance of such technical maps induces a need for the computerization of map making, the need for fully automated algorithms. Typically, labels in technical maps are axis-parallel rectangles of identical sizes. By rescaling one of the axes we can assume that the rectangles are squares. An adequate formalization is as follows:

Problem MAP LABELING

Given n distinct points in the plane. Find the supremum σ_{opt} of all reals σ such that there is a set of n closed squares with side length σ, satisfying the following two properties.

1. Every point is a corner of exactly one square.

2. All squares are pairwise disjoint.

We call σ_{opt} the *optimal size*. A set of non-intersecting squares fulfilling (1) and (2) is called a *valid labeling*, see Figure 1 and 2.

In [FW91] the problem is shown to be $\mathcal{NP}$-hard. The main result of that paper is an approximation algorithm A that finds a valid labeling of at least half the optimal size. In addition, it is shown that no polynomial time approximation algorithm with a better quality guarantee exists, if $\mathcal{P} \neq \mathcal{NP}$. Related results were reported in [AIK89] and [IA86]. The running time of A is in $\mathcal{O}(n \log n)$. In [W94] it is shown that there is a matching lower bound on the running time.

A conceptually works as follows: We start with infinitesimal equally sized squares attached to each point in all four possible positions. Then all squares are expanded uniformly. In order to resolve conflicts between them, we eliminate all those which would contain another point if they were twice as big. It is easy to show that after this process, a point p can not have more than two squares left which overlap other squares.

If we consider p a boolean variable and associate its squares with the values p and $\bar{p}$, we can generate a boolean formula consisting of clauses which encode all conflicts. Suppose square p was overlapping square $\bar{q}$ of a point q, this would give us the clause $\overline{(p \wedge \bar{q})} = (\bar{p} \vee q)$ meaning that we do *not* want p and $\bar{q}$ to be simultanously in the solution. If we join all such clauses with the $\wedge$–operator, the satisfiability of the formula tells us exactly whether there is a solution of the current size. Since all clauses consist of two laterals, the formula is of 2-SAT type, and can be evaluated in time proportional to its length [EIS76].

This works only because we make sure that no point has more than two squares left after the elimination phase. On the other hand, we often eliminate both of two conflict partners, where it would have sufficed to delete one to resolve the conflict. This seems to be the reason for the practically very bad behaviour of A. In fact, A usually produces solutions not much better than 50 percent of the optimum, which makes it nearly useless for practical problems.

On the other hand there is a heuristical approach that shows very convincing results [WW95]. Instead of eliminating the squares as early as possible, it eliminates a square just when it is clear that it cannot be in any solution of the current size. The bad side effect of this is, that some points might have three or four squares left after the elimination phase. In order to handle this, three different methods H, I, and J are suggested to bring their number down to two. Method I is the winner in the experimental contest in [WW95].

In the same paper a hybrid algorithm is suggested that first runs A and then uses its result to control the heuristics in two ways:

1. The approximation size of A gives a lower and an upper bound on the optimal size. These bounds are used to show that there is only a linear number of conflicts between overlapping squares. In addition to that, twice the upper bound is the width of a window which swept across the sites to detect these conflicts. This makes sure that the heuristics' runtime is provably optimal, too.

2. The result of the hybrid algorithm is the maximum of A's and the heuristics' result, thus guaranteeing the optimal approximation quality.

The simplest of the heuristics, H, is used by the City of München for the application mentioned above, by the PTT Research Labs of the Netherlands to produce on-line maps for mobile radio networks, and in a computer system for the automated search for matching constellations in a star catalogue [WKA94] as a tool to label the output on the screen. With a very similar algorithmic approach Formann and Wagner [FW] were able to solve the so-called METAFONT labeling problem posed by Knuth and Raghunathan [KR92].

The main contribution of this paper is the integration of the two parts of the hybrid method into a provably good and efficient algorithm of even better quality:

In order to achieve A's running time efficiency of $\mathcal{O}(n \log n)$, we introduce a new way of detecting conflicts between overlapping squares. It is based on an algorithm to find closest neighbours, which was suggested in [F92]. We maintain A's quality guarantee of 50 percent without being obliged to run A beforehand and use its result. In order to do so, we perform a test whether a solution can still be constructed from a subset of the squares which have not been eliminated so far during a certain phase of the algorithm. We improve this elimination phase with the help of an additional detection rule for squares which are not needed in a valid solution.

We compare A, B, and I in an experimental evaluation using three different classes of random problems and a selection of problems arising in the production of groundwater quality maps by the authorities of the City of München. The best of the three, our new algorithm B, guarantees optimal approximation quality and runtime behaviour, and yields experimental results closer to the optimum than the best heuristic known so far.

2 The Algorithm

Before we can describe the structure of the algorithm we have to give some definitions.

Definition 1 *For a point p in the plane, a real $\sigma \geq 0$, and $i \in \{1, 2, 3, 4\}$, denote by σp_i an axis-parallel square with side length σ and p in its southwest, southeast, northeast respectively northwest corner. The enumeration is chosen like that of quadrants.*

We will call p_i a candidate *of the* site p. *Where the edge length σ is omitted, we refer to a candidate of the current label size.*

A solution of size σ *is a valid labeling with candidates of side length σ.*

For technical reasons, we will from now on consider a candidate an open square, plus the open edges incident to the site. Note that this excludes all corner points, especially the site itself. The idea is that we shrink the squares by a tiny bit, so that an optimal labeling is a valid labeling, too. σ_{opt} then is the size of the maximum valid solution. This is equivalent to the previous definition.

We say that two candidates *overlap* or have a *conflict* if they intersect and neither contains a site. Analogously, two points are in conflict if any of their candidates are. For two candidates we define their *conflict size* as the largest edge length at which they do not intersect.

2.0 Structure

B differs from the hybrid algorithm suggested in [WW95] in the following way: It avoids to run A beforehand by finding the conflict sizes in Step 1 independently of A's result. In Phase II of Step 2, we introduce a new elimination rule that improves the performance substantially. The 2-SAT test of this phase maintains A's quality guarantee of 50 percent. Phase I and II are taken from Heuristic I.

1. Find all *important* conflict sizes.

2. Do a binary search on these conflict sizes. Check for each size you look at, whether there is a solution or not, by going through the following three phases:

    Phase I: Preprocessing.

    Phase II: Eliminate candidates which cannot be part of the solution. Then do a 2-SAT test on a subset of those which have not been eliminated.

    Phase III: For those points which still have two or more candidates left, choose exactly two, and check, whether this remaining problem is solvable by 2-SAT, as described in the introduction.

2.1 Finding conflict sizes

We show that it is sufficient to look at a constant number of closest neighbours of a site p in order to determine all conflict sizes which are not greater than the optimal label size. The reason for this strategy is that the k closest neighbours of all n sites p in any of the four quadrants relative to p can be found efficiently in $\mathcal{O}(kn \log n)$ time with an algorithm described in [F92].

Definition 2 *Let* $\text{Quad}(p_i) = \infty p_i$, *that is the ith quadrant relative to p, $i \in \{1, 2, 3, 4\}$. Note that this includes the border except p.*

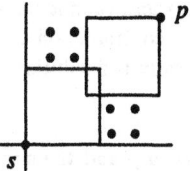

Figure 3: The conflict shown here between s_1 and p_3 is *not* important.

A conflict between a site s and one of its eight closest neighbours in one of the four quadrants relative to s is called important. *See Figure 3 for a counterexample. The size of such a conflict analogously is an* important conflict size.

A label *is the candidate of a site which has been chosen to be in the solution.*

Theorem 3 *All conflict sizes which are not important, are greater than σ_{opt}.*

We do not have to consider conflict sizes greater than σ_{opt}, because there cannot be a solution of that size. There is no need to check any other label size either, since the conflict graph of all candidates does not change inbetween two consequetive conflict sizes. So it is sufficient to do the binary search in Step 2 of the algorithm on important conflict sizes.

It is obvious that the number of important conflict sizes is linear if we only have to look at a constant number of closest sites per candidate. These can be detected efficiently in $\mathcal{O}(n \log n)$ as mentioned above.

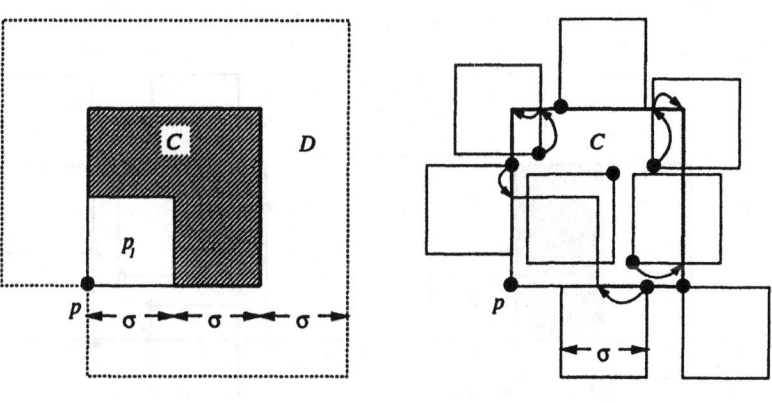

Figure 4:   Figure 5:

Proof. For reasons of symmetry we can focus on conflicts in which some candidate p_1 is involved. First, consider only sites which lie in Quad(p_1). We want to show that the sizes of all conflicts between p_1 and candidates of these sites are greater σ_{opt} if they are not important. Suppose there is a conflict of size $\sigma \leq \sigma_{opt}$ between p_1 and one of the candidates of its kth

closest neighbour s^k. We show that there cannot be a partial solution of size σ for p and $s^1, \ldots, s^k$ if $k > 8$. This would be a contradiction to $\sigma \leq \sigma_{opt}$, because there is always a solution of size σ_{opt}, and it automatically is a solution for all smaller label sizes as well, and certainly for a subset of the sites.

It is clear that the distance between p and the sites with candidates in conflict with p_1 is bounded in the following way: $\sigma \leq \mathrm{d}(p, s^i) \leq 2\sigma \leq 2\sigma_{opt}$ for $i = 1, \ldots, k$, otherwise s^1 would lie in σp_1, or none of the candidates of s^k would have a conflict of size σ with p_1. The area C in which the s^i can therefore lie, is shaded in Figure 4. Look at the area D which is where candidates of the sites in C can lie. The size of D is $15\sigma^2$, but the number of labels which can be placed in D, such that they are not in conflict with each other, is not 15 but 8, because the sites are restricted to C:

The idea is that we find out how many labels intersect the bold part of C's outline, and how many do not. We start with those sites whose labels contain a point of this line. We move them to a point of the outline which has either the same x- or y-coordinate — if the site is not already located on the outline. Since we move them "outwards", into a direction where no other label can lie, we avoid generating conflicts. We can then move sites along the line into its five corners. For those corners which are already contained by a label, take the site to which the label belongs, otherwise the closest site on either side of the outline. Now it is clear that apart from those in the corners not more than two sites can lie on the thick line, and another one in the interior of C, such that their labels do not overlap. Moving these labels on the one hand maximized the space in the interior of C, on the other it minimized the length of the intersection of each of the labels with the bold line. This means that in the original solution, too, not more than seven labels can have intersected the bold line, and only one can possibly have had an empty intersection.

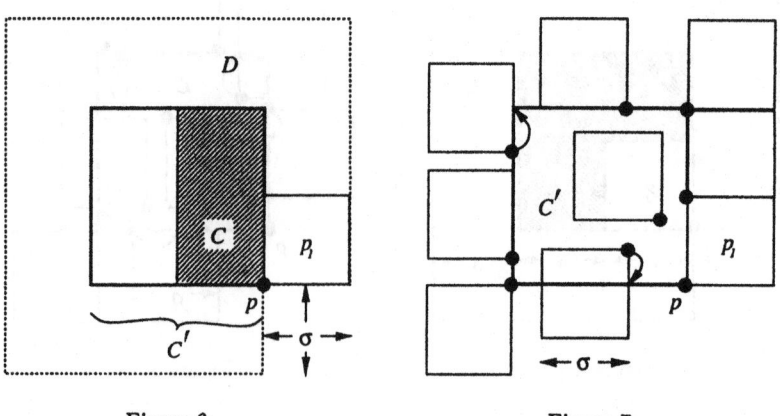

Figure 6:   Figure 7:

Now consider all sites which lie in the second quadrant relative to p, and have candidates in conflict with p_1. Again, let C be the area in which such sites can lie. C is shaded in Figure 6. Look at the area C' in Quad(p_2) of all points which have the same distance to p as points in C.

Push all sites to the outline of C' as above. Three sites can then be moved into the corners of the outline, one on each of its four sides, and again there can be only one site such that its candidate does not contain a point of the outline, see Figure 7.

Exactly the same idea holds for sites in Quad(p_4), since it is symmetric to Quad(p_2) relative to p_1. No candidate of a site in Quad(p_3) can be in conflict with p_1, because it would then automatically contain p. Hence there cannot be a conflict of size $\sigma \leq \sigma_{opt}$ between p_1 and one of the candidates of its kth closest neighbour for $k > 8$. ⊡

2.2 Check whether there is a solution for a given σ

2.2.1 Phase I: Preprocessing

We run through all the candidates p_i. If $d(p_i) < \sigma$ we *eliminate* p_i, i. e. we will not consider it any more because then σp_i contains $\delta(p_i)$. Otherwise we create a new list of overlap information which is an excerpt from p_i's conflict list. We use the fact that two overlapping candidates remain in conflict, until either of them contains a site if they are blown up simultanously. The elements of the new list consist of pointers to the overlap information of those candidates which actually overlap p_i for the given label size σ, and a pointer back to the candidate it belongs to. This can be done in linear time since the sum of the lengths of all conflict lists is linear, confer Section 2.1. For the same reason, the lists of overlap information just need linear space.

2.2.2 Phase II: Making Decisions

We run once through all points p. Heuristic I looks at the following three cases:

- If all candidates of p have been eliminated, we stop and return "no solution" to the program which does the binary search on the conflict list.

- If p has candidates free of intersections with other candidates, we choose an arbitrary one of them (say p_i), and eliminate all other candidates p_j of p. Before a candidate's deletion, we do the following updates: we delete its list of overlap information and the symmetric entries stored with all candidates which overlap it.

- If p has only one candidate p_i left, do the same updates with all candidates q_j which overlap p_i, and then delete them.

For B we add we a new rule. Refer to Figure 13 and 14 to see how this has improved the performance of B as compared to Heuristic I.

- If p has a candidate p_i which overlaps the last two candidates of another site q, then update and eliminate p_i.

While we do this we maintain a stack. On this stack we put all those candidates which now fulfill the same properties as p_i did before, i. e. do not intersect any other squares, or are the last candidates of their sites. Before we look at the next site p, we do all the decisions waiting for us on the stack. Since each of the cases listed above can be detected and handled in constant time, and the number of times they occur is bounded by the number of conflicts, Phase II takes us so far linear time.

Corollary 4 *If there is a solution of the current label size σ, then there is still a subset of the candidates left after Phase II, which forms a solution.*

Proof. This can be shown analogously to the proof of Corollary 5 in [WW95]. □

At the end of this part of Phase II we are done if all sites have exactly one candidate left. Otherwise we know that candidates of sites with several candidates — call them *active* — never intersect with those that are "the last of their breed", i. e. belong to sites with exactly one square left, because then the former ones would have been eliminated. So it is enough to focus on active candidates from now on. The others are already chosen as part of the solution, and do not interfere with the active ones any more.

As a consequence of Corollary 4 we also know that we have not yet returned "no solution" if there is one of size σ. So we can find a solution with the help of 2-SAT as described in the introduction if no site has more than two candidates left. Otherwise we do a test on a subset of the active candidates as follows.

Keeping the Approximation Guarantee

How can we make sure that B, too, keeps the approximation guarantee of 50 percent? Its main difference from the Approximation Algorithm A is that until the end of Phase II it does not destroy any candidate which might be necessary to construct a solution. A on the other hand already eliminates candidates if they contain a site at twice their size. We take advantage of this approach without risking to end up with a practical behaviour similarly bad as A's.

Definition 5 *A candidate p_i is called σ–dead if σp_i contains a site.*

Lemma 6 *Let σ be the current label size. Then after Phase II all sites have at most two candidates left which are not 2σ–dead.*

Proof. Suppose p had three such candidates, w. l. o. g. p_1, p_2, and p_4 as indicated in Figure 8. If $2\sigma p_1$, $2\sigma p_2$, and $2\sigma p_4$ did not contain any other site, then there could clearly be no candidate in conflict with σp_1. This means that p_1 (or eventually p_2 or p_4) should have already been chosen as part of the solution in Phase II, and p's other candidates eliminated. □

Theorem 7 *We can efficiently construct a solution of size σ from the candidates left after Phase II — if $\sigma \leq \sigma_{opt}/2$.*

Proof. If $\sigma \leq \sigma_{opt}/2$, then there must be a solution π of size 2σ. π is also a solution of size σ, since simultanously shrinking all squares reduces the number of conflicts. Now consider all sites p which have more than one candidate left after Phase II. Let $i = \pi(p)$. We want to show that p_i has not yet been eliminated. To see that, we just have to check the conditions under which this could have happened:

- σp_i cannot contain any other site, because $2\sigma p_i$ does not.

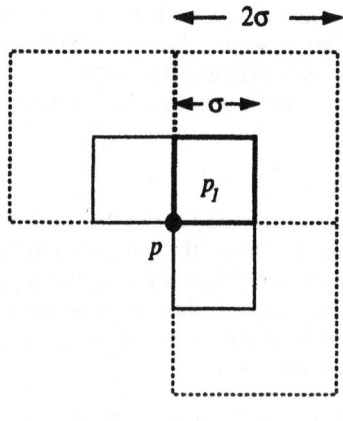

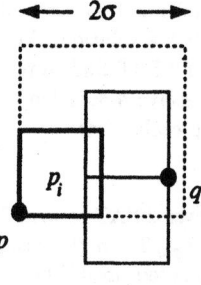

Figure 8:   Figure 9:

- σp_i cannot overlap the last two candidates of some site q, since $2\sigma p_i$ would then contain q (see Figure 9).

- Suppose σp_i overlaps σq_j, and q_j is the last candidate of q (see Figure 10). Then $\pi(q) = k \neq j$, otherwise $2\sigma p_i$ and $2\sigma q_{\pi(q)}$ would overlap. The consequence of this is, that q_k also must have overlapped a candidate which is the last of its site, otherwise q_k and therefore p_i would not have been eliminated. Since the number of sites is finite, there must be a candidate r_m in this chain which overlaps a last candidate we have already looked at, say q_j. Then $2\sigma p_i$ and $2\sigma r_m$ must be disjoint, because they are part of solution π. At the same time σq_j must overlap σp_i and σr_m. This means that one corner of σq_j is contained in σp_i, and another one in σr_m. But then automatically the last two corners of σq_j must lie in $2\sigma p_i$ and $2\sigma r_m$ which are disjoint (see Figure 11). Obviously one of the corners of σq_j must be q. So q lies in either $2\sigma p_i$ or $2\sigma r_m$. This is a contradiction to the assumption that they are part of solution π.

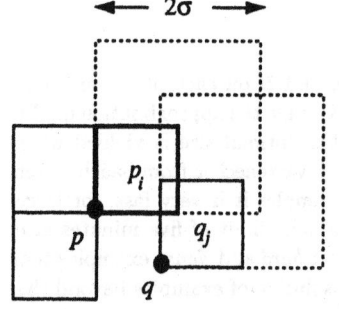

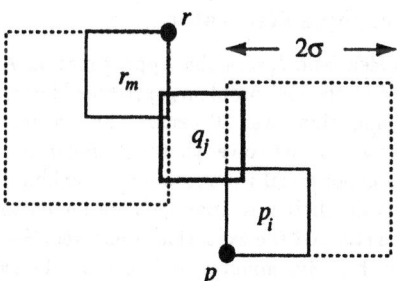

Figure 10:   Figure 11:

We have just shown that if there is a solution of size 2σ, then all sites with active candidates have kept their candidate which is part of this solution all the way through Phase II.

Lemma 6 tells us that after Phase II all sites have at most two squares left which are candidates for such a solution. So we can use 2-SAT on the set of active candidates which are not 2σ–dead. If 2-SAT finds a partial solution for those, we can just add all inactive candidates, and thus receive a solution for all sites, since the sets of active and inactive candidates do not overlap each other.

□

If however 2-SAT returns "no solution", then there cannot be any of size 2σ. That means that $\sigma > \sigma_{opt}/2$. In this case we try to find a solution in Phase III. This 2-SAT test will only be executed until it has returned a negative answer to a problem of size σ_0 to which the heuristics nevertheless found a solution afterwards. We can spare it then, because the binary search for a solution of maximal size continues only on label sizes σ with $\sigma > \sigma_0 > \sigma_{opt}/2$. That means that the test would keep returning negative answers.

Since the first part of Phase II and 2-SAT can be implemented in linear time [EIS76], Phase II has an overall running time of $\mathcal{O}(n)$.

2.2.3 Phase III

We run through all points with active candidates twice. In the first run, we only look at those with four candidates left, eliminate the one with most conflicts, and make all decisions of the type we did in Phase II. During the second run, we do the same for points which still have three active candidates. Then the remaining problem (consisting only of points with exactly two active candidates) is handed over to 2-SAT as descibed in the introduction. This takes linear time.

2.3 Running time analysis

Phase I, II, and III can all be done in linear time, but since we have to look at $\mathcal{O}(\log n)$ conflict sizes during the binary search for the best solution, we need $\mathcal{O}(n \log n)$ time for Step 2. Finding conflicts in Step 1 takes the same time, so the whole algorithm is in $\mathcal{O}(n \log n)$.

3 Experiments

3.1 Example Generators

We run Heuristic I, and the Approximation Algorithms A and B on each of the examples produced by the four problem generators. For every size we average the approximation quality and running time over 30 tests. The information about the optimal size is yielded where possible by an exact solver that was implemented by Erik Schwarzenecker from Saarbrücken. It shows exponential running time behaviour. For small examples it is very fast, for larger ones it is unreliable, so that we were forced to introduce a time limit of five minutes after which we stopped the execution. Only very few of the largest *hard* and *dense* examples took less than these five minutes, and we have observed that the solution of examples beyond that bound then easily takes half an hour or much more.

Random. We choose n points uniformly distributed in a square of size $10n \times 10n$.

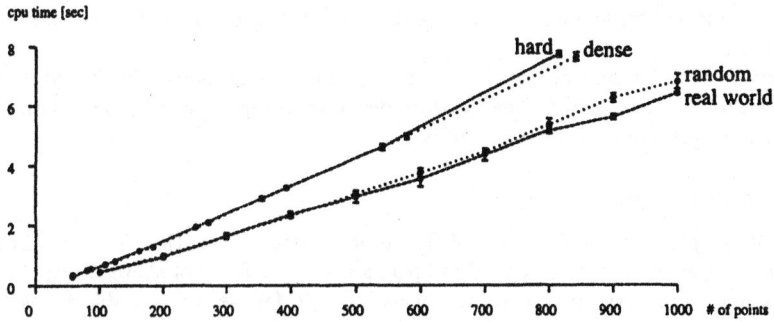

Figure 12: Running time of the Approximation Algorithm B on different example classes

Dense. Here we try to place as many squares as possible of a given size σ on a rectangle. We do this by randomly choosing points p and then checking whether σp_1 intersects with any of the σq_1 chosen before. We stop when we have unsuccessfully tried to place a new square 200 times. In a last step we assign a random corner point to each of the squares we were able to place without intersection, and return its coordinates. This method gives us a lower bound for the label size of the optimal solution. The size of the rectangle on which we place the squares is $\lfloor \alpha\sqrt{n} \rfloor \times \lceil \alpha\sqrt{n} \rceil$. α is a factor chosen such that the number of successfully placed squares is approximately n, the number of sites asked for.

Hard. In principle we use the same method as for Dense, that is, trying to place as many squares as possible into a rectangle. In order to do so, we put a grid of cell size σ on it. In a random order, we try to place a square of edge length σ into each of the cells. This is done by randomly choosing a point within the cell and putting a fixed corner of the square on it. If it overlaps any of those chosen before, we try to place it into the same cell a constant number of times.

Real World. The municipal authorities of Munich provided us with the coordinates of roughly 1200 ground water drill holes within a 10 by 10 kilometer square centred approximately on the city centre. From this list we extract a given number of points being closest to some centre point according to the L_∞-norm, thus getting all those lying in a square around this extraction centre, where the size of the square depends on the number of points asked for. For our tests we chose five different centres; that of the map and those of its four quadrants in order to get results from different areas of the city with strongly varying point density. This is due to the fact that many of the holes were drilled during the construction of subway lines which are concentrated in the city centre, see Figure 15.

3.2 Results

We show the two classical kinds of plots; time and quality. Quality here means the quotient of the solutions of the approximation algorithms and the exact solver. Time is measured in CPU time, which is sufficient since it is closely related to the number of square–square conflicts. This on the other hand determines the number of crucial steps, namely finding all conflicts once, and then extracting those valid for a certain σ in every step of the binary search.

The results both for time and quality are averaged only over those tests the exact solver managed within the time bound. The standard deviation is represented by the length of the vertical bars in each point of the result plots.

3.2.1 Running Time

In Figure 12 we plot the running times of the Approximation Algorithm B on the different example sets, which we measured on a Sun Sparc station 20. The plot shows a rather stable $\mathcal{O}(n)$-behaviour with very small standard deviation. So far we are not able to back the empirically linear running time by a theoretical analysis.

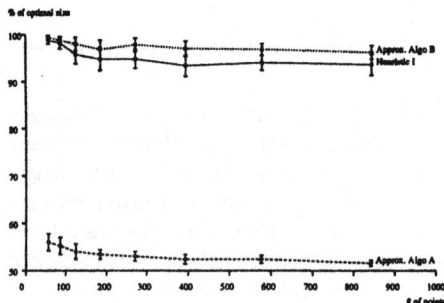

Figure 13: Quality of the algorithms on dense examples

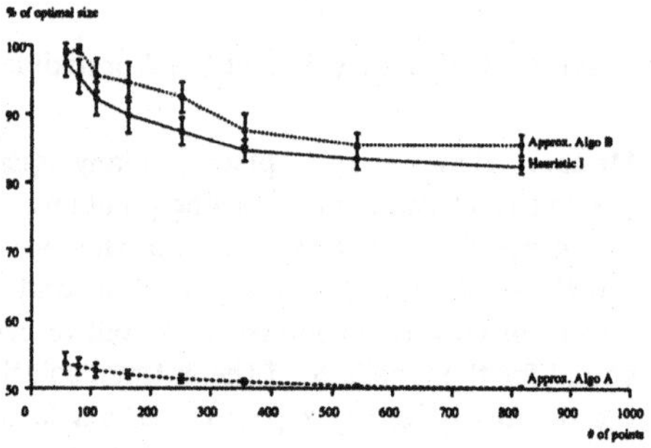

Figure 14: Quality of the algorithms on hard examples

3.2.2 Approximation Quality

The approximation quality is only plotted for dense and hard examples (see Figure 13 and 14), because on random and real world problems the performance of both B and Heuristic I was almost always 100 percent. For an example, see Figure 16 and 17.

A remark on the examples for which X did not give a result within the time bound: As mentioned above we did not include those in the calculation of the quality plots. But using the upper bound provided by the Approximation Algorithm A, and taking into consideration the typical quality of A, we found out that B's behaviour on those examples does not differ significantly from that on the other examples.

4 Implementation

The code was written in C++, and we strongly took advantage of data structures and algorithms provided by LEDA [MN95]. The commands LEDA offers, helped a great deal to shorten and simplify the code. All heuristics, approximation algorithms, exact solvers, and problem generators described here, can be tested on the WWW under URL:
http://www.inf.fu-berlin.de/~awolff/html/labeling.html.

References

[AIK89]   H. AONUMA, H. IMAI, Y. KAMBAYASHI, *A visual system of placing characters appropriatly in multimedia map databases*, Proceedings of the IFIP TC 2/WG 2.6 Working Conference on Visual Database Systems, North Holland (1989) 525–546

[EIS76]   S. EVEN, A. ITAI, A. SHAMIR, *On the complexity of Timetable and Multicommodity Flow Problems*, SIAM Journal on Computing **5** (1976) 691–703

[F92]   M. FORMANN, *Algorithms for Geometric Packing and Scaling Problems*, Dissertation, Fachbereich Mathematik und Informatik, Freie Universität Berlin (1992)

[FW91]   M. FORMANN, F. WAGNER, *A Packing Problem with Applications to Lettering of Maps*, Proceedings of the 7th Annual ACM Symposium on Computational Geometry (1991) 281–288

[FW]   M. FORMANN, F. WAGNER, *An efficient solution to Knuth's METAFONT labeling problem*, Manuscript (1993)

[I75]   E. IMHOF, *Positioning Names on Maps*, The American Cartographer **2** (1975) 128–144

[KR92]   D. E. KNUTH AND A. RAGHUNATHAN, *The Problem of Compatible Representatives*, SIAM Journal on Discrete Mathematics **5** (1992) 422–427

[MN95]   K. MEHLHORN AND S. NÄHER, *LEDA: a platform for combinatorial and geometric computing*, Communications of the ACM **38** (1995) 96–102

[IA86]   H. IMAI, T. ASANO, *Efficient Algorithms for Geometric Graph Search Problems*, SIAM J. Comput. **15** (1986) 478–494

[KMPS93]   L. KUČERA, K. MEHLHORN, B. PREIS, E. SCHWARZENECKER, *Exact Algorithms for a Geometric Packing Problem*, Proceedings of the 10th Annual Symposium on Theoretical Aspects of Computer Science, Lecture Notes in Computer Science **665** (1993) 317–322

[W94]   F. WAGNER *Approximate Map Labeling is in $\Omega(n \log n)$*, Information Processing Letters **52** (1994) 161–165

[WW95]   F. WAGNER, A. WOLFF *Map Labeling Heuristics: Provably Good and Practically Useful*, to appear in: Proceedings of the 11th Annual ACM Symposium on Computational Geometry (1995)

[WKA94]   G. WEBER, L. KNIPPING, H. ALT, *An Application of Point Pattern Matching in Astronautics*, Journal of Symbolic Computation **17** (1994) 321–340

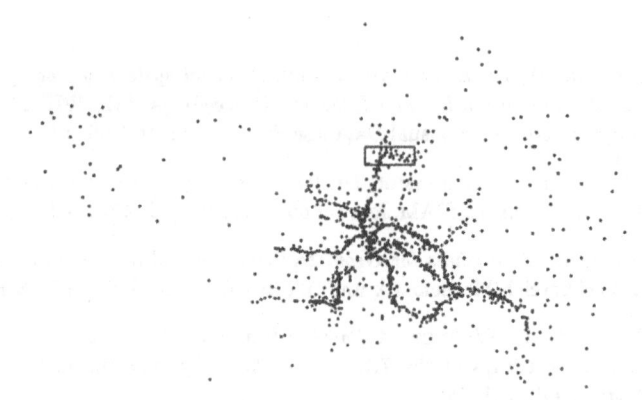

Figure 15: Map showing our sample data of approximately 1200 groundwater drillholes in Munich, and the section tested in Figures 16 and 17. There are no conflicts of interesting size between this section and the rest. The subway lines can be detected easily.

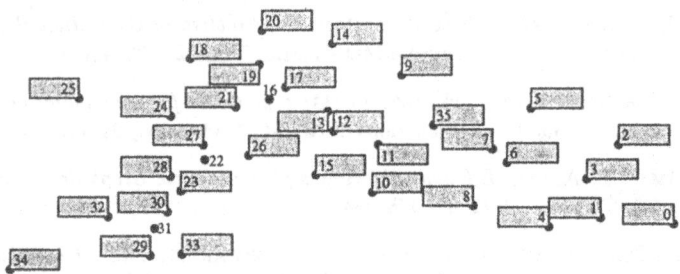

Figure 16: Solution of the program used by the authorities of the City of Munich before (label height 5000, 3 sites not labelled). It tries to maximize the number of sites labelled for a given size.

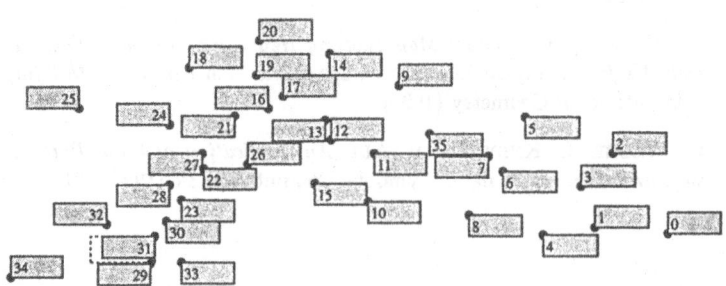

Figure 17: Solution produced by our heuristics (label height 5400, optimal).

Approximating the Bandwidth
for Asteroidal Triple-Free Graphs

T. Kloks[1] *, D. Kratsch and H. Müller[2]

[1] Department of Mathematics and Computing Science
Eindhoven University of Technology
P.O.Box 513, 5600 MB Eindhoven
The Netherlands
[2] Fakultät für Mathematik und Informatik
Friedrich-Schiller-Universität
Universitätshochhaus, 07740 Jena
Germany

Abstract. We show that there is an $O(n^3)$ algorithm to approximate the bandwidth of an AT-free graph with worst case performance ratio 2. Alternatively, at the cost of the approximation factor, we can also obtain an $O(e + n \log n)$ algorithm to approximate the bandwidth of an AT-free graph within a factor 4. For the special cases of permutation graphs and trapezoid graphs we obtain $O(n \log n)$ algorithms with worst case performance ratio 2. For cocomparability graphs we obtain an $O(n^2)$ algorithm with worst case performance ratio 3.

1 Introduction

The BANDWIDTH minimization problem is the problem, given a graph G and an integer k, to map the vertices of G to distinct positive integers, so that no edge of G has its endpoints mapped to integers that differ by more than k. The problem is motivated by the bandwidth minimization problem for matrices, given an $n \times n$ matrix A and an integer k, to find whether there is a permutation matrix P such that $P \cdot A \cdot P^T$ is a matrix with all nonzero entries on the main diagonal or on the k diagonals on either side of this main diagonal. The latter problem is of great importance in many engineering applications. Typically, matrices arising in applications are sparse (even already with many nonzero entries in a small 'band' around the main diagonal). Standard matrix operations like inversion and multiplication as well as Gaussian elimination etc. can be speed up considerably if the matrix A is transformed into a matrix $P \cdot A \cdot P^T$ of 'small bandwidth'. Due to the great interest in the problem, both in the matrix and in the graph version, over the last 30 years a number of heuristics have been proposed (see [4]). We mention that the performance of the Cuthill and McKee heuristic for *random matrices* was investigated in [27]. It is shown that these heuristics produce optimal solutions within a constant factor for random matrices with bandwidth no larger than k. Nevertheless no polynomial time algorithm with a provably good approximation has been found.

---

* Email: ton@win.tue.nl

More precisely, up to our knowledge, there is no approximation algorithm for bandwidth of general graphs with worst case ratio $o(n)$ and the same is the case for trees. We are also not aware of hardness results concerning the approximation of bandwidth. Finally, BANDWIDTH is one of the few well-known NP-complete graph problems that remain NP-complete on trees.

As a first approach in improving the unsatisfactory situation we present efficient approximation algorithms for bandwidth on classes of AT-free graphs with worst case performance ratio two or three. We also show that the BANDWIDTH problem remains NP-complete on complements of bipartite graphs, hence on AT-free graphs.

An independent set of three vertices is called an *asteroidal triple* if between each pair in the triple there exists a path that avoids the neighborhood of the third. A graph is *AT-free* if it does not contain an asteroidal triple. Recently the structure of AT-free graphs has been studied extensively (see [5]). The class of AT-free graphs contains various well-known graph classes, as e.g., interval, permutation, trapezoid and cocomparability graphs. A good reference for more informations on these graph classes is [9].

In this paper we give an efficient algorithm that approximates the bandwidth of AT-free graphs with worst case performance ratio two. Our algorithm can be implemented so that it runs in $O(n^3)$ time. We also present an $O(e + n \log n)$ algorithm with worst case performance ratio four. Here n is the number of vertices and e is the number of edges. For the special case of cocomparability graphs we obtain a simple $O(n^2)$ algorithm that approximates the bandwidth with worst case performance ratio three. For permutation graphs and trapezoid graphs we obtain approximation algorithms with worst case performance ratio two. These algorithms can be implemented to run in $O(n \log n)$ time. For cocomparability graphs of dimension bounded by d, we obtain an approximation algorithm with worst case performance ratio two that can be implemented to run in $O(nd + n \log n)$ time.

Recently, for the class of chain graphs (which is a proper subclass of the bipartite permutation graphs), we obtained an $O(n^2 \log n)$ algorithm to compute the exact bandwidth [14].

2 Preliminaries

2.1 Bandwidth

Let $G = (V, E)$ be a graph. We denote by n the number of vertices of G and by e the number of edges.

Definition 1 *A layout L of G is a 1-1 mapping from V into $\{1, ..., |V|\}$. The width of L is defined as $b(G, L) = \max\{|L(u) - L(v)| \mid (u, v) \in E\}$. The bandwidth of G is*

$$bw(G) = \min\{b(G, L) \mid L \text{ is a layout of } G\}.$$

In general, computing the bandwidth of a graph is NP-complete [20], even for trees having maximum degree three [8].

There are only a few graph classes known for which the bandwidth can be computed efficiently. One such graph class is the class of theta graphs [22]. Another one is the class of caterpillars with hairs of length one and two [2]. However, for caterpillars with hairs of length at most three, the BANDWIDTH problem remains NP-complete [18]. Recently we obtained a polynomial time algorithm computing the bandwidth of chain graphs [14].

It can be checked in linear time whether the bandwidth of a graph is at most two [8]. For general k, there is an $O(n^k)$ algorithm to check whether the bandwidth of a graph is at most k [10]. In some sense, this is best possible, since it was shown in [3] that BANDWIDTH is $W[t]$-hard for all t in the fixed parameter hierarchy. Hence it cannot be expected that there is an $O(n^\alpha)$ algorithm for any fixed α.

There is one other non trivial graph class for which the bandwidth can be computed efficiently. This is the class of interval graphs. It was shown in a series of papers that the bandwidth of an interval graph can be computed efficiently in a greedy manner [16, 12, 26].

Definition 2 *A graph is an* interval graph *if one can associate with each vertex an interval on the real line such that two vertices are adjacent if and only if the corresponding intervals have a nonempty intersection. An interval graph is called* proper *if the intervals can be chosen such that no interval is properly contained in another one.*

In [11] it is shown that the bandwidth of a graph is equal to its *proper pathwidth*. The proper pathwidth of a graph G is defined as the minimum cardinality of a maximum clique of a proper interval supergraph of G decreased by one.

The complements of bipartite graphs, so-called cobipartite graphs, form a subclass of the cocomparability graphs, and hence also of the AT-free graphs.

Theorem 1 *The problem 'Given a cobipartite graph $G = (V, E)$ and a positive integer k, decide whether $bw(G) \leq k$', is* NP-*complete.*

Proof. Let G be a cobipartite graph. Then $\alpha(G) \leq 2$ implies that G is claw-free. Hence G is claw-free and AT-free and this implies $bw(G) = tw(G)$ by corollary 3 on page 10. Finally the problem TREEWIDTH is known to be NP-complete on cobipartite graphs [1] implying that BANDWIDTH on cobipartite graphs is NP-complete. □

Clearly this implies that BANDWIDTH remains NP-complete when restricted to cocomparability or to AT-free graphs.

Most of our approximation algorithms are based on the following theorem. Here G^d is the graph having the same vertex set as G and two vertices u and v are adjacent in G^d if and only if there is a path of length at most d between u and v in G.

Theorem 2 *Let G and H be graphs with $G \subseteq H \subseteq G^d$ or $H \subseteq G \subseteq H^d$ for an integer $d \geq 1$, and let L be an optimal layout for H, i.e., $b(H, L) = bw(H)$. Then L approximates the bandwidth of G by a factor of d, i.e., $b(G, L) \leq d \cdot bw(G)$.*

Proof. First we assume $G \subseteq H \subseteq G^d$. Then

$$
\begin{aligned}
b(G, L) &\leq b(H, L), & &\text{since } G \subseteq H \\
&= bw(H), & &\text{since } L \text{ is optimal for } H \\
&\leq bw(G^d), & &\text{since } H \subseteq G^d \\
&\leq d \cdot bw(G).
\end{aligned}
$$

Now let $H \subseteq G \subseteq H^d$. Then

$$
\begin{aligned}
b(G, L) &\le b(H^d, L), &&\text{since } G \subseteq H^d \\
&\le d \cdot b(H, L), \\
&= d \cdot bw(H), &&\text{since } L \text{ is optimal for } H \\
&\le d \cdot bw(G), &&\text{since } H \subseteq G.
\end{aligned}
$$

This proves the theorem. □

We end this section with a lower bound for the bandwidth of graphs.

Lemma 1 *Let $G = (V, E)$ be a graph. Then for every edge (x, y) of G*

$$
bw(G) \ge \frac{1}{3} \left(|N(x) \cup N(y)| - 1 \right)
$$

Proof. Let L be an optimal layout of G, i.e., $b(G, L) = bw(G)$, and $(x, y) \in E$. W.l.o.g. assume $L(x) < L(y)$. Consider the three sets

$$
\begin{aligned}
S_1 &= \{v \in N(x) \cup N(y) \mid L(v) \le L(x)\} \\
S_2 &= \{v \in N(x) \cup N(y) \mid L(x) \le L(v) \le L(y)\} \\
S_3 &= \{v \in N(x) \cup N(y) \mid L(v) \ge L(y)\}
\end{aligned}
$$

Then $bw(G) = b(G, L) \ge \max_{i \in \{1,2,3\}}(|S_i| - 1)$. Since $\sum_{i=1}^{3} |S_i| = |N(x) \cup N(y)| + 2$ there is an i with $|S_i| - 1 \ge \frac{1}{3} \left(|N(x) \cup N(y)| - 1 \right)$. □

2.2 Preliminaries on triangulations

Definition 3 *A graph is called* chordal *if it does not contain a chordless cycle of length at least four as an induced subgraph.*

Definition 4 *Let $G = (V, E)$ be a graph. A* triangulation *of G is a chordal graph H, with the same vertex set as G such that G is a subgraph of H.*

If $G = (V, E)$ is a graph and A is a subset of edges we denote the graph $(V, E \setminus A)$ by $G - A$. We write $G - e$ if A consists of a single edge e.

Definition 5 *A triangulation H of a graph G is called* minimal *if for any nonempty subset A of edges $H - A$ is not a triangulation of G.*

It was shown in [24] that a triangulation H of G is minimal if and only if for every edge e in H, $H - e$ is not a triangulation of G. In other words, deleting any edge of H which is not an edge of G creates a C_4.

Definition 6 *A subset S of vertices of a graph G is called an a, b-separator for non adjacent vertices a and b in $V \setminus S$ if a and b are in different connected components of $G[V \setminus S]$. If no proper subset of S separates a and b in this way, then S is called a minimal a, b-separator. A subset S is called a* minimal separator *if there exist non adjacent vertices a and b for which S is a minimal a, b-separator.*

It is by now very well known, that a graph is chordal if and only if every minimal separator is a clique (see, e.g., [9]). The concept of an efficient triangulation was introduced and applied for the design of exact treewidth algorithms in [15] (see also [13]).

Definition 7 *Let H be a triangulation of a graph G. Let C be the set of minimal separators of H. The triangulation H is called* efficient *if the following conditions are satisfied.*

1. *If a and b are non adjacent vertices in H then every minimal a, b-separator in H is also a minimal a, b-separator in G.*
2. *If S is a minimal separator in H and C is the vertex set of a connected component of $H[V \setminus S]$, then C induces also a connected component in $G[V \setminus S]$.*
3. *$H = G_C$, which is the graph obtained from G by adding edges between all pairs of vertices contained in the same set C for every $C \in \mathcal{C}$.*

It has been shown in [13, page 16] that any triangulation H of a graph G contains a subgraph that is an efficient triangulation of G. Taking a minimal triangulation H of G this implies that H is also efficient. Hence any minimal triangulation of a graph G is also an efficient triangulation. The next theorem shows that the converse is also true.

Theorem 3 ([14]) *Let $H = (V, F)$ be a triangulation of $G = (V, E)$. Then H is a minimal triangulation of G if and only if H is an efficient triangulation of G.*

Using Theorem 3 we obtain the following algorithm for finding a minimal triangulation of a graph.

Step 1 If G is a clique, then $H = G$ is an efficient triangulation. Stop.

Step 2 Find a minimal separator S. This can be done in linear time, see [15, 13]. Let $C_1, \ldots, C_t$ be the connected components of $G[V \setminus S]$. For $i = 1, \ldots, t$ let S_i be the vertices of S that have a neighbor in C_i.

Step 3 Add edges to make a clique of S. Call the new graph G'.

Step 4 Recursively, add edges, making efficient triangulations of $G'[S_i \cup C_i]$ for $i = 1, \ldots, t$. Let H be the graph obtained in this way. This is an efficient triangulation of G.

It can easily be shown that this algorithm can be implemented to run in $O(n^3)$ time.

Theorem 4 ([14]) *There exists an $O(n^3)$ algorithm which computes an efficient triangulation of a given graph $G = (V, E)$.*

Other algorithms for finding minimal triangulations were obtained in [19, 24, 7].

Definition 8 *A minimal separator S is* good *if for every non adjacent pair x and y in S, x and y have at least one common neighbor in G.*

We introduce a new concept of triangulation that will be useful for the design of efficient algorithms for bandwidth approximation.

Definition 9 *A triangulation H of a graph G is good if for every edge (a, b) in H either (a, b) is also an edge in G or there exists some vertex z such that (a, z) and (b, z) are edges of G.*

Remark 1 *A triangulation H of the graph G is good if and only if H is a subgraph of the graph G^2. Consequently, by Theorem 2 any layout L of a good triangulation H of G with $b(H, L) = bw(H)$ fulfills $b(G, L) \leq 2 \cdot bw(G)$.*

Theorem 5 *If every minimal separator of a graph G is good then every minimal triangulation H of G is good.*

Proof. Let H be an efficient triangulation $H = G_C$. Let (a, b) be an edge of H which is not an edge in G. Then a and b must appear in some minimal separator $S \in C$. Since H is efficient, S is a minimal separator in G, hence S is good. Thus by definition a and b have a common neighbor. □

An important class of graphs for which there exist good triangulations are the graphs without chordless cycle of length at least six.

Lemma 2 *Let G be a graph without induced cycle of length at least six. Then every minimal separator of G is good.*

Proof. Assume there is some minimal separator S containing non adjacent vertices x and y such that $N(x) \cap N(y) = \emptyset$. Assume S is a minimal a, b-separator, and let C_a and C_b be the connected components of $G[V \setminus S]$ containing a and b, respectively. Both x and y have at least one neighbor in both C_a and C_b. Hence we can obtain a chordless cycle containing x and y, one chordless path in C_a and one chordless path in C_b. Since x and y do not have a common neighbor this chordless cycle must have length at least six. □

Theorem 2, Theorem 5 and Lemma 2 imply

Corollary 1 *Let G be a graph without induced cycle of length at least six and let H be a minimal triangulation of G. Then the bandwidth of H is at most twice the bandwidth of G.*

3 Bandwidth of AT-free graphs

Definition 10 *An independent set of three vertices is called an* asteroidal triple *if between each pair in the triple there exists a path that avoids the neighborhood of the third. A graph is* asteroidal triple free *(AT-free) if it contains no asteroidal triple.*

AT-free graphs include interval graphs, cocomparability graphs, permutation graphs, trapezoid graphs etc.

In Section 3.1 we give an $O(e + n \log n)$ approximation algorithm to approximate the bandwidth of an AT-free graph with worst case performance ratio four. In Section 3.2 we present an $O(n^3)$ algorithm with a worst case performance ratio two.

3.1 Factor four approximation

In this section we describe an $O(e + n \log n)$ approximation algorithm to approximate the bandwidth of AT-free graphs with worst case performance ratio four.

We need some results of [6].

Definition 11 *A* dominating pair *is a pair of vertices with the property that every path connecting them is a dominating set.*

In [6] it was shown that every connected AT-free graph has a dominating pair. Their algorithm to find a dominating pair in a connected AT-free graph works as follows. First start a lexicographic breadth-first search from an arbitrary vertex x (see e.g., [24]). Let v be the vertex numbered last by this search. Start a new lexicographic breadth-first search in v. Let $[u = u_1, u_2, \ldots, u_n = v]$ be the ordering found by this lexicographic breadth-first search. In [6] the following theorem was proved.

Theorem 6 ([6]) *The algorithm described above computes in linear time an ordering $M = [u = u_1, \ldots, u_n = v]$ such that for each $i = 1, \ldots, n - 1$, the pair u_i, v is a dominating pair for the subgraph induced by $\{u_i, u_{i+1}, \ldots, u_n\}$.*

A tree $T = (V, E)$ is called a *caterpillar with hairs of length at most k* if there is a set $B \subseteq V$ inducing a path in T (the body of the caterpillar), such that each connected component of $T[V \setminus B]$ is a path with at most k vertices (hair of the caterpillar). Furthermore, we call a caterpillar with hairs of length one simply a *caterpillar*.

Our approximation algorithm first computes a spanning caterpillar T such that G is a subgraph of T^4.

Theorem 7 *Let G be a connected AT-free graph. There exists a caterpillar T such that T is a spanning subgraph of G and such that adjacent vertices in G are at distance at most 4 in T. This caterpillar can be found in $O(n + e)$ time.*

Proof. Consider the breadth-first search tree B determined by the lexicographic breadth-first search that computes the ordering $M = [u = u_1, \ldots, u_n = v]$ described in Theorem 6. We denote L_i as the set of vertices at level i in B, such that the root, $v = u_n$, is at level 1. Notice that since B is a breadth-first search tree, all vertices of L_i can have neighbors only in L_{i-1}, L_i and in L_{i+1}. Assume B has k levels.

Construct a path $P = [u = x_k, x_{k-1}, \ldots, v = x_1]$ in G with vertex x_i from level L_i, $i \in \{1, 2, \ldots, k\}$, as follows. Start with $x_k = u$ in level k. For $i = k, k-1, \ldots, 2$ take the first vertex in the ordering M after x_i which is in level L_{i-1} and which is adjacent adjacent to x_i. This produces a dominating path P from u to v with one vertex from each level.

We claim, that each vertex $z \neq x_i$ in level L_i is either adjacent to x_i or to x_{i-1}. If z is in the last level k, this is immediate. Let z be a vertex of level L_i, $i \in \{1, 2, \ldots, k - 1\}$. If z is *before* x_i in M then z cannot be adjacent to x_{i+1} by the construction of P. If x_i is *before* z in M then the pair x_i, v is a dominating pair for the subgraph induced by $\{x_i, \ldots, u_n\}$ by Theorem 6. Hence, z must be adjacent to x_i or x_{i-1}. This proves the claim.

We can now construct the caterpillar T. The body of the caterpillar is the path P. For each vertex z in level L_i, $z \neq x_i$ make z either adjacent to x_i or to x_{i-1} depending on which edge is present in G.

Let (y, z) be an edge of G. Then either y and z are in the same level of B or in consecutive ones. By the construction of T the neighbors of y and z in the path P are in distance at most two in P. Hence G is a spanning subgraph of T^4. □

Theorem 8 *There exists an $O(e + n \log n)$ approximation algorithm for the bandwidth of AT-free graphs that has worst case performance ratio four.*

Proof. First construct the caterpillar T of Theorem 7 in linear time. Then use the $O(n \log n)$ algorithm of [2] to find an optimal layout L for T. T is a spanning subgraph of G and by Theorem 7 G is a subgraph of T^4. Hence Theorem 2 implies $b(G, L) \leq 4 \cdot bw(G)$ for any optimal layout L of T. □

We describe a simple $O(n)$ time approximation algorithm for the bandwidth of caterpillars with worst case ratio $3/2$. Let $T = (V, E)$ be a caterpillar with body $(b_1, \ldots, b_l)$ and d_i hairs on vertex b_i. We define a layout L of T by $L(b_i) = \lfloor d_i/2 \rfloor + \sum_{j=1}^{i-1} d_j$ for the body and $L(v) \in \{L(b_i) - \lfloor d_i/2 \rfloor, L(b_i) + \lceil d_i/2 \rceil\}$ for the remaining vertices v with neighbor b_i in the body. Then $b(T, L) = \max_i(L(b_{i+1}) - L(b_i)) = \max_i(\lceil d_i/2 \rceil + \lfloor d_{i+1}/2 \rfloor)$. On the other hand we have $bw(T) \geq 1/3 \max_i(d_i + d_{i+1} + 1)$ by Lemma 1. Hence $b(T, L) \leq 3/2 bw(T)$. Using this algorithm one obtains

Theorem 9 *There exists an $O(n + e)$ approximation algorithm for the bandwidth of AT-free graphs with worst case performance ratio six.*

3.2 Factor two approximation

Lemma 3 *Every minimal triangulation of an AT-free graph is good.*

Proof. Notice that an AT-free graph cannot contain a chordless cycle of length at least six. The result follows by Theorem 5 and Lemma 2. □

Theorem 10 ([17]) *Every minimal triangulation of an AT-free graph is an interval graph.*

Using Corollary 1 we obtain the following.

Corollary 2 *If G is AT-free, and H is a minimal triangulation of G, then H is an interval graph, and the bandwidth of H is at most twice the bandwidth of G.*

In [12] an $O(nk)$ algorithm is described which solves the k-bandwidth problem of interval graphs, where n is the number of vertices of G. That is, given an integer k and an interval graph G the algorithm finds a layout L of G with $b(G, L) \leq k$ or outputs that no such layout exists.

We can now obtain an algorithm approximating the bandwidth of AT-free graphs. First the algorithm computes a minimal triangulation H of the given AT-free graph

G. By Theorem 10 the triangulation H is an interval graph. Then the algorithm computes the bandwidth of the interval graph H using the algorithm of [12]. By Corollary 2 we obtain a layout of width at most twice the bandwidth of G. This proves the following.

Theorem 11 *There exists an $O(n^3)$ algorithm to approximate the bandwidth of an AT-free graph with worst case performance ratio two.*

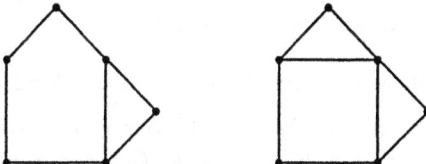

Fig. 1. H_1 and H_2

For all graphs G the parameters bandwidth $bw(G)$, proper pathwidth $ppw(G)$, pathwidth $pw(G)$ and treewidth $tw(G)$ satisfy $bw(G) = ppw(G) \leq pw(G) \leq tw(G)$.

Lemma 4 *If a graph $G = (V, E)$ is claw-free, C_k-free for all $k \geq 6$ and does not contain H_1 or H_2 (see Figure 1) as an induced subgraph, then every minimal triangulation of G is claw-free.*

Proof. Assume G has a minimal triangulation H which has a claw. Any minimal separator S of H is a clique since H is chordal (see [9]). By Theorem 3 the chordal graph H is an efficient triangulation of G. Hence any minimal separator S of H is also a minimal separator of G and the connected components of $H[V \setminus S]$ and $G[V \setminus S]$ coincide. Since G is claw-free $G[V \setminus S]$ has exactly two components.

We call the edges of H which are not edges in G, *new edges*. Since H is an efficient triangulation of G there is a minimal separator of H containing the end vertices of any new edge of H.

Let K be a claw induced by $\{c, x, y, z\}$, with central vertex c, which has a minimal number of new edges among all claws of H. Since G is claw-free there is at least one new edge, say (c, x). Then there is a minimal separator S, containing c and x. Let C_1 and C_2 be the two connected components of $G[V \setminus S]$. If y and z were contained in the same connected component, say C_1, then we could choose in C_2 a vertex x' adjacent to c in G and thus obtain a claw induced by $\{c, x', y, z\}$ in H with one new edge less than K, a contradiction. So we may assume that $y \in C_1$ and $z \in C_2$.

We claim that (c, y) and (c, z) are edges in G. Suppose (c, y) would not be an edge in G. Then there is a minimal separator S' in H containing c and y. Clearly

$x \notin S'$ since $(x, y) \notin E(H)$ and S' is a clique in H. Now we choose a vertex y' adjacent to c in G in the connected component of $G[V \setminus S']$ that does not contain x. Now $x \in S$ implies that in H the vertex x has a neighbor in C_2 and hence there is a path from x to z with all vertices except x belonging to C_2. Furthermore the minimal separator S' is a clique in H containing $y \in C_1$ and therefore S' cannot contain any vertex of C_2. Hence x and z belong to the same component of $H[V \setminus S']$ and therefore $\{c, x, y', z\}$ induces a claw in H with one new edge less than K, a contradiction.

Since S is a minimal separator of G vertex x has neighbors $p \in C_1$ and $q \in C_2$ in G. Since G is C_k-free ($k \geq 6$), c must be adjacent to at least one of p or q in G. Since G is claw-free, the neighbors of c in C_1 and in C_2 form cliques. If c is adjacent to p and to q, we get H_2 as an induced subgraph. In the case that c is adjacent to one of p or q, we get H_2 as an induced subgraph. □

The following Corollary of Lemma 4 was shown independently in [21].

Corollary 3 *A graph G is claw-free and AT-free if and only if any minimal triangulation H of G is a proper interval graph. Consequently, $bw(G) = ppw(G) = pw(G) = tw(G)$ for all claw-free and AT-free graphs G.*

Proof. First assume that G has a claw K. Add all possible edges to G except between vertices of K. In this way we obtain a triangulation H of G with K as induced subgraph. Hence there is also a minimal triangulation H' with a claw. Such a minimal triangulation cannot be a proper interval graph [23, 9].

If G is AT-free then every minimal triangulation is an interval graph by Theorem 10. Hence if G is claw-free and AT-free then, by Lemma 4, every efficient triangulation is a proper interval graph. □

In the next theorem we obtain a characterization of the claw-free and AT-free graphs.

Theorem 12 ([14]) *A connected graph G is claw-free and AT-free if and only if G is either a claw-free cocomparability graph or G is the complement of a triangle-free graph.*

The following lemma shows that the bandwidth of the complements of triangle-free graphs is of little interest with respect to 2-approximations.

Lemma 5 *If G is the complement of a triangle-free graph then the treewidth of G is at least $\frac{n}{2} - 1$.*

Proof. Let k be the treewidth of G. Consider a triangulation H of G with $\omega(H) = k + 1$. If H is a clique, then $k = n - 1 \geq \frac{n}{2} - 1$.

If H is not a clique, then H has at least two non adjacent simplicial vertices, say s_1 and s_2 (see, e.g., [9]). Since G is the complement of a triangle-free graph, every other vertex is adjacent to at least one of s_1 and s_2 in G. Since G is a subgraph of H, every vertex other than s_1 and s_2 is adjacent to at least one of s_1 and s_2 in H. It follows that one of the closed neighborhoods $N[s_1]$ or $N[s_2]$ contains at least $\frac{n}{2}$ vertices. Since H is chordal, these closed neighborhoods are cliques. This proves the lemma. □

4 Bandwidth of permutation graphs

One of the most well-known and well-studied classes of perfect graphs is the class of *permutation graphs*. We refer to [9] for a detailed exposition of properties of permutation graphs. We mention that they can be recognized in time $O(n^2)$ (see [25]).

We think of a permutation π of the numbers $1, \ldots, n$, as the sequence $\pi = [\pi_1, \ldots, \pi_n]$. We use the notation π_i^{-1} for the position of the number i in this sequence.

Definition 12 *If π is a permutation of the numbers $1, \ldots, n$, we can construct an undirected graph $G[\pi] = (V, E)$ with vertex set $V = \{1, \ldots, n\}$, and edge set E:*

$$(i, j) \in E \Leftrightarrow (i - j)(\pi_i^{-1} - \pi_j^{-1}) < 0$$

An undirected graph is called a permutation graph *if there exists a permutation π such that $G \cong G[\pi]$.*

$G[\pi]$ is sometimes called the *inversion graph* of π.

Let π be a permutation of $1, \ldots, n$. The *permutation diagram* of π can be obtained as follows. Write the number $1, \ldots, n$ horizontally from left to right. Underneath, write the numbers $\pi_1, \ldots, \pi_n$, also horizontally from left to right. Draw n straight line segments joining the two 1's, the two 2's, etc. Notice that two vertices i and j of $G[\pi]$ are adjacent if and only if the corresponding line segments in the permutation diagram of π intersect. Geometric intersection models like the permutation diagram are often useful in visualizing certain concepts.

In this section, let $G = G[\pi]$ be a permutation graph with n vertices. We show an algorithm which approximates the bandwidth of a permutation graph.

First we compute a triangulation of the permutation graph. Consider the permutation diagram. For each $j = 1, \ldots, n$ consider the interval given by the left and right endpoints $\ell(j) = \min(\pi_j^{-1}, j)$ and $r(j) = \max(\pi_j^{-1}, j)$. In case $\pi_j^{-1} = j$ we take the *closed* interval (which consists of only the point j). All other intervals are open. Let H be the interval graph obtained in this way.

Lemma 6 *H is a good triangulation of G.*

Proof. It is easy to see that G is a subgraph of H. Let i and j be adjacent in H but not in G. Let t be a point of the intersection of the intervals. Consider a vertical line at position t in the permutation diagram. Then the line segments of i and j cross this vertical line. Without loss of generality we can assume that $i < t < \pi_i^{-1}$ and $j < t < \pi_j^{-1}$. Now notice that the number of elements k with $k < t < \pi_k^{-1}$ is equal to the number of elements k' with $\pi_{k'}^{-1} < t < k'$. Using this fact, it follows that there is a line segment r crossing the vertical line at position t with $\pi_r^{-1} < t < r$. Hence $r \in N(i) \cap N(j)$. □

In [26] an $O(n \log n)$ algorithm is given to compute the bandwidth of an interval graph (when the interval model is given). Using this algorithm we obtain the following.

Theorem 13 *For any permutation graph with given permutation diagram, there exists an $O(n \log n)$ algorithm to approximate the bandwidth with worst case performance ratio two.*

Proof. Use the observations above and Corollary 1. □

In the same manner one can obtain approximation algorithms for finite dimensional cocomparability graphs and for trapezoid graphs. In both cases we assume that a suitable intersection model is part of the input.

Corollary 4 *There exists an $O(nd+n \log n)$ algorithm to approximate the bandwidth of a cocomparability graph of dimension at most d with worst case ratio two. There exists an $O(n \log n)$ algorithm to approximate the bandwidth of a trapezoid graph with worst case ratio two.*

5 Bandwidth of cocomparability graphs

Despite the NP-completeness of the BANDWIDTH problem on cocomparability graphs there is a fairly simple approximation algorithm of worst case performance ratio three.

Theorem 14 *Let $G = (V, E)$ be a cocomparability graph, $P_G = (V, <_{P_G})$ any partially ordered set having $\overline{G}$ as comparability graph and $L = x_1, x_2, \ldots, x_n$ any linear extension of P_G. Then $b(G, L) \leq 3 \cdot bw(G)$.*

Proof. Consider any linear extension $L = x_1, x_2, \ldots, x_n$ of P_G and set $L(x_i) = i$ for all $i \in \{1, 2, \ldots, n\}$. Let x_r and x_s, $r < s$ be two adjacent vertices of G with $|L(x_r) - L(x_s)| = b(G, L)$. Any vertex x_i with $r < i < s$ is adjacent to x_r or x_s, otherwise $x_r <_{P_G} x_i <_{P_G} x_s$ would imply $x_r <_{P_G} x_s$ and $\{x_r, x_s\} \notin E$. Hence $|N(x_r) \cup N(x_s)| \geq s - r + 1$.

By Lemma 1 we get $bw(G) \geq \frac{1}{3} (|N(x_r) \cup N(x_s)| - 1) \geq \frac{1}{3} (s - r) = \frac{1}{3} b(G, L)$. □

Hence any algorithm computing for a given cocomparability graph G a linear extension L of P_G constructs a layout of G with $b(G, L) \leq 3 \cdot bw(G)$. This can simply be done by computing a transitive orientation of $\overline{G}$ and then a topological sort of the resulting directed acyclic graph [9, 25].

Theorem 15 *There is an $O(n^2)$ approximation algorithm for the bandwidth of a cocomparability graph that has worst case performance ratio three.*

6 Conclusions

We have presented an efficient approximation algorithm that approximates the bandwidth of AT-free graphs with worst case performance ratio two. It is likely that our approach is applicable for many other graph classes as well. One interesting open problem in this respect is, to find a characterization of graphs G for which G^2 is an interval graph.

The algorithmic complexity of BANDWIDTH on split graphs and on permutation graphs, respectively, is an open question. We could only show that the bandwidth can be computed in polynomial time for the class of chain graphs, which is a proper subclass of the class of bipartite permutation graphs. The problem remains open for the class of bipartite permutation graphs. Another challenging open problem is to find an approximation algorithm for the bandwidth of trees with constant worst case ratio.

In respect to the celebrated recent results on the hardness of approximation problems it would be very interesting to know whether the approximation of bandwidth on general graphs can be shown to be hard in one of these manners, as e.g., to show that there is an $\epsilon > 0$ such that no polynomial time approximation for the bandwidth minimization problem with worst case ratio n^ϵ exists, unless $P = NP$.

References

1. Arnborg, S., D. G. Corneil and A. Proskurowski, Complexity of finding embeddings in a k-tree, *SIAM Journal on Algebraic and Discrete Methods* 8 (1987), pp. 277–284.

2. Assmann, S. F., G. W. Peck, M. M. Syslo and J. Zak, The bandwidth of caterpillars with hairs of length 1 and 2, *SIAM Journal on Algebraic and Discrete Methods* 2 (1981), pp. 387–393.

3. Bodlaender, H., M. R. Fellows and M. T. Hallet, Beyond NP-completeness for problems of bounded width: Hardness for the W hierarchy, *Proceedings of the 26th STOC*, 1994, pp. 449–458.

4. Chinn, P.Z., J. Chvátalová, A.K. Dewdney, N.E. Gibbs, The bandwidth problem for graphs and matrices — a survey, *Journal of Graph Theory* 6 (1982), pp. 223–254.

5. D.G. Corneil, S. Olariu, and L. Stewart, Asteroidal triple-free graphs, *Proceedings of the 19th International Workshop on Graph-Theoretic Concepts in Computer Science WG'93*, Lecture Notes in Computer Science, Vol. 790, Springer-Verlag, 1994, pp. 211–224.

6. Corneil, D. G., S. Olariu and L. Stewart, Linear time algorithms to compute dominating pairs in asteroidal triple-free graphs. To appear in: *Proceedings ICALP'95*.

7. Dahlhaus, E. and M. Karpinski, An efficient parallel algorithm for the minimal elimination ordering (MEO) of an arbitrary graph, *Proceedings of the 30th FOCS*, 1989, pp. 454–459.

8. Garey, M. R., R. L. Graham, D. S. Johnson and D. E. Knuth, Complexity results for bandwidth minimization, *SIAM Journal on Applied Mathematics* 34 (1978), pp. 477–495.

9. Golumbic, M. C., *Algorithmic graph theory and perfect graphs*, Academic Press, New York, 1980.

10. Gurari, E. and I. H. Sudborough, Improved dynamic programming algorithms for the bandwidth minimization and the mincut linear arrangement problem, *Journal on Algorithms*, 5, (1984), pp. 531–546.

11. Kaplan, H. and R. Shamir, Pathwidth, bandwidth and completion problems to proper interval graphs with small cliques. Manuscript, 1993.

12. Kleitman, D.J. and R.V. Vohra, Computing the bandwidth of interval graphs, *SIAM Journal on Discrete Mathematics* 3 (1990), pp. 373–375.

13. Kloks, T., *Treewidth - Computations and Approximations*, Springer–Verlag, Lecture Notes in Computer Science 842, 1994.

14. Kloks, T., D. Kratsch and H. Müller, Approximating the bandwidth for asteroidal triple-free graphs, Forschungsergebnisse Math/95/6, Friedrich-Schiller-Universität Jena, Germany, 1995.

15. Kloks, T., D. Kratsch and J. Spinrad, Treewidth and pathwidth of cocomparability graphs of bounded dimension, Computing Science Notes 93/46, Eindhoven University of Technology, Eindhoven, The Netherlands, 1993.

16. Mahesh, R., C. Pandu Rangan and Aravind Srinivasan, On finding the minimum bandwidth of interval graphs, *Information and Computation* 95 (1991), pp. 218–224.

17. Möhring, R. H., Triangulating graphs without asteroidal triples. Technical Report 365/1993, Technische Universität Berlin, Fachbereich Mathematik, 1993.

18. Monien, B., The bandwidth minimization problem for caterpillars with hair length 3 is NP-complete, *SIAM Journal on Algebraic and Discrete Methods* 7 (1986), pp. 505–512.

19. Ohtsuki, T., A fast algorithm for finding an optimal ordering for vertex elimination on a graph, *SIAM J. Comput.* 5 (1976), pp. 133–145.

20. Papadimitriou, C.H., The NP-completeness of the bandwidth minimization problem, *Computing* 16 (1976), pp. 263–270.

21. Parra, A. and P. Scheffler, Treewidth equals bandwidth for AT-free claw-free graphs. Technical Report 436/1995, Technische Universität Berlin, Fachbereich Mathematik, 1995.

22. Peck, G. W. and Aditya Shastri, Bandwidth of theta graphs with short paths, *Discrete Mathematics* 103 (1992), 177–187.

23. Roberts, F., S., Indifference graphs, In: *Proof techniques in graph theory*, (F. Harary, ed.), pp. 139–146. Academic Press, New York, (1969).

24. Rose, D. J., R. E. Tarjan and G. Lueker, Algorithmic aspects of vertex elimination of graphs, *SIAM J. Comput.* 5 (1976), pp. 266–283.

25. Spinrad, J., On comparability and permutation graphs, *SIAM J. on Comput.* 14 (1985), pp. 658–670.

26. Sprague, A. P., An $O(n \log n)$ algorithm for bandwidth of interval graphs, *SIAM Journal on Discrete Mathematics* 7 (1994), pp. 213–220.

27. Turner, J. S., On the probable performance of heuristics for bandwidth minimization, *SIAM J. Comput.*, 15, (1986), pp. 561–580.

Near-Optimal Distributed Edge Coloring *

Devdatt Dubhashi[1][**] and Alessandro Panconesi[2][***]

[1] **BRICS**[***], Department of Computer Science, University of Aarhus, Ny
Munkegade, DK-8000 Aarhus C, Denmark
[2] Centrum voor Wiskunde en Informatica 413 Kruislaan, 1098 SJ, Amsterdam,
Holland

Abstract. We give a distributed randomized algorithm to edge color a
network. Given a graph G with n nodes and maximum degree Δ, the
algorithm,
- For any fixed $\lambda > 0$, colours G with $(1+\lambda)\Delta$ colours in time $O(\log n)$.
- For any fixed positive integer s, colours G with $\Delta + \frac{\Delta}{(\log \Delta)^s} = (1 + o(1))\Delta$ colours in time $O(\log n + \log^s \Delta \log \log \Delta)$.

Both results hold with probability arbitrarily close to 1 as long as $\Delta(G) = \Omega(\log^{1+d} n)$, for some $d > 0$. The algorithm is based on the Rödl Nibble, a
probabilistic strategy introduced by Vojtech Rödl. The analysis involves
a certain quasi-random phenomenon involving sets at the vertices of the
graph.

1 Introduction

The edge coloring problem is a basic problem in graph theory and combinatorial
optimization. Its importance in distributed computing, and computer science
generally, stems from the fact that several scheduling and resource allocation
problems can be modeled as edge coloring problems [9, 11, 14, 17]. In this paper,
we give a distributed randomized algorithm that computes a near-optimal edge
coloring in time $O(\log n)$. By "near-optimal" we mean that the number of colors
used is $(1 + o(1))\Delta$ where Δ denotes the maximum degree of the network and
the $o(1)$ term can be as small as $1/\log^s \Delta$, for any $s > 0$. Both performance
guarentees – the running time and the number of colours used – hold with high
probability as long as the maximum degree grows at least logarithmically with

* A preliminary version of this work was presented at the 15th International
Symposium on Mathematical Programming, August 1994, Ann Arbour, Michigan,
USA.
** dubhashi@daimi.aau.dk. Work done partly while at the Max–Planck–Institute für
Informatik supported by the ESPRIT Basic Research Actions Program of the EC
under contract No. 7141 (project ALCOM II).
*** puck@math.tu-berlin.de. Supported by an Ercim postdoctoral fellowship. Current
address: Fachbereich Mathematik, MA 6-1, TU Berlin, Str. des 17 Juni 136, 10623
Berlin.
*** Basic Research in Computer Science,
Centre of the Danish National Research Foundation.

n. Our algorithm can be implemented directly in the PRAM model of computation.

Motivation and Related Work. The edge coloring problem can be used to model certain types of jobshop scheduling, packet routing, and resource allocation problems in a distributed setting. For example, the problem of scheduling I/O operations in some parallel architectures can be modeled as follows [9, 6]. We are given a bipartite graph $G = (\mathcal{P}, \mathcal{R}, E)$ where, intuitively, $\mathcal{P}$ is a set of processes and $\mathcal{R}$ is a set of resources (say, disks). Each processor needs data from a subset of resources $R(p) \subseteq \mathcal{R}$. The edge set is defined to be $E = \{(p, r) : r \in R(p), p \in \mathcal{P}\}$. Due to hardware limitations only one edge at the time can be serviced. Under this constraints it is not hard to see that optimal edge colorings of the bipartite graph correspond to optimal schedules that is, schedules minimizing the overall completion time. Clearly, if a graph G has maximum degree Δ then at least Δ colors are needed to edge color the graph. A classical theorem of Vizing shows that $\Delta + 1$ colors are always sufficient, and the proof is actually a polynomial time algorithm to compute such a coloring (see for example [4]). Interestingly, given a graph G, it is NP-complete to decide whether it is Δ or $\Delta + 1$ edge colorable [8], even for regular graphs [7]. Efforts at parallelizing Vizing's theorem have failed; the best PRAM algorithm known is a randomized algorithm by Karloff & Shmoys that computes an edge coloring using very nearly $\Delta + \sqrt{\Delta} = (1 + o(1))\Delta$ colors. The Karloff & Shmoys algorithm can be derandomized by using standard derandomization techniques [3, 16]. In the distributed setting the previously best known result was a randomized algorithm by Panconesi & Srinivasan that uses roughly $1.58\Delta + \log n$ colors with high probability and runs in $O(\log n)$ time with high probability. For the interesting special case of bipartite graphs Lev, Pippinger & Valiant show that Δ-colorings can be computed in NC, whereas this is provably impossible in the distributed model of computation even if randomness is allowed (see [18]).

Our solution. To state our results precisely, we reproduce below our main theorem:

Theorem 1. *For any fixed $\lambda > 0$, given a graph with n vertices and maximum degee Δ, we can edge colour the graph with $(1 + \lambda)\Delta$ colours in time $O(\log n)$ where n is the number of vertices in the graph. For any fixed positive integer s, we can edge colour it with $\Delta + \Delta/\log^s \Delta = (1 + o(1))\Delta$ colours in time $O((\log \Delta)^s \log \log \Delta + \log n)$. The results hold with failure probability decreasing to 0 faster than any polynomial (in n) provided that $\Delta = \Omega(\log^{1+d} n)$ for some $d > 0$.*

Our algorithm is based on the *Rödl Nibble*, a beautiful probabilistic strategy introduced by Vojtech Rödl to solve a certain covering problem in hypergraphs [2, 20]. The method has subsequently been used very successfully to solve other combinatorial problems such as asymptotically optimal coverings and colorings for hypergraphs [2, 10, 19, 21]. In this paper, we introduce it as a tool for the

design and analysis of randomized algorithms.[4] Although the main component
of our algorithm is the Rödl nibble and the intuition behind it rather compelling,
the algorithm requires a non-trivial probabilistic analysis of a so called quasi-
random process. To explain what this is, it is perhaps best to give a brief outline
of our algorithm. Starting with the input graph G_0 the algorithm generates a
sequence $G_0, G_1, \ldots, G_t$ of graphs. One can view each edge e as possessing a
palette of available colors, starting with the whole set of $[\Delta]$ colours initially. At
an arbitrary stage, a small ϵ fraction of uncolored edges is selected, and each
selected edge chooses a tentative color at random from its current palette. If the
tentative color is not chosen by any neighboring edge it becomes final. Palettes
of the remaining uncolored edges are updated in the obvious fashion– by deleting colors used by neighboring edges. The process is then repeated. Like other
proofs based on the same method our proof hinges on two key features of the
Rödl nibble. The first key idea of the method is that if colors are chosen independently, the probability of color conflict is roughly ϵ^2, a negligible fraction of
all edges attempting coloring at this stage. If the same "efficiency" is maintained
throughout, the overall "wastage" will be very small. The second aspect of the
Rödl nibble is a deeper mathematical phenomenon called *quasi-randomness* (see
[2]). In our context, quasi-randomness means that the palettes of available colors
at the edges at any stage are "essentially" truly independent random subsets of
the original full palette. The crux of the analysis is to show that despite the potential of a complicated interaction regulated by the topology of the underlying
graph, the "nibbling" feature of the coloring process ensures that the palettes
are evolving almost independently of each other. In all applications of the nibble
method, it is the quasi-random aspect which is mathematically challenging and
which usually requires a quite laborious probabilistic analysis.

2 Preliminaries

A *message–passing distributed network* is an undirected graph $G = (V, E)$ where
vertices (or nodes) correspond to processors and edges to bi–directional communication links. Each processor has its unique *id*. The network is *synchronous*, ,
computation takes place in a sequence of *rounds*; in each round, each processor
reads messages sent to it by its neighbors in the graph, does any amount of local
computation, and sends messages back to all of its neighbors. The time complexity of a distributed algorithm, or *protocol*, is given by the number of rounds
needed to compute a given function. If one wants to translate an algorithm for
this model into one for the PRAM then computation locally done by each processor must be charged for. An *edge coloring* of a graph G is an assignment of
colors to edges such that incident edges always have different colors. The edge

---

[4] This research was originally prompted by a conversation that the second author had
with Noga Alon and Joel Spencer, in which they suggested that the nibble approach
should work. Noga Alon has recently informed us that he is already in possession of
a solution with similar performance [1]. However, at the time of writing, a written
manuscript was not available for comparison.

coloring problem is to find an edge coloring with the aim of minimizing the number of colors used. Given that determining an optimal (minimal) coloring is an NP-hard problem this requirement is usually relaxed to consider approximate, hopefully even near-optimal, colorings. The edge coloring problem in a distributed setting is formulated as follows: a distributed network G wants to compute an edge coloring of its own topology. As remarked in the introduction such a coloring might be useful in the context of scheduling and resource allocation. The set $\{1, 2, \ldots, n\}$ will be denoted by $[n]$. Given a graph G and a set of edges F, $G[F]$ denotes the subgraph of G whose edge set is F. In the paper we will use the following approximations repeatedly: $(1 - 1/n)^n \approx e^{-1}$, and $e^\epsilon \approx 1 + \epsilon$ or $e^\epsilon \approx 1 + \epsilon + \epsilon^2/2$, for small values of ϵ. Whenever such an approximation is in effect, we will use the sign $\approx$ in place of the equality sign. We will make use of a slight modification of a well-known vertex coloring algorithm by Luby [13]. Luby's algorithm computes a $(\Delta + 1)$–vertex coloring of a graph in expected time $O(\log n)$, where n is the number of vertices of a graph of maximum degree Δ. The running time of the algorithm is $O(\log n)$ with high probability [12, 13]. When applied to the line graph of G the algorithm computes a $(2\Delta - 1)$–edge coloring. In the original algorithm each vertex is initially given a palette of $\Delta + 1$ colors; it can be easily verified that the algorithm still works in the same fashion if each vertex u is given a palette of $\deg(u) + 1$ colors instead, where $\deg(u)$ is the degree of u. This modification is introduced for explanatory purposes.

3   The Algorithm

The algorithm is in two phases. The first phase is an application of the Rödl nibble and has the goal of coloring most of the edges using a palette of Δ colors. By the end of this phase we will be left with a graph whose maximum degree is at most $\kappa\Delta$ with high probability. In the second phase the modified Luby's algorithm is used to color the remaining graph with at most $2\kappa\Delta$ fresh colors. As we shall see in section 4.1, the number of iterations needed to bring the degree down from Δ to $\kappa\Delta$ is $O(\log(1/\kappa)/\alpha\kappa^2)$, where $\alpha = \epsilon(1 - \epsilon)e^{-4\epsilon}$. Hence, in order to get a $(1 + \lambda)\Delta$, where $\lambda > 0$ is any fixed constant, the first phase takes constant time. To get a $(1 + o(1))\Delta$ coloring takes $O((\log \Delta)^{2s} \log \log \Delta)$ time, where the $o(1)$ term is $1/(\log \Delta)^s$, for any $s > 0$. This holds with high probability. The exact probability of success will be determined in the section devoted to the analysis. We note here that an assumption on the maximum degree of the graph is needed, namely $\Delta(G) = \Omega(\log^{1+d} n)$, for some $d > 0$ (n denotes the number of vertices of G). Phase 2 takes $O(\log n)$ time, with high probability. The basic idea underlying the first phase of the algorithm is for each vertex to select a small "nibble" of edges incident upon it and assign tentative colors to them independently at random. Most of these edges are expected to avoid conflicts with other edges vying for coloring, and get successfully colored at this stage. This is because the nibble keeps the "efficiency" of the coloring close to 1 at each stage. To describe the algorithm more precisely, we introduce some definitions that will also be used later in the analysis. At any stage $k \geq 1$, we have a graph

$G_k(V, E_k)$. Initially, $G_0(V, E_0) := G(V, E)$, the input graph. By Δ_k we denote the maximum degree of the graph G_k (note $\Delta_0 = \Delta(G)$ initially). Each vertex has a palette of available colors, A_k^u with $A_0^u = [\max_{w \in \delta(u \cup \{u\})} \deg(w)]$. (This can be arranged in one round with each vertex communicating its own degree to each of its neighbours.) The set of edges successfully colored at stage k is denoted by C_k. Then, $G_{k+1} := G_k[E - C_k]$ is the graph passed on to the next stage. In the algorithm, $t(\epsilon, \kappa)$ denotes the number of stages needed to bring the maximum degree of the graph from Δ to $\kappa\Delta$ with high probability, and has value

$$t(\epsilon, \kappa) = \left\lceil \frac{\ln(1/\kappa)}{\epsilon(1 - \epsilon)e^{-4\epsilon}\kappa} \right\rceil .$$

The algorithm is more precisely described as follows

Phase 1. RODL NIBBLE

For $k = 1, 2, \ldots, t(\epsilon, \kappa)$ stages repeat the following:

- Each vertex u randomly selects an ϵ fraction of the edges incident on itself, and independently at random assigns them a tentative color from its palette A_k^u of currently available colors. If an edge $e = \{u, v\}$ is selected by both its endpoints, it is simply dropped and not considered for coloring at this stage.
- Let $e = \{u, v\}$ be a selected edge, and $c(e)$ its tentative color. Color $c(e)$ becomes the final color of e unless one of the following two conflict types arises: *i)* some edge incident on e is given the same tentative color, or *ii)* $c(e) \notin A_k^u \cap A_k^v$, the tentative color given to e is not available at the other endpoint of e.
- The graph is updated by setting

$$A_{k+1}^u = A_k^u - \{c : e \text{ incident on } u, c(e) = c \text{ is the final color of } e\}$$

and $G_{k+1} = G_k[E_k - C_k]$, where C_k is the set of edges which got a final color at stage k.

Phase 2.

Color $G_{t(\epsilon, \kappa)}$ with fresh new colors by using the modified Luby's algorithm.

4 Analysis

4.1 Intuitive Outline

Suppose for a start that the graph is Δ–regular. Intuitively, the palettes A_k^u are more–or–less random subsets of the base set $[\Delta]$. Let us assume they are indeed truly random subsets of $[\Delta]$, so precisely, let us assume that the palette of each vertex at stage $k \geq 0$, is a uniformly and independently chosen random subset of $[\Delta]$ of the same size Δ_k. Then, at stage k (with high probability), the size of the common palette between any two vertices is Δ_k^2/Δ. So the probability that a colour chosen by a vertex as a tentative colour for an incident edge is also valid at the other end–point is Δ_k/Δ. Hence, the probability that an edge is successfully coloured at stage k is roughly, $\epsilon\frac{\Delta_k}{\Delta}$ and we have the following recurrence for the vertex degree,

$$\Delta_{k+1} \leq (1 - \epsilon\frac{\Delta_k}{\Delta})\Delta_k \leq \exp(-\epsilon\frac{\Delta_k}{\Delta})\Delta_k$$

This recurrence implies that given a fixed $0 < \lambda < 1$, the vertex degree drops to $\lambda\Delta$ within a constant number of stages, or that for any positive integer $s > 0$, the degree drops to $\Delta/(\log \Delta)^s$ in a poly–logarithmic (in Δ) number of stages. This yields the required time complexity analysis for the algorithm. Unfortunately, neither of the two assumptions above are in fact valid. First, because the graph G can have a very complex, irregular topology, it is not true that vertex degrees and palettes are uniform, at the outset, and they are even less likely to remain so at subsequent stages. In addition, the palettes are not truly independent random subsets either, as they can interact over the stages in a potentially complicated fashion governed by the topology of the graph. However, we show in § 4.2 below, that despite the possibility of a complex interaction in the graph, the "nibbling" feature of the colouring process leads to an essentially local interaction of the palettes. So, while the palettes are not truly random subsets, they behave essentially as such, specifically, with regard to the relative size and composition of the common palettes and the palettes themselves. Given this one simple, but crucial feature of the interaction of the palettes,it follows that the decay law is essentially as given above. To highlight the essential ideas, we start with some simplifying assumptions and progressively, we remove the assumptions and refine the argument. First we give an analysis under the assumption that the initial network is Δ–regular. This will bring out to both the nature of the interaction of the palettes due to the "nibbling" feature of the colouring, and how that determines the decay law. With a high probability analysis using a martingale, we show that the concerned random variables are sharply concentrated around their means. Thus the graph continues to remain almost Δ_k–regular at each stage $k \geq 0$. Finally, we indicate how to remove the assumptions of uniformity made at the outset.

4.2 The Regular Case

Let us start by assuming that the graph is initially Δ–regular, and that it retains symmetry between vertices at each stage. Thus, at each stage $k \geq 0$, each vertex has some degree Δ_k which is also the size of its palette, and the common palette between any two neighbouring vertices also has the same value uniformly, which we denote by Θ_k. The probability that an edge is successfully coloured at stage k is

$$p_k = 2\epsilon(1 - \epsilon)\frac{\Theta_k}{\Delta_k}e^{-4\epsilon} = \alpha\frac{\Theta_k}{\Delta_k}.$$

where we define $\alpha = \alpha(\epsilon) := 2\epsilon(1-\epsilon)e^{-4\epsilon}$. (The factor $2\epsilon(1-\epsilon)$ is the probability that the edge is chosen by exactly one endpoint. The fraction Θ_k/Δ_k is the probability that the tentative color chosen is present at the other endpoint and,

$e^{-4\epsilon}$ is the probability that there is no color conflict.) Hence, we have for $k \geq 0$ and any vertex u,

$$
\begin{aligned}
E[\Delta_{k+1} \mid \Delta_k, \Theta_k] &= \Delta_k - \sum_{w \in N_k(u)} \alpha \frac{\Theta_k}{\Delta_k} \\
&= \Delta_k - \alpha \Theta_k \\
&= \Delta_k (1 - \alpha \frac{\Theta_k}{\Delta_k}) \\
&= \Delta_k (1 - \alpha \eta_k) \\
&\approx \Delta_k e^{-\alpha \eta_k}
\end{aligned}
\tag{1}
$$

where we put $\eta_k := \frac{\Theta_k}{\Delta_k}$.

To compute $E[\Theta_{k+1} \mid \Delta_k, \Theta_k]$ we make use of the "nibbling" feature of the colouring process. For each edge (u, v) and color c,

$$
\begin{aligned}
\Pr[c \in A_{k+1}^u \cap A_{k+1}^v | \Delta_k, \Theta_k] &= 1 - \Pr[c \notin A_{k+1}^u \cap A_{k+1}^v | \Delta_k, \Theta_k] \\
&= 1 - (\Pr[c \notin A_{k+1}^u | \Delta_k, \Theta_k] + \Pr[c \notin A_{k+1}^v | \Delta_k, \Theta_k]) \\
&\quad + \Pr[c \notin A_{k+1}^u, c \notin A_{k+1}^v | \Delta_k, \Theta_k] \\
&\approx 1 - (\Pr[c \notin A_{k+1}^u | \Delta_k, \Theta_k] + \Pr[c \notin A_{k+1}^v | \Delta_k, \Theta_k]) \\
&\approx (1 - \Pr[c \notin A_{k+1}^u | \Delta_k, \Theta_k])(1 - \Pr[c \notin A_{k+1}^v | \Delta_k, \Theta_k]) \\
&= \Pr[c \in A_{k+1}^u | \Delta_k, \Theta_k] \Pr[c \in A_{k+1}^v | \Delta_k, \Theta_k]
\end{aligned}
$$

(since $\Pr[c \notin A_{k+1}^u, c \notin A_{k+1}^v | \Delta_k, \Theta_k] = O(\epsilon^2) = \Pr[c \notin A_{k+1}^u | \Delta_k, \Theta_k] \Pr[c \notin A_{k+1}^v | \Delta_k, \Theta_k]$). Thus, the "nibbling" feature of the colouring process is such that the common palette $A_{k+1}^u \cap A_{k+1}^v$ evolves as if it were the intersection of two palettes evolving independently of each other. Thus for an edge (u, v) at stage $k \geq 0$,

$$
\begin{aligned}
E[\Theta_{k+1} \mid \Delta_k, \Theta_k] &\approx \Theta_k - \sum_{w \in N_k(u)} \alpha (\frac{\Theta_k}{\Delta_k})^2 - \sum_{w \in N_k(v)} \alpha (\frac{\Theta_k}{\Delta_k})^2 \\
&= \Theta_k (1 - 2\alpha \frac{\Theta_k}{\Delta_k}) \\
&= \Theta_k (1 - 2\alpha \eta_k)
\end{aligned}
\tag{2}
$$

It is important to note here the factor $(\frac{\Theta_k}{\Delta_k})^2$ – this arises because when a colour is selected for an edge neighbouring (u, v), it must be in the common palette of *both* edges. Then, ¿from (2) it follows that

$$
\begin{aligned}
E[\Theta_{k+1}] &= E[E[\Theta_{k+1} \mid \Delta_k, \Theta_k]] \\
&= (1 - 2\alpha \eta_k) E[\Theta_k] \\
&\approx e^{-2\alpha \eta_k} E[\Theta_k].
\end{aligned}
\tag{3}
$$

Let us write $\eta_k := \frac{\Theta_k}{\Delta_k} \approx \frac{E[\Theta_k]}{E[\Delta_k]}$; we will justify this shortly by showing that the r.v.s Δ_k and Θ_k are sharply concentrated at their means. Thus from (1) and (3), if follows that [5]

$$\eta_{k+1} = \eta_k e^{-\alpha \eta_k}. \tag{4}$$

This recurrence is well–studied, see for instance [5, § 8.5]. We have that

$$E[\Delta_k] = \Delta \exp(-\alpha \sum_{i \le k} \eta_i)$$

It can be verified that $E[\Delta_k] \le \lambda \Delta$ whenever $k \ge k_0 := (\frac{\log(1/\lambda)}{\alpha \lambda})$. In computing the expectations above, we assumed that the graph was Δ_k–regular at stage k. Even if we assume the initial graph is Δ–regular, it will not remain regular at later stages due to statistical fluctuations. However, we shall now refine the argument by a high–probability analysis and show that the random variables Δ_k^u (denoting the size of the palette of vertex u and also its degree) and $\Theta_k^{u,v}$ (denoting the common palette size $|A_k^u \cap A_k^v|$) are each sharply concentrated around their means computed in the last section. Thus, the graph does remain "almost" regular at each stage. We shall let $E\Theta_k$ and $E\Delta_k$ be the recurrences determined by

$$E\Theta_0 := \Delta, \; E\Theta_{k+1} = e^{-2\alpha\eta_k} E\Theta_k,$$

and

$$E\Delta_0 := \Delta, \; E\Delta_{k+1} = e^{-\alpha\eta_k} E\Delta_k.$$

where, as before, η_k is the sequence determined by the recurrence (4) (with $\eta_0 := 1$). We will show that for each vertex u and each edge (u, v), with high probability,

$$(1 - \delta_k)E\Delta_k \le \Delta_k \le (1 + \delta_k)E\Delta_k,$$

and

$$(1 - \delta_k)E\Theta_k \le \Theta_k \le (1 + \delta_k)E\Theta_k,$$

(for a sequence δ_k to be specified). In this sense, if we start with a graph which is Δ–regular, it remains "almost" regular as we progress through the stages. We shall prove these statements by induction on k; they are trivially true at the start for $k = 0$. The number of edges coloured at any stage around a given vertex or edge is the sum of indicator random variables which are 1 with the probability computed earlier. We would like to use large deviation bounds to show that this sum is sharply concentrated around its mean. However, these random variables are manifestly *not* independent, and we cannot employ the usual Chernoff bound. However due to the nature of the association of the r.v.s in our case, we are able to salvage the Chernoff bound nevertheless. The most efficient way of doing this is to to use the following martingale argument sometimes called "the method of bounded differences" [15]:

---

[5] We use here our approximation that $\frac{1-x}{1-y} \approx 1 + y - x$ and so strictly should write $\approx$. However, since $e^{-(1+\epsilon)x} \le 1 - x \le e^{-x}$ for any $\epsilon > 0$ if x is sufficiently small, we can use $=$ with the tacit understanding that one can substitute these exact inequalities if required.

Proposition 2 (Method of Bounded Differences). *Let* $X := X_1, \ldots, X_m$ *be independent random variables with X_k taking values in a set A_k. Suppose the measurable function $f : \prod_k A_k \to \mathrm{R}$ satisfies*

$$|f(X) - f(X')| \le c_k,$$

whenever X and X' differ only in the kth co-ordinate for each $k \in [m]$ and for some constants $c_1, \ldots, c_m$. Then, for any $t > 0$,

$$\Pr[|f(X) - E[f(X)]| > t] \le 2 \exp(-2t^2 / \sum_k c_k^2).$$

To apply this proposition to compute the palette size of a given vertex or edge after an arbitrary stage $k \ge 0$, we proceed as follows. Let us consider the edge palettes. Fix a certain order of considering the vertices, and think of the random colouring process at stage k as determining the tentative colour assignments to edges in order corresponding to the vertices they are incident on. For an edge e incident on a fixed vertex u, let X_e be the tentative colour it is assigned at this stage, provided it is also available at the other endpoint and suffers no conflict at that endpoint (we can think of each X_e being a special colour $\perp$ at the start, thus the sets A_k in the proposition are each $[\Delta] \cup \{\perp\}$). By the properties of the algorithm, these variables (which are $|A_k^u|$ in number) are indeed independent. The function f, we choose is the size of the resulting edge palette, under these choices at stage k. It can be verified that this function has the "bounded difference" property with each $c_k := 2$. Given Δ_k, Θ_k, we have computed the expectations before. Now, applying the Chernoff bound shows that given Δ_k, Θ_k, we have with high probability (namely that given above), for any $0 < \delta < 1$,

$$e^{-2\alpha} \Theta_k (1 - \delta) \le \Theta_{k+1} \le e^{-2\alpha} \Theta_k (1 + \delta),$$

and inductively assuming high probability bounds on Θ_k, this implies that

$$E\Theta_{k+1} (1 - \delta)(1 - \delta_k) \le \Theta_{k+1} \le E\Theta_{k+1} (1 + \delta)(1 + \delta_k).$$

or,

$$E\Theta_{k+1} e^{-\delta_k - \delta} \le \Theta_{k+1} \le E\Theta_{k+1} e^{\delta_k + \delta}.$$

So, taking $\delta_k := k\delta$ verifies the inductive claim. Similarly for Δ_k, with high probability,

$$E\Delta_k (1 - \delta_k) \le \Delta_k \le E\Delta_k (1 + \delta_k).$$

The analysis is exactly the same for the vertex palettes. The failure probability is pessimistically estimated as $k_0 n$ times the failure probability at a vertex at the last stage k_0. This in turn is given by Proposition 2. We are interested only in vertices which have degree at least $\lambda\Delta$ at this stage. Noting that $\Delta_{k+1} \ge e^{-2\epsilon} \Delta_k$ for any $k \ge 0$, we get that the failure probability is at most

$$2nk_0 \exp(-\lambda^2 \Delta).$$

The statement on the failure probability in Theorem 1 follows ¿from this (recall that we assume $\Delta = \Omega(\log^{1+d} n)$ for some $d > 0$).

4.3 Removing the Regularity Assumption

In this section, we outline how to remove the assumption that the graph is initially Δ-regular. Note that because we want our algorithm to work in a truly distributed fashion, we cannot assume that the maximum degree is known to all vertices. As we shall demonstrate below, the essential feature of the interaction of the palettes (namely the locality) and the decay law obeyed by the palettes continues to hold without the regularity assumption. Let $\eta_k^{\max}$ be determined by the recurrence relations

$$\eta_0^{\max} := \max_{u,v} \frac{\Theta_0^{u,v}}{\Delta_0^u},$$

and

$$\eta_{k+1}^{\max} = \eta_k^{\max} \exp(-\alpha \eta_k^{\max}).$$

Similarly, let $\eta^{\min}$ be determined by the corresponding recurrence with max replaced by min. It is easy to verify by induction that for each $k \geq 0$,

$$1 \leq \eta_k^{\max}/\eta_k^{\min} \leq \eta_0^{\max}/\eta_0^{\min}.$$

Let us now write down the equation corresponding to (2) in the non-regular setting (with Θ_k, Δ_k denoting the vector of the random variables at stage k). Recall that the algorithm sets $\Delta_0^u = \max_{w \in N(u) \cup \{u\}} \deg(w)$ initially and that $\Theta_0^{u,v} = \min(\Delta_0^u, \Delta_0^v)$.

$$E[\Theta_{k+1}^{u,v} \mid \Theta_k, \Delta_k] = \Theta_k^{u,v} - \alpha \sum_{w \in N_k(u)} (\frac{\Theta_k^{u,w}}{\Delta_k^u})^2 + (\frac{\Theta_k^{u,w}}{\Delta_k^w})^2 - \alpha \sum_{w \in N_k(v)} (\frac{\Theta_k^{v,w}}{\Delta_k^v})^2 + (\frac{\Theta_k^{v,w}}{\Delta_k^w})^2$$

$$= \Theta_k^{u,v} - \alpha \sum_{w \in N_k(u)} (\eta_k^{u,w})^2 + (\eta_k^{w,u})^2 - \alpha \sum_{w \in N_k(v)} (\eta^{v,w})^2 + (\eta^{w,v})^2$$

Now, once again, inductively, we have for any two edges (u,v) and (u',v'),

$$\frac{\eta_0^{\min}}{\eta_0^{\max}} \leq \frac{\eta_k^{u,v}}{\eta_k^{u',v'}} \leq \frac{\eta_0^{\max}}{\eta_0^{\min}}.$$

Using this in the previous equation, we get:

$$\Theta_k^{u,v} - \alpha \frac{\eta_0^{\max}}{\eta_0^{\min}} \sum_{w \in N_k(u)} (\eta_k^{u,v})^2 + (\eta_k^{v,u})^2 - \alpha \frac{\eta_0^{\max}}{\eta_0^{\min}} \sum_{w \in N_k(v)} (\eta^{v,u})^2 + (\eta^{u,v})^2 \leq E[\Theta_{k+1}^{u,v} \mid \Delta_k, \Theta_k].$$

and

$$E[\Theta_{k+1}^{u,v} \mid \Delta_k, \Theta_k] \leq \Theta_k^{u,v} - \alpha \frac{\eta_0^{\min}}{\eta_0^{\max}} \sum_{w \in N_k(u)} (\eta_k^{u,v})^2 + (\eta_k^{v,u})^2 - \alpha \frac{\eta_0^{\min}}{\eta_0^{\max}} \sum_{w \in N_k(v)} (\eta^{u,v})^2 + (\eta^{v,u})^2.$$

Thus,

$$\Theta_k^{u,v}(1 - \alpha') \leq E[\Theta_{k+1}^{u,v} \mid \Delta_k, \Theta_k] \leq \Theta_k^{u,v}(1 - \alpha'').$$

We are thus back in essentially the same situation as before and we can refine the calculation of expected values into a high probability argument as before. Finally we have proved:

Theorem 3. *For any fixed $\lambda > 0$, given a graph with n vertices and maximum degee Δ, we can edge colour the graph with $(1 + \lambda)\Delta$ colours in time $O(\log n)$ where n is the number of vertices in the graph. For any fixed positive integer s, we can edge colour it with $\Delta + \Delta/\log^s \Delta = (1 + o(1))\Delta$ colours in time $O((\log \Delta)^s \log \log \Delta + \log n)$. The results hold with failure probability decreasing to 0 faster than any polynomial (in n) provided that $\Delta = \Omega(\log^{1+d} n)$ for some $d > 0$.*

REMARK: It is unlikely that one can improve the above analysis to get a colouring better than the $\Delta + \Delta/(\log \Delta)^s$ bound above, while still retaining a poly-logarithmic running time (in n and Δ). To see this, recall from the intuitive outline in § 4.1, that even if we assume that the initial graph is regular and that the palettes evolve as truly random independent subsets, the decay law has the form

$$\Delta_{k+1} \leq \exp(-\epsilon \frac{\Delta_k}{\Delta})\Delta_k.$$

If η_k is the determined by the recurrence

$$\eta_{k+1} := e^{-\alpha \kappa \eta_k} \eta_k,$$

then one can show (see for instance, [5, § 8.5]) that $\eta_k \geq 1/k$. So, if the shrinking of a vertex degree is governed by an equation of the form $\eta_{k+1} := e^{-\alpha \kappa \eta_k} \eta_k$ the number of iterations needed to bring the degree down to $\Delta/g(\Delta)$ is $k(\Delta) = \Omega(g(\Delta))$.

Acknowledgments

We are indebted to Noga Alon and Joel Spencer for their suggestion that the Rödl nibble should work, and to Noga Alon for pointing out a mistake in a previous draft of this work and for several suggestions. We are grateful to Jirka Matousek who independently also suggested the use of the Rödl Nibble and who gave us a nice explanation of the swift working of the nibble. Kurt Melhorn provided, as usual, insightful remarks. We also thank Sem Borst and Marco Combe for useful discussions. The first author is grateful to the CWI and the Altec project for making possible an extended visit to CWI. The second author acknowledges the generous hospitality of MPI and BRICS.

References

1. N. Alon. Private Communication.
2. N. Alon, J. Spencer, and P. Erdős. *The Probabilistic Method.* Wiley–Interscience Series, John Wiley & Sons, Inc., New York, 1992.
3. B. Berger and J. Rompel. Simulating $(\log^c n)$-wise independence in NC. *J. Assoc. Comput. Mach.*, 38(4):1026–1046, 1991.
4. B. Bollobás. *Graph Theory.* Springer Verlag, New York, 1979.
5. N.G. de Bruijn. *Asymptotic methods in Analysis.* Number 4 in Bibliotheca Mathematics. North Holland Publishing Co., 1958.

6. R. Jain D. Durand and D. Tseytlin. Distributed scheduling algorithms to improve the performance of parallel data transfers. Technical Report 94-38, DIMACS, 1994.

7. Z. Galil and D. Leven. NP-completeness of finding the chromatic index of regular graphs. *J. of Algorithms*, 4:35-44, 1983.

8. I. Holyer. The NP-completeness of edge coloring. *SIAM J. Comp.*, 10:718-720, 1981.

9. R. Jain, K. Somalwar, J. Werth, and J. C. Browne. Scheduling parallel i/o operations in multiple bus systems. *Journal of Parallel and Distributed Computing*, 16(4):352-362, 1992.

10. J. Kahn. Coloring nearly-disjoint hypergraphs with $n + o(n)$ colors. *J. Comb. Theory, Series A*, 59:31-39, 1992.

11. H. J. Karloff and D. B. Shmoys. Efficient parallel algorithms for edge coloring problems. *Journal of Algorithms*, 8:39-52, 1987.

12. R. M. Karp. Probabilistic recurrence relations. In *Proceedings of the ACM Symposium on Theory of Computing*, pages 190-197, 1991.

13. M. Luby. Removing randomness in parallel computation without a processor penalty. In *Proceedings of the IEEE Symposium on Foundations of Computer Science*, pages 162-173, 1988. To appear in a special issue of *Journal of Computer and System Sciences*, devoted to FOCS 1988.

14. N. A. Lynch. Upper bounds for static resource allocation in a distributed system. *Journal of Computer and System Sciences*, 23:254-278, 1981.

15. C. McDiarmid. On the method of bounded differences. In J. Siemons, editor, *Surveys in Combinatorics*, volume 141 of *London Math. Soc. Lecture Notes Series*, pages 148-188. Cambrideg University Press, 1989.

16. R. Motwani, J. Naor, and M. Naor. The probabilistic method yields deterministic parallel algorithms. In *Proceedings of the IEEE Symposium on Foundations of Computer Science*, pages 8-13, 1989.

17. A. Panconesi and A. Srinivasan. Fast randomized algorithms for distributed edge coloring. In *Proceedings of the ACM Symposium on Principles of Distributed Computing*, pages 251-262, 1992.

18. A. Panconesi and A. Srinivasan. Improved distributed algorithms for coloring and network decomposition problems. In *Proceedings of the ACM Symposium on Theory of Computing*, pages 581-592, 1992.

19. N. Pippinger and J. Spencer. Asymptotic behaviour of the chromatic index for hypergraphs. *J. Combinatorial Theory, Series A*, 51:24-42, 1989.

20. V. Rödl. On a packing and covering problem. *European Journal of Combinatorics*, 5:69-78, 1985.

21. J. Spencer. Asymptotically Good Coverings. *Pacific Journal of Mathematics*, 118(2):575-586, 1985.

The Centroid of Points with Approximate Weights

Marshall Bern[1]   David Eppstein[2]   Leonidas Guibas[3]   John Hershberger[4]
Subhash Suri[5]   Jan Wolter[6]

[1] Xerox Palo Alto Research Center, 3333 Coyote Hill Rd., Palo Alto, CA 94304
[2] Dept. of ICS, U. of California, Irvine, CA 92717. Work done in part while visiting
Xerox PARC and supported in part by NSF grant CCR-9258355
[3] Dept. of Computer Science, Stanford University, Stanford, CA 94705
Supported in part by NSF grants CCR-9215219 and IRI-9306544.
[4] Mentor Graphics Corp., 1001 Ridder Park Dr., San Jose, CA 95131
[5] Dept. of Computer Science, Washington University, St. Louis, MO 63130
[6] Dept. of Computer Science, Texas A & M, College Station, TX 77843

Abstract. Let S be a set of points in $\mathbb{R}^d$, each with a weight that is
not known precisely, only known to fall within some range. What is the
locus of the centroid of S? We prove that this locus is a convex polytope,
the projection of a zonotope in $\mathbb{R}^{d+1}$. We derive complexity bounds and
algorithms for the construction of these "centroid polytopes".

1   Introduction

Suppose that $S = \{s_1, s_2, \ldots, s_n\}$ is a set of points in $\mathbb{R}^d$ and that each point s_i
has an unknown nonnegative weight w_i that lies within a known range $[l_i, h_i]$.
The *centroid* of S—also called its "center of mass" or "weighted average"—is
the vector sum $\frac{1}{W} \sum_i w_i s_i$, where $W = \sum_i w_i$. We are interested in $C(S)$, the
set of all possible centroids.

A generalization allows explicit upper and lower bounds on the total weight
W, tighter than the implicit bounds $\sum_i l_i \leq W \leq \sum_i h_i$. Let $C_{L,H}(S)$ denote
the locus of possible centroids of S with the additional constraint that $L \leq W \leq
H$. Thus $C(S) = C_{0,\infty}(S)$. A different generalization, which we defer until the
paper's conclusion, incorporates positional uncertainty for the points of S.

In this paper we characterize the *centroid polytope* $C_{L,H}(S)$ and give algorithms to compute it. Section 2 establishes a connection between centroid polytopes and the classical notion of zonotopes. Section 3 proves that the maximum
complexity of a centroid polytope is $\Theta(n^d)$, but only $\Theta(n^{d-1})$ for the special
case of $C(S)$. Section 4 describes algorithms to compute centroid polytopes. In
two dimensions, our algorithm has running time $O(n \log n + k \log^2 n)$, where k is
the complexity of $C_{L,H}(S)$, the output. Section 5 improves the running time to
$O(n \log n)$ for $C(S)$. This section introduces a new type of semi-dynamic convex
hull data structure. Since $C(S)$ is simply the convex hull of S if each l_i is zero,
this algorithm is asymptotically optimal.

An application of the centroid polytope in two dimensions arises in the mathematical modeling of oil spills [8]. The points are centers of quadtree squares,

with each weight representing the quantity of oil within a square, which is known only approximately. Working top-down in the quadtree, predicted centroids are repeatedly tested against centroid polytopes of observational data in order to set parameters of the model.

Aircraft design provides a natural application in three dimensions: the centroid of a fully loaded airplane depends on the weights of its fuel, cargo, and passengers, and the flight characteristics of the plane in turn depends on the location of its centroid. It is reasonable to add an explicit bound on the total weight W, since a single 300-pound passenger is fairly likely, though not an entire planeload of 300-pound passengers.

2 Connection to Zonotopes

As above, let $S = \{s_1, s_2, \ldots, s_n\}$ be points in Euclidean d-space $\mathbb{R}^d$, each with an unknown weight w_i that lies in the known range $[l_i, h_i]$, where $0 \leq l_i \leq h_i$. Let L and H be bounds on the total weight of S, with $0 \leq L \leq H$. The *centroid polytope* $C_{L,H}(S)$ is

$$\left\{ \frac{1}{W} \sum_{i=1}^{n} w_i s_i \ \middle|\ w_i \in [l_i, h_i] \text{ for } 1 \leq i \leq n, \ W = \sum_{i=1}^{n} w_i, \text{ and } L \leq W \leq H \right\}.$$

Günter Rote (personal communication) points out that a change of variables—"fractional programming"—lets us write the centroid polytope as the feasible region of a linear program, thereby proving that the centroid polytope is indeed a polytope. If we let $t = 1/\sum_i w_i$ and $v_i = w_i t$, then

$$C_{L,H}(S) = \left\{ \sum_{i=1}^{n} v_i s_i \ \middle|\ \sum_{i=1}^{n} v_i = 1, \ l_i t \leq v_i \leq h_i t, \text{ and } Lt \leq 1 \leq Ht \right\}.$$

An interesting special case, the *unconstrained centroid polytope*, is $C_{0,\infty}(S)$, which we also denote by $C(S)$. Figure 1 shows $C(S)$ for $S = \{(0,1), (0,0), (1,0)\}$, assuming that each weight range is simply $[\frac{1}{2}, 1]$. Each of the six sides of $C(S)$ corresponds to fixing one weight to $\frac{1}{2}$, another weight to 1, and letting the third weight vary between these two extremes. The varying weight belongs to the point hit by the continuation of the side. In general, each facet of an unconstrained centroid polytope in $\mathbb{R}^d$ lies in the same hyperplane as (at least) $d - 1$ points of S. The weights of these points vary, whereas other weights are fixed to either their minimum or maximum values, depending on their halfspace.

Let v_i denote $(s_i, 1)$, the vector in $\mathbb{R}^{d+1}$ that agrees with s_i on its first d coordinates and has 1 as its last coordinate. Let

$$Z = \left\{ \sum_{i=1}^{n} w_i v_i \ \middle|\ w_i \in [l_i, h_i] \text{ for } 1 \leq i \leq n \right\}.$$

Thus Z is the Minkowski sum of the n line segments of the form $\{ w_i v_i \mid w_i \in [l_i, h_i] \}$. (The *Minkowski sum* of sets A and B in $\mathbb{R}^{d+1}$ is $\{ p + q \mid p \in A \text{ and } q \in$

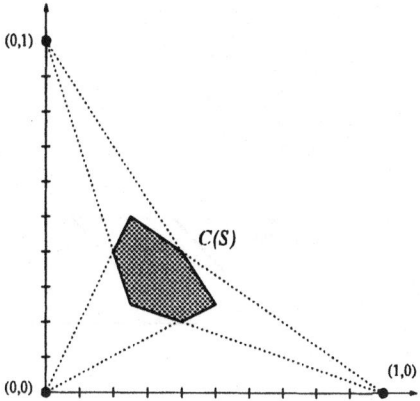

Fig. 1. The centroid polytope $C(S)$ for 3 points, each with weight range $[\frac{1}{2}, 1]$.

B}.) The Minkowski sum of line segments is a special type of convex polytope called a *zonotope*.

Edelsbrunner [3] writes, "it is very instructive to visualize the inductive construction of a zonotope". To add the ith segment z_i to the Minkowski sum, we sweep the current zonotope parallel to z_i and then take the swept volume as the new zonotope. If we assume that the ith segment runs "north-south", this step enlarges the current zonotope in $\mathbb{R}^{d+1}$ by stretching its "equator", a belt of faces of dimensions up to $d-1$, into a "zone" of faces of dimensions up to d. Faces of zonotopes are themselves lower-dimensional zonotopes.

There is a nice connection between zonotopes in $\mathbb{R}^{d+1}$ and arrangements in $\mathbb{R}^d$. We identify $\mathbb{R}^d$ with the space of directions (normal vectors) of hyperplanes tangent to zonotope Z. These directions most naturally live in projective d-space, so we adopt the projective view of an arrangement in $\mathbb{R}^d$, in which we regard antipodal pairs of unbounded cells as single cells (see [3], pp. 46–47). A hyperplane of the arrangement in $\mathbb{R}^d$ corresponds to the directions of hyperplanes tangent to a single zone of Z. A cell of the arrangement corresponds to an antipodal pair of vertices of Z. And, more generally, a k-face of the arrangement corresponds to an antipodal pair of $(d-k)$-faces of Z. Dimensions reverse, because a hyperplane through a $(d-k)$-face in $\mathbb{R}^{d+1}$ has k degrees of freedom.

Returning now to centroid polytopes, we see that $C_{L,H}(S)$ can be rewritten as

$$\left\{ \left(\frac{x_1}{x_{d+1}}, \frac{x_2}{x_{d+1}}, \ldots, \frac{x_d}{x_{d+1}} \right) \;\middle|\; (x_1, x_2, \ldots, x_d, x_{d+1}) \in Z \text{ and } L \leq x_{d+1} \leq H \right\}.$$

Notice that the last coordinate of points in Z measures total weight, that is, for the point $(x_1, x_2, \ldots, x_{d+1}) = \sum_i w_i v_i$, we have $x_{d+1} = \sum_i w_i$. Thus the condition $L \leq x_{d+1} \leq H$ restricts us to points of acceptable total weight, a section of Z lying between two hyperplanes normal to the $(d+1)$-st axis. Dividing the first d coordinates by the $(d+1)$-st then corresponds to a perspective projection of this section. (Only the "silhouette" matters; the interior of the projection is

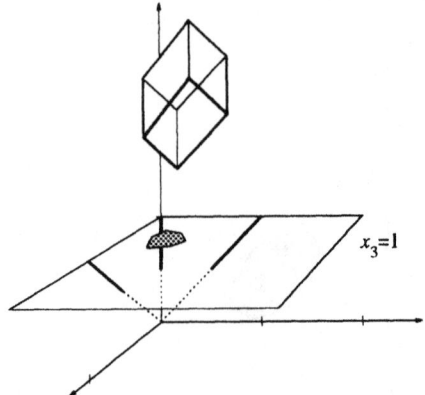

Fig. 2. The centroid polytope $C(S)$ is a perspective projection of a zonotope.

featureless.) In this projection, the viewpoint is the origin, and the "viewing screen" is the hyperplane $x_{d+1} = 1$. Figure 2 shows the zonotope connection for the example from Figure 1. Notice that in the special case of $C(S)$, the section is all of Z. We summarize with the following theorem.

Theorem 1. *The centroid polytope $C_{L,H}(S)$ is a perspective projection of the section of Z lying between the hyperplanes $x_{d+1} = L$ and $x_{d+1} = H$.* $\square$

3   Complexity Bounds

Theorem 1 immediately implies that centroid polytopes are indeed convex polytopes. The zonotope connection also yields bounds on the complexity (total number of faces) of a centroid polytope. We present upper bounds first, then matching lower bounds.

Theorem 2. *The complexity of centroid polytope $C_{L,H}(S)$ is $O(n^d)$.*

Proof. A zonotope formed by n line segments in $\mathbb{R}^{d+1}$ has at most $O(n^d)$ faces [3]. Slicing Z by the hyperplanes $\sum_i w_i = L$ and $\sum_i w_i = H$ increases the complexity by at most a constant factor, and taking a perspective projection cannot increase the complexity, so altogether there are at most $O(n^d)$ faces in $C_{L,H}(S)$. $\square$

Tighter bounds [3] follow from the correspondence between zonotopes in $\mathbb{R}^{d+1}$ and arrangements of hyperplanes in $\mathbb{R}^d$. The number of k-faces of zonotope Z for $0 \le k \le d$ is at most $2 \cdot \sum_{i=0}^{k} \binom{d-i}{k-i}\binom{n}{d-i}$. This implies that the number of facets of $C_{L,H}(S)$ is never more than $2(d+1)\binom{n}{d} + o(n^d)$.

We now turn to the special case of unconstrained centroid polytopes. Curiously, the projection of an entire zonotope has smaller asymptotic complexity than the projection of an arbitrary section. This is due to the following special property of a zonotope: its tangent space is not just any convex decomposition, but rather a hyperplane arrangement.

Theorem 3. *The complexity of the unconstrained centroid polytope $C(S)$ is $O(n^{d-1})$. In particular, the number of facets is at most $2\binom{n}{d-1}$.*

Proof. We exploit the zonotope-arrangement connection described above, in which zonotope Z corresponds to a hyperplane arrangement $\mathcal{A}$ in $\mathbb{R}^d$. Now consider all hyperplanes tangent to Z that pass through the origin in $\mathbb{R}^{d+1}$. The directions of these tangent hyperplanes form a convex body with boundary C in $\mathbb{R}^d$. Furthermore, C is linear within any face of $\mathcal{A}$. (From here on, we think of cells as faces too.) There is a one-to-one, dimension-reversing correspondence between faces of $C(S)$ and faces in the arrangement of piecewise-linear surfaces on C formed by the hyperplanes of $\mathcal{A}$. Thus we can count faces of $C(S)$ by counting the faces in this arrangement. By a standard argument, we can assume that $\mathcal{A}$ is a simple arrangement—meaning hyperplanes in general position—and that C does not pass through any vertex of $\mathcal{A}$; otherwise a small perturbation would increase the complexity.

We count the number of faces on C as follows. Each face is formed as a connected component of an intersection of a face in $\mathcal{A}$ with C. Since C is linear within each face of $\mathcal{A}$, each face of $\mathcal{A}$ gives rise to only one connected component, which necessarily includes a portion of the boundary of the face. Therefore we can "charge" each face of $\mathcal{A}$ cut by C to a lower-dimensional cut face, until we reach 1-faces (edges) of $\mathcal{A}$ that intersect C in single points (corresponding to facets of $C(S)$). Any line in $\mathcal{A}$ crosses C at most twice, and there are $\binom{n}{d-1}$ lines of $\mathcal{A}$, so the arrangement on C has at most $2\binom{n}{d-1}$ vertices and $C(S)$ has at most $2\binom{n}{d-1}$ facets. By general position, any 1-face in $\mathcal{A}$ has exactly $3^{d-1} - 1$ neighboring cells of higher dimension, and hence each facet of $C(S)$ is charged at most 3^{d-1} times. Therefore there are at most $2 \cdot 3^{d-1}\binom{n}{d-1} = O(n^{d-1})$ faces in $C(S)$. □

We now move on to lower bounds, presenting them in reverse order from the upper bounds. The facet bound of Theorem 3 is tight. First consider the case $d = 2$. Let S be the vertices of a regular n-gon, and set all weight ranges to $[\frac{1}{n} - \epsilon, \frac{1}{n} + \epsilon]$ for $\epsilon > 0$ very small. Then $C(S)$ is a small $2n$-gon, with two sides for each s_i in S.

To generalize this bound to higher dimensions, let S be a set of points fairly evenly spaced on a sphere. For ϵ sufficiently small, $C(S)$ behaves like a point, which we may assume is in general position with respect to the points of S. In other words, through any $d - 1$ points of S, some hyperplane misses $C(S)$, and hence two hyperplanes support $C(S)$. The two supporting hyperplanes each contain a facet, on which the weights of the $d - 1$ points vary. Of course, this example also gives an $\Omega(n^{d-1})$ bound for the total complexity.

Figure 3(a) gives an $\Omega(n^2)$ lower bound example for $C_{L,H}(S)$ for the case $d = 2$, matching the bound given by Theorem 2. In this example, there are two rings of $\frac{n}{2}$ points, where $\frac{n}{2}$ is an odd number, forming regular $\frac{n}{2}$-gons. The inner-ring points, denoted S_I, lie just inside the vertices of the central $\frac{n}{2}$-gon formed by the diagonals of the outer-ring points, denoted S_O. Each point of S_O has range $\left[\frac{2}{n} - \epsilon, \frac{2}{n} + \epsilon\right]$, with $\epsilon > 0$ small enough that the centroid polytope $C(S_O)$ lies well inside the inner ring. Each point of S_I has range $[0, \frac{\epsilon}{n}]$. We now assert that $C_{L,H}(S)$, where $S = S_O \cup S_I$ and $L = H = 1 + \epsilon$, is a polygon with $\frac{n(n+1)}{2}$ sides.

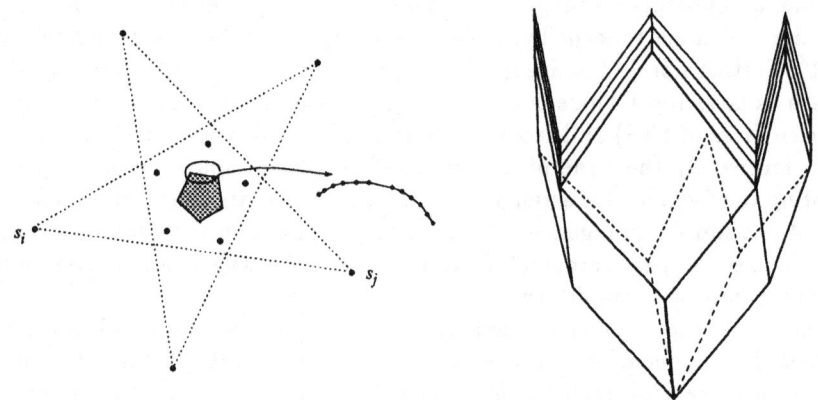

Fig. 3. (a) An $\Omega(n^2)$ example for $C_{L,H}(S)$. (b) Corresponding zonotope.

Why is this assertion true? First consider $C_{L,H}(S_O)$, the centroid polytope of the outer points, with the same restriction on the total weight. This polytope is an $\frac{n}{2}$-gon, shown shaded in Figure 3(a). Each side of $C_{L,H}(S_O)$ is defined by two points of S_O, whose weights vary along that side. For example, the weights of s_i and s_j vary along the side circled in Figure 3(a), while the weights of the top two points are fixed at $\frac{2}{n} + \epsilon$ and the weight of the bottom point is fixed at $\frac{2}{n} - \epsilon$.

Now step up to $C_{L,H}(S)$. The circled side is replaced by a chain of $n + 1$ sides. As we move along the first $\frac{n}{2}$ sides of this chain, weight $\frac{\epsilon}{n}$ transfers from s_i to each of the points of S_I in turn, in clockwise order around s_i. Along the middle side of the chain, weight $\frac{\epsilon}{2}$ transfers from s_i to s_j. Finally, along the last $\frac{n}{2}$ sides, weight transfers from S_I to s_j.

The zonotope corresponding to S looks something like a Ukrainian egg, as shown in Figure 3(b). This zonotope is the Minkowski sum of $\frac{n}{2}$ well-spaced long segments and $\frac{n}{2}$ nearly parallel short segments, and hence contains a zig-zag of $\Omega(n^2)$ faces that can all be cut by a horizontal plane.

By replacing S_O and S_I by points nearly equally spaced around concentric spheres, it should be possible to generalize this construction to give an $\Omega(n^d)$ bound in arbitrary dimension, but we have not worked out the details.

4 Output-Sensitive Algorithms

In this section, we briefly sketch algorithms for computing centroid polytopes. We start by showing how to compute $C_{L,H}(S)$ for $d = 2$. Because this polygon may have complexity anywhere from $O(1)$ to $O(n^2)$, our goal is an *output-sensitive* algorithm, meaning one with running time dependent on k, the number of sides of $C_{L,H}(S)$.

Let z be a rightmost vertex of $C_{L,H}(S)$. The next lemma shows that z is defined by separating the points of S into minimum- and maximum-weight subsets that can be separated by a vertical line ℓ. Line ℓ typically passes through a point of S that has intermediate weight.

Lemma 4. *There is a vertical line ℓ partitioning the plane into open halfplanes h^- and h^+, such that z is a linear combination $\frac{1}{W} \sum_i w_i s_i$ in which $w_i = l_i$ for each s_i in h^- and $w_i = h_i$ for each s_i in h^+. Furthermore, if $z \in h^+$, then the sum of all weights is L, and if $z \in h^-$, then the sum of all weights is H.*

Proof. If s_i is to the left of s_j, with $w_i > l_i$ and $w_j < h_j$, we could increase w_j and decrease w_i causing the centroid to move further to the right. Therefore in the weights for z, the points taking their minimum value must lie to the left of those taking their maximum value. Vertical line ℓ separates these two sets, passing through the point(s) with intermediate weight. If no intermediate-weight point exists, we let ℓ be the separating line closest to z, which passes through either the rightmost minimum-weight or the leftmost maximum-weight point. Thus ℓ is always chosen to go through some particular point s_k. If the conditions on total weight in the last sentence of the lemma were not met, then by changing the weight of of s_k we could move the extremum. □

Lemma 5. *A rightmost vertex z can be found in $O(n)$ time.*

Proof. Assume that the points of S by increasing x-coordinate are $s_1, s_2, \ldots, s_n$. Let x_i denote the x-coordinate of s_i, and let $m = \lfloor \frac{n}{2} \rfloor$.

Using linear-time median finding, we compute x_m. We then assign point s_i weight l_i if $i < m$ and h_i if $i \geq m$, and compute the centroid c with this setting of weights. Let ℓ be the vertical line $x = x_m$. If the total weight is less than L (respectively, greater than H) we know that z lies to the left (right) of ℓ. If the total weight falls within these bounds, it is not hard to show that z lies on the same side of ℓ as c. Now if z lies to the left of ℓ, we collapse the set $S' = \{s_i \mid i \geq m\}$ to a single combined point—the centroid of S' with all weights set to their h_i values—and set its combined weight to $\sum_{i \geq m} h_i$. Similarly if z lies to the right of ℓ, we collapse the points s_i with $i < m$. In either case, we have reduced the problem to one with half as many points. Iterating this procedure gives an overall linear-time algorithm. □

We use a rotational sweep algorithm to compute $C_{L,H}(S)$. At all times, z is a point on the boundary of $C_{L,H}(S)$ that has a tangent parallel to sweep-line ℓ. Initially z is a rightmost vertex of $C_{L,H}(S)$ and ℓ is the line constructed in the proof of Lemma 5. As z moves counterclockwise around the boundary of $C_{L,H}(S)$, we maintain ℓ as the separating line that would be constructed if the plane were rotated to make z rightmost; thus ℓ always passes through a point of S.

In an event in the rotational sweep, line ℓ, pivoting around point s_i, crosses another point s_j. At such an event, we increase or decrease w_j (depending on whether s_j is in h^- or h^+) while making an offsetting change in w_i, until one of w_i and w_j runs into its minimum or maximum value. The one of s_i and s_j left with an intermediate value becomes the new pivot point; if both points reach extrema simultaneously, we break the tie by letting s_j be the new pivot point. If the input is in general position (no collinear triple in S and no "coincidental" sums of w_i's), these weight changes move z to a new vertex of $C_{L,H}(S)$.

Changing weights and recomputing z take only $O(1)$ time, so the overall running time of a step depends on how fast we can detect collisions between ℓ and points of S. One way to detect collisions between ℓ and S is to maintain the convex hull of each of the sets $S \cap h^+$ and $S \cap h^-$. The problem of detecting the next collision then reduces to finding the lines that pass through z and are tangent to these convex hulls. The fully dynamic convex hull data structure of Overmars and van Leeuwen [6] takes $O(\log^2 n)$ time per insertion or deletion, and supports such tangent-finding queries in time $O(\log n)$ apiece.

Theorem 6. *Polytope $C_{L,H}(S)$ in $\mathbb{R}^2$ can be computed in $O(n \log n + k \log^2 n)$ time for inputs in general position, where k denotes the output size.* □

We briefly remark on inputs with degeneracies. Coincidences among w_i's can be handled without increasing the running time, but if S includes many collinear points, ℓ might simultaneously cross many points that give only a single vertex of $C_{L,H}(S)$. This degrades the running time to $O((n + k + n^{2/3}k^{2/3})\log^2 n)$, a bound stemming from the maximum number of incidences in any configuration of n points and k lines [7].

Before we step up to higher dimensions, let us take a dual view of the algorithm just presented. A rotational sweep through S corresponds to moving a point from left to right across the arrangement of lines $\mathcal{A}$ dual to S. When ℓ pivots around a point in the primal plane, its dual point p traces along a segment of $\mathcal{A}$. When ℓ crosses another point of S, p crosses another line of $\mathcal{A}$. Thus we can think of our previous algorithm as tracing a certain path in $\mathcal{A}$. Each successive vertex on the path can be found by a (primal) convex hull data structure, and corresponds to a vertex of the centroid polytope.

Now let us consider the case of arbitrary fixed dimension. It is straightforward to prove generalizations of Lemmas 4 and 5. Any extreme point of $C_{L,H}(S)$ corresponds to splitting S by a hyperplane and giving all points in one open halfspace their minimum weights and all points in the other their maximum weights. Point set S dualizes to a hyperplane arrangement (not the same arrangement as in Theorem 3), and the set of splitting hyperplanes dualizes to a piecewise-linear surface in this arrangement. The features of this surface correspond one-to-one to faces of the centroid polytope; for example, vertices of the surface are the duals of splitting hyperplanes parallel to facets of $C_{L,H}(S)$. Thus we can compute the centroid polytope by visiting each face on the surface.

We use the following query to trace the surface: find all the cells sharing a vertex with a given cell. Moving to one of these neighboring cells corresponds to moving one of the input points from one side of the splitting hyperplane to the other. In the primal space, the query corresponds to searching for the first point hit by a hyperplane as it swings around a $(d-1)$-dimensional pivot. For points in general position, these queries, along with insertions and deletions, can be handled by an algorithm of Agarwal, Eppstein, and Matoušek [1], which takes time $O(n^{1-2/(\lceil d/2 \rceil)+\epsilon})$ per operation for any given $\epsilon > 0$. In particular, each operation takes time $O(n^\epsilon)$ in the case $d = 3$.

Theorem 7. *Polytope $C_{L,H}(S)$ in $\mathbb{R}^d$ can be computed in $O((k+n)n^{1-2/(\lceil d/2 \rceil)+\epsilon})$ time for inputs in general position and $O(kn)$ time for arbitrary inputs.* $\square$

5 Optimal Algorithm for the Unconstrained Polytope

In this section, we show how to improve the algorithm above from $O(n \log^2 n)$ to $O(n \log n)$ for the case of $C(S)$ in $\mathbb{R}^2$. As above, let z be a rightmost vertex of $C(S)$, and let ℓ be a vertical line separating the points taking maximum and minimum weights. For the unconstrained centroid polytope $C(S)$, we may assume that line ℓ passes through z, rather than through a point of S. (We omit the easy proof.)

Now a generic step of the algorithm rotates ℓ counterclockwise about z until it hits a point of S. When ℓ hits a point s_i in half-plane h^+, the algorithm moves s_i from h^+ to h^-, changes its mass from h_i to l_i, and recomputes centroid z. Similarly, when ℓ hits a point $s_i \in h^-$, the algorithm moves s_i to h^+ and changes its mass to h_i. In either case, the new centroid lies on $\ell \cap C(S)$, counterclockwise of z. (Unlike before, special-position inputs do not cause any real problems.)

Smaller Sweeps. The improved algorithm splits the rotational sweep into four $90°$ sweeps. It is not hard to see that within a $90°$ sweep (in fact, even within a $180°$ sweep), no point of S can leave h^+ and subsequently re-enter it. This constraint allows the use of semi-dynamic, rather than fully dynamic, data structures.

Let us consider the $90°$ sweep in which ℓ starts out vertical and tangent to the right side of $C(S)$ and rotates to be horizontal and tangent to the top of $C(S)$. The starting and ending positions of ℓ partition the plane into quadrants, as shown in Figure 4.

During the sweep, the points of S in the four quadrants change sets as follows:

I) All points remain in h^+.
II) All points leave h^- and enter h^+.
III) Some points stay in h^-, but others (shaded) leave h^- and later re-enter it.
IV) All points leave h^+ and enter h^-.

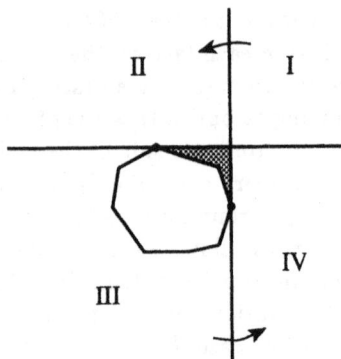

Fig. 4. Quadrants determined by starting and ending positions of ℓ.

Points in quadrants II and IV are easy to handle. For example, when a point s_i in quadrant IV leaves h^+, we delete it from the data structure holding the points of IV and change its weight from h_i to l_i, but we need not insert it into any other dynamic data structure, because it cannot again be crossed by ℓ within this sweep. The deletions-only convex hull data structure of Hershberger and Suri [5] (see also Chazelle [2]) provides a suitable data structure for quadrants II and IV; it supports tangent-finding in $O(\log n)$ time and point deletion in $O(\log n)$ amortized time.

Points in quadrant III, specifically those in the shaded region, pose a difficulty. At any instant of the sweep, the points in quadrant III belong to three sets: (A) points that have not yet left h^-, (B) points now in h^+, and (C) points that have already returned to h^-. Set C does not need to be represented, because ℓ will not cross over these points again in this 90° sweep; set A can be represented using a deletions-only hull; set B, however, requires something more than a deletions-only hull, because this set is subject to both insertions and deletions.

In order to handle the points in set III(B), we modify the deletions-only hull so that it supports insertions at its left end, arbitrary deletions, and queries asking for the point closest to ℓ, each in amortized time $O(\log n)$. The modified data structure no longer supports arbitrary tangent-finding queries, but closest-point queries will suffice for our purposes.

Restricted-Insertion Convex Hull Data Structure. At any instant of the sweep, our data structure will contain the points in the trapezoid bounded by the initial and final positions of ℓ, the current position of ℓ, and the vertical line through z (the point of tangency between ℓ and $C(S)$). The points in this trapezoid are the points of set B that ℓ will hit first as it rotates around z. See Figure 5(a).

Because our new data structure is based on the deletions-only hull, we review the relevant properties of that data structure. Given a set of n points in the plane, the deletions-only structure represents the convex hull of an *active* subset of the points. All the points start out in the active state; they are inactivated by the

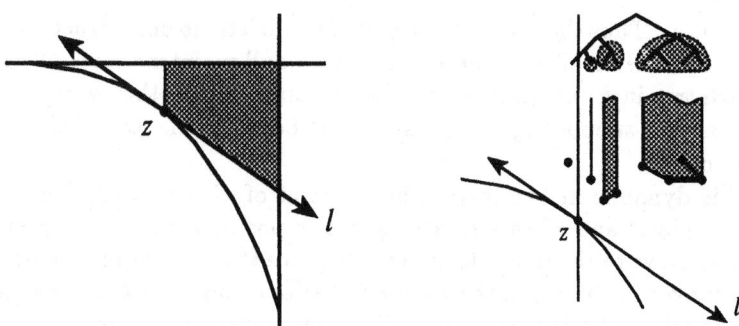

Fig. 5. The restricted-insertion data structure (a) stores the points in the shaded trapezoid and (b) maintains the convex hulls of maximal-subtree subsets.

client of the data structure in an arbitrary order. The data structure supports tangent queries in $O(\log n)$ time apiece and point deletions (inactivations) in $O(\log n)$ amortized time apiece.

The data structure is based on a complete binary tree T that stores all the points (in our case, all the points in quadrant III) at the leaves in x-coordinate order. Each node of T represents the lower convex hull of the active points in its subtree as a linked list. The lower convex hull at a node is the concatenation of two sublists, one from each child node, linked by the lower common tangent of the two child hulls. When a point becomes inactive, the hull is recomputed at its leaf and all of its ancestor nodes. Each active point has a level: $level(s_i)$ is the highest level in T at which s_i appears on a convex hull. The level of an active point increases monotonically. When a point is inactivated (deleted), recomputing convex hulls takes time proportional to the number of point-level changes. Thus the total work of maintaining the convex hull is $O(n \log n)$.

We now modify this deletions-only structure to allow insertions of a restricted sort. The key ideas are the following:

1. Instead of maintaining a single lower convex hull, we maintain hulls for a forest F with $O(\log n)$ connected components, namely the set of maximal subtrees of T to the right of z. The leaves of F correspond to the points of S in quadrant III whose x-coordinates lie between z and the initial position of ℓ. For any active point s_i in this interval, the maximum value of $level(s_i)$ is the height of its containing tree in F. See Figure 5(b).

2. We insert new active points at the left end of the forest F. When the vertical line through z passes a point s_i of quadrant III, above ℓ, F changes: the nodes of T whose leftmost descendant is s_i become nodes in F. We then compute lower convex hulls at these new nodes of F.

   New hulls can be computed in time proportional to the number of point-level changes (see [5]). The basic idea is to compute the common tangents between sub-hulls by marching along their lists, traversing only points whose level is about to increase. A key observation is that $level(s_j)$ does not decrease for any point s_j. Thus the central invariant of the deletions-only structure is preserved, and the total time needed for convex hull maintenance is $O(n \log n)$.

3. For each tree in F, we maintain its closest point to ℓ, in other words, the hull vertex with a supporting line parallel to ℓ. Let E be the set of all $O(\log n)$ closest points.

   Set E is dynamic in two ways: when a point of E is deleted, a new point takes its place; and when ℓ rotates to a new position, the closest point of a tree may move to the right along the hull. Since there are at most n deletions and n moves to the right, the total number of changes to E during the 90° sweep is $O(n)$. We now show that these two types of changes can each be handled in $O(\log n)$ amortized time.

   When a point $s_j \in E$ is deleted (inactivated), a new closest point must be computed. Let s_i and s_k be the neighbors of s_j on its convex hull. After s_j is deleted, the new closest point lies on the new hull between s_i and s_k. The points strictly between s_i and s_k are those whose level has increased, so the cost of finding the new extremum is dominated by the cost of recomputing the convex hull.

   When ℓ rotates to a new position ℓ', we identify the point $s_i \in E$ that has the minimum-slope (most vertical) edge to its right. There are only $O(\log n)$ candidates, so this step clearly takes only $O(\log n)$ time. If the slope to s_i's right is greater than that of ℓ', then E does not change. Otherwise we replace s_i in E by its right neighbor and repeat the process until we reach the slope of ℓ'.

The Complete Algorithm. We now describe how to use these data structures to implement one step of the 90° sweep. Of course, the complete algorithm consists of four 90° sweeps. We use deletions-only convex hull data structures for the points in quadrants II, IV, and III(A) and a restricted-insertion convex hull data structure for III(B). One step has the following substeps:

1. Find the first point s_i in II, IV, or III(A) that is hit by ℓ as it rotates around z. Let ℓ' be the line through z and s_i.

2. Rotate ℓ toward ℓ' while maintaining E. The next change in E occurs when ℓ's slope equals that of the minimum slope to the right of a point in E. Before making the change to E, check whether ℓ would cross over any point of the current set E before reaching this slope. If not, change E and continue rotating ℓ. If so, reset s_i to be the point of E that ℓ hits and and stop rotating ℓ at the slope determined by z and s_i.

3. Delete s_i from its data structure, move it from h^+ to h^- or vice versa, and recompute z. Update the forest F in the restricted-insertion data structure to include all the points in III whose x-coordinates lie between the old and the new positions of z.

We have now proved the following result.

Theorem 8. *The planar, unconstrained centroid polytope $C(S)$ can be computed in $O(n \log n)$ time.* □

6   Conclusions

In this paper, we have characterized and given algorithms for centroid polytopes. Centroid polytopes simultaneously generalize the centroid and the convex hull of a point set.

We can generalize the centroid polytope to allow positional uncertainty as well as weight uncertainty. Suppose we do not know the exact locations of the points of S, but just bounding boxes (or other zonotopes of fixed complexity). This change adds more segments to the zonotope Z, but does not affect our complexity bounds.

Yet another interpretation of the centroid polytope $C_{L,H}(S)$ in $\mathbb{R}^d$ involves the following system of homogeneous linear equations:

$$\begin{pmatrix} x_{11} & x_{12} & \cdots & x_{n1} & z_1 \\ x_{12} & x_{22} & \cdots & x_{n2} & z_2 \\ \vdots & \vdots & & \vdots & \vdots \\ x_{1d} & x_{2d} & \cdots & x_{nd} & z_d \\ 1 & 1 & \cdots & 1 & 1 \end{pmatrix} \begin{pmatrix} w_1 \\ w_2 \\ \vdots \\ w_n \\ -W \end{pmatrix} = \begin{pmatrix} 0 \\ 0 \\ \vdots \\ 0 \\ 0 \end{pmatrix}$$

Now if each $s_i = (x_{i1}, x_{i2}, \ldots, x_{id})$, then the centroid polytope is exactly the set of $z = (z_1, z_2, \ldots, z_d)$ that leaves a solvable system under the linear constraints $l_i \le w_i \le h_i$ and $L \le W \le H$. With this non-geometric formulation, it makes sense to generalize centroid polytopes to allow negative weights. If the total weight of S can equal zero, the centroid polytope becomes unbounded. In this case, the centroid polytope is best computed by dividing it into pieces containing centroids of positive and negative total weight.

References

1. P.K. Agarwal, D. Eppstein, and J. Matoušek. Dynamic half-space reporting, geometric optimization, and minimum spanning trees. In *Proc. 33rd IEEE Foundations of Computer Science*, Pittsburgh, 1992, 80–89.
2. B. Chazelle. On the convex layers of a planar set. *IEEE Trans. Inf. Theory*, IT-31 (1985), 509–517.
3. H. Edelsbrunner. *Algorithms in Combinatorial Geometry*, Springer-Verlag, 1987.
4. H. Edelsbrunner and E.P. Mücke. Simulation of simplicity, a technique to cope with degenerate cases in geometric computations. *ACM Trans. Graphics* 9 (1990) 66–104.
5. J. Hershberger and S. Suri. Applications of a semi-dynamic convex hull algorithm. *BIT* 31 (1992), 249–267.
6. M. Overmars and J. van Leeuwen. Maintenance of configurations in the plane. *J. Comput. Syst. Sci.* 23 (1981), 166–204.
7. E. Szemerédi and W.T. Trotter. Extremal problems in discrete geometry. *Combinatorica* 3 (1983), 381–392.
8. J. Tsao, J. Wolter, and H. Wang. Model-based understanding of uncertain observational data for oil spill tracking. In *Proc. 3rd Int. Conf. on Industrial Fuzzy Control & Intelligent Systems*, Houston, 1993, 149–154.

0/1–Integer Programming:
Optimization and Augmentation are Equivalent

Andreas S. Schulz[1]*, Robert Weismantel[2], and Günter M. Ziegler[3]**

[1] Technische Universität Berlin, Fachbereich Mathematik (MA 6–1),
Straße des 17. Juni 136, D–10623 Berlin, Germany,
schulz@math.tu–berlin.de
[2] Konrad–Zuse–Zentrum für Informationstechnik Berlin,
Heilbronner Straße 10, D–10711 Berlin, Germany,
weismantel@zib–berlin.de
[3] Technische Universität Berlin, Fachbereich Mathematik (MA 6–1),
Straße des 17. Juni 136, D–10623 Berlin, Germany,
ziegler@math.tu–berlin.de

Abstract. For every family of sets $\mathcal{F} \subseteq \{0,1\}^n$ the following problems are strongly polynomial time equivalent: given a feasible point $x^0 \in \mathcal{F}$ and a linear objective function $c \in \mathbb{Z}^n$,

- find a feasible point $x^* \in \mathcal{F}$ that maximizes $c\,x$ (Optimization),
- find a feasible point $x^{\text{new}} \in \mathcal{F}$ with $c\,x^{\text{new}} > c\,x^0$ (Augmentation), and
- find a feasible point $x^{\text{new}} \in \mathcal{F}$ with $c\,x^{\text{new}} > c\,x^0$ such that $x^{\text{new}} - x^0$ is "irreducible" (Irreducible Augmentation).

This generalizes results and techniques that are well known for 0/1–integer programming problems that arise from various classes of combinatorial optimization problems.

1 Introduction

For every family of sets $\mathcal{F} \subseteq \{0,1\}^n$ of *feasible* 0/1–points, we show that optimization and (irreducible) augmentation with respect to linear objective functions are strongly polynomial time equivalent. For example, $\mathcal{F}$ can be given as $\mathcal{F} = \{x \in \{0,1\}^n : A\,x \le b\}$ for a matrix $A \in \mathbb{Z}^{m \times n}$ and a vector $b \in \mathbb{Z}^m$, but such an explicit representation need not be given for the following. However, we assume throughout that some *feasible* solution $x^0 \in \mathcal{F}$ is known in advance. The optimization problem for $\mathcal{F}$ is the following task.

> **The Optimization Problem (OPT)**
> Given a vector $c \in \mathbb{Z}^n$ and a point $x^0 \in \mathcal{F}$, find a point $x^* \in \mathcal{F}$ that maximizes $c\,x$ on $\mathcal{F}$.

* Andreas S. Schulz has been supported by the graduate school "Algorithmische Diskrete Mathematik". The graduate school "Algorithmische Diskrete Mathematik" is supported by the Deutsche Forschungsgemeinschaft (DFG), grant We 1265/2–1.
** Günter M. Ziegler acknowledges support by a DFG Gerhard–Hess–Forschungs-förderungspreis.

Many combinatorial optimization problems, such as the maximum weight independent set and the traveling salesman problem as well as unit capacity network flow and matching problems, can be described in this form. One can also model linear optimization problems for integral points $x \in \mathbb{Z}^n$ in this way, as long as there are bounds $0 \leq x_j \leq u_j$ for the variables $x \in \mathbb{Z}^n$ that are polynomially bounded in the input size. For this, just replace x_j by $x_{j1} + x_{j2} + \cdots + x_{ju_j}$, where the x_{ji} are 0/1–variables.

Several known polynomial time algorithms for solving special 0/1–optimization problems are of primal nature. That is, given a feasible solution, i. e., a point $x^0 \in \mathcal{F}$, one successively produces new feasible solutions $x^1, x^2, \ldots$ with $cx^0 < cx^1 < cx^2 < \cdots$ until an optimal solution is reached. From an abstract point of view an *augmentation problem* is solved in each iteration: for given x^k find an *augmentation vector* z such that $cz > 0$, and $x^k + z$ is feasible.

The Augmentation Problem (AUG)
Given a vector $c \in \mathbb{Z}^n$ and a point $x^{\text{old}} \in \mathcal{F}$, find a point $x^{\text{new}} \in \mathcal{F}$ such that $cx^{\text{new}} > cx^{\text{old}}$, or assert that no such x^{new} exists.

Most primal algorithms, however, need to make a special choice of the augmentation vector $x^{\text{new}} - x^{\text{old}}$ in each iteration, in order to come up with a polynomial number of overall iterations. In cycle–canceling algorithms for the min cost flow problem, for instance (see, e.g., [AMO93]), in each iteration flow has to be augmented along a negative cycle with minimum (mean) cost.

We show that (OPT) can be solved by calling a strongly polynomial number of times an oracle that solves (AUG). This has also been shown by Grötschel and Lovász [GL]. Notice that this holds for arbitrary 0/1–programs and that it is sufficient to be able to find, for any integral objective function, an arbitrary augmentation vector: we do not need a "best" one in any respect.

Observe that there is another difference to cycle–canceling algorithms. In these algorithms one restricts to simple cycles (irreducible augmentation vectors), while in (AUG) arbitrary augmentation vectors are allowed for. By supp(x) we denote the support of a vector x, i. e., supp$(x) = \{j : x_j \neq 0\}$. We call an augmentation vector z *reducible* if there exist two vectors v, w such that

- $v + w = z$,
- $v, w \neq 0$,
- supp$(v) \subseteq$ supp(z),
- $x^{\text{old}} + v, x^{\text{old}} + w \in \mathcal{F}$.

Observe that this implies that supp$(v) \subset$ supp(z) and supp$(w) \subset$ supp(z). In this situation v or w is a "smaller" augmentation vector that can be applied instead of z at the point x^{old}. In case z is not reducible, it is called *irreducible*.

The Irreducible Augmentation Problem (IRR–AUG)
Given a vector $c \in \mathbb{Z}^n$ and a point $x^{\text{old}} \in \mathcal{F}$, find a point $x^{\text{new}} \in \mathcal{F}$ such that $x^{\text{new}} - x^{\text{old}}$ is irreducible and $cx^{\text{new}} > cx^{\text{old}}$, or assert that no such x^{new} exists.

The restriction to irreducible augmentation vectors does not affect the existence of a strongly oracle–polynomial time algorithm solving (OPT) when $\mathcal{F}$ is given by an oracle for (IRR–AUG). In fact, whereas (AUG) and (IRR–AUG) have the same input, each solution of (IRR–AUG) is a solution of (AUG), but not vice versa. It is by no means trivial to solve the irreducible augmentation problem given an oracle for solving (OPT), or, equivalently, (AUG).

Theorem 1. *Any one of the following three problems:*

- *optimization* (OPT),
- *augmentation* (AUG),
- *irreducible augmentation* (IRR–AUG),

can be solved in strongly oracle–polynomial time for any set $\mathcal{F} \subseteq \{0,1\}^n$ given by an oracle for any of the other two problems.

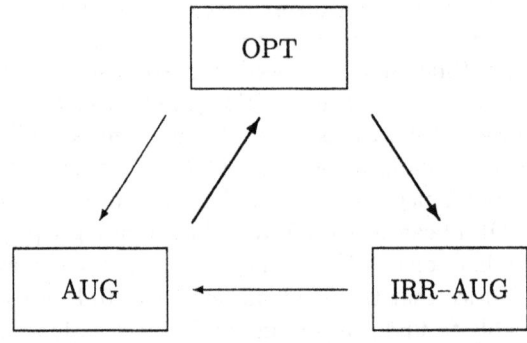

Figure 1

Figure 1 indicates the trivial relations between the three problems by thin arrows. Here, an arrow means that the problem at the head of the arrow can be solved in strongly oracle–polynomial time given an oracle for the problem at the tail of the arrow. The two thick arrows represent our main results that are presented in Sections 2 and 3, respectively. In both cases we first derive a polynomial time algorithm, and then use the "preprocessing algorithm" of Frank and Tardos [FT87] to turn it into a strongly polynomial procedure. The two missing arrows follow then from transitivity.

For a thorough introduction to oracles and oracle–polynomial time algorithms as well as strongly polynomial time algorithms we refer to Grötschel, Lovász, and Schrijver [GLS88], see also Lovász [Lov86].

2 An Oracle–Polynomial Time Algorithm for (OPT)

In this section we state and analyze an oracle–polynomial time algorithm that solves (OPT), assuming an oracle for (AUG) is given. We first concentrate on nonnegative objective function vectors but shall conclude with a discussion of the transformation needed to allow arbitrary objectives. Finally, we point out how to turn this into a strongly polynomial time algorithm.

The essential idea underlying the algorithm is *bit-scaling* (see [EK72]). Thus we have to represent data by binary numbers. For $\alpha \in \mathbb{N}$ and a given number K that is at least as big as the number of bits needed to encode α, i. e., $K \geq \lceil \log(\alpha + 1) \rceil$, we represent α as a K-bit binary number, adding leading zeros if necessary. We denote by $\alpha(k)$ the number obtained by considering the k leading bits only. With ${}^k\alpha$ we refer to the k-th bit of α. Thus, $\alpha(k) = ({}^1\alpha, {}^2\alpha, \ldots, {}^k\alpha) = \sum_{i=1}^{k} {}^i\alpha \, 2^{k-i}$, and $\alpha(K) = \alpha$.

Scaling methods have extensively been used to derive polynomial time algorithms for a wide variety of network and combinatorial optimization problems (see, e.g., [AMO93]). In this section we use bit-scaling of costs to derive an oracle-polynomial time algorithm for optimizing 0/1-integer programs. This technique has been used earlier by Röck [Rö80] and Gabow [Gab85] for solving minimum cost flow and shortest path problems, respectively.

As already mentioned we assume first that all coefficients of the objective function vector are nonnegative.

The Optimization Problem (OPT)$^{\geq 0}$
Given a vector $c \in \mathbb{N}^n$ and a point $x^0 \in \mathcal{F}$, find a point $x^* \in \mathcal{F}$ that maximizes cx on $\mathcal{F}$.

Restricting the optimization problem in this way, it seems to be reasonable to do the same with the augmentation problem. That is, we restrict it to nonnegative input vectors c, too.

The Augmentation Problem (AUG)$^{\geq 0}$
Given a vector $c \in \mathbb{N}^n$ and a point $x^{\text{old}} \in \mathcal{F}$, find a point $x^{\text{new}} \in \mathcal{F}$ such that $c\,x^{\text{new}} > c\,x^{\text{old}}$, or assert that no such x^{new} exists.

Theorem 2. *There exists an algorithm that solves* (OPT)$^{\geq 0}$ *by* $\mathcal{O}(n \log C)$ *calls of an oracle that solves* (AUG)$^{\geq 0}$, *for* $C = \max\{c_j : j = 1, \ldots, n\} + 1$.

Proof. Let $K = \lceil \log C \rceil$. We present a bit-scaling algorithm solving a sequence of problems $(P_1), (P_2), \ldots, (P_K)$. The objective of (P_1) consists of the most significant bit only, the one of (P_2) of the first two most significant bits, and so on. Problem (P_K) will be the original problem to be solved.

For $k = 1, \ldots, K$ we define (P_k) as

$$\max \ c(k)\,x$$
$$\text{s. t. } \ x \in \mathcal{F}$$

Here, $c(k) \in \mathbb{N}^n$ denotes the vector obtained from c by restricting each component to the k leading bits, $c(k) = (c_1(k), \ldots, c_n(k))$.

Bit–Scaling Algorithm for Solving $(\text{OPT})^{\geq 0}$

      1. let x^0 be a feasible solution;

      2. $k := 1$;

      3. while $k \leq K$ do

      4.   solve (P_k) by iterated use of

            $(\text{AUG})^{\geq 0}$, with initial solution x^{k-1};

      5.   let x^k be the optimal solution of (P_k);

      6. $k := k + 1$;

      7. end.

Step 4 of this algorithm needs further explanation. The general idea is to start with a feasible solution y^0, say, to call the augmentation oracle with y^0, and to obtain a better solution y^1 that serves as the new input, and so forth. Since (P_k) is bounded, the procedure is finite. To keep the number of these inner iterations small it is important to use x^{k-1} as the starting solution.

Since (P_K) coincides with the original optimization problem the algorithm is correct. It remains to be shown that the number of calls of $(\text{AUG})^{\geq 0}$ in Step 4 to determine x^k starting from x^{k-1} is polynomially bounded in the input size. The following calculation shows that in Step 4 for each problem (P_k) the oracle for $(\text{AUG})^{\geq 0}$ is called at most n times:

$$c(k)\,(x^k - x^{k-1}) = 2\,c(k-1)\,(x^k - x^{k-1}) + {}^k c\,(x^k - x^{k-1}) \leq 0 + n.$$

The inequality follows from the optimality of x^{k-1} for (P_{k-1}), from ${}^k c \in \{0,1\}^n$, and from $x^k, x^{k-1} \in \{0,1\}^n$. (We define $c(0)$ to be zero.) $\square$

The last obstacle on our way to an oracle–polynomial time algorithm for $0/1$–programming is to get rid of the assumption on the objective function vectors. We made this nonnegativity assumption in order to simplify the bit–scaling. We shall use a standard transformation from (OPT) to $(\text{OPT})^{\geq 0}$ as well as from $(\text{AUG})^{\geq 0}$ to (AUG). Given an instance of (OPT),

$$\max\ cx$$
$$\text{s. t. } x \in \mathcal{F}$$

with $c \in \mathbb{Z}^n$, we define an instance of $(\text{OPT})^{\geq 0}$ as follows. Let $\tilde{c} \in \mathbb{N}^n$ be the vector with coefficients

$$\tilde{c}_j = \begin{cases} c_j, & \text{if } c_j \geq 0, \\ -c_j, & \text{otherwise,} \end{cases}$$

and let for $x \in \mathcal{F}$

$$\tilde{x}_j = \begin{cases} x_j, & \text{if } c_j \geq 0, \\ 1 - x_j, & \text{otherwise.} \end{cases}$$

With $\tilde{\mathcal{F}} := \{\tilde{x} : x \in \mathcal{F}\}$ the following defines an instance of $(\text{OPT})^{\geq 0}$:

$$\max \ \tilde{c}\,\tilde{x}$$
$$\text{s. t.} \ \tilde{x} \in \tilde{\mathcal{F}}$$

Then $x \in \mathcal{F}$ is optimal with respect to c if and only if $\tilde{x} \in \tilde{\mathcal{F}}$ is optimal with respect to $\tilde{c}$.

From the discussion above, we know that $(\text{OPT})^{\geq 0}$ can be solved in polynomial time assuming an oracle for solving $(\text{AUG})^{\geq 0}$ is given. Since $(\text{AUG})^{\geq 0}$ for $\tilde{\mathcal{F}}$ can be solved by calling the (AUG) oracle for $\mathcal{F}$, we obtain a polynomial time algorithm for optimization in terms of an augmentation oracle, as follows.

Corollary 3. *There exists an algorithm that solves* (OPT) *by* $\mathcal{O}(n \log C)$ *calls of an oracle that solves* (AUG), *for* $C = \max\{\,|c_j| : j = 1, \ldots, n\} + 1$.

Using the "preprocessing algorithm" of Frank and Tardos [FT87, Theorem 3.3] that relies on simultaneous Diophantine approximation, we turn this into a strongly polynomial algorithm. The input and output of their algorithm is as follows.

INPUT: $c \in \mathbb{Q}^n$, $N \in \mathbb{N}$

OUTPUT: $\bar{c} \in \mathbb{Z}^n$ such that $\overline{C} \leq 2^{4n^3} N^{n(n+2)}$,

and $\text{sign}(c\,z) = \text{sign}(\bar{c}\,z)$ for all $z \in \mathbb{Z}^n$ with $|z|_1 \leq N - 1$

Here $|z|_1$ denotes the L_1-norm of z. By taking the quality parameter N to be $N = n+1$, we replace the original objective function c by a new integral objective function $\bar{c}$ whose size is polynomially bounded in n, such that a point $x \in \mathcal{F}$ is optimal with respect to c if and only if it is optimal with respect to $\bar{c}$. Then the algorithm just described is run for the new objective function $\bar{c}$.

This completes the proof of the implication $(\text{AUG}) \longrightarrow (\text{OPT})$ of Theorem 1. In particular, there exists an algorithm that solves (OPT) by $\mathcal{O}(n^4)$ calls of an oracle for (AUG), and by performing $\mathcal{O}(n^8)$ additional arithmetic steps to compute $\bar{c}$.

3 An Oracle–Polynomial Algorithm for (IRR–AUG)

This section is devoted to showing that (IRR–AUG) can be solved in (strongly) polynomial time, assuming an oracle that solves (OPT) is given. We divide this problem into three parts. First, we show how to determine a maximum mean augmentation vector, i. e., an augmentation vector with the maximum ratio of improvement to cardinality of support. Since there exists an irreducible augmentation vector attaining this optimum value, we shall then provide an algorithm to determine such an augmentation vector. Finally, we use again the preprocessing algorithm of Frank and Tardos to turn this into a strongly polynomial time algorithm.

One remark shall be given in advance. In the formulation of (OPT) we assumed the objective function vector c to be integer valued since (starting from rational numbers) this can always be achieved by appropriate scaling. In the following we will construct objective functions that are not integer valued. This is only done for simplifying the presentation. In these cases we always assume the scaling to be performed implicitly. One should observe, however, that this scaling can indeed be done without affecting the magnitude of the size of the objective function.

3.1 The Maximum Mean Augmentation Problem

One step towards solving the irreducible augmentation problem via a polynomial number of calls of the optimization oracle is to find a maximum mean augmentation vector. This question is addressed in this section. The technique that we use is quite similar to the one used for the *minimum cost–to–time ratio cycle problem* (see [AMO93, pp. 150–152]).

The Maximum Mean Augmentation Problem (MMA)
Given a vector $c \in \mathbb{Z}^n$ and a point $x^{\text{old}} \in \mathcal{F}$, find a point $x^{\text{new}} \in \mathcal{F}$ such that $c\,x^{\text{new}} > c\,x^{\text{old}}$ and x^{new} maximizes $\frac{c\,x - c\,x^{\text{old}}}{|x - x^{\text{old}}|_1}$ over $\mathcal{F} \setminus \{x^{\text{old}}\}$, or assert that no such x^{new} exists.

Here, $|z|_1$ denotes again the L_1–norm of z, which coincides with the cardinality of the support of an augmentation vector. Throughout this section we assume that $c \in \mathbb{Z}^n$ and $x^{\text{old}} \in \mathcal{F}$ are given. By S^{old} we denote the support of the vector x^{old}, i. e., $S^{\text{old}} := \{j : x^{\text{old}}_j = 1\}$.

In order to find a maximum mean augmentation vector we proceed as follows. We first call (OPT) with the linear functional c. In case that x^{old} is optimal, there does not exist an augmentation vector. Thus, in the following we assume that x^{old} is not optimal with respect to c. Let μ^* denote the optimal objective function value of (MMA). For any arbitrary value $0 < \mu \le C$ we define an objective function c^μ as follows:

$$c^\mu_j := \begin{cases} c_j + \mu, & \text{if } j \in S^{\text{old}}, \\ c_j - \mu, & \text{otherwise.} \end{cases}$$

Calling (OPT) with c^μ as input, we distinguish three different situations. Let x^μ denote the output of (OPT).

Case 1. $x^\mu = x^{\text{old}}$.
  In this case, $c^\mu x \le c^\mu x^{\text{old}}$ for every $x \in \mathcal{F} \setminus \{x^{\text{old}}\}$. Alternatively,

$$\frac{c\,(x - x^{\text{old}})}{|x - x^{\text{old}}|_1} \le \mu \ .$$

Therefore, μ is an upper bound on μ^*.

Case 2. $x^\mu \neq x^{\text{old}}$ and $c^\mu x^\mu = c^\mu x^{\text{old}}$.

As in the previous case we obtain

$$\frac{c\,(x - x^{\text{old}})}{|x - x^{\text{old}}|_1} \leq \mu, \quad \text{for all } x \in \mathcal{F} \setminus \{x^{\text{old}}\}.$$

Since x^μ satisfies this inequality with equality, $\mu = \mu^*$.

Case 3. $x^\mu \neq x^{\text{old}}$ and $c^\mu x^\mu > c^\mu x^{\text{old}}$.

In this case,

$$\mu^* \geq \frac{c\,(x^\mu - x^{\text{old}})}{|x^\mu - x^{\text{old}}|_1} > \mu \ .$$

Consequently, μ is a strict lower bound on μ^*.

Based on the preceding case analysis we can determine μ^* by binary search. Observe that μ^* lies in the interval $(0, C]$. Thus, we start with the lower bound $\underline{\mu} = 0$ and the upper bound $\overline{\mu} = C$. At every iteration we consider $\mu = (\underline{\mu} + \overline{\mu})/2$ and call (OPT) with input c^μ. If we encounter Case 1 or 3 we reset $\overline{\mu} = \mu$ and $\underline{\mu} = \mu$, respectively, whereas in Case 2 we are done. At every iteration, we halve the length of the search interval. Given any two augmentation vectors with distinct ratios the absolute difference between their ratios is at least $1/n^2$. Hence, if $\overline{\mu} - \underline{\mu} < 1/n^2$, the current interval $[\underline{\mu}, \overline{\mu}]$ contains at most one ratio of the form $\frac{cz}{|z|_1}$ where z is an augmentation vector. Consequently, calling (OPT) again with $\underline{\mu}$ we either obtain $\mu^* = \underline{\mu}$ but $x^{\mu^*} = x^{\text{old}}$ (Case 1), or we obtain $\mu^* = \underline{\mu}$ and x^{μ^*} immediately (Case 2), or, in Case 3, we obtain x^{μ^*} from which we can also compute μ^*. In the first case, we still have to determine $x^{\text{new}} \neq x^{\text{old}}$ such that

$$\frac{c\,(x^{\text{new}} - x^{\text{old}})}{|x^{\text{new}} - x^{\text{old}}|_1} = \mu^* \ .$$

This can be done as follows. Set $M := 2nC + 1$. For every $i \in \{1, \ldots, n\}$, we define an objective function d^i by setting

$$d^i_j := \begin{cases} c^{\mu^*}_j + (-1)^k M, & \text{if } i = j, \\ c^{\mu^*}_j, & \text{otherwise}, \end{cases}$$

with $k = 1$ if $i \in S^{\text{old}}$, and $k = 2$, otherwise. Then we call (OPT) with objective function d^i, $i = 1, \ldots, n$, and denote by y^i the point that is returned by (OPT). Among all these points y^i, $i \in \{1, \ldots, n\}$, there is one y^*, say, such that $y^* \neq x^{\text{old}}$ and $\frac{cy^* - c x^{\text{old}}}{|y^* - x^{\text{old}}|_1} = \mu^*$, because there exists $y \neq x^{\text{old}}$ such that $\frac{cy - c x^{\text{old}}}{|y - x^{\text{old}}|_1} = \mu^*$, and $y \neq x^{\text{old}}$ implies that there exists an index i^* with $y_{i^*} \neq x^{\text{old}}_{i^*}$.

Corollary 4. *There exists an algorithm that solves* (MMA) *by* $\mathcal{O}(n + \log nC)$ *calls of an oracle that solves* (OPT).

3.2 Determining an Irreducible Augmentation Vector

In the preceding section we showed how to determine a maximum mean augmentation vector, assuming an optimization oracle is given. We shall now show that there exists an irreducible augmentation vector sharing this property. This will enable us to determine such a vector. This completes the proof that (IRR–AUG) can be solved by calling an oracle for (OPT) a polynomial number of times. We continue using the notation introduced in the previous section.

Proposition 5. *Assume that $x \in \mathcal{F} \setminus \{x^{\text{old}}\}$ is a point such that $\frac{c\,x - c\,x^{\text{old}}}{|x - x^{\text{old}}|_1} = \mu^*$. If $z := x - x^{\text{old}}$ is reducible, $z = v + w$, say, then $\frac{c\,v}{|v|_1} = \frac{c\,w}{|w|_1} = \mu^*$.*

Proof. Since z is reducible by v and w, we obtain $|v + w|_1 = |v|_1 + |w|_1$. Together with $\frac{c\,(v+w)}{|v+w|_1} \geq \frac{c\,v}{|v|_1}$ and $\frac{c\,(v+w)}{|v+w|_1} \geq \frac{c\,w}{|w|_1}$, this implies $\frac{c\,v}{|v|_1} = \frac{c\,w}{|w|_1}$. We conclude that $\frac{c\,(v+w)}{|v+w|_1} = \frac{c\,v}{|v|_1} = \frac{c\,w}{|w|_1}$. $\square$

We use Proposition 5 to compute an irreducible maximum mean augmentation vector: if z is reducible by v and w, the support of both v and w is properly contained in the support of z. We exploit this by appropriate perturbation of the objective function.

Let x^{μ^*} still denote the output of (MMA), and let $z := x^{\mu^*} - x^{\text{old}}$ be the associated augmentation vector. Observe that there always exists an index $i^* \in \{1, \ldots, n\}$ such that $z_{i^*} = (-1)^k$ for some $k \in \{1, 2\}$, and x^{μ^*} is optimal with respect to the objective function d^{i^*} whereas x^{old} is not. If z is reducible, $z = v + w$, we may assume that $v_{i^*} = (-1)^k$. Therefore, Proposition 5 implies that $x^{\text{old}} + v$ is also optimal with respect to d^{i^*}. We now describe the appropriate perturbation of d^{i^*} in order to end up (by calling (OPT) with the perturbed function) with an irreducible augmentation vector. For two points $y, y' \in \mathcal{F}$ that have two different objective function values the difference $|d^{i^*} y - d^{i^*} y'|$ is at least $\frac{1}{n}$ because μ^* has denominator at most n. Setting $\epsilon := \frac{1}{1 + 2n^2}$, we define the perturbed objective function $\tilde{d}^{i^*}$ as follows:

$$\tilde{d}^{i^*}_j := \begin{cases} d^{i^*}_j + \epsilon & \text{if } j \in S^{\text{old}}, \\ d^{i^*}_j - \epsilon & \text{if } j \notin S^{\text{old}}. \end{cases}$$

For every point $y \in \mathcal{F}$ we have

$$\tilde{d}^{i^*} y = d^{i^*} y + \epsilon \left(|S^{\text{old}}| - |\{j \in S^{\text{old}} : y_j = 0\}| - |\{j \notin S^{\text{old}} : y_j = 1\}| \right)$$
$$= d^{i^*} y + \epsilon |S^{\text{old}}| - \epsilon |y - x^{\text{old}}|_1$$
$$= c^{\mu^*} y + (-1)^k M y_{i^*} + \epsilon |S^{\text{old}}| - \epsilon |y - x^{\text{old}}|_1 \ .$$

Now, observe that by the choice of ϵ every point $y \in \mathcal{F}$ that is optimal with respect to $\tilde{d}^{i^*}$ is also optimal with respect to d^{i^*}. Therefore, (OPT) will return a vector $x^{\text{old}} + v$ with $\frac{c\,v}{|v|_1} = \mu^*$ and $v_{i^*} = (-1)^k$. The vector v is irreducible, for if not, then there would exist a vector w such that $x^{\text{old}} + w \in \mathcal{F}$, $\frac{c\,w}{|w|_1} = \mu^*$, $w_{i^*} = (-1)^k$, and $|w|_1 < |v|_1$. The latter fact would imply that $\tilde{d}^{i^*} w > \tilde{d}^{i^*} v$, a contradiction.

The following theorem summarizes what we have obtained so far.

Theorem 6. *There exists an algorithm that solves* (IRR–AUG) *by* $\mathcal{O}(n+\log nC)$ *calls of an oracle that solves* (OPT).

Again, using the preprocessing algorithm of Frank and Tardos [FT87], we turn this into a strongly polynomial time algorithm.

This time, the "quality" of the preprocessing is chosen to be good enough to guarantee that $\text{sign}(cv) = \text{sign}(\bar{c}v)$ holds for all vectors $v \in \mathbb{Z}^n$ with $|v_j| \leq n$. That is we take $N = n^2 + 1$. This implies that not only a vector $v \in \{+1, 0, -1\}^n$ is an augmentation vector with respect to the original objective function c if and only if it is an augmentation vector with respect to $\bar{c}$, but also that the maximum mean augmentation vectors with respect to c coincide with those for $\bar{c}$. Namely, x provides a better mean augmentation than x' if and only if

$$c\left((x - x^{\text{old}}) \, |x' - x^{\text{old}}|_1\right) \;>\; c\left((x' - x^{\text{old}}) \, |x - x^{\text{old}}|_1\right),$$

and by Frank and Tardos this is equivalent to

$$\bar{c}\left((x - x^{\text{old}}) \, |x' - x^{\text{old}}|_1\right) \;>\; \bar{c}\left((x' - x^{\text{old}}) \, |x - x^{\text{old}}|_1\right).$$

Now we can run the polynomial time algorithm described above on the objective function $\bar{c}$, and obtain a strongly polynomial time procedure for the implication (OPT)$\longrightarrow$(IRR–AUG) of Theorem 1. The output is, in particular, a maximum mean augmentation vector with respect to the original objective function c.

4 Concluding Remarks

It is clearly an interesting question to what extent the results of this paper can be extended to general integer programs. In this case we have established that the optimization problem (OPT) and the maximum mean augmentation problem (MMA) are equivalent, and that an (MMA) oracle can be used to solve the irreducible augmentation problem (IRR-AUG) in polynomial time. It is not clear to us whether an augmentation oracle (AUG) can be used to simulate any of the others.

For the 0/1–case we have here shown the equivalence between the optimization and the augmentation problem with respect to strongly polynomial time solvability. The algorithm presented for solving (OPT) given an oracle for (AUG) does not only generalize some known algorithms for special combinatorial problems, but is also a new one for treating some of them. Its essence is that a problem is (strongly) polynomially tractable if and only if, for every objective function and every nonoptimal feasible solution, we can find an improving solution in (strongly) polynomial time.

On the other hand, we cannot expect to solve (AUG) in polynomial time if the corresponding optimization problem is $\mathcal{NP}$–hard, unless $\mathcal{P}=\mathcal{NP}$. We may hope however, to be able to solve (AUG) efficiently if we restrict ourselves to special augmentation vectors. Then, of course, by the algorithm described above we will not necessarily obtain an optimal solution, but maybe a good one. The equivalence of (AUG) and (IRR–AUG) suggests to search for irreducible augmentation vectors.

Finally, let us point out the relation to test sets in integer programming (see, e.g., [Sch86]). Given an oracle that solves (AUG) we have access to a test set for the given integer program, for any objective function. Similarly, with any oracle for (IRR–AUG) we are implicitly given an *irreducible* test set. From this point of view, the results presented in this paper in particular imply that for 0/1–programming problems optimization and augmentation by use of (irreducible) test sets are equivalent in terms of computational complexity.

Acknowledgments. Thanks to Lex Schrijver for an important augmentation vector for this paper. Thanks also to Maurice Queyranne who pointed out reference [GL] to us after reading an earlier version of this paper.

References

[AMO93] Ravindra K. Ahuja, Thomas L. Magnanti, and James B. Orlin: *Network Flows: Theory, Algorithms, and Applications*, Prentice Hall, Englewood Cliffs NJ, 1993.

[EK72] Jack Edmonds and Richard M. Karp: Theoretical improvements in algorithmic efficiency for network flow problems, *Journal of the Association for Computing Machinery* 19 (1972), 248–264.

[FT87] András Frank and Éva Tardos: An application of simultaneous Diophantine approximation in combinatorial optimization, *Combinatorica* 7 (1987), 49–65.

[Gab85] Harold N. Gabow: Scaling algorithms for network problems, *Journal of Computer and System Sciences* 31 (1985), 148–168.

[GL] Martin Grötschel and László Lovász: *Combinatorial Optimization*, Chapter 28 of the Handbook on Combinatorics, R. Graham, M. Grötschel, and L. Lovász (eds.), to appear.

[GLS88] Martin Grötschel, László Lovász, and Alexander Schrijver: *Geometric Algorithms and Combinatorial Optimization*, Algorithms and Combinatorics 2, Springer, Berlin, 1988; Second edition 1993.

[Lov86] László Lovász: *An Algorithmic Theory of Numbers, Graphs and Convexity*, CBMS–NSF Regional Conference Series in Applied Mathematics 50, Society for Industrial and Applied Mathematics (SIAM), Philadelphia, 1986.

[Rö80] Hans Röck: Scaling techniques for minimal cost network flows, in: V. Page (ed.), *Discrete Structures and Algorithms*, Carl Hanser, Munich, 1980, pp. 181–191.

[Sch86] Alexander Schrijver: *Theory of Linear and Integer Programming*, John Wiley & Sons, Chichester, 1986.

The Online Transportation Problem

Bala Kalyanasundaram[*]   Kirk R. Pruhs[†]

Abstract

We study the online transportation problem under the weakened adversary model. More specifically, we assume that the adversary has only half at many servers at each site as the online algorithm. We show that the halfopt-competitive ratio for the greedy algorithm is $\Theta(\min(m, \log C))$, where m is the number of server sites, and C is the total number of servers. We then present an algorithm BALANCE, which is a simple modification of the greedy algorithm, that has a halfopt-competitive ratio of $O(1)$.

1 Introduction

We consider the natural online version of the well-known transportation problem [3, 6]. The initial setting consists of a collection $S = \{s_1, \ldots, s_m\}$ of server sites in a metric space $\mathcal{M}$. Each server site s_j has a positive integral capacity c_j. The online algorithm $\mathcal{A}$ sees over time a sequence $R = \{r_1, \ldots r_n\}$ of requests for service, with each request being a point in $\mathcal{M}$. In response to the request r_i, $\mathcal{A}$ must select a site $s_{\sigma(i)}$ to service r_i. The cost for this assignment is the distance $d(s_{\sigma(i)}, r_i)$ in the metric space between $s_{\sigma(i)}$ and r_i. Each site s_j can service at most c_j requests. The classic dilemma faced by the online algorithm $\mathcal{A}$ is that, at the time of the request r_i, $\mathcal{A}$ is not aware of the location of the future requests. The goal for the online algorithm is to minimize $\frac{1}{n}\sum_{i=1}^{n} d(s_{\sigma(i)}, r_i)$, the average cost to service the requests.

For concreteness, consider the following two examples of online transportation problems. In the *fire station problem*, the site s_j is a fire station that contains c_j fire crews. Each request is a fire that must be handled by a fire crew. The problem is to assign the crews to the fire so as to minimize the average distance traveled to get to a fire. In the *school assignment problem*, the site s_j is a school that can has a capacity of c_j students. Each request is a new student who moves into the school district. The problem is to assign the children to a school so as to minimize the average distance traveled by the children to reach their schools.

---

[*] *kalyan@cs.pitt.edu*, Computer Science Dept., University of Pittsburgh, Pittsburgh, PA 15260, Supported in part by NSF under grant CCR-9202158.

[†] *kirk@cs.pitt.edu*, Computer Science Dept., University of Pittsburgh, Pittsburgh, PA 15260, Supported in part by NSF under grant CCR-9209283.

The standard measure of "goodness" of an online algorithm is the competitive ratio. For the online transportation problem, the competitive ratio for an online algorithm $\mathcal{A}$ is the supremum over all possible instances I, of the cost incurred when the assignment is made by $\mathcal{A}$ on I divided by the minimum possible cost of any assignment for instance I. We can assume that the minimum possible assignment is computed by an offline algorithm that has prior knowledge of I. Note that the instance I specifies the metric space as well as the values of each s_j, c_j, and r_i.

In [1, 4] the online assignment problem, a special case of online transportation in which each capacity $c_i = 1$, was studied. In [1], it was shown that the competitive ratio of the intuitively appealing greedy algorithm, which assigns the nearest available server site to the new request, has a competitive ratio of $2^m - 1$. In [1, 4], it was shown that the optimal deterministic competitive ratio is $2m - 1$. The algorithm that achieves this competitive ratio requires a shortest augmenting path computation for each request. These results illustrate some criticisms of the adversarial nature of the definition of competitive ratio, namely:

1. The achievable competitive ratios often grow quickly with input size, and would seem to overly pessimistic for "normal" inputs.

2. The algorithm that achieves the optimal competitive ratio is probably unnecessarily complicated for "normal" inputs.

3. The worst-case analysis of the intuitive greedy algorithm does not reflect the fact that it may perform reasonably well on "normal" inputs.

In situations where the traditional competitive ratio is not very informative, it is important to find alternate ways to to identify what works well in practice. In this paper we adopt a modified version of competitive analysis that we call the *weak adversary model*. Generally speaking, in this model the offline algorithm is given slightly less resources than the online algorithm. The intuition is that for "normal" inputs, the performance of an **offline** algorithm will not degrade significantly if its resources are reduced slightly. Hence, if we can prove that an online algorithm is competitive against an offline algorithm with slightly less resources, then we can argue that the online algorithm will be competitive against an equivalently equipped offline algorithm on "normal" inputs. One can also view this weakening of adversary as measuring the additional resources required by the online algorithm to offset the decrease in performance due to the online nature of the problem.

In the case of the transportation problem we compare the online algorithm with c_i servers at s_i to an offline line algorithm with $a_i = c_i/2$ servers at s_i (we assume for simplicity that c_i is even). Given an instance I of the online transportation problem with $n = \sum_{i=1}^{m} c_i/2$ requests, we let I' be the same instance with each capacity c_i replaced by a_i. We then say the *halfopt-competitive ratio* of an online algorithm $\mathcal{A}$ is the supremum over all instances I with $n = \sum_{i=1}^{m} a_i$

requests of the cost incurred when the assignment is made by $\mathcal{A}$ on I divided by the minimum possible cost of any assignment for instance I'.

In this paper we present the following results. In section 3, we show that the halfopt-competitive ratio of the greedy algorithm is $\Theta(\min(m, \log C))$, where $C = \sum_{i=1}^{m} c_i$ is the sum of the capacities. If the server capacity of each site is constant, then the halfopt-competitive ratio is logarithmic in m, a significant improvement over the exponential bound on the traditional competitive ratio. In section 4, we present an online algorithm, BALANCE, that is a simple modification of the greedy algorithm. Intuitively, if there are two server sites s_1 and s_2 that are roughly equally close to a request, and if s_1 has already moved a server but s_2 hasn't, then BALANCE will move a server from s_2. We show that BALANCE has a halfopt-competitive ratio of $O(1)$. Recall that the traditional competitive ratio of every online algorithm is $\Omega(m)$.

We now summarize related results. As far as we can tell, the weakened adversary model was introduced in [7] in the context of studying paging. This model has also been used to study variants of the k-server problem, a generalization of the paging problem (see for example [9]). References to other ways that have been suggested to modify the definition of competitiveness can be found in [5]. Further ancillary results on online assignment, which are not directly related to the results in this paper, can be found in [1]. In [8] the average competitive ratio for the greedy algorithm in the online assignment problem is studied under the assumption that the metric space is the Euclidean plane and the points are uniformly distributed in a unit square. The offline transportation problem can be solved in polynomial-time[3, 6].

2 Preliminaries

In this section we introduce some definitions and concepts that are common to the remaining sections. We generally begin by assuming the simplifying condition that the online capacity c_i of each server site is two. We will think of s_i as containing two *online servers* s_i^1 and s_i^2 that move to service requests. We also think of s_i as containing one *adversary server* s_i^a. In proving upper bounds on the halfopt-competitive ratio of an algorithm we imagine that the online algorithm is playing a game against an all powerful adversary that specifies the requests and services them in the optimal way. We assume, without loss of generality, that the adversary services request r_i with s_i^a. We use $s_{\sigma(i)}$ to denote the site that the online algorithm uses to service request r_i. We define a weighted bipartite graph $\mathcal{G} = (S \cup R, E)$, which we call the *response graph*, by including an *online edge* $(r_i, s_{\sigma(i)})$, and an *adversary edge* (r_i, s_i) for each request r_i. The weight of each edge (r_i, s_j) in $\mathcal{G}$ is the distance $d(r_i, s_j)$ between the r_i and s_j in the underlying metric space. The following lemmas reveal useful facts about the response graph.

Lemma 1 Each connected component of the response graph $\mathcal{G}$ contains at most one cycle.

487

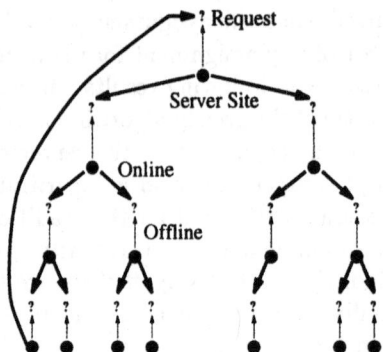

Figure 1: A Connected Component of the Response Graph

Lemma 2 For each r_i, the connected component of $\mathcal{G} - (s_{\sigma(i)}, r_i)$ that contains r_i is a tree.

In each section we will consider a tree $\mathcal{T}$ rooted at request r_i in $\mathcal{G} - (s_{\sigma(i)}, r_i)$. The tree $\mathcal{T}$ is defined differently for each algorithm, and need not include all of the connected component of $\mathcal{G} - (s_{\sigma(i)}, r_i)$. Each leaf of $\mathcal{T}$ will be a server site.

Lemma 3 For each $r_j \in \mathcal{T}$ the server site s_j is the one and only child of r_j in $\mathcal{T}$, and if r_j is not the root, then the server site $s_{\sigma(j)}$ is the parent of r_j in $\mathcal{T}$.

We denote the cumulative weight of the adversary edges in $\mathcal{T}$ by $OPT(\mathcal{T})$. Analogously, we denote the cumulative cost of the online edges that service a request in $\mathcal{T}$ by $ON(\mathcal{T})$. Note that the online edge servicing the root r_i of $\mathcal{T}$ will not be in $\mathcal{T}$. For a node in $\mathcal{T}$, we define $ld(x)$ to be the shortest path from x to any leaf in $\mathcal{T}$. If r_i is a request in $\mathcal{T}$, we define $\mathcal{T}_i$ to the the subtree of $\mathcal{T}$ rooted at r_i.

3 Analysis of Greedy

We begin with the upper bound on the competitive ratio for GREEDY. GREEDY uses the nearest available server to service each request. We first assume that the online capacity of each node is two, and then show how to extend the proof to the general case.

We now wish to divide the response graph into edge disjoint trees, $\mathcal{T}^1, \ldots, \mathcal{T}^l$. We then prove the competitive bound separately for each tree. Each tree $\mathcal{T}^k$ will satisfy the following conditions:

1. The root of $\mathcal{T}^k$ is a request $r_{\alpha(k)}$ with the property that $r_{\alpha(k)}$ is the latest request not in one of the previously constructed trees, $\mathcal{T}^1, \ldots, \mathcal{T}^{k-1}$.

2. Each nonleaf server site s_i in $\mathcal{T}^k$ has two children in $\mathcal{T}^k$.

3. Each leaf of T^k is a server site s_i such that either:

   (a) s_i has zero or one online edges incident to it in $\mathcal{G}$, or

   (b) one of the online edges incident to s_i in $\mathcal{G}$ is of the form $(s_i, r_{\alpha(j)})$ for some $j \leq k$, i.e. s_i is connected to the current root or to a previous root.

We construct the tree T^k in the following manner: Let $r_{\alpha(k)}$ be the most recent request site in $\mathcal{G}$ that has not been included in any previously constructed trees. The online edge incident to $r_{\alpha(k)}$ is then removed from $\mathcal{G}$. To construct T^k we can then perform a graph search from $r_{\alpha(k)}$ stopping at a server site if it satisfies either condition 3(a), or condition 3(b). Lemma 2 implies that no cycle will be found in this search. The edges and request vertices in T^k are then removed from G, and the procedure is repeated.

Lemma 4 For all the requests r_i in T^k and for all the leaf servers s_j in T^k, it is the case that GREEDY had a server available at the site s_j at the time of the request r_i.

Proof Sketch: If s_j has degree zero or one in $\mathcal{G}$ the result follows immediately. If s_j has an edge leading to a root $r_{\alpha(j)}$, with $j \leq k$, then the result follows since the roots are selected in reverse chronological order. $\blacksquare$

We now fix a particular tree, say $T = T^k$.

Lemma 5 For each request r_i, $ld(r_i) \leq OPT(T_i)$, and $d(s_{\sigma(i)}, r_i) \leq OPT(T_i)$.

Proof Sketch: The proof is by induction on the number k of request nodes in the induced tree T_i. If $k = 1$ then the child s_i of r_i is a leaf, and T_i consists of one adversary edge with weight $OPT(T_i)$. The first claim follows by the triangle inequality, and the second claim follows by the definition of the greedy algorithm.

Suppose $k > 1$. If s_i is not a leaf then it has two children in T_i, say r_a and r_b. By induction, it must be the case that $ld(r_a) \leq OPT(T_a)$, $d(s_i, r_a) \leq OPT(T_a)$, $ld(r_b) \leq OPT(T_b)$, and $d(s_i, r_b) \leq OPT(T_b)$. Therefore, by the triangle inequality

$$ld(r_i) \leq \min(d(r_i, s_i) + d(s_i, r_a) + OPT(T_a), d(r_i, s_i) + d(s_i, r_b) + OPT(T_b))$$

By induction $ld(r_i) \leq d(r_i, s_i) + 2 \min(OPT(T_a), OPT(T_b))$. Hence, $ld(r_i) \leq d(r_i, s_i) + OPT(T_a) + OPT(T_b) = OPT(T_i)$. The second claim follows by the definition of the greedy algorithm. $\blacksquare$

In the next theorem, log means the logarithm base 2.

Lemma 6 For each tree T_i, with k request vertices, it is the case that $ON(T_i) \leq 2 \cdot \max(1, \log k) \cdot OPT(T_i)$.

Proof Sketch: The proof is by induction on k. For $k = 1$, $ON(T_i) \leq OPT(T_i)$ by the definition of GREEDY. By the definition of T_i, it is not possible for $k = 2$. The case $k = 3$ can be verified by using lemma 5.

Now consider the case $k > 3$. Once again let the two children of s_i in T_i be r_a and r_b. Notice that $ON(T_i) \leq ON(T_a) + ON(T_b) + d(s_{\sigma(i)}, r_i)$. In the proof of lemma 5 it was shown that $d(s_{\sigma(i)}, r_i) \leq d(r_i, s_i) + 2\min(OPT(T_a), OPT(T_b))$. Hence, $ON(T_i) \leq ON(T_a) + ON(T_b) + d(r_i, s_i) + 2\min(OPT(T_a), OPT(T_b))$. Let $w = OPT(T_a) + OPT(T_b)$, and $x = OPT(T_a)$. Assume, without loss of generality, that $OPT(T_a) \leq OPT(T_b)$. Hence, $x \leq w/2$. Also let y be the number of request nodes in T_a. We now break the proof into cases.

In the first case we assume that both T_a and T_b consist of more than one request vertex. Hence, $3 \leq y \leq k - 4$. By induction $ON(T_a) \leq 2x \log y$ and $ON(T_b) \leq 2(w - x) \log(k - 1 - y)$. Hence, $ON(T_i) \leq 2x \log y + 2(w - x) \log(k - 1 - y) + 2x + d(r_i, s_i)$. We wish to show that the right hand quantity is less that $2OPT(T_i) \log k$. Since $OPT(T_i) = w + d(r_i, s_i)$, it is sufficient to show $f(x, y) = x \log y + (w - x) \log(k - 1 - y) + x \leq w \log k$. Notice that $f(x, y)$ is linear in x. Hence the maximum of $f(x, y)$ must occur on the boundary at $x = 0$ or $x = w/2$. The inequality follows immediately for $x = 0$. For $x = w/2$ we have to find the maximum $f(w/2, y) = (w/2)(\log y + \log(k - 1 - y) + 1)$. By taking derivative of $f(w/2, y)$ with respect to y we can see that the maximum occurs at $y = (k - 1)/2$. We now must show that $f(w/2, (k - 1)/2) = (w/2)(2\log((k - 1)/2) + 1) \leq w \log k$. This holds if $2((k - 1)/2)^2 \leq k^2$. One can see that this holds for $k \geq 3$.

We now consider the case that T_a contains only one request. So $y = 1$. By induction $ON(T_a) \leq 2x$ and $ON(T_b) \leq 2(w - x) \log(k - 2)$. As before it is sufficient to verify that $x + (w - x) \log(k - 2) + x \leq w \log k$. Once again since the left hand side is linear is x we need only consider $x = 0$ and $x = w/2$. The case $x = 0$ follows immediately. If $x = w/2$ we are left with $(w/2) \log(k - 2) + w \leq w \log k$. This will be true if $2^2(k - 2) \leq k^2$. One can verify that this holds for $k \geq 3$.

We now consider the case that T_b contains only one request. So $y = k - 2$. As before it is sufficient to show $x \log(k - 2) + (w - x) + x \leq w \log k$. The proof is almost identical to the previous case. ∎

Theorem 7 The halfopt-competitive ratio of GREEDY for online transportation, in the case that each online capacity is two, is at most $2 \log m$, for $m \geq 3$.

We now discuss how to extend the result to the case of $c_i > 2$. Recall that $C = \sum_{i=1}^{m} c_i$ is the total online capacity.

Theorem 8 The halfopt-competitive ratio of GREEDY for online transportation is $O(\min(m, \log C))$.

Proof Sketch: We pair an arbitrary pair of online servers at s_i with one adversary server at s_i, and then consider s_i as consisting of $a_i = c_i/2$ different sites. The upper bound of $O(\log C)$ follows immediately. To see the $O(m)$ bound we want to show that we can construct the T^k's so that the height of each tree is at most m, and hence the logarithm of the number of nodes in the tree will be $O(m)$.

To see this assume that there is some root to leaf path that first passes through s_i and then passes through s_j, where s_i and s_j were originally the same site. By pairing s_j^1 and s_j^2 with s_i^a, and pairing s_i^1 and s_i^2 with s_j^a we can reduce the redundancy in T^k. By repeating this process the desired bound on the height of each T^k follows. ∎

We now prove an asymptotically matching lower bound for the halfopt-competitive ratio for GREEDY.

Theorem 9 The halfopt-competitive ratio of GREEDY for the transportation problem is $\Omega(\min(m, \log C))$.

Proof Sketch: Assume without loss of generality that $m = \Theta(\log C)$. We embed m server sites on the real line. The server site s_1 is located at the point -1. The server site s_i $(2 \leq i \leq m)$ is located at $2^{i-1} - 1$. The online algorithm has $c_i = 2^{m-i+1}$ servers at s_i, while the adversary has $a_i = 2^{m-i}$ servers at s_i. Thus the online algorithm has a total of $C = 2^{m+1} - 2$ servers, while the adversary has a total of $A = 2^m - 1$ servers. The requests occur in m batches. The first batch consists of 2^{m-1} requests at the point 0. The ith batch $(2 \leq i \leq m)$ contains 2^{m-i} requests that occur at $2^{i-1} - 1$, the location of s_i. GREEDY responds to batch i $(i < m)$ by answering each request in batch i with a server at site s_{i+1}, thus depleting site s_{i+1}. GREEDY responds to the mth batch by moving one server from s_1. Thus the total online cost is $m2^{m-1}$. By using the servers in s_i to handle batch i $(1 \leq i \leq m)$ it is possible to obtain a total cost of 2^{m-1}. ∎

4 The Algorithm Balance

In this section we present an algorithm, BALANCE with a halfopt-competitive ratio of $O(1)$. Let us once again first assume that the online capacity of each server site is two.

Algorithm BALANCE: At each site s_i we classify the server s_i^1 as *primary* and the server s_i^2 as *secondary*. Let $c > 1$ be some constant that we define later. For each new request r_i, find the closest available primary server s_j^1 and the closest available secondary server s_k^2. Use the server s_j^1 if $d(r_i, s_j^1) \leq c \cdot d(r_i, s_k^2)$. Otherwise use the server s_k^2.

We now wish to divide the response graph into edge disjoint trees, $T^1, \ldots, T^l$. Each tree T^k will satisfy the following conditions:

1. The root of T^k is a request $r_{\alpha(k)}$ with the property that $r_{\alpha(k)}$ is the latest request not in one of the previously constructed trees, $T^1, \ldots, T^{k-1}$.

2. Each leaf of T^k is a server site s_i such that either every online edge incident to s_i is of the form $(s_i, r_{\alpha(j)})$ for some $j \leq k$, i.e. each online edge leads to the current root or a previous root. Note this condition is trivially satisfied if s_i has no children.

T^k can be constructed as follows. Let $r_{\alpha(k)}$ be the most recent request not in a previous tree. Remove the online edge $(s_{\sigma(\alpha(k))}, r_{\alpha(k)})$ from $\mathcal{G}$. Let T^k be the connected component of the remaining graph that contains $r_{\alpha(k)}$. We then remove all the edges and request sites in T^k from $\mathcal{G}$ and repeat this process. We now restrict our attention to one tree, say $T = T^k$.

Lemma 10 Let r_i be a request in T. If s_j is a leaf server site in T^k, then BALANCE has both s_j^1 and s_j^2 available at the time of the request r_i. If s_j is a server site with only one child in T, then BALANCE has s_j^2 available at the time of the request r_i.

Lemma 11 If BALANCE uses a primary server s_j^1 to handle a request r_i, then $d(s_j, r_i) \le d(r_i, s_i) + ld(s_i)$. Hence, $ld(s_j) \le 2(d(r_i, s_i) + ld(s_i))$.

Lemma 12 If BALANCE uses a secondary server s_j^2 to handle a request r_i, then $d(s_j, r_i) \le (1/c)(d(r_i, s_i) + ld(s_i))$. Hence, $ld(s_j) \le (1 + 1/c)(d(r_i, s_i) + ld(s_i))$.

Lemma 13 Assume that BALANCE uses a server at s_j to handle a request r_i and that s_i has at most one child in T. Then $d(s_j, r_i) \le c \cdot d(r_i, s_i)$, and $ld(s_j) \le (c+1)d(r_i, s_i) + ld(s_i)$.

Lemma 14 Assume that BALANCE uses a server at s_j to handle a request r_i, that s_i has two children r_a and r_b, and that s_a has at most one child in T. Then $d(s_j, r_i) \le d(r_i, s_i) + (c+1)d(r_a, s_a)$, and $ld(s_j) \le 2d(r_i, s_i) + (c+1)d(r_a, s_a) + ld(s_i)$.

Lemma 15 For all nonnegative reals x and y, $\min(2x, (1 + 1/c)y) \le \frac{2c+2}{3c+1}(x + y)$.

Lemma 16 Assume that BALANCE uses a server at s_j to handle a request r_i, and that s_i has two children r_a and r_b. Further assume that s_a has two children r_c and r_d, and that s_b has two children r_e and r_f. Then

$$
\begin{aligned}
ld(s_j) \le {}& 2d(r_i, s_i) + \\
& (2 \cdot \frac{2c+2}{3c+1}(d(r_a, s_a) + d(r_b, s_b)) + \\
& (2(\frac{2c+2}{3c+1})^2)(d(r_c, s_c) + d(r_d, s_d) + d(r_e, s_e) + d(r_f, s_f)) + \\
& (2(\frac{2c+2}{3c+1})^2)(ld(s_c) + ld(s_d) + ld(s_e) + ld(s_f))
\end{aligned}
$$

Proof Sketch: First $ld(s_a) \le 2(d(r_c, s_c) + ld(s_c))$ by lemma 11, and $ld(s_a) \le (1 + 1/c)(d(r_d, s_d) + ld(s_d))$ by lemma 12. By lemma 15

$$
ld(s_a) \le (\frac{2c+2}{3c+1})(d(r_c, s_c) + d(r_d, s_d) + ld(s_c) + ld(s_d))
$$

Similarly,

$$
ld(s_b) \le (\frac{2c+2}{3c+1})(d(r_e, s_e) + d(r_f, s_f) + ld(s_e) + ld(s_f))
$$

Applying the same idea to s_i we get $ld(s_i) \leq 2(d(r_a, s_a) + ld(s_a))$ by lemma 11, $ld(s_i) \leq (1 + 1/c)(d(r_b, s_b) + ld(s_b))$ by lemma 12, and by lemma 15

$$ld(s_i) \leq (\frac{2c+2}{3c+1})(d(r_a, s_a) + d(r_b, s_b) + ld(s_a) + ld(s_b))$$

By combining these inequalities we get

$$
\begin{aligned}
ld(s_i) \leq{} & (\frac{2c+2}{3c+1})(d(r_a, s_a) + d(r_b, s_b)) + \\
& (\frac{2c+2}{3c+1})^2(d(r_c, s_c) + d(r_d, s_d) + d(r_e, s_e) + d(r_f, s_f)) + \\
& (\frac{2c+2}{3c+1})^2(ld(s_c) + ld(s_d) + ld(s_e) + ld(s_f))
\end{aligned}
$$

Now $d(s_j, r_i) \leq d(r_i, s_i) + ld(s_i)$ and $ld(s_j) \leq d(s_j, r_i) + d(r_i, s_i) + ld(s_i)$. Hence by substitution,

$$
\begin{aligned}
ld(s_j) \leq{} & 2d(r_i, s_i) + \\
& (2 \cdot \frac{2c+2}{3c+1})(d(r_a, s_a) + d(r_b, s_b)) + \\
& (2(\frac{2c+2}{3c+1})^2)(d(r_c, s_c) + d(r_d, s_d) + d(r_e, s_e) + d(r_f, s_f)) + \\
& (2(\frac{2c+2}{3c+1})^2)(ld(s_c) + ld(s_d) + ld(s_e) + ld(s_f))
\end{aligned}
$$

∎

Theorem 17 The halfopt-competitive factor of BALANCE is $O(1)$.

Proof Sketch: By lemmas 11 and 12 it is sufficient to show that

$$\sum_{r_i \in T} ld(s_i) = O(OPT(T))$$

First notice that every request r_i meets the conditions of one of lemmas 13, 14 or 16. By expanding the recurrences in these lemmas, and by commutitivity of addition, we get

$$\sum_{r_i \in T} ld(s_i) \leq \sum_{r_i \in T} d(r_i, s_i) \sum_{k=1}^{\infty} 3(2(\frac{2c+2}{3c+1})^2))^k (2(c+1))$$

If c is large enough, so that $2(\frac{2c+2}{3c+1})^2 < 1$, this geometric series converges, and the result follows. ∎

To construct a constant competitive algorithm for the general case of $c_i = a_i/2$ we need only apply the technique in lemma 9.

5 Conclusion

The most obvious avenue for further investigation is to determine the competitive ratio in the weakened adversary model when the adversary capacity is more than half of the online capacity. It seems that some new techniques will be needed in this case since the response graph no longer has the treelike property that was so critical in our proofs. The obvious generalizations of BALANCE are no longer constant competitive for the general case where the online capacity of each server site is k and the adversary capacity is l, with $l \leq k$. In contrast, we can not rule out the possibility of a constant competitive algorithm even for the case $l = k - 1$.

We feel that the weakened adversary model will likely be useful in other online problems. Subsequent to this investigation, we have used this model to study online scheduling problems [2].

References

[1] B. Kalyanasundaram, and K. Pruhs, "Online weighted matching", *Journal of Algorithms*, 14, 478–488, 1993.

[2] B. Kalyanasundaram, and K. Pruhs, "Speed is as powerful as clairvoyance", Technical Report, Computer Science Dept., University of Pittsburgh, 1995.

[3] Jeff Kennington, and Richard Helgason, *Algorithms for Network Programming*, John Wiley and Sons, 1980.

[4] S. Khuller, S. Mitchell, and V. Vazirani, "On-line algorithms for weighted matchings and stable marriages", *Theoretical Computer Science*, 127(2), 255–267, 1994.

[5] Elias Koutsoupias, and Christos Papadimitriou, "Beyond competitive analysis", *FOCS*, 394–400, 1994.

[6] Eugene Lawler, *Combinatorial Optimization: Networks and Matroids*, Sanders College Publishing, 1976.

[7] Daniel Sleator and Robert Tarjan, "Amortized efficiency of list update and paging rules", *Communications of the ACM*, 28, 202–208, 1985.

[8] Ying Teh Tsai, Chuan Yi Tang and Yunn Yen Chen, "Average performance of a greedy algorithm for the on-line minimum matching problem on Euclidean space", *Information Processing Letters*, 1994.

[9] N. Young, "The k-server dual and loose competitiveness for paging," *Algorithmica*, 11, 525–541, 1994.

A Polyhedral Approach to Planar Augmentation and Related Problems

Petra Mutzel

Max-Planck-Institut für Informatik, Im Stadtwald, D-66123 Saarbrücken,
Germany

Abstract. Given a planar graph G, the planar (biconnectivity) augmentation problem is to add the minimum number of edges to G such that the resulting graph is still planar and biconnected. Given a nonplanar and biconnected graph, the maximum planar biconnected subgraph problem consists of removing the minimum number of edges so that planarity is achieved and biconnectivity is maintained. Both problems are important in Automatic Graph Drawing. In [JM95], the minimum planarizing k-augmentation problem has been introduced, that links the planarization step and the augmentation step together. Here, we are given a graph which is not necessarily planar and not necessarily k-connected, and we want to delete some set of edges D and to add some set of edges A such that $|D| + |A|$ is minimized and the resulting graph is planar, k-connected and spanning. For all three problems, we have given a polyhedral formulation by defining three different linear objective functions over the same polytope, namely the 2-node connected planar spanning subgraph polytope $2\text{-}\mathcal{NCPLS}(K_n)$. We investigate the facial structure of this polytope for $k = 2$, which we will make use of in a branch and cut algorithm. Here, we give the dimension of the planar, biconnected, spanning subgraph polytope for $G = K_n$ and we show that all facets of the planar subgraph polytope $\mathcal{PLS}(K_n)$ are also facets of the new polytope $2\text{-}\mathcal{NCPLS}(K_n)$. Furthermore, we show that the node-cut constraints arising in the biconnectivity spanning subgraph polytope, are facet-defining inequalities for $2\text{-}\mathcal{NCPLS}(K_n)$. We give first computational results for all three problems, the planar 2-augmentation problem, the minimum planarizing 2-augmentation problem and the maximum planar biconnected (spanning) subgraph problem. This is the first time that instances of any of these three problems can be solved to optimality.

1 Introduction

Many algorithms work only for biconnected graphs. A graph that does not satisfy this condition, has to be augmented by adding a set of edges such that the resulting graph is biconnected. In general, the problem of augmenting a graph by a minimum number of edges so that it meets certain edge or node connectivity requirements given by $k \in \mathbb{Z}$ is called *augmentation problem*. We will consider node connectivity requirements with $k = 2$.

The (node-)connectivity $\kappa(G)$ of a graph G is the minimum number of nodes whose removal together with its incident edges results in a disconnected or trivial graph. A graph is said to be *k-node-connected*, or *k-connected*, if $\kappa(G) \geq k$. The problem of finding the minimum cost k-node connected spanning subgraph in a general graph is proven to be NP-hard even for $k = 2$ and uniform edge costs [GJ79]. Garg, Santosh and Singla have given a 3/2-approximation algorithm for $k = 2$ and uniform edge costs [GSS93], whereas Ravi and Williamson have given a $2\mathcal{H}(k)$-approximation algorithm for general k and general costs, where $\mathcal{H}(k) = 1 + \frac{1}{2} + \ldots + \frac{1}{k}$ [RW95]. For $G = K_n$, the complete graph on n vertices, Harary has given a polynomial time algorithm for constructing a minimum k-vertex connected spanning subgraph for $k \in \mathbb{N}$ [Har62].

Another type of connectivity problems are *augmentation problems*. Here, the given graph has to be augmented in a minimum way to a k-connected graph. The problem of augmenting a given graph by the minimum number of edges in order to obtain a k-node connected graph seems still to be open, whereas there are polynomial time algorithms for $k = 2$ and $k = 3$ [RG77,HR91,WN93]. In the case that the augmented edges have general edge costs, the problem is NP-hard even for $k = 2$ [ET76].

The *planar (k-)augmentation problem* has been brought up by Kant, and consists of adding a minimum number of edges to a planar graph in order to obtain a k-connected graph, which is still planar. Kant showed that this problem is NP-hard for $k = 2$, and gives a linear time approximation algorithm which adds at most 2 times the minimum required number of edges [Kan93].

The planar augmentation problem has a wide application in Automatic Graph Drawing. Here, many algorithms for drawing planar graphs work only for biconnected or even triconnected graphs. In order to use these algorithms for a simple connected graph there are two possibilities. Either to draw the biconnected parts of the graph separately, or to augment it to biconnectivity and draw the resulting graph while suppressing the augmented edges. In general, the second approach leads to nicer drawings.

Since there are many graph drawing algorithms for planar graphs and only a few algorithms for nonplanar graphs, Tamassia, Batini and Di Battista suggested a method using planarization [TBB88]. Here, the maximum planar subgraph P of a given nonplanar graph is determined, and either P is drawn and the deleted edges are inserted again, or the deleted edges are reinserted before the drawing step, and the produced crossings are substituted by artificial nodes in order to obtain a planar graph. In the former case, it is often required that the found maximum planar subgraph is biconnected. Requiring this condition and the condition that the subgraph should be spanning leads to the *maximum planar biconnected (spanning) subgraph problem*. Goldschmidt and Takvorian showed that even finding a biconnected spanning planar subgraph of a biconnected nonplanar graph is NP-hard [GT94].

For the application in Automatic Graph Drawing it is, indeed, advantageous to link the planarization step and the augmentation step together (see [JM95]). This leads to the *minimum planarizing k-augmentation problem*. Given a graph $G = (V, E)$, we like to delete a set of edges $D \subseteq E$ and to add a set of edges

$A \subseteq (V \times V) \setminus E$ such that $|D| + |A|$ is minimum and the resulting graph is planar, spanning and k-connected.

All three problems, the planar k-augmentation problem, the maximum planar k-connected spanning subgraph problem and the minimum planarizing k-augmentation problem can be formulated as an optimization problem of a linear objective function over a single polytope. The *planar k-connected (spanning) subgraph polytope* k-$\mathcal{NCPLS}(G_0)$ is defined to be the convex hull over all incidence vectors of planar, k-connected and spanning subgraphs of a graph G_0.

For a given graph $G = (V, E)$, we choose $G_0 = (V_0, E_0)$ to be the complete graph on the vertex set $V_0 = V$ and consider the optimization problem

$$\max\{w^T x \mid x \in k\text{-}\mathcal{NCPLS}(G_0)\}.$$

Depending on the value of the vector $w \in R^{E_0}$ we can formulate all three problems.

In order to solve the planar k-augmentation problem we set $w_e = -1$ for $e \in E_0 \setminus E$ and $w_e = M$ for $e \in E$, where $M = |E_0 \setminus E| + 1$. By setting $w_e = 1$ for $e \in E$ and $w_e = -M'$ for $e \in E_0 \setminus E$ with $M' = |E| + 1$, we can formulate the maximum planar k-connected spanning subgraph problem. The natural setting of c occurs for the minimum planarizing k-augmentation problem. Here, $w_e = 1$ for $e \in E$ and $w_e = -1$ for $e \in E_0 \setminus E$. Maximizing $w^T x$ leads to taking as many edges of G as possible and as few edges of $G_0 - G$ as possible. This way, the difference between the given graph G and the new graph will be minimized.

Hence, we are interested in the facial structure of the *planar k-connected (spanning) subgraph polytope* k-$\mathcal{NCPLS}(K_n)$. This polytope is the intersection of two well studied polytopes, namely the planar subgraph polytope $\mathcal{PLS}(G)$ for the maximum planar subgraph problem [JM95] and the k-node-connected spanning subgraph polytope k-$\mathcal{NCS}(G)$ for the network-survivability problem [Sto92]. Since the polyhedral approach for both problems seems to be promising, i.e., the optimum solution can be found by branch and cut algorithms within a few seconds for graphs of moderate sizes, our hope is to obtain similar results for all three problems using the polyhedral approach.

The outline of the paper is as follows. In Section 2 we give some mathematical background in polyhedral combinatorics and recall the most important classes of facet-defining inequalities for the planar subgraph polytope. A short introduction into the k-connected subgraph polytope is given in Section 3. New results of the structure of the planar biconnected spanning subgraph polytope are contained in Section 4. Recall that we restrict our consideration to $k = 2$, since many applications require biconnectivity. Moreover, in the second part of Section 4 we restrict ourselves to k-$\mathcal{NCPLS}(G_0)$ for $G_0 = K_n$, the complete graph on n vertices, which is not really a restriction, since all three problems can be formulated over k-$\mathcal{NCPLS}(K_n)$. In Section 5 we show how these theoretical results can be useful in practice. We describe our branch and cut algorithm which has been used in our computational experiments given in Section 6.

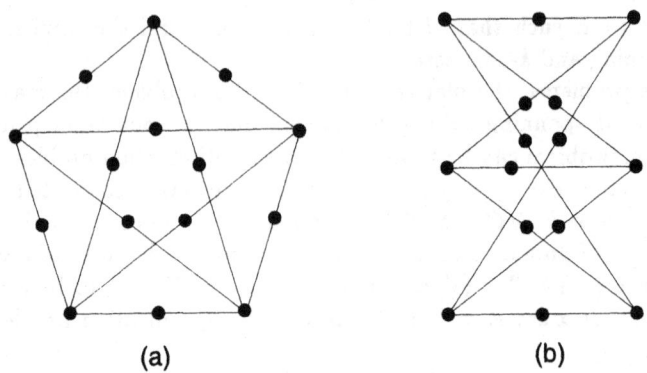

Fig. 1. shows some Kuratowski graphs, i.e., a subdivision of K_5 and K_{33}

2 The Planar Subgraph Polytope

Given a graph $G = (V, E)$ with edge weights $w_e \in \mathrm{R}$ for all $e \in E$, let $\mathcal{P_G}$ be the set of all planar subgraphs of G. For each planar subgraph $P = (V', F) \in \mathcal{P_G}$, we define its incidence vector $\chi^F \in \mathrm{R}^E$ by setting $\chi_e^F = 1$ if $e \in F$ and $\chi_e^F = 0$ if $e \notin F$. The *planar subgraph polytope* $\mathcal{PLS}(G)$ of G is defined as the convex hull over all incidence vectors of planar subgraphs of G. The problem of finding a planar subgraph $P = (V', F)$ of G with weight $w(P) = \sum_{e \in F} w_e$ as large as possible can be written as the linear program $\max\{w^T x \mid x \in \mathcal{PLS}(G)\}$, since the vertices of the polytope $\mathcal{PLS}(G)$ are exactly the incidence vectors of the planar subgraphs of G. Kuratowski characterized the minimal nonplanar graphs to be exactly the subdivisions of K_5 and $K_{3,3}$. Hence we get the following integer programming formulation for the maximum planar subgraph problem:

$$\text{maximize } w^T x$$
$$\begin{aligned}
\text{subject to } & 0 \le x_e \le 1, && \text{for all } e \in E, && (1) \\
& x(K) \le |K| - 1, && \text{for all Kuratoski subgraphs } (V', K),\ K \subseteq E && (2) \\
& x_e \text{ integral}, && \text{for all } e \in E && (3)
\end{aligned}$$

Since integer programming is NP-hard, we drop the integer constraints. In order to apply linear programming techniques to solve this linear program one has to represent $\mathcal{PLS}(G)$ as the solution of an inequality system. Due to the NP-hardness of our problem, we cannot expect to be able to find a complete description of $\mathcal{PLS}(G)$ by linear inequalities. But even a partial description of the facial structure of $\mathcal{PLS}(G)$ by linear inequalities is useful for the design of a "branch and cut"-algorithm, because such a description defines a relaxation of the original problem. Such relaxations can be solved within a branch and bound framework via cutting plane techniques and linear programming in order to produce tight bounds. An irredundant description of $\mathcal{PLS}(G)$ by linear inequalities contains only inequalities which describe proper faces of maximal dimension of $\mathcal{PLS}(G)$, so-called facet-defining inequalities.

For efficiency, also in a partial description by inequalities, we concentrate on those valid inequalities for $\mathcal{PLS}(G)$ which are facet-defining. For ease of notation, we define $x(F) = \sum_{e \in F} x_e$ for $F \subseteq E$. In [JM94] we state the following

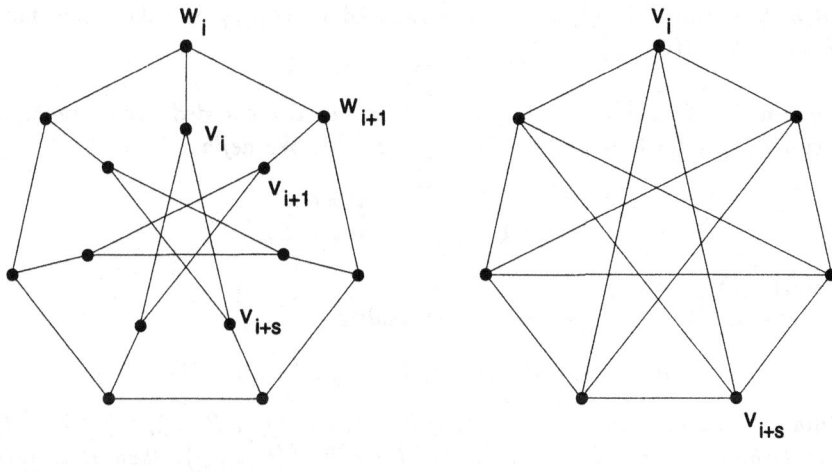

(a)   (b)

Fig. 2.(a) The generalized Petersen graph $P(7,3)$ and the (b) s-chorded cycle graph $G_{(n,s,t)}$ for $n = 7$ and $s = 3$

Theorem 1 [JM94]. *The dimension of the planar subgraph polytope $\mathcal{PLS}(G)$ of $G = (V, E)$ is $|E|$, so it is full dimensional.*
For all edges $e \in E$ the inequalities $x_e \geq 0$ and $x_e \leq 1$ define facets of $\mathcal{PLS}(G)$.
For all subdivisions $G' = (V', K)$ of K_5 or $K_{3,3}$ contained in G the inequality $x(K) \leq |K| - 1$ defines a facet of $\mathcal{PLS}(G)$.
For all cliques (V', F) (or complete bipartite subgraphs) contained in G, the Euler inequalities $x(F) \leq 3|V'| - 6$ (or $x(F) \leq 2|V'| - 4$, respectively) are facet-defining for $\mathcal{PLS}(G)$.

The following class of graphs, called s-chorded cycle graphs, has been introduced in [Mut94b]. This class of graphs can be derived from the generalized Petersen graphs by the contraction of certain edges (see Figure 2). The s-chorded cycle graphs give rise to huge classes of inequalities generalizing the Kuratowski inequalities.

    For the rest of this section all sums of integers representing nodes of graphs $G = (V, E)$, which are greater than $n = |V|$ are to be taken modulo n.

Definition 1 For $s, n, r, t \in \mathbb{N}$, $s \geq 2$, $n = st + r$, $0 \leq r < s$, the s-chorded cycle graph $G_{(n,s,t)} = (V, E)$ is defined via

$$V = \{1, 2, \ldots, n\} \quad \text{and}$$
$$E = C_n \cup D_n, \quad \text{where}$$

C_n is a cycle of length n, $C_n = \{(i, i+1) \mid i = 1, \ldots, n\}$ and
D_n is the set of s-chords of C_n, $D_n = \{(i, i+s) \mid i = 1, \ldots, n\}$.

An s-chorded cycle graph gives rise to the definition of the corresponding s-chorded cycle inequality. In [Mut94b], it is investigated for which values of s,

t and n the s-chorded cycle inequality induced by $G_{(n,s,t)}$ is valid, resp. facet-defining for $\mathcal{PLS}(G_{(n,s,t)})$.

Theorem 2 [Mut94b]. *Let* $G_{(n,s,t)} = (V, E)$ *be a s-chorded cycle graph, i.e.* $n = st + r$ *for* $s, t, r \in \mathbb{N}$ *and* $r \in \{0, 1, \ldots, s - 1\}$. *We define*

$$c_e = \begin{cases} n - 2t - s, & \text{if } x \in C_n \\ 1, & \text{if } x \in D_n \end{cases}$$

and $c(E) := \sum_{e \in E} c_e$.
If $n - 2t - s > 0$, *the s-chorded cycle inequality*

$$(n - 2t - s)x(C_n) + x(D_n) \leq c(E) - (n - 2t)$$

is valid for $\mathcal{PLS}(G_{(n,s,t)})$ *if and only if* $(t \geq 3, s \geq 3)$ *or* $(t = 2, s \geq 3, r \geq 2)$.
If the s-chorded cycle inequality is valid for $\mathcal{PLS}(G_{(n,s,t)})$, *then it is facet-defining for* $\mathcal{PLS}(G_{(n,s,t)})$ *if* $r > 0$.
Let $G_{(n,s,t)}$ *be an s-chorded cycle graph, which is a subgraph of G. If* $t = 2$, *or* $s = 3$, *or* $r \geq \lfloor \frac{s}{2} \rfloor$ *and the corresponding s-chorded cycle inequality corresponding to* $G_{(n,s,t)}$ *is facet-defining for* $\mathcal{PLS}(G_{(n,s,t)})$, *then the s-chorded cycle inequality is facet-defining for* $\mathcal{PLS}(G)$.

For $n = 5$ the 2-chorded cycle inequality is identical to the Kuratowski inequality for K_5. So, the general s-chorded cycle inequalities are generalizations of the Kuratowski inequality for K_5. For the special case that $n = 2k + 1$ and $s = k$, the s-chorded cycle graphs give rise to another kind of inequality, the odd n-ladder inequality, which is an alternative generalization of the K_5-inequality.

Theorem 3 [Mut94b]. *If G contains the k-chorded cycle graph* $G_{(2k+1,k,2)} = (V, E)$ *on $2k + 1$ nodes with* $E = C_{2k+1} \cup D_{2k+1}$, $k \in \mathbb{N}$, $k \geq 2$, *then the odd n-ladder inequality*

$$(2k - 3)\, x(C_{2k+1}) + x(D_{2k+1}) \leq (2k - 1)^2$$

is facet-defining for $\mathcal{PLS}(G_{(2k+1,k,2)})$ *and for* $\mathcal{PLS}(G)$.

In the special case $s = k$ and $n = 2k$, the k-chorded cycle graphs $G_{(2k,k,2)}$ contain multiple edges. If we take each diagonal only once, we obtain a Möbius-ladder, which gives rise to an inequality which is a generalization of Kuratowski's $K_{3,3}$ inequality.

Definition 2 For $k \in \mathbb{N}$, $k \geq 3$, we define the (even) Möbius-ladder to be the graph $G_M = (V, E)$ with

$$V = \{1, 2, \ldots, 2k\} \quad \text{and}$$
$$E = C_{2k} \cup D_k, \quad \text{where}$$

C_{2k} is a cycle of length $2k$, $C_{2k} = \{(i, i+1) \mid i = 1, \ldots, 2k\}$,
D_k is the set of longest chords of C_{2k}, $D_k = \{(i, i+k) \mid i = 1, 2, \ldots, k\}$.

Theorem 4 [Mut94b]. *For the Möbius-ladder graph $G_M = (V, E)$ with $E = C_{2k} \cup D_k$ and $k \geq 3$ the Möbius-ladder inequality*

$$(k-2)\ x(C_{2k}) + x(D_k) \leq 2(k-1)^2$$

is facet-defining for $\mathcal{PLS}(G_M)$. Moreover, the Möbius ladder inequality is facet-defining for $\mathcal{PLS}(G)$ whenever G_M is a subgraph of G.

A complete overview of the currently known structure of the planar subgraph polytope can be found in [Mut94a].

3 The k-connected Subgraph Polytope k-$\mathcal{NCS}(G)$

Given a k-connected graph $G = (V, E)$, we are interested in the set of all k-connected spanning subgraphs of G. For each k-connected spanning subgraph $K = (V, F)$ of G we define its incidence vector $\chi^F \in \mathbb{R}^E$ by setting $\chi_e^F = 1$ if $e \in F$ and $\chi_e^F = 0$ if $e \notin F$. The k-connected (spanning) subgraph polytope k-$\mathcal{NCS}(G)$ is defined as the convex hull over all incidence vectors of k-connected spanning subgraphs of G. In order to solve the minimum k-connected subgraph problem for a given k-connected graph $G = (V, E)$, we define the weight w_e for an edge $e \in E$ to be 1. The problem of determining a k-connected subgraph of G with the minimum number of edges, can be formulated as the linear program $\min\{w^T x \mid x \in k\text{-}\mathcal{NCS}(G)\}$. If we like to solve the k-augmentation problem for a given graph G, we choose $G_0 = (V, E_0) := K_n$, define the weight w_e for an edge to be $-M$, if $e \in E$ and 1 if $e \notin E$, where $M = |E_0 \setminus E| + 1$, and solve $\min\{w^T x \mid x \in k\text{-}\mathcal{NCS}(G_0)\}$.

The k-connected subgraph polytope was already studied by Stoer in a more general form [Sto92]. For $k = 2$, the computational results in [Sto92] are promising.

Theorem 5 [Sto92]. *The integer points of k-$\mathcal{NCS}(G)$ are characterized by the following system of inequalities:*

$$\begin{aligned}
0 \leq x_e \leq 1, & \quad \text{for all } e \in E & (4)\\
x(\delta_{G-Y}(W)) \geq 1, & \quad \text{for all } Y \subseteq V, |Y| = k-1, W \subseteq V \setminus Y & (5)\\
x_e \text{ integral}, & \quad \text{for all } e \in E & (6)
\end{aligned}$$

The inequalities (5) essentially say that if a node set $Y \subseteq V$ of size $k-1$ is removed, the resulting graph must still be connected.

4 Intersecting $\mathcal{PLS}(G)$ and k-$\mathcal{NCS}(G)$

In this section we are interested in the integer points contained in the polytope

$$k\text{-}\mathcal{NCPLS}(G) = k\text{-}\mathcal{NCS}(G) \cap \mathcal{PLS}(G)$$

for $k = 2$. We already gave a system of inequalities characterizing the integer points for both polytopes. So, the integer points of the new polytope k-$\mathcal{NCPLS}(G)$ are defined by the system of inequalities given by (1), (2), (3) and (5).

One of the first questions occuring in connection to a polytope is its dimension. For general graphs G, the dimension of 2-$\mathcal{NCPLS}(G)$ is unlikely to be determined. Even for biconnected graphs 2-$\mathcal{NCPLS}(G)$ may be empty, like, for example, for the Kuratowski graph shown in Figure 1(a). If we restrict our attention to $G = K_n$, the complete graph on n nodes, we will see that 2-$\mathcal{NCPLS}(K_n)$ has full dimension.

Theorem 6 *The dimension of the planar biconnected spanning subgraph polytope 2-$\mathcal{NCPLS}(K_n)$ for $K_n = (V, E)$, the complete graph on $n = |V|$ vertices, is $|E|$.*

Proof. We will show that there are $|E| + 1$ affinely independent elements in R^E that are incidence vectors of planar, biconnected and spanning subgraphs of K_n. For simplicity let the vertices be numbered by $1, 2, \ldots, n$. Let P consist of the Hamiltonian cycle $C = ((1, 2), (2, 3), \ldots, (n-1, n), (n, 1))$ together with the edge $e = (1, 3)$. Adding any of the edges $e_i \in E \setminus (C \cup \{e\})$ gives P^i, $i = 1, 2, \ldots, |E| - n - 1$. Removing the edge $(u, u + 1)$, resp. $(n, 1)$, and adding the edges (u, v), $(v, u + 1)$, resp. (n, v), $(v, 1)$, for any $v \notin \{u, u + 1\}$ gives P_j, $j = 1, 2, \ldots, n$. Let P' be equal to $P \setminus \{(1, 3)\}$. Then P, P^i for $i = 1, 2, \ldots, |E| - n - 1$, P_j for $j = 1, 2, \ldots, n$, and P' induce planar, biconnected and spanning subgraphs of K_n. Moreover, all the $|E| + 1$ incidence vectors χ^P, χ^{P^i} for $i = 1, 2, \ldots, |E| - n - 1$, χ^{P_j} for $j = 1, 2, \ldots, n$, and $\chi^{P'}$ are linearly independent. □

The natural question occurs whether known facets of the planar subgraph polytope are also facets for the planar biconnected spanning subgraph polytope. Suppose $c^T x \leq c_0$ is a facet of $\mathcal{PLS}(G)$. It will also be a facet of 2-$\mathcal{NCPLS}(G)$ if $d = \dim(2\text{-}\mathcal{NCPLS}(G))$ affinely independent incidence vectors of subgraphs $P \subseteq G$ exist that are planar, biconnected, spanning and satisfy the equation $c^T \chi^P = c_0$. The following theorem answers the question whether the support graph of a facet-defining inequality for $\mathcal{PLS}(K_n)$ is biconnected. The *support graph* of an inequality $c^T x \leq c_0$, resp. $c^T x \geq c_0$, is induced by the edge set $\{e \in E \mid c_e \neq 0\}$.

Theorem 7 *Let $c^T x \leq c_0$ be a facet of the planar subgraph polytope $\mathcal{PLS}(K_n)$. Then either the inequality $c^T x \leq c_0$ is identical with $x_e \geq 0$ or $x_e \leq 1$, or the support graph of $c^T x \leq c_0$ is biconnected.*

Proof. Let S_G denote the edge set of the support graph of $c^T x \leq c_0$, and suppose that $G[S_G]$ is not biconnected with $|S_G| \geq 2$. Then, let a be an articulation point of S_G that separates the blocks induced by K_1 from $K_2 := S_G \setminus K_1$. At least one of the blocks induced by K_1 and K_2 is nonplanar (say K_1), since S_G is nonplanar. We have $\{x \in \mathcal{PLS}(G) \mid c^T x = c_0\} = \{x \in \mathcal{PLS}(K_n) \mid c_1^T x + c_2^T x = c_1^0 + c_2^0\}$, where c_i denotes the vector of coefficients for block K_i for $i = 1, 2$. We have that $\{x \in \mathcal{PLS}(K_n) \mid c^T x = c_0\} \subset \{x \in \mathcal{PLS}(K_n) \mid c_1^T x = c_1^0\}$, since otherwise we would have $c_2^T x > c_2^0$ or $c_1^T x > c_1^0$, both of which are nonvalid inequalities for $\mathcal{PLS}(K_n)$. Moreover, we have $\{x \in \mathcal{PLS}(K_n) \mid c^T x = c_0\} \neq \{x \in \mathcal{PLS}(K_n) \mid c_1^T x = c_1^0\}$. Hence the inequality $c^T x \leq c_0$ can not be facet-defining for $\mathcal{PLS}(K_n)$, which is a contradiction. $\qquad\qquad\square$

In general, biconnectivity of the support graph of a facet-defining inequality $c^T x \leq c_0$ for $\mathcal{PLS}(G)$ is not sufficient to guarantee that it is also a facet for $2\text{-}\mathcal{NCPLS}(G)$. Consider, for example, the subdivision S of K_5 shown in Figure 1(a). The Kuratowski inequalities that are facets for $\mathcal{PLS}(S)$, are not facets for $2\text{-}\mathcal{NCPLS}(S)$, since removing any edge of S leads to a graph that is not biconnected. Fortunately, for complete graphs the situation is promising. We can show the following theorem.

Theorem 8 *Let (7) $c^T x \leq c_0$ be a facet of the planar subgraph polytope $\mathcal{PLS}(K_n)$. Then $c^T x \leq c_0$ is also a facet of the planar, biconnected and spanning subgraph polytope $2\text{-}\mathcal{NCPLS}(K_n)$.*

Proof. Let $K_n = (V, E)$. From Theorem 7 we know that either the inequality $c^T x \leq c_0$ is identical with $x_e \geq 0$ or $x_e \leq 1$, or the support graph of $c^T x \leq c_0$ is biconnected. For the inequalities $x_e \geq 0$ and $x_e \leq 1$ we give a direct proof similar to the one for Theorem 6. For simplicity the nodes are numbered by $1, 2, \ldots, n$. We first show that $x_e \leq 1$ is a facet-defining inequality for $2\text{-}\mathcal{NCPLS}(K_n)$. Without any restriction let $e = (1, 3)$. Consider the graphs induced by the edge sets P, P^i for $i = 1, 2, \ldots |E| - n - 1$ and P_j for $j = 1, 2, \ldots, n$ defined in the proof to Theorem 6. All of them are planar, biconnected and spanning, satisfy $x_e = 1$ and their incidence vectors are linearly independent. Next, removing the edge $e = (1, 3)$ from P, P^i for $i = 1, 2, \ldots |E| - n - 1$ and P_j for $j = 1, 2, \ldots, n$ gives us still $|E|$ planar, biconnected and spanning subgraphs of K_n, satisfying $x_e = 0$. Moreover, their incidence vectors are affinely independent.

Now we will show the facet-defining properties of those inequalities $c^T x \leq c_0$ which are facet-defining for $\mathcal{PLS}(K_n)$ and have biconnected support. There are sufficiently many planar subgraphs satisfying (7) with equality that can be used in an indirect proof to show the facet-defining property of (7) for $\mathcal{PLS}(K_n)$. That is, suppose there exists an inequality $a^T x \leq a_0$ with $\{x \in \mathcal{PLS}(K_n) \mid c^T x = c_0\} \subseteq \{x \in \mathcal{PLS}(K_n) \mid a^T x = a_0\}$. From this it follows that $c^T = a^T$ and $c_0 = a_0$. If all these planar subgraphs are biconnected and spanning, we have already found enough affinely independent points of $2\text{-}\mathcal{NCPLS}(K_n)$ in $\{x \in 2\text{-}\mathcal{NCPLS}(K_n) \mid c^T x = c_0\}$. Suppose, some of these subgraphs are not biconnected and spanning. Then we claim that we can augment them to planar, biconnected and spanning subgraphs of K_n by adding a set $F \subseteq E$, with $c(F) = 0$ and $a(F) = 0$. With the new set of independent points in $\{x \in 2\text{-}\mathcal{NCPLS}(K_n) \mid c^T x = c_0\}$ we can proceed as in the indirect proof for the facet-defining property of (7) for $\mathcal{PLS}(G)$.

Let P be the edge set of a planar subgraph used in the indirect proof for $\mathcal{PLS}(K_n)$, that is not biconnected nor spanning with $c^T \chi^P = c_0$. Let a be an articulation point of P separating the blocks induced by K_1 from $K_2 := P \setminus K_1$. Let v_i be a vertex next to a at the outer face of a planar embedding of block K_i, $i = 1, 2$. Adding the edge $e = (v_1, v_2)$ to P still maintains a planar subgraph P_1, where a does not anymore separate component K_1 from K_2. Moreover, we have $c_0 \geq c^T \chi^{P_1} = c^T \chi^P + c_e = c_0 + c_e$, hence $c_e = 0$ that implies $a_e = 0$. Repeating this iteration subsequently for all articulation points in P leads to a biconnected planar subgraph P_k with $P_k = P \cup F$ with $c(F) = 0$ and $a(F) = 0$. If P_k is not spanning, we add the edges (z, u), (z, w) for $(u, w) \in P_k$ and all vertices z that are not contained in P_k. Now the new edge set $P'_k = P_k \cup F'$ is spanning, while planarity and biconnectivity is maintained. Using the same arguments as before gives $c(F') = a(F') = 0$. $\qquad\qquad\square$

Our investigations concerning the facet-defining inequalities arising from biconnectivity conditions lead to the following theorem. For $k = 2$, inequalities (5) reduce to the ones given in Theorem 9.

Theorem 9 *Let $K_n = (V, E)$ be the complete graph on n nodes. Furthermore, let $z \in V$, and a set $W \subseteq V \setminus \{z\}$ with $\emptyset \neq W \neq V \setminus \{z\}$. The node-cut constraint*

$$x(\delta_{G-\{z\}}(W)) \geq 1 \qquad (8)$$

defines a facet of $2\text{-}\mathcal{NCPLS}(K_n)$.

Proof. We will prove it indirect. Let $c^T x \geq c_0$ denote inequality (8). Suppose there exists an inequality $a^T x \geq a_0$ with $\{x \in 2\text{-}\mathcal{NCPLS}(K_n) \mid c^T x = c_0\} \subseteq \{x \in 2\text{-}\mathcal{NCPLS}(K_n) \mid a^T x = a_0\}$. We will show that then $c^T = \lambda a^T$ and $c_0 = \lambda a_0$ for $\lambda > 0$. For any fixed $z \in V$ and fixed set $W \subseteq V \setminus \{z\}$, $\emptyset \neq W \neq V \setminus \{z\}$ let $U = V \setminus \{W \cup z\}$. We will construct a planar, biconnected and spanning subgraph of G induced by the edge set $P \subseteq E$. P consists of the edge sets of an Hamiltonian cycle H_W in W and H_U in U, resp. an Hamiltonian Path H_W if $|W| = 2$ or H_U if $|U| = 2$, the edge (u, w) for $u \in U$, $w \in W$, and the edges (u, z), (u', z), (z, w) and (z, w') for $u \neq u' \in U$, if $|U| \geq 2$, $w \neq w' \in W$ if $|W| \geq 2$. Obviously, P is the edge set of a planar, biconnected and spanning subgraph of G satisfying inequality (8) by equality, hence also $a^T \chi^P = a_0$.

Substituting in P the edge (u, w) by any of the edges (u'', w''), $u'' \in U$, $w'' \in W$ still yields a planar, biconnected spanning subgraph P_1 satisfying $c^T \chi^{P_1} \geq c_0$ with equality, hence also $a^T \chi^{P_1} = a_0$. We have $0 = c^T \chi^P - c^T \chi^{P_1} = a^T \chi^P - a^T \chi^{P_1} = a_{(u,w)} - a_{(u'',w'')}$, hence $a_{(u,w)} = a_{(u'',w'')}$ for all $u, u'' \in U$, $w, w'' \in W$.

By adding any edge contained in the complete graph induced by W to H_W the required properties are still satisfied, hence $a_e = 0$ for all $e \in G[W \setminus H_W]$. If

$|W| \geq 4$, then by choosing a different Hamiltonian cycle in $G[W]$, we get $a_e = 0$ for all $e \in G[W]$. In the case that $|W| = 3$, we can remove the edge $e = (w, w')$ from P without loosing biconnectivity, hence we get $a_e = 0$ for all $e \in G[W]$. The same arguments hold for the set U, so we have $a_e = 0$ for all $e \in G[U]$.

Moreover, adding any of the edges (z, w''), $w'' \in W \setminus \{w, w'\}$ to P still yields a planar, biconnected spanning subgraph of G satisfying inequality (8) with equality. Hence $a_{(z,w)} = 0$ for $w \in W$ and, because of symmetry reasons, $a_{(z,u)} = 0$ for $u \in U$. Hence we have shown $a_e = \lambda c_e$, and $a_0 = \lambda c_0$ for all $e \in E$. □

5 Algorithm

In [JM94] we give a branch and cut algorithm for the maximum planar subgraph problem using facet-defining inequalities for $\mathcal{PLS}(G)$ as cutting planes. We have to change and to add only a few routines in order to get a branch and cut algorithm for solving the three problems, the minimum planarizing 2-augmentation problem, the planar 2-augmentation problem and the maximum planar biconnected spanning subgraph problem.

In a cutting plane algorithm, a sequence of relaxations is solved by linear programming. After the solution x of some relaxation is found, we must be able to check whether x is the incidence vector of a planar, biconnected and spanning subgraph (in which case we have solved the problem) or whether any of the known facet-defining inequalities are violated by x. If no such inequalities can be found, we cannot tighten the relaxation and have to resort to branching, otherwise we tighten the relaxation by all facet-defining inequalities violated by x which we can find. Then the new relaxation is solved, etc. The process of finding violated inequalities (if possible) is called "separation" or "cutting plane generation".

The cutting plane generation as well as the lower bound heuristic for the planarity part are based on a planarity testing algorithm of Hopcroft and Tarjan [HT74]. At the beginning we solve the Linear Program (LP) consisting of the trivial inequalities $x_e \geq 0$, $x_e \leq 1$ and the inequality $x(E) \leq 3|V| - 6$. Let x be an LP-solution produced in the cutting plane procedure applied in some node of the enumeration tree. For $0 \leq \varepsilon \leq 1$ we define $E_\varepsilon = \{e \in E \mid x_e \geq 1 - \varepsilon\}$ and consider $G_\varepsilon = (V, E_\varepsilon)$. For the unweighted graph G_ε the linear planarity testing algorithm of Hopcroft and Tarjan is called. The algorithm stops if it finds an edge set F which is not planar. In case the inequality $x(F) \leq |F| - 1$ is violated, we reduce it to a facet-defining inequality before we add it to the constraints of the current LP. We also use a heuristic which searches for violated Euler-inequalities and inequalities given by some classes of s-chorded cycle graphs.

The cutting plane generation for achieving biconnectivity makes use of the facet-defining inequalities given in (5). For $k = 2$, the inequalities reduce to the ones given in (8). Given an LP-solution x produced in the cutting plane procedure, we are able to give a node $z \in V$ and a set W, $\emptyset \neq W \subset V \setminus \{z\}$ violating inequality (8) or guaranteeing that all the inequalities in (8) are satisfied

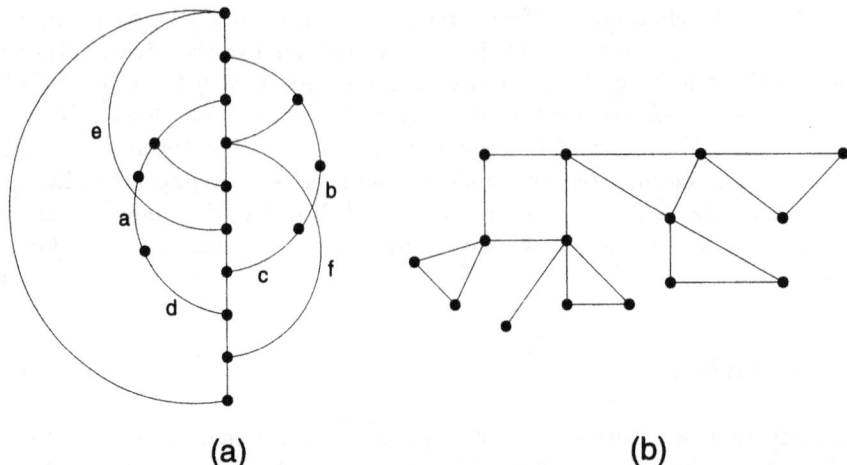

(a)   (b)

Fig. 3. shows two of the instances used for the computations. For the graph shown in (a) the approach, first to determine a maximum planar subgraph may lead to removing the edges a and b. In order to augment the graph to biconnectivity, we have to add at least three edges, whereas the optimum solution of the minimum planarizing 2-augmentation problem removes the edges e and f. The graph shown in (b) has been given by Kant. The heuristic he suggests adds 5 new edges to the graph in order to obtain a planar biconnected graph, whereas the optimum solution is to add 3 edges.

by x. This can be done in polynomial time by the following separation routine. For all nodes $z \in V$ construct the graph $G' := G - \{z\} = (V', E')$ and search for the minimum cut in G' with edge values x_e for all $e \in E'$. Let y be the value of this minimum cut. If $y \geq 1$, all inequalities of the type (8) for the specific node z are satisfied. Otherwise, the inequality $x(\delta_{G-\{z\}}(W)) \geq 1$ is violated for the set W determined by the minimum cut $\delta(W)$.

Although the vectors x coming up as solutions of LP-relaxations in the above outlined process have fractional components in general, they are often useful to obtain information on how a high-valued planar subgraph might look like. We exploit this idea with a greedy type heuristic with respect to the solution values of the edges. Starting from the empty graph, a planar subgraph is constructed by adding the edges in order of decreasing values if they do not destroy planarity. In a second phase this planar subgraph is augmented to a biconnected, spanning and planar subgraph by a trivial heuristic. So, in addition to the upper bounds $w^T x$ on the value of a maximum planar biconnected spanning subgraph, we also obtain a lower bound $w^T \bar{x}$ from the incidence vector $\bar{x}$ of the planar, biconnected and spanning subgraph derived heuristically from x.

6 Computational Experiments

The computational experiments were run on a Solbourne 5E/702. For our "first" experiments we used the above described algorithm. Our current preliminary implementation does not yet contain a procedure that tries to find good feasible solutions whose objective function values constitute lower bounds. However, in all examples discussed here, the optimum solution was found by cutting

Table 1. Computational Results for graphs from the literature

Problemname	$|V|$	$|E|$	#Var	Del1	Add1	Tim1	Del2	Add2	Tim2	Del3	Tim3
Jaya.10.22	10	22	45	2	0	1	2	0	0	2	1
Beck.11.21	11	21	55	1	1	0	1	1	0	-	-
JueMut.14.20	14	20	91	2	0	1	2	0	0	2	1
JueMut.16.22	16	22	120	2	0	1	2	3	2	2	1
Kant.15.20	15	20	105	0	3	0	0	3	0	-	-
EadMar.20.30	20	30	190	2	0	10	2	0	1	2	10
Harel.20.35	20	35	190	3	0	11	2	1	0	3	4
Martin.30.56	30	56	435	3	3	19	3	3	6	-	-
Hims.34.45	34	45	561	2	6	335	2	6	192	-	-
Kant.45.85	45	85	990	3	0	231	3	0	15	3	421
Hims.46.64	46	64	1035	2	1	64	2	2	23	-	-
Cim.60.166	60	160	1770	1	0	17	1	0	17	1	32

planes, i.e. the LP-solution x turned out to be the incidence vector of a planar, biconnected and spanning subgraph.

Table 1 shows computational results for some graphs in the [LAYOUT-LIB], a library for benchmark sets in Automatic Graph Drawing. The columns show the problem name, the number of nodes and edges of the given graph G, and the number of variables. Moreover, it shows the solution of the minimum planarizing 2-augmentation problem, i.e. the number of edges removed from G and added to G, and the solution time in seconds. Columns 8-10 show the solution, when first a maximum planar subgraph is determined and then the planar 2-augmentation problem is been solved. Surprisingly, the total time for solving both problems, the maximum planar subgraph problem and the planar 2-augmentation problem, is in many cases much less than the time for solving the minimum planarizing 2-augmentation problem. Moreover, the solutions found by the two different methods are in most cases identical. We do not have a code for augmenting a nonplanar graph to biconnectivity. Hence, we only give the results of the third approach, first augmenting the graph to biconnectivity and then solving the maximum planar biconnected subgraph problem, for the graphs that are already biconnected.

Consider the graph shown in Figure 3 (JueMut.16.22). In the run for solving the minimum planarizing 2-augmentation problem, 14 violated node-cut constraints have been found, 14 Kuratowski-inequalities and 1 Euler-inequality. It took 17 LP's with 4 branch and cut nodes in 1.48 seconds. At the end, the two edges e and f have been deleted. This solution is identical to the solution for the maximum planar biconnected subgraph problem. Solving first the maximum planar subgraph problem leads to removing the edges a and b (using 10 Kuratowski inequalities and 1 Euler inequality, 6 LP's in the root node in 0.54 seconds). Solving the planar 2-augmentation problem on the obtained planar subgraph leads to adding three new edges. In this case 8 violated node-cut inequalities were detected, 34 Kuratowski inequalities and 1 Euler inequality. The computation took 31 LP's and 8 branching nodes in 2.33 seconds.

This is the first time that these problems have been solved to optimality. We think that we will be able to solve even bigger instances in the near future.

507

References

[ET76]   Eswaran, K.P., and R.E. Tarjan: Augmentation problems. SIAM Journal on Computing 5 (1976) 653–665

[Fra92]   Frank, A.: Augmenting graphs to meet edge-connectivity requirements. SIAM J. Discr. Math. 5 (1992) 25–53

[GJ79]   Garey, M.R., and D.S. Johnson: Computers and Intractability: A Guide to the Theory of NP-completeness. Freeman, San Franc. (1979)

[GSS93]   Garg, N., V.S. Santosh, and A. Singla: Improved Approximation Algorithms for Biconnected Subgraphs via Better Lower Bounding Techniques. SODA (1993), 103–111

[GT94]   Goldschmidt, O. and A. Takvorian: Finding a Biconnected Spanning Planar Subgraph of a Nonplanar Graph. Technical Report ORP93-03, University of Texas, Austin, USA (1994)

[Har62]   Harary, F.: The maximum connectivity of a graph Proceed. of the National Academy of Sciences. USA 48 (1962) 1142–1146

[Him93]   Himsolt, M.: Konzeption und Implementierung von Grapheneditoren. Dissertation, Universität Passau (1993)

[HT74]   Hopcroft, J., and R.E. Tarjan: Efficient planarity testing. J. ACM 21 (1974) 549–568

[JM94]   Jünger, M. and P. Mutzel: Maximum planar subgraphs and nice embeddings: Practical layout tools. to appear in Algorithmica, special issue on Graph Drawing, Ed. G. Di Battista und R. Tamassia (1994)

[JM95]   Jünger, M. and P. Mutzel: The Polyhedral Approach to the Maximum Planar Subgraph Problem: New Chances for Related Problems. Lect. Notes in Comp. Sci 894, Proc. DIMACS GD'94, Ed. R. Tamassia and I.G. Tollis, Springer-Verlag, Princeton (1995) 119–130

[HR91]   Hsu, T.-S. and V. Ramachandran: A linear time algorithm for triconnectivity augmentation. Proc. 32th Annual Symp. on Found. of Comp. Science, Puerto Rico (1991) 548–559

[Kan93]   Kant, G.: Algorithms for Drawing Planar Graphs. Ph.D.-Thesis, Utrecht University (1993)

[LAYOUT-LIB] per ftp available at layoutlib@informatik.uni-koeln.de (1995)

[Mut94a]   Mutzel, P.: The Maximum Planar Subgraph Problem. Dissertation, Universität Köln (1994)

[Mut94b]   Mutzel, P.: s-Chorded Cycle Graphs and their Relation to the Planar Subgraph Polytope. Technical Report No. 94-161, Angewandte Mathematik und Informatik, Universität zu Köln (1994)

[RW95]   Ravi, R. and D.P. Williamson: An Approximation Algorithm for Minimum-Cost Vertex-Connectivity Problems. Proc. 6th Annual ACM-SIAM Symposium on Discrete Algorithms, San Francisco (1995)

[RG77]   Rosenthal, A. and A. Goldner: Smallest augmentation to biconnect a graph. SIAM J. on Computing 6 (1977) 55–66

[Sto92]   Stoer, M.: Design of Survivable Networks. Lecture Notes in Mathematics, Springer-Verlag, Berlin (1992)

[TBB88]   Tamassia, R., G. Di Battista, and C. Batini: Automatic graph drawing and readability of diagrams. IEEE Transactions on Systems, Man and Cybernetics 18 (1988) 61–79

[WN93]   Watanabe, T., and A. Nakamura: A minimum 3-connectivity augmentation of a graph. J. Comp. and Sys. Sci. 46 (1993) 91–128

Optimal Layouts on a Chain ATM Network

(Extended Abstract)

Ornan Gerstel[1], Avishai Wool[2] and Shmuel Zaks[1]

[1] Department of Computer Science, Technion, Haifa 32000, Israel,
E-mail: orig@csc.cs.technion.ac.il, zaks@cs.technion.ac.il.

[2] Department of Applied Mathematics and Computer Science, The Weizmann
Institute, Rehovot 76100, Israel, E-mail: yash@wisdom.weizmann.ac.il.

Abstract. We study a routing problem which occurs in high-speed
(ATM) networks, termed the "one-to-many virtual path layout problem" on chain networks. This problem is essentially a tree embedding
problem on a chain host graph. We present four performance measures
to the quality of such an embedding which have practical implications,
and find optimal solutions for each of them. We first show that the search
can be restricted to the class of layouts with no crossovers. Given bounds
on the load ℓ and number of hops h in a layout, we then present a family
of ordered trees $T(\ell, h)$, within which an optimal solution can be found
(if one exists at all); this holds for either the worst-case or average-case
measures, and for a chain of length n, with $n \leq \binom{l+h}{l}$. For the worst-
case measures these trees are used in characterizing, constructing, and
proving the optimality of the solutions . For the average-case measures
polynomial dynamic programming algorithms are presented which find
optimal solutions for all cases.

1 Introduction

1.1 Motivation

The advent of fiber optic media has dramatically changed the classical views on
the role and structure of digital communication networks. Specifically, the sharp
distinction between telephone networks, cable television networks, and computer
networks, has been replaced by a unified approach.

The most prevalent solution for this new networking challenge is called *Asynchronous Transfer Mode* (ATM for short), and is thoroughly described in the
literature [ITU90, HH91, CS94]. ATM is based on relatively small fixed-size
packets termed *cells*. Each cell is routed independently, based on two small routing fields at the cell header, called *virtual channel index* (VCI) and *virtual path
index* (VPI). At each intermediate switch, these fields serve as indices to two
routing tables (one for each field), and the routing is done in accordance to the
predetermined information in the appropriate entries.

Routing in ATM is hierarchical in the sense that the VCI of a cell is ignored
as long as its VPI is not null. This algorithm effectively creates two types of

predetermined simple routes in the network - namely routes which are based on
VPIs (called *virtual paths* or VPs) and routes based on VCIs and VPIs (called
virtual channels or VCs). VCs are used for connecting network users (e.g., a
telephone call); VPs are used for simplifying network management — routing of
VCs in particular. Thus the route of a VC may be viewed as a concatenation of
complete VPs.

As far as the mathematical model is concerned, given a communication network, the VPs form a virtual network on top of the physical one which we term
the *virtual path layout* (VPL for short), on the same vertices, but with a different
set of edges (typically a superset of the original edges). Each VC is a simple path
in this virtual network.

The VP layout must satisfy certain conditions to guarantee important performance aspects of the network. In particular, there are restrictions on: (1)
the number of virtual edges that share any physical edge (termed the *load*), (2)
the diameter of the virtual graph (termed the *hop count*), and (3) the connection between shortest paths in the physical and virtual graphs (see [GZ94] for a
justification of the model for ATM networks).

In many works (e.g., [ABNLP89, AP92, GZ94, CGZ94]), a general routing
problem is solved using a simpler sub-problem as a building block; In this subproblem it is required to enable routing between all vertices to a single vertex
(rather than between any pair of vertices). This restricted problem for the ATM
VP layout problem is termed the *one-to-many VPL* problem [GZ94] and is the
focus of the present work.

1.2 Related Works

A problem which is related to ours is that of keeping small routing tables for
routing in conventional computer networks. This problem was widely studied
[ABNLP89, AP92, FJ86, KK80, PU88] and yielded interesting graph decompositions and structures, but it differs from ours in some major aspects which
deemed most of these solutions impractical for our purposes. The main difference stems from the fact that in our case there is no flexibility as to the routing
scheme itself since it is determined by the ATM standard [ITU90].

A few works have tackled the VP layout problem, some using empirical techniques [ATTD94, LC93], and some using theoretical analysis [GZ94, CGZ94];
However, none of these works has attempted to combinatorially characterize the
optimal solution, and achieve a tight upper bound for the problem. In addition,
most of these works have considered only one of the relevant performance measures, namely the worst case load measure, while we solve the problem for several
others, equally important, performance measures too.

Of particular practical interest is the weighted hop count measure, since it
determines the expected time for setting up a connection between a pair of users,
given the relative frequency of connection requests between network vertices. A
similar problem was empirically handled in [GS95]. To the best of our knowledge,
the present work is the first to analytically tackle this problem.

Our problem is essentially a graph-embedding problem, in which it is required to find a tree T and embed it in a given chain, so that the congestion is low, and so that the height of T is kept low. However, all of the embedding problems we are aware of focus on embedding a *given graph* in a host graph, while here we are allowed to *choose* an embedded graph with minimal diameter. In addition, due to practical considerations, we restrict the discussion to paths in the embedded tree which map to shortest paths on the chain, a restriction which does not appear in embedding problems, and substantially affects the solutions. Thus these problems are not applicable for this case.

A related criterion, minimizing the *number* of VPs to achieve a required maximal hop count, is discussed in [BTS94].

1.3 Summary of Results

In the paper we consider the problem of constructing a VP layout on a network with chain topology. This simple topology enables us to study the problem in greater depth than previous works. As proven in the sequel (Theorem 9), this topology is the worst case with respect to all the relevant performance measures, and thus our upper bound results serve as an upper bound for more complex topologies as well.

We also restrict the discussion to "one-to-many" layouts, in which it is required to enable connections between all vertices to a single vertex called the *root* (rather than to all vertices). As mentioned earlier, these layouts have been shown to be a useful tool in constructing more complex layouts.

We consider four performance measures, and achieve optimal solutions for each measure:

- Given an upper bound on the maximum number of hops, minimize the maximum load ($\mathcal{L}_{\max}$),
- Given an upper bound on the maximum load, minimize the maximum number of hops ($\mathcal{H}_{\max}$),
- Given an upper bound on the maximum number of hops, minimize the average load ($\mathcal{L}_{\mathrm{avg}}$), and
- Given an upper bound on the maximum load, and vertex weights (representing the frequency of connection requests between the vertex and the root), minimize the average number of hops ($\mathcal{H}_{\mathrm{avg}}^{w}$).

After defining the model and measures (in Section 2), we show (in Section 3) that it is sufficient to consider layouts of a canonic form (Lemma 10 and Theorem 12), and consider the number of such different layouts (Catalan number C_{n-1} for a chain of n vertices). Next we focus (in Section 4) on the maximal load and hops measures and define a new class $\mathcal{T}(\ell, h)$ of ordered trees which includes as subtrees all feasible layouts that satisfy given load and hops constraints. This tree helps in characterizing tight bounds for both the $\mathcal{L}_{\max}$ and $\mathcal{H}_{\max}$ measures, namely, given a chain with n vertices such that $\binom{\ell+h-1}{\ell} < n \leq \binom{\ell+h}{\ell}$ then $\mathcal{H}_{\max} = h$ and if $\binom{\ell+h-1}{h} < n \leq \binom{\ell+h}{h}$ then $\mathcal{L}_{\max} = \ell$.

In Section 5 we study the average load measure, for which we obtain an $O(n^2h)$ algorithm for finding the optimal layout, based on dynamic programming, and achieve a similar $O(n^2\ell)$ algorithm for the unweighted average hops measure (i.e., when all weights are 1). Finally, we study the more complex weighted average hops measure and present an $O(n^3\ell)$ optimal algorithm for it. We conclude and list a few of the remaining open problems in Section 6. Most proofs are either omitted or only sketched in this Extended Abstract.

2 The Model

We model the underlying communication network as an undirected graph $G = (V, E)$, where V corresponds to the set of switches and E to the set of physical links between them.

Definition 1. A *virtual path layout* (VPL^{1-m} *for short*) Ψ is a collection of simple paths in G and a vertex $r \in V$ termed the *root* of the layout (denoted $root(\Psi)$)[3].

Definition 2. The *load* $\mathcal{L}(e)$ of an edge $e \in E$ in a VPL^{1-m} Ψ is the number of VPs $\psi \in \Psi$ that include e. The *maximal load* $\mathcal{L}_{\max}(\Psi)$ of a VPL^{1-m} Ψ is $\max_{e \in E} \mathcal{L}(e)$. The *average load* of a VPL^{1-m} Ψ is $\mathcal{L}_{\text{avg}}(\Psi) \equiv \frac{1}{|E|} \sum_{e \in E} \mathcal{L}(e)$.

Definition 3. The *hop count* $\mathcal{H}(v)$ of a vertex $v \in V$ in a VPL^{1-m} Ψ is the minimum number of VPs whose concatenation forms a shortest path in G from v to $root(\Psi)$. If no such VPs exist, define $\mathcal{H}(v) \equiv \infty$. The *maximal hop count* of a VPL^{1-m} Ψ is $\mathcal{H}_{\max}(\Psi) \equiv \max_{v \in V} \{\mathcal{H}(v)\}$.

Definition 4. Let $w(v)$ be non-negative weights assigned to the vertices $v \in V$ and let $W = \sum_{v \in V} w(v)$. The *weighted total hop count* of a VPL^{1-m} Ψ is $\mathcal{H}_{\text{tot}}^w(\Psi) \equiv \sum_{v \in V} w(v)\mathcal{H}(v)$, and the *weighted average hop count* is $\mathcal{H}_{\text{avg}}^w(\Psi) \equiv \mathcal{H}_{\text{tot}}^w(\Psi)/W$. When the weights are all $w(v) = 1$ then we denote the total hop count by $\mathcal{H}_{\text{tot}}(\Psi)$, and the *average hop count* is $\mathcal{H}_{\text{avg}}(\Psi) \equiv \frac{\mathcal{H}_{\text{tot}}(\Psi)}{n-1}$.

In the rest of this paper we assume that the underlying network is a *chain*. Therefore w.l.o.g. we can assume that the root of every VPL we consider is the leftmost vertex of the chain. For simplicity we denote the vertices $1, 2, \ldots, n$ and the root is always vertex 1. In a chain the path between two vertices is unique, so we can denote a VP $\psi \in \Psi$ between vertices u and v by the names of its endpoints, i.e., $\psi \equiv (u, v)$.

Definition 5. Let $\psi = (u, v)$ be a VP. Then the *length* of ψ, denoted $|\psi|$, is the number of physical links that ψ traverses, $|\psi| = v - u$. Let Ψ be a VPL^{1-m}, then the *total load* of Ψ is $\mathcal{L}_{\text{tot}}(\Psi) \equiv \sum_{\psi \in \Psi} |\psi|$.

---

[3] The "1-m" superscript is taken from [GZ94] and denotes the fact that we refer to "one-to-many" layouts. This fact is taken into account in Definition 3.

Lemma 6. *For any* VPL^{1-m} Ψ *on a chain,* $\mathcal{L}_{\mathrm{avg}}(\Psi) \equiv \frac{\mathcal{L}_{\mathrm{tot}}(\Psi)}{n-1}$.

To minimize the load, one can use a VPL^{1-m} Ψ which has a VP on each physical link, i.e., $\mathcal{L}_{\mathrm{max}}(\Psi) = 1$, however such a layout has a hop count of $n - 1$. The other extreme is connecting a direct VP from the root to each other vertex, yielding $\mathcal{H}_{\mathrm{max}} = 1$ but $\mathcal{L}_{\mathrm{max}} = n - 1$. For the intermediate cases we need the following definitions.

Definition 7. Let $\mathcal{H}_{\mathrm{opt}}(n, \ell)$ denote the *optimal hop count* of any VPL^{1-m} Ψ on a chain of n vertices such that $\mathcal{L}_{\mathrm{max}}(\Psi) \leq \ell$, i.e.,

$$\mathcal{H}_{\mathrm{opt}}(n, \ell) \equiv \min_{\Psi}\{\mathcal{H}_{\mathrm{max}}(\Psi) : \mathcal{L}_{\mathrm{max}}(\Psi) \leq \ell\}.$$

Definition 8. Let $\mathcal{L}_{\mathrm{opt}}(n, h)$ denote the *optimal load* of any VPL^{1-m} Ψ on a chain of n vertices such that $\mathcal{H}_{\mathrm{max}}(\Psi) \leq h$, i.e.,

$$\mathcal{L}_{\mathrm{opt}}(n, h) \equiv \min_{\Psi}\{\mathcal{L}_{\mathrm{max}}(\Psi) : \mathcal{H}_{\mathrm{max}}(\Psi) \leq h\}.$$

3 The Structure of an Optimal VPL^{1-m}

We first show that the chain network serves as an upper bound on VPL^{1-m} for any other tree topology, in any of the relevant performance measures.

Theorem 9. *Given a chain network with n vertices, a VPL^{1-m} Ψ on it, and a tree network T with n vertices and a root r, then there exists a VPL^{1-m} Ψ' for T with r as its root such that $\mathcal{L}_{\mathrm{max}}(\Psi) \geq \mathcal{L}_{\mathrm{max}}(\Psi')$, $\mathcal{L}_{\mathrm{avg}}(\Psi) \geq \mathcal{L}_{\mathrm{avg}}(\Psi')$, $\mathcal{H}_{\mathrm{max}}(\Psi) \geq \mathcal{H}_{\mathrm{max}}(\Psi')$, and $\mathcal{H}_{\mathrm{avg}}^{w}(\Psi) \geq \mathcal{H}_{\mathrm{avg}}^{w}(\Psi')$.*

Sketch of proof. Given a chain C, we gradually transform it into T using a series of intermediate trees $C = T_1, T_2, ..., T_k = T$, and modify Ψ accordingly: We first split C into sub-chains according to the sizes of the root's subtrees in T, and connect each sub-chain directly to vertex 1 (which is Ψ's root). At the same time we modify each VP $(l, r) \in \Psi$ that has endpoints in two different sub-chains to connect the root directly (i.e., $(1, r)$ using the new edge between the sub-chain and vertex 1). This transformation may only decrease the load on edges, and the hop counts of some vertices, and thus can only improve the layout w.r.t. any of the above measures. We continue this process recursively in each sub-chain. □

We now establish a canonic form of a VPL^{1-m}, which will simplify the rest of the discussion.

Lemma 10. *Every vertex $v \geq 2$ on a chain is the right-most endpoint of a single VP in a VPL^{1-m} which is optimal under one of the $\mathcal{L}_{\mathrm{max}}$, $\mathcal{L}_{\mathrm{avg}}$, $\mathcal{H}_{\mathrm{max}}$, or $\mathcal{H}_{\mathrm{avg}}^{w}$ measures. In other words, an optimal VPL^{1-m} Ψ induces a tree rooted at vertex 1 with the VPs corresponding to the tree edges.*

Definition 11. Two VPs denoted (l_1, r_1) and (l_2, r_2) constitute a *crossing* if $l_1 < l_2 < r_1 < r_2$. A VPL^{1-m} is called *crossing-free* if no pair of VPs constitute a crossing.

Theorem 12. *For each performance measure ($\mathcal{L}_{\max}$, $\mathcal{H}_{\max}$, $\mathcal{L}_{\text{avg}}$, and $\mathcal{H}^w_{\text{avg}}$) there exists an optimal VPL^{1-m} which is crossing-free.*

Sketch of proof. For the sake of brevity, we shall present only the proofs for the $\mathcal{H}_{\max}$ and $\mathcal{H}^w_{\text{avg}}$ cases, the proof for the other measures follows along the same lines. Assume there is a pair of crossing VPs (l_1, r_1) and (l_2, r_2) in Ψ.

(1) The $\mathcal{H}_{\max}$ case: Denote by a_i ($i \in \{1, 2\}$) the minimal number of hops from l_i to the root (1), denote by b_i the maximum number of hops between r_i to a vertex $v_i > l_i$ (see Figure 1(a)). Due to Lemma 10, there is a single VP path from v_i to the root, which must traverse (l_i, r_i), thus $a_i + 1 + b_i \leq h$ to satisfy $\mathcal{H}_{\max}(\Psi) \leq h$. If $b_1 > b_2$ then replace the VP (l_2, r_2) by (r_1, r_2) in Ψ (see Figure 1(b)). It is easy to see that the maximum load does not increase: in the segment (l_2, r_1) the load is decreased by one, and the rest of the chain is not affected. As to the maximum hops, the only vertices affected by the change are in the subtree of r_2 in the VPL "tree", where in the worst case

$$\mathcal{H}(v_2) = b_2 + 1 + 1 + a_1 \leq b_1 + 1 + a_1 \leq h.$$

In case $b_1 \leq b_2$, we replace (l_1, r_1) in Ψ by (l_2, r_1) (see Figure 1(c)). Again, both the maximum load and hops are not increased. The load has changed only in the segment (l_1, l_2), in which it has decreased by one, and the hops have changed only in the subtree of r_1 in the VPL, where in the worst case

$$\mathcal{H}(v_1) = b_1 + 1 + a_2 \leq b_2 + 1 + a_2 \leq h.$$

Note that both transformations reduce the number of crossings by at least one, and do not increase the maximum load and hops. Hence all the crossings may be eliminated by iterating the transformation a finite number of times.

(2) The $\mathcal{H}^w_{\text{avg}}$ case: Refer again to Figure 1(a), but define b_i to be the sum of weights of all vertices in the VPL which are in the subtree to r_i (a_i remains the number of hops from l_i to the root). Let c_i denote the weighted sum of hops of all vertices except those in the subtree of r_i in Ψ, and d_i denote the weighted sum of hops of in r_i's subtree up to r_i (i.e., *not* all the way to vertex 1). Thus initially we have

$$\mathcal{H}^w_{\text{tot}}(\Psi) = c_i + d_i + b_i(1 + a_i)$$

where the third component of the sum is the additional cost of vertices in r_i's subtree, from r_i to vertex 1.

If $a_1 < a_2$ then apply the transformation of Figure 1(b), After which $\mathcal{H}^w_{\text{tot}}(\Psi') = c_2 + d_2 + b_2(1 + 1 + a_1)$ so we get $\mathcal{H}^w_{\text{tot}}(\Psi') \leq \mathcal{H}^w_{\text{tot}}(\Psi)$, and therefore $\mathcal{H}^w_{\text{avg}}(\Psi') \leq \mathcal{H}^w_{\text{avg}}(\Psi)$. On the other hand, if $a_1 \geq a_2$ then we use the transformation of Figure 1(c). In this case $\mathcal{H}^w_{\text{tot}}(\Psi) \geq c_1 + d_1 + b_1(1 + a_2) = \mathcal{H}^w_{\text{tot}}(\Psi')$. Again, in both cases the number of crossings has decreased by at least one. □

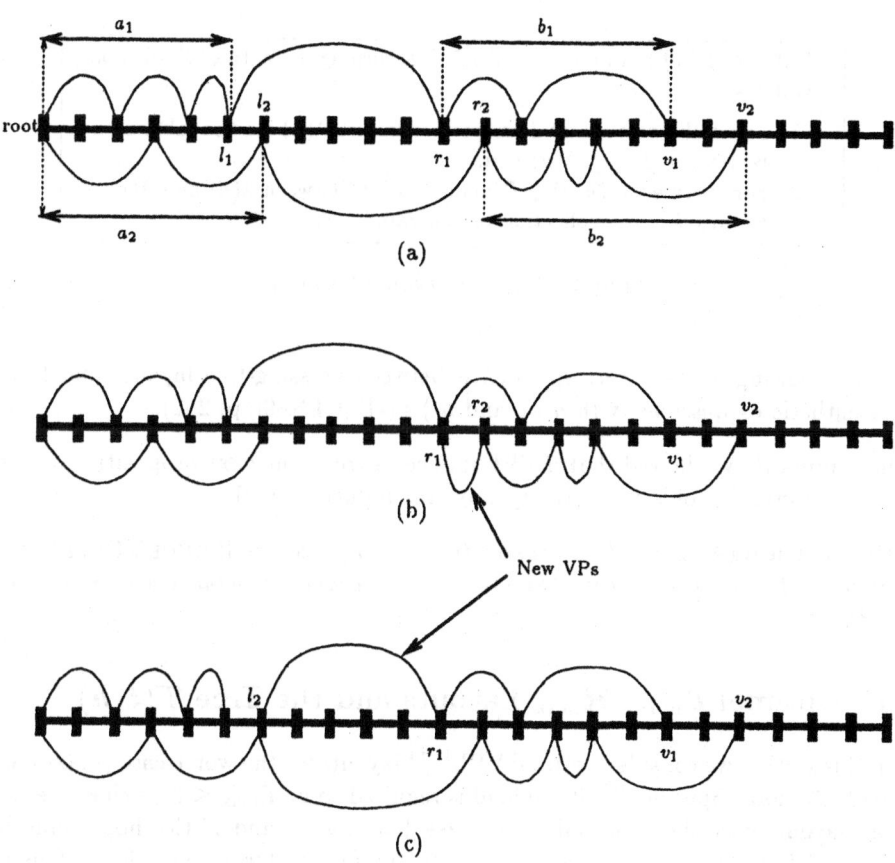

Fig. 1. Crossing elimination transformations

In the rest of the paper we restrict ourselves to VPLs satisfying Lemma 10 and Theorem 12. The next Lemma characterizes the number of such canonic VPLs.

Lemma 13. *The number $\gamma(n)$ of crossing-free VPL^{1-m} constructions on a chain with n vertices is a Catalan number,*

$$\gamma(n) = C_{n-1} = \frac{1}{n}\binom{2n-2}{n-1} = \Omega\left(\frac{4^n}{n^{3/2}}\right).$$

Sketch of proof. The lemma can be proven using the following recurrence (based on the structure of $T(\ell, h)$ — see next section):

$$\gamma(n) = \begin{cases} 1, & n = 1, \\ \sum_{d=1}^{n-1} \gamma(d)\gamma(n-d), & n \geq 2. \end{cases}$$

INDUCEVPL(T): Induce a VPL^{1-m} according to a tree T with n vertices.

1. Label the vertices of T in pre-order. Let $\lambda(u)$ be the label of a vertex $u \in T$, $1 \leq \lambda(u) \leq n$.
2. For every edge $(u, v) \in T$ connect a VP between $\lambda(u)$ and $\lambda(v)$.
3. Return Ψ_T, the collection of generated VPs.

Fig. 2. Procedure INDUCEVPL(T).

or by showing a $1 - 1$ correspondence between crossing-free layouts and legal parenthetic expressions with n ('s and n)'s (cf. [CLR89], p. 262). $\quad\square$

In Lemma 10 we showed that a VPL induces a tree. The next proposition shows that the inverse holds too, namely, any tree induces a VPL.

Proposition 14. *Let T be a ordered tree. Then procedure* INDUCEVPL(T) *(see Figure 2 for the pseudo-code, Figure 3 for an example) induces a crossing-free* VPL^{1-m}.

4 Optimal $\mathcal{L}_{\max}$, $\mathcal{H}_{\max}$ Layouts and the Tree $\mathcal{T}(\ell, h)$

In this section we consider optimal VPL^{1-m} layouts for the worst-case (maximal) load and hops. Specifically, if the load is required to be $\mathcal{L}_{\max} \leq \ell$ we characterize the layout with the minimal worst case hop count, and if the hop count is $\mathcal{H}_{\max} \leq h$ we characterize the layout with the minimal worst-case load. This is done using a new class of trees $\mathcal{T}(\ell, h)$ (see next definition). These trees contain all VPL^{1-m}s on a chain that satisfy the above load and hop constraints.

Definition 15. The ordered tree $\mathcal{T}(\ell, h)$ is defined recursively as follows. The root r has ℓ children, numbered $1, \ldots, \ell$ from left to right. The ith child is the root of a $\mathcal{T}(i, h - 1)$ subtree. A tree $\mathcal{T}(\ell, 0)$ or $\mathcal{T}(0, h)$ is a single vertex (see Figure 3 for an example).

Note that an internal vertex of $\mathcal{T}(\ell, h)$ which is the ith child (from the left) of its parent has i children. The tree $\mathcal{T}(1, h)$ is a rooted chain of $h + 1$ vertices. Also note that $\mathcal{T}(\ell, h)$ has height h and maximum degree ℓ.

Lemma 16. *The tree $\mathcal{T}(\ell, h)$ contains $\binom{\ell+h}{h}$ vertices.*

Proof. Let $N(\ell, h)$ denote the number of vertices in $\mathcal{T}(\ell, h)$. Then $N(\ell, h)$ satisfies the recurrence $N(\ell, h) = 1 + N(1, h - 1) + N(2, h - 1) + \cdots + N(\ell, h - 1)$. Since $N(0, h) = 1$ we can write

$$N(\ell, h) = \begin{cases} 1, & \ell = 0 \text{ or } h = 0, \\ \sum_{j=0}^{\ell} N(j, h - 1), & \text{otherwise.} \end{cases}$$

One can prove by induction (cf. [GKP89], p. 174) that $N(\ell, h) = \binom{\ell+h}{h}$. $\quad\square$

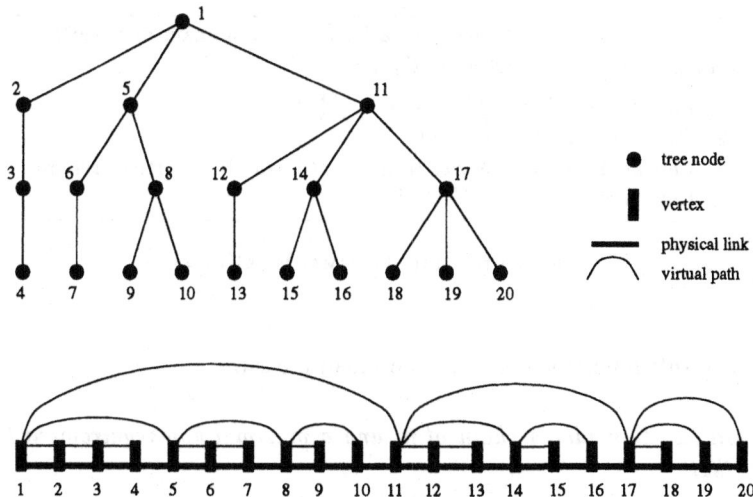

Fig. 3. The tree $T(3,3)$ and its induced VPL.

Definition 17. An ordered tree T is *embedded* in $T(\ell, h)$ if its root is the root of $T(\ell, h)$ and the subtrees of the root's children in T are (recursively) embedded in the subtrees of a subset of the children of the root in $T(\ell, h)$.

Proposition 18. *Let T be an ordered tree that is embedded in $T(\ell, h)$, and let Ψ_T be the output of Procedure* INDUCEVPL(T).
Then $\mathcal{L}_{\max}(\Psi_T) \leq \ell$ and $\mathcal{H}_{\max}(\Psi_T) \leq h$.

Proposition 19. *For every crossing-free VPL^{1-m} Ψ, with $\mathcal{L}_{\max}(\Psi) \leq \ell$ and $\mathcal{H}_{\max}(\Psi) \leq h$ there exists a tree T which is embedded in $T(\ell, h)$ such that $\Psi =$* INDUCEVPL(T).

Theorem 20. *Consider a chain of n, and a maximal load requirement ℓ. Let h be such that*
$$\binom{\ell + h - 1}{\ell} < n \leq \binom{\ell + h}{\ell}.$$
Then $\mathcal{H}_{\mathrm{opt}}(n, \ell) = h$.

Sketch of proof. It is easy to verify that h hops are sufficient (i.e., there exists a VPL Ψ such that $\mathcal{H}_{\max}(\Psi) = \mathcal{H}_{\mathrm{opt}}(n, \ell)$): use Procedure OPTIMALMAXH(n, ℓ) (see Figure 4). To show that there exists no VPL with less hops, assume (to obtain a contradiction) that there exists a VPL Ψ' with $\mathcal{L}(\Psi') \leq \ell$ and $\mathcal{H}(\Psi') \leq h - 1$. Then by Proposition 18 there exists a tree T' embedded in $T(\ell, h-1)$ which induces Ψ', but

$$|T'| \leq |T(\ell, h-1)| = N(\ell, h-1) = \binom{\ell + h - 1}{h - 1} = \binom{\ell + h - 1}{\ell} < n,$$

a contradiction. □

OPTIMALMAXH(n, ℓ): Construct a VPL^{1-m} on a chain of n vertices with minimal $\mathcal{H}_{\max}$ such that $\mathcal{L}_{\max} \leq \ell$.

1. Find h such that $\binom{\ell+h-1}{h-1} < n \leq \binom{\ell+h}{h}$.
2. Construct the tree $T = \mathcal{T}(\ell, h)$.
3. Repeatedly remove leaves from T until exactly n vertices remain.
4. Return $\Psi_T = \text{INDUCEVPL}(T)$.

Fig. 4. Procedure OPTIMALMAXH(n, ℓ).

A similar result holds for the maximum load measure:

Theorem 21. *Consider a chain of n, and a maximal hop requirement h. Let ℓ be such that*

$$\binom{\ell + h - 1}{h} < n \leq \binom{\ell + h}{h}.$$

Then $\mathcal{L}_{\text{opt}}(n, h) = \ell$.

Proposition 22. *Given a chain with $n = N(\ell, h)$ there exists a unique VPL^{1-m} with $\mathcal{L}_{\max}(\Psi) \leq \ell$ and $\mathcal{H}_{\max}(\Psi) \leq h$. As is evident from the OPTIMALMAXH(n, ℓ) procedure (Figure 4) there exist several such VPLs for smaller chains (i.e., $n < N(\ell, h)$).*

Remark. The above results relate to results in [CGZ94, GZ94] in the following way:

- Theorem 20 precisely characterizes the optimal solution on a chain. In addition we prove that if $n = N(\ell, h)$ then $(h!n)^{1/h} - \frac{h+1}{2} \leq \ell \leq (h!n)^{1/h} - 1$. This is an improvement to the upper bound $\ell \leq hn^{1/h}$ of [CGZ94], since $(h!)^{1/h} \leq h$.
- In [GZ94], a greedy optimal algorithm for finding a VPL^{1-m} on tree networks is presented. The algorithm does not give insight into the structure of the obtained VPL^{1-m}, in particular no upper bound is easily derived from it. It is interesting to note that the resulting VPL^{1-m} of the algorithm[4] for a chain with $n = N(\ell, h)$ is identical to the one obtained by our construction, as indeed predicted by Proposition 22.

---

[4] The model in [GZ94] differs from ours in that the load is measured on the vertices rather on the edges of the network. Therefore the greedy algorithm should be modified to optimize the edge-load. We refer to this modified algorithm in this comparison.

5 Minimizing the average case

5.1 The average load

In this section we consider the case where the maximal number of hops is limited to h, and we wish to find the layout with the smallest average load. Recalling Lemma 6, we observe that a layout Ψ_{opt} that minimizes $\mathcal{L}_{\mathrm{avg}}$ also minimizes its total load $\mathcal{L}_{\mathrm{tot}}(\Psi_{\mathrm{opt}})$. Hence the following definition.

Definition 23. Let $\mathcal{L}_{\mathrm{tot}}(n, h)$ denote the minimal total load of any VPL^{1-m} on n vertices with at most h hops, namely

$$\mathcal{L}_{\mathrm{tot}}(n, h) \equiv \min_{\Psi}\{\mathcal{L}_{\mathrm{tot}}(\Psi) : \mathcal{H}_{\max}(\Psi) \leq h\}.$$

The following example shows that we cannot simply use the constructions of the previous section for the average cases.

Example 1. Consider the layouts in Figure 5, on a chain of $n = 7$ vertices, all of which satisfy the worst-case constraints $h \leq 2$ and $\mathcal{L}_{\max} = 3$. However layout (a) has the optimal total load, $\mathcal{L}_{\mathrm{tot}}(\Psi_a) = 11$, while layouts (b) and (c) have total loads of 12 and 14 respectively.

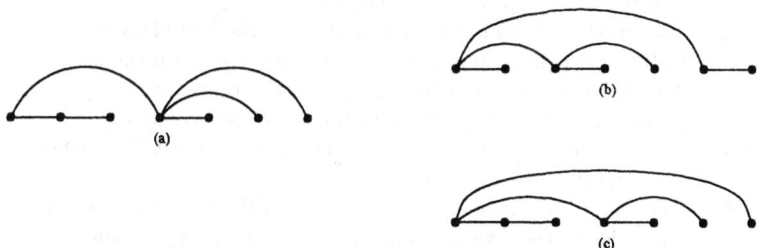

Fig. 5. Three layouts on $n = 7$ vertices, $h \leq 2$ hops, and $\mathcal{L}_{\max} = 3$.

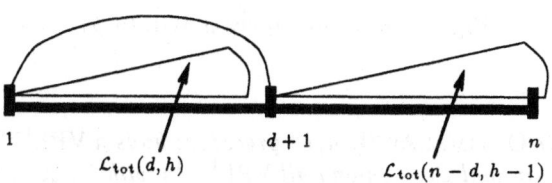

Fig. 6. The optimal average load layout

Next we describe the rationale behind our dynamic programming algorithm for finding optimal $\mathcal{L}_{\text{tot}}$ ($\mathcal{L}_{\text{avg}}$) layouts. Let Ψ_{opt} be the optimal VPL^{1-m} (that achieves $\mathcal{L}_{\text{tot}}(\Psi_{\text{opt}}) = \mathcal{L}_{\text{tot}}(n,h)$). Let $(1, d+1)$ be the longest VP connected to the root (see Figure 6). Since, by Theorem 12, we can assume that Ψ_{opt} is crossing-free, it follows that no VP of ψ connects a vertex $i \leq d$ with a vertex $j > d+1$, thus Ψ_{opt} splits into two disjoint optimal layouts, one on vertices $1, \ldots, d$ and the other on $d+1, \ldots, n$. However, the second layout (rooted at $d+1$) may use only $h-1$ hops since one hop is used to traverse the VP $(1, d+1)$. Thus if d is known, then $\mathcal{L}_{\text{tot}}$ satisfies the recurrence

$$\mathcal{L}_{\text{tot}}(n,h) = d + \mathcal{L}_{\text{tot}}(d,h) + \mathcal{L}_{\text{tot}}(n-d, h-1),$$

so clearly

$$\mathcal{L}_{\text{tot}}(n,h) = \min_{1 \leq d \leq n-1} \{d + \mathcal{L}_{\text{tot}}(d,h) + \mathcal{L}_{\text{tot}}(n-d, h-1)\}.$$

There are two simple "boundary" cases: (1) If $1 \leq n \leq h+1$ then clearly $\mathcal{L}_{\text{tot}}(n,h) = n-1$, (2) If $h = 1$ then we must connect a direct VP to each vertex, so $\mathcal{L}_{\text{tot}}(n,1) = n(n-1)/2$. The above argument leads to the dynamic programming algorithm in Figure 7 for finding an optimal VPL.

OPTIMALAVGL(n,h): Construct a VPL^{1-m} on a chain of n vertices with minimal $\mathcal{L}_{\text{avg}}$ such that $\mathcal{H}_{\max} \leq h$.

1. Maintain an $n \times h$ table A in which the values of $\mathcal{L}_{\text{tot}}(i,j)$ will be stored in $A[i,j]$, for increasing values of i, j.
2. Initialize the relevant entries according to the boundary cases.
3. Gradually calculate new values of A[i,j] by relying on the already calculated ones (to calculate $A[i,j]$ use the values of $A[k,j]$ and $A[k, j-1]$ for $k < i$ according to the above recurrence).
4. For each calculated A[i,j], store d (the length of the VP for which the minimum sum was obtained).
5. After $A[n,h]$ is calculated, traverse the VPL from the root, recalling the above d values: First recall d for $A[n,h]$, producing a VP $(1, d+1)$, next recall the d values for each of the two segments $[1, d]$ and $[d+1, n]$ and produce the relevant VPs (namely the values d', d'' for $A[d,h]$ and $A[n-d, h-1]$ producing the VPs $(1, d')$ and $(d+1, d+1+d''+1)$ respectively). Proceed recursively in each segment.
6. Return the resulting layout.

Fig. 7. Procedure OPTIMALAVGL(n,h).

Lemma 24. *The* OPTIMALAVGL(n,h) *procedure finds a* VPL^{1-m} *on a chain of* n *vertices with minimal* $\mathcal{L}_{\text{avg}}$ *among all* VPL^{1-m}s *with* $\mathcal{H}_{\max} \leq h$.

Lemma 25. *The time complexity of* OPTIMALAVGL(n,h) *is* $O(n^2 h)$.

5.2 The unweighted average hops measure

We now turn to the average hops measure, given a maximum bound ℓ on the load. We start with the simpler unweighted average case, which can be solved by an algorithm similar to that of the average load.

Definition 26. Consider a crossing-free optimal VPL Ψ_{opt} for a chain with n vertices and maximum load ℓ (which achieves minimum $\mathcal{H}_{tot}(\Psi)$). Recalling Definition 4, define $\mathcal{H}_{tot}(n, \ell) \equiv \mathcal{H}_{tot}(\Psi_{opt})$.

Let $(1, d+1)$ be the longest VP connected to the root. Again, it follows that there exists no VP connecting the vertices $1, ..., d$ to the vertices $d + 2, ..., n$, and thus the layouts in these two segments are independent and should both be optimal in Ψ_{opt}. In this case however, the layout on the vertices $1, ..., d$ should not exceed the load $\ell - 1$ (since together with the VP $(1, d)$ the load should not exceed ℓ). By the above discussion, it is evident that

$$\mathcal{H}_{tot}(n, \ell) = min_{1 \le d \le n-1}\{\mathcal{H}_{tot}(d, \ell - 1) + (n - d) + \mathcal{H}_{tot}(n - d, \ell)\}.$$

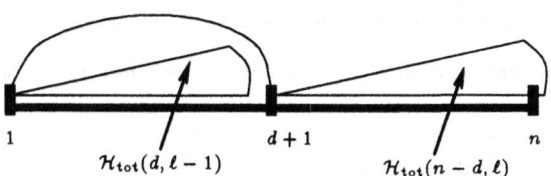

Fig. 8. The optimal average hops layout

The first and third components of the sum are the values of $\mathcal{H}_{tot}$ in the two separate segments, and the second component is the cost of an additional hop incurred by all vertices in the segment $d + 1, ..., n$ (see Figure 8). The boundary cases here are $\mathcal{H}_{tot}(n, 1) = (n - 2)(n - 1)/2$ (if the maximum load is 1 then the only possible VPs are identical to the network edges), and if $n \le \ell$ then $\mathcal{H}_{tot}(n, \ell) = n - 1$ (since we can afford to construct direct VPs from all vertices to the root).

In this case too a similar dynamic programming algorithm can be devised with time complexity $O(n^2 \ell)$. The algorithm is omitted for the sake of brevity.

5.3 The weighted average hops measure

This case is based on similar consideration as for the unweighted case, but the resulting algorithm turns out to have a higher time complexity since the values for $\mathcal{H}_{tot}^w$ depend not only on the size of the chain but on the specific weights assignment.

In [GZ94] an optimal solution for the $\mathcal{H}_{\max}$ measure is presented, based on a greedy algorithm from leaves towards the root of a tree. The next lemma indicates that no such greedy algorithm is possible for the weighted average case.

Lemma 27. *There exists a chain and a weights assignment to its vertices for which an algorithm that determines the VPL^{1-m} based on subtrees from the leaves towards the root exclusively, cannot find an optimal VPL^{1-m}.*

Definition 28. Given a chain with n vertices and weights assignment $w(v)$, let $W(i,j) = \sum_{v=i}^{j} w(v)$. Define $\mathcal{H}_{\text{tot}}^w(i,j,\ell)$ to be the optimal value for $\mathcal{H}_{\text{tot}}^w$ for a chain rooted at i and ending at $j \geq i$, with maximum load ℓ.

We prove that

Lemma 29. *Given any weight assignment W, the minimal weighted total load $\mathcal{H}_{\text{tot}}^w$ satisfies*

$$\mathcal{H}_{\text{tot}}^w(i,j,\ell) = \min_{i \leq d \leq j-1}\{\mathcal{H}_{\text{tot}}^w(i,d,\ell-1) + W(d+1,j) + \mathcal{H}_{\text{tot}}^w(d+1,j,\ell)\}$$

Based on this lemma, we present a dynamic programming algorithm for this problem. This algorithm maintains a three dimensional table A with dimensions $n \times n \times \ell$ where $A[i,j,k]$ contains the value of $\mathcal{H}_{\text{tot}}^w(i,j,k)$. We show that

Lemma 30. *The time complexity of the above algorithm for calculating $\mathcal{H}_{\text{tot}}^w(1,n,\ell)$ is $O(n^3\ell)$.*

6 Summary and Open Problems

We have considered a routing problem termed the "one-to-many VP layout problem" that arises in ATM networks. In the work we mainly focused on the worst case topology for this problem — a chain network (and proven it is indeed the worst case). We have identified four performance measures for a solutions to the above problem and presented optimal solutions with respect to each measure. We characterized the structure of the VP layout for the worst-case measures $\mathcal{L}_{\max}$ and $\mathcal{H}_{\max}$. We also presented dynamic programming polynomial algorithms for the average case measures $\mathcal{L}_{\text{avg}}$, $\mathcal{H}_{\text{avg}}$, and $\mathcal{H}_{\text{avg}}^w$.

The most immediate open problem is to generalize these results for arbitrary trees, a task which seems non-trivial, as far as the dynamic programming algorithms are concerned, due to the additional structural information that is attached to each subtree (which does not exist in chains).

Another open problem which is of practical interest is to use our VPL^{1-m}s to construct an optimal many-to-many VPL even for chains. Of particular practical interest is such a construction for the $\mathcal{H}_{\text{avg}}^w$ measure.

Acknowledgment

We would like to thank Nicola Santoro for stimulating discussions.

References

[ABNLP89] B. Awerbuch, A. Bar-Noy, N. Linial, and D. Peleg. Compact distributed data structures for adaptive routing. In *21st Symp. on Theory of Computing*, pages 479–489, 1989.

[AP92] B. Awerbuch and D. Peleg. Routing with polynomial communication-space tradeoff. *SIAM Journal on Discrete Math*, 5(2):151–162, May 1992.

[ATTD94] S. Ahn, R.P. Tsang, S.R. Tong, and D.H.C. Du. Virtual path layout design on ATM networks. In *IEEE Infocom'94*, pages 192–200, 1994.

[BTS94] H.L. Bodlaender, G. Tel, and N. Santoro. Trade-offs in non-reversing diameter. *Nordic Journal of Computing*, 1:111–134, 1994.

[CGZ94] I. Cidon, O. Gerstel, and S. Zaks. A scalable approach to routing in ATM networks. In G. Tel and P.M.B. Vitányi, editors, *The 8th International Workshop on Distributed Algorithms (LNCS 857)*, pages 209–222, Terschelling, The Netherlands, October 1994.

[CLR89] T. H. Cormen, C. E. Leiserson, and R. L. Rivest. *Introduction to Algorithms*. MIT Press, 1989.

[CS94] R. Cohen and A. Segall. Connection management and rerouting in ATM networks. In *IEEE Infocom'94*, pages 184–191, 1994.

[FJ86] G.N. Frederickson and R. Janardan. Separator-based strategies for efficient message routing. In *27th Symp. on Foundations of Computer Science*, pages 428–437, 1986.

[GKP89] R. L. Graham, D. E. Knuth, and O. Patashnik. *Concrete Mathematics*. Addison-Wesley, 1989.

[GS95] O. Gerstel and A. Segall. Dynamic maintenance of the virtual path layout. In *IEEE Infocom'95*, pages 330–337, April 1995.

[GZ94] O. Gerstel and S. Zaks. The virtual path layout problem in fast networks. In *The 13th ACM Symp. on Principles of Distributed Computing*, pages 235–243, Los Angeles, USA, August 1994.

[HH91] R. Händler and M.N. Huber. *Integrated Broadband Networks: an introduction to ATM-based networks*. Addison-Wesley, 1991.

[ITU90] ITU recommendation. I series (B-ISDN), Blue Book, November 1990.

[KK80] L. Kleinrock and F. Kamoun. Optimal clustering structures for hierarchical topological design of large computer networks. *Networks*, 10:221–248, 1980.

[LC93] F.Y.S. Lin and K.T. Cheng. Virtual path assignment and virtual circuit routing in ATM networks. In *IEEE Globecom'93*, pages 436–441, 1993.

[PU88] D. Peleg and E. Upfal. A tradeoff between space and efficiency for routing tables. In *20th Symp. on Theory of Computing*, pages 43–52, 1988.

Efficient Dynamic-Resharing "Verifiable Secret Sharing" Against Mobile Adversary

Noga Alon[1][*] and Zvi Galil[2][**] and Moti Yung[3]

[1] Tel-Aviv University
[2] Columbia University and Tel-Aviv University
[3] IBM T. J. Watson Research Center, Yorktown Heights, NY

Abstract. We present the notion of "Dynamic Re-sharing Verifiable Secret Sharing" (VSS) where the dealing of shares is dynamically and randomly refreshed (without changing or corrupting the secret). It works against the threat of the recently considered *mobile adversary* that may control all the trustees, but only a bounded number thereof at any time period.

VSS enables a dealer to distribute its secret to a set of trustees, so that they are assured that the sharing is valid and that they can open it later, and further no small group of trustees can open it prematurely. Recently, such sharing of cryptographic tools gained much attention, e.g., in the context of "key escrow cryptography" where a user enables a group of trustees to potentially open its information (when authorized by the court). Our dynamic-sharing VSS allows for mobile adversary attacking different sets of trustees at different time periods (modeling, e.g., network viruses that get spread as well as get eliminated).

Technically, we concentrate on simple direct methods that are combinatorial and number-theoretic in nature, and employ only simple public-key functions. (All previous schemes withstanding linear number of faults embedded secrets in polynomials which we do not do). In addition, our protocols are constant round.

The work is a sequence of reductions. We reduce $t(t < n(1/2-\epsilon))$ out-of n VSS from n out-of n one (assuming ex-or homomorphic encryption), then we reduce dynamic resharing (by the dealer) VSS from the static VSS, finally we reduce proactive VSS (dynamic VSS with no dealer presence after the initial sharing) from our dynamic resharing VSS.

1 Introduction

Verifiable secret sharing (VSS), introduced by Chor et al. [9], is an important primitive for sharing cryptographic data and control as well as a tool for other

---

[*] Supported in part by a USA-Israel BSF grant and by the Fund for Basic Research administered by the Israeli Academy of Sciences

[**] Supported in part by NSF grant CCR-93-16209 and CISE Institutional Infrastructure grant CDA-90-24735

applications in distributed computing protocols (a recent application is key escrow systems such as fair cryptosystems [20]). VSS enables a dealer to transmit a secret to a set of n trustees (share-holders), so that up to a threshold of t trustees (t out-of n) cannot get the secret and later a threshold of trustees (or more) can reconstruct the secret. These are the properties of secret sharing [24, 5]. Further, the sharing is verified: after the secret is shared the trustees are assured that reconstruction is possible, otherwise they reject the sharing.

In applications like key escrow (e.g., the Clipper encryption chip, but with voluntary key choice as in [20]), the dealer keeps on storing the secret but the trustees may open it upon request (this is exactly the escrow-agent scenario where a court asks the agents to open the secret). The verifiability of the sharing is crucial in such applications since the trustees when required (by court order) to open the shared secret, must be able to do so. Thus, we have to force the dealer to perform a valid sharing, and otherwise reject its actions.

Two basic methods for realizing VSS are known to withstand a *linear number of faults*. The first is based on general zero-knowledge techniques or simulation of private channel and the embedding the secret in a polynomial function; these methods are general, tolerate up to $n/2$ faults, but also generally inefficient (e.g., [15, 4, 8, 25]. The second method is a direct (and efficient) but requires, in addition to public key system, homomorphic functions (Exponentiations over finite fields) that encrypt a polynomial function nicely and make VSS easy [12, 22] (polynomial interpolation is a traditional secret sharing method [24]). In contrast, it is interesting to note that, technically speaking, our method will not rely on such polynomial-embedding in an encryption method (we use different encryption assumption not known before to be sufficient for efficient highly fault-tolerance polynomial-time VSS). We are, in fact, able to tolerate linear faults (say, $n/3$ faults which are the upper bound for, say, broadcast or agreement protocols). We essentially build such VSS from the simpler n-out-of-n VSS.

VSS, as defined, assumes that the secret is long-lived (like a cryptographic key that is used for a long period of time). Yet the models assume that the adversary is restricted to a sub-set of up to t trustees for the entire life of the secret. Given the fact that networks are dynamically attacked by hackers and viruses, a new model was suggested recently by Ostrovsky and Yung [21]. It assumes a bound on corrupted trustees within a time period, but lets the adversary be mobile and be able to change the set of corrupted trustees from period to period. This models the fact that viruses are detected and cleaned up, while the attacker keeps its efforts to break into various new machines. We consider here a dynamic-resharing VSS, where every time period the dealer is able to verifiably refresh the shares of the trustees, so that they are uncorrelated with previous shares. This forces the adversary to corrupt the required threshold of trustees within a time-period bound which is a harder task than the one without the time period limitation. This enables us to consider security while assuming mobility of the adversary (other recent works in this model are in [21, 7, 18, 19].)

In more detail, the dynamic-resharing VSS assumes that we have VSS and there are time-periods in which the sharing is refreshed. As long as at each time period the adversary is limited to up to t trustees, it cannot learn the secret prematurely, and it cannot prevent the trustees from opening the secret whenever required. Further, the dealer cannot corrupt the secret, the protocol either updates the shares or detects the cheating attempt of the dealer in which case the shares are not getting updated to prevent their potential corruption. (This last property extends the verifiability of the initial VSS to the update stages and is important in key escrow application where the dealer may want to corrupt data).

In proactive VSS, the community maintains the secret and refreshes it, rather than the dealer. We remark that dynamic-resharing VSS is a different protocol from mobile-adversary proactive Secret Sharing as presented in [21, 18] (which employs polynomial-based sharing methods). In the former, the dealer has a special role where it keeps on having the secret (as in escrow systems applications [20]). The goal is to keep the refreshing of secret verifiable. In the later, the secret was given to the community and forgotten by the dealer and the refresh is done by the community itself (so the set of applications of such a system is different, e.g. maintaining a secure data-base by the community). We show connections between the two and show under what conditions given dynamic-resharing VSS one can construct proactive VSS.

Our Results: The results achieved are four-fold:

1. We introduce "dynamic-sharing VSS."
2. It is constructed from a VSS protocol which uses simple probabilistic encryption tools yet tolerates a linear number of faults. This protocol is a reduction of t out of n VSS from n out of n one using exor (or addition) homomorphic encryption.
3. Further, the VSS protocol is constant round (and is non-interactive in case of honest parties). We reduce dynamic-sharing VSS from our VSS.
4. Finally we show how proactive VSS is reduced from the dynamic resharing VSS.

2 Preliminaries

2.1 Crytptographic Background

We employ public keys [11, 23] and in particular Goldwasser and Micali's probabilistic encryption method [16]. It is based on composite numbers $N = PQ$ where N is a Blum integer, that is $P = Q = 3(\bmod 4)$. This assures that -1 does not have a square root while its Jacobi symbol is $+1$. That is, take any element x from the multiplicative group mod N, $x \in Z_N^*$, square it get the set $QR = \{x^2(\bmod N)\}$ this gives the set of quadratic-residues (q.r.'s), take these numbers and multiply each by -1 to get the set $QNR = \{-x^2(\bmod N)\}$, this is the set of quadratic non-residue (q.n.r.'s).

It is assumed that it is hard to distinguish a random element from QR from such an element from QNR, in the sense that any circuit that is able to distinguish between the two sets with a non-negligible (say inverse polynomial) probability, must be super-polynomial in size. (In general inverse polynomial probability is considered substantial while smaller than any super polynomial is negligible). On the other hand, this task becomes easy when one knows the factorization of N (P, Q). In [16] the two sets QR and QNR are used to encrypt a bit: squares (residues) to encrypt 0, and non-squares (non-residues) as above to encrypt 1. For a bit b, we denote $enc(b)$ an element from QR in case $b = 0$ and an element from QNR otherwise. This is the original probabilistic encryption method.

In our protocols, the dealer uses a Blum integer N as its key, the same key throughout the execution of the protocol.

We will employ the following simple properties of the probabilistic encryption of [16]. Note that when one multiply a q.r. with a q.r. the result is a q.r. and similarly multiplying a q.n.r. with a q.n.r. results in a q.r. On the other hand, multiplication of a q.r. and a q.n.r. (or q.n.r. with a q.r.) gives a q.n.r. as a result. Note therefore, that this multiplication rules of the ciphertexts of probabilistic encryptions corresponds to ex-or-ing (sum modulo 2) of the cleartext.

We comment that we could have used r-th residuosity and its additive homomorphism property [10, 3].

2.2 Model and Problem Definition

The parties in our protocols are communicating polynomial-time Turing machines [17]. These machines flip *on-line* random coins in their computation, read communication tapes and internal (private) tapes, and write messages and internal tapes. The unpredictable on-line nature of coin-flipping is important to enable the ability to cope with mobile adversaries.

We assume a system of n trustees $\mathcal{A} = \{P_1, P_2, \ldots, P_n\}$ and a dealer D that will share secret bits among the trustees (we will assume describe only one bit secret b). The goal of the scheme is to prevent the adversary from learning x. At the same time, the adversary cannot prevent the trustees $\mathcal{A}$ from reconstructing x themselves when they need to, after a turning point in the protocol.

Communication Model (Network) the dealer D and each trustee in $\mathcal{A}$ is connected to a common broadcast medium. We assume for simplicity (to avoid dealing with authenticity of users) that each message origin is known (that is, each user has a Bulletin Board [10]). The property is that when a message is sent from any party, it immediately reaches all the others and the identity of its sender gets globally known. We assume the system is synchronized and rounds are well defined. In each round all parties act based on information available from previous rounds and local coin-flips.

For now, we assume that the adversary can corrupt at the start of the protocol up to $t = n(1/2 - \epsilon)$ trustees, and it may or may not corrupt the dealer. Corrupting a trustee means potentially, both, learning its memory and making it behave arbitrary.

2.3 The Problem

A Verifiable Secret Sharing (VSS) is a protocol between a dealer and a set of trustees, all are polynomial-time probabilistic communicating machines. The VSS is a two part protocol.

The first part: **sharing**, where a dealer has input a secret (bit) and at the end of this stage each of the n trustees has a private value and decides to accept or reject the sharing.

The second part: **reconstruction**, where the trustees pull their values together and reconstruct an output value.

We would like the following properties to hold:

1. Secrecy: Till reconstruction, no set of up to $t(t < n(1/2 - \epsilon))$ trustees can compute any information about the secret. (Our notion is computational: no such set will be able to get a non-negligible computational advantage in computing the secret).
2. Reconstructability: After the first part, if validity of sharing is assumed, at any moment any group of at least $n - t$ trustees can produce the output via the second part protocol and with overwhelming probability the result is the secret.
3. Verifiability: After the first part, the trustees can decide whether the part is successful or not. If successful, the trustees will decide that the sharing is valid and accept, otherwise they will reject the sharing. A honest dealer D can always cause the trustees to decide "valid" (in the presence of up to t faults), while a dishonest dealer that executes a sharing protocol part that fails reconstructability, will be rejected by the majority.

The dynamic resharing has additional properties which are in the spirit of the above. We require that a misbehaving dealer will not be able to corrupt its shares throughout. More details about this variant are in section 6.

3 The Construction Ideas

We describe below how the dealer sends its bit b to the set of trustees. (For many bits the protocol is run concurrently).

3.1 The Assignment Idea:

The first idea is random *families - of - committees assignment* (assignment for short) which generalizes the idea of partitioning the set of trustees. The network will be organized by an assignment, which consists of a number of *families*, each family consists of a number of *committees*. The latter are subsets of processors (families and committees may overlap).

We require some properties from an assignment. It is good (called "ϵ-terrific") if for any set of bad guys (trustees) of size $(1/2 - \epsilon)n$: (1) in each family there is a committee of only good guys and (2) in most (say, 90 percent) of the families no committee consists of bad guys only. (A good random assignment which we will employ is analogous to but more involved than "Bracha assignment" [6].) For number of bad guys $t = (1/2 - \epsilon)n$, a good polynomial-size assignment exists. Furthermore, a random assignment is good with overwhelming (exponentially small) probability $1 - 2^{(-K)}$ where K is the security parameter.

Remark: In fact, we can make the random assignment good with probability as high as the probability that the underlying public keys are secure.

We first assume an assignment is given, then we will show how to generate one by the processors themselves.

3.2 The Secret Encoding

Given an assignment, the encoding of a bit will be as follows: Let m be the number of families and r the number of committees per family. The bit b will be encrypted m times, we call these repetitions $b_1, ..., b_m$.

Then, each time we will use the following "partitioned encryption": The bit b will be distributed to r bits ($r - 1$ of which are random) $b_{i,1}, b_{i,2}, ...b_{i,r}$, where

$$\sum_{j=1}^{r} b_{i,j} = b(\bmod 2)$$

(namely, the sum modulo 2 of these bits is b). The bit $b_{i,j}$ will be "sent encrypted" to the j-th committee of the i-th family.

Let $enc(b_{i,j}) = $ a random q.r. if $b_{i,j} = 0$, and random q.n.r. otherwise. Recall that when $\sum_{j=1}^{r} b_{i,j} = b(\bmod 2)$ then

$$\prod_{j=1}^{r} enc(b_{i,j}) = enc(b)$$

and for all i, let $ENC(b) = ENC(b_i)$ and (abusing notation) we represent it as a vector:

$$ENC(b_i) =< enc(b_{i,1}), ...enc(b_{i,r}) >$$

For each vector $ENC(b_i)$ (n-out-of-n sharing) the bad guys will not get b since one committee is good in each family and therefore one $b_{i,j}$ is not known ahead of time (since the XOR of the other possible known bits $b_{i,1}, ..b_{i,r}$ is just a random bit independent of the partial opened bits of other families).

4 The Protocol

Assume we have an assignment of processors, and each committee has a committee public key with corresponding private key known to its members. We next show that:

Theorem 1. *(QR is hard) There is a (single-round sharing) protocol for VSS in a broadcast network with up to any subset of $((1/2 - \epsilon)n)$ being faulty processors (given good "assignment" into committees).*

THE VSS PROTOCOL

Part 1: Distribution of the secret

- 1a. The dealer encodes its bit b (as described above) as $ENC(b)$ (a value it is committed to). It broadcasts ENC(b). For each family it independently encodes this value as a vector encoding: $ENC(b_i), i = 1,..,m$ for each family F_i. It broadcasts the vector encodings.
- 1b.The dealer also sends its private random bits used in the encryption of $enc(b_{i,j})$ To the j-th committee of the i-th family using the committee's public key.
- 2a. Each trustee verifies that $ENC(b_i)$ is the same value for all i, by verifying that for all i:

$$\prod_{j=1}^{r} enc(b_{i,j}) = ENC(b)$$

  and all elements has Jacobi symbol $+1$; this shows that the "vector encodings" represent the same bit b that the dealer committed itself to.
- 2b. If in one of the rounds a trustee catches the dealer cheating in its publicized encryptions, it announces this (in fact all honest trustees will complain). The dealing is cancelled if there is a protest by a majority.
- 2c. Also, each party that gets random bits from the dealer that are different from the bits used in the encryption, protests. The parties can verify the protest and reject the dealing if protest justified, or ignore the protest. This is one round of protest.
  When this single round is over this is the **TURNING POINT** of the protocol, from which on the dealer is committed to a specific value.

Part 2. Reconstruction of the secret

- 1. Each member (in each committee for all families) opens the bits of the encodings. The parties calculate the resulting bits
  Note that bad committees can simply disappear or all present a false random bits (assuming they collaborated with the bad dealer, etc.) or behave in any devious way. As long as there is a family in which each committee bit is correctly opened, the bit is reconstructed.

Let us sketch the proof of the theorem above and check the properties of VSS. honest. Note that in each family there is a bit which is in the hand of a good committee that has the information about the bit encrypted randomly using semantically secure system [16]. The random bits used in the encryption were encrypted by the committee's public key, which is a randomly chosen key (distributed, in turn, encrypted under randomly chosen keys of individual honest participants)– so the dealer hides these bits semantically.

The formal proof will show that an advantage in guessing the bit b can be translated to the fact that one of the keys of either the dealer, or the ones of committee good members, or the committee's key, is not semantically secure (a contradiction).

Note that a honest dealer will be able to get into the turning point successfully as only up to t (dishonest) trustees may unjustifyingly complain at any time.

On the other hand, note that 0.9 of the families have a good participant in any committee and these bits will be opened, so a successful cheating implies that the dealer has to give these honest participants bits not corresponding to the committee encrypted bit, but then they will successfully protest publicly.

In reconstruction, we are assured that a family exists which all its committees' bits are opened, this will generate the witness for the encryption of b.

5 Preprocessing

Next we describe the assignment and the preprocessing protocol and its properties. We show that:

Theorem 2. *There exists a constant-round preprocessing protocol that in a network with up to any subset of $((1/2 - \epsilon)n)$ being faulty processors, assigns the processor to an "ϵ-terrific" assignment, each committee with a public key, where good committees have secure keys.*

5.1 The Assignment

Let $N = \{1, 2, ..., n\}$ be a set of n elements. An (n, m, r, s)-assignment A is a collection $(F_1, F_2, ..., F_m)$, where each F_i is family of r subsets $\{S_{i,j}\}, j = 1, ..., r$ of N that satisfy $|S_{i,j}| \leq s$ for $i = 1, ..., m, j = 1, ..., r$.

For a subset T of N, and a subset S of N, S is T-bad if S is a subset of T, S is T-good if S is a subset of $N - T$.

A family $F = (F_1, F_2, ..., F_r)$ of subsets of N is T-reasonable if at least one of the Fj's is T-good. F is T-wonderful if no F_j is T-bad.

Finally, an (n, m, r, s)-assignment $A = (F_1, F_2, ..., F_m)$ is T-terrific if each Fi is T-reasonable and at least $0.9m$ of the F_i's are T-wonderful.

The assignment A is ϵ-terrific if it is T-terrific for every subset T of N that satisfies

$$|N - T|/|T| >= 1 + \epsilon$$

The next proposition shows that for every n and every $\epsilon > 0$, there exists an ϵ-terrific (n,m,r,s)- assignment A, where m, r, s are polynomials (of degree $O(1/\epsilon)$ in n. Given n and $\epsilon > 0$, let k satisfy

1. $k \geq \max\{2n + 2, 100\}$, and
2. define s, r, and m by: $s = \lceil\{2\log k/\log(1 + \epsilon)\}\rceil$, $r = ((2 + \epsilon)^s)/k$, $m = 10k$.

Notice that $s = O(\log k/\epsilon)$ and $r = k^{O(1/\epsilon)}$.

Proposition 3. *For every integer n and $\epsilon > 0$, there exists an ϵ- terrific-assignment for every m, r, s that satisfy (1) and (2). Moreover, if k, m, r, s satisfy (1) and (2) and we construct randomly an (n, m, r, s)-assignment $A = (F_1, ..., F_m)$ where $F_i = (S_{i,j}), j = 1, ..., r$ and each $S_{i,j}$ is chosen by picking s random elements (with repetitions) from N, then the probability that A is ϵ-terrific is at least $1 - 2^{(-k/2)}$.*

The proof uses some standard probabilistic arguments including Chernoff's Inequality [1]. (Remark: we can also make k a function of a security parameter for small n by assuming a large n and assigning many ID's to each participant).

5.2 Preprocessing Protocol

We assume each trustee has a public key published and the dealer has one based on a Blum integer. (This fact can be certified earlier, the need for a zero-knowledge certification was shown in [13]).

THE PREPROCESSING PROTOCOL

- Choosing a random assignment which is represented by :
  1. Each trustee commits to $m\ r\ s \log n$ bits which represent an assignment.
  2. Each trustee opens it commitment. Then the bits of trustees that fully open their commitments are bitwise ex-ored together to result in an assignment.
- Let the trustee with a smallest ID in a committee be the leader of the committee. Each leader chooses a public key for the committee, publishes it (with the committee tag $\{i, j\}$) and then transfers its corresponding secret key to the committee members, using probabilistic encryption of each committee member.

The following propositions imply Theorem 2.

Proposition 4. *The assignment is ϵ-terrific with probability $1 - 2^{-k}$ for any subset of misbehaving parties of size up to $(1/2 - \epsilon)n$.*

Assume the dishonest trustees are committed and they have a hidden random string. They commit and then the string is given. Now they may decide how to open their strings (which subset to fully open) and ex-or this subset of strings with the random string. The idea of the proof is that the set of faulty trustees has an exponential number of ways to behave and determine outputs (after commitment). However the density of bad assignments is exponentially smaller than the number of strings the dishonest trustees can open (after they see the hidden string). This means that a set of faulty guys will not be able to hit an assignment which does not qualify as being good – but (still) with exponentially vanishing probability.

This is true when they start from a random string. They actually start from a string that is the ex-or of the honest guys' commitments. If they can bias it now much (inverse polynomially) better than when against a random unknown string, then the encryptions of the commitment by honest trustees can be broken (a contradiction).

Proposition 5. *The committee key of any good committee is secure with respect to the adversary.*

This is implied by the fact that a good trustee chooses with very high probability a good key and probabilistically encrypts it with good keys and the key is stored at secure memory (good trustees' memory).

6 Mobile Adversary and Dynamic-Resharing

Now we partition the computation into time periods at the end of which a resharing protocol is executed.

We will show that:

Theorem 6. *(QR is hard) There is a constant-round dynamic-resharing VSS protocol in a broadcast network with up to any subset of size $t = ((1/2 - \epsilon)n)$ being faulty processors at each time period.*

6.1 The Mobile Adversary Model

We now change our underlying paradigm so that an adversary does not control a server forever. Instead, an adversary can succeed in breaking-in for limited periods of time and move on to attack other servers. In other words, the adversary is mobile as was initially suggested in [21]. This assumption is motivated by two facts: first, intrusions in networks (such as software modification, viruses, etc.) are eventually exposed and cleared, and second the fact that machines are easily accessible through the network for an adversary to attack. The main goal behind dynamic protocols solution is to make the information gained by the adversary during a break-in useless during future break-ins (to the same or other servers), thus neutralizing the power given by the mobility property.

We assume the computation is divided into periods of time, at the end of a period of time there is an update phase. If a server is corrupted during an update phase, we consider the server as corrupted during both periods adjacent to that update phase. We assume that the adversary corrupts *no more* than $t \leq n(1/2 - \epsilon)$ out of n parties. (Note that there are enough "good parties" at any time period). If at the time period prior to the update there are t_1 corruptions and t_2 corruptions at the time-period after the update and t_3 corruption during the update then $t_1 + t_3 \leq t$ and $t_2 + t_3 \leq t$. When a trustee is cleaned, it waits till the next resharing step and starts afresh (announcing its status, first). We require that as long as the dealer is not corrupted throughout, its secret remains secure, and that the adversary is unable to prevent reconstruction of the message at any point.

6.2 The problem

A Dynamic-Sharing Verifiable Secret Sharing (VSS) is a protocol between a dealer and a set of trustees. Its first part is sharing as in VSS, its last part is reconstruction as in VSS. Between the two stages it has time periods, in each time period at most t trustees are controlled by the adversaries (as explained above). At the end of a period, a resharing protocol is performed, with the following properties: (1) At the end of the resharing the trustees either reject and hold the old shares, or accept and update the shares they hold; (2) reconstruction is maintained throughout. We would like the following security properties to hold:

1. Secrecy: As long as at each period the adversary has up to $t, t < n(1/2 - \epsilon)$ trustees, it cannot compute any information about the secret.
2. Reconstructability: If the original sharing was valid, no further resharing can reduce the probability of reconstruction but with negligible probability.
3. Verifiability: The trustees can decide whether the resharing is successful or not. If successful, the trustees will decide to update, otherwise they will reject the re-sharing. A honest dealer D can always cause the trustees to decide "valid" (in the presence of up to t faults), while a dishonest dealer that executes a resharing protocol part that fails reconstructability, will be rejected by at least $n - t$ trustees.

Note that for secrecy, the dealer has to be kept honest (otherwise the secret is easily learned). For reconstructability, any behavior of the dealer cannot hurt this property.

6.3 The Protocol

In the protocol to follow we have to refresh the public-keys at each update (resharing) so that previous keys learned are not valid anymore (are independent of the current values).

We then let the dealer refresh the shares. To do this we use the fact that an encryption when multiplied with a q.r. does not change its cleartext value. Let us review how it looks in the cleartext domain. So a family will get a random vector of cleartext bits with sum modulo 2 equals 0, each committee will ex-or its new bit with its old bit. As a result each committee now has a newly random bit independent of any previous partial bit pattern. At the same time the value of the family bit (the "family value") remains the same. At any period the adversary is only able to get partial bit encryptions earlier and partial updates or partial update and new shares which are less than the required threshold. Recall that the assignment is such that any subset of size t cannot "cover" any family in full.

THE DYNAMIC RESHARING PROTOCOL (STAGE l)

- 0. Each trustee i sends a new public key $E_{i,l}$. Then, each head of a committee sends a new committee key.
- 1. The dealer encodes the bit 0 (as described above) as $ENC^l(0), i = 1, .., m$ for each family F_i.
- 2. The dealer proves that $ENC^l(0)$ indeed encodes the same bit 0. It computes, $ENC^l(0) = e^l = e^l i = enc^l(bi, 1) * enc^l(bi, 2) * ... * enc^l(bi, r)$, for each family i. It shows that $ENC^l(0)$ is a quadratic residue using the protocol of [2] (parallel constant-round proofs to all trustees can be performed in zero-knowledge).
- 2.a If in one of the rounds a trustee catches the dealer cheating, it announces this fact and reveals its secret bits and convinces the other trustees.
- 3. All trustees compute the accumulated vector encryption of every committee which is the encryption of its bit at stage l

$$enc^l(bi, j) = enc^{l-1}(b_i, j) * enc^{l-1}(0, j)$$

which is

$$enc^l(b_i, j) * enc^{l-1}(0, j) * ... * enc^1(0, j) * enc(b_i, j)$$

(This is the initial encrypted bits encryption multiplied by all the updates). Each trustee sends a list of all the accumulated public encryptions.

3.a. Each trustee takes the majority of echoed encryptions as the right encryption.

3.b. The dealer computes majority and decrypts the correct current bits and sends to the trustees. (It sends the random bits used in the encryption of $enc(b_{i,j})$ To the j-th committee of the i-th family using the committee's public key.)

3.c. Committees complain about the bits they receive, complaints are verifiable, if a complaint is valid the dealer is dishonest, his resharing is ignored (each trustee restores its state prior to the protocol). Otherwise, updates takes place, and old states are erased.

When this is over this is the **START NEW PERIOD POINT**.

Note that any choice of t faulty guys, will keep the assignment such that every family has a good committee, and the majority of families will be able (if necessary) to open the right encryption.

7 Proactive VSS from Dynamic-Resharing VSS

Next we will discuss how we can use the dynamic-resharing to have a proactive secret sharing (a secret maintained by the community). In such a protocol, after the first sharing, the dealer will forget its secret and the community will maintain the secret (secrecy and reconstructability is verifiably kept).

7.1 The Result

We will show that:

Theorem 7. *(QR is hard) There is a proactive VSS protocol in a broadcast network with up to any subset of $((1/2 - \epsilon)n)$ processors being faulty at each time period.*

We only sketch the ideas below.

The first idea is that the prover in the dynamic-resharing actually needs to know the square of the encryption rather than the factorization of the key N. Thus, every user can serve as a dealer of an encryption of a zero and prove the validity of its actions– being the actual dealer does not help!

The second idea is a transformation of a dynamic resharing protocol where anyone can refresh the shares, into a proactive VSS.

In fact, we will have the following initialization steps: (1) A Sharing stage protocol. (2) A sharing of the shares, using the same protocol each user will share its shares. At this point the secret can be erased.

Then we maintain the shares. At the end of each time period we have a refresh protocol. Each dynamic resharing is done by all users (each user, or at least $t + 1$ of them to make sure one is honest, re-randomizes the shares, each acting as a dealer). All shares-of-shares get refreshed. We have the following two refresh stage:

- (1) Before refreshing, users that have recovered in the period must have a chance to rejoin. So they should get from holders of shares of their share to send them secretly their share for reconstruction, so they can rejoin gracefully.

  In fact, we can have a global "secret echoing" step, where a user sends a new public key first, and the parties send to him secretly the random secret bits of the shares of its share encrypted with this new key, and the parties also publish the current encrypted share-of-share. This gives the party a "reconstruction of its share". Now we are ready for refresh.
- (2) Each party (or at least $t + 1$ of them) is refreshing (re-randomizing) all the shares-of-shares. This is done via the dynamic-resharing protocol above. Note that this can be viewed as maintaining a memory against mobile adversary. Share-of-share ideas have been used in various ways in the secure distributed computing literature, and were first introduced in [14]. (As in the dynamic resharing, only updates which are valid are taken into accounts and other resharings are ignored).

A few remarks: First note that maintaining only the shares directly is not enough as a share is held by a committee and the adversary can erase it– slowly erasing all committees' bits. But, shares of shares enable global maintenance via on-going intermediate reconstructions and recoveries when needed. A misbehaving processor that sends a weak key is controlled by the adversary which can anyway learn its secret to start with (and does not need the other processors to open its share– in any case at most t shares get opened).

8 Conclusions

We have presented a new method that relies only on probabilistic encryption [16] with its ex-or homomorphism and still gives a VSS protocol with linear faults. The fact that a new complexity assumption is sufficient for VSS is interesting by itself and makes the first result incomparable with previous solutions. We have then considered the new concept of "mobile adversary" and showed new constructions in this areas based on reductions. We introduced and implemented the notion of "dynamic re-sharing" by a dealer which fits settings like key escrow, and can be used to achieve proactive VSS.

References

1. N. Alon, and J. Spencer, *The Probabilistic Method*, (1991), John Wiley and Sons.
2. M. Bellare, S. Micali, and R. Ostrovsky, *Perfect Zero Knowledge in Constant Rounds*, Proceedings of the 22th Annual Symposium on the Theory of Computing, 1990, pp. 482–493.
3. J. C. Benaloh and M. Yung, *Distributing the Power of a Government to Enhance the Privacy of Voters*, Proc. of the 5th ACM Symposium on the Principles in Distributed Computing, 1986, pp. 52-62.
4. M. Ben-Or, S. Goldwasser, and A. Wigderson, "Completeness theorems for non-cryptographic fault-tolerant distributed computation," ACM STOC 1988, 1–9.
5. G.R. Blakley, *Safeguarding Cryptographic Keys*, AFIPS Con. Proc (v. 48), 1979, pp 313–317.
6. G. Bracha, *An $O(\log N)$ Expected Round Randomized Byzantine Generals Protocol*, Proc. of ACM STOC 1985.
7. R. Canetti and A. Herzberg, Maintaining Security in the Presence of Transient Faults, Crypto 94.
8. D. Chaum, C. Crépeau, and I. Damgård, "Multiparty unconditionally secure protocols," ACM STOC 1988, 11–19.
9. B. Chor, S. Goldwasser, S. Micali and B. Awerbuch, *Verifiable Secret Sharing and Achieving Simultaneous Broadcast*, Proc. of IEEE Focs 1985, pp. 335-344.
10. J. Cohen and M. Fischer, *A robust and verifiable cryptographically secure election scheme*, Proc. 26th Annual Symposium on the Foundations of Computer Science, 1985, pp 372–382.
11. W. Diffie and M. Hellman, *New Directions in Cryptography*, IEEE Trans. on Information Theory 22 (6), 1976, pp. 644-654.

12. P. Feldman, *A Practical Scheme for Non-Interactive Verifiable Secret Sharing*, Proc. of the 28th IEEE Symposium on the Foundations of Computer Science, 1987, 427-437

13. Z. Galil, S. Haber and M. Yung, *Minimum-Knowledge Interactive Proof for Decision Problems*, SIAM J. Comp., 18, 1989, pp 711–739.

14. Z. Galil, S. Haber, and M. Yung, "Cryptographic computation: secure fault-tolerant protocols and the public-key model," Crypto 87.

15. O. Goldreich, S. Micali, and A. Wigderson, *How to play any mental game*, Proceedings of the Nineteenth annual ACM Symp. Theory of Computing, 1987, pp 218–229.

16. S. Goldwasser and S. Micali, *Probabilistic Encryption*, J. Com. Sys. Sci. 28 (1984), pp 270-299.

17. S. Goldwasser, S. Micali and C. Rackoff, *The Knowledge Complexity of Interactive Proof-Systems*, Siam J. on Computing, 18(1) (1989), pp 186-208.

18. A. Herzberg, S. Jarecki, H. Krawczyk, M. Yung, *Proactive Secret Sharing, or how to cope with perpetual leakage*, Crypto 95.

19. A. Herzberg, M. Jakobsson, S. Jarecki, H. Krawczyk, M. Yung, *Proactive Public-Key and Signature Schemes*, Manuscript.

20. S. Micali, *Fair public-key cryptosystems*, Crypto '92.

21. R. Ostrovsky and M Yung, *How to withstand mobile virus attacks*, Proc. of the 10th ACM Symposium on the Principles in Distributed Computing, 1991, pp. 51-61.

22. T. P. Pedersen, Distributed Provers with Applications to Undeniable Signature, Eurocrypt '91. 1991.

23. R. Rivest, A. Shamir and L. Adleman, *A Method for Obtaining Digital Signature and Public Key Cryptosystems*, Comm. of ACM, 21 (1978), pp 120-126.

24. A. Shamir. *How to share a secret*, Commun. ACM, 22 (1979), pp 612-613.

25. T. Rabin and M. Ben-Or, *Verifiable Secret Sharing and Multiparty Protocols with Honest Majority*, STOC 1989, ACM, pp. 73-85.

26. G. J. Simmons. *An introduction to shared secret and/or shared control schemes and their application*, In G. J. Simmons, editor, *Contemporary Cryptology*, pp. 441–497. IEEE Press, 1992.

Adaptive Video on Demand

Sudhanshu Aggarwal[1]   Juan A. Garay[2]   Amir Herzberg[2]

[1] Lab. for Computer Science, MIT, 545 Technology Square, Cambridge, MA 02139.
[2] IBM T. J. Watson Research Center, P.O. Box 704, Yorktown Heights, NY 10598.

Abstract. In this paper we formulate the problem of Video on Demand (VOD) from a resource allocation perspective. In particular, we introduce the *decision element* into a movie vending environment, which complements the current approaches. In contrast with more the traditional resource allocation problems (such as machine scheduling and call control), the problem possesses the distinctive *batching* property, which stands for the feasibility of several requests being served by one resource (channel). We investigate the problem in an *on-line* fashion, namely, having to accept or reject a request for a movie without the knowledge of future requests. We show upper and lower bounds on the *competitive ratio* of deterministic on-line movie scheduling algorithms for a variety of scenarios (an algorithm is called *competitive* if it performs, up to a constant factor, as well as its off-line, clairvoyant counterparts for the same problem). In particular, for the natural case of *refusal by choice* with *delayed notification*, we present a class of algorithms that exhibit, under certain conditions, an asymptotically optimal behavior.

We also compare the performances of the different algorithms under various distributions of requests over time, and evaluate the effect of various heuristics.

1 Introduction

The area of interactive home video entertainment is actively developing. For example, hybrid networks that enable multi-media connections are being studied as a step toward all-digital video networks [11], as is the problem of bandwidth allocation strategies for combined analog/digital transmission of data over a CATV system [9]. Experiences and services trials are being performed, mostly for *near* VOD, which consists of broadcasting movies at fixed time invervals (e.g., every 15 min) [13, 14]. Also, special hardware and switches are been built for this purpose.

Most published works and on-going projects seem to share a common view on the basic architecture for VOD services. The movies are stored in a central *video server*, which may be connected to other servers over a high bandwidth WAN. The *video server* is connected by a high-capacity fiber link to local distribution centers (hubs), from which coax cables are used to broadcast to the households.

In this architecture, there appear to be two major bottlenecks:

- The limited number of broadcast channels available on the coax cable (shared by many households).[3]

- The number of movies which the server is able to transmit concurrently (see, in particular, [10]).

Previous works and current implementations attempt to achieve, under these constraints, *Full Video on Demand* and/or *Near Video on Demand*:

- **Full Video on Demand:** Whenever a user's request for a movie arrives, it is immediately served, provided there is capacity (a channel) available; otherwise the request is rejected.
- **Near Video on Demand:** A fixed set of movies is played regularly, at fixed time intervals.

Much attention has been given to the hardware requirements for such designs, and the quality of service, due to the bottlenecks mentioned above. It is immediate that either policy may poorly utilize the available resources. Namely, a considerable number of feasible distribution of requests may cause the first approach to tie up the distribution resources, while the latter provides—by definition—a limited service, and a bad decision directly translates into resource wastefulness. The two approaches are shown in Figure 1. The purpose of this paper is to in-

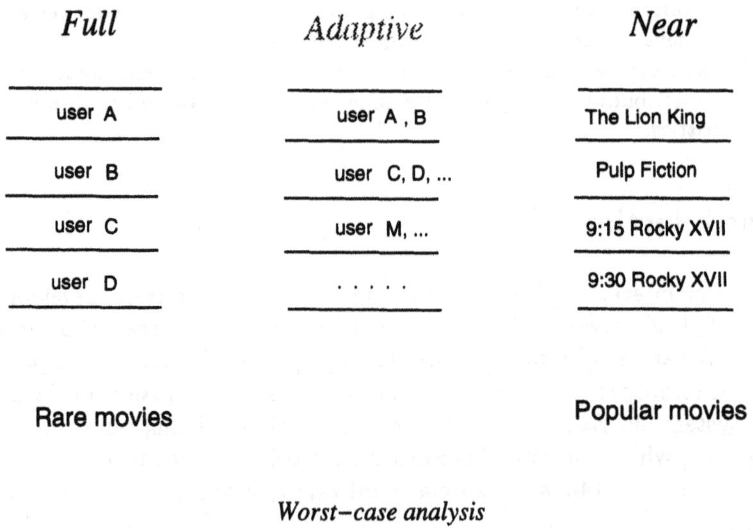

Full	*Adaptive*	*Near*
user A	user A , B	The Lion King
user B	user C, D, ...	Pulp Fiction
user C	user M, ...	9:15 Rocky XVII
user D	. . . . .	9:30 Rocky XVII

| Rare movies | | Popular movies |

Worst–case analysis

Fig. 1. VOD approaches.

vestigate the intermediate terrain that lies between the two extreme policies.

---

[3] It appears that, for economical reasons, it may be desirable to put a much larger number of homes on the same broadcast cable. For example, the number of channels per coax may be about 100, and one may expect up to, say, 1000 concurrent users.

Indeed, we introduce the *decision element* into the video service architecture, which leads to the concept of:

- **Adaptive Video on Demand:** Upon arrival, a movie request is accepted and served (possibly with some delay), or rejected. The decision as to whether accept or reject (and the amount of delay) is made by a scheduling algorithm.

Note that the scheduling algorithm might specify that the request be rejected, even though there might be channels that are currently not being used.

The approach is also depicted in Figure 1. To the best of our knowledge, this is the first attempt so far to tackle VOD from an optimization perspective. Namely, the purpose of this paper is to optimize the decisions on which movie requests to accept, when to deliver them, and which requests to reject. We therefore try to maximize the utilization of a (limited) resource being shared by a number of users. Our motivation is to complement approaches based on predicting demand for movies (i.e., Near VOD), which are best suited for popular movies, but inappropriate for more unusual selections. This leads us to investigate the problem in an *on-line* fashion, as explained below.

The type of problem we have just described is not unrelated to the more "classical" problems of (parallel) machine scheduling (e.g., [6, 16]) call control [4, 7, 8], and virtual circuit routing [2, 3]. However, our problem has the distinctive property of allowing the movie scheduling algorithm, under certain circumstances, to serve several users simultaneously on the same channel.

The general nature of the case of unusual selections is such that it is always possible to "second guess" decisions made in the past. In other words, a decision made previously to accept a request, may have been wrong from the service provider point of view, because it may cause a channel to be used up, thus preventing a subsequent more "valuable" set of requests (for example, for a more popular movie) to be rejected. Thus, the *on-line* nature of the problem, i.e., the fact that decisions are to be made when requests for movies arrive into the system without knowledge of future requests, might lead to significantly lower efficiency than would have been possible with full *off-line* knowledge of the entire pattern of request arrivals. This motivates the emphasis we put on this issue. Namely, in this paper we investigate movie scheduling mechanisms in an *on-line* fashion. A second motivation for tackling a new problem with this type of analysis is the hope of achieving a deeper, meaningful understanding of the problem. We evaluate on-line algorithms in terms of their *competitiveness* [15]. An algorithm is said to be κ-competitive, if its performance on *any* sequence of requests is within a factor κ of the performance of any other algorithm on the same sequence, including the *off-line* algorithms for the problem. A measure of performance of movie-scheduling algorithms is the *revenue* they produce. (We can assume that the operating costs are fixed, so the profit is a function of the revenue only.) Let c denote the number of available channels; m the number of movies; and u the number of users. Also let $\mu \stackrel{\text{def}}{=} u/c$. Our results are summarized in Table 1.

541

| COMPETITIVE | No refusal by | Refusal by choice | |
RATIO	choice	Immediate Notification	Delayed Notification
Lower bound	μ	$\frac{u}{c+1} \approx \mu$	$\Omega(\ln \mu)$
Algorithm	μ	μ	$O(\ln \mu)^{\dagger}$

Table 1. Our results

† For adequate values of c, and notification time equal to service time.

Regarding the terminology of Table 1, one important property of Adaptive VOD is the possibility of a request being rejected even though there is available capacity to serve it. We call this property *refusal by choice*. We show that if refusal by choice is prohibited, then the competitive ratio of any Adaptive VOD algorithm cannot improve compared to the Full VOD approach. Another relevant parameter is the maximum time the system is allowed to spend (respectively, the time a user can be made to wait) before the user is informed of the decision on his/her request. We thus distinguish between systems with *immediate* or *delayed* notification. For the more natural case of refusal by choice with delayed notification, we show a lower bound of $\Omega(\ln \mu)$ on the performance of any algorithm, and introduce a class of algorithms (which we call Pre-partition) that, for a big enough number of channels and notification time, exhibit an asymptotically optimal behavior. In particular, we analyze the *Harmonic* version, which outperforms the more straightforward exponential algorithms by a constant factor in the worst case.

We have also compared the performances of the different algorithms under various distributions of requests over time, and evaluated the effect of various heuristics. Our studies of these more realistic scenarios show an even stronger performance advantage of Harmonic compared to Exponential, Near, and variants of Full VOD. Some of these results are discussed at the end of the paper. Finally, we remark that we consider the special case of *non-preemptive* video scheduling algorithms, and work exclusively with *flat* charging policies, although many of our results can be generalized to hold for other policies.

The assignment of users to resources, as embodied by the VOD problem, has a natural graph-theoretic formulation. This has been recently done by Awerbuch, Azar, and Fiat [1], who consider an on-line version of the *set cover* problem. For the case of immediate notification with refusal, they show that the linear bound can be improved to (super)logarithmic when randomization is allowed. They also generalize the problem to a setting in which every customer discloses a list of alternative movie titles he/she wants to watch (*non-disjoint max cover*).

The remainder of the paper is organized as follows. In Section 2 we provide a more technical description of the model and necessary definitions. Our lower bounds and algorithms for the on-line case are presented in Section 3. We conclude with some results for other scenarios and directions for future work.

2 Model and Definitions

The model we consider in this work is represented in Figure 2. This seems to be a reasonable model for the two likely bottlenecks for VOD identified in the previous section. The model considers finite, fixed sets of movies, users and channels, denoted respectively M, U and C. Denote the cardinality of these sets by the corresponding lower case letter, i.e. m, u and c.

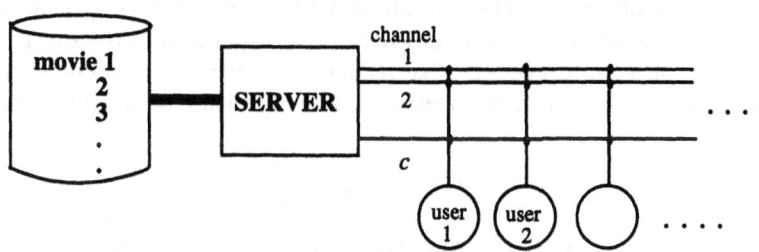

Fig. 2. Model for Video On Demand.

A *request* is a tuple (*time, user, movie*) where *time* is a positive number, *user* $\in U$ and *movie* $\in M$. Similarly, a response is a tuple (*time, user, movie, channel, servetime*) where *channel* $\in C$ and *servetime* is a real number. If *servetime* is negative, we say that the response is negative (or a refusal).

The system has three positive parameters related to time:

- T – the duration of a movie. (For simplicity, we assume it is the same for all movies. We also assume that no user can place more than one request in each interval of length T.)
- τ – the maximal delay between a request and the start of the movie (if it is accepted).
- ν – the maximal delay between a request and its response (notification).

We assume $\nu \leq \tau \leq T$. A sequence of requests (t_i, u_i, m_i) is *valid* if the time values t_i are monotonously increasing and there is at least time T between any two subsequent requests of the same user.

Intuitively, the movie scheduling algorithm determines the responses at any moment, based on the requests up to that moment. More formally, a *scheduling algorithm* is a function from a time, a state, and a sequence of requests (all before the given time), to a time (for next wakeup, if no requests would occur before), a new state, and possibly one or more responses.

We consider different classes of movie scheduling algorithms, based on the definitions above. An algorithm $\mathcal{A}$ is said to *refuse by choice* if in some run it has a response $(t_i, u_i, m_i, c_i, s_i)$ with $s_i < 0$, while there is some free channel at t_i. A channel c_k is *free* at time t if there is no response $(t_k, u_k, m_k, c_k, s_k)$ with $t_k \leq t$ and for which $[s_k, s_k + T]$ intersects with $[t, t + T]$. An algorithm $\mathcal{A}$ is said to perform *immediate notification* if each request must be "decided upon" as soon as it is presented. (Note that if a request is accepted, $\mathcal{A}$ must serve it within time τ.) Algorithms that do not always notify immediately are said to perform *delayed notification*, and in this case they have to notify within the parameter ν defined above.

We now give a more technical definition of competitive ratio. A measure of performance of movie schedulers is the *revenue* they produce. A *revenue function* is a function that maps integers (delay) to the rationals (income). The revenue of a response $(t_i, u_i, m_i, c_i, s_i)$ corresponding to a request (t_j, u_i, m_i) is $rev(s_i - t_j)$ if $s_i \geq 0$, and 0 otherwise. The revenue of a run α, denoted $rev(\alpha)$, is the sum of the revenues of all the requests in the run. Let $\mathcal{A}(\rho)$ denote the run produced by algorithm $\mathcal{A}$ when presented with request sequence ρ (we assume that $\mathcal{A}$ is deterministic). The *competitive ratio* of online algorithm $\mathcal{A}$ is defined as

$$\phi(\mathcal{A}) = \sup_{\mathcal{A}^*, \rho} \frac{rev(\mathcal{A}^*(\rho))}{rev(\mathcal{A}(\rho))}$$

where $\mathcal{A}^*$ is an offline algorithm and ρ is a sequence of requests. It is sometimes convenient for the analysis to describe things in terms of a game between a player (the online algorithm) and an *adversary* (the offline algorithm), whose goal is to produce a request sequence that would force the player to perform poorly.

3   Our Results

We now show upper and lower bounds on the competitive ratio $\phi(\mathcal{A})$ of online movie scheduling algorithms $\mathcal{A}$ (see Table 1). Intuitively, there must exist algorithms with a competitive ratio of μ, because by "bunching up" requests, the adversary can serve at most u requests for every c requests the online algorithm serves. We first formalize this argument and show that there indeed exists a simple algorithm $\mathcal{A}_1$ with a competitive ratio of μ. This algorithm is essentially the same as the Full VOD approach.

We then show that any algorithm $\mathcal{A}$ which is *not allowed refusal by choice* must have a competitive ratio of at least μ. Thus, no online strategy can, in the worst case, "gain" anything by bunching up requests with respect to the simple strategy $\mathcal{A}_1$, unless it is allowed to reject requests. Next, we show that even refusal by choice, by itself, is not enough to improve worst-case performance. The timing of *notification* also determines how well an algorithm can do. Specifically, we show that an algorithm that performs *immediate notification* must have a competitive ratio of at least $\frac{u}{c+1}$ (which is nearly equal to μ). Thus, such algorithms do not have any advantage, in the worst case, over those which do not refuse.

Finally, we consider the case of algorithms which are allowed to refuse by choice together with *delayed notification*. As expected, they may perform substantially better than those with immediate notification. More specifically, we show a lower bound of $\Omega(\ln \mu)$ on the performance of such algorithms, and present a simple algorithm ($\mathcal{H}$, for *Harmonic*) which matches this bound in the case $\nu = \tau$.

3.1 No refusal

In this section we show that a competitive ratio of μ is easy to achieve by giving a simple "naive" algorithm that achieves this ratio (and which is not allowed to refuse requests by choice). For simplicity, we consider the special case of the maximum service delay τ being 0; the proof is easily extended to any arbitrary delay τ. Consider the following algorithm $\mathcal{A}_1$: Given request sequence $\rho = r_1, r_2, r_3, ..., r_n$ (as defined in the previous section), $\mathcal{A}_1$ assigns r_i to any channel that is free at time t_i; otherwise r_i is rejected.

Theorem 1. *Algorithm $\mathcal{A}_1$ is μ-competitive.*

Proof. Each request in ρ is either accepted or rejected. Let n_A and n_R be the number of accepted and rejected requests, respectively. Thus, $n = n_A + n_R$.

If request r_i is rejected, it is rejected because exactly c previously accepted requests are being served at that instant. For each of these c requests r', we say r_i *is rejected because of r'*. Define the "rejection graph" to be the graph whose vertices consist of all n requests, and for any pair of vertices r_a and r_b, there is a link connecting them if and only if r_a is rejected because of r_b, or vice-versa. (Note that each rejected request must have exactly c incident links, and that the total number of links incident on all accepted requests must equal the total number of links incident on all rejected requests.)

Now we claim that each accepted request cannot have more than $u-c$ incident links. For suppose there exists an accepted request r_i which has $u-c+d$ incident links, i.e., $u - c + d$ requests (comprising, say, set R) are rejected because of r_i. Consider the latest request in set R, say, request r_l at time t_l. Note that $t_l < t_i + T$. Thus, there were exactly $u - c + d$ rejections in the time interval $[t_i, t_l]$. Further, since r_l was rejected, there were exactly c acceptances in the time interval $(t_l - T, t_l]$. It follows that there were at least $(u - c + d) + c = (u + d)$ requests in $(t_l - T, t_l]$, which contradicts the fact that each user of the u users may have at most one request during $[t_l - T, t_l]$.

Thus, the total number of links incident on accepted requests is at most $n_A(u - c)$, and the number incident on rejected requests is exactly $n_R c$. As a consequence, we have $n_A(u - c) \geq n_R c$, or $n_A u \geq nc$, which implies $\frac{n}{n_A} = \phi(\mathcal{A}_1) \leq \frac{u}{c}$. □

The following theorem shows that, under the circumstances, this is the best that can be done.

Theorem 2. *Let $T > \tau$. Any algorithm $\mathcal{A}$ that is not allowed to refuse by choice must have a competitive ratio $\phi(\mathcal{A}) \geq \mu$.*

Proof. Let $\mathcal{A}$ be an algorithm that does not refuse by choice. We construct request sequences ρ that would necessitate a competitive ratio of at least $\mu = u/c$.

Let ρ be a request sequence for $\mathcal{A}$, which begins with a set of c requests for c different movies, presented to $\mathcal{A}$ at time 0. Assume there is no additional request until time ν. Algorithm $\mathcal{A}$ must schedule all of the first c requests until time ν, since it does not refuse by choice. Furthermore, the schedules should all start before time τ.

We now have two mutually exclusive cases:

1. Suppose there at least one movie is scheduled before τ; let m_i be such a movie. Then ρ would contain $u - c$ requests for the same movie m_i, all arriving exactly at time τ. However, at this time all channels are already scheduled, and since we do not allow preemption, the online algorithm would have to refuse all of these movies.

    Thus, A honors a total of c requests in its run. On the other hand, obviously there exists an offline algorithm A^* which can serve all the requests. Thus, $\phi(A) = u/c$.

2. Suppose that A schedules each of the c movies to begin at time τ. The request sequence ρ would continue with $u - c$ requests at time $T - \frac{\tau}{2}$ for some movie. Of course, algorithm A cannot honor any of these requests since all channels are still occupied at that time. Thus, A honors a total of c requests in this run.

    On the other hand, there exists an offline algorithm A^* that honors all requests by honoring the first set immediately and the second set at time T (within the permitted delay). Thus, again, $\phi = u/c$. □

3.2 Refusal by choice and immediate notification

One might wonder whether an algorithm, when allowed to reject movies, could perform substantially better. As the following theorem shows, that is not the case: Refusal by choice, by itself, is not enough to improve worst-case performance. The timing of notification also determines how well an algorithm can do.

Theorem 3. *Any algorithm A capable of refusal by choice, but which is required to notify immediately, must have a competitive ratio $\phi(A) \geq \frac{u}{c+1}$.*

Proof. Let A be an algorithm that is allowed to refuse requests by choice, but which always notifies immediately. Let it have a competitive ratio $\phi(A) = \kappa$. The proof is by contradiction. It is based on the fact that if A has a lower competitive ratio, the adversary can construct a request sequence that makes A violate the ratio. The adversary first presents a series of "baits" such that A is forced to fill up its channels in order to maintain its competitive ratio. Once A is filled up, the adversary then presents as many requests as possible; thus, the adversay can take all of these while the A can take none.

Specifically, the adversary first presents a request sequence $\rho' = r_1, r_2, ...r_l$. Each request r_i is a request for movie m_i made at time $i\epsilon$, where ϵ is some arbitrarily small interval of time. The adversary constructs the request sequence according to the following rules:

1. If r_i is rejected by A, $m_{i+1} = m_i$;

2. if r_i is accepted by $\mathcal{A}$, $m_{i+1} \notin \{m_1, m_2, ...m_i\}$; and
3. the total number of different movies requested in ρ' is c.

Thus, the adversary presents individual requests for movie m (at instants separated by ϵ) until one of these is accepted by $\mathcal{A}$, then it presents a group of individual requests for another movie m' until one is accepted, and so on, for a total of c request groups for c different movies. The total number of requests accepted by $\mathcal{A}$ is c (the last one in each of the c groups).

On the other hand, by "bunching up" the requests in each group and transmitting a single movie to serve an entire group, the adversary can accept each one of these j requests. Since $\mathcal{A}$ must maintain a competitive ratio of κ, $\mathcal{A}$'s ith acceptance must occur by request $r_{i\kappa}$. Thus, the total number of requests is $l \leq c\kappa$. Note that since there are c groups of requests of total size $l \leq c\kappa$, there must exist a group G of size $g \leq \kappa$.

Having "filled up" all the channels of $\mathcal{A}$, the adversary now presents a request sequence $\rho'' = r_{l+1}, r_{l+2}, ...r_u$ for any one of the movies not requested so far. It accepts all the requests from the earlier sequence ρ', except for the $g \leq \kappa$ requests in group G. Thus, at this point the adversay has one channel free; it accepts all the requests in ρ'' and serves them on this channel.

The online algorithm $\mathcal{A}$ has served exactly c requests, while the adversary has served $(l-g)$ requests from ρ' and $(u-l)$ from ρ'', for a total of $(u-g) \geq (u-\kappa)$. Thus, to preserve a competitive ratio of κ, a necessary condition is $\kappa \geq \frac{u-\kappa}{c}$, which reduces to $\kappa \geq \frac{u}{c+1}$. □

Remark. The above lower bound can be readily extended to non-zero (but small) values of ν, namely, for all $\nu < \frac{\tau}{c\kappa}$.

3.3 Refusal by choice and delayed notification

In this section we give bounds on the competitive ratio of movie scheduling algorithms that are allowed refusal by choice and delayed notification. For the case of the upper bound we also assume big, but yet reasonable, values of c. We first show a lower bound of $\ln \mu/2$ on the competitive ratio of such algorithms for the case of short notification times (specifically, $\nu < \frac{2\tau}{\mu}$).

Let $\mathcal{A}$ be an online algorithm. We prove the result using a particular type of game between $\mathcal{A}$ and its adversary $\mathcal{A}^*$. The structure of the game will be as follows. The adversary will present $\mathcal{A}$ with several sets $R_1, R_2, ...$ of requests in quick succession. When presented with each set of requests, $\mathcal{A}$ has to schedule just enough to maintain a competitive ratio of κ. However, the adversary chooses the request sets in such a manner that it is able to force $\mathcal{A}$ to accept "cheap" requests, and thereby block its channels to further, more remunerative requests, which are served by the adversary. If the adversary $\mathcal{A}^*$ ends up serving at least κ times more requests than $\mathcal{A}$, we will have shown that $\mathcal{A}$ cannot have a competitive ratio of κ. This allows us to state:

Theorem 4. *Let $\nu < \frac{2\tau}{\mu}$. Any algorithm $\mathcal{A}$ capable of refusal by choice and delayed notification must have a competitive ratio $\phi(\mathcal{A}) > \ln \frac{\mu}{2}$.*

We first describe the game in more detail, and establish some technical facts. Suppose there is a κ-competitive online movie scheduler algorithm $\mathcal{A}$. Consider the following sequence of requests. At time 0, request set R_1 is presented, which contains one request for each of c movies. $\mathcal{A}$ must eventually serve at least $\frac{c}{\kappa}$ of them (in fact, $\lceil \frac{c}{\kappa} \rceil$) to maintain a competitive ratio of κ, otherwise $\mathcal{A}^*$ could serve all of them and have a ratio greater than κ. We say that $\mathcal{A}$ has *committed* to at least $\frac{c}{\kappa}$ requests. Note that $\mathcal{A}$ must do so by time ν.

At time ν, two requests for each of c movies are presented which are different from the $\frac{c}{\kappa}$ already committed. More precisely, for each of the $c - \frac{c}{\kappa}$ movies not committed there is an additional request, plus two new requests for $\frac{c}{\kappa}$ movies which were not requested before. Let the additional requests presented at time ν comprise set R_2. Again, $\mathcal{A}$ must serve at least $\frac{c}{\kappa} \cdot \frac{1}{2}$ movies (each with 2 requests), otherwise $\mathcal{A}^*$ could satisfy all and accrue $2c$. In other words, the adversary would have gotten revenue of $2c$ while $\mathcal{A}$'s revenue would be $\frac{c}{\kappa}$ from the movies taken by time ν plus *less than* $\frac{c}{\kappa}$ from the additional movies taken by time 2ν (two movies per channel, but less than $\frac{c}{2\kappa}$ channels).

This generalizes to request set R_i at time $(i-1)\nu$ in the following way: At time $(i-1)\nu$ there are i requests for each of c movies which are not yet being served. Again, old outstanding requests are "reused." After set R_{i-1} at time $(i-2)\nu$ is presented, $\mathcal{A}$ has to serve $\frac{c}{(i-1)\kappa}$ movies, each with $i-1$ requests. There are $c - \frac{c}{(i-1)\kappa}$ movies left, with $i-1$ outstanding requests; the adversary will just add one additional request for each one of these movies. For the remaining $\frac{c}{(i-1)\kappa}$ channels, the adversary will produce i requests for new movies. Thus, the total number of new requests is $(c - \frac{c}{(i-1)\kappa}) + i \cdot \frac{c}{(i-1)\kappa} = c + \frac{c}{\kappa}$. Again, $\mathcal{A}$ must satisfy at least $\frac{c}{\kappa} \cdot \frac{1}{i}$ of these requests, otherwise the adversary could satisfy them all and beat the competitive ratio. If this was the case, the adversary would have gotten revenue of $i \cdot c$, while $\mathcal{A}$'s revenues would be $\frac{(i-1)c}{\kappa}$ from the movies taken before time $(i-1)\nu$, plus *less than* $\frac{c}{\kappa}$ from the movies taken from R_i at time $(i-1)\nu$ (i movies per channel but less than $\frac{c}{i \cdot \kappa}$ channels). Thus, the adversary would achieve a ratio greater than κ.

This process continues until $\mathcal{A}$ exhausts its supply of channels. This happens after the kth set when $\sum_{i=1}^{k} \frac{c}{i \cdot \kappa} \geq c$, hence

$$H_k = \sum_{i=1}^{k} \frac{1}{i} \geq \kappa , \qquad (1)$$

where H_k is the kth harmonic number.

We also need to make sure for our argument to hold that the number of requests is not too large to exhaust the supply of users in a period of length T. The size of the first request set is c, while the size of each of the $k-1$ subsequent sets is $(c + \frac{c}{\kappa})$. Thus, the total number of requests needed by this strategy is $c + (k-1)(c + \frac{c}{\kappa})$, which must be less than u. This imposes the following condition on k:

$$k + \frac{k-1}{\kappa} \leq \mu . \qquad (2)$$

Once the adversary has exhausted the online algorithm's supply of channels after presenting R_k, it can serve all the remaining users on one channel. In order for the adversary to take advantage of the situation by accepting requests that $\mathcal{A}$ cannot, we also require that the acceptance/rejection notifications for the kth set of requests be issued *before* time T. This will require

$$k\nu < \tau . \tag{3}$$

Just after set R_k is presented, there exist kc unsatisfied requests (k requests for each of c movies). $\mathcal{A}$ has already committed to a total of $\frac{(k-1)c}{\kappa}$ requests. $\mathcal{A}$ serves $\frac{c}{\kappa}$ more requests at the kth round, for a total of $\frac{kc}{\kappa}$ requests. On the other hand, $\mathcal{A}^*$ serves all the kc unfulfilled requests. In addition, it presents set R_{k+1} at time $k\nu$, which contains as many requests as possible (to make a total of u requests), all for one of the movies that are not been served by $\mathcal{A}$. At this point, the only requests not served by $\mathcal{A}^*$ are the $\frac{(k-1)c}{\kappa}$ requests that $\mathcal{A}$ committed to. Thus, the adversay is able to serve a total of $u - \frac{(k-1)c}{\kappa}$ requests.

In order for algorithm $\mathcal{A}$ to preserve a competitive ratio of κ, we must have

$$\kappa \geq \frac{u - (k-1)\frac{c}{\kappa}}{\frac{kc}{\kappa}} = \frac{u\kappa - (k-1)c}{kc} = \frac{\kappa}{k}\mu - (1 - \frac{1}{k}) ,$$

which implies $(1 - \frac{1}{k}) \geq \kappa(\frac{\mu}{k} - 1)$, or

$$\kappa \leq \frac{1 - 1/k}{\mu/k - 1} . \tag{4}$$

We now proceed to prove the bounds on the competitive ratio.

Proof. Suppose there exists an online algorithm $\mathcal{A}$ with competitive ratio $\phi(\mathcal{A}) = \kappa \leq \ln \mu/2$. The adversary chooses $k = \lfloor \mu/2 \rfloor$. Since $\forall k$, $\ln(k+1) \leq H_k$, we have $\ln\lceil \mu/2 \rceil \leq H_k$. Thus, $\kappa \leq H_k$, and constraint (1) is satisfied.

Regarding constraint (2), since $\kappa \geq 1$ (by definition), we have $(k + \frac{k-1}{\kappa}) \leq 2k$, always. If $k = \lfloor \mu/2 \rfloor$, we obtain $2k \leq \mu$, so constraint (2) is also satisfied. Thus the "game" can be completed. Finally, the assumption (stated earlier) that $\nu < 2\tau/\mu$ allows constraint (3) to be satisfied. We now check whether (4) is satisfied. The RHS of (4) is now

$$\frac{1 - \frac{1}{\lfloor \mu/2 \rfloor}}{\frac{\mu}{\lfloor \mu/2 \rfloor} - 1} \leq \frac{1 - \frac{1}{\mu/2}}{\frac{\mu}{\mu/2} - 1} = 1 - \frac{2}{\mu} .$$

Thus, the RHS is always smaller than 1. Since the LHS of (4), κ, is larger than 1, (4) fails to hold, and therefore no algorithm can have a competitive ratio of $\kappa \leq \ln \mu/2$. $\square$

Remark. The $\Omega(\ln \mu)$ bound can be extended to hold for all $\nu \leq \tau$, at the expense of a bigger user space. (Of course, the bound can also be readily strengthened to μ for small values of c.)

3.4 Prepartition algorithms

In this section we present algorithm $\mathcal{H}$ which, when allowed delayed notification and refusal by choice, and given a large enough (but reasonable) number of channels, matches (asymptotically) the lower bound of the previous section. More specifically, $\phi(\mathcal{H}) \leq 1 + 8H_\mu$, where H_μ is the μth harmonic number.

Roughly, $\mathcal{H}$ divides the c channels into μ sets $S_1, S_2, ...S_\mu$. Each set S_i contains $\frac{c}{iH_\mu}$ channels (in fact, $|S_i| = \lfloor \frac{c}{iH_\mu} \rfloor$). Thus, the total cardinality of the sets is $\frac{c}{H_\mu} \cdot (1 + 1/2 + ... + 1/\mu) = c$. The algorithm stipulates that any transmission made on a channel in set S_i must serve at least i requests (see Figure 3).

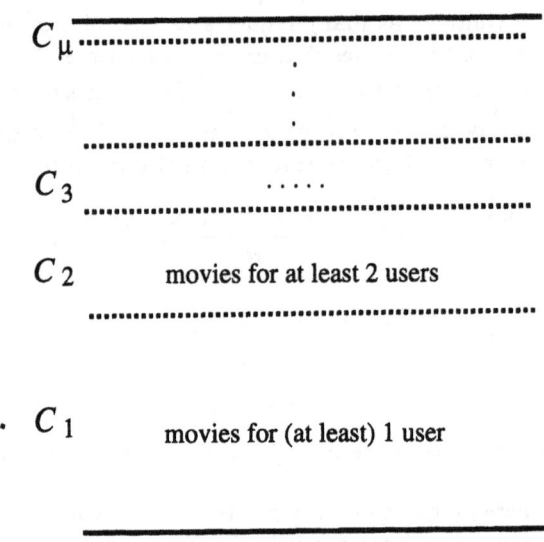

Fig. 3. The *Harmonic Prepartition* algorithm.

$\mathcal{H}$ schedules the arriving requests as follows: Each movie m_j is associated with a distinct *queue* Q_j of requests for that movie. Initially, $Q_j = nil$ for all j. When a request for m_j is made, it is immediately inserted into Q_j. The purpose of Q_j is to accumulate as many requests as possible for movie m_j, so that it can serve all of them with a single transmission at the end. Each queue Q_j is associated with a "start time" $start(Q_j)$, representing the time at which the first request in the queue was made.

At time $start(Q_j) + \nu$, the algorithm decides whether to serve the requests in Q_j (say k in number). If there exists a free channel in a set S_i with index $i \leq k$, it immediately notifies and begins a transmission for m_j on one such channel, thus serving all k requests, and Q_j is re-initialized to nil. If, on the other hand, there exists no such channel, $\mathcal{H}$ rejects only those requests in Q_j made at time $start(Q_j)$, deletes them from Q_j, and resets $start(Q_j)$ to the time of the earliest request now in Q_j. We are able to state the following.

Theorem 5. *Let* $\nu = \tau$ *and* $c \geq (\mu + 1)H_\mu$, *where* H_μ *is the* μ*th harmonic number. Then* $\phi(\mathcal{H}) = O(H_\mu)$.

Proof. We define the *saturation level* at instant t to be the highest i such that *all* channels in sets $S_1, S_2, ...S_i$ are occupied at time t. The saturation level of an *interval* of time is the *highest* saturation level achieved during the interval.

Consider any execution of $\mathcal{H}$ in response to a request sequence ρ, and a corresponding execution of an offline algorithm $\mathcal{A}^*$ in response to the same request sequence. We divide the entire execution into *intervals* $I_0, I_1, ...I_l$ of length T. Interval I_j spans the half-open time interval $[jT, (j + 1)T)$.

For all $j = 0, 1, ...l$, let $A(j)$ be the number of requests accepted by $\mathcal{H}$ in interval I_j, and $R(j)$ the number of requests that were *rejected by* $\mathcal{H}$ *in* I_j *but accepted by* $\mathcal{A}^*$ *in its execution*. Let σ_j be the saturation level of interval I_j. Thus, there exists some instant $t_j \in I_j$ in which all channel sets $S_1, S_2, ...S_{\sigma_j}$ are "full." Since such requests must have been accepted no earlier than T before t_j, we have, $\forall j = 1, 2, ..., l$,

$$A(j - 1) + A(j) \geq \frac{\sigma_j c}{H_\mu} - \frac{\sigma_j(\sigma_j + 1)}{2} \geq \frac{\sigma_j c}{2H_\mu} , \qquad (5)$$

given our assumption on c. Along similar lines, $A(0) \geq \frac{\sigma_0 c}{2H_\mu}$.

Now consider $R(j)$. By definition, all requests in $R(j)$ were rejected by $\mathcal{H}$ during I_j, so each such request was *made* in the interval $[jT - \nu, (j + 1)T - \nu)$. The adversary $\mathcal{A}^*$ could thus serve these requests anytime during the interval $[jT - \nu, (j + 1)T)$. Hence the offline can utilize upto $2c$ transmissions to serve these requests (≤ 2 on each channel).

Assume first $\sigma_j < \mu$. Each such transmission by $\mathcal{A}^*$ must contain $\leq \sigma_j$ requests from $R(j)$. For if there existed an offline transmission serving more than σ_j requests from $R(j)$, $\mathcal{H}$ too would be in a position to accept these requests, since its saturation level in that interval was σ_j. Hence we have, $\forall j$,

$$R(j) \leq 2\sigma_j c . \qquad (6)$$

From (5) and (6) above, we have, $\forall j$,

$$4H_\mu(A(j - 1) + A(j)) \geq R(j) . \qquad (7)$$

The case $\sigma_j = \mu$ is handled similarly, yielding the same result. Let now A be the total number of requests accepted by $\mathcal{H}$, and let R be the total number of requests rejected by $\mathcal{H}$ but accepted by $\mathcal{A}^*$. Note that $A = A(0)+A(1)+...+A(l)$, and similarly, $R = R(0) + ... + R(l)$. Summing up the $(l + 1)$ inequalities of the form (7) (and making the necessary adjustments), we obtain $4H_\mu \cdot (2A) \geq R$, or

$$R \leq 8H_\mu A . \qquad (8)$$

From the definitions, $rev(\mathcal{A}^*) \leq A + R$, while $rev(\mathcal{H}) = A$. Hence, from (8), $\phi(\mathcal{H}) \leq 1 + \frac{R}{A} \leq 1 + 8H_\mu$. $\qquad \square$

4 Discussion

In this paper we have formulated the problem of Video on Demand (VOD) from a resource allocation perspective. We have investigated the problem in an *on-line* fashion, and established upper and lower bounds on the competitive ratio of on-line movie scheduling algorithms for a variety of scenarios.

In the previous section we have analyzed VOD from a worst-case perspective. We have also compared the performances of the different algorithms under various distributions of requests over time, and evaluated the effect of various heuristics. For example, it is expected that the number of requests for movies in a VOD system would peak during "prime time" hours. Define the *load* of such a system as the number of incoming requests in a given time interval. Alternatively, this can be expressed as $x \cdot c$, where x is the *loading factor*. Figure 4 depicts one plausible load variation over a period of 8τ.

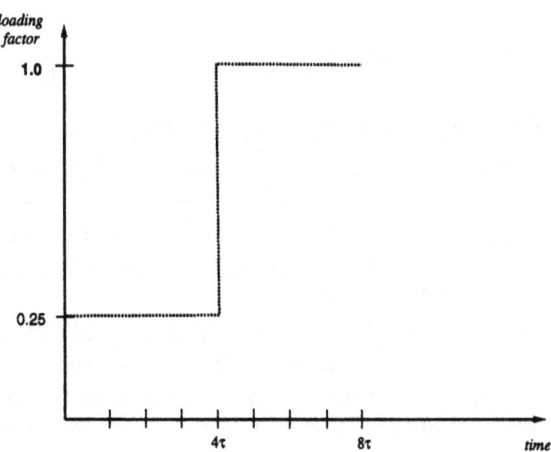

Fig. 4. Load variation in a VOD system.

Let π denote the probability that a user will place a request for a movie, given that the movie is to be shown in some channel. Since we are concerned with unusual selections/rare movies, π would typically be a small number. Figure 5 shows the performance (i.e., expected number of users served) as a function of π achieved by the different approaches: Near, Full, Full$^+$ VOD, and Harmonic from the previous section. (Full$^+$ is the extension of Full that waits for time τ before serving a request.) In the example, $c = 1000$ and $u = 10000$. The comparison is done under the load conditions of Figure 4. For the purpose of Harmonic, we model the probability of i users showing interest for a (rare) movie to be shown in some channel as $\frac{\pi^i}{i!}$. The poor performance of Near VOD (i.e., $\pi \cdot c$) is expected, as the method is aimed at popular movies. Harmonic also outperforms the more straightforward Exponential algorithm (not shown).

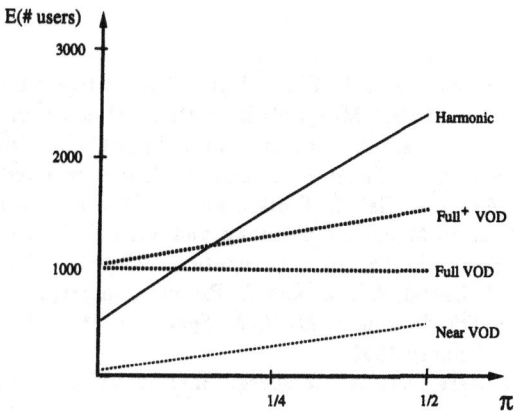

Fig. 5. Experimental comparison of VOD approaches.

The basic teachings of our on-line analysis of VOD are summarized as follows:

- From the adversarial setting: the offline algorithm capitalizes by serving what the online "just missed;"
- from lower bounds' reasoning: long notification times are needed; and
- from the experimental evaluation and "data:" low π for most movies – need long periods to collect requests.

We have used our findings to come up with a VOD scheme which we believe has the potential of practical development [5].

Finally, we remark that in this work we have implicitly assumed that all requests for movies are made in "real time," i.e., ideally the customer would like the request to be honored "as soon as possible." A natural extension is to allow for *reservations*, i.e., requests which are done substantially ahead of the time desired for the delivery of the movie. Some additional natural extensions would be to investigate requests (movies) of different length (cf. [12]), different capacity requirements (e.g., HDTV vs. regular TV quality), dynamically changing capacity requirements (e.g., multi-media documents), and multiple stages of video servers and capacity-limited channels.

Acknowledgements

The authors thank Ernst Biersack, Anna Karlin, and Orli Waarts for helpful comments on a preliminary version of the manuscript. Work of the first author was performed while he was visiting IBM.

553

References

1. B. Awerbuch, Y. Azar and A. Fiat. Tight Competitive ALgorithms for On-Line Max Cover or How to Beat Murphy's Law. Personal communication.
2. B. Awerbuch, Y. Azar and S. Plotkin. Throughput-Competitive On-Line Routing. *Proc. 34th Annual IEEE Symp. on Found. of Computer Science*, 32–40, 1993.
3. J. Aspnes, Y. Azar, A. Fiat, S. Plotkin and O. Waarts. On-Line Load Balancing with Applications to Machine Scheduling and Virtual Circuit Routing. *Proc. 25th Annual ACM Symp. on Theory of Computing*, 623–631, 1993.
4. B. Awerbuch, Y. Bartal, A. Fiat and A. Rosen. Competitive Non-Preemptive Call Control. *Proc. 5th Annual ACM-SIAM Symp. on Discrete Algorithms*, 312–320, Arlington, VA, January 1993.
5. A. Bar-Noy, J. Garay and A. Herzberg. Sharing Video on Demand. Unpublished manuscript.
6. R.L. Graham. Bounds on multiprocessing anomalies. *SIAM Journal of Applied Mathematics*, 17:263–269, 1969.
7. J. A. Garay and I. S. Gopal. Call Preemption in Communication Networks. *Proc. INFOCOM '92*, 1043–1050, Florence, Italy, May 1992.
8. J.A. Garay, I. Gopal, S. Kutten, Y. Mansour and M. Yung. Efficient On-line Call Control Algorithms. *Proc. 2nd Israeli Symp. on Theory of Computing and Systems*, Netania, Israel, 285–293, June 1993.
9. Colin J. Horton. 110 channels without boundaries: why restrict options? *Communications Engineering and Design*, 19/1:29–30, January 1993.
10. Steven M. McCarthy. Integrating Telco Interoffice Fiber Transport with Coaxial Distribution. *Proc. SPIE - Int. Soc. Opt. Eng.*, 1786:23–33, November 1993.
11. John Koegel and Andrzej Syta. Routing of multi-media connections in hybrid networks. *Proc. SPIE - Int. Soc. Opt. Eng.*, 1786:2–10, November 1993.
12. Richard J. Lipton and Andrew Tomkins. Online Interval Scheduling. *Proc. 5th Annual ACM-SIAM Symp. on Discrete Algorithms*, Arlington, VA, 302–311, January 1994.
13. Kenneth Metz. Next generation catv networks. *Proc. SPIE - Int. Soc. Opt. Eng.*, 1786:184–189, November 1993.
14. Chris Sell. Video on demand internal trial. *Proc. SPIE - Int. Soc. Opt. Eng.*, 1786:168–175, November 1993.
15. Daniel D. Sleator and Robert E. Tarjan. Amortized efficiency of list update and paging rules. *Communications of the ACM*, 32:652–686, 1985.
16. D. Shmoys and J. Wein and D. Williamson. Scheduling Parallel Machines On-line. *Proc. 32nd IEEE Symp. on Foundtations of Computer Science*, 131–140, October 1991.

The Binomial Transform and its Application to the Analysis of Skip Lists*

Patricio V. Poblete[1], J. Ian Munro[2] and T. Papadakis[3]

[1] Department of Computer Science, University of Chile, Casilla 2777, Santiago, Chile (E-mail: ppoblete@dcc.uchile.CL)
[2] Department of Computer Science, University of Waterloo, Waterloo, Ontario, Canada N2L 3G1 (E-mail: imunro@watsol.uwaterloo.ca)
[3] Department of Computer Science, York University, 4700 Keele Street, North York, Ontario, Canada M3J 1P3 (E-mail: tom@cs.yorku.ca)

Abstract. To any sequence of real numbers $\langle a_n \rangle_{n \geq 0}$, we can associate another sequence $\langle \widehat{a}_s \rangle_{s \geq 0}$, called its *binomial transform*. This transform is defined through the rule

$$\widehat{a}_s = B_s a_n = \sum_n (-1)^n \binom{s}{n} a_n.$$

We study the properties of this transform, obtaining rules for its manipulation and a table of transforms, that allow us to invert many transforms by inspection.

We use these methods to perform a detailed analysis of *skip lists*, a probabilistic data structure introduced by W. Pugh as an alternative to balanced trees. In particular, we obtain the mean and variance for the cost of searching for the first or the last element in the list (confirming results obtained previously by other methods), and also for the cost of searching for a random element (whose variance was not known).

We obtain exact (albeit sometimes complicated) expressions for all $n \geq 0$, and from them we find the corresponding asymptotic expressions.

1 Introduction

The inversion formula

$$a_n = \sum_k (-1)^k \binom{n}{k} b_k \iff b_n = \sum_k (-1)^k \binom{n}{k} a_k \tag{1}$$

plays an important rôle in the analysis of some algorithms and data structures, and in the solution of many combinatorial problems [18, 8].

In [12] Knuth used this relation to define a *transform*, mapping sequences of real numbers onto sequences of real numbers. Even though the experience with

* This research was supported in part by the Natural Sciences and Engineering Research Council of Canada under grant No. A-8237, the Information Technology Research Centre of Ontario, and FONDECYT(Chile) under grant 1940271. Part of this work was done while the first author was on sabbatical and the third author a post doc at the University of Waterloo.

other similar transforms indicates that they can be powerful tools, in this case the concept does not seem to have been developed, and the literature is lacking in tables of binomial transforms, rules for their manipulation, etc. The relation of the binomial transform to other methods, and some generalizations of it, have been studied by Prodinger [10].

In this paper, we develop the theory of the binomial transform, and show how it can be applied to analyze the performance of *skip lists*, a probabilistic data structure introduced by W. Pugh [17, 14].

Using the binomial transform, we give alternative derivations for some known results about skip lists, and solve some problems that were still open. Our results give exact expressions for the relevant performance measures, as functions of n (the number of keys in the data structure). Having exact expressions is useful, because it allows us to check the consistency of our solutions, by comparing them to the values obtained by numerical evaluation of the corresponding recurrence equations for small values of n. However, these expressions do not always have simple closed forms, but from them we can obtain asymptotic expressions. As a possible drawback, the manipulations required for the analysis tend to be complicated, and usually require the use of a symbolic algebra system.

2 Notation

We generally follow the notational conventions of [8]. In particular, for any boolean expression C, $[C]$ denotes a function that has the value 1 if C is true, and 0 otherwise. If $\Gamma(x)$ is the Gamma function, $\psi(x)$ is its logarithmic derivative $(\psi(x) = \Gamma'(x)/\Gamma(x))$. The harmonic numbers H_n are defined as $H_n = \sum_{1 \le k \le n} \frac{1}{k}$. Equivalently, $H_n = \psi(n+1) + \gamma$, where $\gamma = 0.5772\ldots$ is Euler's constant. Asymptotically, $H_n = \ln n + \gamma + O(\frac{1}{n})$. Harmonic numbers can be generalized to non-integer arguments by the formula $H_x = \sum_{k \ge 1}(\frac{1}{k} - \frac{1}{k+x})$, and it can be shown that $H_x = \psi(x+1) + \gamma$. The second order harmonic numbers $H_n^{(2)}$ are defined as $H_n^{(2)} = \sum_{1 \le k \le n} \frac{1}{k^2}$, and their limiting value is $H_\infty^{(2)} = \frac{\pi^2}{6}$. The rising and falling powers are defined as $x^{\overline{n}} = x(x+1)\cdots(x+n-1)$ and $x^{\underline{n}} = x(x-1)\cdots(x-n+1)$ respectively. They can be extended to negative exponents by means of the relation $x^{\overline{-n}} = 1/(x-1)^{\underline{n}}$. In addition to the standard binomial coefficients, we also use Comtet's notation for the symmetric binomial coefficients $(i,j) = \binom{i+j}{i} = \binom{i+j}{j}$ [2]. The equivalent formula $(x,n) = \frac{(x+1)^{\overline{n}}}{n!}$ extends this to the case of one non-integer argument. The notation $Q_r(m,n)$ denotes the Q functions [12], defined as $Q_r(m,n) = \sum_{0 \le j \le n}(j,r)\frac{n^{\underline{j}}}{m^j}$, for $r \ge 0$. The notation $B(x,y)$ denotes the Beta function $(B(x,y) = \Gamma(x)\Gamma(y)/\Gamma(x+y))$. The operator $\mathbf{D}_x$ denotes a partial derivative with respect to x. Whenever the variables p and q are used, it will be assumed that $p+q = 1$. Finally, throughout the paper, each occurrence of the symbol $\chi(\cdot)$ will denote a (possibly different) periodic function of very small amplitude (see Section 3.7).

3 The Binomial Transform

3.1 Definition

For any sequence of real numbers $\langle a_n \rangle_{n \geq 0}$ we define its *binomial transform* as the sequence $\langle \widehat{a}_s \rangle_{s \geq 0}$, where

$$\widehat{a}_s = \mathcal{B}_s a_n = \sum_n (-1)^n \binom{s}{n} a_n. \tag{2}$$

Note that, since s is a nonnegative integer, the binomial coefficients vanish for $n < 0$ or $n > s$. Therefore the actual range of the summation is really $0 \leq n \leq s$, but we prefer to write it as an unconstrained sum to simplify notation. Furthermore, we will always assume that $a_n = 0$ for all $n < 0$, and $\widehat{a}_s = 0$ for all $s < 0$. For example, when we write "$a_n = 1$" we actually mean "$a_n = [n \geq 0]$."

3.2 Inverse Transform

Because of the inversion formula (1), it is easy to see that we can recover a_n from $\widehat{a}_s$ by using the same formula (2), with the rôles of n and s reversed. Thus,

$$a_n = \mathcal{B}_n \widehat{a}_s = \sum_s (-1)^s \binom{n}{s} \widehat{a}_s. \tag{3}$$

3.3 Relation with other transforms

For any sequence of real numbers $\langle a_n \rangle_{n \geq 0}$, its Poisson Transform [7, 16] can be defined as

$$\mathcal{P}_x a_n = e^{-x} \sum_{n \geq 0} a_n \frac{x^n}{n!}$$

(this corresponds to the case $m = 1$ in [7, 16]). For any function $f(x)$, its Mellin Transform [13, 5] is defined as

$$\mathcal{M}_s f(x) = \int_0^\infty f(x) x^{s-1} dx.$$

It is not hard to prove that the following formal identity holds between these two transforms and the binomial transform:

$$\mathcal{B}_s a_n = \frac{1}{\Gamma(-s)} \mathcal{M}_{-s} \mathcal{P}_x a_n$$

There is also an interesting and useful relation linking the Binomial Transform to exponential generating functions (EGFs). If the EGF of $\langle a_n \rangle$ is $A(z)$, then the EGF of $\langle \widehat{a}_s \rangle$ is $e^z A(-z)$ [10].

3.4 Rules for the Manipulation of Binomial Transforms

The following is a list of useful rules for the manipulation of binomial transforms:[4]

R1. $\mathcal{B}_s(\alpha a_n + \beta b_n) = \alpha \mathcal{B}_s a_n + \beta \mathcal{B}_s b_n$ (Linearity)

R2. $\mathcal{B}_s \dfrac{a_{n+1}}{n+1} = -\dfrac{\widehat{a}_{s+1}}{s+1}$

R3. $\mathcal{B}_s \displaystyle\sum_{0 \le k < n} a_k = -\widehat{a}_{s-1}$ $\qquad \mathcal{B}_s \displaystyle\sum_{0 \le k \le n} a_k = \widehat{a}_s - \widehat{a}_{s-1}$

$\mathcal{B}_s a_{n-1} = -\displaystyle\sum_{0 \le t < s} \widehat{a}_t \qquad \mathcal{B}_s(a_n - a_{n-1}) = \displaystyle\sum_{0 \le t \le s} \widehat{a}_t$

R4. $\mathcal{B}_s \dfrac{1}{n+1} \displaystyle\sum_{0 \le k \le n} a_k = \dfrac{\widehat{a}_s}{s+1} \qquad \mathcal{B}_s \dfrac{a_n}{n+1} = \dfrac{1}{s+1} \displaystyle\sum_{0 \le t \le s} \widehat{a}_t$

R5. $\mathcal{B}_s(a_{n+1} - a_n) = -\widehat{a}_{s+1} \qquad \mathcal{B}_s a_{n+1} = \widehat{a}_s - \widehat{a}_{s+1}$

R6. $\mathcal{B}_s \dbinom{n}{k} a_{n-k} = (-1)^k \dbinom{s}{k} \widehat{a}_{s-k}$ (or equivalently, $\mathcal{B}_s n^{\underline{k}} a_{n-k} = (-1)^k s^{\underline{k}} \widehat{a}_{s-k}$)

R7. $\mathcal{B}_s n a_n = s(\widehat{a}_s - \widehat{a}_{s-1}) \qquad \mathcal{B}_s n(a_n - a_{n-1}) = s\widehat{a}_s$

For the following three properties, recall our convention that $p + q = 1$:

R8. $\mathcal{B}_s \displaystyle\sum_k \dbinom{n}{k} p^k q^{n-k} a_k = p^s \widehat{a}_s$

R9. $\mathcal{B}_s \displaystyle\sum_k \dbinom{n}{k} p^k q^{n-k} a_k b_{n-k} = \displaystyle\sum_k \dbinom{s}{k} p^k q^{s-k} \widehat{a}_k \widehat{b}_{s-k}$

R10. $\mathcal{B}_s \displaystyle\sum_k \dbinom{n+1}{k+1} p^k q^{n-k} a_k = p^s \widehat{a}_s - p^{s-1} q \displaystyle\sum_{0 \le t < s} \widehat{a}_t = p^{s-1} \widehat{a}_s - p^{s-1} q \displaystyle\sum_{0 \le t \le s} \widehat{a}_t$

R11. $\mathcal{B}_s(a_n - a_0) = \mathcal{B}_s(a_n - a_0)[n > 0] = \widehat{a}_s[s > 0]$

3.5 Table of Transforms

Because the binomial transform and the inverse binomial transform are defined by the same summation, with the rôles of n and s interchanged, we use the following format for our table of transforms: Table 1 lists pairs of functions $a_n \rightleftharpoons b_n$ that are related by the transform; when applying the binomial transform, the table can be used to find $a_s = \mathcal{B}_s b_n$ and $b_s = \mathcal{B}_s a_n$, and when inverting the transform, $a_n = \mathcal{B}_n b_s$ and $b_n = \mathcal{B}_n a_s$.

Transforms T6–T10 will be useful for inverting transforms expressed as partial fraction expansions. Transforms T11 and T12 are not used in the rest of the paper, but we include them for the sake of completeness, as the functions involved appear in similar analyses of other algorithms.

---

[4] The proofs of these rules and other properties of binomial transforms have been omitted for reasons of space.

	a_n	$\rightleftharpoons$	b_n
T1.	1	$\rightleftharpoons$	$[n = 0]$
T2.	$[n > 0]$	$\rightleftharpoons$	$-[n > 0]$
T3.	p^n	$\rightleftharpoons$	q^n
T4.	$\binom{n}{k}$	$\rightleftharpoons$	$(-1)^k [n = k]$
T5.	$\binom{n}{k} p^{n-k}$	$\rightleftharpoons$	$(-1)^k \binom{n}{k} q^{n-k}$
T6.	$\dfrac{1}{n+1}$	$\rightleftharpoons$	$\dfrac{1}{n+1}$
T7.	H_n	$\rightleftharpoons$	$-\dfrac{[n > 0]}{n}$
T8.	$\dfrac{1}{2}(H_n^2 + H_n^{(2)})$	$\rightleftharpoons$	$-\dfrac{[n > 0]}{n^2}$
T9.	$\dfrac{1}{n+x}$	$\rightleftharpoons$	$\dfrac{n!}{x^{n+1}} = B(x, n+1)$
T10.	$\dfrac{1}{(n+x)^2}$	$\rightleftharpoons$	$B(x, n+1)(\psi(x+n+1) - \psi(x))$ $= B(x, n+1)(H_{x+n} - H_{x-1})$
T11.	$n(H_n - 1)$	$\rightleftharpoons$	$-\dfrac{[n > 1]}{n-1}$
T12.	$Q_r(m, n)$	$\rightleftharpoons$	$(-1)^n (n, r) \dfrac{n!}{m^n}$

Table 1. Pairs of binomial transforms

3.6 Two-dimensional binomial transforms

If a_{n_1, n_2} is a sequence indexed by two nonnegative integer variables, we can compute its binomial transform on the first variable ($\hat{a}_{s_1, n_2}$), on the second one ($\grave{a}_{n_1, s_2}$), or on both ($\hat{a}_{s_1, s_2}$). Note that in the latter case, the result is independent of the order in which the transforms are applied, as these transforms are finite summations.

All the known properties apply to each variable considered in isolation, but there are some interesting results about the two variables together. We will use the two following properties:

P1. If a_{n_1, n_2} is a two-dimensional sequence, and b_n a sequence on one variable, then

$$a_{n_1, n_2} = b_{n_1 + n_2} \forall n_1, n_2 \geq 0 \iff \hat{a}_{s_1, s_2} = \hat{b}_{s_1 + s_2} \quad \forall s_1, s_2 \geq 0.$$

P2. If a_{n_1, n_2} is a two-dimensional sequence, and b_n a sequence on one variable, then

$$b_n = \frac{1}{n+1} \sum_{0 \leq k \leq n} a_{k, n-k} \quad \forall n \geq 0 \iff \hat{b}_s = \frac{1}{s+1} \sum_{0 \leq t \leq s} \hat{a}_{t, s-t} \quad \forall s \geq 0.$$

3.7 Oscillating functions

Consider the function $\frac{1}{s+\sigma}$ where σ is an imaginary parameter. From transform T9[5], we know that its inverse is $B(\sigma, n+1) = \frac{n!}{\sigma^{n+1}}$, which is asymptotically a periodic function of $\log n$, since $B(\sigma, n+1) \sim \Gamma(\sigma)e^{-\sigma \ln n}$.

In our applications, we will encounter functions of the form

$$F_n = \frac{1}{\ln p} \sum_{k \neq 0} B(\sigma_k, n+1) = B_n \frac{1}{\ln p} \sum_{k \neq 0} \frac{1}{s + \sigma_k}$$

where $\sigma_k = \frac{2\pi i k}{\ln p}$, p is a real parameter $(0 < p < 1)$, and the summation ranges over all negative and positive integers. Besides this basic function, several others related to it will appear, and it will be convenient to define a family of functions

$$F_n^{[r]} = \frac{1}{\ln p} \sum_{k \neq 0} \sigma_k^{\bar{r}} B(\sigma_k, n+1)$$

such that

$$\widehat{F}_s^{[r]} = \frac{1}{\ln p} \sum_{k \neq 0} \frac{\sigma_k^{\bar{r}}}{s + \sigma_k}.$$

The F functions are asymptotic to periodic functions of $\log n$. If p is not too small (say, $p > .3$)[6], these are functions of very small amplitude (less than 10^{-3}). Using the notation of [15], we have that $F_n^{[r]} \sim -f_{r,1/p}(n)$, where $f_{r,1/p}(n) = -\frac{2}{\ln p} \sum_{k \geq 1} \Re[\Gamma(r - \sigma_k)e^{-\sigma_k \ln n}]$. See [14, 15] for further discussion of the f functions, and for a table of values for different values of p.

Consider now the effect of differentiating the equation $B_n \frac{\sigma^{\bar{r}}}{s+\sigma} = \sigma^{\bar{r}} B(\sigma, n+1)$ with respect to σ. Using the differential operator $\mathbf{D}_\sigma$ (derivative with respect to σ), we have that

$$\mathbf{D}_\sigma B_n \frac{\sigma^{\bar{r}}}{s+\sigma} = -\sigma^{\bar{r}} B(\sigma, n+1)(H_{\sigma+n} - H_{\sigma+r-1})$$

$$= -\sigma^{\bar{r}} B(\sigma, n+1)H_n + \sigma^{\bar{r}} B(\sigma, n+1)(H_n - H_{\sigma+n} + H_{\sigma+r-1}).$$

Since $H_n - H_{\sigma+n} = O(\frac{1}{n})$, the second term is another asymptotically periodic function of $\log n$. Based on it, we define a new family of functions

$$G_n^{[r]} = \frac{1}{\ln^2 p} \sum_{k \neq 0} \sigma_k^{\bar{r}} B(\sigma_k, n+1)(H_n - H_{\sigma_k+n} + H_{\sigma_k+r-1}).$$

These functions are also asymptotically periodic functions of $\log n$, and in the range of values of interest for p they are of very small amplitude.

---

[5] Actually, to do this we must first extend the binomial transform to the complex numbers, but this presents no problem.

[6] The case of very small p is not very interesting for skip lists, as the data structure degenerates into a linear linked list.

Note: As we have stated, this paper will focus on the application of these techniques to the analysis of algorithms, and in particular to the analysis of skip lists. The statistics associated with these algorithms usually involve terms consisting of linear combinations or squares of the oscillating functions studied in this section. For reasonable values of p, the contribution of these terms decreases dramatically as n increases, and asymptotically they are periodic functions of $\log n$ of very small amplitude (but nonvanishing). In our theorems, we give the exact form for all these terms, and additionally we provide asymptotic expressions. In these asymptotic expressions, we lump the contribution of all the oscillatory terms into a term we denote as $\chi(\cdot)$, in a manner analogous to the $O(\cdot)$ notation. Each occurrence of the χ symbol will denote a possibly different periodic function of very small amplitude.

4 Analysis of Skip Lists

4.1 Review of the data structure and known results

A *(probabilistic) skip list* [17] is a generalization of a linked list. Elements are stored in order of increasing key value, and the records are linked into several lists (see Figure 1).

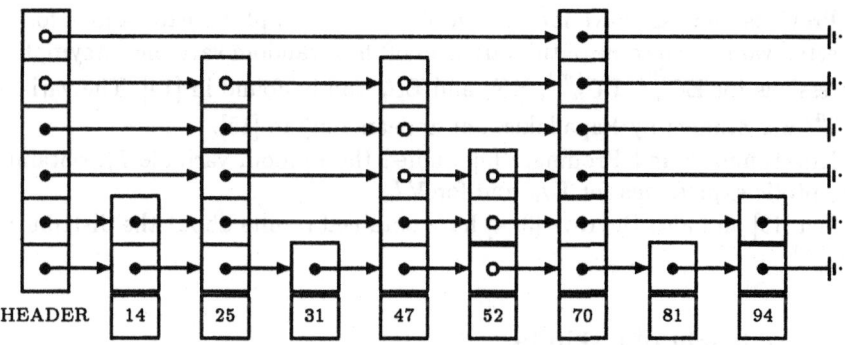

Fig. 1. A skip list, showing the search path for $x \in (52, 70]$

The bottom such list links all elements, and each successive list links a subset of the elements in the list below it. For each element in a given list, the decision to link it into the next level list is made probabilistically: a (possibly biased) coin is flipped, and if it lands "heads" the element is included in the next level list. In our analyses we will assume that the coin has probability p of landing "heads," and probability $q = 1 - p$ of landing "tails."

To search for a given element x, we start at the header of the top list, and compare x to the key of the next record. If x is less than or equal to that key, we move down to the pointer one level below. If not, we advance to the next pointer to the right. We perform a final comparison for equality only when we have reached the bottom level. Figure 1 shows (in white) the pointers inspected to search for any element en the range $(52, 70]$. We will use the number of pointers inspected as the measure of the cost of the search[7].

Note that, to simplify the description of the algorithm (and possibly its implementation), we assume the nil pointers point to a "trailer" record whose key is $+\infty$ (not shown in the figures).

We now define some random variables of interest for the analysis of skip lists. Let us assume that the skip list contains n keys

$$K_1 < K_2 < \cdots < K_n$$

(plus two fictitious keys $K_0 = -\infty$, $K_{n+1} = +\infty$). Then, we will write $C_n^{(m)}$ to denote the random variable "cost of searching for an element in the range $(K_m, K_{m+1}]$," S_n for the random variable "cost of searching successfully for a random element" (assuming all n keys are equally likely), and U_n for the random variable "cost of searching unsuccessfully for a random element" (assuming all $n + 1$ intervals are equally likely). Note that $S_n = C_n^{(M_{n-1})}$ and $U_n = C_n^{(M_n)}$, where M_k is a uniform random variable taking integer values over the range $[0, k]$. Another random variable of interest is the total cost of searching for each of the n elements is a skip list, denoted I_n. (This last quantity is the equivalent for skip lists of the internal path length for binary search trees.)

Previous analyses have focused on obtaining asymptotic expressions for the expected values and some of the variances of these random variables. Asymptotic expressions for $\mathbf{E}C_n^{(0)}$, $\mathbf{E}C_n^{(m)}$, $\mathbf{E}S_n$ and $\mathbf{E}U_n$ can be found in [14]. The variance $\mathbf{V}C_n^{(n)}$ was studied by Papadakis, but appears only in [15].

Kirschenhofer and Prodinger [9] studied the random variable I_n, obtaining asymptotic expressions for $\mathbf{E}I_n$ and for $\mathbf{V}I_n$.

Sen [19] and also Devroye [3, 4] have obtained results about the distribution of $C_n^{(m)}$.

4.2 The height of a skip list

We begin our analyses by studying the height of a skip list, i.e. the number of pointers in the header of the data structure. The asymptotic behavior of this random variable is well known. It is basically the maximum of n independent geometric random variables, and it appears in the analysis of many algorithms. (See for example the analysis of tries in [12], and also [20].) In our case, it will serve to illustrate the use of the binomial transform, and we will use it as a "warm up" before tackling more complicated problems.

Equivalently, the height of a skip list is the number of pointers of the tallest element in the skip list, and it is also equal to the cost of searching for any

---

[7] Note that our definition for the search cost gives a value 1 greater than the search cost defined by Pugh in [17].

element less than or equal to the first element in the list. Figure 2 shows the path followed by this kind of search. To simplify, only the "profile" of the skip list is shown (i.e. the stack of pointers for each element), and not the key values nor the arrows.

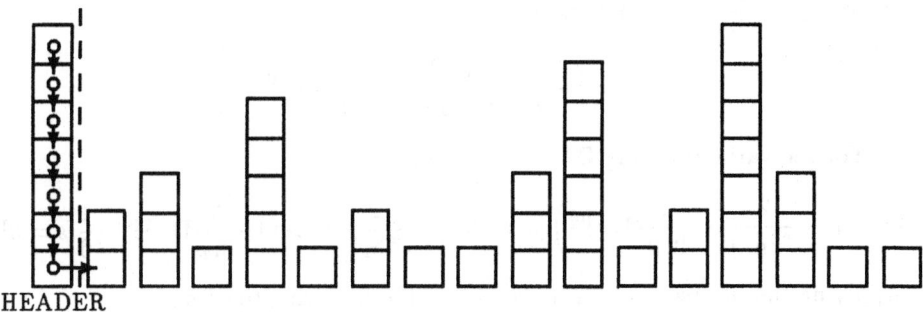

Fig. 2. The height of a skip list

Let $P_n(z)$ be the probability generating function for the random variable "height of the skip list." Since p is the probability that we keep adding pointers when inserting an element, then if we remove the bottom layer of pointers the probability that i elements will still have a nonempty stack of pointers, while j are left with no pointers is $(i,j)p^i q^j$, where $i + j = n$. Therefore, noting that $P_0(z) = 1$, we have that

$$P_n(z) = z \sum_{\substack{i,j \geq 0 \\ i+j=n}} (i,j)p^i q^j P_i(z) + (1-z)[n=0] \quad \forall n \geq 0.$$

Applying the binomial transform, we get

$$(1 - zp^s)\widehat{P}_s(z) = (1 - z) \tag{4}$$

Now, to compute the average and the variance we have to differentiate equation (4) and evaluate at $z = 1$. Let us introduce the notation $a_n = P'_n(1)$, $b_n = P''_n(1)$. Note that the average is given by a_n, and the variance by $b_n + a_n - a_n^2$. Observing that $\widehat{P}_s(1) = [s = 0]$ (because $P_n(1) = 1$), we obtain

$$\widehat{a}_s = \frac{[s > 0]}{p^s - 1}$$

and

$$\widehat{b}_s = \frac{2\widehat{a}_s}{1 - p^s} = -\frac{2p^s}{(p^s - 1)^2}[s > 0].$$

To invert these transforms the following expansions are useful:

E1. $\dfrac{1}{p^s - 1} = -\dfrac{1}{2} + \dfrac{1}{s \ln p} + \dfrac{1}{\ln p} \displaystyle\sum_{k \neq 0} \dfrac{1}{s + \sigma_k} = -\dfrac{1}{2} + \dfrac{1}{s \ln p} + \widehat{F}_s^{[0]}$

E2. $\dfrac{p^s}{(p^s - 1)^2} = \dfrac{1}{s^2 \ln^2 p} + \dfrac{1}{\ln^2 p} \displaystyle\sum_{k \neq 0} \dfrac{1}{(s + \sigma_k)^2} = \dfrac{1}{s^2 \ln^2 p} - \dfrac{\mathbf{D}_s \widehat{F}_s^{[0]}}{\ln p}$

where $\sigma_k = \frac{2\pi i k}{\ln p}$. Using **E1**, we have

$$\widehat{a}_s = \left(-\dfrac{1}{2} + \dfrac{1}{s \ln p} \right) [s > 0] + \widehat{F}_s^{[0]}.$$

For the variance, using **E2**, we have that

$$\widehat{b}_s = \left(-\dfrac{2}{s^2 \ln^2 p} + \dfrac{2}{\ln p} \mathbf{D}_s \widehat{F}_s^{[0]} \right) [s > 0] = -\dfrac{2}{s^2 \ln^2 p}[s > 0] + \dfrac{2}{\ln p}\mathbf{D}_s \widehat{F}_s^{[0]} - \dfrac{1}{6}[s = 0].$$

Applying the inverse transform, we obtain the following theorem:

Theorem 1. *The expected value of the height of a skip list is*

$$\mathbf{E}C_n^{(0)} = -\dfrac{H_n}{\ln p} + \dfrac{1}{2}[n > 0] + F_n^{[0]} \quad \text{for all } n \geq 0$$

$$= -\log_p n + \dfrac{1}{2} - \dfrac{\gamma}{\ln p} + \chi(\log n) + O(\dfrac{1}{n}),$$

and its variance is equal to

$$\mathbf{V}C_n^{(0)} = \dfrac{H_n^{(2)}}{\ln^2 p} + \dfrac{1}{12} - \dfrac{1}{4}[n = 0] - (F_n^{[0]})^2 + 2G_n^{[0]} \quad \text{for all } n \geq 0$$

$$= \dfrac{\pi^2}{6 \ln^2 p} + \dfrac{1}{12} + \chi(\log n) + \chi^2(\log n) + O(\dfrac{1}{n}).$$

4.3 Searching for $+\infty$

Let us now turn our attention to the opposite end of the data structure. A search for "$+\infty$" is (on the average) the most expensive one. We analyze now the cost of that search.

We will use the same notation as before, only the random variable we are studying now is "number of pointers inspected when searching for $+\infty$." (We redefine P_n, a_n and b_n accordingly.) Figure 3 shows the path followed in this case.

Consider the search path from right to left. At the rightmost end, there is some number, say k, of elements of height 1, all of whose bottom-level pointers had to be traversed during the search. In addition, the bottom-level pointer belonging to the element immediately preceding them was also traversed. Note that this last element must necessarily be of height > 1. If we now erase the bottom layer of pointers, all k rightmost elements will disappear. Of the other $n - k$ elements, some of them, say i, were of height > 1, and the remaining ones,

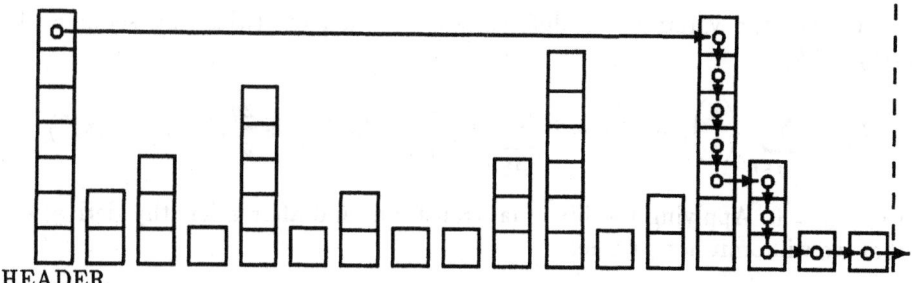

Fig. 3. Searching for $+\infty$ in a skip list

say j, were of height 1, so they also disappear. These i and j elements may be intermixed in any of $(i-1,j)$ possible orders, because one of the i "tall" elements is constrained to be in the rightmost position.

Therefore, the generating function for the search cost is

$$P_n(z) = \sum_{\substack{i,j,k \geq 0 \\ i+j+k=n}} (i-1,j)p^i q^{j+k} P_i(z)z^{k+1} + (1-z)[n=0] \quad \forall n > 0. \quad (5)$$

After the substitution $h = i - 1 + j$, we can obtain the following somewhat simpler form:

$$P_n(z) = \sum_{0 \leq i \leq n} p^i q^{n-i} P_i(z)g_{i,n}(z) + (1-z)[n=0] \quad \forall n \geq 0, \quad (6)$$

where $g_{i,n}(z) = \sum_{i-1 \leq h \leq n-1} \binom{h}{i-1} z^{n-h}$.

The summation defining the function $g_{i,n}(z)$ does not seem to have a simple closed form, but its derivatives at $z = 1$ are not hard to find. It can be shown by induction that $g_{i,n}(1) = \binom{n}{i}$ and $g_{i,n}^{(k)}(1) = k!\binom{n+1}{i+k} \quad \forall k \geq 1$.

Differentiating equation (6) to find the average $a_n = P_n'(1)$, we get

$$a_n = \sum_{0 \leq i \leq n} p^i q^{n-i} \left(\binom{n}{i} a_i + \binom{n+1}{i+1} \right) - [n=0]$$

$$= \sum_{0 \leq i \leq n} \binom{n}{i} p^i q^{n-i} a_i + \frac{1-q^{n+1}}{p} - [n=0],$$

whose transform is

$$\widehat{a}_s = \left(\frac{q}{p} + \frac{1}{p(p^s-1)} \right)[s>0] = \frac{[s>0]}{sp\ln p} + (\frac{1}{2p}-1)[s>0] + \frac{\widehat{F}_s^{[0]}}{p}. \quad (7)$$

To find the variance, we differentiate equation (6) twice and set $z = 1$, obtainig b_n:

$$b_n = \sum_{0 \le i \le n} \binom{n}{i} p^i q^{n-i} b_i + 2 \sum_{0 \le i \le n} \binom{n+1}{i+1} p^i q^{n-i} a_i + \frac{2q}{p^2}(1 - q^n - npq^n)$$

with $b_0 = 0$. Applying the binomial transform, and after a lengthy derivation that we will omit here, we get

$$\hat{b}_s = \left(-\frac{2p^s}{p^2(p^s - 1)^2} - \frac{2qs}{p^2}\left(1 + \frac{1}{p^s - 1}\right) + \frac{2q}{p^2}\beta(p) + \frac{2q}{p^2}\hat{d}_s \right) [s > 0].$$

where $\beta(p) = \sum_{k \ge 1} \frac{k}{p^{-k}-1}$ and $\hat{d}_s = -\sum\sum_{1 \le j \le k}(1 - p^{js})/(p^{-k} - 1)$. Note that $\beta(p)$ is a constant, and it is possible to prove that $d_n = O(\frac{1}{n})$.

Applying the inverse transform, we obtain the following theorem:

Theorem 2. *The expected value of the cost of searching for $+\infty$ in a skip list is*

$$\mathbf{EC}_n^{(n)} = -\frac{H_n}{p \ln p} + \left(1 - \frac{1}{2p}\right) [n > 0] + \frac{F_n^{[0]}}{p} \quad \text{for all } n \ge 0$$

$$= -\frac{1}{p} \log_p n + 1 - \frac{1}{2p} - \frac{\gamma}{p \ln p} + \chi(\log n) + O(\frac{1}{n}),$$

and its variance is

$$\mathbf{VC}_n^{(n)} = -\frac{q}{p^2 \ln p} H_n$$

$$+ \frac{q}{p^2}[n = 1] + \frac{H_n^{(2)}}{p^2 \ln^2 p} + \left(\frac{1}{2p} - \frac{5}{12p^2} + \frac{2q}{p^2 \ln p} - \frac{2q}{p^2}\beta(p)\right) [n > 0]$$

$$- \frac{(F_n^{[0]})^2}{p^2} + \frac{2q}{p^2} F_n^{[1]}[n > 0] + \frac{2}{p^2} G_n^{[0]} + \frac{2q}{p^2} d_n \quad \text{for all } n \ge 0$$

$$= -\frac{q}{p^2} \log_p n$$

$$+ \frac{\pi^2}{6p^2 \ln^2 p} + \frac{1}{2p} - \frac{5}{12p^2} + \frac{q(2 - \gamma)}{p^2 \ln p} - \frac{2q}{p^2}\beta(p)$$

$$+ \chi(\log n) + \chi^2(\log n) + O(\frac{1}{n}),$$

where $\beta(p) = \sum_{k \ge 1} \frac{k}{p^{-k}-1}$ and $d_n = \mathcal{B}_n[-\sum\sum_{1 \le j \le k} \frac{1-p^{js}}{p^{-k}-1}]$.

4.4   Searching for an arbitrary element

We consider now a more general case. Suppose the element we are looking for is not necessarily the first or the last one in the list, but an arbitrary one (see Figure 4). If we assume that the rank of the element is known to us, we can view the list as partitioned in two sections: the left part, of size n_1 contains

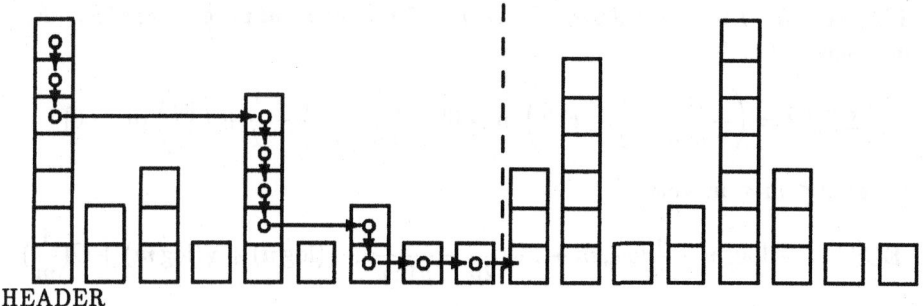

HEADER

Fig. 4. Searching for an arbitrary element

all elements smaller than the given one; the right part, of size n_2, contains all elements greater than or equal to the given one. (Of course, $n_1 + n_2 = n$.)

Generalizing our notation, we write $P_{n_1,n_2}(z)$ for the probability generating function of the search cost. In a similar way, we write a_{n_1,n_2} and b_{n_1,n_2}.

Note that we have already studied the cases $n_1 = 0$ (height of the skip list, or cost of searching for the first element) and $n_2 = 0$ (cost of searching for $+\infty$). Also note that this means that we already know the results for the case $s_1 = 0$ and for the case $s_2 = 0$ in the transformed domain.

The recurrence equation for $P_{n_1,n_2}(z)$ is

$$P_{n_1,n_2}(z) = \sum_{\substack{0 \le i \le n_1 \\ 0 \le \ell \le n_2}} p^i q^{n_1-i} g_{i,n_1}(z) \binom{n_2}{\ell} p^\ell q^{n_2-\ell} P_{i,\ell}(z) + (1-z)[n_1 = 0][n_2 = 0].$$

As before, we apply the binomial transform and differentiate. Omitting the details of the derivations, we finally obtain

$$\hat{a}_{s_1,s_2} = \frac{q}{p}\left(1 + \frac{1}{p^{s_1} - 1}\right)[s_1 > 0][s_2 = 0] + \frac{1}{p^{s_1+s_2} - 1}[s_1 + s_2 > 0].$$

and

$$\hat{b}_{s_1,s_2} = q\,\hat{b}_{s_1,0}[s_2 = 0] + p\,\hat{b}_{s_1+s_2,0} - \frac{2q}{p}\frac{p^{s_1+s_2}}{p^{s_1+s_2} - 1}\sum_{1 \le t < s_2}\frac{1}{p^t - 1}[s_1 + s_2 > 0].$$

From this expression, we can obtain an explicit formula for a_{n_1,n_2} and also for b_{n_1,n_2} (and therefore for the variance). However, we have been unable to obtain the latter in closed form. (Surprisingly, the expression we have found here for $\hat{b}_{s_1,s_2}$ will enable us in the next section to find a closed form expression for the variance of the cost of searching for a random element.)

Therefore, we state only our results for the expected cost:

Theorem 3. *The expected cost of searching for an element in the range $(K_m, K_{m+1}]$ in a skip list is*

$$\mathbf{EC}_n^{(m)} = \left(-\frac{H_n}{\ln p} + \frac{1}{2} + F_n^{[0]}\right)[n > 0] + \frac{q}{p}\left(-\frac{H_m}{\ln p} - \frac{1}{2} + F_m^{[0]}\right)[m > 0]$$

for all $0 \le m \le n$, and

$$\mathbf{EC}_n^{(m)} = -\log_p n - \frac{q}{p}\log_p m + 1 - \frac{1}{2p} - \frac{\gamma}{p\ln p} + \chi(\log n) + \chi(\log m) + O\left(\frac{1}{m}\right)$$

for $m, n \to \infty$.

4.5 Searching for a random element

We consider the case of an unsuccessful search for a random element. When the search stops, the list has been split in (n_1, n_2) elements as in the previous section, and the $n + 1$ possible splits are equally likely.

If we write $\bar{P}_n(z)$ to denote the probability generating function for the search cost, we have that

$$\bar{P}_n(z) = \frac{1}{n+1}\sum_{0 \le k \le n} P_{k,n-k}(z)$$

where the P in the right hand side is from the previous section. Using property **P2** of two-dimensional binomial transforms, we have that a similar relationship holds between their transforms:

$$\widehat{\bar{P}}_s(z) = \frac{1}{s+1}\sum_{0 \le t \le s} \widehat{P}_{t,s-t}(z).$$

Similar equations hold for $\bar{a}_n$ and $\bar{b}_n$.

Using this and the results from the previous section we can analyze the cost of this search. The details are lengthy and we omit them in this paper. The results are stated in the following theorem:

Theorem 4. *The expected value of the cost of an unsuccessful search in a skip list, for all $n \ge 0$, is*

$$\mathbf{EU}_n = -\frac{H_n}{p\ln p} + 1 - \frac{1}{2p} + \frac{q}{p\ln p} - \frac{1}{2}[n = 0] + F_n^{[0]} - \frac{q}{p}F_n^{[-1]} + \frac{1}{p(n+1)}$$

$$= -\frac{1}{p}\log_p n + 1 - \frac{1}{2p} + \frac{q - \gamma}{p\ln p} + \chi(\log n) + O\left(\frac{1}{n}\right),$$

and the variance is

$$\mathbf{VU}_n = -\frac{q}{p^2\ln p}H_n$$

$$+ \frac{H_n^{(2)}}{p^3\ln^2 p} + \frac{1}{2p} - \frac{5}{12p^2} + \frac{3q}{p^2\ln p} + \frac{q(1+p)}{p^2\ln^2 p} - \frac{2q\alpha(p)}{p\ln p} - \frac{2q(1+p)\beta(p)}{p^2} - \frac{2q\gamma(p)}{p}$$

$$+ \left(\frac{q(p-2)}{p^2} - \frac{2q}{p\ln p} + \frac{2q\alpha(p)}{p}\right)F_n^{[0]} + \left(\frac{q(p-3)}{p^2} + \frac{2q^2}{p^2\ln p} + \frac{2q\alpha(p)}{p}\right)F_n^{[-1]}$$

$$-(F_n^{[0]} - \tfrac{q}{p}F_n^{[-1]})^2 + \tfrac{q}{p}F_n^{[1]}[n > 0] + \tfrac{2}{p}G_n^{[0]} - \tfrac{2q}{p^2}G_n^{[-1]}$$

$$+ \left(-\tfrac{(q+1)^2}{4p^2} - \tfrac{q}{p\ln p} + \tfrac{q\alpha(p)}{p} + \tfrac{2q\beta(p)}{p}\right)[n = 0] + \tfrac{q}{2p}[n = 1]$$

$$+ \left(\tfrac{p^2 - 6p + 3}{qp^2} - \tfrac{2q}{p^2\ln p} - \tfrac{2\alpha(p)}{p} + \tfrac{2q\beta(p)}{p^2} + \tfrac{2q\gamma(p)}{p} - \tfrac{2}{p}F_n^{[0]} + \tfrac{2q}{p^2}F_n^{[-1]}\right)\tfrac{1}{n+1}$$

$$+ \tfrac{2}{p^2\ln p}\tfrac{H_n}{n+1} + \tfrac{2q}{p}d_n + \tfrac{2q}{p^2}e_n + \tfrac{2q}{p}f_n - \tfrac{1}{p^2(n+1)^2} \quad \text{for all } n \geq 0$$

$$= -\tfrac{q}{p^2}\log_p n$$

$$+ \tfrac{\pi^2}{6p^2\ln^2 p} + \tfrac{1}{2p} - \tfrac{5}{12p^2} + \tfrac{q(3-\gamma)}{p^2\ln p} + \tfrac{q(1+p)}{p^2\ln^2 p} - \tfrac{2q\alpha(p)}{p\ln p} - \tfrac{2q(1+p)\beta(p)}{p^2} - \tfrac{2q\gamma(p)}{p}$$

$$+ \chi(\log n) + \chi^2(\log n) + O(\tfrac{\log n}{n}),$$

where $\alpha(p) = \sum_{k \geq 1} \frac{1}{p^{-k} - 1}$, $\beta(p) = \sum_{k \geq 1} \frac{k}{p^{-k} - 1}$, $\gamma(p) = \sum_{k \geq 1} \frac{k}{(p^{-k} - 1)^2}$, $d_n = \mathcal{B}_n[-\sum_{1 \leq j \leq k} \frac{1 - p^{js}}{p^{-k} - 1}]$, $e_n = \mathcal{B}_n[-\frac{1}{s+1}\sum\sum_{1 \leq j \leq k} \frac{1 - p^{js}}{p^{-k} - 1}]$, and $f_n = \mathcal{B}_n[-\frac{1}{s+1}\sum\sum_{1 \leq j \leq k} \frac{1 - p^{js}}{(p^{-k} - 1)^2}]$.

Note that the first line in the huge expression for the variance is the leading term, that coincides with that of the variance of the cost of searching for $+\infty$. The second line shows the constant term, while lines 3–4 contain the oscillating part. We already know that $d_n = O(\frac{1}{n})$. A similar analysis shows that $e_n = O(\frac{\log n}{n})$ and $f_n = O(\frac{1}{n})$. Therefore, the three final lines contain terms that are either transient, or can be bounded by $O(\frac{\log n}{n})$.

5 Conclusions and Further Work

We believe the main contribution of this paper is the development of a theory of the binomial transform, that brings it to a level comparable to that of other, better known, tools for solving recurrence equations. We have shown its usefulness by deriving a number of results for the behavior of (probabilistic) skip lists. These results confirm previously known asymptotic results. In one case (variance of the search cost for a random element) we have solved a problem that was still open. While we do not claim that this method is always better than alternative ones, there are certainly cases where the binomial transform should be the method of choice. Since no method is uniformly best, the availability of a new tool should enhance our chances of solving a given problem.

There are several ways in which this work can be continued. One obvious line of development is the application of this transform to the analysis of other problems. Some preliminary results indicate that some hashing algorithms can be analyzed using the binomial transform. Another interesting line of research is to try to find ways of writing down the equations in the domain of the binomial transform (the "s domain") directly from the algorithm being analyzed, without going through the intermediate step of writing down recurrence equations. This is the approach that P. Flajolet and his group have been investigating for several kinds of generating functions, with great success (for instance, see [6]).

Acknowledgements

We thank Hosam Mahmoud for many illuminating discussions on the topic of this paper and related issues, and also for his very useful comments on a preliminary draft.

References

1. M. Abramowitz and I.A. Stegun, *Handbook of Mathematical Functions*, Dover, 1972.
2. L. Comtet, *Advanced Combinatorics*, D. Reidel, Dordrecht, 1974.
3. L. Devroye, *Expected Time Analysis of Skip Lists*, Technical Report, School of Computer Science, McGill University, Montreal, 1990.
4. L. Devroye, "A Limit Theory for Random Skip Lists," *Annals of Applied Probability*, **2**(3), August 1992, 597-609.
5. P. Flajolet, M. Régnier and R. Sedgewick, "Some Uses of the Mellin Integral Transform in the Analysis of Algorithms," in *NATO Advanced Research Workshop on Combinatorial Algorithms on Words*, A. Apostolico and Z. Galil (Eds.), Maratea, Italy, June 1984, 241-254.
6. P. Flajolet, B. Salvi and P. Zimmermann, "Automatic average-case analysis of algorithms," *Theoretical Computer Science* **79**(1), 1991, 37-109.
7. G.H. Gonnet and J.I. Munro, "The Analysis of a Linear Probing Sort by the Use of a New Mathematical Transform," *Journal of Algorithms* **5**(4), Dec. 1984, 451-470.
8. R.L. Graham, D.E. Knuth and O. Patashnik, *Concrete Mathematics*, Addison-Wesley, 1989.
9. P. Kirschenhofer and H. Prodinger, "The Path Length of Random Skip Lists," *Acta Informatica* **31** (8), November 1994, 775-792.
10. H. Prodinger, "Some Information about the Binomial Transform," *The Fibonacci Quarterly* **32** (5), November 1994, 412-415.
11. D.E. Knuth, *The Art of Computer Programming*, vol. 1: *Fundamental Algorithms*, Addison-Wesley, 1973.
12. D.E. Knuth, *The Art of Computer Programming*, vol. 3: *Sorting and Searching*, Addison-Wesley, 1973.
13. H. Mellin, "Über den Zusammenhang Zwischen den Linearen Differential- und Differenzengleichungen," *Acta Mathematica* **25**, 1902, 139-164.
14. T. Papadakis, J.I. Munro and P.V. Poblete, "Average Search and Update Costs in Skip Lists," *BIT* **32**, 1992, 316-332.
15. T. Papadakis, *Skip Lists and Probabilistic Analysis of Algorithms*, Ph.D. Thesis, University of Waterloo, Waterloo, Ontario, Canada (May 1993). [Available as Technical Report CS-93-28.]
16. P.V. Poblete, "Approximating Functions by their Poisson Transform," *Information Processing Letters* **23**(3), Oct. 1986, 127-130.
17. W. Pugh, "Skip Lists: A Probabilistic Alternative to Balanced Trees," *Comm. ACM* **33**(6), June 1990, 668-676.
18. J. Riordan, *Combinatorial Identities*, John Wiley & Sons, 1968.
19. S. Sen, "Some Observations on Skip Lists," *Information Processing Letters* **39**(4), August 1991, 173-176.
20. W. Szpankowski and V. Rego, "Yet Another Application of a Binomial Recurrence. Order Statistics," *Computing* **43**, 1990, 401-410.

An Optimal Parallel Algorithm for Digital Curve Segmentation Using Hough Polygons and Monotone Function Search

Peter Damaschke

FernUniversität, Theoretische Informatik II
58084 Hagen, Germany
peter.damaschke@fernuni-hagen.de

Abstract

The problem of finding all the digital line segments in a digital curve is interesting in pattern recognition and image processing. We develop an algorithm solving this problem in $O(\log n)$ time using $O(n)$ work on a CREW PRAM. This beats all known algorithmic results about digital curve segmentation. Moreover, our algorithm solves a more general segmentation problem which can be formulated in terms of semigroups.

Our algorithm makes use of two prerequisites, possibly interesting for their own: (a) certain parameter polygons with few vertices, called Hough polygons, which we assign to every digital line segment, and (b) the parallel search for monotone integer functions in $O(\log n)$ time by $O(n)$ threshold queries.

1 Introduction

Recognizing simple geometric shapes in digital images is one of the central tasks in image processing, being particularly important for data compression and automatic image understanding. Here we consider the problem of finding all the inclusion-maximal digital line segments (DLS for short) in a given digital curve. First we outline the known algorithmic results and the contributions of this paper. The necessary definitions are given afterwards.

In the following, n always means the number of pixels in the given digital curve. The complexity of algorithms is measured by the running time on a RAM, or by the parallel execution time and the work (i.e. the total number of arithmetic operations) on a CREW PRAM. We will not further mention these models of computation. For basics on parallel algorithms we refer to [11].

Several authors [8] [7] provided $O(n)$ time algorithms for determining one specific partition of a digital curve into a minimum number of DLS. Such a segmentation is not unique, actually it depends on where the segmentation algorithm starts. This arbitrariness might be unsatisfactory for some applications, therefore we consider the more general problem stated above. Once the family of all maximal DLS is determined, one can easily construct also a minimum segmentation by standard techniques.

In [5] we presented an algorithm that finds all the maximal DLS in $O(\log n)$ time using $O(n \log^2 n)$ work. The idea was to exploit the characterization of DLS by their convex hulls (cf. e.g. [1]) and the efficient parallel fusion of convex polygons due to [4]. The complexity bound ensued from the $O(\log n)$ bound for the vertex numbers of convex hulls of DLS. Since this bound is tight [14] [5], we could not simply reduce our complexity estimation. Now we are able to improve it to $O(n)$ work.

Our present algorithm uses the convex hulls of DLS in a dual way. Instead of working with them directly we compute related polygons in a parameter plane. Due to the similarity to Hough transformation we call them Hough polygons. Concatenating two DLS then corresponds to intersecting their Hough polygons. The nice property is that their vertex numbers are bounded by a constant, thus we can drop one $\log n$ factor.

In [6] we have already successfully used parameter polytope intersection to get an $O(n)$ time recognition algorithm for digital circular and elliptical arcs, by an $O(n)$ time reduction to linear programming with three or five variables [13]. In view of the immense literature about the inner structure of DLS, Hough polygons may also be of interest for their own. Dualization was also used for line and curve detection in digital images in [2] [3].

The second $\log n$ factor is saved by a more economical search strategy than the brute-force search in [5]. The key to this is the parallel search for a monotone discrete function in $O(\log n)$ steps using a total of $O(n)$ threshold queries of kind "$f(x) \geq y$?". This result is related to similar search problems such as studied in [10] [9].

Altogether we improve the known results on the digital curve segmentation topic in several directions, and the methods of Sections 4–6 may have applications also in other branches where maximal intervals with a certain hereditary property shall be found.

2   Basic Definitions

A pixel is an element of the grid $\mathcal{Z}^2$, that is, a point in the plane with integer coordinates. The 8-neighbors of a pixel (p, q) are the eight pixels whose abscissa and ordinates differ by at most 1 from p and q, respectively. A digital curve (or 8-connected curve) is a sequence of pixels such that any neighbors in the sequence are 8-neighbors in the grid. A digital curve containing at most one pixel (p, q) for every p is called a digital function curve (DFC), for it is an 8-connected curve of an integer function.

The digitalization dig(E) of a planar point set $E \subseteq \mathcal{R}^2$ is defined to be the set of all pixels (p, q) such that there exists a point $(x, y) \in E$ satisfying:

$$(x = p \wedge q - 0.5 < y \leq q + 0.5) \vee (y = q \wedge p - 0.5 < x \leq p + 0.5).$$

That means $(p, q) \in \text{dig}(E)$ iff E meets the axis-parallel cross with center (p, q) and four bars of length 0.5 where both the northern and eastern bars include their endpoints.

If E is an euclidean straight line segment then $D = \text{dig}(E)$ is called a digital line segment (DLS). In the following we consider only flat lines with slopes from -1 to 1. For steep lines with slopes outside the interval $[-1, 1]$ one must switch the roles of x and y. According to this we speak of flat and steep DLS.

As well-known, any flat DLS can be described as the set of all (p, q) satisfying $q = \lfloor \frac{ap - \mu}{b} \rfloor$, $l \leq p \leq r$, where a, b, μ, l, r are integers, $|a| \leq b$, and a, b are relatively prime. This is just the digitalization of a segment of the line $y = \frac{a}{b} x - \frac{\mu}{b} - 0.5$.

Note that any flat DLS is a DFC. On the other hand, any DLS contained in a DFC is flat. So we can restrict our problem to DFC rather than arbitrary digital curves. This often simplifies notations. It is a straightforward exercise to decompose a digital curve into maximal DFC (with x or y as the independent variable) in $O(\log n)$ time using $O(n)$ work.

3 Hough Polygons of DLS

Let (p, q) be a pixel. The set of all lines $y = sx + t$ such that $q = \lfloor sp + t \rfloor$ is described by $-ps + q \leq t < -ps + q + 1$, hence their parameters form a stripe in the s, t-plane. Due to the similarity to Hough transformation we call it the Hough stripe of (p, q). For a fixed set D of pixels, let H denote the intersection of the Hough stripes of all pixels from D. We call H the Hough polygon of D. Trivially, a DFC D is a (flat) DLS iff the intersection of H and the stripe $-1 \leq s \leq 1$ is nonempty. Moreover, since any flat DLS is the digitalization of some flat line segment, we have a simpler equivalence:

Lemma 1 *A DFC D is a DLS iff its Hough polygon H is nonempty.*

H is a convex polygon, in general not including all of its vertices and edges. Let $\bar{H}$ be the intersection of the topological closures of of Hough stripes of the pixels from D. If $\bar{H}$ is two-dimensional then it is exactly the closure of H.

In the following, D is assumed to be a flat DLS. Our next goal is to characterize the vertices of $\bar{H}$. Let C be the convex hull of D. The boundary of C consists of the upper and lower envelope, denoted by U and L, respectively. Let L' be the lower envelope, shifted by one length unit upwards, i.e. in positive y direction. Since the vertical breadth of C is smaller than 1, L' lies completely above U.

For an euclidean line E with equation $y = sx + t$ we have $(s,t) \in \bar{H}$ iff passes through the funnel between U and L', possibly being tangent to them. So $\bar{H}$ can be considered as the intersection of two unbounded convex polygons $\bar{H}_U$ and $\bar{H}_L$ containing the parameter points of all lines above U and below L', respectively. Due to a well-known geometric duality there is a 1-1 correspondence between the edges of U and the vertices of $\bar{H}_U$, and similarly between the edges of L' and the vertices of $\bar{H}_L$. Since $\bar{H} = \bar{H}_U \cap \bar{H}_L$, there may be two vertices of $\bar{H}$ being vertices of neither $\bar{H}_U$ nor $\bar{H}_L$. We summarize our consideration:

Lemma 2 *Every vertex of $\bar{H}$, possibly except two vertices, corresponds to an edge e of U or L' such that the line through e passes through the funnel bounded by U and L'.*

We want to estimate the number of edges satisfying the condition of Lemma 2. For this we transform the problem into an equivalent one which is easier to overlook. Let D be represented by $q = \lfloor \frac{ap - \mu}{b} \rfloor$, $l \leq p \leq r$, as introduced in Section 2, and let Δ be the mapping

$$\Delta \begin{pmatrix} x \\ y \end{pmatrix} = \begin{pmatrix} 1 & 0 \\ a & -b \end{pmatrix} \begin{pmatrix} x \\ y \end{pmatrix} - \begin{pmatrix} 0 \\ \mu \end{pmatrix}.$$

Applying Δ to $x = p$, $y = \lfloor \frac{ap - \mu}{b} \rfloor$, we get for the second component:

$$
\begin{aligned}
ax - by - \mu &= ap - b\lfloor \frac{ap - \mu}{b} \rfloor - \mu \\
&= ap - (ap - \mu) + (ap - \mu) \bmod b - \mu \\
&= (ap - \mu) \bmod b.
\end{aligned}
$$

Thus we have:

$$\Delta(D) = \{(x,y) : x, y \text{ integer} \wedge y = (ax - \mu) \bmod b \wedge l \leq x \leq r\}.$$

Since Δ is an affine mapping, vertices and edges of C are mapped onto vertices and edges of the convex hull $\Delta(C)$ of $\Delta(D)$.

Consider an edge e of U. Since Δ stretches all vertical distances by a factor b, one easily verifies equivalence of the following conditions:

- The line including e passes through the funnel between U and L'.

- The line through $\Delta(e)$ has vertical distance at most b to every pixel of $\Delta(D)$.

Note that $\Delta(e)$ is an edge of the lower envelope of $\Delta(C)$. We refer to such edges as lower edges in the following.

We call the rectangle determined by $l \leq x \leq r$ and $0 \leq y \leq b - 1$ the frame of $\Delta(D)$. Let e_0 be the leftmost lower edge with non-negative slope, and $e_1, e_2, \ldots$ the next lower edges to the right of e_0. The left endpoint of e_k is denoted by (x_k, y_k).

The next periodicity lemma is simple but crucial, since it yields strong restrictions for the lower edges.

Lemma 3 *Let (x, y) and (u, v) be pixels of $\Delta(D)$ such that $(2u - x, 2v - y)$ is inside the frame. Then the latter pixel also belongs to $\Delta(D)$.*

The proof is straightforward.

Lemma 4 *For $k \geq 0$ we have $x_{k+1} - x_k > r - x_{k+1}$.*

Proof. Assume on the contrary $2x_{k+1} - x_k \leq r$. Since the e_i have increasing slopes, we also have $2y_{k+1} - y_k < b$, otherwise one easily finds that $\Delta(D)$ cannot contain a pixel with $x = r$. Hence $(2x_{k+1} - x_k, 2y_{k+1} - y_k)$ is inside the frame. So by Lemma 3 it belongs to $\Delta(D)$. Due to collinearity, (x_{k+1}, y_{k+1}) cannot be a vertex of $\Delta(C)$, a contradiction. $\square$

Now we turn to the central lemma of this section.

Lemma 5 *For each $k \geq 2$ there exists a pixel in $\Delta(D)$ whose vertical distance to the line through e_k is larger than b.*

Proof. Lemma 4 yields $x_1 - x_0 > r - x_1 \geq x_2 - x_1$, hence $2x_1 - x_2 \geq x_0$. Consequently, if $2y_1 - y_2 \geq 0$ then $(2x_1 - x_2, 2y_1 - y_2)$ is inside the frame. By Lemma 3 it is a member of $\Delta(D)$, hence (x_1, y_1) cannot be a vertex, a contradiction. This shows $2y_1 - y_2 < 0$. From this we immediately conclude $(2x_1 - x_2, 2y_1 - y_2 + b) \in \Delta(D)$.

Since the line through e_k passes below e_1 and has larger slope than e_1, it includes the point $(2x_1 - x_2, y)$ where $y < 2y_1 - y_2$. Hence its vertical distance to some pixel from $\Delta(D)$ is larger than b. $\square$

Thus at most two lower edges of $\Delta(C)$ with non-negative slope, namely e_0 and e_1, can satisfy the condition equivalent to that of Lemma 2 for $\Delta^{-1}(e_k)$. Similarly we find at most two such lower edges with negative slopes and at most four such upper edges. Together with the two exceptional vertices from Lemma 2 we obtain:

Theorem 6 *The (closure of the) Hough polygon of a flat DLS has at most ten vertices.*

The only important thing for our purpose is that the vertex numbers of Hough polygons are bounded by some constant.

Corollary 7 *The intersection of two Hough polygons can be computed in $O(1)$ time.*

A first consequence is another incremental $O(n)$ algorithm for partitioning a DFC into a minimum number of DLS: Scan the DFC pixel by pixel, and always intersect the Hough stripe of the actual pixel with the Hough polygon of the previously scanned portion of the curve. Whenever the intersection becomes empty, start a new DLS at the pixel considered last.

Later we shall prove stronger algorithmic results.

4 Parallel Search for a Monotone Function by Threshold Queries

Let us be given an unknown function $f : M \longrightarrow N$ where M, N are totally ordered sets with m and n elements, respectively, and $x \leq z$ always implies $f(x) \leq f(z)$. Assume that, for any pair $(x, y) \in M \times N$, we can ask whether $f(x) \geq y$. Our aim is to determine f by such threshold queries.

First let us agree on a few suppositions which can be done without loss of generality. Let $M = \{1, \ldots, m\}$ and $N = \{1, \ldots, n\}$. Due to a symmetry of the problem we assume $m \leq n$. Otherwise we can search for the discrete inverse function $f^{-1}(y) = \min\{x : f(x) \geq y\}$. Note that f is uniquely determined by f^{-1} and $f(x) \geq y$ is equivalent to $f^{-1}(y) \leq x$. For simplicity we finally assume $m = 2^k - 1$. Otherwise we extend M and N such that m reaches the next power of 2 and n remains larger than m, and we add dummy pairs $(x, f(x))$ with $f(x) = n$. These augmentations remain within constant factors, so they do not affect our asymptotic result.

Lemma 8 Let $1 \leq b_i \leq t_i \leq n$ for $1 \leq i \leq j$. We define $b_0' = 1$, $b_i' = b_i$, $t_i' = t_{i+1}$, $t_j' = n$. Then we have

$$\sum_{i=0}^{j}(t_i' - b_i' + 1) = n + j + \sum_{i=1}^{j}(t_i - b_i).$$

Obviously we have only rearranged the sum. The lemma is suitable in the proof of:

Theorem 9 *With the above assumptions, the monotone function f can be found in $O(\log n)$ time by a total of $O(n)$ threshold queries.*

Proof. We give a parallel search algorithm. A procedure ask(x, b_x, t_x) is defined as follows:

```
begin
    y := ⌈(bₓ+tₓ)/2⌉;
    if f(x) ≥ y then bₓ := y else tₓ := y − 1
end;
```

Provided that initially $b_x \leq f(x) \leq t_x$, successive calls of $\text{ask}(x, b_x, t_x)$ will compute $f(x)$ by binary search. The main program is the following. As usual, "pardo" stands for "do in parallel".

```
begin
    b₀ := 1; b_{m+1} := n;
    for i := k − 1 downto 0 do
    begin
        for all x with x mod 2^{i+1} = 2^i pardo
            begin b_x := b_{x−2^i}; t_x := t_{x+2^i} end;
        for all x divisible by 2^i pardo
            begin ask(x, b_x, t_x); ask(x, b_x, t_x) end;
    end;
    for all x pardo
        begin repeat ask(x, b_x, t_x) until b_x = t_x; write(x, b_x) end
end.
```

Since f is monotone, one can easily show by induction that $b_x \leq f(x) \leq t_x$ when the search interval for argument x is initialized. From this the correctness of the algorithm follows.

After $O(\log m)$ steps the search has been started for each $x \in M$, and every binary search terminates after $O(\log n)$ steps. Thus the running time is $O(\log n)$.

We have 2^{k-i-1} search processes at arguments x divisible by 2^i but not by 2^{i+1}. These search processes run over a total range of $s(i) := \sum_x (t_x - b_x + 1)$ where the sume is taken over all such x, and the b_x, t_x are the initial values here. Since the procedure "ask" is called two times for every activated x before the next set of search intervals is initialized, the already existing search intervals are reduced by a factor 4 in each round. Now Lemma 8 yields the recurrence $s(k-1) = n$, $s(j) \leq 2n + \sum_{i>j} 4^{j-i} s(i)$. Downwards induction on j shows $s(j) \leq 3n$.

The total number Q of queries is bounded by

$$Q \leq \sum_{i=0}^{k-1} \max\{ \sum_{j=1}^{2^{k-j-1}} \log s_{ij} \},$$

where for each i the s_{ij} are integers satisfying

$$\sum_{j=1}^{2^{k-j-1}} s_{ij} = s(i).$$

(By convention let be $\log 0 = 0$.) Since the log function is concave, the inner sum becomes maximal if all s_{ij} are equal. Hence

$$Q \leq \sum_{i=0}^{k-1} 2^{k-i-1} \log \frac{s(i)}{2^{k-i-1}} = \sum_{i=0}^{k-1} 2^i \log \frac{s(k-i-1)}{2^i}.$$

Together with $m = 2^k - 1 \leq n$ and $s(i) = O(n)$ we finally obtain

$$Q \leq \sum_{i=0}^{\log n} 2^i \log \frac{O(n)}{2^i} = O(n).$$

□

Of course, the algorithm can also run in $O(n)$ sequential time.

5   Finding all the Maximal DLS in Linear Time

Before we give our final result, we show that all maximal DLS in a digital curve can be found in $O(n)$ sequential time. The algorithm already delivers also the high-level description of the parallel algorithm. So we can postpone the details of the parallelization.

First we introduce some further denotations. D is the given DFC, consisting of $n = r - l + 1$ pixels (x, y) with $l \leq x \leq r$. The subcurve of D in the vertical stripe $l' \leq x \leq r'$ is denoted by $D[l', r']$, and $H[l', r']$ is its Hough polygon. A partition of $D = D[l, r]$ is called a segmentation if it has the form $D[l_1, r_1], D[l_2, r_2], \ldots, D[l_k, r_k]$ where $l_1 = l$, $r_k = r$, $l_{i+1} = r_i + 1$, and each $D[l_i, r_i]$ is a DLS whereas $D = D[l_i, r_{i+1}]$ is not. We do not demand k to be minimal!

It follows that every DLS in D is contained in at most three contiguous segments of the partition. Hence the following algorithm correctly computes all maximal DLS in D.

```
begin
     construct an arbitrary segmentation D[l₁, r₁], D[l₂, r₂], ..., D[l_k, r_k];
     for i := 1 to k do
          begin
               for j := l_i to r_{i+1} do compute H[l_i, j];
               for j := r_{i+1} downto l_i do compute H[j, r_i]
          end;
     for i := 1 to k − 2 do compute f_i : {l_i, ..., r_i} ⟶ {r_i, ..., r_{i+2}}
          where f_i(x) = max{y : H[x, y] ≠ ∅};
     remove all non-maximal DLS among the D[x, f_i(x)]
end.
```

Theorem 10 *The algorithm finds all maximal DLS in $O(n)$ time.*

Proof. As earlier mentioned, we can construct a segmentation in $O(n)$ time. The Hough polygons in the segments are computed incrementally in $O(n)$ time. The monotone functions f_i are computed by $O(n)$ threshold queries using Theorem 9. Note that, by Lemma 1, we have $H[x, y] \neq \emptyset$ iff $D[x, y]$ is a DLS iff $f_i(x) \geq y$. By Corollary 7, each threshold query can be answered in $O(1)$ time: We only have to intersect two of the previously computed Hough polygons. (At this point we need the segmentation from the first line!) The final step is trivially executable in $O(n)$ time by pairwise comparisons: $D[x, f(x)]$ is non-maximal iff $D[x - 1, f(x)]$ is also a DLS. $\square$

6 The Parallel Algorithm

The parallelization of our $O(n)$ time sequential algorithm from Section 5 yields the main result. The next lemma settles the most difficult first step.

Lemma 11 *A segmentation of a DFC $D = D[l, r]$ can be found in $O(\log n)$ time using $O(n)$ work.*

Proof. Consider the following algorithm. W.l.o.g. we assume n to be a power of 2, otherwise we add suitable dummy pixels to D. Further it is no restriction to assume $l = 0$ and $r = n - 1$.

```
begin
    for i := 0 to n - 1 pardo
        begin compute H[i, i]; label D[i, i] "good" end;
    for k := 0 to log₂ n - 1 do
        for all multiples i of 2^(k+1) pardo
            begin
                H[i, i + 2^(k+1) - 1] := H[i, i + 2^k - 1] ∩ H[i + 2^k, i + 2^(k+1) - 1];
                if H[i, i + 2^(k+1) - 1] ≠ ∅
                then begin
                label D[i, i + 2^(k+1) - 1] "good";
                label D[i, i + 2^k - 1] "bad";
                label D[i + 2^k, i + 2^(k+1) - 1] "bad"
                    end
            end
end.
```

Obviously, this algorithm computes, within the asserted bounds, in a tree-like fashion a partition of D into DLS, finally labeled "good". But this is in general not a segementation of D yet: The union of several neighbored good DLS may form a larger DLS if none of them are siblings in that complete binary tree. We shall now construct a segmentation each segment of which is a union of consecutive good DLS.

An abscissa value $i + 2^k$ such that 2^{k+1} divides i is called a break if both $D[i, i + 2^k - 1]$ and $D[i + 2^k, i + 2^{k+1} - 1]$ are good. In other words, a break separates two good siblings. Note that in this case $D[i, i + 2^{k+1} - 1]$ is surely not a DLS, hence also no other DLS can include this segment. Consequently, it suffices to consider the portions of D between any two neighbored breaks, independently from each other.

So let $D_0, D_1, D_2, \ldots$ be consecutive good DLS without any break between them, and assume D_0 to be horizontally longer than D_1. Then D_2 cannot be longer than D_1, according to the tree structure of the partition. Now assume that D_1 and D_2 have equal length. Since the longer segment D_0 stands immediately before them, D_1 and D_2 must be siblings in the tree, but then they are separated by a break which contradicts our suppositions. So D_1 must be longer than D_2, and we inductively conclude that D_k is longer than D_{k+1}. Clearly, the lengths decrease by a factor at least 2. Similarly we conclude for a sequence of good DLS directed to the left. Finally, there cannot exist three consecutive good DLS of equal length – two of them would be siblings.

This shows that a portion of D between two breaks contains $O(\log n)$ good DLS. Hence a segmentation of each portion can even be found in $O(\log n)$ sequential time using the previously computed Hough polygons. Altogether this yields a segmentation of the complete curve. $\square$

Theorem 12 *All maximal DLS in a DFC with n pixels can be found in $O(\log n)$ time using $O(n)$ work.*

Proof. Now we are ready to parallelize the $O(n)$ algorithm from Section 5. The first line is replaced by the method of Lemma 11. Since intersection is an associative operation, the Hough polygons are obtained by parallel prefix computations. The complexity bounds follow from Corollary 7 and [12]; see also [11]. The f_i can be computed independently, using Theorem 9 for each function. Let us remark that this is the only point where we need concurrent read operations; an unbounded number of threshold queries may access the same Hough polygon at the same time. The final removal of the non-maximal DLS is trivial. $\square$

7 Concluding Remarks

The methods of the last three sections do not rely on special geometric properties of DLS. The technique would also work for all problems of the following kind: Given a sequence of symbols, we want to find all subsequences which share some hereditary property. Assume that there is a homomorphism h of the semigroup of these sequences into another semigroup with zero element 0, and a sequence w has the interesting property iff $h(w) \neq 0$. Then the problem is solvable in $O(\log n)$ time using $O(n)$ operations of the second semigroup in the PRAM model. It would be interesting to know of other meaningful instances of this abstract problem.

Recently we found a way to avoid concurrent read operations in Theorem 12. For this we have to search for monotone functions in such a way that simultaneous threshold queries "$f(x) \geq y$?" are allowed only for mutually distinct x and y, respectively. Our new search algorithm is somewhat more tricky than the present algorithm in Section 4, particularly it uses accelerated cascading. It shall appear in the full version of the paper. So our segmentation algorithm runs even on an EREW PRAM in $O(\log n)$ time using $O(n)$ work.

References

[1] T.A.Anderson, C.E.Kim: Representation of digital line segments and their preimages, *Computer Vision, Graphics, and Image Processing* 30 (1985), 279-288

[2] T.Asano, N.Katoh: Number theory helps line detection in digital images – an extended abstract, *Int. Symposium on Algorithms and Computation ISAAC'93, Lecture Notes in Computer Science* 762, Springer 1993, 313-322

[3] T.Asano, N.Katoh, T.Tokuyama: A unified scheme for detecting fundamental curves in binary edge images, *2nd European Symposium on Algorithms ESA'94*, Utrecht 1994, *Lecture Notes in Computer Science* 855, Springer 1994, 215-226

[4] M.J.Atallah, M.T.Goodrich: Parallel algorithms for some functions of two convex polygons, *Algorithmica* 3 (1988), 535-548

[5] P.Damaschke: Line segmentation of digital curves in parallel, *12th Symposium on Theoretical Aspects of Computer Science STACS'95*, Munich 1995, *Lecture Notes in Computer Science* 900, Springer 1995, 539-549

[6] P.Damaschke: The linear time recognition of digital arcs, *Pattern Recognition Letters*, to appear

[7] I.Debled-Rennesson, J.P.Reveillès: A linear algorithm for segmentation of digital curves, *3rd Int. Workshop on Parallel Image Analysis*, College Park/MD 1994

[8] L.Dorst, A.W.M.Smeulders: Decomposition of discrete curves into piecewise segments in linear time, *Contemporary Math.* 119 (1991), 169-195

[9] F.Gao, L.J.Guibas, D.G.Kirkpatrick, W.T.Laaser, J.Saxe: Finding extrema with unary predicates, *Algorithmica* 9 (1993), 591-600

[10] R.Hassin, N.Megiddo: An optimal algorithm for finding all the jumps of a monotone step function, *Journal of Algorithms* 6 (1985), 265-274

[11] J.JáJá: *An Introduction to Parallel Algorithms*, Addison-Wesley 1992

[12] R.E.Ladner, M.J.Fischer: Parallel prefix computations, *Journal of the ACM* 27 (1980), 831-838

[13] N.Megiddo: Linear programming in linear time when the dimension is fixed, *Journal of the ACM* 31 (1984), 114-127

[14] J.P.Reveillès: Géométrie discrète, calcul en nombres entiers et algorithmique, Thèse d'Etat, Univ. Louis Pasteur, Strasbourg 1991

Fast Skeleton Construction

Rolf Klein
FernUniversität Hagen

Andrzej Lingas
Lund University

Abstract

For a polygon P, the skeleton of P is a partition of P into regions assigned to the edges of P. A point p inside P belongs to the region of an edge e if and only if e is the closest edge of P. We present a randomized algorithm that builds the skeleton of a simple polygon in linear expected time. We also observe that the Delaunay triangulation (equivalently, the Voronoi diagram) of a planar point set can be computed from its connected spanning subgraph in linear expected time.

1 Introduction

The Voronoi diagram and its dual, the Delaunay triangulation, belong to the most useful structures in computational geometry, due to the variety of problems they help to solve; see [2, 20, 21] for surveys. Computing the diagram of n points or straight-line segments in the Euclidean plane is well known to be a problem of complexity $\Theta(n \log n)$. But since the Voronoi diagram is so useful it is natural to ask if faster algorithms are available if more information about the sites is given.

In 1987, Aggarwal, Guibas, Saxe, and Shor [1] obtained a linear-time upper bound for the problem of computing the Voronoi diagram of the vertices or edges of a convex polygon. In [8, 15], it was shown that the method of Aggarwal et al. yields also a linear-time upper bound on computing the Voronoi diagram of more general site configurations, e.g., a monotone point sequence, or more generally, a point sequence for which a special Hamiltonian path in the dual of the Voronoi diagram is known in advance [15]. Also, Yap considered the generalization of the linear-time deterministic method due to Aggarwal et al. to include a convex set of points and circular segments [27]. On the other hand, already in 1986 Chew presented an extremely simple linear-time randomized algorithm for the Voronoi diagram of a convex polygon [4, 23] (for its generalization to include a monotone sequence of line segments see [14]). Devillers used the random sampling method and Seidel's acceleration technique to derive an $O(n \log^* n)$-time randomized algorithm for the Voronoi diagram of the edges of a simple polygon [7]. Also, he showed that if a bounded-degree, connected spanning subgraph of the Delaunay triangulation of a planar point set (e.g., the Euclidean minimum spanning tree) is known in advance then the diagram can be computed in $O(n \log^* n)$ expected-time [7].

The Voronoi diagram of a set L of straight-line segments in the Euclidean plane has been termed as *the skeleton of L* by Kirkpatrick [11]. If L forms a simple polygon, the skeleton inside L is also called the *medial axis* of L [2, 21].

Skeletons have numerous applications in biology, pattern recognition, geography and computer graphics (see [2, 11, 19, 20] for more details). They are also very useful in planning collision-free motions of figures in the presence of obstacles [2, 27]. The medial axis is frequently used in pocket machining; it helps to plan offset motions for the milling tool, see Held [10].

In this paper, we improve the $O(n \log^* n)$ time-bound on simple-polygon skeleton construction due to Devillers [7], providing a new randomized algorithm for this problem running in linear expected time. A basic subroutine in this algorithm is a linear-time randomized algorithm for the skeleton within a monotone polygon. Similarly as the linear-time algorithm for the Voronoi diagram of the vertices of a convex polygon due to Chew [4], and the linear-time algorithm for the so called bounded Voronoi diagram of a special case of a monotone polygon called histogram [13], it constructs the skeleton incrementally in a random order [†]. In spite that our subroutine is valid for a larger class of polygons it is much more natural than that for the bounded Voronoi diagram of a histogram from [13].

We also observe that the other aforementioned $O(n \log^* n)$ expected-time bound due to Devillers can be improved in the following aspects using [13]. First, the time can be reduced to linear. Second, the degree bound is not needed. Thus, we can compute the Delaunay triangulation (equivalently, the Voronoi diagram) of a planar point set from its connected spanning subgraph in linear expected time.

The organization of this paper is as follows.

In Section 2 we state some basic properties of the skeleton of a simple polygon. Then we show key lemmata on computing the skeleton on "the other side of a straight-line" and merging skeletons of subpolygons within linear time.

In Section 3 we present and analyze, by means of backward analysis [23], a linear time randomized algorithm for constructing the skeleton of a monotone polygon. Then, in Section 4, we apply the method of decomposing a simple polygon into *pseudo-histograms* within linear time given in [13]. Using a merging technique based on our key lemmata we are able to compute, in linear time, the skeleton of the polygon from the diagrams of the pseudo-histograms in the decomposition. In Section 5 we improve the other upper time-bound due to Devillers.

2   Skeletons

We start with formal definitions of the *bisector* of two straight-line segments, the skeleton of a set of straight-line segments, and the skeleton of a simple polygon.

Definition 2.1 A *bisector* of two disjoint straight-line (open or half-open, or closed) segments is a continuous curve that separates the two segments, dividing the plane in two regions, each consisting of points that are closer to the segment contained in it than to the other segment.

Kirkpatrick has observed that a bisector of two straight-line segments consists of a finite number (at most seven) of either straight or parabolic segments (see p. 19 in [11]).

[†]The $\Omega(n \log^* n)$ bound for a game related to accelerated randomized incremental constructions due to Clarkson *et al.* [6] doesn't hold in any of the three specific cases since the assumption about a logarithmic localization cost for a new object is not fulfilled.

Definition 2.2 Let L be a set of disjoint straight-line segments. For each line segment l in L, the *region* $R(l)$ of l consists of all points p in the plane for which l is a closest segment of L. The *skeleton of* L, $SK(L)$, is the union of the boundaries of the bounded regions. The maximal segments of bisectors of two segments in $SK(L)$ are called *skeleton edges*. The endpoints of the skeleton edges are called *skeleton vertices*.

If L has n vertices then $SK(L)$ is of size $O(n)$ [11].

Further, to make polygon edges disjoint we shall identify a polygon edge leading from a vertex v and to a vertex w in clockwise order with the half-open segment $[v, w)$.

Definition 2.3 Let P be a simple polygon. The *skeleton of* P, $SK(P)$, is the skeleton of the set of (the half-open segments in one-to-one correspondence with) the edges of P. The *inner skeleton of* P, $ISK(P)$, is $SK(P)$ restricted to the area of P. The *outer skeleton of* P is defined analogously.

Now we state a useful lemma on skeleton computing. In similar form, it has been used in [14] first.

Definition 2.4 Let L be a finite sequence of disjoint line segments in the plane. A *search supporting function for* L is a function f satisfying the following condition for any subsequence U of L with at least k elements (k is a constant):
For at least a constant fraction of u in U, $f(U, u)$ is a skeleton edge of $SK(U - \{u\})$ at least a fragment of which doesn't occur in $SK(U)$. Otherwise, $f(U, u) = blank$.

Note that whenever $f(U, u)$ is not *blank*, we can start to construct the region of u in $SK(U - \{u\})$ from the point of intersection of $f(U, u)$ with the bisector of u and any of the two sites defining $f(U, u)$.

Lemma 2.5 *Let L be a sequence of n disjoint line segments in the plane, and let f be a search supporting function for L. Suppose that for any subsequence U of L, and u in U, the time of computing $f(U, u)$ is not greater than $t(n)$. $SK(L)$ can be computed in $O(t(n)n)$ expected time by a simple randomized algorithm.*

Proof: Consider the following algorithm.

Input: A set L of n disjoint line segments in the plane, a procedure for computing a search supporting function f for L.
Output: $SK(L)$.

1. If L has no more than k line segments then compute $SK(L)$ directly;

2. If L has more than k line segments then pick a segment u in L randomly;

3. Compute $f(L, u)$; if $f(L, u) = blank$ go to Step 2;

4. Set L' to $L - \{u\}$ and recursively compute $SK(L')$;

5. Transform $SK(L')$ to $SK(L)$.

As in Step 5 one edge of $SK(L')$ is known that will not fully belong to $SK(L)$, we are able to construct the region of u in $SK(L)$ in time proportional to $n(u)$, the number of neighboring regions (see [12], Section 4).

By the planarity of $SK(L)$ and the random choice of u, $n(u)$ has expected value $O(1)$. Now to obtain the thesis it remains to observe that the expected number of trials for finding an u for which $f(U, u) \neq blank$ is $O(1)$ by our assumptions. $\square$

The following lemma easily follows from Lemma 2.5

Lemma 2.6 *Let K be a straight line dividing the plane into half-planes H_1 and H_2. Suppose that for a set L of disjoint straight line segments in H_1, the intersection $SK(L) \cap K$ is given. The skeleton of L within H_2, i.e., $SK(L) \cap H_2$, can be computed in linear expected time.*

Proof: We may assume w.l.o.g that K is a horizontal straight-line. Scan $\mathrm{SK}(L) \cap K$ from the left to the right. For each encountered intersection d of the region of a line segment b with K, compute the point d_l on the closure of b closests to the left endpoint of d and the point d_r on the closure of b closests to the right endpoint of d. Set b_d to $b \cap [d_l, d_r]$. Let L' denote the sequence of subsegments b_d of L produced in this way (in linear time). Note that $\mathrm{SK}(L') \cap H_2 = \mathrm{SK}(L) \cap H_2$. Thus, it is sufficient to show that $\mathrm{SK}(L')$ can be computed in linear expected time by Lemma 2.5. To achieve this, observe that by the definition of L', the insides of the vertical projections of the segments in L' on K don't overlap and occur in the same order as the segments in L'. In case of overlapping endpoints we can always make some of the segments open at this point. It follows that any two consecutive segments in any subsequence of L' have to share a skeleton edge intersecting K in the skeleton of the subsequence. Therefore, for any subsequence U of L', and any internal element u of U, we can set $f(U, u)$ to the skeleton edge separating the immediate predecessor from the immediate successor of u in U, and intersecting K. As at least a fragment of this skeleton edge has to disappear in $\mathrm{SK}(U)$, f satisfies the requirements of Lemma 2.5. Since it can be computed in constant time, the lemma follows by Lemma 2.5. $\square$

The following, so called Merging Lemma will be used in Section 4. Here, a *subpolygon* of a simple polygon P is a simple polygon whose boundary consists of edges and (inner) diagonals of P.

Lemma 2.7 *Let d be a diagonal of the subpolygon Q of P. For $i = 1, 2$ let Q_i (resp. P_i) be the two subpolygons resulting from splitting Q (resp. P) along d. Finally, assume the following notation:*
(i) for $i = 1, 2$, k_i is the number of regions in $\mathrm{ISK}(P_i - \{d\})$ adjacent to d ("support cost"),
(ii) m is the number of edges in $\mathrm{ISK}(P) \cap Q$ separating regions of edges placed on opposite sides of d in P ("merge line cost").
Then, given $\mathrm{ISK}(P_1 - \{d\}) \cap Q_1$ and $\mathrm{ISK}(P_2 - \{d\}) \cap Q_2$, the skeleton $\mathrm{ISK}(P) \cap Q$ can be computed in time $O(k_1 + k_2 + m)$.

Proof: relies on Lemma 2.6 and is similar to that of an analogous lemma for bounded Voronoi diagrams from [13]. $\square$

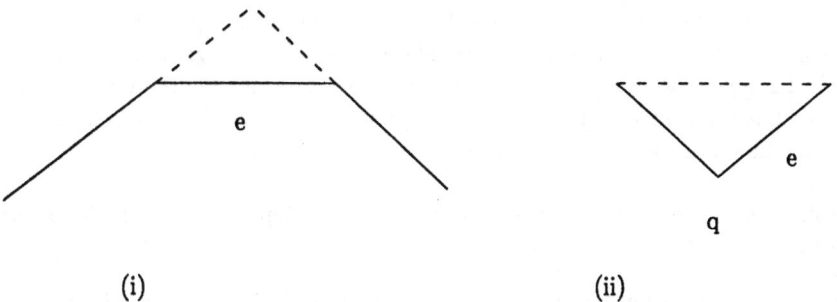

(i)   (ii)

Figure 1: The transformation of P into P': convex case (i), and reflex case (ii).

3   Monotone polygons

We present a natural linear-time randomized algorithm for the skeleton of a monotone polygon in this section. A simple polygon is called *monotone* if there is a direction such that any line perpendicular to it intersects the inside of the polygon at most once. Note that our definition of a monotone polygon allows edges perpendicular to the monotonicity direction and therefore it is more general than that in [21].

Algorithm 1

Input: A monotone polygon P.
Output: ISK(P).

1. If P has only five vertices then construct $SK(P)$ directly.

2. If P has more than five vertices then randomly pick an edge e of P such that the sum of the angles within P at the endpoints of e is greater than 180°. If both endpoints of e are convex vertices (i. e., ones whose internal angle is not greater than 180°) then set P' to the polygon resulting from deleting e and extending the two incident edges up to their intersection. Otherwise, set q to a reflex endpoint of e, and set P' to the polygon resulting from cutting off the triangle induced by the two edges incident to q in P.

3. Recursively compute ISK(P').

4. Transform ISK(P') to ISK(P).

Clearly, the transformation of P into P' neither cause self-intersections nor unboundness, since P is monotone and the sum of the two angles incident to e is greater than 180°. Consequently, P' is also monotone. The following lemma will facilitate Step 4 of Algorithm 1.

Lemma 3.1 *Let x be a point of P that lies in the region of an edge d in ISK(P'). If x cannot see the closest point on d in P then it belongs to the ISK(P) region of e or the edge sharing the reflex corner q with e.*

Proof: In Case (i) of Figure 1, the closest point on d must belong to the extensions of the edges incident to e. Hence, x is trivially in the ISK(P) region of e. Consider Case (ii) of Figure 1 where q is a reflex vertex. Since x cannot see the closest point v on d, the line segment $\overline{vx}$ intersects both edges incident to q. Thus, x is closer to a point in the union of these two edges than to v. □

To estimate the complexity of the transformation in Step 4 we need the following auxiliary lemma.

Lemma 3.2 *The transformation of ISK(P') into ISK(P) in Step 4 in Algorithm 1 can be done in time $O(k)$, where k is the number of regions adjacent to the regions of the edges of $P - P'$ in ISK(P).*

Proof: **Case 1:** Both endpoints of e are convex. Inserting the edge e and chopping off the extensions of incident edges from P' doesn't change region membership within P except for those points that are going to be in the region of e anyway. Thus, we only need to construct the skeleton region of e, e.g, starting from the bisector of e and one of the incident edges. We construct the region of e in a standard way, traversing only the region boundaries of ISK(P') within the region of e. By planarity, this can be done within time proportional to the number of regions in ISK(P) sharing boundary with the region of e.

Case 2: At least one endpoint of e is reflex. Lemma 3.1 implies that inserting the edges e and the other edge, say d, incident to the reflex endpoint q of e does not change region membership except for those points that are going to be in the regions of these two edges anyway. Again, we only need to construct the skeleton regions of e, d in a standard way, starting the construction from the bisector of e (or d, respectively) and the incident edge in $P \cap P'$, and traversing the region boundaries of ISK(P') within the regions of e, and d. By planarity, it takes $O(k)$ time. □

Lemma 3.3 *If e is randomly chosen among the edges of P for which the sum of the incident angles is greater than 180° then the expected value of k is $O(1)$.*

Proof: Suppose that $P = (q_0, q_1, ..., q_{n-1})$, and consider the randomly picked edge $e = (q_j, q_{j+1 \bmod n})$ where $0 \le j < n$ and e satisfies the angle requirements. The value of k is bounded from above by the total number of regions sharing boundary with the regions of the edges $(q_{j-1 \bmod n}, q_j)$, $(q_j, q_{j+1 \bmod n})$, $(q_{j+1 \bmod n}, q_{j+2 \bmod n})$ in ISK(P). Note also that a monotone polygon on at least five vertices can have at most a constant fraction of edges for which the sum of incident angles is bounded by 180°. Hence, by the linear size of ISK(P), the expected value of k is $O(1)$. □

Theorem 3.4 *Algorithm 1 constructs the inner skeleton of a monotone polygon in linear expected time.*

Proof: The correctness of Algorithm 1 follows from the fact that if P is respectively a monotone polygon then P' also has this property.

Step 1 takes constant time. To implement Step 2 in a constant time we keep a linear array A indexed by vertex numbers. An entry $A[i]$ is marked as passive if the vertex v_i doesn't occur in the current polygon P'. Otherwise two pointers to the two vertices adjacent to v_i in P' are kept in $A[i]$. (In the convex case of the reduction of P to P' we change the coordinates of, say, the left endpoint of e to those of the intersection point of the extensions of the incident edges, set the pointers appropriately and make the other endpoint of e passive). Assume for a moment that no more than half of the elements of A are passive. To choose randomly, say, the left endpoint of e, a logarithmic number of random bits is used. If the corresponding entry is passive or the angle condition isn't fulfiled, a new logarithmic sequence of random bits is generated *etc.* By our temporary assumption, the specification of Algorithm 1 and the fact that at most a constant fraction of edges in P doesn't fulfil the angle condition, the probability that a randomly choosen entry of the array corresponds to a "good" active vertex is bounded from below by a positive constant c. Consequently, the expected number of trials in Step 2 until an appropriate e is found is $\leq \sum_{i=1}^{\infty} ci(1-c)^{i-1} = O(1)$. Thus, Step 2 can be implemented in constant expected time. When half of the entries in A become passive, they are removed and the array is shrunk. The remaining "active" entries become reindexed appropriately. The above operation takes time linear in the array size, and totally throughout the algorithm $\sum_{i=1}^{\infty} O(\frac{i}{2^i}) = O(n)$ time is used. Step 4 can be done in constant expected time by Lemmata 3.2 and 3.3. Since all steps in Algorithm 1 but for the recursive computation of $\text{ISK}(P')$ take constant expected time and P' has one less vertex than P the whole algorithm runs in linear expected time. $\quad\square$

Definition 3.5 A simple polygon H is called a *histogram* if there is an edge e of H such that any line perpendicular to e either intersects e and the inside of H exactly once or doesn't intersect the inside of H at all. Such a distinguished edge is called the *base* of H and its endpoints are called the *base vertices* of H. A *pseudo-histogram H with base e* is a simple polygon H which, by adding at most two right triangles flush with e, can be transformed into a histogram whose base is the extension of e by the co-linear edges of the triangles. The edges of H which are incident to e and are also edges of the augmenting triangles are called *pseudo-base edges* of H.

Corollary 3.6 *The inner skeleton of a pseudo-histogram can be constructed in linear expected time.*

4 Polygons

Klein and Lingas presented a useful method of partitioning simple polygons into pseudo-histograms in [13].

Definition 4.1 Let P be a simple polygon with a distinguished edge e. The *partition of P into the set $SH(P, e)$ of pseudo-histograms* is defined as follows. Let H be the largest pseudo-histogram within P which has as its base e and as pseudo-bases these of the two edges of P incident to e that form an angle between 90° and 180° with e inside P. If $H = P$ then $SH(P, e) = \{H\}$. Otherwise, let $P_1, P_2, \ldots, P_k$ be the subpolygons of P into which the rest of P is partitioned by H, and let e_i be the edge of H adjacent to P_i. Then $SH(P, e)$ is the union of $\{H\}$ with the sets $SH(P_i, e_i), i = 1, \ldots, k$; see Figure 2 for an example.

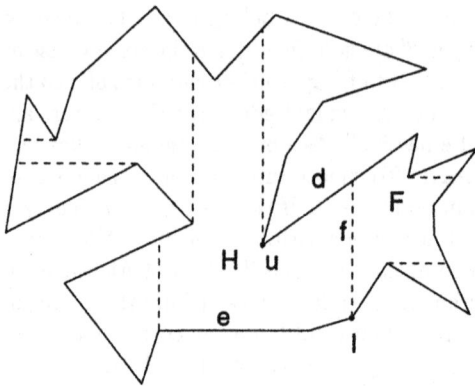

Figure 2: The decomposition of a polygon into pseudo-histograms.

The following fact has been proved in [13].

Lemma 4.2 *Let P be a simple polygon on n vertices, and let e be a vertical or horizontal edge of P. The total number of vertices in the pseudo-histograms in $SH(P, e)$ is $O(n)$. All reflex vertices in the pseudo-histograms are original reflex vertices of P. Given the vertical and horizontal trapezoidations of P and an arbitrary edge e of P, $SH(P, e)$ can be computed in linear time.*

The main idea of our algorithm for polygon skeleton will be as follows. First the input polygon will be decomposed into pseudo-histograms according to Lemma 4.2. Next, for each of the resulting pseudo-histograms the inner skeleton with the regions of artificial edges partitioned into the regions of remaining edges will be computed. Finally, the histogram skeletons will be locally merged. To specify and analyze the algorithm we need the following definitions and lemma.

Definition 4.3 Let $H \in SH(P, e)$. The *open inner skeleton* of H (OISK(H) for short) is the diagram resulting from deleting the open edges d of H that are not on the perimeter of P and repartitioning their regions $R(d)$ in ISK(H) between the regions of the remaining edges of H. A point p in $R(d)$ will be now in the region of the closest edge on the perimeter of P that p can see (i.e., be connected by a straight-line segment) within H.

Definition 4.4 Assume $H \in SH(P, e)$ has a horizontal base. A maximal chain of non-base edges of H is called a *left wall* of H if it is illuminated from the right by the vertical light beams that emanate from the base and pseudo-bases of H. Symmetrically a *right wall* is defined. A child F of H is a *left neighbor* of H if its base lies on the left wall of H. Otherwise, it is a *right neighbor*. The base of a left (respectively, right) neighbor of H is called a *left* (respectively, *right) door* of H. The pseudo-histogram H itself is called the *base neighbor* of F.

To show that for $H \in SH(P, e)$, OISK(H) can be computed in linear expected time, we need the following lemmata. In the first obvious one, we say that a skeleton region is *horizontally convex* if no horizontal line intersects its perimeter in more than two points.

Lemma 4.5 *Let d be a left (vertical) door in H. The region of d in $ISK(H)$ is horizontally convex and lies between the bisector of d and the edge including its top endpoint, and the bisector of d and its bottom endpoint. Thus, the region doesn't share boundary with the region of any other left door in H.*

Lemma 4.6 *The region $R(d)$ of a left door d in $ISK(H)$ can be eliminated in linear (in the number of adjacent regions) expected time.*

Proof: To start with, we partition the region of the half-open segment corresponding to d into the region of the bottom endpoint of d and the region $R(d)$ of the open segment d. By Lemma 4.5 $R(d)$ is horizontally convex and lies within the vertical strip bounded by the horizontal straight-lines passing through the endpoints of d. In $ISK(H)$, $R(d)$ is also vertically convex. Suppose otherwise. Then, a vertical segment s with endpoints inside $R(d)$ and the middle of s inside the region of a site on a right wall of H exists. We obtain a contradiction since the site on the right wall or a higher site on this wall is closer to the top endpoint of s than d. Consequently, the boundary of $R(d)$ is formed of a non-increasing upper chain C_1 starting from the top endpoint of d and a non-decreasing bottom chain C_2 starting from the bottom endpoint of d. Only the bottom endpoint of d and possibly the base and pseudo-bases have their bisectors with d on C_2. We can trivially compute the skeleton of these few sites in constant time.

Let L be the sequence of the sites whose bisectors with d form C_1 (in the order of the bisector fragments along C_1). Since the sites in L are disjoint from the base and pseudo-bases of H their vertical projections don't overlap. Also, we may assume w.l.o.g that they don't extend below C_1 (we can always cut the last site in L to its maximal essential for C_1 part, see the proof of Lemma 2.6).

Consider any subsequence U of L including the first and the last element of L. By the properties of L, the sites in U don't extend below C_1 and their vertical projections don't overlap. Hence, if a site u precedes a site v in U then v lies to the right of u. Consider $SK(U)$. Clearly, for any site u in U, the fragment $c(u)$ of the bisector between d and u on C_1 lies within the region of u in $SK(U)$. Let u, v be two consecutive sites in U. Suppose that a fragment b of C_1 between $c(u)$ and $c(v)$ belongs to another site w in U. Then, by the monotonicity of C_1 we obtain a contradiction. Simply, if w was to the left of u then u would be closer to b than w. Otherwise, if w was to the right of v then v would be closer to b than w. Consequently, any two consecutive sites in U have a common boundary in $SK(U)$ intersecting C_1. Let u be any inner element of U. Since $U - \{u\}$ is also a subsequence of L, we analogously infer that the predecessor and the successor of u in U have a common boundary in $SK(U - \{u\})$ intersecting C_1. Hence, we can set $f(U, u)$ to the bisector of the predecessor and the successor in order to derive a linear expected-time bound on the construction of $SK(L)$ from Lemma 2.5.

Now it is sufficient to merge $SK(L)$ with the skeleton of the few sites having their bisectors with d on C_2 in linear time by using the technique from [11], and to cut off the resulting diagram along the boundary of $R(d)$ to eliminate $R(d)$. $\qquad\square$

By combining Lemmata 4.5, 4.6 with Corollary 3.6, we obtain:

Theorem 4.7 *Let $H \in SH(P, e)$. $OISK(H)$ can be computed in linear expected time.*

Proof: We start from constructing ISK(H) in linear expected time by running Algorithm 1. Note that Algorithm 1 can be trivially modified to compute the inner skeleton of H where the base is not treated as a site. Therefore, in case the base of H is different from the edge e we may assume that its region is already partitioned into the neighboring regions.

By Lemmata 4.5, 4.6, we can eliminate the regions of left doors in ISK(H) in total linear expected time. Then, analogously, we can eliminate the regions of the right doors. □

The following definition and lemma lie behind the idea of local merging of the open skeletons used in our algorithm.

Definition 4.8 Let F be a child of $H \in SH(P, e)$. By $s(F)$ we denote the part of P illuminated by light beams emanating perpendicularly from the base of F through H. Actually, $s(F)$ is a histogram but we shall rather call it *the strip of F*, to avoid confusion with the histograms of the decomposition.

It has been observed in [13] that if F and F' are two left neighbors of H (or, right neighbors of H, respectively) then $s(F) \cap s(F') = \emptyset$. This observation leads to the following lemma.

Lemma 4.9 *Let H be the base neighbor of F in $SH(P, e)$, and let d be an edge site of F (with subtracted base vertices of F in case any of them originally overlaps with d). Then the region, U, of d in ISK(P) is contained in the union of F, the left and right neighbors of F, and the strip $s(F)$ that passes through H and possibly enters some neighbors of H opposite to F.*

Proof: Suppose the base of F is horizontal and a point x of U lies in the base neighbor, H, of F or in some neighbor of H opposite to F but outside the strip. Let v be the point on the closure of d closest to x. The segment $\overline{vx}$ belongs to U, but the point of intersection with the vertical line through the lower base vertex, l, of F is closer to l than to v. We obtain a contradiction by triangle inequality since the lower base vertex is always in an original edge site of P. Similarly, if a point x of U lies in a grandchild F'' of F in the multiway tree $SH(P, e)$ then the point of the intersection of the vertical line through the lower base vertex l'' of F'' with $\overline{vx}$ is closer to l'' than to v. We obtain again a contradiction. □

Algorithm 2

Input: A simple polygon P with horizontal edge e.
Output: The inner skeleton ISK(P).

1. Compute the pseudo-histogram decomposition SH(P, e).

2. For each H compute OISK(H).

3. For each H merge OISK(H) with the diagrams OISK(F) of the left neighbors F of H into the diagram LOISK(H). Also, merge OISK(H) with the diagrams of its right neighbors into ROISK(H).

4. For each H merge LOISK(H) with the diagrams OISK(F) of the right neighbors F of H into the diagram RLOISK(H).

5. For each H merge RLOISK(H) with

   (a) the diagram ROISK(F), if H is a left neighbor of F,
   (b) the diagram LOISK(F), if H is a right neighbor of F,

   of the base neighbor F of H into the diagram $\text{OISK}_1(H)$.

6. Glue together all diagrams $\text{OISK}_1(H)$ into the diagram ISK_1.

7. Output the resulting diagram.

Theorem 4.10 *Algorithm 2 produces the inner skeleton of P in linear expected time.*

Proof: The correctness of Algorithm 2 follows from the correctness of the merging steps (3-6). The latter is due to Lemma 4.9.

A linear worst-case time implementation of Step 1 is provided by Chazelles's algorithm [3] combined with Lemma 4.2. Step 2 can be carried out in linear expected time, due to Theorem 4.7.

Further, let $|F|$ denote the total number of vertices (original and extra ones) of pseudo-histogram F in SH(P, e).

In Step 3, let F_i with base f_i, $1 \leq i \leq m$, be the left neighbors of H. For each i we apply the Merging Lemma 2.7 with its notations to $P_1 = Q_1 = F_i$, $P_2 = Q_2 = H$, and $d = f_i$. Trivially, $k_1 \leq |F_i|$ holds. Due to Lemma 4.9 we have $k_2 \leq |s(F_i) \cap H|$. Since the strips $s(F_i)$ are disjoint, the total supporting cost, taken over all i, is bounded by $|H| + \sum_{i=1}^m |F_i|$. Moreover, these m merge operations do not interfere and correctly result in ISK($H \cup F_1 \cup \ldots \cup F_m$). Since each of the m merge lines is part of this diagram, their total size (= merge line cost) is also bounded by the number of vertices of H and of all its left neighbors F_i. Summarizing over all pseudo-histograms H, we still obtain a linear bound because each F can be a left neighbor of at most one H—its base neighbor.

In Step 4, let F_i, $1 \leq i \leq m$, denote the right neighbors of H. In applying the Merging Lemma we extend Q_1 and Q_2 to the union of F_i and H with all their left neighbors, respectively. During these m merge steps we compute, within $H \cup F_1 \cup \ldots \cup F_m$, the skeleton within H, the F_i, and all their left neighbors, and the total cost is bounded by the sum of the number of vertices of these pseudo-histograms. Again, taking the sum over all H leads to a linear bound.

In Step 5, (a), Q_1 is the union of F with its right neighbors while Q_2 consists of H and both the left and the right neighbors of H. The reason for using ROISK(F) rather than RLOISK(F) is to avoid an "echo" from H. Altogether, we have, in Step 5, to merge the diagrams of each F into the diagrams of its left (a) and right (b) neighbors. The total cost is still linear in $|P|$.

Steps 6, 7 can be clearly done in linear time. $\qquad\square$

5   The Delaunay triangulation

We adhere to the standard definitions of a *planar straight-line graph*, the Voronoi diagram of a planar point set, its straight-line dual called *Delaunay triangulation* and the *convex hull* of a planar figure (see [21]). In the following definition of the *generalized Delaunay triangulation* of a planar straight-line graph, two points in the plane are mutually *visible* iff their connecting line segment doesn't cross properly any edge of the graph.

Definition 5.1 *The generalized Delaunay triangulation of a planar straight-line graph is a triangulation where no circumcircle of a triangular face contains a different vertex visible from all of the three vertices of the face.*

The generalized Delaunay triangulation of a planar straight-line graph on n vertices can be built in $O(n \log n)$ time [22, 25]. Also, if the graph is a set of isolated points then it is simply the Delaunay triangulation of the point set. The following fact has been proved in [13].

Lemma 5.2 *[13]: The generalized Delaunay triangulation within a simple polygon can be built in linear expected time.*

It enables us to improve the $O(n \log^* n)$ upper time-bound on the construction of the Voronoi given a bounded-degree spanning subgraph of its straight-line dual due to Devillers [7] in two aspects. Firstly, the upper bound is reduced to a linear one. Secondly, the requirement on bounded degree is not needed.

Theorem 5.3 *The Delaunay triangulation (equivalently, the Voronoi diagram) of a planar point set S can be computed from a connected spanning subgraph of the Delaunay triangulation of S (e.g., the Euclidean minimum spanning tree) in linear expected time.*

Proof: Let W be a connected spanning subgraph of the Delaunay triangulation of S. Compute a spanning tree T of W, e.g., by depth first search in linear time. Observe that T includes as its vertices all points in S. Also, T can be regarded as a (degenerate) simple polygon after doubling its edges. Therefore, we can apply a known linear-time algorithm for the construction of the convex hull of a simple polygon [9, 16] to construct the convex hull of T. The hull together with T partition the area of the hull into a collection of simple polygons whose vertices are exactly the points in S. The partition can be easily found in linear time by traversing T and the hull. The generalized Delaunay triangulations within all of these polygons can be computed in linear expected time by 5.2. As the edges of the polygons are in the Delaunay triangulation the union of the generalized Delaunay triangulations within the polygons is simply the Delaunay triangulation of S.   □

Acknowledgements

Thanks go to an unknown referee for valuable comments.

References

[1] A. Aggarwal, L.J. Guibas, J. Saxe, and P.W. Shor. A Linear-Time Algorithm for Computing the Voronoi Diagram of a Convex Polygon. Discrete and Computational Geometry 4, 1987.

[2] F. Aurenhammer. Voronoi Diagrams—A Survey of a Fundamental Geometric Data Structure. ACM Computing Surveys 23, 1991, pp. 345–405.

[3] B. Chazelle. Triangulating a Simple Polygon in Linear Time. Discrete and Computational Geometry 6, 1991, pp. 485–524.

[4] P. Chew. Building Voronoi Diagrams for Convex Polygons in Linear Expected Time. Manuscript, 1986.

[5] P. Chew. Constrained Delaunay Triangulations Proc. 3rd ACM Symposium on Computational Geometry, 1987, pp. 215–222.

[6] K.L. Clarkson, K. Mehlhorn and R. Seidel. Four results on randomized incremental constructions. Computational Geometry: Theory and Applications 3 (1993), pp. 185-212.

[7] O. Devillers. Randomization yields simple $O(n \log^* n)$ algorithms for difficult $\Omega(n)$ problems. International Journal of Computational Geometry and Applications, Vol2, No1 (1992), pp. 97-111

[8] H. Djidjev and A. Lingas. On Computing the Voronoi Diagram for Restricted Planar Figures. Proc. WADS'91, LNCS 519, pp. 54-64, Springer Verlag. To appear in International Journal of Computational Geometry and Applications.

[9] R.L. Graham and F.F. Yao. Finding the convex hull of a simple polygon. J. Algorithms 4(4), pp. 324-331, 1983.

[10] M. Held. On the Computational Geometry of Pocket Machining. LNCS 500, Springer-Verlag, 1991.

[11] D.G. Kirkpatrick. Efficient computation of continuous skeletons. Proceedings of the 20th FOCS, pp. 18-27.

[12] R.Klein, K. Mehlhorn, and S. Meiser. Randomized Incremental Construction of Abstract Voronoi Diagrams. Computational Geometry: Theory and Applications 3 (1993), pp. 157–184.

[13] R. Klein and A. Lingas. A Linear-Time Randomized Algorithm for the Bounded Voronoi Diagram of a Simple Polygon. 9th ACM Symposium on Computational Geometry, San Diego, U.S.A., 1993, pp. 124–132.

[14] R. Klein and A. Lingas. A Note on Generalizations of Chew's Algorithm for the Voronoi Diagram of a Convex Polygon. Proc. 5CCCG, pp. 370-374, 1993.

[15] R. Klein and A. Lingas. Hamiltonian Abstract Voronoi Diagrams in Linear Time. Proc. ISAAC'94, LNCS, Springer Verlag.

[16] D.T. Lee. On finding the convex hull of a simple polygon. Int'l J. Comput. and Infor. Sci. 12(2), 87-98, 1983.

[17] D.T. Lee and A. Lin. Generalized Delaunay Triangulations for Planar Graphs. Discrete and Computational Geometry 1, 1986, pp. 201-217.

[18] A. Lingas. Voronoi Diagrams with Barriers and the Shortest Diagonal Problem. Information Processing Letters 32, 1989, pp. 191–198.

[19] U. Montanari. Continuous Skeletons from Digitized Images. J. ACM 16 (1969) pp. 564-549.

[20] A. Okabe, B. Boots, and K. Sugihara. Spatial Tessellations, Concepts and Applications of Voronoi Diagrams. John Wiley & Sons, West Sussex, 1992.

[21] F.P. Preparata and M.I. Shamos. Computational Geometry: An Introduction. Texts and Monographs in Theoretical Computer Science, Springer Verlag, New York, 1985.

[22] R. Seidel. Constrained Delaunay triangulations and Voronoi diagrams with obstacles. In Rep. 260, IIG-TU Graz, Austria, pp. 178-191.

[23] R. Seidel. Backwards Analysis of Randomized Geometric Algorithms. New Trends in Discrete and Computational Geometry, Janos Pach (ed.), Springer-Verlag, 1993.

[24] R. Seidel. A simple and fast incremental and randomized algorithms for computing trapezoidal decompositions and for triangulating polygons. Computational Geometry: Theory and Applications, vol. 1, no. 1, 1991.

[25] C. Wang and L. Schubert. An Optimal Algorithm for Constructing the Delaunay Triangulation of a Set of Line Segments. Proc. 3rd ACM Symposium on Computational Geometry, Waterloo, pp. 223–232, 1987.

[26] C. Yap. An $O(n \log n)$ algorithm for the Voronoi diagram of a set of simple curve segments. Discrete Computational Geometry 2 (1987), pp. 365-393.

[27] C. Yap and H. Alt Motion Planning in the CL-Environment. Proc. WADS'89, Ottawa, Canada, LNCS 382, pp. 373-380.

Authors' Index

Lecture Notes in Computer Science

For information about Vols. 1–903

please contact your bookseller or Springer-Verlag